AF616018

ENERGETIC PHENOMENA ON THE SUN

ASTROPHYSICS AND SPACE SCIENCE LIBRARY

A SERIES OF BOOKS ON THE RECENT DEVELOPMENTS
OF SPACE SCIENCE AND OF GENERAL GEOPHYSICS AND ASTROPHYSICS
PUBLISHED IN CONNECTION WITH THE JOURNAL
SPACE SCIENCE REVIEWS

VOLUME 153
CURRENT RESEARCH

ENERGETIC PHENOMENA ON THE SUN

Edited by

M. R. KUNDU
Astronomy Program, University of Maryland at College Park, Maryland, U.S.A.

B. WOODGATE
NASA/Goddard Space Flight Center, Greenbelt, Maryland, U.S.A.

and

E. J. SCHMAHL
Astronomy Program, University of Maryland at College Park, Maryland, U.S.A.

KLUWER ACADEMIC PUBLISHERS
DORDRECHT / BOSTON / LONDON

Library of Congress Cataloging in Publication Data

Energetic phenomena on the sun / edited by Mukul R. Kundu and Bruce Woodgate, and E.J. Schmahl.
p. cm. -- (Astrophysics and space science library ; 153)
"A result of three meetings ... held at the Goddard Space Flight Center on January 24-28, 1983, June 8-14, 1983, and February 13-17, 1984"--Foreword.
Includes bibliographies and index.
ISBN 0-7923-0172-2
1. Solar flares--Congresses. I. Kundu, Mukul Ranjan, 1930- II. Woodgate, Bruce. III. Schmahl, E. J. IV. Series: Astrophysics and space science library ; v. 153.
QB526.F6E52 1989
523.7'5--dc19 89-31037

ISBN 0-7923-0172-2

Published by Kluwer Academic Publishers,
P.O. Box 17, 3300 AA Dordrecht, The Netherlands.

Kluwer Academic Publishers incorporates
the publishing programmes of
D. Reidel, Martinus Nijhoff, Dr W. Junk and MTP Press.

Sold and distributed in the U.S.A. and Canada
by Kluwer Academic Publishers,
101 Philip Drive, Norwell, MA 02061, U.S.A.

In all other countries, sold and distributed
by Kluwer Academic Publishers Group,
P.O. Box 322, 3300 AH Dordrecht, The Netherlands.

printed on acid free paper

Printed in The Netherlands

TABLE OF CONTENTS

FOREWORD

This publication is a result of three meetings, each 5 days long, held at the Goddard Space Flight Center on January 24-28, 1983, June 8-14, 1983, and February 13-17, 1984.

The meetings were held in the interim between the full operations of the Solar Maximum Mission (SMM) in 1980, and the renewed operations after its repair in orbit in April 1984. Their general objectives were as follows:

o Synthesize flare studies after three years of SMM data analysis. Many analyses of individual flares and individual phenomena, often jointly across many data sources had been published, but a need existed for a broader synthesis and updating of our understanding of solar flares since the Skylab Flare Workshops held several years earlier.

o Encourage a broader participation in the SMM data anlysis and combine this more fully with theory and other data sources--data obtained with other spacecraft such as the HINOTORI, P78-1, and ISEE-3 spacecrafts, and with the Very Large Array (VLA) and many other ground-based instruments. Many coordinated data sets, unprecedented in their breadth of coverage and multiplicity of sources, had been obtained within the structure of the Solar Maximum Year (SMY).

o Stimulate joint studies, and publication in the general scientific literature. The intended primary benefit was for informal collaborations to be started or broadened at the Workshops with subsequent publications.

o Provide a special publication resulting from this Workshop.

o Provide a starting point of understanding for planning renewed full observations with the repaired SMM.

The widespread interest in the Workshop topics drew over 150 people from 17 countries, which enabled the desired broad participation of people, data sources and ideas to be achieved. It also led to fragmentation of closely related subjects into topic groups in order to allow reasonably sized groups for discussion, and to allow more time for detailed discussion of each topic. The results of the studies of each group form the chapters of this book. Table 1 shows the seven topics, the topic leaders, and the sub-topics expressing the scope of each topic, shown as a set of questions. The Workshop participants, are shown at the back of the book. Some integration of topics occurred by having joint meetings between topic groups and plenary sessions.

This volume is the product of research carried out by all the workshop participants, and in particular of the dedicated efforts of all the team leaders. We express our sincere thanks to Drs. E. J. Schmahl, S. M. White and N. Gopalswamy for their generous help in various aspects of the production of this volume; to Ms. Gloria Wharen of GSFC and Ms. Betty J. Stevenson of U. MD. for the preparation of the manuscript; to Dr. David Bohlin of NASA for providing financial support for the workshops; to Goddard Space Flight Center, and the University of Maryland for various administrative support services.

M. R. Kundu

B. Woodgate

Table 1 SMM Workshop Topic Groups

Chapter	Topic	Sub Topics
1	Preflare Activity Leaders: E. Priest and V. Gaizauskas	What changes in physical conditions can lead to instabilities? What are typical pre-flare phenomena: Photosphere: Velocity and magnetic shears? Emerging flux? Chromosphere, T.R.: UV fluctuations, downflows? UV upflows? Filament eruptions? Corona: Soft X-ray increases-mechanism? Early radio events-arch destabilizers?
2	Impulsive Phase Acceleration of Particles Leader: L. Vlahos	What instabilities will occur first in plausible physical situations? Where is the primary energy release located? What are the physical conditions in the loop implied by gamma ray lines and continuum? What are accelerating mechanisms of the electrons and ions that are allowed by the observations?
3	Impulsive Phase Transport Processes Leader: R. Canfield	Are hard X-ray and UV bursts the result of electron beams or conduction fronts? What observational constraints exist on impulsive phase models? What is the distribution of thermal and non-thermal plasma?
4	Chromospheric Explosions Leader: G. Doschek	What are the observed motions during and after the impulsive phase? What is the evidence for a high pressure region at this time? How do thermal and mechanical energies compare, e.g., using soft X-ray intensities and line shifts? Can the total energy be contained?
5	Flare Energetics Leader: S-T. Wu	Is the maximum energy in the thermal phase greater than the energy input by the impulsive phase? Is there theoretical need or observational evidence for post-impulsive energy input? Is there a time inconsistency between the end of the impulsive phase and the maximum of the thermal phase?
6	Coronal Mass Ejections and Coronal Structures Leader: E. Hildner	What is the relationship between coronal mass ejections and flares and filament eruptions? How can metric radio phenomena be integrated with other pre-flare, flare and post-flare phenomena and observed coronal structure? What is the detailed structure of coronal mass ejections in the corona and how do they appear in the interplanetary medium? How are they initiated and related to other solar phenomena? How well do models and observations of coronal transients compare? How did the global corona evolve in the post-maximum epoch of SMM?
7	Numerical Analysis Leader: R. Kopp	Do the results of different flare modelling codes agree when applied to the same initial conditions?

ENERGETIC PHENOMENA ON THE SUN

INTRODUCTION

Mukul R. Kundu[1] and Bruce Woodgate[2]

[1]University of Maryland, College Park, Maryland
[2]Goddard Space Flight Center, Greenbelt, Maryland

PREVIOUS HISTORY

The SKYLAB Workshop on Solar Flares was a focal point for much of our recent understanding on the origin of solar flares and their after effects. That workshop was concerned with observations of soar flares obtained with the first full-scale, manned, astronomical observatory in space, supplemented by observations in other spectral domains such as optical and radio from ground-based observatories. The SKYLAB carried instruments covering the wavelength range from 2 to 7000A encompassing the soft X-ray, ultraviolet and visible light portions of the electromagnetic spectrum. It was the first observatory in space that produced daily X-ray pictures of active regions and flares, and provided unprecedented images of magnetic loops, flare plasmas, coronal holes and bright points. These observations provided fresh insight into the physics of active region emissions and flare processes. Prior to SKYLAB, great improvement in our understanding of active regions and flares was achieved from the series of spacecrafts known as Orbiting Solar Observatories (OSO), which made possible a systematic study of the structure and variability of the sun's atmosphere in spectral domains (EUV) inaccessible with ground-based instruments.

One of the major emphases of the Solar Maximum Mission (SMM) program was the realization, right from the beginning, that in order to understand the flare phenomenon fully, one has to have simultaneous observations over the entire electromagnetic spectrum--from shortest wavelength gamma rays all the way to the long radio meter-decameter wavelengths. This concept of simultaneous coverage was encouraged by the SMM Investigators and the NASA administration via the SMM Guest Investigator Program and other means. The result was a comprehensive coverage of simultaneous solar flare observations by SMM instruments as well as large ground-based instruments in both optical and radio domains. The SMM was designed to make simultaneous observations of as many aspects of solar flares as possible. The emphasis with SMM instrumentation was on comprehensive coverage. The observations obtained with SMM also benefited from improvements in sensitivity, spectral and temporal resolution, particularly for the highest energy electromagnetic radiation. A major advance achieved with SMM was the capability of imaging X-rays with energies up to 30 keV. Subsequently, HINOTORI, a Japanese solar satellite, extended the energy range of hard X-ray imaging of solar flares up to 40 keV. The results obtained with this spacecraft as well as the X-ray and coronagraph results obtained with the P78-1 satellite are also included in this workshop report. Among the ground-based instruments we must note the contributions made by the Sacramento Peak Vacuum Tower Telescope, routine optical observations in Hα, helium D3, the green and red coronal lines, and white light by Sacramento Peak, Big Bear Solar Observatory (BBSO), Ottawa River Solar Observatory and the U.S. Air Force Solar Optical Observatory Network (SOON), the magnetograms taken by Kitt Peak, BBSO and MSFC; high spatial resolution (a few arc seconds) radio observations made by the Very Large Array (VLA), Owens Valley (OVRO) and Westerbork (WSRT) interferometers, the Clark Lake Radio Observatory (CLRO) radopheliograph, the Nancay and Culgoora radioheliographs, Toyokawa and Nobeyama interferometers, Berne and Zurich high time resolution radio telescopes at decimeter and centimeter wavelengths, the Air Force RSTN radio patrol telescopes, and the Itapetinga (Brazil) millimeter telescope with high time resolution.

In Table 2 is given a list of the seven experiments that are on board SMM. Of these seven experiments, only the first six are relevant for our studies of active regions and flares, although the Active Cavity Radiometer Irradiance Monitor (ACRIM) has provided useful upper limits on the total radiated flux from flares (see Chapter 5).

THE SOLAR MAXIMUM MISSION

A detailed description of the SMM program, the spacecraft, the instruments and their objectives appeared in a special issue of Solar Physics Volume 65, (1980). Examples of initial coordinated results appeared in a special issue of Astrophys. J. Letters Volume 244, L113, 1981. We provide a brief summary of the spacecraft and instrument capabilities here.

OBJECTIVES

Observations of flares prior to SMM suffered from several major problems: Observations of different aspects of flares were usually taken from different flares or at different times in a flare. Coordinated observations of the same flare at the same time by instruments with different energy ranges were very rare, so that comparisons of energy structure and timing could only be made in the broad statistical sense across many dissimilar flares. Observations of the early stage of flares, the pre-impulsive and impulsive phases were rare. The time resolution of observations was generally inadequate for studying the impulsive phase.

SMM was designed to provide substantial improvements in these areas. A set of instruments was selected to provide broad energy coverage from gamma rays to the visible, all operating at the same time, and able to point at the same area of the Sun, so that simultaneous co-spatial studies could be carried out across this broad energy range. Preflare and impulsive phase studies could occur using automatic observation sequences over long periods of time waiting for the flare, using a free-flying spacecraft and photoelectric detectors (no fear of using up all the film or having no flares that week). A rapid response to impulsive phase phenomena was achieved across the instrument set without waiting for human intervention by providing an automatic flare flag from the Hard X-ray Imaging Spectrometer.

Table 2 Instruments on Solar Maximum Mission

Experiment	Spectral Range	Spectral Resolution	Spatial Resolution	P.I.	Organization
GRS (Gamma Ray Spectrometer)	0.1–17 MeV 10-160 MeV	7.5% (at 0.66 MeV)	Full Sun	E.L. Chupp	U. New Hampshire (Max Planck Institute, Garching and NRL)
HXRBS (Hard X-ray Burst Spectrometer)	25-500 keV	15 channels (128 ms time)	Full Sun	K. Frost	Goddard Space Flight Center
HXIS (Hard X-ray Imaging Spectrometer)	3.5-30 keV	6 channels	8″ FWHM, 32″ FWHM	C. deJager	SRL Utrecht (Birmingham)
XRP (X-ray Polychromator) —Includes BCS (Bent Crystal Spectrometer) and FCS (Flat Crystal Spectrometer)	1.76-3.23 Å 1.44-22.4 Å	0.14-0.62 mÅ 0.2-20 mÅ	6′ FWHM 14″ FWHM	L.W. Acton J.L. Culhane A.H. Gabriel	Lockheed Mullard Space Science Lab. Appleton Lab.
UVSP (UV Spectrometer and Polarimeter)	1100-3300 Å	0.02 Å	Variable > 2″	E. Tandberg-Hanssen	Marshall Space Flight Center Goddard Space Flight Center
C/P (Coronagraph/Polarimeter)	4435-6583 Å	200 Å ±20 Å, ±2 Å	6.4″ or 12.8″	R. MacQueen	High Altitude Observatory
ACRIM (Active Cavity Radiometer Irradiance Monitor)	UV-IR	—	Full Sun	R.C. Wilson	Jet Propulsion Laboratory

INSTRUMENTS

The Gamma Ray Spectrometer (GRS) was designed to study the acceleration of nucleons and electrons in solar flares and cosmic sources, to detect gamma rays between 0.3 and 9 MeV with high spectral resolution, measure nuclear lines and the continuum, and to measure 10-100 MeV gamma rays with lower spectral resolution. It also detects neutrons above 10 MeV. These modes have 2 s time resolution. Two small 10-140 keV X-ray detectors have 1 s time resolution.

The Hard X-ray Burst Spectrometer (HXRBS) measures X-ray spectra between 25 and 500 keV in 15 channels with a time resolution of 0.128 s. It also records the total count rate in the same energy range with a time resolution of 1-10 ms during the larger flares. The goals include determination of electron acceleration, propagation and deposition processes primarily in the impulsive phase of flares, particularly by comparison of the X-ray energetics and timing with microwave and ultraviolet emissions.

The Hard X-ray Imaging Spectrometer (HXIS) was designed to measure the position, structure, and coarse spectrum of hard X-rays in flares. It obtained images in six energy bands between 3.5 and 30 keV, in a fine field of view of 2'40" diameter with 8 arc sec resolution, and in a coarse field of view of 6'24" diameter with 32 arc sec resolution. Its time resolution was 1.5-9 s depending on the mode. It generated a flare flag showing the flare location where the X-ray flux reached a threshold and sent this flag to the spacecraft computer for use by the other experiments.

The X-ray Polychromator (XRP) consists of two instruments, the Bent Crystal Spectrometer (BCS) and the Flat Crystal Spectrometer (FCS). Both instruments measure the individual emission lines which dominate spectra from the solar corona and flare X-ray plasmas, to diagnose their temperatures, densities, velocities, abundances, and non-equilibrium states of all phases of flares, and active regions. The BCS was designed for high time resolution studies in lines of FeI-FeXXVI and CaXIX. It covers its entire spectral range at once, and integrates spatially over an entire active region with a field of view of 6 arc min FWHM. The FCS provides mapping of flares and active regions in the resonance lines of 7 ions between O VIII and Fe XXV with 14 arc sec resolution and field of view up to 7 arc min. It can also spectrally scan the 1.4-22.5 Å region.

The Ultraviolet Spectrometer and Polarimeter (UVSP) was designed to measure the lower temperature plasmas (5×10^3 to 2×10^5 K) in flares, active regions and the quiet Sun. The wavelength range was 1170-3500 Å and a spectral scanning capability was included. Spatial resolution is 2-3 arc sec, with a spatial range up to 256 arc sec square for the raster imaging. Dopplergrams were possible using two detectors, one on either side of a spectral line. Time resolution can be as fast as 64 ms. Linear and circular polarimetry is possible. Any combination of timing, slit size, spatial step size, image size, polarimetry and wavelength scanning in any order was possible. (All except the wavelength scanning is still possible.)

The Coronagraph/Polarimeter (C/P) observes the corona between 1.6 and 6 solar radii with 10 arc sec resolution, one quadrant at a time, using an external occulter to block out the solar disc. Seven wavebands in the visible and a linear polarimetry capability are available. The goals include observations of coronal mass ejections from flares and other phenomena, large scale structure of the solar corona over long periods of time and its evolution.

The Active Cavity Radiometer Irradiance Monitor (ACRIM) measures the total irradiance of the Sun to a short term stability of better than 0.01% and long term stability of calibration better than 0.1%. Its goals are to measure the long term variability of the Sun for climatological effects, and to measure irradiance variations due to sunspots and active regions, and oscillations.

The spacecraft is three-axis stabilized, with controllable roll normally set so that one axis of the imaging instruments aligns with the solar North-South direction. The spacecraft aspect control system is able to point the instrument at a chosen feature to about 20 arc sec, with a stability normally better than 1 arc sec. It has a programmable solar feature tracking capability to compensate for solar rotation.

OPERATIONS

The scientific operations of the SMM are centered at the Experimenters Operations Facility (EOF) at the Goddard Space Flight Center. Daily solar forecasts are provided by the National Oceanic and Atmospheric Administration (NOAA) personnel resident at the EOF, with data obtained through the network of observatories centered on the Solar Forecast Center in Boulder.

Experimenter team representatives select an orbit-by-orbit observing plan for the following day, based on the match between the solar forecast, the current SMM objectives, guest observer plans and engineering considerations. Current SMM objectives are reviewed every 2-4 weeks by the Investigators' Working Group (now the Experimenters' Forum). Experiment personnel then convert their scientific plans to command loads which are uplinked to the spacecraft overnight. Similarly the spacecraft pointing plan commands are uplinked ready for the next day's operation. Data are sent down to the EOF and collected and reduced by each experimenter team, mostly within 1-2 days of the observations being taken. This allows both scientific and operational feedback, to optimize experimental sequences, react to different solar conditions, and to begin joint analyses of interesting phenomena. Early review of data can lead to reformulation of scientific questions and corresponding experimental sequences. The daily scientific plan is sent out via the NOAA network to collaborating observatories around the world, according to joint agreements or as part of the SMY arrangements.

The SMM was launched on February 14, 1980, to an orbit with an altitude of 574 km, an inclination of 28°.6, and a period of 96 min. All instruments operated successfully for 7 months, after which the coronagraph electronics failed; after 91/2 months the spacecraft fine pointing system failed. From November 1980 until April 1984, coarse pointing only was possible with the instruments offset by sometimes as much as 15° from the Sun. This did allow, however, the GRS, HXRBS and ACRIM experiments to continue obtaining useful data for most of this time. The HXIS instrument failed during this time. In April 1984, the SMM was repaired in orbit by the Shuttle astronauts. The fine pointing system and the Coronagraph were repaired, but not HXIS. Since then the Observatory has been operating well. The UVSP wavelength drive operated for 7 months during this period but since November 1985 has not been working, and some of the XRP detectors are not working. Otherwise, SMM is fully operational.

SUMMARY OF WORKSHOP RESULTS

In what follows we summarize briefly some of the main results obtained during the workshops by different teams.

PREFLARE ACTIVITY

Theoretical mechanisms which could lead to the initiation of flares through a variety of instabilities were discussed. In particular the combination of tearing modes with other instabilities which can greatly increase the rate of energy release were investigated. The tearing mode may switch over to the radiative mode which is a hundred times faster at high magnetic Reynold's numbers. Alternatively the tearing mode can initiate fast non-linear Petscheck-Sonnerup reconnection. This can give rise to broad slow shock waves, hot fast jets, and thinner fast mode shocks. These shocks could be effective particle accelerators. Cases arising from tearing of a sheared magnetic field, and from an outside driver such as emerging flux were discussed.

Non-linear tearing may saturate, or mode couple producing helical structures, in which turbulence can produce rapid small scale heating. It also produces multiple islands, which may be attracted to each other creating a coalescence instability which can be very rapid. Flux pile-up can occur between the islands, lengthening the sheet, which becomes unstable and enters an impulsive bursty regime.

The role of emerging flux was discussed. Reconnection crossing a microturbulence threshold was suggested for small flares, and destabilization of a filament in an arcade for large flares.

Equilibrium conditions for an arcade by balancing tension and buoyancy, relating height and footpoint separation show that instability to spontaneous eruption occurs above a critical height. The equilibrium is affected by an external magnetic field and by twist. The effect of line tying on stabilizing a filament is dependent on whether flow occurs parallel to the magnetic field at the loop footpoints.

Preflare magnetic and velocity fields were studied to identify the conditions in an active region in which flares are more likely to occur. It was found that a preflare active region is more dynamic than average; more surges and more evidence of heating occurred. Computations of coronal magnetic fields deduced from photospheric magnetic field patterns showed evidence in one case that the fields above a flaring active region were not force-free (the electric currents were not field-aligned), whereas the fields above a non-flaring region were force-free. However, the clearest finding was the association between flare occurrence and magnetic shear, deduced from the angle of the transverse magnetic field with respect to the neutral line of the longitudinal (line of sight) magnetic field. Velocity shear was also evident in the association of transition region horizontal flow direction reversal with the neutral line. The existence of the magnetic shear was associated with motions of sunspots. Computations from the vector magnetic fields showed large scale persistent vertical electric currents. This increased interest in current-driven instabilities. The electric currents were correlated with the location of UV preflare brightenings. The connection between flare occurrence and emerging flux was found to be uncertain; cases were found where flares had evidence of phenomena associated with emerging flux, and other well-observed cases were found where there was no evidence of emerging flux.

A set of selected flares were studied in which precursor phenomena were observed by a wide range of instruments, and relationships between several phenomena were noted. Filament eruptions were confirmed to be a frequent preflare phenomena, evidence for heating and upflow in structures beneath the filament were found. In a flare at the limb, with extensive preflare X-ray brightenings, a preflare rising compact loop was seen in Hα and UV which erupted at its top during the impulsive phase. Associations between radio type III bursts and microwave preflare bursts have been observed, demonstrating their nonthermal character. Soft X-ray preflare brightenings near the flare location were coincident in time with the projection of coronal mass ejections back to the surface. Microwave observations have shown preflare step increases in intensity and changes in polarization. Microwave active region maps made with arc second resolution using the VLA have demonstrated that the brightness temperature enhancement of an active region is not a sufficient condition for the occurrence of a flare. The appearance of a new microwave region and its interaction with the preexisting region, as suggested by the reversal of polarization or change of orientation of the neutral plane in the corona, a few minutes (< 15 minutes) before the flare appears to be necessary for the onset of the impulsive flare.

PARTICLE ACCELERATION

The primary goal of this group was the study of the acceleration of electrons and ions to high energies before, during and after the impulsive phase of flares.

The SMM and HINOTORI satellites provided several exciting new results that changed our focus on the study of particle acceleration in solar flares. The most important new information includes the following:

Hard X-ray imaging has, for the first time, provided evidence for discrete isolated footpoints during the impulsive phase. The hard X-ray source gradually evolves and forms a single source of hot plasma in the decay phase of some flares. Some other flares never develop footpoints and the thermal plasma dominates their evolution. These results seem to have provoked a vigorous debate on the role of "thermal" plasma vs "nonthermal" beams as the main product of a solar flare. Current theoretical understanding of the interaction of "hot" and "cold" plasma combined with the above mentioned data suggest that "thermal" and "nonthermal" plasma are in fact always present and they are interrelated, e.g., a locally heated plasma in the corona can be the source of nonthermal particles for the chromosphere and these particles are quickly thermalized. X-ray imaging has provided input for this interconnection and the simplistic division of flares into two artificial classes has been abandoned.

High spatial resolution microwave images have led to important advances in our understanding of the evolution of the magnetic topologies that lead to a flare. Interacting loops and bipolar structures commonly observed suggest that loop-like structures are the elementary components of the flare process. The isolated flaring loop, a very popular concept during and after the SKYLAB, should be replaced by more complex topologies (e.g., interacting loops, emerging flux or even a catastrophic interaction of many loops). The loop, however, remains as the elementary structure that participates in these interactions. The radio maps at meter wavelengths and detailed studies of meter wave/X-ray correlations suggest that the region of acceleration must encompass open and closed field lines, located in the low corona.

An outstanding contribution made by this group was the realization that "microflares" observed with high resolution X-ray detectors continuously occur in the corona and accelerate particles to energies up to tens of keV. In other words, coronal heating and flaring may not be two distinctly different processes and "flares" have no energy threshold. The number of flares appears to increase as the total energy per flare decreases. The main contribution of this result to particle acceleration theory is that the ambient velocity distribution may not be a "Maxwellian"; non-thermal particles may be present at all times in the corona.

The last few years have seen a surge of theoretical and observational activity on the understanding of the intense, coherent, polarized microwave pulses. We now believe that these "spikes" may be the result of the conversion of kinetic energy of the precipitating electrons to coherent electromagntic radiation through the convergence of the magnetic field lines and the formation of unstable distributions. Considerable discussion was devoted to the interaction of these intense electromagnetic pulses with the solar atmosphere and to the acceleration of electrons outside the flaring volume.

The most dramatic result regarding the acceleration of particles in solar flares came from the gamma-ray detectors aboard the SMM. The conventional two-phase acceleration--a prompt first phase acceleration of mildly relativistic electrons followed by a slower (delayed by several minutes from the first phase) second phase acceleration that energizes the ions and further accelerates the electrons to relativistic energies--had to be abandoned, because in many flares fast synchronous (or nearly synchronous) pulses occur simultaneously in a wide range of energies. Hence, the concept of prompt acceleration of both electrons and ions to all energies in a few seconds is a new important discovery, which has changed dramatically our thinking on the processes responsible for particle acceleration in flares. More recent work on the interaction of 100 keV ions with an oblique turbulent shock which developed in the vicinity of the flaring volume has shown that ions reach energies as high as 50 MeV in less than 10 milliseconds. Studies of the interaction of the hot flaring plasma with the "cold" environment outside the flaring volume using a hybrid numerical code have shown that a piston driven perpendicular shock is formed in a few microseconds. In other words, shock formation and particle acceleration can occur in 10-20 msec, below the resolution of current instruments.

This group discussed at great length the observed time fluctuations in solar bursts. Fast pulses with duration that sometimes reaches the instrumental resolution and delays between pulses in different wavelengths are among the new results. Some of the observed pulses are clearly the result of pulsations in the acceleration source, in other cases the pulses could be the result of the radiation mechanism. Our theoretical understanding on the time development of the acceleration source is currently weak but the workshop raised many concrete questions for future study.

IMPULSIVE PHASE TRANSPORT

The work of this group was concerned with how the energy released in a solar flare is transported through the solar atmosphere before escaping in the form of radiant and mechanical energy.

The mechanisms suggested for dissipation of magnetic energy in flares suggest that the actual sites of reconnection must involve scale lengths well below currently achievable spatial resolution. Consequently, observational evidence in support of a flare theory is necessarily indirect. Indirect evidence may be of several kinds. For example, high spatial resolution observations may indicate the geometry, on a larger scale, of the magnetic environment in which the mechanism operates. This may permit distinction between, for example, emerging flux models and twisted arch models or between mechanisms driven by currents parallel to the magnetic field, as opposed to perpendicular. High time resolution observations, on the other hand, may set limits on instability growth rates, imply the occurrence of repetitive or multiple dissipation, indicate the production sequence of the various flare manifestations and set an upper limit to the size of the primary dissipation site. Transport of energy away from the primary sites, and its ultimate thermalization, depends not only on the primary mechanism itself but also on the larger-scale structure of the active region atmosphere in which the transport occurs. Consequently the study of energy transport as a diagnostic of flare mechanisms involves extensive theoretical modelling of the transport processes, as well as observational input, to provide the framework for interpreting the observations.

The topic most thoroughly studied is that of the transport of nonthermal electrons. The thick-target electron beam model, in which electrons are presumed to be accelerated in the corona and typically thermalized in the chromosphere and photosphere, is supported by observations. At the higher energies, the anisotropy of gamma-ray emission above 10 MeV clearly indicates that these photons are emitted by anisotropically-directed particles. The timing of this high-energy gamma radiation with respect to lower-energy hard X-radiation implies that the energetic particles have short lifetimes. For collisional energy loss, this means that they are stopped in the chromosphere or below. Stereoscopic observations at hard X-ray energies (up to 350 keV) imply that these lower-energy nonthermal electrons are also stopped deep in the chromosphere. Hard X-ray images show that, in spatially resolved flares whose radiation consists of impulsive bursts, the impulsive phase starts with X-radiation that often comes mostly from the footpoints of coronal loops and microwave radiation from their tops. Combined hard and soft X-ray spectra suggest the presence of nonthermal electrons early in the impulsive phase. The thick-target electron-beam model accounts for the close temporal coincidence of UV and hard X-ray intensity. White-light emission from the largest flares is closely related temporally and energetically to the nonthermal electrons that produce deka-keV X-rays. However, white-light spectra imply that commonly such emission originates in the upper photosphere or temperature minimum region while deka-keV electrons are stopped in the upper chromosphere. Both timing and spectra of Hα emission imply that energetic nonthermal electrons heat the chromosphere during the impulsive phase. Finally, the interpretation of microwave spike bursts requires a level of microwave maser emission that dramatically enhances the precipitation of thick-target electrons.

Although the thick-target nonthermal-electron model meets with some success, it has significant problems. For example, the thick-target model appears to predict much more EUV emission, in proportion to hard X-ray emission, than is observed. In some flares, white-light emission is observed well beyond the end of the impulsive phase, which appears inexplicable in the thick-target model; such emission appears to correlate more strongly with the thermal X-ray emission of the hot ($T > 2 \times 10^7$ K) thermal phase.

CHROMOSPHERIC EXPLOSIONS

The work of this team addressed the question of the response and relationship of the flare chromosphere and transition region to the hot coronal loops that reach temperatures of about 10^7 K and higher. Flare related phenomena such as surges and sprays were also discussed. The team members debated three main topics:1) whether the blue-shifted components of X-ray spectral lines are signatures of "chromospheric evaporation" (evaporated plasma is defined as cool chromospheric plasma that is heated to multimillion degree temperatures and therefore contributes to the X-ray emission of the soft X-ray flare. In the process of heating it moves upward from the chromosphere into the coronal flux tubes); 2) whether the excess line broadening of UV and X-ray lines is accounted for by "convective velocity distribution" in evaporation; and 3) whether most chromospheric heating is driven by electron beams. These debates illustrated the strengths and weaknesses of our current observations and theories.

Observations with crystal spectrometers on SMM, P78-1, and HINOTORI have shown that during the rise phase of flares the profiles of X-ray lines exhibit a blue-shifted component, characteristic of velocities between 100 and 500 km s^{-1}. The intensity of the blue-shifted component is usually much less than the intensity of the non-Doppler shifted, or stationary component, throughout the rise phase of flares. The two most important questions regarding the blue-shifted component are concerned with its origin, and its relationship to the stationary SXR emission. Based on the presence of the blue-shifted emission during the rise phase of many of the larger flares observed with BCS, the blue-shifted emission was interpreted as chromospheric evaporation, with the primary flare energy release occurring in coronal flux tubes. Part of this energy would be subsequently deposited by either conduction or transport by high energy particles into the chromosphere. This would result in heating the chromosphere to multimillion degree temperatures. The heated plasma would move upward into the flaring flux tubes, and the Doppler effect would cause the X-ray emission from this upwelling plasma to be blue-shifted relative to stationary plasma already present in the flux tubes. In this picture, the ablated or evaporated plasma is the main cause of the increase in emission measure of the soft X-ray flare. This point of view was not shared by some team members who found discrepancies between the observed characteristics of the blue-shifted emission and what is predicted by numerical simulations of chromospheric evaporation. Also, certain previous observations from SKYLAB are not in apparent agreement with predictions. The debate focused, among other things, on the timing and magnitude of the blue-shifted emission relative to emission from stationary plasma.

The second topic of debate is also centered around the idea of chromospheric evaporation, and is based upon the finding that the soft X-ray spectral lines from plasma at temperatures between 6×10^6 and 3×10^7 K are broadened during the rise phase of SXR flares. The origin of this broadening of spectral lines is unknown, but is reasonably assumed to be due to mass motions and not an ion temperature that is much higher than the electron temperature. This nonthermal broadening could be caused by true plasma turbulence or by spatially non-uniform heating in flux tubes which produces non-uniform velocity fields over the regions of spectral line formation. This latter process was referred to as convective evaporation, although the term is not very precise. The team debated the question of whether convective evaporation was a viable line broadening mechanism. The debate focused on single loop and multiple loop numerical simulations and the effects of the velocity fields on X-ray and UV line profiles.

The third topic debated is concerned with the relative importance of electron beams and conduction fronts in heating the flare chromosphere. There are significant differences between the two types of flare heating, and it is not clear whether conductive heating or beam heating provides the closest correspondence to observational data. Is the energy released in the chromosphere by a beam sufficient to drive chromospheric evaporation and produce the initial emission measure growth of the SXR flare? The debate focused on such issues as the differences in appearance of Hα line profiles produced by conduction front and beam heating, and the results of numerical simulations that assume beam heating and conduction heating. One interesting result of the numerical simulation is that beam heating simulations usually produced much higher evaporative velocities than conduction front simulations. The effects of beam and conduction fronts on the transition region and chromosphere are of fundamental importance to the understanding of energy transport in flares.

FLARE ENERGETICS

The work of this team was concerned with the investigation of flare energetics, identifying the major sources and sinks of flare energy, together with the mechanisms by which the energy flows from one form to another. To achieve this goal, the team chose five flares of different types that were well observed with instruments on the SMM, and with other space-borne and ground-based instruments. The team sought to study several key questions: 1) the characteristics of the impulsive phase and whether all flares have an impulsive phase; 2) the total energy content in the impulsive phase, and the relative proportion of energy in the impulsive and gradual phases of a flare; 3) the relative importance of thermal and nonthermal components in the impulsive phase; and 4) the need for providing continual energy input to post-flare loops.

1) The classical definition of the impulsive phase depends on the presence of spiky (time scales < 10 s) hard X-ray bursts, temporally correlated with microwave bursts. The microwave post-burst increase is equivalent to the X-ray gradual phase and its source is identified with the thermal soft X-ray source. It is clear that this thermal source coexists with the energetic electrons of the impulsive phase. It appears that most flares have an impulsive phase with the qualification that the absence of hard X-rays may be no more than a threshold effect within the bounds of the observations.

2) The total energy in electrons above 25 keV calculated from the hard X-ray spectra assuming thick-target interactions ranges from 10^{29} to 10^{31} ergs; alternatively, thermal energy at the time of the peak hard X-ray flux calculated from the hard X-ray spectrum assuming a source with a temperature $> 10^8$ K is $\sim 10^{29} - 10^{30}$ ergs. Thermal energy at the time of the peak soft X-ray flux computed from a multi-thermal analysis with $10^7 < T < 5 \times 10^7$ K and the HXIS source area ranges from $\sim 10^{29}$ to 5×10^{30} ergs. The increase in the thermal and turbulent energy of the plasma plus the total radiated and conducted energy losses up until the time of the peak in soft X-ray flux ranges from $\sim 10^{30}$ to 10^{31} ergs. The peak energy of the turbulent plasma motions is $\sim 10^{29}$ ergs, and the radiant energy in Hα integrated over the duration of the impulsive phase is $\sim 10^{29}$ ergs. The computation of the relative proportion of energy in the impulsive and gradual phase is model-dependent because of the ambiguity of the thermal/nonthermal question and missing information in each case. In the nonthermal case the major problem is the determination of the low energy cut-off in the electron spectrum. In the thermal case it is the determination of the density. In both cases, the uncertainty in the filling factor limits the precision with which the energy in the soft X-ray plasma can be determined. With these uncertainties, the team gave two answers based on the thermal and nonthermal impulsive phase energies and the gradual phase thermal energies. In the thermal case, the impulsive phase tends to be relatively unimportant and the main flare energy release occurs gradually throughout the duration of the flare. In the nonthermal case, it is possible that the impulsive phase contains a large fraction of the total flare energy.

3) At this stage of our knowledge it does not seem possible to determine the relative importance of the thermal and nonthermal components of the impulsive phase of a flare, except to say that if the hard X-rays are produced in thick target interactions by nonthermal electrons accelerated during the impulsive phase, the electrons have a dominant energetic role, and that the thermal sources of soft X-rays are only one of several subordinate effects produced by this inherently non-thermal energy release.

4) In large, two-ribbon flares such as the 1980 May 21 flare, the reduction of the radiative cooling time due to filamentary fine structure would worsen the discrepancy between the observed long cooling time and the predicted shorter time. Thus, the SMM data confirm the need for continued energy input in the late phases of such flares. The microwave data also appear to support continual energy input.

CORONAL MASS EJECTION AND CORONAL STRUCTURES

The work of this team was concerned with modelling of post-flare arches, the reconnection theory of flares, the slow variation of coronal structure, and the coronal and interplanetary detection, evolution, and consequences of mass ejections. Post-flare arches are a newly-discovered phenomenon, wherein a very large coronal loop appears to undergo energization after a flare, thus allowing it to shine more brightly in soft X-rays. Some post-flare arches seem to be re-energized in nearly homologous fashion after each of a sequence of flares, without suffering significant disruption. Analysis of coronal mass ejections (CMEs) involved: case studies of individual events, in which it was attempted to provide the most complete descriptions possible, using correlative observations at diverse wavelengths; statistical studies of the properties of CMEs and their associated activity; interplanetary observations of associated shocks and energetic particles--even observations of CMEs traversing interplanetary space; and synoptic charts which show to what degree mass ejections affect the background corona and how rapidly the corona recovers its pre-disturbance form. Theoretical investigations of CMEs involved attempts to simulate CMEs by introducing pressure pulses at the bases of isothermal coronas permeated by a variety of potental magnetic field configurations, and analytic, self-similar descriptions of propagating ejections. These theoretical attempts provide insights into the ways ejections might evolve as they ascended and the stability of pre-event coronal structures.

A number of interesting coronal transient events observed by C/P were studied by the team. The observations range from the inner corona as low as 1.2 R_0 with the Mauna Loa (MLO) K-coronameter to reconstructions of transient brightness distributions at 0.3 AU with the Helios spacecraft. Two events observed with C/P showed unusual features that might offer insight into the physical structure and processes occurring in transients. One of these is a "disconnection" event that was interpreted as a pinching-off of a transient loop, so that the magnetic fields threading the transient no longer connected to the Sun. The other showed features that expanded in a self-similar fashion, as predicted by Low (1982). Five other events were studied because of the existence of meter-wave radio observations which indicate the presence of energetic electrons and/or shock waves in the same range of coronal heights as traversed by the transient.

The case studies include an example of a dark or depletion transient (at 1.2 R_0) which acquired a bright rim as it rose through the corona. Thus, this depletion transient (perhaps the ascent of the cavity often seen around a prominence at the limb) became a bright coronal mass ejection in an orbiting coronagraph's field-of-view. In this case, it was clear that the excess mass ejected was not raised from the chromosphere but, rather, started in the corona. This was predicted from the SKYLAB observations. As mentioned above, in one event there is a strong suggestion of magnetic disconnection from the Sun; that is, there is a rising, outwardly concave, intensity-enhanced structure. Similar structures were seen in perhaps as many as 10 percent of all C/P events. Another CME was studied which possibly resulted from the eruption of an arcade rather than a single loop. In this case, a moving Type IV radio burst was associated with a denser blob of rising plasma. The soft X-ray emission enhancement commenced 17 minutes before the earliest Hα activity and a forerunner similar to those reported for SKYLAB-era events surrounded the ejection. In some type II/type IV associated CME's, the type II source occurs below the top of coronal white light loops in the early stages and travels faster than the transient. Interplanetary magnetic clouds appear to be the likely manifestations of CMEs at 1 AU.

Some coronal mass ejections events--with or without accompanying flares--appear to start at a time coincident with weak, soft X-ray bursts. Such "precursor bursts" occur 10 to 20 minutes before the associated flare onsets. Rising, X-ray emitting counterparts of white light coronal mass ejections were detected at low heights (e.g., 1.14 R_0) for some events, also in association with the "precursor bursts".

Coronal mass ejections associated with flares show rapid initial acceleration, usually followed by a constant speed or deceleration. By contrast, coronal mass ejections associated with prominence eruptions tend to move more slowly and often are still accelerating at great heights in orbiting coronagraphs' field-of-view (6-10 solar radii).

Numerical model calculatons show that it is not possible to simulate a realistic coronal mass ejection by calculating the response of an atmosphere in hydrostatic equilibrium, permeated by an initially potential magnetic field, to a perturbing pressure pulse at its base. Self-similar solutions of the MHD equations were suggested as a good way to model the asymptotic behavior of coronal mass ejections far from their initiating sites.

Observations with the Helios satellites' zodiacal light photometers permitted "stereoscopic" views of three coronal mass ejections to be obtained. The observations showed that coronal mass ejections typically have a complicated 3-dimensional structure, a result supported by the SOLWIND coronagraph observations of "halo" CMEs which sometimes appear to surround the occulting disk completely. The values of speed, mass, and energy of CMEs observed during solar maximum (P78-1 data) appear to be similar to those observed during solar minimum (SKYLAB data), although the latitude of occurrences spread much higher.

Of 80 interplanetary shocks observed when Helios was within 30° of a solar limb and when the SOLWIND coronagraph was observing, 40 had good associations with coronal mass ejections, and another 19 had possible coronal mass ejections. The mass ejections associated with the shocks at Helios were generally faster, brighter, and at lower latitude than the typical CME; they tended to fill an arc of heliocentric latitude in the corona which encompassed the Helios-Sun line. Of 27 prompt proton events with Hα flares which were observed, 26 had associated coronal mass ejections. The 27th event appears to be a member of a new class of events associated with short-lived flares having gamma-ray and radio Type III bursts but no CME. By contrast, energetic particle events which were rich in ^{3}He(^{3}He/^{4}He > 0.2 at ~ 1.5 MeV/nucleon) were not well associated with radio bursts or coronal mass ejections.

Studies of coronal evolution showed that sometimes there were permanent changes in coronal structures due to coronal mass ejections; sometimes, coronal structures endured for more than a day, even in the presence of repeated flaring below them; often, there were changes which occurred over a period of hours. Type III storms tended to occur where the coronal structure appeared to be bundles of small, discrete rays; generally, the corona was brighter after a radio noise storm than before; it was suggested that the eruption of new flux at the base of the corona created the proper conditions for the noise storm and carried relatively denser plasma laterally, and upward to greater heights as well, thereby brightening the corona.

INTERCOMPARISON OF NUMERICAL MODELS OF FLARING CORONAL LOOPS

The numerical modelling group was concerned with computations relevant to the problem of the hydrodynamic and radiative response of a single magnetic flux tube to a sudden release of energy in it. The group initially considered a simple "Benchmark Model" in which the physics of real loops--radiation, thermal conduction, compressible hydrodynamics, gravity and nonthermal heating could be incorporated with some degree of realism. The proposed Benchmark Model consisted of a one-dimensional magnetic flux tube containing a low-beta plasma in which the field strength was so large that the plasma could move only along the flux tube and whose shape remained invariant with time. The flux tube cross section was taken to be constant over its entire length. The flux tube was considered to have a semi-circular shape, symmetric about its midpoint $s = s_{max}$ and intersecting the chromosphere-corona interface (CCI) perpendicularly at each footpoint. The arc length from the loop apex to the CCI was 10,000 km. The flux tube extended an additional 2000 km below the CCI to include the chromosphere, which initially had a uniform temperature of 8000 K. The temperature at the top of the loop was fixed initially at 2×10^6 K. The plasma was assumed to be a perfect gas (γ = 5/3), consisting of pure hydrogen, fully ionized at all temperatures; and the electron and ion temperatures were taken to be everywhere equal at all times.

The anticipated dynamical response of the loop atmosphere to a transient heating function showed that nearly all of the flare energy is deposited in the corona. This leads to a rapid rise in the coronal temperature from its initial value on a time scale given by $\tau_F = 3N_ekT/E \cong 0.02$ s, which is much less than the acoustic transit time for the loop: $\tau_a = s_{max}/c \cong 50$ s, where c is the velocity of sound. Thus, much of the heating takes place before substantial mass motions can occur. The temperature rise is most rapid near the loop top (where the heating is strongest), and this drives a supersonic thermal wave downward along the loop. When this conduction front reaches the top of the chromosphere, the resultant sudden heating of the cool plasma there causes an expansion in both directions along the flux tube. The downward-propagating pressure wave rapidly steepens to form a shock, which ultimately overtakes the thermal wave as both move deeper into the chromosphere. At the same time, the upward-moving (evaporated) chromospheric plasma pushes a weaker pressure wave ahead of it into the corona. This wave may or may not have time to steepen into a shock before reaching the top of the loop. The loop soon becomes filled with hot and dense matter.

The primary goal of the Benchmark Model was to intercompare code calculations on a standardized, although hypothetical, problem, rather than to establish the best possible physical model. Although the group did not converge upon a unique Benchmark Model, it succeeded in defining a well-posed Benchmark Problem, and this was considered a necessary first step. The most valuable exercise of the group was the general recognition within the flare-modeling community of certain difficulties associated with trying to model the energetic flare process numerically. In particular, the importance of confronting directly the difficult numerical problems associated with the rapid motion of a very steep thermal wave front through the chromosphere, as well as that of the extremely thin, dense compression wave that runs ahead of it was recognized.

CHAPTER 1: PREFLARE ACTIVITY

CHAPTER 1: PREFLARE ACTIVITY

TABLE OF CONTENTS

E.R. Priest, V. Gaizauskas, M.J. Hagyard, E.J. Schmahl
and D.F. Webb

Page

TABLE OF CONTENTS (continued)

TABLE OF CONTENTS (continued)

CHAPTER 1: PREFLARE ACTIVITY

E.R.Priest, V. Gaizauskas, M.J. Hagyard, E.J. Schmahl
and D.F. Webb

1.1 INTRODUCTION

The Preflare Activity Team at the Solar Maximum Mission Workshops was split into three groups. The first group, concerned with Magnetohydrodynamic Stability (see Section 1.2), was led by E.R. Priest. Its members were A. Aydemir, F. Brunel, P. Cargill, T. Forbes, A. Hood, J. Melville, B. Schmieder, R. Steinolfson, and G. Van Hoven. The second group discussed Preflare Magnetic and Velocity Fields (see Section 1.3). It was led by M. Hagyard and included G. Chapman, A. de Loach, F. Drago, A. Gary, B. Haisch, W. Henze, V. Gaizauskas, H. Jones, J. Karpen, M. Martres, J. Porter, E. Reichmann, B. Schmieder, G. Simon, J. Smith, Jr., and J. Toomre. The third group considered Coronal Manifestations of Preflare Activity (see Section 1.4) under the leadership of E. Schmahl and D. Webb. Its members were R. Bentley, F. Drago, S. Enome, V. Gaizauskas, R. Harrison, G. Hurford, B. Jackson, T. Kosugi, M. Kundu, K. Lang, A. Magun, P. Martens, G. Pneuman, E. Reichmann, A. Schadee, J. Schrijver, B. Schmieder, G. Simon, J. Smith, Jr., K. Strong, P. Waggett and B. Woodgate.

This long list of participants with very diverse interests underscores the subtleties in the ways the Sun prepares for a major flare. Nevertheless, they worked constructively together and have all learnt a great deal about the complexities of preflare solar activity. The authors of this chapter are extemely grateful to their group members for helping in its construction.

The main features of preflare activity such as soft X-ray heating and filament activation are well-known, but we are nowhere near understanding them fully. We do not know the necessary and sufficient conditions for a flare and still cannot predict them with much confidence. Yet this must be our aim -- just what is it that gives rise to this bewildering event? Great progress has certainly been made recently. For example, ten years ago it was felt that a flare was due to a magnetic instability of some vague kind and that a current concentration such as a current sheet was needed to release the energy fast enough, since the diffusion and tearing times characteristic of an active region as a whole are years and days, respectively. Now, however, the detailed stability of a loop and an arcade have been calculated (Section 1.2.6) and the ways that magnetic fields reconnect at current sheets have been studied analytically and numerically (Section 1.2.1). Much has been learned empirically about the role of velocity fields which stress the lower solar atmosphere during the birth and growth of active regions

(Section 1.3). Measurements made with the vector magnetograph now permit the calculation of electric currents in the stressed areas (Section 1.3). Preflare changes in magnetic fields at coronal heights are directly observable as polarization changes in spatially resolved microwave structures (Section 1.4). The precise location in the corona of the initial flare outbursts can now be determined from VLA microwave maps and from HXIS maps in hard X-rays. And during flare onset, there are indications of preflare energy released by non-thermal as well as by thermal processes (Section 1.4).

1.1.1 The Preflare State -- a Review of Previous Results

A comprehensive summary of our earlier understanding of the preflare state was given by Van Hoven et al. (1980) at the Skylab Workshops [see also Svestka (1976) and Priest (1981)]. The basic philosophy in preflare studies presumes that free energy is stored in stressed magnetic fields, although the electric currents can be extrapolated from photospheric measurements of the magnetic field rather than being directly observable. The energy released during a flare was assumed to be by magnetic reconnection of the coronal fields in narrow current sheets; with Skylab, the coronal loops and arcades were seen for the first time. It was well-known that flares occur usually near a polarity inversion line when an active region is changing rapidly. The flare start was defined as the beginning of the flash phase (the rapidly rising or impulsive part of the event). Theoretical studies had concentrated on the storage of energy in the chromosphere and corona in a twisted loop or a sheared arcade in response to the slow passive evolution of footpoints. The cause of the preflare heating was unknown, as was the means of destabilising the magnetic configuration, although it had been suggested that instability might occur when the shear reaches a critical value. The necessity to demonstrate that the marginal state is meta-stable and the instability explosive was appreciated. In a parallel development to the Skylab studies, the Emerging Flux Model (Heyvaerts, Priest and Rust, 1977) had been proposed to explain a class of flare-related erupting filaments. It suggested that energy for large flares was stored in a sheared arcade and that its release was triggered by emerging flux.

Ground-based observations had long ago given evidence for preflare magnetic changes. Evolving magnetic features (emerging flux or satellite sunspots) were often seen near a polarity inversion line. Photospheric velocity measurements showed that flares occur near reversals in the line-of-sight velocity and sometimes near vortex patterns. Also, the presence of strong shear (which tends to promote filament formation) and of large spot motions were common elements in the preparation for a large flare. Another precursor was the expansion of arches as seen at 5303Å. The eruption of an active region filament was recognized as an important feature of a two-ribbon flare. First, there is an increase in absorption in the wings and centre of Hα, with the filament crest rising at a few km s^{-1}. Then Hα changes from absorption to emission. The filament splits into fragments and (partially) disappears, with plasma falling down along its legs. Filament activation can begin as much as 3 hours before or as short as 15 minutes before flare start; filament acceleration is largest between 0 and 10 minutes before Hα onset.

Spacecraft observations had revealed a preflare soft X-ray rise in 80% of flares for a few minutes before flare onset accompanying a filament activation, although in 30% of cases the soft X-rays were thought to begin after Hα. A preflare EUV enhancement had been found 30 minutes before flare start from a small point near some emerging flux, while soft X-rays were emitted along the whole length of a filament.

At the Skylab Workshops Van Hoven et al. considered various flare precursors. Coronal transients had forerunners containing 10-20% of its mass which could be traced back to an origin before any possibly associated flare. Also type III radio bursts were twice as common as normal in the last few hours before flare onset. Fluctuating UV bright knots seen by the astronauts were thought to be loop footpoints. In one example a flare occurred after flux emergence below a pre-existing EUV loop, while in another a flare took place in a pre-existing X-ray loop. Sometimes EUV bright points were present. Events with no preflare X-ray rise were often small and there was little evidence for a preflare rise in subflares. Both X-rays and EUV showed preflare brightenings for 2-20 minutes before flare start in two out of three cases, although often displaced from the flare site. Thus a preflare coronal enhancement was common although not always in the later dominant loop. However, it was not possible to relate the character of the slow X-ray rise (its duration or magnitude) to that of the subsequent flare.

Van Hoven et al. studied the flare of 5 September 1973 in particular detail. Several satellite spots emerged over a few days around the main sunspot, producing a series of small flares. Before the biggest flare of the series, the spot rotated with a speed of 20-100 ms^{-1} producing a possible energy increase of 2×10^{31} erg. A small filament developed at the subsequent flare site. It darkened an hour before the flare and ascended with a speed of up to 14 km s^{-1}. Downflow was seen in its legs and a soft X-ray enhancement at its crest. In Hα and Ca II a faint arc propagated away perhaps representing the projection on the disc of a coronal transient. In EUV some compact bright knots smaller than 5 arc sec appeared and disappeared on time scales of minutes. The flare began in an EUV loop which became visible in X-rays and was subsequently replaced by a loop that was higher and less inclined to the arcade axis.

The theorists in Van Hoven's team studied non-linear force-free solutions for coronal arcades and found that there are no nonlinear static solutions (equilibrium states) when a parameter λ (related to the axial magnetic field) exceeds a critical value. They also set up models for the thermal and magnetic structure of coronal loops. These exhibit nonequilibrium or magnetic instability when the loop is twisted too much.

There have been many advances in our preflare knowledge since Skylab days. In MHD stability much work has been done on the theory of ideal (non-resistive) and resistive modes. In X-rays the HXIS and XRP instruments, by virtue of spatial and spectral resolution, have revolutionised our understanding of the flare. At radio wavelengths we have seen active regions resolved in great detail with the VLA. In the transition region the details of the UV flows have been studied and magnetic fields (of strength 1000 Gauss) have been measured for the first time with UVSP. For filaments, detailed velocity maps have been constructed and associated EUV arcades have been seen a few hours preflare. Also, the preflare heating of filaments, and vector magnetic field have been measured. At the photosphere and chromosphere the importance of velocity and magnetic shears and of emerging flux has been underlined.

1.1.2 Some Questions

During the present workshop series many questions have been raised, and some are at least partially answered in the rest of this chapter. They include the following:

MHD instability - Is the preflare equilibrim structure a loop or an arcade? What changes in physical conditions lead to instability? Is the basic instability ideal or resistive? What is the threshold? Can preflare changes occur such that repeated "tries" are needed to exceed the threshold? What is the best way to model photospheric line tying? Is the filament just a tracer of magnetic field lines or crucial to the instability? How do we model an active region when linear force-free fields are inappropriate?

Soft X-rays - What causes the preflare rise in soft X-rays? Is it due to a thermal instability, joule heating, magnetic reconnection, or a response to a non-thermal mechanism? Does one type of process first energize the corona with another responsible for the flare phenomenon? Is the preflare gradual phase energized in the same way as the postflare phase?

Radio - What are the preflare signatures? How are they related to those of other wavelengths? Are there preflare non-thermal bursts? What are the coronal magnetic fields and what are their relevant preflare changes? What is the β of the coronal plasma? How does the corona respond to emerging flux?

Ultra-violet - What are the properties of the brightness fluctuations and mass flows? What causes them? How are they related to changes in Hα, emerging flux, and/or footpoint motions of filaments or loops?

Filaments - How do they form? What is their magnetic structure? What drives the observed flows? What happens during a filament activation? What are the causes of filament eruption? What is the nature of the coronal environment around the eruptive filament? Is there a difference between filaments -- those activated in the presence of new magnetic flux and those which are not? Are there two distinct classes of precursor -- with and without active filaments?

Velocity and Magnetic Shear - How are they related to each other and to flare productivity? What is the role of the resulting electric currents? Is there a critical value of shear for eruptive instability?

What are the preflare characteristics of the velocity field and how do they evolve?

Emerging Flux - What are the necessary and sufficient conditions for it to trigger a large flare? What is different in the numerous flux emergences which produce no large flares? What are typical velocities, magnetic field strengths and rates of growth? Are they related to particle acceleration?

1.2 MAGNETOHYDRODYNAMIC INSTABILITY

E.R. Priest, P. Cargill, T.G. Forbes, A.W. Hood
and R.S. Steinolfson

1.2.1 Magnetic Reconnection

Our basic understanding of magnetic reconnection has changed recently due to the beginning of detailed numerical experiments on various aspects of the process (Priest 1984a and 1984b). These have linked the two previous strands of reconnection theory, namely tearing mode instability and Petschek-Sonnerup reconnection (as described below), and have presented us with new surprises (Section 1.2.2, 1.2.3).

1.2.1.1 Linear Tearing Modes

A current sheet of width d is spontaneously unstable to the linear tearing mode (Furth et al., 1963), which creates long thin magnetic islands by reconnection on a time-scale

$$\tau \sim (\tau_d \tau_A)^{1/2} \qquad (1.2.1)$$

where $\tau_d = d^2\eta$ is the resistive diffusion time and $\tau_A = d/v_A$ is the Alfven time in terms of the Alfven speed (v_A). For the active-region corona with global length-scales(d) of typically 10^3-10^4 km, the tearing mode growth-time (1.2.1) is days to weeks and is therefore much too long to explain a flare although it may well be important for normal coronal heating, (Heyvaerts and Priest 1985; Parker 1984).

Tearing may also take place in a sheared magnetic field such as a flux tube. However, in solar coronal applications it is important to incorporate the stabilising effect of photospheric line tying, since the footpoints of coronal magnetic field lines are anchored in the dense photosphere. This has led to suggestions that the resistive modes be completely stable in a loop (Mok and Van Hoven 1982) or in an arcade (Hood 1984a, Migliuolo and Cargill 1983) unless there is a reversal in the axial (loop) or azimuthal (arcade) field component.

An important new development is the discovery of a much faster radiative tearing mode (Van Hoven et al., 1982, Steinolfson 1983, 1984a, 1984b). Steinolfson and Van Hoven have solved the normal incompressible resistive MHD equations but they have allowed the magnetic diffusivity to depend on temperature $\eta = \eta_0 T^{-3/2}$), which introduces a coupling to the energy equation

$$\frac{nk_B}{\gamma-1}\frac{dT}{dt} = \nabla \cdot (\kappa \mathbf{B}\ \mathbf{B}\cdot\nabla T) - R_\rho{}^2 T^{-\alpha} + \mu_0 \eta(T) J^2, \qquad (1.2.2)$$

and produces resistive field changes on energy-transport time scales.

The predicted growth rate of linearly unstable modes is shown in Figure 1.2.1. At coronal values of the magnetic Reynolds number S (typically $S_s = 10^{10.6}$ for $T = 10^6$ K, $n = 10^{17}$ m^{-3}, B = 100 G, a = 100 km) there are two distinct modes, namely the tearing mode and the radiative mode, which is typically a hundred times faster. However, both modes are modified significantly. Local cooling at the X-points increases the magnetic diffusivity and so enhances the reconnection for the tearing mode. In the radiative mode a considerable amount of reconnection is present (Figure 1.2.2): the island width is typically 30% of that produced by the tearing mode and the perturbed magnetic energy is typically five times the perturbed thermal energy.

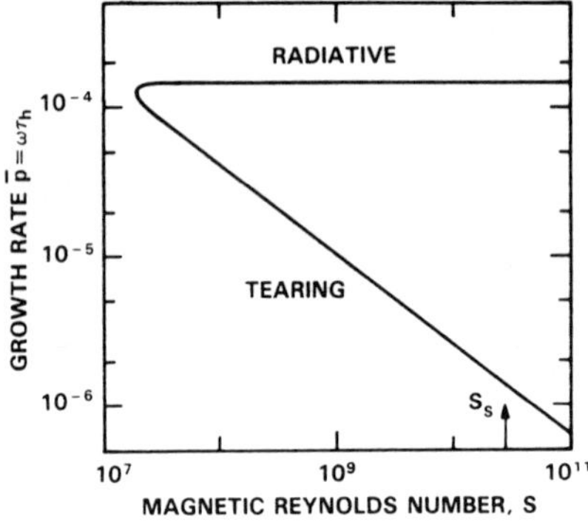

Figure 1.2.1 Growth rate(ω) for radiative tearing instability in units of the Alfven travel time ($\tau_A = a/v_A$) as a function of the magnetic Reynolds number ($S = \tau_d/\tau_h$), where $\tau_d = a^2/\eta$ is the diffusion time. Here the dimensionless wavenumber (ka) is 0.1, corresponding to a wavelength of 10π times the shear length (a), (From Steinolfson and Van Hoven 1984).

The inclusion of compressibility is found to be unimportant for solar coronal conditions. It inverts the tearing temperature at very long wavelengths and increases the radiative rate by typically a factor of five at very short wavelengths. Steinolfson (1984) includes perpendicular thermal conduction which introduces spatial temperature oscillations normal to the tearing surface and on a scale comparable with the width of the resistive tearing layer. Such thermal ripples create velocity oscillations but don't affect the magnetic field.

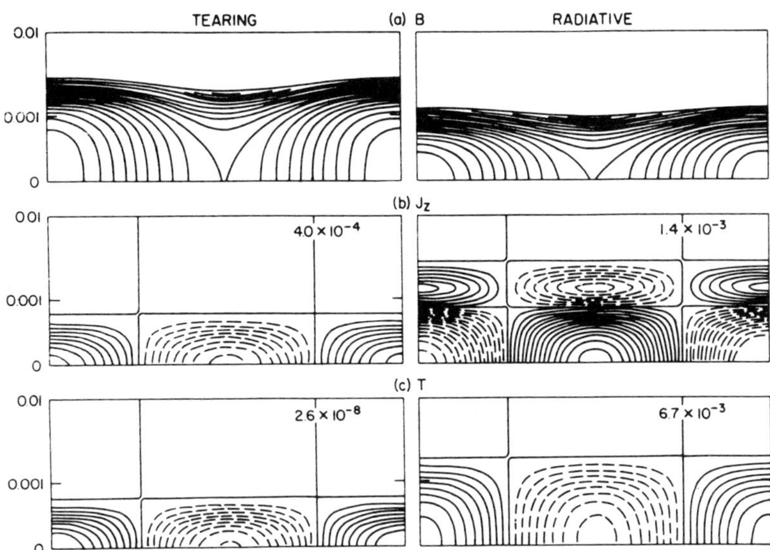

Figure 1.2.2 A comparison of the magnetic field (B), electric current (J_z), temperature (T) and ohmic dissipation ($\mu_0 \eta j^2$) for the tearing and radiative modes at $S = S_s$. Contours are normalised with respect to the maximum values indicated in the top right of each box, and the y scale is expanded. Dashed (and solid) contours represent values that are below (and above) the initial values. (From Steinolfson and Van Hoven, 1984).

Steinolfson (1983) derives analytical expressions for the growth-rates in the constant-ψ and long-wavelength approximations. He finds that for S smaller than about 10^6 the modes are generally stabilised. Also, he discusses the Joule heating instability which is present in the absence of radiation.

1.2.1.2 Petschek-Sonnerup Reconnection

The second main theme of reconnection theory has been the fast nonlinear mode of steady Petschek-Sonnerup reconnection (Petschek 1964, Sonnerup 1970), which has been reviewed many times (e.g., Priest 1984b, Vasyliunas 1975). In this fast nonlinear state the current sheet bifurcates into two pairs of slow shock waves, which exist because the inflow plasma speed exceeds the slow magnetoacoustic wave speed. The shocks are standing in the flow and, as the plasma and magnetic field lines pass through them, they have the effect of transferring inflowing magnetic energy into the heat and kinetic energy of hot fast jets.

In general one would expect the external boundary conditions at the sources of the inflowing plasma to produce a hybrid Petschek-Sonnerup regime. Particular forms of those boundary conditions (namely free or fixed corner conditions) may, however, produce the pure Petschek or Sonnerup extremes, respectively (Vasyliunas 1975). The difference between the two extremes is as follows. The Petschek mode has a pure fast magnetoacoustic expansion in the inflow regions upstream of the slow shocks, such that the flow converges and the magnetic field strength decreases as the central diffusion region is approached. By contrast, the inflow region for the Sonnerup mode consists of a slow mode expansion with the flow diverging and the field strength increasing. Although in Sonnerup's original analysis the slow mode expansion fan was very thin and generated at a single point in the inflow, Sonnerup reconnection here refers also to the more general situation with a wide fan and generation across a substantial part of the inflow region.

When the reconnection develops locally from the tearing of a sheared magnetic field, such as in the simulation of the Kopp-Pneuman model for main phase reconnection (Section 1.2.5), the nonlinear steady state is expected to be Petschek-like (Figure 1.2.9). When the reconnection is driven from outside, as in a simulation of emerging flux (Section 1.2.4), the nonlinear state can be closer to the Sonnerup regime (Figure 1.2.7).

Recent numerical experiments by Forbes and Biskamp have produced two main surprises. They have demonstrated that the tearing mode can develop in its nonlinear phase into the fast Petschek-Sonnerup mode (Section 1.2.2). They have also revealed some new regimes of fast nonlinear unsteady reconnection when the Petschek-Sonnerup mode breaks down or goes unstable (Section 1.2.3).

1.2.2 Nonlinear Tearing

The nonlinear development of the tearing mode is far from simple and not yet completely understood. Several pathways along which the instability may develop appear to be possible, depending on the geometry and the parameter regime, as outlined below.

1.2.2.1 Saturation

The first possibility is that the mode may saturate at an extremely low amplitude, when the island width has only grown equal to the resistive layer width (Rutherford 1973). This benign outcome with an extremely small energy release has been the most commonly expected development in laboratory devices. However, some recent calculations have been performed by Steinolfson and Van Hoven at large values of S ($\approx 10^6$) and at long wavelengths, conditions much more appropriate to solar applications than previous attempts (Steinolfson and Van Hoven 1983). At a wavelength of only twice the shear length (a) the reconnection is indeed found to slow down drastically, as in the Rutherford regime. But at wavelengths of 20a the nonlinear reconnection rate is ten times faster and the island grows enormously up to a width of 2a (see also Section 1.2.3.2).

1.2.2.2 Mode Coupling

In a magnetic flux tube, surfaces at different radii are unstable to modes with different values of m. m = 1 represents a simple kinking of the tube near the surface, with the cross-section remaining circular. For higher m values the cross-section becomes distorted: for instance, m = 2 perturbs the tube to a double-helix shape and m = 3 to a triple helix.

Normally, one expects several such modes to be present and, when they grow to a large enough amplitude, modes on neighbouring surfaces may couple to one another (Waddell et al., 1978).

Aydemir and Barnes (1984) have performed some numerical studies for a reversed field pinch which may be of relevance to coronal structures. Such a toroidal laboratory device possesses a toroidal field component (B_S) and a toroidal current (I_S) which produces a poloidal field component (B_θ) of the same order of magnitude.

Experimentally, an initially turbulent state leads to a spontaneous reversal of the field near the axis followed by a long quiescent phase. In this state the magnetic field is near to a constant-α force-free field which, according to Taylor's hypothesis (1974), minimises the magnetic energy subject to toroidal-flux and magnetic-helicity conservation. Heyvaerts and Priest (1984) have generalised the hypothesis and applied it to the corona in order to deduce the coronal heating and mini-flarings that are produced by tearing turbulence.

Aydemir has studied the relaxation and sustenance of the quiescent state using an incompressible 3D MHD code, with $S \approx 5 \times 10^3$. The torus is approximated by a periodic cylinder of length $2\pi R_O$, and so the variables are Fourier expanded in the axial and azimuthal directions.

Starting from an unstable equilibrium the nonlinear evolution is followed, with the dominant modes having $m = 1$ and $n = 2$ and 3. The system is driven in the sense that the toroidal current is maintained against resistive diffusion by an external source. For a 2D single helicity calculation, in which only those modes with a given ratio m/n are retained, a steady helical magnetic field is maintained in a laminar manner. For a 3D multiple helicity calculation, when two modes such as $m/n = 1/2$ and $1/3$ are perturbed, many other modes are generated by nonlinear coupling. Below a critical current (such that $\alpha a \approx 3$) the field evolves to a steady state, and above that value a quasi-steady state is reached with fluctuations about a mean value maintained by a turbulent dynamo for at least 1000 τ_A. Also, for some currents a bifurcation or frequency doubling is observed. Another important effect is that the magnetic flux surfaces break up and the field lines become stochastic, which produces a rapid increase in heat transport across the magnetic field.

1.2.2.3 Coalescence

In the linear regime the fastest growing tearing mode has a very long wavelength ($\approx S^{1/4}a$), and so in many cases only one magnetic island will form. Sometimes, however, the structure may be long enough for several islands to grow, and then, in the nonlinear regime, neighbouring islands may be attracted towards one another by an ideal mode known as the coalescence instability (Finn and Kaw, 1977, Pritchett and Wu, 1979). Being an ideal instability, unlike the tearing mode, this mode grows extremely rapidly on Alfvenic times.

The results of numerical simulation by Bhattacharjee, Brunel and Tajima (1983) are shown in Figure 1.2.3b for a plasma β of 0.02 and a magnetic Reynolds number of 10^3. They begin with two magnetic islands in equilibrium, which are assumed to have been created by tearing (first

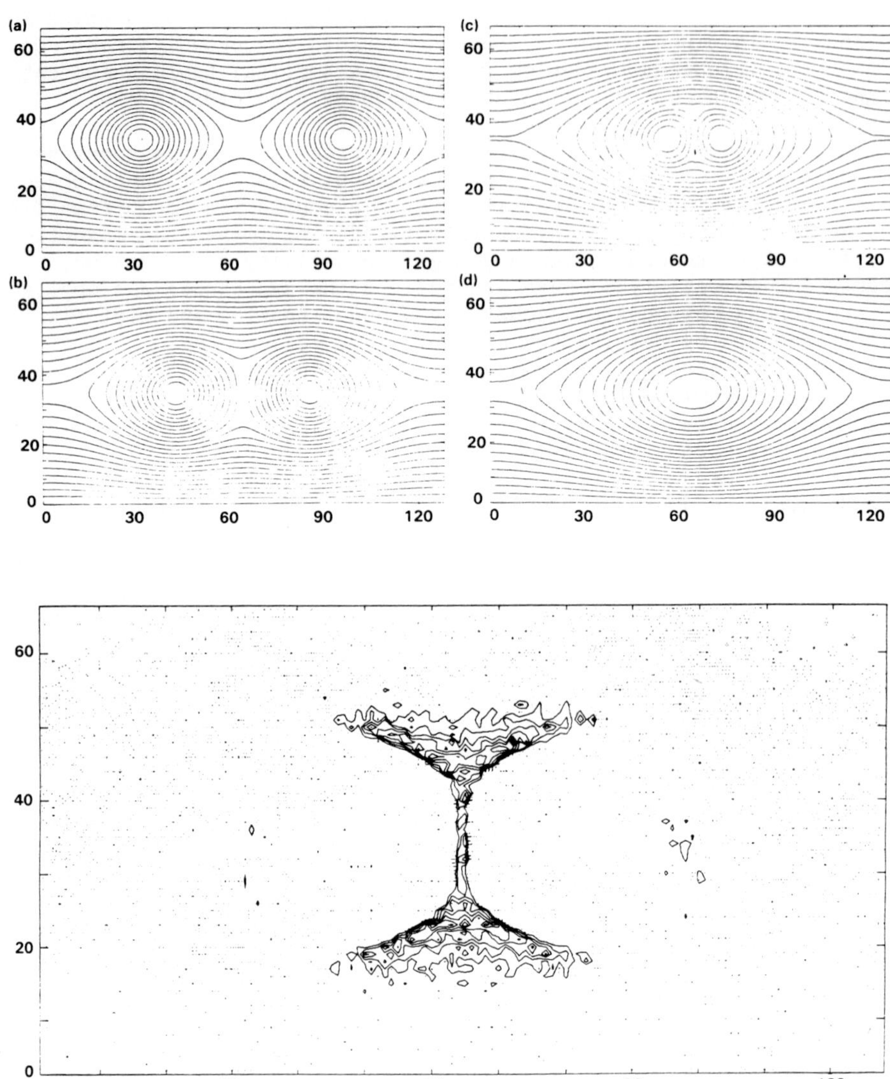

Figure 1.2.3 A numerical simulation of the coalescence instability showing (a) magnetic field lines at several times (b) plasma density contours. (From Bhattacharjee *et al.*, 1983).

frame). The two islands rapidly approach one aother and create an intense current sheet at the interface between them as the coalescence instability saturates (second frame). Then the two islands reconnect (third frame) and coalesce to form a single island (fourth frame), which oscillates in response to its violent birth. Plasma density contours at the beginning of the reconnecting phase (Figure 1.2.3b) at $t = 1.6L/v_A$, where $L = 128$ is the length of the system, suggest the presence of two pairs of slow shocks propagating from the ends of the central current sheet. Also, the large length of the central current sheet and the high speed of approach of the two islands suggest that this may represent a flux pile-up regime (Section 1.2.3.1).

1.2.2.4 Petschek-Sonnerup Reconnection

When the outflow boundary conditions are free enough and the inhibiting effect of the large tokamak axial field is absent, it is possible for the tearing mode to evolve nonlinearly into the fast steady state of Petschek-Sonnerup reconnection (Forbes and Priest, 1982, 1983a), as described in Section 1.2.1.2. A new discovery by Forbes is that fast-mode shocks may be present in the outflowing hot jets (Forbes and Priest, 1983a, 1983b). These have the effect of degrading the kinetic energy into heat and may be much more efficient at accelerating fast particles than the much thicker slow shocks. The steady Petschek-Sonnerup mode is possible when the inflow speed (v) of plasma at large distances is less than a maximum value, $v < v_{max}$, which depends on the magnetic Reynolds number and also on the external boundary conditions. For pure Petschek reconnection it is typically 0.01 v_A, but for pure Sonnerup reconnection it is roughly the Alfven speed (v_A).

1.2.3 Nonlinear Reconnection Experiments

1.2.3.1 New Regimes of Fast Reconnection

Recent numerical experiments at high magnetic Reynolds number by Forbes (Forbes and Priest, 1982, 1983a, 1983b) and by Biskamp (1982a, 1982b, 1982c) have revealed two new regimes of fast unsteady reconnection when the Petschek-Sonnerup mode breaks down (Figure 1.2.4).

The flux pile-up regime occurs when the inflow of plasma is so fast that ($v > v_{max}$) is violated, as for instance when reconnection is driven by an ideal instability such as coalescence or kinking. In this case the flux cannot reconnect as fast as it is brought in and so it piles up outside the central diffusion region and causes it to grow in length.

The impulsive bursty regime occurs when the length (L) of the central diffusion region in either the Petschek-Sonnerup or flux pile-up regime becomes too great

$$L > L_{max} \tag{1.2.3}$$

In this case the central sheet goes unstable to secondary tearing (on the tearing mode time-scale) and coalescence (on the Alfven time-scale). The result is a more rapid energy release in a series of bursts as the islands coalesce. This could be extremely important for particle acceleration and the impulsive energy release that is often seen in flares.

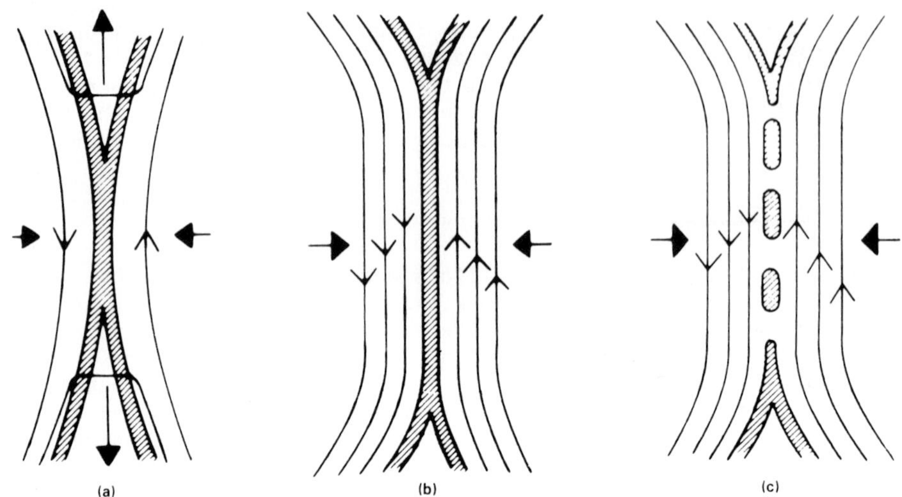

Figure 1.2.4 Regimes of Fast Reconnection: (a) Petschek-Sonnerup, (b) Flux Pile-up, (c) Impulsive Burst. (From Priest, 1984).

1.2.3.2 Nonlinear Tearing at High S

Steinolfson and Van Hoven (1984b) have investigated numerically the nonlinear evolution of the tearing mode at values of Lundquist number (S), (normally equal to the magnetic Reynolds number) between 10^2 and 10^6 and dimensionless wavenumber (α = ka) between 0.042 and 0.5. They find that the growth slows considerably from the linear rate, and at least 80% of the stored magnetic energy is converted into thermal energy for the long wavelength modes with ka < 0.5. Also the maximum electric fields are about three orders of magnitude smaller than the Dreicer field. However, other features depend on the k-S parameter regime, as follows.

The incompressible resistive MHD equations are solved with temperature (T) and magnetic diffusivity (η) assumed constant. The initial state is taken to be one isolated wavelength of a linear oscillatory mode extending from the centre (x < 0) of one island to the adjacent X-point (x_{max}) and from the tearing surface (y = 0) to a large distance (y_{max}) such that the perturbation is negligible and is decaying exponentially with y. Symmetry boundary conditions are applied at x = 0, x = x_{max}, y = 0, and a non-uniform grid is used in the y-direction with a concentration of grid-points near y = 0 in order to resolve the resistive layer.

Figure 1.2.5 plots the reduction $\Delta E_m = E_{mo} - E_m(t)$ in the magnetic energy $E_m(t)$ stored in the shear layer, where $E_{mo} = E_m(0)$. Except for the constant ψ solution, the energy that has been released by the end of the computation is between 8% and 27%. It can also be seen that the longer

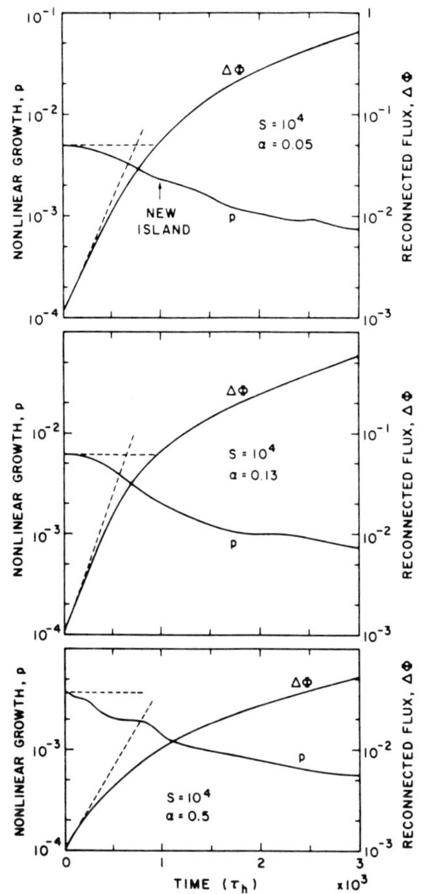

Figure 1.2.5 a) Nonlinear growth rate (p) and reconnected flux ($\Delta\phi$) as functions of time in units of the Alfven time ($\tau_h = a/V_A$) for the tearing instability of one island at a Lundquist number (s) of 10^4 and several values of the dimensionless wavenumber ($\alpha = ka$).

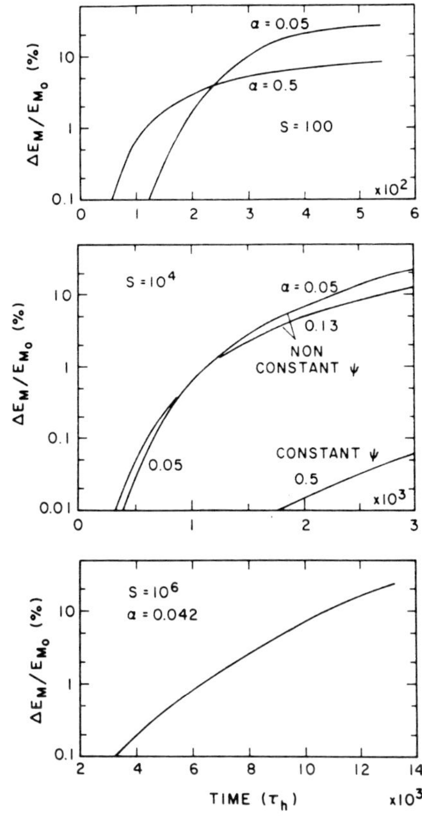

Figure 1.2.5 b) Temporal behavior of the percentage of magnetic energy removed from the shear layers (from Steinolfson and van Hoven 1984).

the wavelength ($\lambda = 2\pi a/\alpha$), the more magnetic energy is ultimately converted, even though the conversion rate is slower at first.

Figure 1.2.6 shows the formation near the X-point of a secondary flow vortex in the opposite sense to the initial linear vortex. When S and the wavelength are large enough the secondary flow can create a second magnetic island near the original X-point. Also, one finds intense current filaments and electric fields near the X-points. At the same time for long wavelengths ($\alpha = ka = 0.05$) the width of the magnetic island grows to more than twice the initial shear layer width (a) by the end of the computation, even though nonlinear saturation has not yet occurred.

1.2.4. Emerging Flux and Moving Satellite Sunspots

1.2.4.1 Their Two Roles

Small regions of emerging flux and small satellite sunspots are often observed before flares. They signify the interaction of separate magnetic flux systems, in the first case by means of a vertical motion and in the second case via a horizontal motion, but in either case the effect is similar, namely the pressing of one flux system against another and the creation of a current sheet at the interface at some height h.

The first role of such flux evolution is to create small flares when the current sheet reaches a critical height such that the current density exceeds the threshold for the onset of microturbulence. This has been estimated by solving the energy balance equation within the sheet and so deducing the resistivity from the temperature (Heyvaerts, Priest and Rust 1977, Milne and Priest 1981).

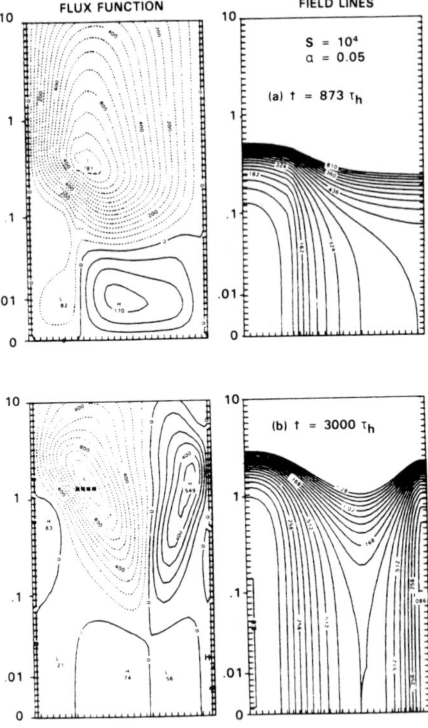

Figure 1.2.6 Stream function (left) and magnetic field lines (right) at two times during nonlinear tearing, illustrating the formation of a new magnetic island. Dashed curves represent a clockwise flow (the modified linear vortex) and solid curves an anticlockwise flow (the new vortex).

The second role of emerging flux and moving satellites is thought of as a trigger of large flares by initiating energy release in a much more extensive overlying field. In particular, emerging flux may push up against a magnetic arcade containing an active-region filament until it goes magnetically unstable of its own accord (Section 1.2.6). Alternatively, it may tear away some of the overlying field lines that are helping to stabilise the arcade, or it may cause a large-scale reconnection by creating a small region of enhanced resitivity.

1.2.4.2 Numerical Experiment

Forbes (Forbes and Priest, 1984 submitted) has recently conducted the first numerical experiment of emerging magnetic flux by solving the resistive MHD equation with a code that is especially designed to treat shock waves well. The initial state consists of a uniform horizontal magnetic field in a numerical box with free-floating conditions on the top and sides. New oppositely directed flux is forced in through the base rather rapidly at a speed of $v_A/8$, and the magnetic Reynolds number is 2000. The resulting magnetic and flow patterns are shown in Figure 1.2.7. In the first three panels the flux emerges and reconnects, with internal energy being converted to the kinetic energy of fast jets of

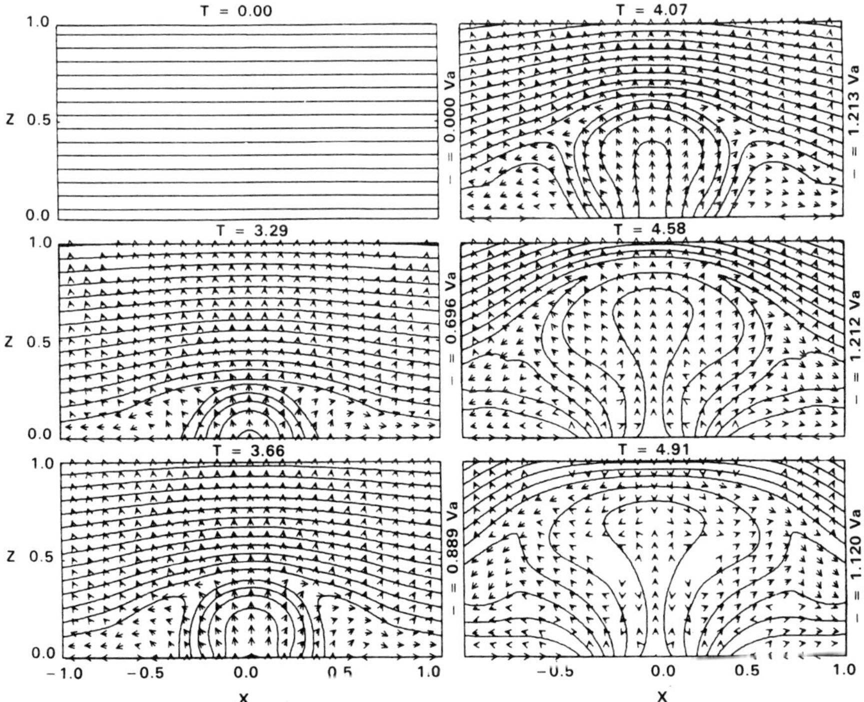

Figure 1.2.7 Magnetic field lines and plasma flow vectors for an emerging flux numerical experiment. At the top of each panel the time is indicated in units of the Alfven travel time across unit distance, and at the right the maximum flow speed is given. (From Forbes and Priest, 1984).

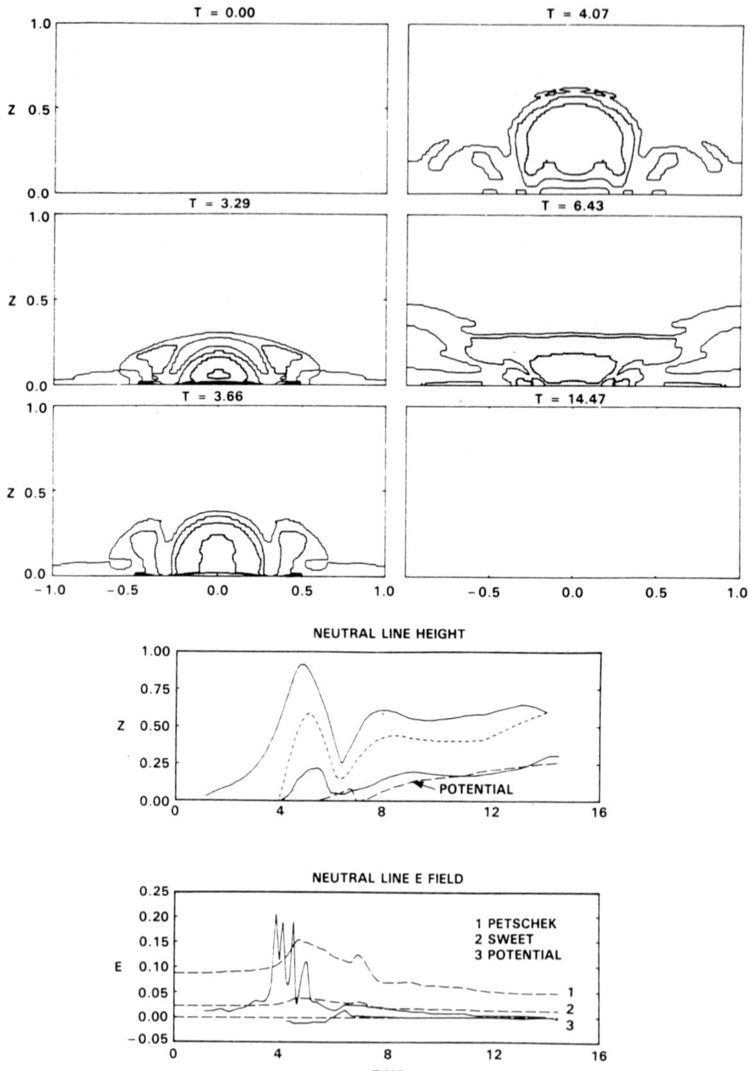

Figure 1.2.8 (a) Mass density contours. (b) Height and electric field of the neutral lines. (From Forbes and Priest, 1984).

plasma. At t = 4 the emergence of new flux through the base is halted, but the flux continues to rise and enters a highly dynamic stage as it loses quasi-equilibrium. The magnetic field lines pinch off near the base and for a plasmoid, which ultimately disappears as the field reduces to a potenital state.

Figure 1.2.8a presents a time-development of the mass density contours, showing the dense emerging flux region and, especially at t = 3.29, the two regions compressed by the shock pairs extending from the central current sheet. This sheet is longer than expected for steady Petschek-Sonnerup reconnection, indicating that reconnection is taking place at this time in a flux pile-up regime (Section 1.2.3.1). The top panel of Figure 1.2.8b gives the heights of the neutral lines as functions of time, with continuous and dashed curves referring to X- and O-points, respectively. It can be seen that at t = 4 a pair of such points is created and at t = 14 a pair is annihilated. The lower panel indicates the electric field as a function of time at the X-points. Just before t = 4 it shows the onset of an impulsive bursty regime (Section 1.2.3.1), with the electric field having impulsive spikes in excess of the steady Petschek value.

Since the electric field at the X-line is a direct measure of the reconnection rate (i.e. the rate at which closed flux in the emerging region is converted into open flux), the variation of E_0 in Figure 1.2.8b gives an indication of the rate at which magnetic energy is converted into kinetic and thermal energy. During the impulsive bursty regime from t = 3.7 to 4.5 there is a rapid release of magnetic energy over a period of about an Alfven scale-time, while before and after this period there is a slower release on the order of the tearing-mode time-scale. That this sequence is suggestive of the flare cycle of pre-flare, impulsive, and main phases is possibly not a coincidence. Instead it seems more likely that structures which incorporate both ideal MHD and resistive instabilities (or, alternatively, non-equilibrium states) will show fluctuations in the reconnection rate on both Alfven and tearing-mode time scales.

1.2.4.3 Current-Driven Instabilities

At the workshop, J. Karpen presented an investigation of the potential role of current-driven instabilities in producing brightenings and general preheating of the plasma in the preflare coronal loops (Karpen and Boris, 1985, submitted). This analytical work calculates the temporal evolution of the ambient plasma characteristics in an emerging, expanding flux tube with an axial magnetic field and electric current. In lieu of focussing on a specific current-driven instability (CDI), they constructed a "generic" CDI with a typical on/off cycle and associated heating. The magnetic-field configuration of an emerging flux tube is simulated by moving two point sources of magnetic flux away from a common origin with constant velocity. The density evolution is derived by considering the mass influx required, but not necessarily attained, for equilibrium in the expanding loop; thus, a range of mass flow rates into the tube is allowed, the zero mass flow yielding complete mass conservation. The temperature evolution s determined by the energy equation, which includes the effects of volumetric heating, intermittent heating due to the CDI, radiation, thermal conduction, and adiabatic cooling. The threshold criterion for the onset of the CDI heating depends solely on the characteristics of the beam current and ambient plasma, and operates under the assumption of marginal stability (e.g., Manheimer and Boris, 1977). Calculations were performed with fast and slow footpoint-separation rates, and with high, intermediate, and low mass-flow rates. The results show a variety of temperature behaviors: in particular, some cases include a single episode of excess CDI heating, lasting for tens of seconds, while others manifest temperature oscillations throughout a large fraction of the simulation period. These types of activity are reminiscent of the patterns of preflare brightenings often observed in the EUV, soft X-ray, and microwave regimes before flare onset (Section 1.3). The authors plan to use the NRL Dynamic Flux Tube Model (cf. Mariska et al., 1982) to obtain more detailed calculations of the effects of the CDI on the dynamics and energetics in the expanding loop.

1.2.5 Main Phase Reconnection in Two-ribbon Flares

The overall magnetic behaviour during a large two-ribbon event is believed to be as follows. Throughout the preflare phase a large flux tube (containing an active-region filament) and its overlying magnetic arcade rise slowly. The rise may be caused by an ideal eruptive instability when the twist in the flux tube or its height become too great (Hood and Priest, 1980, Hood 1984a). Alternatively, it may be due to magnetic nonequilibrium when the equilibrium of a curved tube ceases to exist (Parker 1979, Browning and Priest 1975, Section 1.2.6.1) or it may be triggered by emerging flux.

The onset of the flare itself coincides with the start of the much more rapid eruption of the filament. It probably occurs because the magnetic field lines of the stretched out arcade start to reconnect below the filament (Priest 1981a, 1981b). The linear tearing of the field lines leads on to the fast Petschek and impulsive bursty regimes of reconnection, as described below (Figure 1.2.9). During the main phase the reconnection is thought to continue and create hot "post'-flare loops with Hα ribbons at their footpoints as the field closes down. The source of the immense mass of plasma that is subsequently seen to be falling down along cool "post'-flare loops below the hot loops is an upflow of plasma from the chromosphere along the open field lines before they reconnect (Kopp and Pneuman 1976). The cause of the upflow may well be evaporation driven by thermal conduction or by fast particles that are accelerated at the shocks associated with the reconnection process. Furthermore, it has now been shown that these slow magnetoacoustic shocks can heat the upflowing plasma to the temperatures observed in the hot loops of up to twenty million degrees (Cargill and Priest 1982).

A numerical experiment on the line-tied reconnection that takes place below the erupting filament has been undertaken by Forbes (1982, 1983a, 1983b). He starts with open, stretched out and oppositely directed magnetic field lines in equilibrium and solves the 2D resistive MHD equations for the subsequent development of the right-hand half of the structure. The base of the numrical box is line-tied. Its lefthand edge is an axis of symmetry, and free-floating conditions are imposed on the other two sides.

First of all, the sheet tears near the base and the magnetic field lines start to close down with the X-type neutral point rising and a plasmoid being ejected from the top of the box. In the nonlinear development, reconnection enters a quasi-steady Petschek regime, which is shown in detail in Figure 1.2.9. The decrease of magnetic field strength and convergence of the flow vectors as the reconnection point is

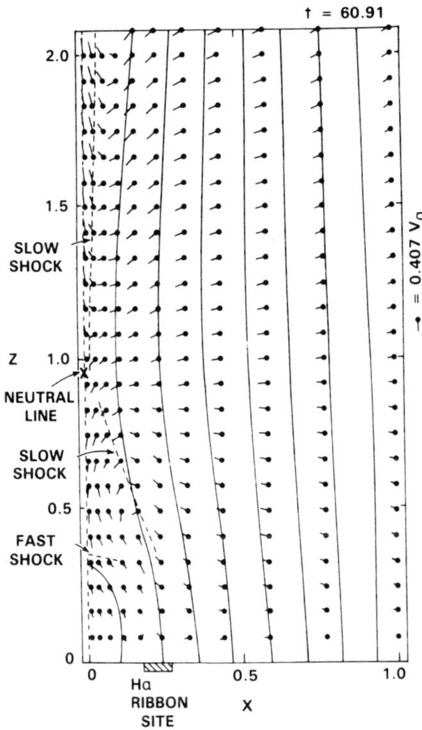

Figure 1.2.9 Magnetic field lines and flow velocity vectors at a quarter of the grid points during the Petschek phase of quasi-steady line-tied reconnection. (Forbes and Priest, 1984).

approached are characteristic of a fast-mode expansion associated with a Petschek-type of regime (Section 1.2.1.2). Also, the fast shock in the downflowing jet may be important for particle acceleration. In the subsequent development the sheet thins and the Petschek mode goes unstable, with the reconnection entering an impulsive bursty regime. Secondary tearing creates a new pair of O and X points, and reconnection at the upper X dominates so that the O is moved down and coalesces with the lower X. Meanwhile, a new pair appears and the process of creation and annihilation of neutral point pairs is repeated. The energy release in this process is faster than the steady Petschek rate, and it occurs in the impulsive manner that is observed in many flares.

1.2.6 Magnetic Instability Responsible for Filament Eruption in Two-Ribbon Flares

1.2.6.1 Loop Configuration

Many people have modelled the preflare magnetic configuration by a single loop and have investigated its stability, with applications to both small simple-loop flares and large two-ribbon flares in mind. Many such stability analyses have been undertaken neglecting for simplicity the curvature of the loop and regarding it as a straight cylinder (e.g., Raadu 1972, Hood and Priest 1979, Einaudi and Van Hoven 1981). Line tying of the ends of the loop in the dense photosphere is an important stabilising effect which makes the perturbation (ξ) vanish there. It keeps the loop stable until the amount of twist in the loop exceeds a critical value, typically 2π or more, depending on the particular equilibrium. The most complete analyses of this type have so far been carried out by Hood and Priest (1981) and Einaudi and Van Hoven (1983). The perturbed equation of motion is solved numerically to give the threshold twist for instability.

The effect of curvature on the equilibrium of an isolated slender coronal loop has also been considered in a simple model (Parker 1979, Browning and Priest 1985) which balances tension and buoyancy. One finds that the variation of the height H of the loop summit with the footpoint separation W is given by

$$\tan^2 \frac{W}{2\Lambda} = e^{H/\Lambda} - 1, \tag{1.2.4}$$

where Λ is the gravitational scale height. Thus, as the foot-points move apart (W increases) the summit rises (H increases) until, as W approaches $\pi\Lambda$, the loop summit floats up indefinitely. For large footpoint separations there is no equilibrium at all. Including an external magnetic field lowers the buoyancy force and therefore the summit height, but it doesn't change the critical footpoint separation. Including a twist in the loop lowers the magnetic tension and so increases the summit height. It also lowers the critical width and changes its nature to a nonequilibrium point.

1.2.6.2 Arcade Configuration

For two-ribbon flares the preflare magnetic configuration has been modelled more accurately by a coronal arcade. In particular, the effect of line tying has been included in models of force-free arcades (Migliuolo and Cargill 1983, Hood and Priest 1980, Birn and Schindler 1981, Ray and Van Hoven 1982, Hood 1983a, Cargill et al., 1984). The original analysis (Hood and Priest 1981) considered various classes of ideal perturbations and found that a simple arcade with its magnetic axis below the photosphere is always stable to those classes. It also appears to be stable to resistive modes usually (Hood 1984, Migliuolo and Cargill 1983). However, arcades with their magnetic axis a distance d above the photosphere are more interesting, since they are more likely to represent configurations within which an active-region (or plage) filament can form (Hood and Priest 1979). Such filaments are quite different from the large quiescent filaments and form along flux tubes. They are indicators of a highly sheared field and very often erupt before two-ribbon flares, slowly at first and then much more rapidly at flare onset. This type of coronal arcade (whose cross-section contains a magnetic island) is found to become unstable when either the height of the magnetic axis (and therefore of the filament), or the amount of twist become too great (Hood and Priest 1981). This suggests that the eruption of the arcade may be caused by a spontaneous eruptive instability when the filament height or the magnetic shear become too great.

Recently, attention has been focussed on magnetohydrostatic arcades with a force balance between the Lorentz force, a pressure gradient and gravity. For a two-dimensional isothermal arcade in which the variables are independent of the direction z along the arcade, the magnetic field components can be found, especially in the limit as H approaches infinity such that the gravitational force is negligible (Low 1979, Heyvaerts et al., 1982, Priest and Milne 1980, Zweibel and Hundhausen 1982, Melville et al., 1983, 1984).

Having obtained the equilibria for magnetostatic arcades, it is important to analyse their stability, since an arcade must be stable if it is to store magnetic energy prior to flares. On the other hand, it is also necessary that this energy can be released by an instability when somecritical threshold is reached. The stability of arcades can be studied either by the energy method or by solving the equations of motion. Using the energy method Schindler et al. (1983) and Hood (1984a, 1984b) independently obtained a sufficient condition for stability. For free-flow boundary conditions (see below) this condition also becomes necessary when there is no axial field ($B_z = 0$).

The strong stabilising influence of the dense photosphere, known as line-tying, has been modelled in two different ways, either by setting the perturbation ($\xi_\perp$) perpendicular to the magnetic field at the photospheric footpoints equal to zero or by making the total perturbation ($\xi_\perp + \xi_\parallel$) vanish there. The physical argument is that the high density (and low temperature) does not allow the photosphere to move in response to disturbances that propagate from the corona. For example, assuming the ratio of photospheric to coronal wave speed to be 10^3, 99.6% of the energy of a non-resonant MHD wave propagating from the corona should be reflected back and only 0.4% transmitted. It is generally agreed that line tying makes perturbations that are perpendicular to the magnetic field vanish at the photosphere for perfect reflection. However, the condition on perturbations parallel to the magnetic field is more controversial (Cargill et al., 1985 submitted). The two main choices are to regard the ends as being rigid and set $\xi = 0$ (e.g., Hood and Priest 1979) or to allow free flow through the ends ($\nabla\cdot\{\xi \exp -y/H\} = 0$ for an isothermal plasma). Many results in the literature differ because of the choice of this parallel boundary condition as well as the choice of equilibrium.

By solving the equations of motion with free-flow boundary conditions parallel to the magnetic field Migliuolo et al. (1984), have demonstrated that arcades with $B_z = 0$ are unstable to interchange modes with very short wavelength (λ_z) along the arcade. This instability may be important for the small-scale structure of the corona rather than for global flare instability. They also showed that, if B_z is non-zero, the arcade becomes unstable when the pressure gradient is large enough, a result which may account for the second stage of a double impulsive flare, in which the second part of the flare occurs after plasma has been evaporated up to the corona by the first part.

More recently, the work has been extended to compare the stability thresholds that result from free-flow and rigid end conditions (Cargill et al., 1985, submitted). For cylindrically symmetric equilibria the presence of a rigid boundary gives rise to substantial differences in the stability thresholds. Equilibria with $B_z = 0$ may be either stable or unstable, depending upon the exact details of the equilibrium and the ratio of the specific heats (γ). Inclusion of shear (B_z) is stabilising, and for the equilibria considered a small amount of shear is sufficient to stabilise all the equilibria. Physically, the rigid conditions do not permit incompressible modes, and so there is an increase in the potential energy due to compression of the plasma. Clearly, the difference in the results from two sets of boundary conditions makes it important to understand the real nature of line tying, and to model it adequately (Cargill et al., 1985,submitted).

Hood (1983b) has considered the arcade equilibrium

$$A = A_o \cos\left(\frac{x}{4H}\frac{\sin Y}{Y^{1/2}}\right) \tag{1.2.5}$$

where

$$Y = 2\beta H \exp(-1/2\ y/H) \tag{1.2.6}$$

and $(2\beta H)^2$ is the plasma beta. This is a special case of the class considered by Zweibel and Hundhausen (1982), and the field lines are shown in Figure 1.2.10a. As the base pressure (and therefore β) increases, so the magnetic field lines bow outwards, and eventually for $2\beta H > 1.15$ a magnetic island appears (Figure 1.2.10b). When the pressure is so large that $2\beta H > \pi$ the upper field lines become detached from the photosphere and the configuration ceases to be physically realistic. When the magnetic island is present, Zweibel (1981) has shown that such fields tend to be unstable. Hood (1983b) has extended her analysis to include the effect of magnetic tension, which makes the field stable for small β.

It should be stressed that stability analyses of the above type may be used to estimate the amount of magnetic energy that may be stored in the corona in the stable state. The equilibria that are considered are certainly not accurate representations of active-region fields (see Section 1.3), but they do typify their expected properties.

1.2.6.3 Prominence Models

Recently, Malherbe (Malherbe and Priest 1983, Malherbe et al., 1983) set up some new current sheet models for quiescent prominences using complex variable theory. Figure 1.2.11a,b shows two models of the Kippenhahn-Schluter type, while Figures 1.2.11c,d indicate some of the Kuperus-Raadu type with the magnetic field lines crossing through the prominence in the opposite direction. Leroy, Bommier, and Sahal-Brechot

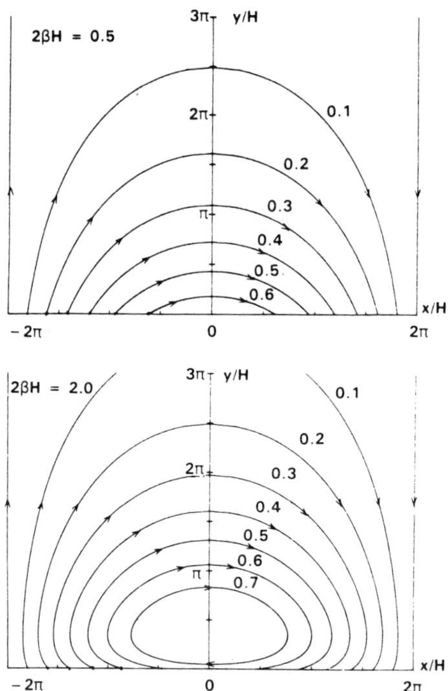

Figure 1.2.10 The magnetic field lines of the equilibria (1.2.5) have the form (a) when the base pressure is so small that $2\beta H < 1.15$, (b) when $1.15 < 2\beta H < \pi$. (From Hood, 1983b).

(1984) find observationally that those of Kuperus-Raadu type are, somewhat surprisingly, twice as common. Also, observations of slow steady upflows in prominences when seen on the disc can be explained by a dynamic prominence model in which the magnetic field evolves in response to photospheric motions. The large quiescent prominences are mainly of Kuperus-Raadu type and the footpoint motions would need to be convergent, which suggests that such prominences lie at the boundaries of large-scale giant cells. By comparison, the plage filaments have a Kippenhahn-Schluter field orientation and would need divergent footpoint motions. At present, Malherbe, Forbes and Priest are studying numerically the formation of prominences in current sheets by radiative tearing. They have adopted the previous radiative code of Forbes and Priest (1982, 1983a, 1983b) to include an energy equation with Joule heating, coronal heating and radiative cooling. In particular, the cases when the cooling time is a factor of between 0.1 and 10 times the tearing times are being investigated.

1.2.7 Conclusion

During the Workshop there have been major advances in the theory of magnetic reconnection and of magnetic instability, with important

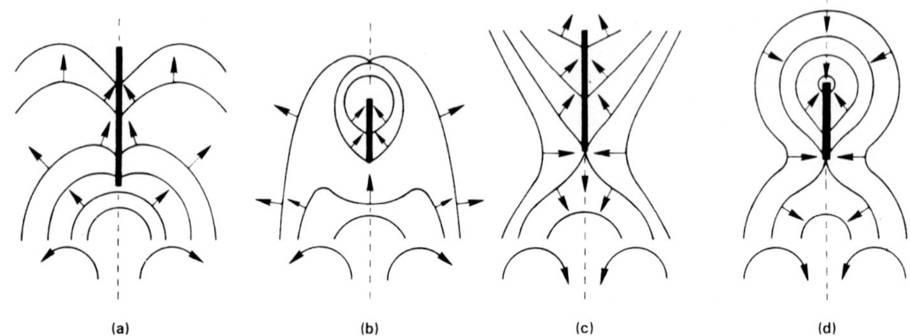

Figure 1.2.11 Current sheet models of prominences.

implications for the observations, as follows:

1. Fast and slow magnetic shock waves are produced by the magnetohydrodynamics of reconnection and are potential particle accelerators.

2. The impulsive bursty regime of reconnection gives a rapid release of magnetic energy in a series of bursts.

3. The radiative tearing mode creates cool filamentary structures in the reconnection process.

4. The stability analyses imply that an arcade can become unstable when either its height or twist or plasma pressure become too great.

1.3 PREFLARE MAGNETIC AND VELOCITY FIELDS

M.J. Hagyard, V. Gaizauskas, G.A. Chapman, A.C. deLoach, G.A. Gary, H.P. Jones, J.T. Karpen, M.-J. Martres, J.G. Porter, B. Schmieder, J.B. Smith, Jr., and J. Toomre

A description of the structure, dynamics and energetics of the preflare state depends on our ability to characterize the magnetic and velocity fields of the preflare active region. In this SMM Workshop, we fortunately had at our disposal many sets of coordinated SMM and ground-based observations of magnetic and velocity fields from the photosphere, chromosphere, transition region and corona to aid in this characterization. At the outset we decided that several aspects of these fields are of special interest to the preflare state: configurations in the magnetic and velocity fields that seem peculiar to flaring active regions; the existence of shears (in both the magnetic and velocity fields); the occurrence of emerging flux. Some questions naturally arise concerning these topics. Do flares occur in active regions where the magnetic field is force-free (currents are field-aligned), non-force-free or both? If it is force-free, can it be specified by a constant-alpha? [Alpha is the ratio between current density and field strength]. Is magnetic shear correlated with the occurrence of flares and, if so, is there a critical value of this shear? What is the role of the resulting electric currents? What are the preflare characteristics of the velocity field and how do they evolve? What is the spatial and physical correlation between sheared velocity and magnetic fields? What are the conditions necessary for emerging flux to trigger a flare? What growth rates of flux are significant? How does the flux emerge into the corona?

Although we did not find answers to all these questions, we made significant progress in many areas. We found that the preflare active region is very dynamic, exhibiting recurrent mass surges and intermittent heating events at many sites. In one case, that of an active region not particularly productive of large flares, the structure of the magnetic field was best represented by a nonlinear force-free field; for a particularly flare-productive region, there were indications that, subject to certain restrictions on the boundary conditions, the field was non-force-free, exhibiting a measurable Lorentz force. We also found that both the magnetic and velocity fields are sheared in flaring regions; the shear of the magnetic field attained maximum values at the sites of flare onset, whereas the velocity field sometimes exhibited an unusual vortical structure at these sites. These sheared magnetic fields produced persistent, large-scale concentrations of electric currents at the flare sites; numerical values for the magnitudes of these currents provided input to models describing preflare brightenings based on joule heating or current-driven instabilities. Finally, we found the role of emerging flux in flares to be ambivalent, providing an obvious triggering of some classes of flare while having no role in the flare process in others.

In describing these various results the material has been arranged as follows. We begin with a characterization of the preflare magnetic field, using theoretical models of force-free fields together with observed field structure to determine the general morphology. We then present direct observational evidence for sheared magnetic fields. The role of this magnetic shear in the flare process is considered within the context of an MHD model that describes the buildup of magnetic energy, and the concept of a critical value of shear is explored. The related subject of electric currents in the preflare state is discussed next, with emphasis on new insights provided by direct calculations of the vertical electric current density from vector magnetograph data and on the role of these currents in producing preflare brightenings. Next we discuss results from our investigations concerning velocity fields in flaring active regions, describing observations and analyses of preflare ejecta, sheared velocities, and vortical motions near flaring sites. This is followed by a critical review of prevalent concepts concerning the association of flux emergence with flares.

1.3.1 General Morphology of the Preflare Magnetic Field

It is generally accepted that magnetic fields are the ultimate source of the energy released in a flare (e.g., Svestka, 1976) and that this energy is stored in an active region prior to the flare as a result of the stressing of these fields into non-potential configurations. We have accumulated observational evidence for such stressed fields, both on large and small scales, and studied the stressing processes which result in the eruption of a flare. We first discuss our studies of the general morphology of the preflare magnetic field.

A. Gary endeavored to classify the non-potential character of magnetic fields in active regions assuming that the fields are force-free, i.e., that the following relation is valid:

$$\nabla \times \underset{\sim}{B} = \mu_o \underset{\sim}{J} = \alpha \underset{\sim}{B}. \tag{1.3.1}$$

Several active regions observed during SMM were modeled using the force-free formulation developed by Nakagawa and Raadu (1972), who assumed that the parameter α is spatially invariant. One of these regions, AR2684, was of particular interest since it was observed by instruments on a Lockheed rocket flight at 20:30 UT on September 23, 1980, as well as by SMM and ground-based instruments. Although the flare activity in the region was relatively minor, several C- and M-class flares occurred on the 23rd and 24th, the largest being a 1B/M1 event at 07:28 UT on the 24th.

This coordinated observational program produced UVSP, XRP AND HXIS data from SMM, Lyman-alpha and 1600Å spectroheliograms from the rocket experiments, and magnetic field, H-alpha and He 10830Å data from the ground-based observatories; B. Haisch assembled these data for the workshop. It was hoped that chromospheric and coronal filamentary structure would delineate both footpoints and field lines of the magnetic field of the entire active region. Then, using these inferred structures, the most appropriate value of the parameter alpha could be selected, i.e., the one that matched calculated field lines with the observed ones. A similar method was used extensively in analyses of Skylab data to model the magnetic field of active regions. The basic observational data used in the analyses are shown in Figure 1.3.1a. The bright points observed in the 1600Å spectroheliograms were assumed to be footpoints of the Lyman-alpha loops an H-alpha fibrils and were thus used as the initial coordinates of the field line calculations for various constant-alpha computations. As expected, Gary found that the field lines calculated for a single value of alpha would not fit all of the observed structures. In Figure 1.3.1b, areas are indicated where field lines calculated for different alphas showed good agreement with the observed fibrilar orientations. Although there are areas of positive, negative and zero alpha, a dominance of positive alpha is indicated. Because the "footpoints" used in the calculations were generally outside the sunspots, this "alpha map" applies only to the weaker field areas.

To investigate the non-potential character of the region's magnetic field without resorting to an assumption of constant alpha, Gary compared the direction (azimuth) of the observed transverse field in the photosphere with that of a potential field distribution. The use of the azimuth of the transverse field as an indicator of the non-potential character of the field is based on the observation that the projected field lines from the force-free calculations were parallel to the observed azimuthal directions; the only exceptions were the field lines that rose very high into the corona and whose lengths were characteristic of the scale of the magnetogram's field-of-view. In his analysis, Gary found the deviation of the observed azimuth from a potential orientation at each grid point, and assigned to these points a positive or negative sign depending on the sense of the observed deviation or "twist"; the results are shown in Figure 1.3.1c. Since this "alpha map" only pertains to the regions in and near the sunspots where the transverse field is above 200 G, it is difficult to relate it to the alpha map in Figure 1.3.1b for the areas of the weaker fields. However, where there is some overlap, the two methods give consistent results, and confirm that a constant alpha force-free field cannot characterize the magnetic topology of this active region.

A second region of interest, AR 2372 (very flare-productive on the solar disk in early April 1980), also proved difficult to model, but for another reason. As will be discussed in Section 1.3.2.2, this region exhibited a large degree of shear in its magnetic field in the area of a magnetic δ-configuration (umbrae of opposite polarity within the same penumbra). This extreme shear could not be reproduced with the linear force-free computation of Nakagawa and Raadu because of the limitations on the maximum value of alpha (i.e., twist or shear). In their formulation, alpha must be less than $2\pi/L$, where L is the scale of the magnetogram. However, an analysis by Krall (private communication) produced some evidence that the magnetic field in the area of the δ-configuration was non-force-free. Using transverse field measurements obtained with the MSFC vector magnetograph, he calculated the resultant Lorentz force in the region of the delta from a formulation derived by Molodensky (1974). Krall found this force had a non-zero horizontal component that was consistent with the observed sunspot motions in that area.

These attempts to model the magnetic fields indicate that, for a moderately active region, the structure of its field was fairly well represented by a nonlinear force-free field. On the other hand, calculations based on the observed field of a highly flare-productive region resulted in a non-zero Lorentz force. Until analyses of other regions are available, these results must be regarded as very preliminary.

1.3.2 Magnetic Field Shear

1.3.2.1 Evidence for Sheared Magnetic Fields

Storage of flare energy in stressed magnetic fields arises from the increasing deformation of the magnetic field from a potential

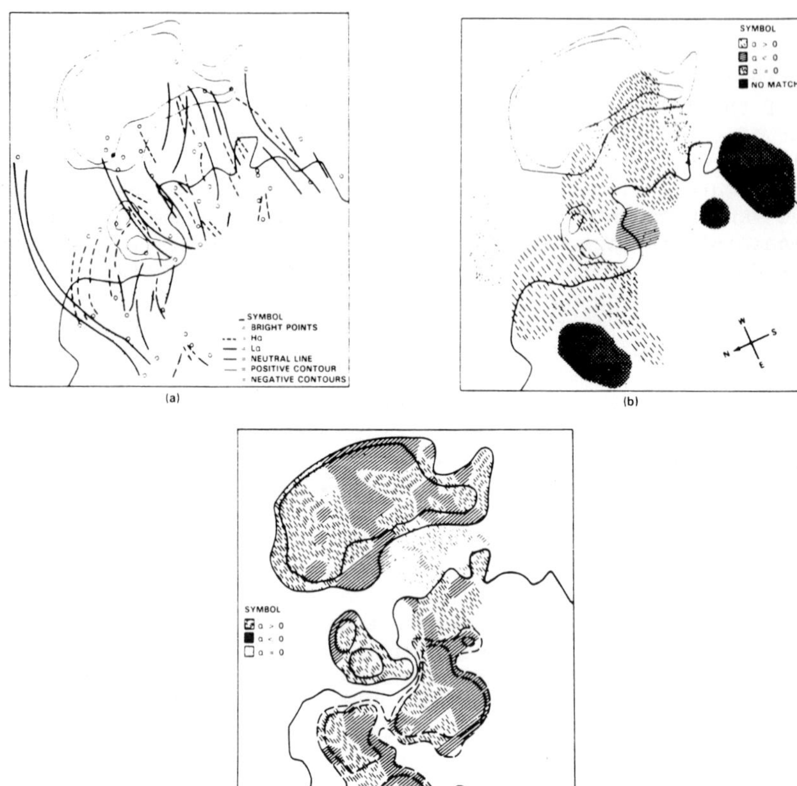

Figure 1.3.1 Force-free field modeling of AR 2684 for September 23, 1980. (a) The filamentary structures inferred from H-α and Lyman-α spectroheliograms are shown superposed on a plot of the photospheric line-of-sight magnetic field of the region, which was located at N18W22. Also shown are the bright points observed in the 1600 A spectroheliograms. The solid (dashed) magnetic field contours represent positive (negative) field levels of 250 and 500 G; the solid curve separating positive and negative fields is the magnetic neutral line. The field-of-view is 2.5′ × 2.5′ and the solar orientation is as shown in panel b. (b) Regions are specified where the fibrilar structures of panel a are matched (or not matched) by force-free field lines calculated with positive, negative or zero (potential field) values of the parameter alpha. The symbols are shown in the upper right corner of the panel; cross-hatched areas specify regions where the fibrils and field lines could not be matched to within 45 degrees. The contours shown are identical to those in panel a. (c) Map of the differences between the orientations of the observed transverse field and a potential field, where the potential field fits the boundary conditions imposed by the observed line-of-sight field. Areas of positive, negative and zero deviations or "twist" are shown according to the legend for the parameter alpha since the force-free parameter alpha is also a measure of the twist of the field. Only areas for which the line-of-sight field is greater than 250 G are indicated, except for one region in which the transverse field was above 200 G. The contours shown are again those designated in panel a.

configuration. This deformation can occur, for instance, through the shearing of magnetic loops as a result of footpoint translations, or through the twisting of individual loops rooted in sunspots which rotate. Some of our indirect evidence for preflare energy storage in stressed fields comes from the geometry of fibrils and structures within filaments in the vicinity of flares; these fibrils and filamentary structures presumably delineate the chromospheric magnetic field. For example, in a detailed study of the August 1972 flares, Zirin and Tanaka (1973) inferred the presence of strongly-sheared, transverse magnetic fields from the twisted appearance of penumbral filaments. In more recent work using both SMM and ground-based observations, Athay et al. (1984) determined the broad features of the magnetic field geometry from chromospheric and transition region Dopplergrams, assuming that the fluid flow follows magnetic lines of force. H-alpha filament orientation and motion, and the relationship of the filaments to sunspots provided additional information on the field geometry. From these data, they deduced that pronounced magnetic shear was present at transition region heights over the entire length of a prominent segment of the polarity-inversion line. This shear remained relatively steady for periods of several days except for temporary local disruptions due to emerging flux regions (see Section 1.3.4.2).

Based on such indirect indications of sheared magnetic fields, much of the previous modeling of preflare magnetic fields during the Skylab series of flare workshops was performed with the assumption that configurations of sheared magnetic fields did indeed exist in preflare active regions. Now, however, direct evidence for sheared magnetic loops comes from measurements of transverse magnetic fields in the photosphere near the magnetic neutral line. At the neutral line, the direction of the transverse component of the field will indicate the orientation of low-lying field loops which connect footpoints on opposite polarity sides. Initial observations of transverse field directions which appeared sheared relative to the neutral line were reported by Smith et al. (1979) using data from the NASA/Marshall Space Flight Center (MSFC) vector magnetograph (Hagyard et al., 1983), which observes in the FeI 5250 A line originating in the photosphere. During SMM, many subsequent examples of sheared magnetic fields in a number of active regions, most of which produced flare activity, were reported by the MSFC group (Krall et al., 1982; Patty and Hagyard, 1984; Smith, 1984 [private communication]).

1.3.2.2 Correlation with Flare Activity

These direct measurements of magnetic shear provide compelling evidence for linking magnetic shear with the incidence of flares. For example, in a statistical study J.B. Smith, Jr. (private communication) found a distinct preference for high flare productivity and major flares in ares with significant magnetic shear. In evolving regions, an increase in shear clearly accompanies an increase in flare frequency and magnitude, while decreasing shear is commonly accompanied by a decrease in flare production. To evaluate the magnetic shear, Smith used the vector MSFC magnetograms which depict the transverse fields at the photosphere as line segments. Their length and orientation give the strength

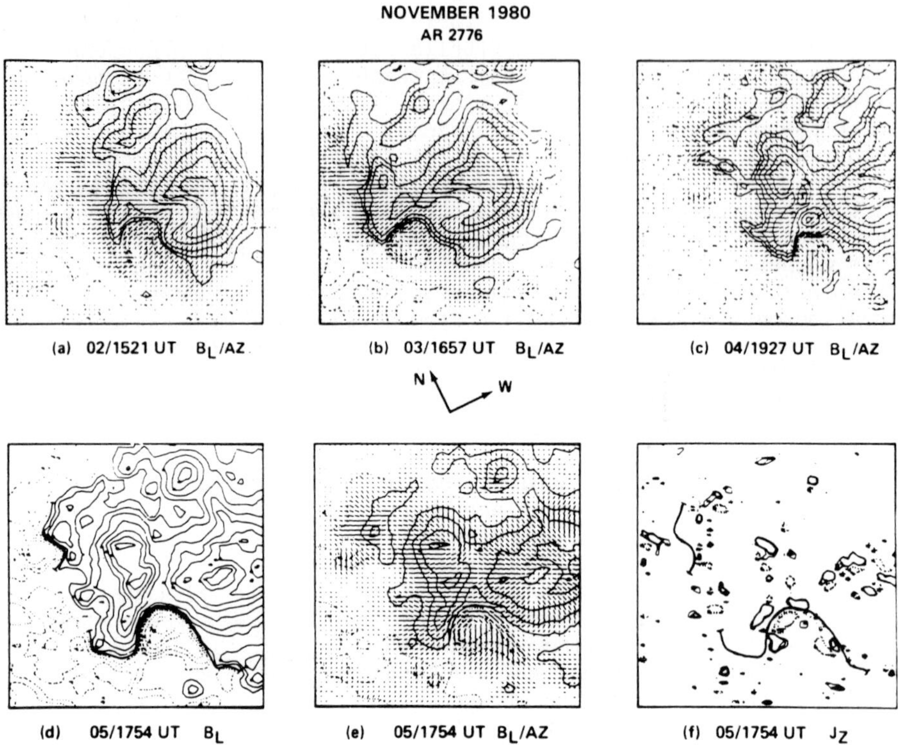

Figure 1.3.2 Magnetic evolution of AR 2776 over the period November 2-5, 1980. The field-of-view in all panels is approximately 2.1′ × 2.1′. In all line-of-sight magnetic field maps (B_L), solid (dashed) contours represent positive (negative) fields of 100, 250, 500, 1000, 1500, 2000 and 2500 G. In overlaid transverse field plots (AZ), the transverse field strength and direction are indicated by the length and orientation of the line segments. (a) Overlay of the transverse magnetic field (AZ) on contours of B_L from observations on November 2. A magnetic delta configuration is formed by the intrusion of the negative-polarity sunspot into the positive-polarity field just to the east of the large positive-polarity spot. Analysis of the line segments representing the transverse field that are overlaid on the contours of B_L reveals a generally potential-appearing field, aligned more or less directly from the positive center to the negative portion of the delta. (b) Overlay of AZ on contours of B_L from observations on November 3. Note the growth of the positive fields to the north of the delta. (c) AZ/B overlays for November 4. Growth of the fields continues, but the field orientation remains generally potential in appearance. (d) B_L field on November 5. (e) Overlay of AZ and B_L for November 5. Only the high B_L field contours from panel d have been depicted in order to make more visible the highly-sheared transverse field along the neutral line. The increase in field strengths and gradients can be seen from comparisons of panels d with panels a, b and c, where the contour levels are all the same. (f) Contours of the vertical component of the electric current density (Jz) for November 5 (see Section 1.3.3.1.). Positive (negative) Jz values of 150, 200 and 250 × 10^{-4} A m^{-2} are depicted by solid (dashed) curves. The neutral lines of panel d have been superposed to aid in orientation.

and direction of the transverse field at each point in the field-of-view. The "angle of shear" along the neutral line can be determined by comparing the directions of the line segments to the orientation of the neutral line. This interpretation of shear assumes that line segments of a potential field cross the neutral line orthogonally.

Smith qualitatively evaluated several regions for correlation between this "angle of shear" and flare production. In two cases, a pair of regions were simultaneously visible, one with measurable shear and the other with fields that appeared more potential. In both cases, the region with observable shear produced flares while the other was essentially quiet. Perhaps the most notable example occurred in April 1980 when AR 2370, a large region promising significant activity but producing little of note, rotated onto the visible disk a few days before the birth nearby of AR 2372 early on the 4th of April. Following rapid development of the main spots of AR 2372, pronounced photospheric shearing motions were observed between the 5th and 7th of April for the sunspots of opposite polarities inside this region. During this epoch of spot motions, the transverse field directions indicated the presence of strong shear along the neutral line in relatively strong magnetic fields; flares, some major, were frequent (Krall et al., 1982). A decrease in the shear of AR 2372 after the 7th was followed by a sharp decrease in flare production. Several other regions were analyzed with similar results: those with strongly sheared fields were flare productive while those with essentially potential fields (or weakly sheared fields) had only minor activity.

Smith also studied the development and evolution of shear within AR 2776, during November 1980 during SMM. The evolution of its vector magnetic field over the period November 2-5 is shown in Figure 1.3.2. On November 2nd (Figure 1.3.2a), some magnetic complexity was evident in the presence of a δ configuration, although the surrounding magnetic gradients were moderate, the fields only moderately strong, and the field alignments generally appeared to be potential. This situation held also on the 3rd and 4th (Figures 1.3.2b,c) but with some complexity added by the building of the positive fields to the north. Still, the observed shear was not extensive. However, the changes between the 4th and 5th were striking, particularly in the pronounced shear seen in the near alignment of the transverse field with the entire length of the neutral line in the area of the delta on the 5th (Figures 1.3.2d,e). In addition, field strengths and gradients increased markedly from the 4th to the 5th. Smith examined the X-ray flares that occurred during the period November 1-12. He found that energetic soft X-ray flares (class M1 or greater) were infrequent until the 5th, when both frequency and magnitude rapidly increased and several major flares followed. Again,the correspondence of increased shear with increased flare activity is borne out by this study.

Smith also analyzed the magnetic shear in other active regions: AR 2522 (June 1980), AR 2544 (June/July 1980) and AR 2725 (October 1980). In Figure 1.3.3, the line-of-sight (BL) and transverse (BT) components of the magnetic fields of these three regions are shown along with the calculated vertical electric current densities, Jz. Varying degrees of magnetic complexity were reflected in the level of flare activity for two of the three regions; the third region was somewhat of an anomaly. AR

2725 (columns c and d in Figure 1.3.3) was the most magnetically complex and also the most flare productive. Examination of the transverse magnetic field data revealed significant shear and moderate field strengths along that portion of the neutral line to the left of center in the magnetograms where the flare of October 11 at 17:41 UT (classified as 1B/C7) occurred, as determined from SMM soft X-ray data. Although significant flares were infrequent, a few class M X-ray flares were observed and a major flare (3B/X3) occurred on October 14. AR 2522 (column a in Figure 1.3.3) has the complexity of a convoluted neutral line and an isolated island of positive polarity, but examination of the transverse field reveals only weak shear along the convoluted neutral line. The region produced numerous minor flares, with a few class M events during its disk passage.

But the third region AR 2544 (column b) proved to be anomalous. It produced only a few flares, but there was a rash of activity on June 30th, including an M1, and a major flare (X2) on July 1st, the day of the illustrated magnetogram. From the magnetic data, there appears to be only the minor complexity of the isolated positive spot intruding into the southern portion of the leading negative polarity, and there is clearly no evidence of strong or extensive shear.

In another anomalous example, the active region complex composed of NOAA active regions 2516, 2517 and 2519 produced very little flare activity along the segments of the magnetic neutral line which showed the most evidence of shear in the transition region (Athay et al., 1984). The flares that did occur were mainly associated with emerging flux regions. The magnetic shear was either "rather stable" or did not generate sufficient free energy to fuel a flare.

However, an important factor may be the overall configuration of the magnetic field from the photosphere through the transition region. For the relatively inactive June complex, Athay et al. (1984) inferred the configuration of the magnetic field from observations of fluid flows in the transition region. They found that the extreme velocity shears in the transition region diminished greatly at the photosphere. If the fluid flow follows magnetic lines of force, this result seems to indicate that the magnetic shear also decreased from the transition region into the photosphere (see Section 1.3.4.2). Such a configuration appears to be exactly opposite to that reported by Krall et al. (1982) for a flare-productive active region where the flares occurred at the site of greatest shear (Hagyard et al., 1984a) as deduced from photospheric observations with the MSFC vector magnetograph. Investigating the field configuration at higher levels, Krall et al. found that the short fibril seen in Hα aligned with the sheared photospheric field, whereas the orientation of the longer, and presumably higher, fibrils was more or less normal to the magnetic neutral line, a configuration indicative of less shear. Thus, the overall structure of the magnetic field in this more active region was one of decreasing shear in going from the photosphere up into the lower corona.

1.3.2.3 Formation of Magnetic Shear

There is good observational evidence that magnetic shear forms as

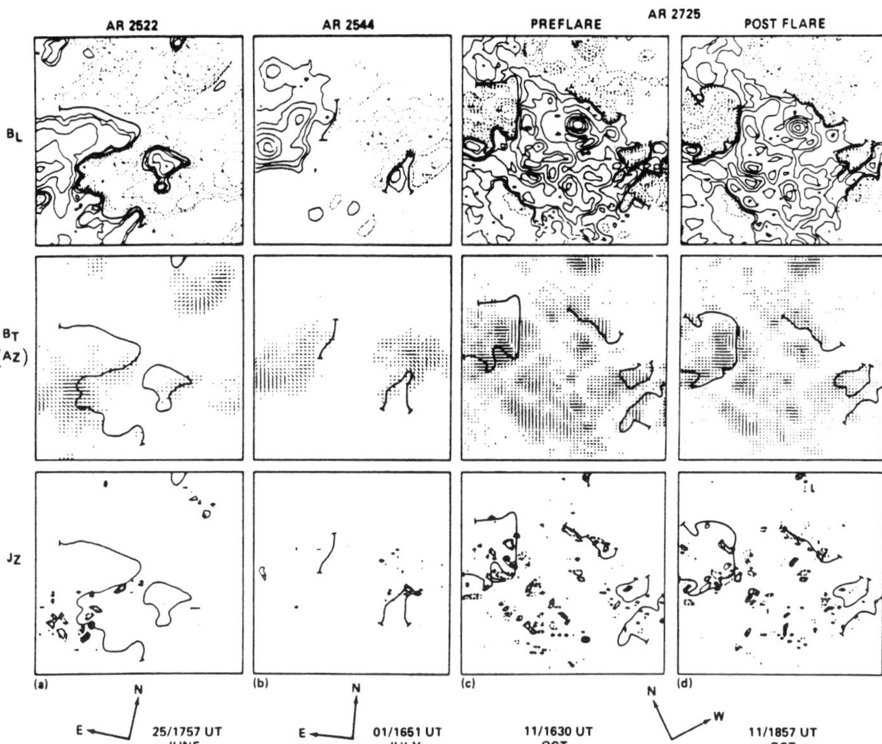

Figure 1.3.3 Vector magnetic fields and electric currents for active regions selected for study by the Preflare Group. The panels in the top row depict the longitudinal magnetic field B_L wherein positive (negative) fields are represented by solid (dashed) contours; contour levels are 100, 250, 500, 1000, 1500 and 2000 G. Panels in the center row show the transverse field B_T depicted as line segments whose length and direction represent the strength and orientation of the transverse component; the magnetic neutral lines are superposed. The bottom row depicts the corresponding vertical electric current densities with the neutral line overlaid in bold lines; contour levels are 150, 200, and 250 $\times$ 10^{-4} A m^{-2}. (a) Data for AR 2522 on June 25, 1980. Only weak shear is seen along the convoluted neutral line in the area of the southern boundary of the large intruding peninsula of positive polarity near the center of the field-of-view (2.1′ × 2.1′). (b) Data for AR 2544 on July 1, 1980. This region was born near central meridian on June 28, but it developed only minor magnetic complexity in the area of the isolated positive spot to the south of the leading negative spot. Field-of-view is 2.1′ × 2.1′. (c) Preflare data for AR 2725 on October 11, 1980. Significant magnetic shear is seen along the neutral line that lies to the left of the center of the magnetogram, with additional, weaker shear found along the neutral line lying above and to the right of center. The field-of-view is 2.5′ × 2.5′. (d) Postflare data for AR 2725 on October 11, 1980. The IB/C7 flare, which occurred about midway in time between the two megnetograms (c and d), coincided with the area of most significant shear. At 17:42 UT, near frame maximum, the X-ray loops appeared centered over the north-south portion of the neutral line to the left of center. The field-of-view is the same as in c.

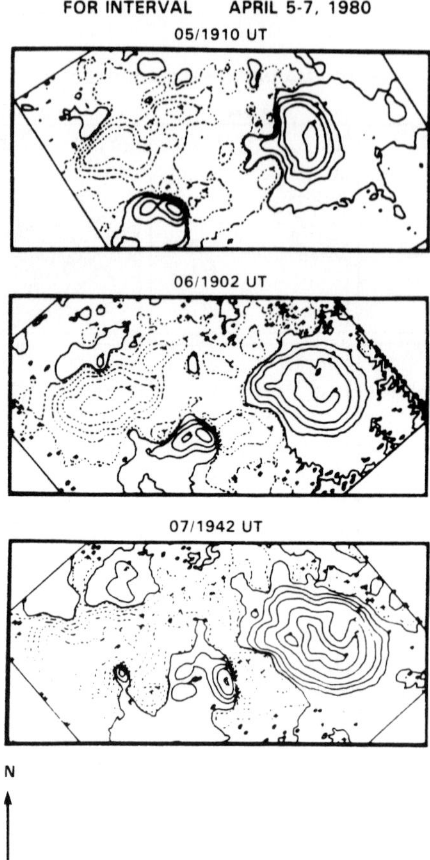

Figure 1.3.4 Evolution of the line-of-sight magnetic field in AR 2372 for April 5-7. In all panels the contours represent positive (solid curves) and negative (dashed curves) magnetic field polarities. The panels are portions of 5′ × 5′ fields-of-view. Top panel: the sunspot motions occur in the area of the internal magnetic dipole lying to the south of the large following-polarity (negative) sunspot that is located to the left of the center in this panel. Middle panel: from the 5th to the 6th, the isolated positive-polarity spot moved westward toward the large, leading-polarity (positive) spot at $\approx$ 160 m s^{-1}, while the adjacent negative spot, initially observed just to the north of the isolated positive spot, moved eastward toward the large, following-polarity spot at $\approx$ 60 m s^{-1}. Bottom panel: the field configuration on the 7th after the major spot motions had ceased.

the result of sunspot motions. Two particularly good examples correlating sunspot motions with the development of photospheric magnetic shear and subsequent flare activity were unveiled in the course of the workshop. The first was the active region of early April 1980, Boulder number 2372. Born on the solar disk early on April 4, it produced many flares during its period of growth and development. Observations made at the Yunnan Observatory on April 5 from 00:50 to 09:15 UT showed the rapid development of three sunspots through the coalescence of several smaller spots (Hoyng et al., 1982). The motions of the smaller sunspots were grouped into three sectors, with the spots in each sector converging and coalescing into one of the three major sunspots seen on the 6th. Since these motions involved spots of different magnetic polarities, stretching and/or shearing of the inter-connecting fields probably occurred, with energy buildup taking place in the process. Sunspot motions occurred through the 6th, as inferred from MSFC white-light photographs and the magnetic field changes seen in Figure 1.3.4. These motions continued until 19:00 UT on the 7th, whereafter no significant motions were observed. In an extensive study of this active region, Krall et al. (1982) found that these spot motions produced significant shear in the magnetic field with resulting flare activity in the region. This can be seen in Figure 1.3.5 which shows the first large flare on the 5th, and the most intense one, on the 6th, both occurring at the locations of the isolated positive spot and the eastward-moving negative spot. Krall et al. related the formation of magnetic shear to the spot motions using the observed orientations of the transverse fields as shown in Figure 1.3.6. These observations confirm that the transverse magnetic field evolved from a slightly sheared configuration on the 5th (Figure 1.3.6a) into a strongly sheared one on the 6th (Figure 1.3.6b), which then relaxed on the 7th (Figure 1.3.6c) as the sunspot motions ceased. Following this apparent relaxation of the field, the high frequency of flares which occurred through the 7th ceased, and little significant flaring was produced on the 8th and 9th.

The second example was provided by G. Chapman, who presented filtergrams, spectroheliograms and magnetograms showing the buildup of stressed magnetic fields through both rotational and translational motions of a large sunspot in conjunction with a satellite spot of opposite polarity in close proximity to the main spot. Observations of AR 2530 were obtained at the San Fernando Observatory (SFO) for approximately 9 1/2 hours on June 24, 1980.

They showed that during this interval, the leading sunspot of this region rotated and deformed substantially from a round to a U-shape. The satellite spot of opposite polarity was adjacent to one edge of the evoving leader spot; the satellite remained intact until the following day despite the drastic changes of its larger companion. The deformations increased the magnetic gradients in the area of the satellite spot. Major flaring took place while the compression and twisting of the magnetic fields were occurring, rather than after the maximum deformation had been reached. This suggests that the rate of change in the stressing of the magnetic field was more a factor in the flare activity than the sheared topology of the field. In addition, the persistence of the satellite spot following the large flare suggests that

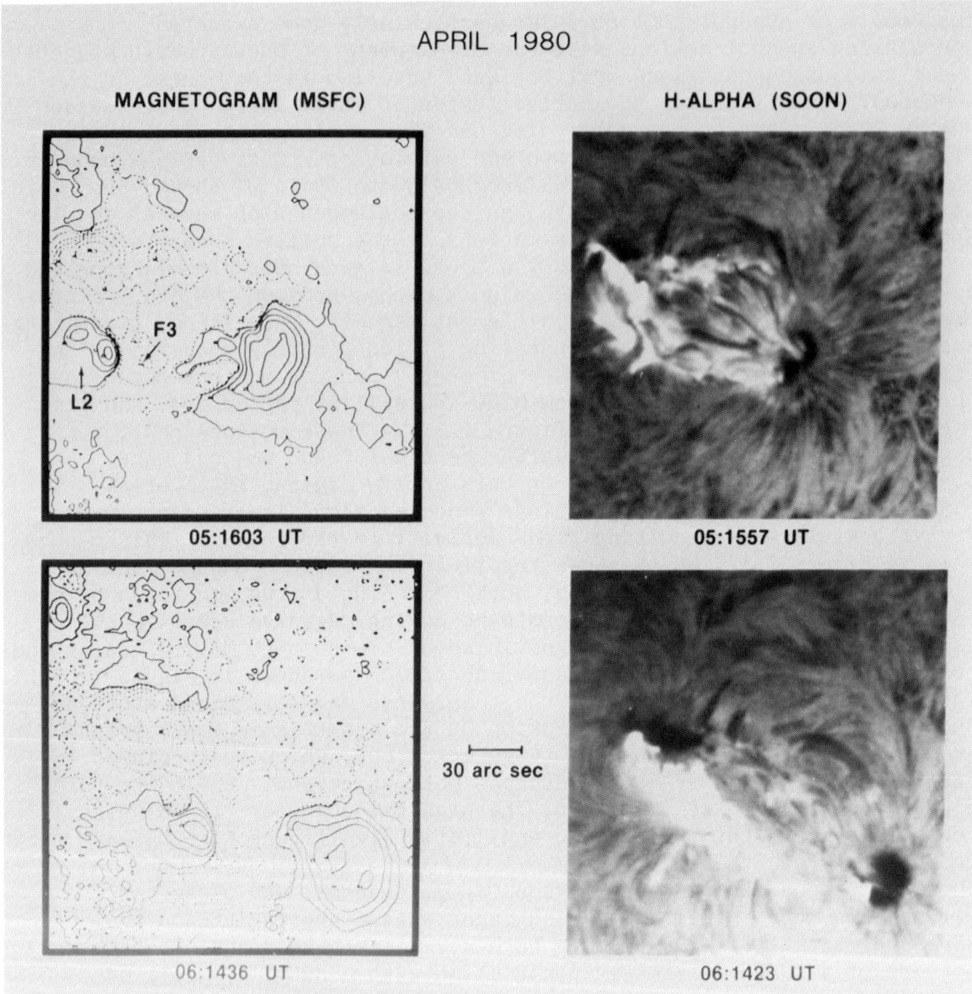

Figure 1.3.5 Magnetograms and H-alpha images for two flares in AR 2372. The left-hand panels depict the observed line-of-sight magnetic field, using the format of figure 3.4. The right-hand panels show H-alpha images from the SOON system for the flares of April 5th (a IB/M5 at 15:57 UT) and April 6th (a 1B/X2 at 14:23). Comparisons of the flare locations with the magnetograms show that both flares occurred in the area of the bipole where significant spot motions were taking place. The fields-of-view are 5′ × 5′.

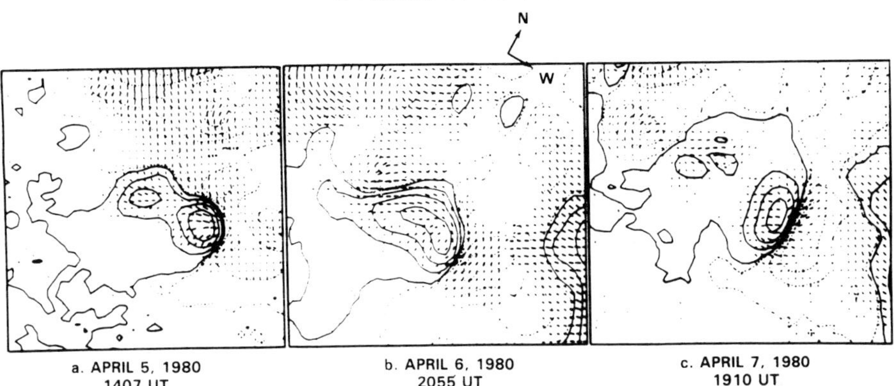

Figure 1.3.6 Evolution of the transverse magnetic field of AR 2372. All panels show the observed line-of-sight magnetic field as solid (positive) and dashed (negative) contours with the transverse field superposed as line segments whose length and direction indicate the strength and orientation of the transverse field. The fields-of-view are 167′ × 167′ and represent blowups of the bipolar region in Figure 3.5. (a) At 14:07 UT on April 5 (shortly before the flare shown in Figure 3.5), the transverse field east of the isolated positive spot was oriented perpendicular to the neutral line. However, to the north and west of this spot, some alignment with the neutral line was seen, implying the presence of shear in the field. (b) By 20:55 UT on the 6th, during the period of spot motion, the strong transverse fields were sheared along most of the neutral line, indicating significant energy storage. (c) Following cessation of spot motion, the shear in the transverse field was less pronounced as observed on the 7th at 19:10 UT.

magnetic shear between the satellite and main sunspot, rather than the high field gradient associated with the satellite spot, was the more important factor for the flare.

There are mechanisms other than spot motions that might produce sheared magnetic fields. Newly-emerged flux can produce such complex magnetic configurations as "kinky" neutral lines, satellite spots, and δ-configurations, all recognized as correlating positively with the frequent occurrence of flares. The statistical studies of J.B. Smith, Jr. (private communication) and Patty and Hagyard (1984a) show that both kinky neutral lines and δ-spots associated with flare activity are areas of sheared magnetic fields. Sturrock (1983, FBS Study Work Group on magnetic shear, Big Bear Solar Observatory) has suggested that in the process of flux emergence the upflows that bring the field to the surface may shear the emerging field due to the coriolis force. Athay et al. (1984) proposed that the sheared magnetic and velocity configurations they observed in the transition region might be produced by two coherent masses of gas of opposing magnetic polarity converging in the stably stratified layers of the solar atmosphere. This suggestion that shear is produced by two converging eddies is based on analogy with shear in the terrestrial atmosphere. Tang (1983) argues that shear also is produced when originally unconnected sunspots of opposite polarity move past each other. Flux cancellation and submergence probably occurs at the neutral lines in active regions (Rabin et al., 1984), and processes may contribute in the formation of sheared magnetic fields.

1.3.2.4 The Role of Magnetic Shear in the Flare Process

However magnetic shear is produced, its existence and association with flares have been directly demonstrated. Questions then arise as to the exact nature of this association, the role of magnetic shear in the buildup of flare energy, and the triggering and eruption of the flare when a critical value of shear is attained.

In a study of magnetic shear, Wu et al. (1984) used a self-consistent magnetohydrodynamic (MHD) model of shearing magnetic loops to investigate the magnetic energy buildup in AR 2372 over the period April 5-7. Wu et al. argued that the evolution of the field observed between the opposing poles of the bipolar region was consistent with a gradual, relative displacement of the bipolar footpoints which occurred during the period April 5-6 (see Figure 1.3.6). The separation of the footpoints of the loops, the maximum footpoint field strength, and the average separation speed of the two spots were all determined from the observational data and used as the initial boundary conditions for an MHD model of an arcade of magnetic loops whose footpoints undergo shearing motions in opposite directions.

Calculations were performed for two different initial configurations of this field: a potential and a force-free field. The photospheric shearing motion of the footpoints of the magnetic arcade was simulated by imposing antiparallel motions on the footpoints on opposite sides of the neutral line. The magnitude of the velocities varied sinusoidally with distance from the neutral line (axis of the arcade). The self-consistent solutions from this model provided numerical values for the magnetic field, velocity, density, temperature, and pressure as functions of two spatial dimensions and time.

Figure 1.3.7 shows a typical result from this calculation; the different energy modes are shown as functions of time as the photospheric shearing motions proceed for the case of an initial potential magnetic field. The magnetic energy buildup clearly dominates, and its growth rate becomes constant after a short interval, while the interaction among the other modes of energy becomes insignificant. Results obtained for different initial values of the parameters are summarized in Table 1.3.1, which gives the energy growth rates in erg day^{-1} per km of arcade length. For parameter values typical of the observed conditions, where magnetic energy density dominated the plasma thermal energy density at the photosphere by a factor of 10, i.e., $\beta_o = 0.1$, and where the maximum shearing velocity was 0.1 km s^{-1}, the rates of energy buildup are 2 x 10^{30-31} and 1 x 10^{31-32} erg day^{-1} for initially potential and force-free fields, respectively, taking arcade lengths of 10^{4-5} km. These values are consistent with the observed flare output rates that were estimated by Krall et al. (1982) to be 2 x 10^{31} erg day^{-1}. Examination of the spatial distributions of magnetic energy showed that the highest concentration of magnetic energy was located near the neutral line, with the concentration being more pronounced in the case of the pre-sheared (force-free) field configuration.

Critical Value of Shear -- The preceding studies demonstrate that magnetic energy sufficient to fuel solar flares is accumulated as a result of increased shear in the magnetic field, with the growth rate of magnetic energy roughly proportional to the shearing speed, to the length of the affected neutral line, and approximately inversely proportional to the initial value of the plasma parameter β_0. The next problem, then, is

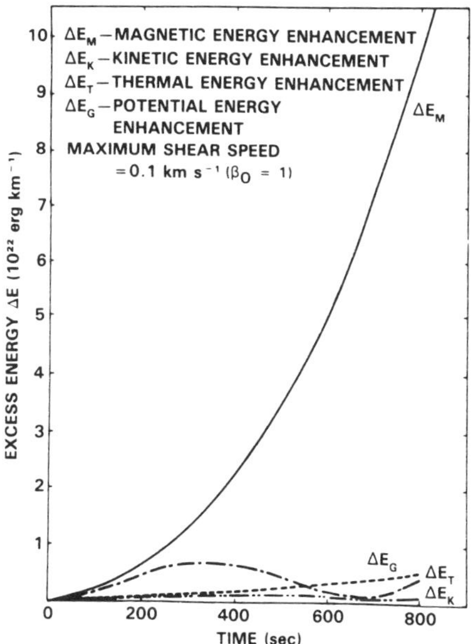

Figure 1.3.7 Magnetic energy buildup through shearing of magnetic fields: results of an MHD model calculation. The calculation models an arcade of magnetic loops whose footpoints on opposite sides of a magnetic neutral line undergo shearing motions in opposite directions. The excess (above that at time t = 0) magnetic, kinetic, thermal, and potential energies are shown as functions of time for an initially unsheared magnetic field configuration. Note that the buildup of magnetic energy dominates and is approximately an order of magnitude greater than the kinetic, thermal and potential energy modes.

to determine the mechanism by which this free energy is suddenly released in the form of a solar flare. In the context of the smooth buildup of magnetic energy through increasingly greater shear in the field, it is tempting to think in terms of a "critical" value of this shear which, when exceeded, triggers the release of energy at these sites of excess shear. Such a concept has been proposed by several theorists (Barnes and Sturrock, 1972; Low, 1977a, 1977b; Birn et al., 1978; Hood and Priest, 1980). The shear in their static models is successively increased until a critical value is reached, after which there are no equilibrium solutions; this critical point has been interpreted as the threshold for the onset of a flare.

Recently, observational evidence for the existence of such a critical shear has been reported by Hagyard et al. (1984a). In their study, these authors analyzed the degree of shear along the neutral line

Table 1.3.1 Energy growth in erg day^{-1} per km depth

Shearing Velocity km s^{-1}	Initially Untwisted Magnetic Field		Initially Twisted Magnetic Field (A 40° Twist)
	$\beta_o = 0.1$	$\beta_o = 1.0$	$\beta_o = 0.1$
0.1	1.94×10^{26}	2.07×10^{25}	1.08×10^{27}
1.0	1.3×10^{28}	2.27×10^{27}	1.12×10^{29}
20	2.66×10^{30}	3.24×10^{29}	

of AR 2372 at a time midway through its early period of flare activity, April 5-7, 1980. They defined the degree of shear, $\Delta\phi$, to be given by the difference at the photosphere between the azimuths of a potential field and the observed field, where the potential field satisfies the boundary conditions provided by the observed line-of-sight field; these fields are depicted in Figure 1.3.8. Using these data, the parameter $\Delta\phi$ was evaluated at 55 points along the neutral line; points 1 and 50 are designated in Figure 1.3.8b. Figures 1.3.8c and 1.3.8d show the variations of the magnitude of the transverse magnetic field (B_T) and $\Delta\phi$, respectively, along the neutral line. In Figure 1.3.8d, the asterisks mark points for which B_T is less than or about 100 G; there probably are substantial errors in the observed azimuth at these points, so the corresponding values of $\Delta\phi$ should be regarded with skepticism. Excluding these points, one can see in Figure 1.3.8d that the degree of shear is non-uniform along the neutral line, and has two negative maxima of values -85^0 and -80^0 along the segments marked A and B, respectively. In addition, from comparisons of B_T with $\Delta\phi$, these very large shears seem to occur preferentially at locations of maximum values in the transverse field strength. The authors argued that the configuration of the magnetic field at the time of the observations (21:10 on April 6) represented the most sheared state attained by the field in the period April 5-8. During the period 14:00 to 21:00 on the 6th, there were no significant changes in the observed azimuth, even though four major flares then occurred.

The sites of flare onset for the 1B/X2 flare at 14:18 UT on April 6, as inferred from the most intense chromospheric flare emissions observed in off-band Hα, were located on either side of the magnetic neutral line along the segments corresponding to A and B in Figure 1.3.8d, that is, at the sites of maximum photospheric shear. Furthermore, in a later flare for which spatially-resolved X-ray data were available (00:48 UT on April 7), the locations of soft X-ray onset were placed at A. Because the flares occurring in AR 2372 in this time period were homologous, the sites of flare onset for these two observed flares are probably representative of the sites for most of the flares during this period.

Based on these results, the authors proposed a scenario for these flares wherein continued magnetic evolution caused the field's maximum shear to exceed a critical value (> 80 - 85°), resulting in a flare at and above the site of maximum photospheric shear. The flare signaled a

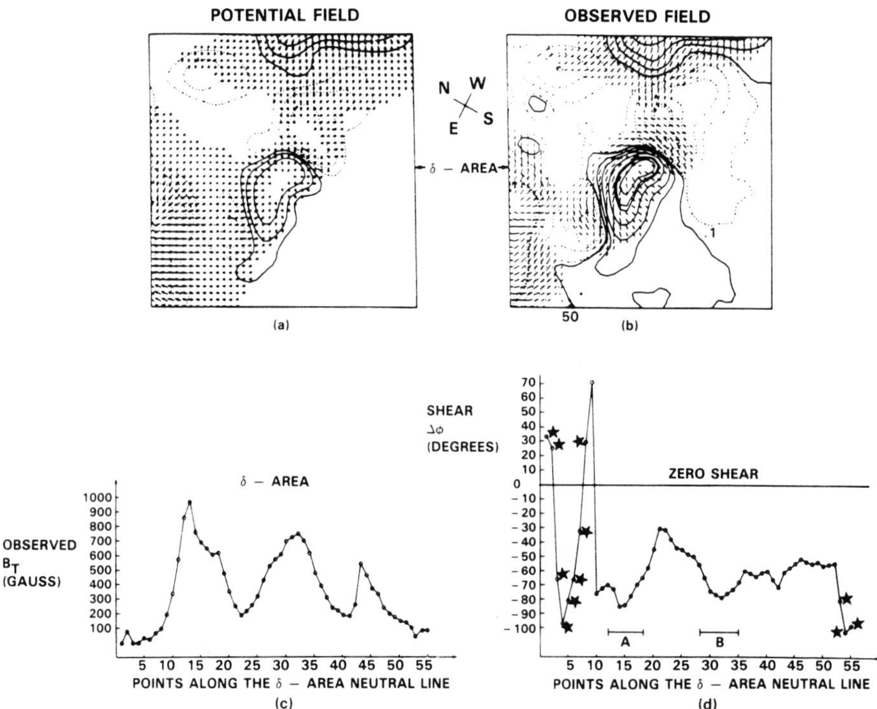

Figure 1.3.8 A quantitative evaluation of magnetic shear in AR 2372 on April 6, 1980. In the region of the small bipole lying to the southeast of the large leader sunspot in that flare-productive active region (see Figure 3.5b), the observed field is compared with a potential field in order to determine their deviations near the magnetic neutral line. (a) The potential magnetic field in the area of the bipole. The potential field was calculated from the observed line-of-sight field. Solid (dashed) curves represent positive (negative) contours of the line-of-sight magnetic field. The directed line segments represent the strength (length) and orientation (direction) of the transverse field. (b) A similar image for the observed magnetic field. In both panels the field-of-view is 1.67′ × 1.67′. (c) Variation of the magnitude of the observed transverse field along the magnetic neutral line. (d) Variation of the degree of shear along the neutral line. Points marked with an asterisk occurred in areas of very weak transverse field and thus are unreliable. The degree of shear exhibits two maxima at points "A" and "B" along the neutral line. These maxima coincide with increased values of the strength of the transverse field.

relaxation of the shear to a value somewhat smaller than the critical value, with further evolution increasing the shear above threshold, another flare, and so on. This scenario, based on observational evidence of persistently sheared fields throughout a flaring epoch, argues against the idea that relaxation of the magnetic field, due to the release of energy in the form of a flare, must proceed until the local shear is negligible. This argument gains support from the preceding analysis of Wu et al. (1984) who showed that energy is more efficiently stored in a field that is already significantly deformed, and that this energy is more concentrated near the neutral line for such a deformed field.

1.3.3 ELECTRIC CURRENTS IN THE PREFLARE ACTIVE REGION

The presence of sheared, photospheric magnetic fields in the pre-flare state implies the existence of electric currents in the atmosphere above the photosphere in which superpotential energy is stored and subsequently released in the outbreak of a solar flare. This release of energy is generally considered to result from resistive MHD instabilities that involve currents flowing either parallel to the magnetic field or perpendicular to the field as in the case of an X-type neutral point. Thus, the concept of a critical value of magnetic shear may have its counterpart in "critical" values for these currents.

Observational data that yield information on the magnitudes and distributions of these currents provide useful constraints on their models of solar flares. The most direct measure of solar currents available comes from observations of the vector magnetic field at the photosphere, from which we can derive the vertical component (Jz) of the electric current density passing through the photosphere, using the relation $(\nabla \times B)_z = \mu_o J_z$. Quantitative values of J_z have been estimated with an uncertainty of 25×10^{-4} Am^{-2} which is inherent in the measurement of the magnitude and azimuth of the transverse component of the magnetic field.

1.3.3.1 J_z Concentrations in Flaring Active Regions

The study by J.B. Smith, Jr. (Section 1.3.2.2) of sheared magnetic fields in active regions was extended by deLoach to include the vertical component of the photospheric electric current density, J_z. This study utilized the magnetic field data for the regions and times listed in Table 1.3.2 with notable flares. The J_z patterns shown in Figure 1.3.2f and in the bottom panels of Figure 1.3.3 correspond to the regions listed in that table. In a majority of these regions, concentrations of J_z were seen along the magnetic neutral line. Indications of additional strong currents often appeared well away from the neutral line; but most of these latter J_z features are probably artifacts of the computational techniques used to resolve the 180° ambiguity in the direction of the magnetic field's transverse component or of the small signal-to-noise ratio in umbral areas. There is a greater level of confidence in the J_z patterns calculated from the sheared transverse magnetic fields in the vicinity of the neutral line where these complications are not a factor.

Table 1.3.2 Date and Times (UT) of MSFC Magnetograms (B) and SMM Events (E) Flare Precursor Matrix Study

AR 2522 JUNE 1980	AR 2544 JUNE/JULY 1980	AR 2725 OCTOBER 1980	AR 2776 NOVEMBER 1980
19 — 1839 E	30 — 2044 B	9 — 2116 B	5 — 1754 B
20 — 1722 B	30 — 2135 B	10 — 1504 B	5 — 2223 E
20 — 2025 B	30 — 2312 B	10 — 1809 B	6 — 1445 B
25 — 1550 E	1 — 1628 E	11 — 1529 B	
25 — 1757 B	1 — 1651 B	11 — 1630 B	
25 — 2156 B	1 — 2052 B	11 — 1741 E	
26 — 1410 B	1 — 2235 B	11 — 1857 B	
26 — 2130 B	2 — 1434 B	12 — 1423 B	
27 — 1402 B			
29 — 0235 E			
29 — 1041 E			

With these caveats in mind, examination of the data showed that J_z concentrations exist in neutral-line regions where flaring occurs; the strength of these currents depends upon the degree of shear present in the vicinity. As discussed in Section 1.3.2.2, a notable increase in field strength and gradient took place in AR 2776 on November 5, and the transverse field directions were closely aligned along the principal neutral line. The J_z calculations for that region show strong currents in that same area, as seen in Figure 1.3.2f. In the case of AR 2725 (Figures 1.3.4c and d), in which the transverse field is highly sheared along the north-south portion of the principal neutral line, strong currents are present in those locations as well. It was also noted that the X-ray flare which took place in this region on the 11th at 17:41 UT had a loop structure that crossed, and was centered over, this area of strong shear and J_z.

The example shown for AR 2522 in Figure 1.3.4a reveals that an area of weak shear lies along the neutral line to the east. Once again the strongest currents lie in the vicinity of these same areas of the neutral line. Finally, in the example of AR 2544, a rather active region of very little magnetic complexity, the amount of shear present was weak and located near the top of the loop-shaped neutral line at the right of Figure 1.3.4b; that is also the site of the only area of significant J_z.

Examination of these and other data for J_z reveals that the existence and general behavior of the photospheric currents in flare-productive active regions are consistent with the degree of persistence and amount of shear exhibited by the transverse magnetic field along the magnetic neutral line. We infer from these observations that the somewhat stable nature of the sheared field configuration throughout the active lives of these regions applies to J_z as well, and that the continued presence of magnetic shear and electric currents at and near flare sites indicates that further activity is likely to occur.

1.3.3.2 Correlations of J_z Concentrations with Sites of Flares

For one active region studied in detail, Hagyard et al. (1984b) found a strong correlation between the sites of flare knots and concentrations of J_z, thus confirming the previous work of Moreton and Severny (1968). The observations were carried out on April 6, 1980, at 21:10 UT in active region 2372 (see also section 1.3.2.4); the results are shown in Figure 1.3.9 which shows the observed magnetic field and derived electric currents. In Figure 1.3.9b, one can pick out seven compact areas of maximum J_z with $J_z \approx 0.025$ A m^{-2}; these maxima are designated by the numbered labels indicated in Figure 1.3.9b. Since the maxima at areas 1, 3, and 7 occur for only one data pixel, they should be viewed with some skepticism in comparison with the other four maxima.

These locations of maximum Jz were compared with the spatial distribution of flare intensities observed for the 1B/X2 flare that began at 14:18 UT on the 6th, about 7 hours prior (see Section 1.3.3.3 following) to the time of the J_z data. In Figure 1.3.9, the locations of the most intense Hα emissions for that flare are shown with respect to the magnetic neutral line derived from an observation of the magnetic field near the time of the flare. In addition, "A" indicates the location

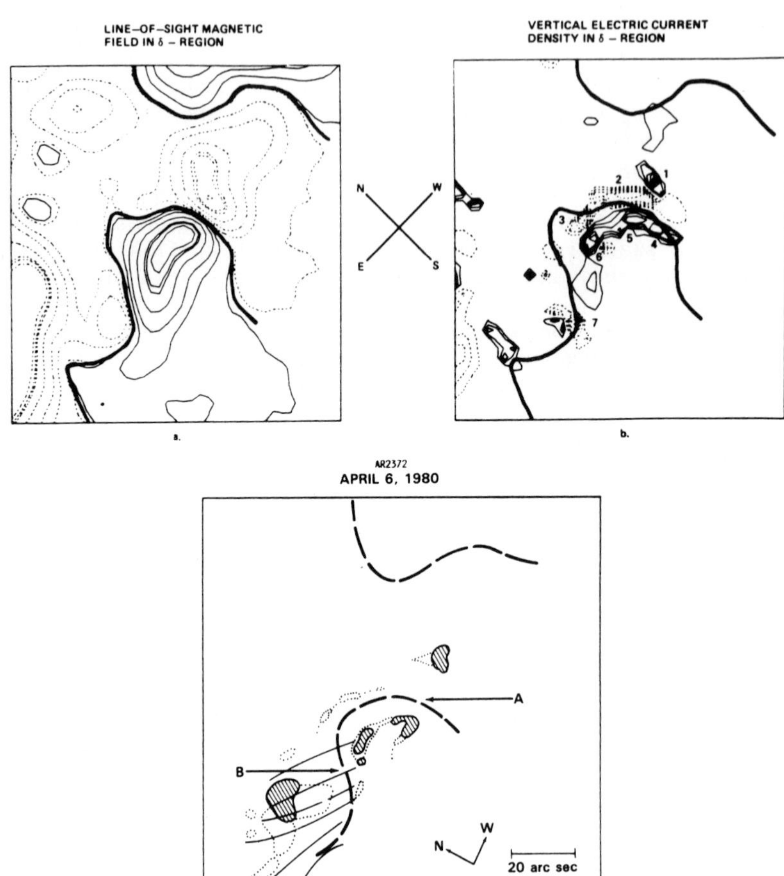

Figure 1.3.9 Concentrations of electric currents at flare sites in AR 2372 on April 6, 1980. (a) The line-of-sight magnetic field observed at 21:10 UT in the area of the magnetic bipole lying to the southeast of the leader spot (see also Figure 3.5b). This area was the site of most of the early flares in AR 2372. Solid (dashed) curves represent positive (negative) contours of the field. (b) Calculated vertical electric current densities (J_z) for the same field-of-view (1.67′ × 1.67′) as panel a. The heavy solid curve delineates the magnetic neutral line in panel a to aid in orientation. Five contour levels of J_z are shown in the panel: 50, 100, 150, 200, and 250 $\times\ 10^{-4}$ A m^{-2}; solid (dashed contours) represent positive (negative) values of J_z, with "positive" indicating a J_z flowing upward out of the photosphere. Note the seven areas of concentrated maxima of J_z. (c) Locations of flare intensities in AR 2372 for the IB/X2 flare at 14:18 UT on April 6, 1980. The 2′ × 2′ field-of-view of this figure is centered on the bipole area of Figure 1.3.14a, and the heavy dashed curves locate the two segments of the neutral line corresponding to those shown in Figure 1.3.14. The hatched regions show the areas of most intense off-band emission, while the dots outline areas of fainter emission. The loops sketched at "B" are inferred from the emission seen at line center and ± 0.4 Å in Hα. The area designated by "A" is explained in the text.

of onset for a flare at 00:48 UT on the 7th (4 hours after the J_z observations) as seen in soft X-ray observations from SMM (Machado et al., 1983). Comparison of Figures 1.3.9b and 1.3.9c shows that the areas of enhanced J_z were approximately cospatial with the sites of flare onset.

1.3.3.3 Evolution of J_z Patterns

The sheared configuration of region AR 2372 persisted during the period of these flares on the 6th and early 7th, with no perceptible relaxation taking place following the flares (Hagyard et al., 1984a). From this observation, one infers that the pattern of J_z concentrations also showed no significant changes during this particular period. However, during most of the interval April 5-7, intense and rapid magnetic evolution took place in the area of the δ configuration with significant flare eruptions. Thus, one suspects that the J_z configuration also was undergoing significant changes during this interval. To investigate this further, deLoach has studied the emergence and evolution of the J_z patterns in the δ-area of AR 2372 from April 5-7. His aim was to determine whether the patterns changed rapidly in strength and/or location, or were in fact maintained over periods long in comparison with the observed flare activity. The region history has been discussed in a previous section (1.3.2.3) and in Krall et al. (1982). Although the region emerged very early on 1-58 April 4, the first vector magnetic field data were not obtained until about 14:00 UT onthe 5th. Thus it was not possible to see the initial emergence of a J_z pattern as the region appeared and grew on the 4th.

For the study of the evolution on the 5th, vector magnetograms obtained at five intervals between 14:00 and 19:00 UT were selected for J_z calculations; the results are summarized in Figure 1.3.10. The sheared nature of the magnetic field along the neutral line is evident in each of the panels depicting the transverse field (B_T): the azimuths are closely aligned with the neutral line. This nonpotential field configuration implies a non-zero curl in these areas and, hence, that electric currents are present. This expectation is borne out in the maps of J_z shown in the bottom panels of Figure 1.3.10. In that figure obvious changes appear in the magnitudes of the line-of-sight and transverse components of the magnetic field as well as in the azimuths near the neutral line. These might be expected because of the evolving magnetic complexity during this period, although some of the subtler changes are seeing effects. The most significant change in the transverse field occurred between the magnetograms obtained at 14:07 and 16:03 UT. This observation is especially interesting since a flare classified as 1B/M5 took place in the region at 15:54 UT. Despite its restructuring the field remained highly sheared at the neutral line. Furthermore, the site of the strongest current density, which was seen in the earliest observation at 14:07 UT, persisted in its location relative to the neutral line throughout the series of observations, that is, just above the "dip" in the neutral line and extending along the left section of that part of the neutral line.

To extend the study of the evolution of J_z in this active region, deLoach used the maps of Jz generated by Krall et al. (1982) for April 6

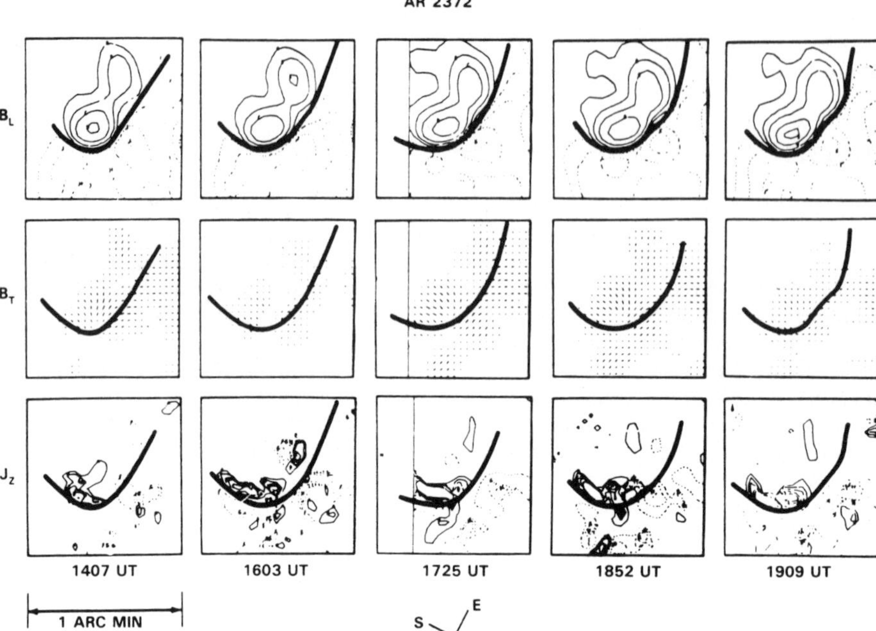

Figure 1.3.10 Evolution of magnetic shear and electric currents in AR 2372 on April 5, 1980. The top row of panels depicts the line-of-sight magnetic field (B_L) at different times in the area of the small bipolar region lying to the southeast of the main leader spot of AR 2372 (see Figure 1.3.5a). Positive (negative) fields are depicted by solid (dashed) countours; the heavy, solid curve denotes the magnetic neutral line. The middle panels show the observed transverse field (B_T) with the neutral line superposed. The line segments represent in length and direction the strength and orientation of the transverse magnetic field. The bottom row of panels shows the derived vertical electric current densities J_z, again with the magnetic neutral line superposed. The contour levels denote positive (solid lines) and negative (dashed lines) values of J_z from 50 to 250 $\times$ 10^{-4} A m^{-2}, in increments of 50. All panels are approximately 1′ $\times$ 1′ in size.

and 7; these show that the sites of strongest current density on the 6th remained along the neutral line, with the primary maximum still located in its western "dip" but migrated westward along with the isolated positive polarity. On April 7th, although still present, the J_z maxima are less clearly defined above the background noise.

Throughout the period studied (April 5-7), the persistent positive J_z concentration is situated in approximately the same location relative to the neutral line. However, the more notable negative J_z areas seem to shift. On the 5th, the primary negative concentration of J_z is located north of the neutral line, coincident with the negative polarity region seen in the panels of the line-of-sight magnetic field (B_L) in Figure 1.3.10 (April 5th). This suggests that the electric currents are flowing along force-free like paths that connect the positive and negative magnetic polarities. From the 5th to the 7th, significant proper motion of the spots associated with these two fields was observed. The changes in the current distributions are consistent with these magnetic changes;

as the fields and currents restructure, the currents still flow across the neutral line.

1.3.3.4 J_z Correlations with Preflare Brightenings

Observations with the SMM/UVSP instrument have revealed numerous intensity enhancements in various bright points in an active region prior to a flare (Cheng et al., 1982). Studies show that these preflare bright points are sometimes, but not always, associated with the UV flare kernels. For example, for the flare on April 8, 1980, at 03:03 UT in AR 2372, Cheng et al. inferred that four UV bright points observed prior to the flare were footpoints of two loops crossing the magnetic neutral line; the ensuing flare occurred in one of these preexisting loops. To investigate the source of these preflare UV brightenings deLoach et al. (1984) have compared distributions of vertical electric current density, J_z, with UV spectroheliograms. If ohmic heating from the electric current is a significant contributor to these brightenings, these comparisons should show that sites of maximum J_z underlie areas of UV enhancements.

The active region chosen for the study was, again, AR 2372, for the period April 6-7, 1980. Two series of UV spectroheliograms were used, one in Lyα (1216Å) and the other in N V (1239Å). The complete series of Lyα and N V spectroheliograms indicate that the region showed persistent internal structure with preferentially bright and dark areas. In addition, the area of strongest J_z maintained its pattern within the δ-region during the period covered by the UV data. When the J_z and UV data were spatially registered, the maximum concentration of J_z fell in an area that was persistently enhanced in the Lyα/N V series.

While this result encourages the view that these brightenings are due to ohmic heating, there is no simple relation between the measured J_z and the heating. For example, there were other areas of enhanced UV emission with no cospatial current at the level of the lowest J_z value measurable. Moreover, the measured J_z maximum of 0.01 A m^{-2} would supply only about 0.1% of the average Lyα flux ($\approx 10^6$ erg cm^{-1} sec^{-1}) radiated from active regions. deLoach et al. concluded that although resistive heating may be important in the transition region, the currents responsible for the heating are largely unresolved in the measurements of their study (≈ 5 arc sec resolution). This conclusion is substantiated by recent work of Rabin and Moore (1984), who suggest that the lower transition region is heated by filamentary, fine-scale electric currents flowing along the magnetic field.

1.3.3.5 Stochastic Joule Heating

In addition to these preflare UV brightenings, it appears that a broad range of smaller-amplitude brightenings are present in most active regions, according to the results of a study by Porter, Toomre and Gebbie (1984) reported at the workshop. Using observations obtained with the UVSP instrument, they found frequent and rapid fluctuations in Si IV and O IV line emission at sites of enhanced intensity within an active region. These brightenings were smaller in amplitude than the UV bursts that have been studied in some detail through observations from the OSO-8 satellite in the CIV and SiIV spectral lines (e.g., Lites and Hansen, 1977, Athay et al., 1980).

The observations reported by Porter et al. (1984) were carried out during seven consecutive orbits of the SMM satellite on October 27 and 28, 1980, in Active Region 2744. Large spatial rasters were performed intermittently to locate the brightest pixel in the region. Subsequent to this, the Si IV and O IV counts in the brightest pixel were measured 1500 times with 0.08 s temporal resolution. 3" x 3" and 4" x 4" entrance slits were used. A total of 67 such sequences was obtained, covering several bright points in the region.

Though the nature of the experiment was to wander from brightest point to brightest point throughout the active region, the pointing did return a number of times to a few consistently bright points. In Figure 1.3.11, the data are shown for one of the brightest sites during the first two orbits on the 27th; this point was perhaps a footpoint of a loop that flared at 02:24 UT on the 28th. Many of the more rapid intensity variations in time were due to small-amplitude jitter in the satellite pointing. By comparing their observations with satellite-pointing data, the authors attempted to remove these effects. The relatively smooth curves drawn through the data points represent their

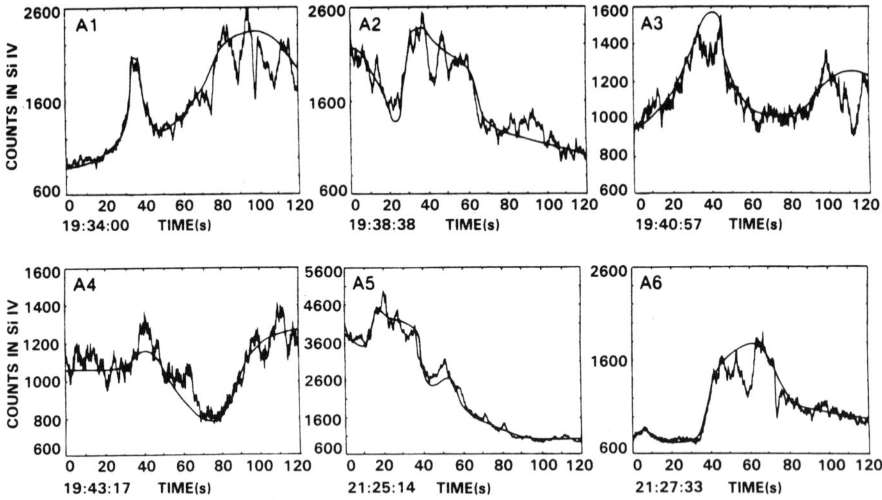

Figure 1.3.11 Rapid variations in time of Si IV λ1394 line emission observed at a persistently bright point in the active region AR 2744. Shown are results from six observing intervals each lasting 120 s; indicated are counts per sampling interval of 0.08 s. Starting times of each sequence on October 27, 1980, are indicated at the lower left. Smoother curves represent estimated true solar output; correction has been made for effects of satellite pointing variations on measured line intensity.

best estimates of the actual solar output. From these and similar high time-resolution observations of bright points in this active region, the authors showed that significant increases in Si IV intensity occurred almost continually on time scales of 10 to 60 s. The intensity enhancement during the brightenings was commonly 20 to 100%, and sometimes larger. These brightenings were present throughout the period of observation, and were so prevalent as to be found in about two-thirds of the selected observing intervals. The analysis of pointing errors leads to an estimate of the possible size of the bright elements as 10 or less.

Evidently, the spectrum of heating responses in the transition region extends from flares and the large amplitude bursts down to the smaller, and usually short-lived (20 s to 60 s), brightenings reported in this study. The transition region is apparently subjected to a variety of heating events on a broad range of spatial and temporal scales. The timing for these smaller events is compatible with an almost instantaneous local heating within the transition region followed by radiative cooling.

1.3.4 Characterization of the Preflare Velocity Field

Of equal importance to sunspot motions in characterizing the preflare state are the Doppler velocity patterns that are observed in active regions. These cover the spectrum from filament activation and eruption, surges, and preflare mass ejections to the predominantly horizontal shearing patterns that are suggestive of cyclonic motions in the photospheric areas where flares occur. For example, some of the most easily recognizable flare precursors are the distinct Doppler patterns observed in the ascending phase of erupting filaments (Martin, 1980). Furthermore, detailed studies of photospheric velocity fields in active regions show that flares tend to be associated with complex velocity patterns (e.g., Martres et al., 1971). Harvey and Harvey (1976) reported that flares seem to occur in areas of strong horizontal velocity shear along the magnetic polarity-inversion line. They concluded that the velocity field is at least as significant as the magnetic field in the flare-buildup process, and urged that equal emphasis be placed on the study of the velocity field. During the SMM observing period in 1980, this counsel initiated many coordinated observing programs in which ground-based Doppler measurements were obtained in conjunction with observations of transition-region velocities using the UVSP instrument. In the following sections we report on results of those observations that pertain to the preflare active region.

1.3.4.1 Preflare Ejecta

During SMM, the Meudon group, represented in the Preflare Group by Schmieder and Martres, observed many surges in their investigation of the flare-buildup process. They were specifically interested in determining whether surges are produced by the same mechanism(s) as flares. They carried out several programs in conjunction with UVSP observations to define the geometrical, thermal and dynamical characteristics of surges (Schmieder et al., 1983).

In one of these programs, they obtained simultaneous ground-based Hα and UVSP C IV observations of recurrent surging in an active region prior to a period of flare activity. The active region (AR 1646) was observed on September 1, 1980, during the time period 12:30-14:08 UT, near disk center. This region developed as a new active center on August 28; there followed a complex evolution over the next few days with a new dipole emerging to the east of the initial one on the 29th (Figure 1.3.12). The new preceding spot increased in size during the 31st and began to move toward the first preceding spot with a velocity of 0.1 km s^{-1}. This motion led to a compression of the following (inverse polarity) spots between the two preceding spots and caused recurrent ejections of matter on September 1 with time intervals of about 10 minutes. Frequent but minor flaring also occurred on the 1st, beginning with a SN/C3 event at 14:08 UT and ending at 22:06.

The Multi-Channel Subtractive Double Pass-Spectrograph (MSDP) in the solar tower of Meudon provided two-dimensional observations of velocities and intensities in the Hα spectral line with a spatial resolution of 1" x 1" and a repetition rate of one minute. The UVSP provided a dopplergram and an intensity map in the C IV (1548 A) line over a 4' x 4' field, followed by a set of 24 intensity maps with a reduced field of 2' x 2' with spatial resolution of 3" x 3" which focussed on the surge event.

The intensity maps in both Hα and C IV showed recurrent maxima in brightness of the facula, but the brightenings in C IV tended to occur a few minutes prior to those in Hα. Three surge events were observed during the time period 12:32-13:15 UT. In Figure 1.3.13, the Hα velocities and intensities are shown for times near the beginning and end of the first event. The surging is clearly located between the two sunspots, along an axis perpendicular to the line joining these spots. The intensity and velocity in the transition region near the beginning of the first surge show that the C IV brightness is elongated in the same direction as the Hα emission but with a slightly greater extension. Only one UV velocity map was obtained for the first surge, at 12:32 UT. This showed upward velocities of 30 km s^{-1} and a more extended (> 2') velocity structure than the one seen in Hα (< 1'). The C IV velociy structure also is more extended than the emission structure seen in C IV.

The time evolution of the surge was obtained from the sequence of MSDP Hα data. The brightening of the facula occurred a few minutes before the maximum extension and velocity of the absorbing feature at 12:36 UT. The redshifted area, which initially corresponded to the emission feature, progressively enveloped all of the absorbing material. The horizontal velocity along the axis of the surge was derived from the position of the absorbing matter and found to be 60 km s^{-1}.

In order to interpret the Hα data in terms of radial velocities, the Hα profiles must be carefully studied. In their interpretation, Schmieder et al. (1984) proposed a "cloud model" to represent the absorbing feature and considered the measured line profiles to be a convolution of two profiles corresponding to the cloud overlying the chromosphere. The observed temporal behavior of the spectra was consistent with a cloud whose velocity reversed direction. Radial velocities were evaluated to be 25-30 km s^{-1} in the ascending phase of the ejection, and 40 km s^{-1} in the descending phase.

Analysis of the complete set of data indicated that recurrent ejecta lasting 10 minutes occurred on September 1 with intervals of about 20 minutes. The observations were consistent with the following scenario. A loop emerging from the brightened facula was compressed by the two merging sunspots. In the first phase, the cold material visible in Hα followed the field lines of this loop with motions upward from the feet. However, the kinetic energy of the material was not sufficient to propel

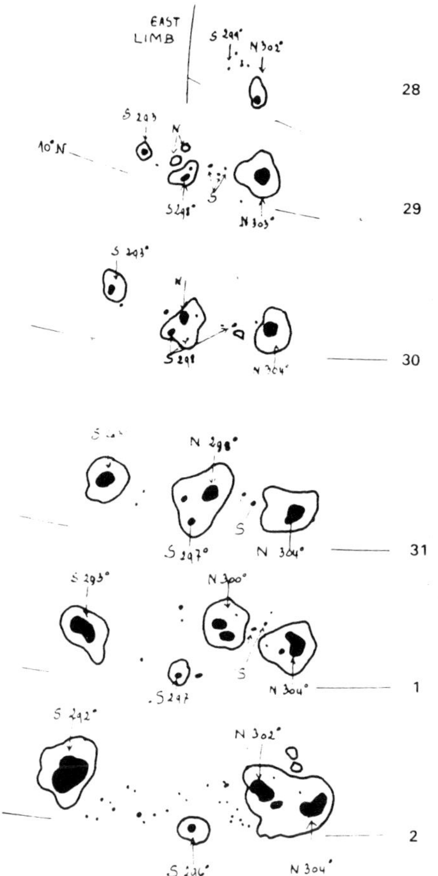

Figure 1.3.12 Evolution of AR 2646 from August 28 to September 2, 1980. The small south (S) polarity spots that were compressed between the larger, north (N) polarity spots can be seen just to the left of the leading (rightward) N-spot on the 31st.

the matter along the whole length of the arch. Instead, the matter fell back along the loop, and the process was repeated periodically. Because of this periodic reorganization, it was conjectured that energy sufficient to trigger a flare could not be built up.

1.3.4.2 Observations of Velocity and Magnetic Shears in Flaring Regions

Evidence of sustained magnetic and velocity shears in an active region, reported by Athay et al. (1984), was discussed at the SMM Workshop by H. P. Jones. The authors compared UVSP Dopplergrams and spectroheliograms in C IV, Si IV, and other UV lines for Hale region 16918 (Boulder region 2517) with ground-based magnetograms, Dopplergrams, and spectroheliograms taken at Kitt Peak, and with Hα filtergrams taken at Big Bear Solar Observatory. The active region was in a complex which passed across the solar disk in the latter half of June 1980. The region was characterized by an unusually long magnetic neutral line that ran more or less eastward from its western extremity for some 300", and then curved rather sharply in a southeastward direction for about 600". A set

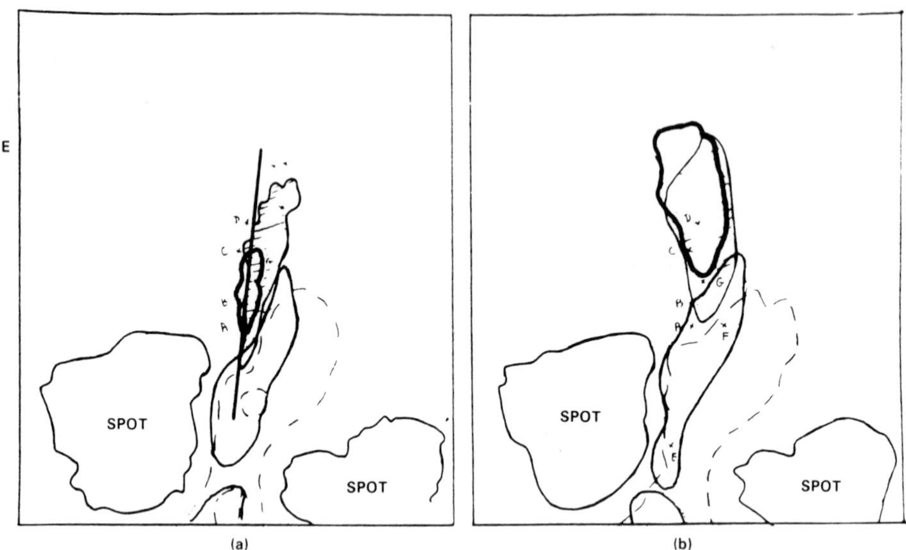

Figure 1.3.13 Evolution of Hα intensity and velocity during the surge event at 12:32 UT on September 1, 1980. The two spots indicated are the N-polarity spots shown in Figure 1.3.12 that compressed the smaller, S-polarity spots, causing recurrent surging on this date. The shaded regions indicate areas of diminished Hα intensity whereas the dashed curve outlines areas of enhanced intensity. The solid contours outline Doppler-shifted areas with the darker (lighter) contour representing upward (downward) velocities. (a) 12:33:20 UT. (b) 12:36:20 UT. The field-of-view in both frames is 1′ × 1′. The projection of the surge axis on the solar surface is depicted in (a) by the solid line.

of six Kitt Peak photospheric magnetograms is shown in Figure 1.3.14; the two segments of the neutral line are quite evident on the magnetogram for June 19. A complex, time-varying system of Hα filaments paralleled the magnetic neutral line; several filaments with footpoints anchored in regions of opposite polarity were spaced at wide intervals along the neutral line. Along the east-west segment, the filaments tended to be long, gently curved, and frequently quite broad in their north-south dimension. Filaments along the second segment of the neutral line were shorter, more curved and of shorter lifetime than those in the east-west segment.

UVSP Dopplergrams were made simultaneously in Si II and C IV several times per day during the period of observations of this region. Dopplergrams in C II were made during one orbit on June 17, and a series of Dopplergrams were made in Fe XII on June 17,18 and 19. Figure 1.3.15 exhibits examples of the Dopplergrams made in these four ions. Ground-based Dopplergrams were made in the Fe I (8688 A) and Ca II (8542 A) lines on June 18, 19 and 20. The authors found similar velocity patterns in C IV, C II and Ca II, although the areal extent of the observed pattern decreased in the order given. Examination of the magnetic and velocity data showed an inversion line in the line-of-sight velocity fields in the transition-region and chromosphere which conformed to the extended, stable, magnetic inversion line. The lateral gradients in both magnetic and velocity fields across this "neutral" line were strong and sharp; this structure maintained its general character for well over a week. From the C IV Dopplergrams, they found velocity differences between adjacent red and blue shifted bands either side of the velocity inversion line that frequently exceeded 20 km s^{-1}. The average velocity gradient across the inversion line was approximately 2 km s^{-1} $arcsec^{-1}$, and this steep gradient often persisted for distances of 9". In the chromospheric Ca II data, these values were reduced: the maximum velocity gradient was 0.5 km s^{-1} $arcsec^{-1}$, and the maximum velocity difference across the inversion line was about 1.8 km s^{-1}. The authors inferred that there was a strong horizontal component of the velocity field since the flow pattern weakened as the region approached central meridian and changed sense after meridian passage. This effect is illustrated by the three C IV dopplergrams for June 17, 18, and 19 in Figure 1.3.15. The authors identified the orientation of the fluid flow in the region with the orientation of the magnetic field, assuming that the fluid flow follows the magnetic field.

The resultant picture of the magnetic field near the neutral line was one of extreme shear at the level of the transition region where C IV is formed, diminishing to greatly reduced shear at the photosphere where the magnetic field was observed. Moreover, the simplest velocity pattern consistent with the data was that of material flow diverging from the tops of low-lying, sheared loops that were closed over the neutral line. The data did not provide obvious clues regarding the supply of material to these loops, but the authors suggested that the sheared magnetic and velocity configuration might be sustained by an underlying, large-scale circulation pattern such as the slow convergence of two giant cells of opposing magnetic polarity.

Numerous small flares and emerging flux regions were reported around the neutral line by Martin et al. (1983), and the observed transition-region line intensities showed continual brightenings. However, the segments of the magnetic neutral line that showed the most evidence of shear produced very little flare activity; the majority of the activity was mainly associated with emerging flux regions. Evidently, this configuration, with strong magnetic shear in the transition region diminishing to reduced shear in the photosphere, is eiher rather stable or does not contain an abundance of free energy.

1.3.4.3 Vortical Velocities Near Flaring Sites

The Solar Department at the Paris-Meudon Observatory carries out simultaneous observations with ≈ 2" spatial resolution in Hα and photospheric spectral lines to study chromospheric structures, line-of-sight magnetic and velocity fields, and flare locations. These studies led to the recognition of large-scale (15-20") structures in both the magnetic and velocity fields: evolving magnetic structures (e.g.,

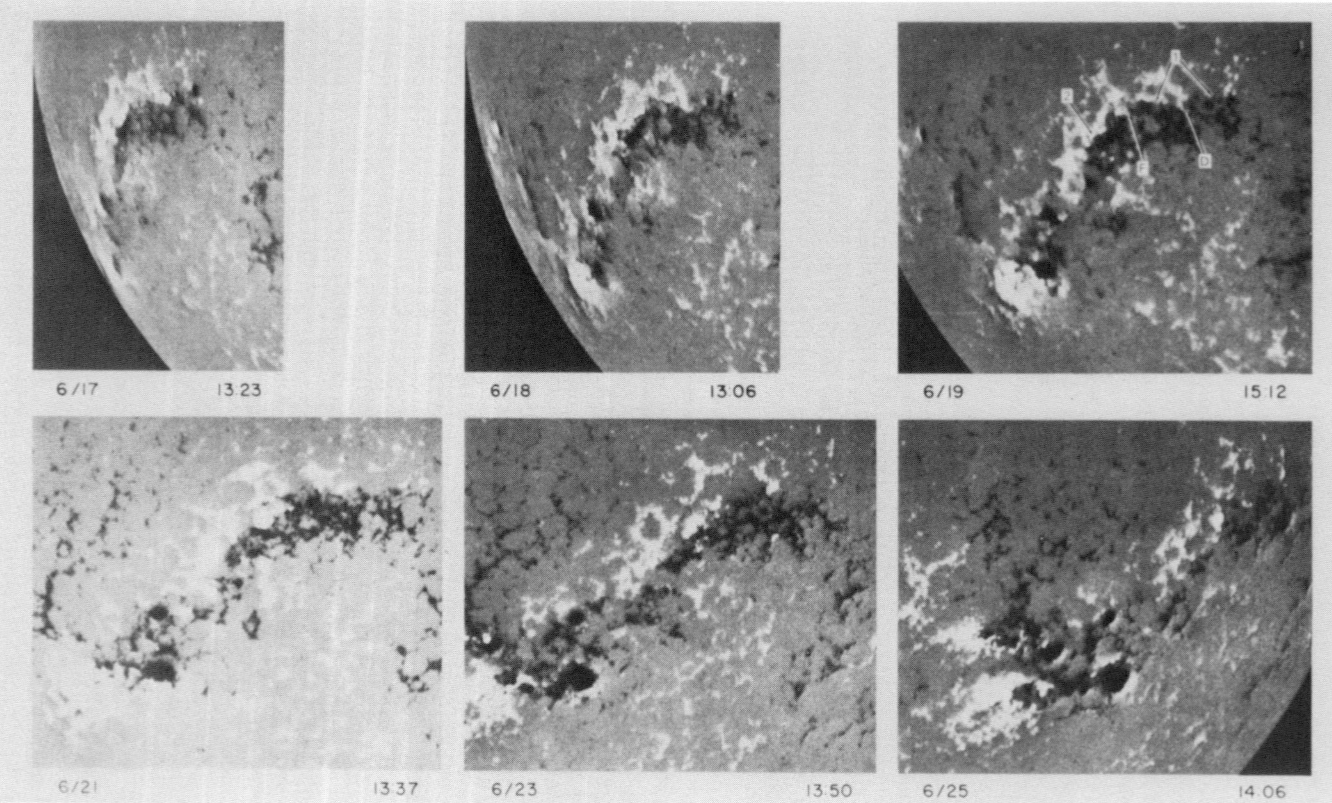

Figure 1.3.14 Magnetic evolution of AR 2517 for the period June 17-24, 1980. The individual frames are portions of six full-disk photospheric magnetograms taken at Kitt Peak. White and dark areas denote positive and negative line-of-sight magnetic fields, respectively.

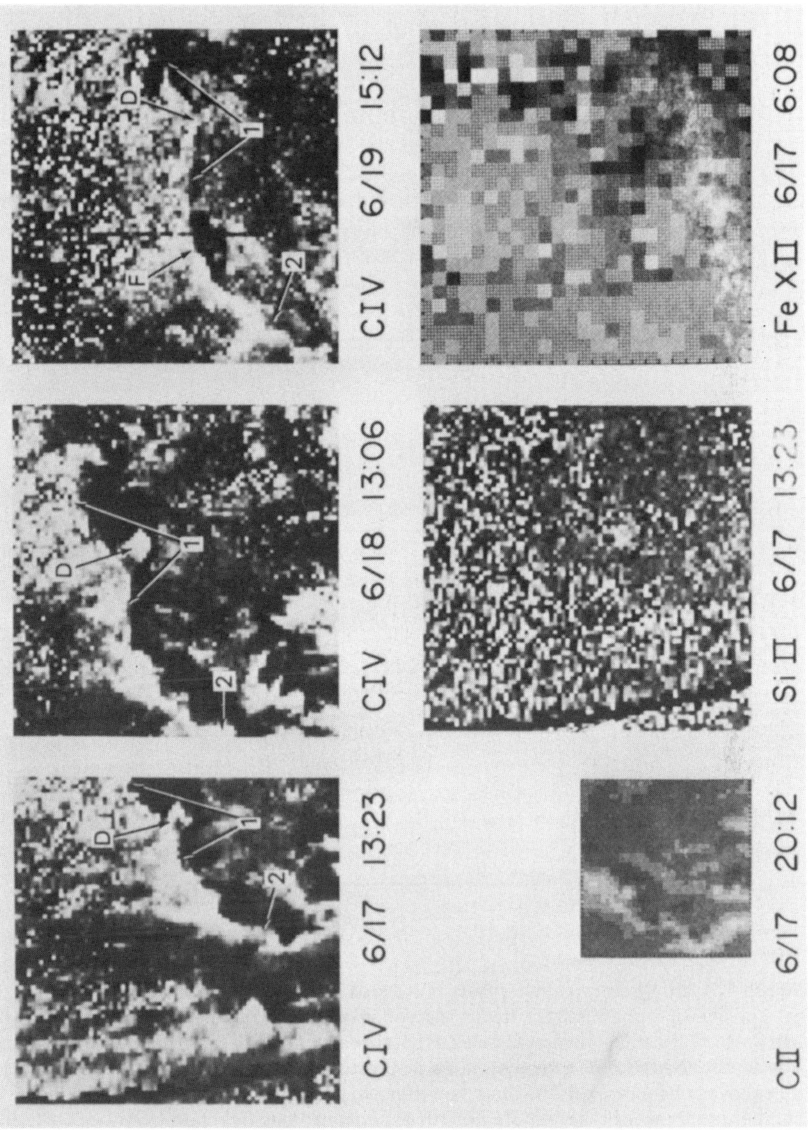

Figure 1.3.15 UV Dopplergrams of AR 2517 for the period June 17–19, 1980. All frames are to the same scale with fields-of-view of 4′ × 4′ for the larger frames. Red shifts are shown in white. The C IV and Si II Dopplergrams are corrected for solar rotation to mimic the observing times of the corresponding magnetograms in Figure 1.3.14. Note the absence of any clear pattern in the Doppler signals for the Si II and Fe XII lines.

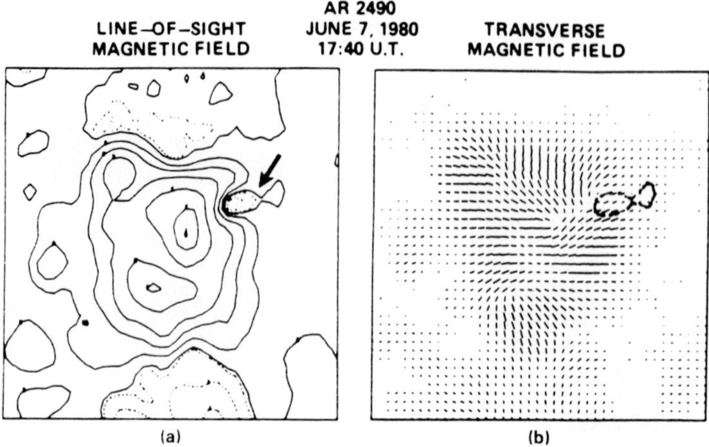

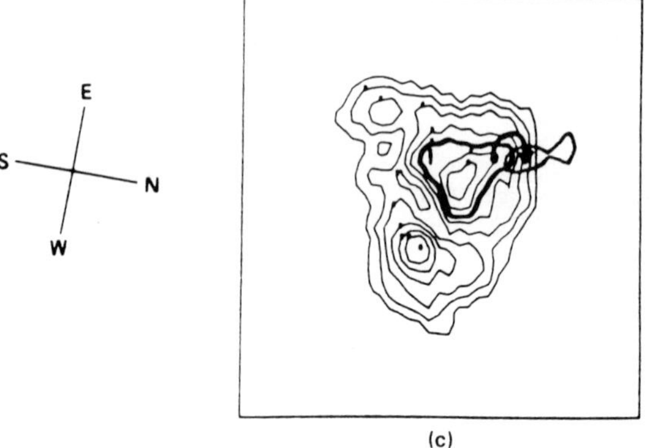

Figure 1.3.16 Vector magnetic field in the area of a vortical velocity cell. (a) The line-of-sight magnetic field observed in AR 2490 on June 7, 1980 at 17:40 UT. Positive (negative) fields are outlined by solid (dashed) curves. The arrow points to the region of parasitic polarity where a vortical velocity cell was observed. (b) The observed transverse magnetic field. The line segments represent the strength (length of segment) and direction (orientation of segment) of the field. The region of parasitic polarity has been superposed to aid in orientation. (c) The observed sunspot and vortex cell. Contours represent levels of intensity of the sunspot. The heaviest contour shows the umbral boundary. Again, the region of parasitic polarity has been superposed on this panel. The arrow indicates the location and sense of circulation of the vortex cell that was observed in this region at 8:49 UT. All fields-of-view are 1.67′ × 1.67′.

Martres et al., 1968) and star-shaped and vortical velocity patterns (Martres et al., 1973). The appearance of combinations of these patterns has been related to magnetic evolution and solar flares (Martres et al., 1974, 1977).

During the workshop series, the Preflare Group tried to identify flare events for which there were observations of vortical motions and transverse magnetic fields. Only one active region was found, AR 2490, which was discussed previously to illustrate the correspondence between the occurrence of vortex motions and flares on June 7 and 8, 1980. Figure 1.3.16a shows the line-of-sight field as measured by the MSFC vector magnetograph on the 7th at 17:40 UT; an area of parasitic polarity is indicated by the arrow. The corresponding transverse magnetic field is seen in Figure 1.3.16b; the area of parasitic (opposite) polarity has been superposed on this image to aid in orientation. The field direction appears sheared along the eastern and western sections of the magnetic neutral line separating the parasitic polarity from the field of the large sunspots. However, it is difficult to interpret these data because the area is so small. The location and sense of rotation of the vortex pattern observed by the Meudon group at 8:49 UT on the 7th is indicated on a plot of the sunspot's intensity in Figure 1.3.16c. The observation of the vortex motion took place during a flare (8:45 - 8:52 UT) in the vicinity of the vortex cell; later observations at 12:09 UT on the 7th showed no signature of a vortex cell in this area. It is interesting to note that this vortical pattern occurs in an area of large horizontal gradients of the transverse field: just to the south of the parasitic polarity in Figure 1.3.16b the transverse field is very weak, whereas to the west of the neutral line the transverse component is larger than in any other area of the field-of-view. Since the vector field was observed almost 9 hours after the detection of such correlations. Clearly, future coordinated observations are needed to study these interesting phenomena.

A preliminary simulation of the short-term evolution of vortex cells by Rayrole and Berton was outlined at the workshop by Martres. The study involved the observation and analysis of magnetic and velocity data for AR 2438 on May 10, 1980. Observations of the velocity field in this region are shown in the top row of Figure 1.3.17; they show the evolution of the velocity during the interval 13:20 to 14:00 UT.

To interpret these observed patterns, Rayrole and Berton performed numerical simulations that combined different vortical flows with typical Evershed motions. The parameters for the simulations were (1) the profiles of the vertical and horizontal components of velocity, V_Z (r) and V_H (r), where r is the distance from the origin of the coordinate system chosen for the structure, (2) the profile of ϕ (r) = $\tan^{-1}$ (V_{HT}/V_{HR}), where V_{HT} and V_{HR} are the tangential and radial components of V_H (r), and (3) the maximum values V_Z, V_H and ϕ of the respective velocity profiles. The assumed horizontal velocity patterns are shown along the bottom row of Figure 1.3.17. Initially, the velocity pattern was a pure Evershed flow. This was followed by the development of a central vortex cell and then an outer vortex cell. The middle row of Figure 1.3.17 depicts the resulting "Dopplergrams" that would be observed as a result of these configurations of velocity. Comparisons of the

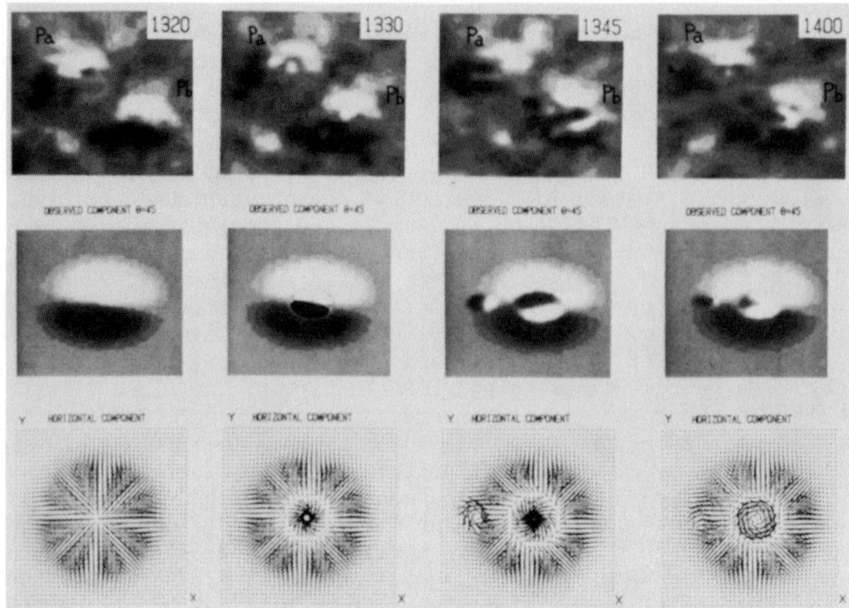

Figure 1.3.17 Comparison between observed velocity patterns in AR 2438 and numerical simulation. Top row: Dopplergrams observed on May 10, 1980, during the period 13:20 to 14:00 UT. Blueshifted (redshifted) velocities are represented by white (dark) shadings. The arrows point to features that are modeled in the numerical simulation. Middle row: ''Dopplergrams'' generated by numerical simulation. The shadings are similar to those shown in the observed Dopplergrams. Bottom row: the velocity patterns assumed in the numerical simulations. The velocity parameters used in the simulations for each panel are the following: 13:20 UT a pure Evershed flow with a small torsion ($V_Z = \Phi$, $V_H = 1500$ m s^{-1}, $\Phi = 10°$); 13:30 UT the same Evershed flow but with a central vortex cell with $V_Z' = \Phi$, $V_H' = 1800$ m s^{-1} and $\Phi' = 20°$; 13:45 UT: the Evershed flow (unchanged), the central vortex but now with a reversed horizontal component ($V_Z' = \Phi$, $V_H' = -2000$ m s^{-1}, $\Phi' = -20°$), and an outer vortex cell with components $V_Z'' = -400$ m s^{-1} (downward), $V_H' = 1800$ m s^{-1} and $\Phi'' = 100°$; 1400 UT: the same Evershed flow, the central vortex with $V_H' = 2000$ m s^{-1} (reversed from the previous time) and $\Phi' = 100°$, and now with a strong upward component $V_Z' = 1500$ m s^{-1}, and the outer vortex with a decreased horizontal component, $V_H'' = 400$ m s^{-1}, the same torsion ($\Phi'' = 100°$), and an increased downward flow ($V_Z'' = 1600$ m s^{-1}). Note the appearance of the ''horseshoe-shaped'' structure in the Dopplergram at the site of the central vortex in the simulation for 14:00 UT. This is a result of the nearly equal horizontal and vertical velocities: $V_Z'/V_H' \approx 1$.

middle and upper panels of Figure 1.3.17 show that the numerical simulations show similarities with parts of the observed Dopplergrams.

The appearance prior to flares of these intriguing vortex patterns in the vicinity of the flaring areas raises the question of their relationship to the sites of maximum shear observed in the magnetic field at flare onset. Certainly, observations imply that the magnetic shear is a fairly persistent feature throughout a flare epoch, whereas the vortical flows only appear 10-60 min before the flare. Thus, the available observations do not support a scenario wherein the vortex motions produce the increased shear at the flare sites. However, more coordinated observations of the vortical velocity patterns and the transverse magnetic field should be carried out to investigate this further. Perhaps the magnetic shear is caused by vortex flows of smaller magnitude than can be detected at the present time. To resolve weak shear motions, it is necessary to average a number of Dopplergrams to overcome the effects of atmospheric seeing, short-lived vertical motions and the 5 min oscillations.

1.3.5 Emerging Flux

The long-established association (Giovanelli, 1939) between flares and changing, especially growing, magnetic fields has been confirmed and extended by many subsequent analyses (Martres et al., 1968, Smith and Howard 1968, Rust 1972). Observations of flare-associated filament eruptions over growing pores (Rust and Roy 1975, Rust et al., 1975) led to the proposal of a specific mechanism to produce two-ribbon flares: the reconnection of newly emerging flux with an overlying filament (Canfield et al., 1974). This mechanism is central to the Emerging Flux Model developed by Heyvaerts et al. (1977). Now emerging flux figures in several models of the preflare state as either continuously driving a flare or triggering it from a metastable state in which magnetic energy has accumulated.

Although the Emerging Flux Model is rooted in observations, its empirical verification in the form pictured by Heyvaerts et al. (1977) is by no means simple. A basic problem lies in the profusion, over a wide spatial range, of magnetic changes. Magnetic flux frequently appears as intensely concentrated bundles in the presence of existing flux. Such changes arise, for example: at the birth of new active regions in the "old" chromospheric network (Bumba and Howard 1965, Born 1974) in already growing active regions at the rate of one or two pairs of bipolar "points" per hour (Schoolman 1973); as "satellite spots" (Rust 1968); as ephemeral regions scattered over the entire solar surface, of which hundreds form and disappear per day (Harvey and Martin 1973); and as "complexes of ativity" (Gaizauskas et al., 1983). With enough spatial resolution and magnetic sensitivity, one should not be surprised to find, during high solar activity, some changing magnetic fields conveniently close to any flare.

How then to discriminate between those changes in magnetic flux whose proximity to a flare is coincidental and those which could conceivably initiate reconnection (e.g., Priest, 1984a, 1984b) or some other process of destabilization (e.g., Hood and Priest, 1980; Kuperus and van Tend, 1981)? The evolution of the flux and the thermal history of its environs must be followed continuously from its first appearance in order to test its relevance to any associated flare. Ideally, we need to measure: the rate, duration, magnetic field intensity and total flux of the new magnetic fields; the magnetic topologies of the emerging flux and of its surroundings; the relative orientation of new and pre-existing magnetic fields; the relative motions of interacting magnetic field patterns, including internal shears; and the changing radiative output with time from photospheric to coronal heights above the increasing magnetic flux. The opportunities to make such measurements existed during the coordinated observations of the Solar Maximum Year. The practical difficulties are such, however, that these ideals were not fulfilled for a single flare.

Evidence that at least some flares are produced by reconnecting "new" and "old" magnetic flux is confused because important details are obscured or still beyond our grasp. Basic concepts about preflare process thus remain unsettled. We shall discuss specific flares associated with the appearance of magnetic flux, which were discussed during the SMM Workshop. But first we examine the background of basic empirical facts concerning flux emergence; for more details, the reader is referred to reviews by Zwaan (1978, 1981).

1.3.5.1 Signatures of Emerging Flux

(i) Chromospheric. The emerging flux model presupposes that magnetic flux rises vertically from beneath the photosphere as bundles of flux tubes in the shape of loops. This fundamental concept is firmly rooted in the customary pattern of growth of bipolar concentrations of magnetic flux ranging in size from ephemeral regions to sunspot groups (Zwaan 1978). While an active region is in its phase of rapid growth, it is referred to as an Emerging Flux Region (EFR, after Zirin, 1970, 1972). The EFR usually has a characteristic signature in the chromosphere which is an important diagnostic for locating new magnetic flux and for tracing its evolution, the Arch Filament System (AFS, after Bruzek, 1967). An AFS consists of a succession of parallel, low-lying arches which bridge the dividing line between opposite polarities in a newly forming group of sunspots. These arches have a distinctive velocity pattern: an ascending motion at the top (< 10 km s^{-1}) and stronger flows (≈ 50 km s^{-1}) down each branch to the footpoints (Bruzek 1969, Roberts 1970). The arches are embedded in conspicuously bright and amorphous Hα plage which extends along the entire length of the arches while an EFR is evolving with its greatest vigour. An EFR will sometimes be resolved into tightly-packed clusters of Ellerman bombs ("moustaches") and their attendant surges when the region is viewed at Hα $\pm$ 1.0Å (Bruzek 1972). The pronounced flow patterns, the intense (almost subflare-bright) plage, and the underlying Ellerman bombs, plainly distinguish AFS from other low-lying arches, the Field Transition arches (Zirin, 1974), which sometimes replace an AFS in its later evolution. The AFS lasts for 3-4 days in the case of an EFR which matures into bipolar spots with penumbrae and with typical life-times of 2-3 weeks (for details, see Weart, 1970; Frazier, 1972; Zirin, 1974; Zwaan, 1978). For ephemeral regions and short-lived active regions, the AFS may last from a few hours to about a day.

Observations at high spatial and temporal resolution of the pre-AFS phase are still so sparse that few key properties of this critical phase of the growth of active regions can be stated with assurance. For example, Martin (1983) claims the existence of an earlier state of development of an active region: a succession of very small flares and associated surges seen one or more hours before the appearance of an AFS. The relation has yet to be determined between the onset of this dynamic stage and the first appearance of a new bipolar field at the photosphere. There is general agreement however that the first chromospheric response to an emerging bipole is the conspicuous brightening of chromospheric faculae observed in Hα and Ca II K (Bumba and Howard 1965, Born 1974, Glackin 1975, Kawaguchi and Kitai 1976). But for an estimation of the delay between the detection of a bipole at the photosphere and the chromospheric response, we must appeal to studies of the more numerous ephemeral regions which are believed to be indistinguishable from active regions at their onsets. It takes about 15 min to be sure that a new ephemeral region bipole has formed, depending upon spatial resolution and sensitivity of the magnetograph; the Hα brightening is detectable within another 30 min (Harvey and Martin 1973). The first arch of an AFS then appears from an hour (Glackin 1975) to about 1.5 hour later (Kawaguchi and Kitai 1976).

Once an AFS forms, the lifetime of individual flat arches is about 20 min; they fade without individually changing length and are continually renewed as long as the AFS is active (Bruzek 1967). On the assumption that every Hα arch traces a magnetic loop, Born (1974) estimated that each brings $\approx 10^{19}$ Mx to the surface. Few ephemeral regions form an AFS; there is some suggestion that a threshold of $\approx 1.5 \times 10^{20}$ Mx for the total flux of an ephemeral region must be exceeded in order to support an AFS (Harvey and Martin 1973).

(ii) Coronal. The morphological description given in the review by Sheeley (1981) of the multithermal coronal plasma associated with emerging flux is based on XUV spectroheliograms obtained with the NRL Skylab/ATM slitless spectrograph. No examples have yet been published of simultaneous chromospheric and coronal observations at high temporal resolution sustained over the birth of an EFR. Most of the following summary, taken from Sheeley (1981; 1980), pertains to the coronal geometries of partially evolved EFR's.

The low temperature EFR plasma (at the 0.5×10^6 K temperature of Ne VII) is confined to the footpoints and legs of magnetic field lines; it rarely forms complete loops. The most striking Ne VII features are long spikes which diverge from the outer ends of an EFR, i.e. from the outer edges of the spots which are normally growing at each end of a bipolar EFR. The spiky structures project 10^4-10^5 km from their footpoints; each lives on the order of 30 min. The high temperature plasma (at the 2.0×10^6 K temperature of Fe XV), on the other hand, is confined to relatively diffuse loops or systems of unresolved loops which join opposite poles within the same EFR or between adjacent active regions, presumably along closed magnetic field lines. Unlike the low-temperature features, the high-temperature loops fade out toward the footpoints. Individual high-temperature structures evolve on a time scale of roughly 6 hr while their collective patterns endure for several days or more.

Sheeley and Golub (1979) were able to establish variability on the scale of ≈ 6 min in the multiple, small, elongated, high-temperature (Fe XV) structures in both the active region and a coronal "bright point" (possibly associated with an ephemeral region). Their observations suggest that the life history of a single "bright point" consists of continuous sequence of miniature loops which evolve rapidly and independently of each other.

Coordinated observations of ephemeral regions and coronal bright points were performed briefly during the Solar Maximum Mission (Tang et al., 1983). These limited data show that more ephemeral regions are found at the photosphere than are their counterparts at higher levels in the atmosphere. There is no indication that UV bright points (observed in the 1548Å line of CIV, characteristic of the transition zone) are enhanced before their associated ephemeral regions are born in the photosphere. Partial UV light curves are available for only two bright points which can be identified with specific ephemeral regions. Brightness of these points maximizes 1/2 and 1 hr after their corresponding ephemeral regions are detected; the brightness drops drastically in the final hour of these two ephemeral regions. (These times are strongly influenced by the sensitivity and spatial resolution of the magnetograph used for their estimation).

At the time of this Workshop, no account existed of the evolution from birth of the spatially-resolved microwave emission from ephemeral regions or EFRs.

(iii) Photospheric. The onset of flux emergence is defined with respect to the photospheric level - specifically by the first detection (e.g., by a magnetograph) of elements paired as a bipolar ephemeral region or a new EFR. Discussion of this aspect of emerging flux has been deferred in order to emphasize the following point: the photospheric feature so commonly cited in the literature on solar flares, the growing pore, is preceded by other well-defined phenomena in both the photosphere and chromosphere.

Thus Bray and Loughhead (1964) found anomalous alignments and darkenings of intergranular lanes which preceded the appearance of a pore by 3 hr. Strong downdrafts, amounting to 1-2 km s^{-1} or more as measured with photospheric lines (Kawaguchi and Kitai 1976, Bumba 1967, Brants et al., 1981) occur near protopores. The downdrafts are localized in small patches ($\approx$ 2") beside, not inside, a protopore (Zwaan et al., 1984). The downflow lasts for at least an hour and stops after the initial detection of the associated photospheric dipole (Harvey and Martin 1973), or after the appearance of the first arches in an AFS (Born 1974).

An existing pore in an EFR grows by adding new flux at the edge facing the centre of the growing active region (Brants 1983). This behaviour is consistent with the later developments in an active region (Zirin 1974): sunspots invariably form at the outer ends of an active region while new flux is added near the middle. The leading and trailing umbrae grow by coalescence of pores of the same polarity (Vrabec 1974, McIntosh 1981). A major region will attain its maximum flux of $\approx 3 \times 10^{22}$ Mx in several days. A new group of spots spreads apart in longitude at a typical rate of 0.1 km s^{-1} which can be sustained for 5-6 days (Kiepenheuer 1953). At birth, the velocity of separation within an EFR can range from 1 to 2 km s^{-1}; some hours later it drops to 0.5 km s^{-1} or less (Born 1974).

In the simpler circumstances of the more abundant ephemeral regions a conspicuous spreading, at 5 km s^{-1}, is evident in the first minutes after the bipolar pair of magnetic elements are detected (Harvey and Martin 1973, Martin 1984). In about 30 min, the expansion rate of an ephemeral region drops by an order of magnitude and remains roughly steady for typically 6 hr (Harvey and Martin 1973).

In summary, the emergence of new magnetic flux is marked in the photosphere by rapidly spreading bipolar fragments, in the chromosphere by conspicuous brightening of chromospheric faculae followed by the formulation of an AFS with distinctive strong downflow at its roots. Pores appear several hours later.

We do not yet have the observations to relate fine-scale structures in the corona with their individual counterparts in the lower atmosphere during these initial few hours. We specifically exclude, as evidence for emerging flux at coronal heights, rapidly expanding structures such as: coronal transients; coronal loops or arches seen in the λ5303Å line; filaments and other mass ejected from a polarity inversion line. Phenomena such as these may rise from other causes than emerging flux, such as for example, reconnection.

1.3.5.2 Flare-Associated Emerging Flux

The association between flares and EFR is strong. But the numerous flares which erupt within an isolated growing region while it is still in the AFS stage are minor ones (Bruzek 1967, Weart and Zirin 1969). More intense flaring is observed when new flux appears within an existing active region, especially if the emergence places following polarity ahead of normally preceding polarity (Zirin 1970, Vorpahl 1973). The importance of interactions between adjacent flux patterns prevails as well as the scale of ephemeral regions. Miniature flare-like events in Hα occur when a spreading bipolar ephemeral region interacts with neighbouring elements in the network (Marsh 1978).

For the larger, more interesting flares, we therefore look at the magnetic changes which can be thought of as flux emerging in a well-developed active region. These can be placed in two broad categories: changes associated with new flux bearing the AFS trademark or changes within already developed patterns of flux without AFS. We consider as well the influence of observed magnetic changes in an active region on the way a filament erupts preceding a flare.

(i) Development of Magnetic Complexity by Emerging Flux. In dealing with the first category of changes, we note that active regions do not form at random. They show a remarkable tendency to cluster in space and time (Gaizauskas et al., 1983) as complexes* of active regions which last for many solar rotations. Liggett and Zirin (1984) measured a rate of flux emergence 27 times higher within active regions than in quiet background areas at the same latitudes. If the packing density of new regions within a complex becomes high enough, we can expect greater magnetic complexity also, and therefore enhanced flare productivity (Giovanelli 1939, Smith and Howard 1968, Bell and Glazer 1959). This is not a common occurrence. Present knowledge indicates that complexity can

*"Complex" used in this sense and without a modifier refers exclusively to spatial and not magnetic complexity.

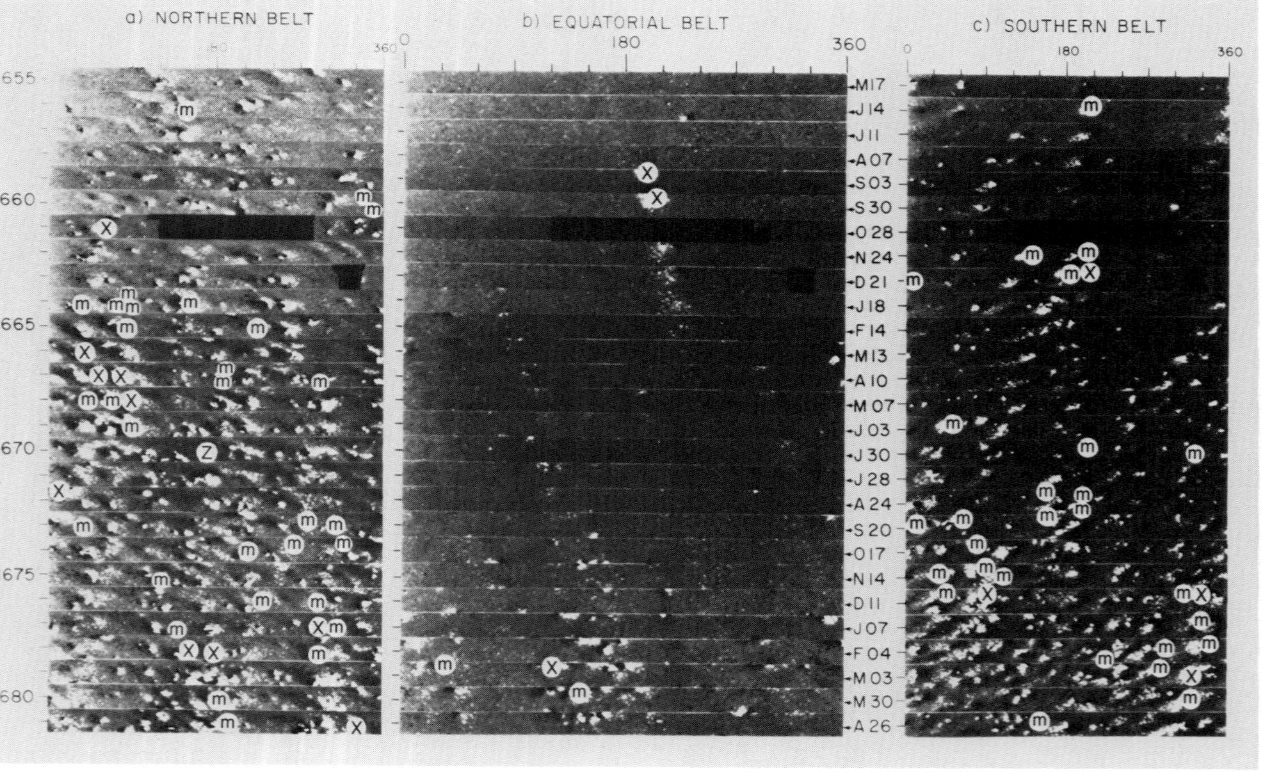

Figure 1.3.18 Chronological arrangement of KPNO synoptic magnetograms of magnetic flux density adapted from Gaizauskas *et al.* (1983). The Carrington rotation number for each strip is shown in the extreme left column. (a) Latitude belt 10°-40° N for each strip of 0-360° in longtitude. (b) Equatorial belt of latitudes, 10°S-10°N for each strip of 0-360° in longitude. (c) Latitude belt 10°-40°S for each strip 0-360° in longitude. The regions in which they occur are designated as "m", "X", or "Z" according as they contain at least one flare classed, respectively, from M_1 to M_9, X_1 to X_9, or X_{10}, in the SESC system.

develop in several ways when two or more bipolar EFR overlap: either beginning with their simultaneous appearance as AFS (Weart 1970), or from the later intrusion of an EFR into a mature, closely-spaced active region (Zirin and Tanaka 1973, Tang 1983, Zirin 1983), or even from the expansion and interpenetration of adjacent bipolar active regions at the same latitude (Tang 1983). The time in the evolution of the region and speed of superposition are probably key factors.

For the SMM Workshop, Gaizauskas and McIntosh (1984) investigated how the rejuvenation of magnetic flux in complexes of activity affects flare productivity. They compared two sets of homogeneous data for the same 27 solar rotations between 1977 and 1979; the set of flares classed M_1 or stronger from among all 1900 X-ray flares recorded in that period by the SMS-GOES satellites; and the set of synoptic maps of the photospheric magnetic field produced by the Kitt Peak National Observatory (Harvey et al., 1980). The subset of 384 flares so defined is not distributed randomly among the 934 active regions enumerated during the same interval. One-half of the flare subset is accounted for by only 12 regions, the other half by another 65; a tenth of the subset erupted in one hyperactive region, McMath 15403 (CMP on 15 July 1978).

The 77 regions with strong flares, clustered in 37 complexes of activity, are marked according to the class of their strongest flares on the chronological arrays shown in Figure 1.3.18 of active-belt strips from the KPNO synotic maps (cf. Figures 2 to 4, Gaizauskas et al., 1983). For half of the complexes, flaring continues above the M1 threshold for 2 or more rotations. In extreme cases, powerful flares do occur on the fifth (in an X-class region) or even on the ninth rotation (for an M-class region). But in at least 7 complexes, flares in the X-category occur without an episode of even M-class flares during the preceding rotation. The most conspicuous example of an immediate output of very intense flares in just one rotation is McMath 15403. Its behaviour is compared in Figure 1.3.19 with another flare-rich complex, the so-called "great complex" (Gaizauskas et al., 1983).

For the great complex, Figure 1.3.19a, the magnetic flux jumps suddenly from a low initial level to a high level which remains roughly steady over 6 rotations before it subsides quickly. But the incidence of strong flares in this same complex is more erratic as indicated by the X-ray Flare Index (XFI, hatched bars). The XFI is both high and low during the strong outbreak of flux sustained for 6 rotations. When the XFI is high, the great complex consists of extended clusters of large spots with many EFR and strong shears; later, when the XFI is low, he great complex contains only one large spot, some EFR and considerable flux distributed throughout the photospheric network. In contrast, hyperactive region McMath 15403 reaches a relatively high flux level and very high XFI in just one rotation (Figure 1.3.19b). During its visible passage, it is a compact and formidably complicated δ-configuration.

Another example of rapid, highly localized development of magnetic complexity occurs during the evolution of the complex activity containing Hale Regions 16862, -3, -4 from 1980 (Gaizauskas 1983) to June 1980 (Martin et al., 1983). Only one new region formed during the first passage of this complex: a very compact δ-configuration which formed in less than a day and produced the only energetic flares for that entire

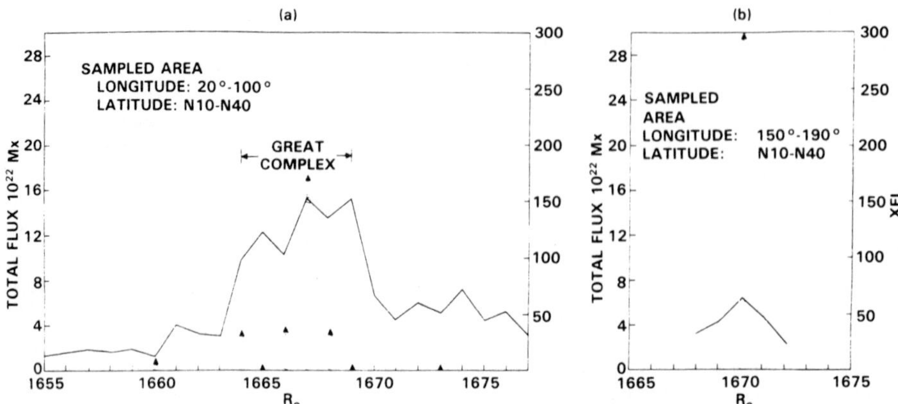

Figure 1.3.19 Episodes of strong X-ray flare activity during the evolution of complexes of activity: (a) in the latitude-longitude limits enclosing the great complex of activity, April-July 1978; (b) in the latitude-longitude limits enclosing the hyperactive McMath Region 15403 (one-half the area of (a)). Abscissa is the Carrington rotation number; left ordinate is the total flux density in the enclosed area; right ordinate is the X-ray Flare Index (XFI) in both panels. XFI = $\Sigma j(M) + 10\Sigma j(X)$ where j(K) is the multiplier used in X-ray class K (K = either M or X). For an X3 flare J(X) = 3 so that its XFI is 10 times greater than for a single M3 flare.

disk passage. On its second passage, 17 new bipolar regions were identified within the evolved complex in an 8-day interval. The great majority of flares, mostly subflares, then erupted within EFR or on their boundaries (Martin et al., 1983), or in locations where new flux appeared adjacent to recently evolved flux from the same rotation (Schmahl 1983).

The build-up of severe complexity is thus more likely to involve new patterns of flux appearing within a few days rather than patterns differing in age by as much as one rotation of the Sun. This finding is consistent with the rapid disappearance, in situ, of large quantities of magnetic flux and its rapid replacement by new, differently arranged flux in a still-active complex (Gaizauskas et al., 1983). The months-long survival of complexes of activity is not in itself sufficient to build up magnetic complexity as supposed by Bumba and Obridko (1969).

Currents flowing in the strongly sheared and twisted magnetic fields in magnetically complex regions are presumed to build up a reservoir of "free energy" which is then released in flares (Nakagawa and Raadu 1972). These concepts encounter difficulty, however, with the phenomenon of flare homology. Experience during the SMY indicates that homology is not at all rare (Woodgate et al., 1983) and must be included among the constraints on flare models. If prior storage of energy is a necessary condition, it must be released by the same process in the same spatial domain in order to reproduce the geometries and time profiles of the flare emissions in a chain of homologues. Now, if flares are "released" by a mechanism which operates as a simple safety valve attached to a constantly stressed reservoir, the output energy ought to remain the same for each flare in a homologous chain. On the other hand, if the release mechanism is not continuously operative, the output energy per flare ought to be proportional to the build-up time between activations of the release mechanism.

Neither steadiness nor proportionality of output were evident for the chain of 5 homologous flares observed within 13 hours, 28-29 May 1980 in a compact δ-region (Gaizauskas 1983). The compactness makes it difficult to distinguish with certainty between relative motions of pores (Nagy 1983) and their appearance or disappearance in a few hours, with new flux possibly appearing in nearly the same place. Yet it seems implausible that new flux, if it were to be acting as the flare-release mechanism, would re-emerge in precisely the same location and in the same manner so many times in succession. The variation of output energy per flare, despite the close homology in this instance, suggests that flares could be driven from the source whose energy supply determines the ultimate flare energy. For example, this chain of flares could have arisen from the shearing action produced during the rapid relative motions of adjacent pores in such a way that the deformation of the magnetic field was not uniform with time.

At some point in the growth of a bipolar active region, the emergence of its flux must stop. Other new bipoles may continue to appear nearby as part of a process for sustaining a complex of activity. Their signatures are easily recognized (Section 1.3.5.1) and correspond, to the best of our knowledge, to flux which emerges from well below the photosphere. But other prominent changes in magnetic flux happen in well-developed active regions without these signatures; they too figure in flares. The origin of these features being uncertain, they are designated here as "appearances" rather than "emergences".

For example, magnetic flux outflow (Vrabec 1974) appears around the edges of some large-spot penumbrae as a rim of reversed polarity moving radially away from the spot with steady speeds ≈ 0.2 km s^{-1} and greater (Sheeley 1969, Harvey and Harvey 1973). It appears without forming AFS and without enhanced Hα emission even though its average flux may be $\approx 10^{19}$ Mx (Harvey and Harvey 1973). Magnetic flux outflow can produce many small surges and even small subflares (Roy and Michalitsianos 1974). Its relationship to other persistent flaring sites on the periphery of sunspots, the so-called "satellite sunspots" (Rust 1968) is uncertain. During observations coordinated with SMM in May 1980, a location of intermittent magnetic flux outflow activity on the rim of the leading spot in Hale 16863 was found to persist for several days as a favoured site for miniature surges and subflares (Gaizauskas 1983). Magnetic flux outflow is believed to be part of a process whereby a sunspot dissolves through fragmentation of its flux tubes (Harvey and Harvey 1973). In that sense, magnetic flux outflow is conceived as a redistribution of already emerged flux, rather than an emergence from a great depth.

Even less understood than magnetic flux outflow is the appearance of pores and small spots without AFS in the middle of mature bipolar active regions. Pores can be found in profusion around the polarity inversion lines underlying the field-transition arches in large, slowly spreading regions like Hale 16863, -4 (Gaizauskas 1983). At the SMM Workshop, Schadee provided examples of transient miniature X-ray events in the lowest energy (3.5-5.5 keV) channel of the HXIS. The low background noise in this instrument allows the detection of very weak X-ray sources. These sources are abundant in field-transition arches and in general along polarity-inversion lines of mature active regions (cf. Gaizauskas 1983,

Figure 10). Some sources enact the 2-ribbon flare scenario in miniature, such as the tiny subflares in mature region Hale 16850 which erupted adjacent to the very long filament which separates still-evolving network of opposite polarities (Schadee and Gaizauskas 1984). VLA observations at 6 cm of other bipolar regions show that radio sources over similar areas of transverse magnetic fields are brighter than those associated with sunspots (McConnell and Kundu 1984).

No systematic study has yet been made to explain why a bipolar AFS transforms into a bipolar region with field-transition arches, why coronal emission is enhanced and eruptive along these polarity inversions, or whether the pores appearing at this stage bear any special relation to the field-transition arches. We may speculate that some of the rapid appearances of new photospheric structures in the middle of mature active regions represent a concentration of superficial flux rather than the emergence of deep-rooted flux. Whatever the origin of these new concentrations, it is in just these circumstances of late development in the active regions that the 2-ribbon flares, advanced in support of the emerging flux model, have occurred (Rust and Roy 1975, Rust et al., 1975, Rust and Bridges 1975, Canfield and Fisher 1976, Hoyng et al., 1981, Simon et al., 1984, Moore et al., 1984).

(ii) Filament Activations and Emerging Flux. Strong circumstantial evidence links the appearance of new pores with flares for the events just cited. But an attempt to extract from them the quantities needed for a rigorous test of the Emerging Flux Model is soon frustrated by the lack of essential pieces of information. Missing most frequently from the examples cited above are the velocity fields associated with emerging flux and the morphologies of fine structures which would yield the relative orientations of new and pre-existing magnetic fields. Breaks in the data at critical periods further amplify the uncertainties. Conclusions drawn from changing patterns of magnetic fields remain inferences based on the premise that rising flux creates the changes.

The small pores of opposite polarity which are believed to figure in the destabilization of a filament as described by Simon et al. (1984) approach each other, contrary to the normal spreading action in an EFR. No direct evidence is provided that these pores are linked magnetically to each other or to the filament; their presence and proper motions are suggestive but enigmatic. For the well-studied class 2B flare of 1980 May 21, Hoyng et al. (1981) attribute the destabilization of a long filament over an extended polarity-inversion line to the emergence nearby of a bipolar region containing a new pore. Subsequent analysis of magnetograms by Harvey (1983) suggests that the pore formed not by emergence but by the compression of existing flux at the surface. New flux did appear nearby as patches of polarity opposite to their unipolar surroundings and in such a way that the shape of the polarity-inversion line was sharply altered. The net flux at a location directly beneath the activated filament actually decreased. These changes may have destroyed the eqilibrium between the filament and its surroundings without recourse to reconnection.

An example of preflare filament eruption without associated merging flux is the 1980 June 25 class 1B flare at 1552 UT studied by Kundu et al. (1985) and discussed at length during the SMM Workshop. The filament was located in the trailing part of AR 2522 (Hale 16931). This region contained two sites of emerging flux which produced several subflares on that same day: on the northern rim of the large leading spot, and in a small magnetically complex region immediately SE of the erupting filament. An adjacent region of comparable size, AR 2530 (Hale 16923) had even more extensive emerging flux in its mid-section. The flare-associated filament was only 40,000 km long; its mid-point passed close to a compact cluster of pores which had no associated chromospheric activity. A close examination of KPNO magnetograms taken hours before and after the flare shows minor, subtle changes in the magnetic flux of photospheric structures adjacent to the filament (Kundu et al., 1985). Nothing observed in 7 hours of rarely interrupted wavelength-sweeping across Hα at high spatial resolution with the photoheliograph at the Ottawa River Solar Observatory can be suspected as adjacent emerging flux which might trigger a filament eruption.

Strong proper motion of the leading sunspot in AR 2530 ($\approx$ 150 m s^{-1}) implies strong shearing near the inter-region boundary which intersects the active filament at its western footpoint (Schmahl 1983). There were no significant proper motions of the spots in which the filament terminated or of the pores near its mid-point. The regions were sufficiently far from the center of the disk that perspective aids the determination of the velocity structure of the filament. Its lateral displacement can be followed with respect to the fixed pattern of photospheric features. Doppler shifts indicated by off-band filtergrams give the sense and an estimate of the line-of-sight velocity. The combined measurements show that the center of the filament began rising slowly ($<$ 1 km s^{-1}) about 3 hours before the flare and accelerated steadily but not uniformly right into the eruptive stage. Matter drained down each end of the filament. About 2 hours pre-flare, the downflow at the eastern footpoint gained a sudden impetus coincident with the activation of a huge boundary filament on the southern side of AR 2522. This larger structure was co-terminal with the flare-associated filament inside the large, trailing sunspot.

At 20 and 10 min preflare, the Hα blue shifts towards the eastern half of the structure increased substantially while at the same time the lateral movement of the filament ceased entirely. This indicates that the suddenly enhanced Hα blue shifts observed during this period had to arise from axial flow rather than an accelerated upheaval. For these two episodes of enhanced flows, filtergrams at Hα ± 0.6Å were subtracted in pairs. They reveal that the filament was rapidly untwisting in the fashion described by Rust et al. (1975) for an event in which new spots strengthened or first appeared adjacent to a filament during its eruption and a subsequent flare. Yet there were no comparable flux changes in the 1980 June 25 event. Bright Hα kernels and an additional short velcity structure appeared behind (and presumably beneath) the most twisted portion of the rising filament for just several minutes during the last episode of enhanced axial flows. These transient features are co-spatial, in projection, with briefly enhanced blue-shifted C IV emission ($\sim$ 10^5 K plasma) detected by the UVSP on board SMM. We may speculate that these transients are counterparts of the short-lived velocity feature also recorded by Canfield and Fisher (1976) for another erupting filament.

Emerging flux did not figure as a direct trigger of the flare of 1980 June 25. This is not to deny that flux emerging elsewhere in AR 2522 and AR 2530 may have played an indirect role by modifying the global magnetic structure. Or the growth of instability in the filament may have been initiated by a major disturbance in a magnetically connected structure in the same active region.

The emergence of flux is clearly necessary to set the stage for subsequent flare activity on a major scale. That stage-setting is very important and takes place over many hours or even a few days. Of all the flares studied so far, the 1980 May 21 event comes closest to providing direct evidence that further emergence of a small amount of flux triggers release of a vast amount of energy stored in a sheared filament. But the details of that particular flux emergence are subtle (Harvey 1983); it is by no means clear how the interaction between the filament and emerging flux should be conceived when the adjacent net photospheric flux decreases during the emergence. The experience with the 1980 June 25 flare, where adjacent emergent flux could not be found, should caution us that the flare triggering process is still elusive.

1.3.5.3 Summary and Recommendations for Studies of Emerging Flux

The vigorous advance of theory (Priest 1984a, 1984b) has brought into sharp focus the observational requirements to test the Emerging Flux Model. From an observational perspective, however, even the conceptual role of emerging flux in the flare process is clouded. Growth of magnetic flux is a necessary pre-condition for flares: small flares are common during the AFS stage of an active region: large flares often have their initial kernels rooted in new, rapidly growing flux. But the vast bulk of magnetic flux appears at the surface without producing flares as strong as a M1 event in X-rays. The published cases of flare-associated filament eruptions lack key facts which are needed either to validate the reconnection inherent in the Emerging Flux Model or to constrain the model in terms of our understanding of flux emergence in the absence of flares. One study of a flare-associated filament eruption on 1980 June 25, observed in detail for many hours at heights in the photosphere, chromosphere, transition zone and corona, rules out local emerging flux as either a driver or a trigger of the activation of that particular flare.

An important new result is the association of Cancelling Magnetic Features with Flares (Martin, 1984). These may have a similar role to emerging flux in triggering flares (Priest, 1985), since what is important is the interaction of flux, whether through material vertical or horizontal motions.

A major advance towards clarifying this situation would come from coronal observations aimed specifically at the problem of emerging flux. Jackson and Sheridan (1979) found general increases in activity of Type III radio bursts prior to flares, which imply that energy, originating in the emergence of new flux, is entering the corona on a time scale of many hours. We badly need to supplement the detailed chromospheric and photospheric observations of emerging flux, now available, with simultaneous multi-wavelength coronal observations of comparable spatial resolution (≈ 1") and comparable duration (many hours, even days, preceding the emergence). Target regions of emerging flux need to be followed long enough at all levels of the atmosphere to come to grips with the formation of AFS and field-transition arches, and in contrasting emergences of flux accompanied by their well-established signatures from simple appearances of new flux without those signatures. More than semantics are at stake; our concepts of the magnetic interconnections in the latter situation are woefully inadequate. Finally, we need to clarify the association between ephemeral regions and coronal bright points on an individual basis. In so doing, we should gain insight into the dissipative mechanisms which seem to occur with great frequency on a basic scale, and which might be applicable to ordinary flares.

1.4 CORONAL MANIFESTATIONS OF PREFLARE ACTIVITY

E.J. Schmahl, D.F. Webb, B. Woodgate, P. Waggett, R. Bentley, G. Hurford, A. Schadee, J. Schrijver, R. Harrison and P. Martens

1.4.1 INTRODUCTION

Recent observations confirm the view that the initial release of flare energy occurs in the corona, with subsequent emissions arising from the interchange of mass and energy between different levels of the solar atmosphere. Knowledge of coronal preflare conditions is essential to understanding how energy is stored and then released in flares. Observational evidence for storage is, however, difficult to interpret owing to our inability to observe the three dimensional structure of the magnetic field and to the lack of coordinated observations with high resolution in space and in time.

More than sufficient energy to power flares can be stored in local magnetic fields on time-scales of hours. Long-term changes include emerging and evolving magnetic flux regions, satellite sunspots, sunspot motions, and velocity patterns (Martin, 1980; Section 1.3 these proceedings). Although such evolutionary changes are considered necessary for the storage of energy leading to especially large flares, it is very difficult to relate specific long-term changes to particular flares since similar changes occur in their absence.

More rapid changes can occur within minutes or a few hours preceding a flare, and they can be more unambiguously interpreted as flare precursors. Clearly, this distinction is arbitrary, but it does provide a useful operational definition of preflare patterns. These more rapid changes, especially in the corona, were the subject of the third subgroup of the Preflare Activity Team.

1.4.1.1 Review of Previous Studies of Coronal Precursors

Earlier searches for rapid flare precursors involving coronal phenomena recognized the physical importance of the corona for storage and release of flare energy. Reported coronal precursors have included X-ray brightenings associated with filament activations (Rust et al., 1975), expanding and brightening green-line arches (Bruzek and DeMastus, 190), gradual enhancements and spectral hardening of soft X-ray and microwave flux (Webb, 1983), "forerunners" of white-light transients (Jackson and Hildner, 1978), changes in circular polarization and intensity at centimeter wavelengths (Lang, 1974; Kundu et al., 1982; Willson, 1983), pre-burst activity at 1.8 cm (Kai et al., 1983) and preflare type III burst activity at meter wavelengths (Jackson and Sheridan, 1979).

Filament activations and associated manifestations, which are frequently observed with two-ribbon flares, have been the most readily observed and most studied forms of rapid flare precursors (Smith and Ramsey, 1964; Martin and Ramsey, 1972). The enhanced darkenings, organized motions and reconfigurations which constitute an "activation" were summarized by Smith and Ramsey (1964) and more recently by Martin (1980). The prevalence of the phenomenon is evident in the statistic (Martin and Ramsey, 1972) from a sample of 297 flares (importance > Class 1) that about half the flares in that sample exhibited preflare filament activity. Prior to or during a filament activation, changes in certain photospheric and chromospheric structures occur, which have been taken as evidence of evolving or emerging magnetic flux (e.g., Rust, 1976).

Preflare observations in soft X-rays have been used in a number of studies. Culhane and Phillips (1970) observed 7 precursor events at 1-12 A, one occurring 15 min before flare onset, using an OSO-4 full-sun detector. Thomas and Teske (1971) performed a statistical study using a full-sun detector on OSO-3 and found a tendency for the onsets of X-ray events to precede those reported in Hα. For a small number of events Roy and Tang (1975) found specific enhancements in full-sun X-ray flux to be associated with different stages of preflare filament activity.

With better spatial resolution (20 arc-sec) Rust et al. (1975) identified OSO-7 EUV and soft X-ray enhancements with a filament activation 30 minutes before flare onset. Van Hoven et al. (1980) studied the preflare phase of a set of 12 flares observed by the same OSO-7 detectors with one-minute time resolution. Eight of the 12 showed definite enhancements in both X-ray and EUV 2-20 min prior to the onset. Interestingly, in 6 of these 8 cases the enhancements were observed simultaneously in both cool He II and hot Fe XXIV lines. Although the Skylab experiments had excellent spatial resolution (arc-sec), the operational modes limited the availability of preflare data to a few specific observations of EUV and soft X-ray precursors (see Van Hoven 1980 for details). Petrasso et al. (1975) and Levine (1978) observed pre-existing coronal loops to brighten 5-10 min before they flared. The XREA full-sun X-ray detector typically detected preflare enhancements 2-20 min before the impulsive phase. There was evidence for slight temperature increases in these events, and an increasing tendency for large flares to have associated precursors.

In more comprehensive, statistical studies, Vorpahl et al. (1975) found many cases where X-rays from the flare regions were enhanced prior to onset, but Kahler and Buratti (1976) and Kahler (1979) found that there were no systematic preflare X-ray brightenings at the locations of subsequent small flares, and therefore no requirement for coronal preflare heating of the flare loops. However, coronal preflare brightenings were observed in the Skylab X-ray data in areas of the active region adjacent to the flare site.

Recently, Webb (1983) studied similar sets of the AS&E Skylab X-ray data with the goal of determining whether X-ray precursors systematically occurred within the flare active region and what their characteristics were. The study differentiated between observations relating to the preheating of flare structures, and precursors which might have time and spatial scales and locations different from that of the flare. High time-resolution Hα and daily photospheric magnetograms were also used. A majority of the flares studied had preflare X-ray features, but typically not at the flare site, occurring within 30 minutes prior to onset. The X-ray precursors consisted of one to three brightened loops or kernels per interval, with Hα emission at the feet of the loops or cospatial with kernels. Electron pressures of a few dyne cm^{-2} were derived for several typical coronal features. In half of the cases the X-ray precursors were associated with preflare Hα filament activity. The preflare and flare events occurred on or near the main active-region neutral line.

Using moderately resolved (arc-min) OSO-8 X-ray observations, Mosher and Acton (1980) and Wolfson (1982) reported no systematic enhancement in active regions in 20-minute intervals preceding flare onsets. But their detector was less sensitive to the lower energy, cooler precursors reported earlier from Skylab.

Radio observations provide important data on coronal emission and changing magnetic fields before flares. Individual observations of microwave preflare activity in the form of increased intensity and changing polarization have been reported in the past. With the increasing sensitivity and spatial resolution of such instruments as the VLA, these observations have become better defined, as discussed in Sections 1.4.2 and 1.4.5.

Green line (5303Å) observations above the solar limb showed acceleration and expansion of coronal arches up to one hour before two flares (Bruzek and DeMastus 1970). Skylab observations of white light mass-ejection "forerunners" (Jackson and Hildner 1978) indicated that such activity might precede Hα flare onset. Recent SMM results, together with improved metric radio and lower altitude K-coronameter data (Wagner 1982) support the overall picture that a large volume of the corona can become activated up to an hour or so before a flare.

1.4.1.2 Objectives

Our objectives in studying preflare coronal phenomena were threefold: to select a suitable data set, to determine appropriate physical parameters, and to search for associations among events so as to identify the relevant physics in preflare phenomena.

A wealth of new information about active regions and preflare activity is now available from the coordinated observations conducted during the Solar Maximum Year by the Solar Maximum Mission satellite, by other spacecraft, and by ground-based observatories. The wavelengths accessible to a study of the preflare coronal condition range from centimeter-wavelength microwaves to hard X-rays. Table 1.4.1 summarizes the data that were used in our study by wavelength and instrument, and the references to publications of events covered in this report. Previous multi-wavelength studies of this sort (Martin, 980; Van Hoven et al., 1980; Webb, Krieger and Rust, 1976; Rust, Nakagawa and Neupert, 1975; Webb, 1983; Kahler and Buratti, 1976) were more limited because of sporadic or slower image cadences, lower sensitivity, poorer resolution, or fewer wavelengths available. In addition to a broader range of data with better coverage, we also had the advantage of observing the sun at its maximum level of activity, with flares occurring six times more frequently than during the Skylab period.

Our approach was to assemble all available preflare data for a number of well-observed events and the results of several "cross-sectional" studies in specific wavelength ranges. We have selected good simultaneous observations at as many levels of the solar atmosphere as possible, from the photosphere through the chromosphere and transition region to the lower and middle corona. We concentrated on data with time resolution ranging from tens of seconds to minutes, collected over a time interval ranging from about one hour before the flare up to impulsive onset as defined in hard X-rays by HXRBS. In very few cases were observations at all levels of equally high quality, but a sufficiently large set of well-covered events was available for comparative and statistical analyses. The study of this large set of data continues in order to test the associations noted here between various preflare signatures.

Among the questions guiding this study were:

- Where is the flare trigger located?
- Is flux emerging at the photospheric level a necessary condition for preflare coronal activity?
- Do flares "try" to start, fail, and "try" again?
- Do flare precursors have both thermal and non-thermal components?
- Are there two distinct classes of precursor, some associated with filament activation and some not?

We have organized material leading up to a discussion of these questions as follows. In the next section we define the onset phase and precursors and explain how we distilled our preflare data set. In Section 1.4.3 are presented several key events which illustrate the connections that we discovered among their preflare phenomena. In Section 1.4.4 we describe an important comparison of the location of preflare activity in FCS and UVSP images. In Sections 1.4.5 and 1.4.6 are reviewed the observations of certain radio precursors which are taken as evidence favouring preheating and non-thermal particle acceleration. Finally Sections 1.4.7 and 1.4.8 describe HXIS observations of X-ray precursors.

1.4.2 Defining the Preflare Regime

We follow Svestka (1976) and Sturrock 1980, (especially p.413) in defining the "onset" of the flare as the time of the first rise in emission at the site of the flare itself. We adopt a somewhat more general definition of "precursor" than used in the Skylab studies (Sturrock, 1980). We take a "precursor" to be a transient event preceding the impulsive phase, possibly even before the onset and not necessarily at the precise site of the flare itself.

The initial sample of preflare events included 54 flares selected by Woodgate to have sufficient coverage in time by UVSP and BCS before the impulsive hard X-ray burst recorded by HXRBS. When multiple bursts occurred, the largest was taken to be the primary flare. In this respect, the study was similar to that of Webb (1983), in that "minor" flaring was included as "preflare" activity. From a combination of Woodgate's sample with 12 other events which showed interesting preflare activity in microwaves, Hα, white light and/or in X-rays, we selected 26 events which had the best overall coverage in data for concentrated study. These events along with key precursor observations and references to publications are summarized in Table 1.4.4.

Table 1.4.1a

Wavelength	Instrument	References Discussed
Microwave (spatially resolved)	Very Large Array: 2 cm, 6 cm, 20 cm	9, 14, 15, 18
	Owens Valley Radio Observatory	10
	Nobeyama Interferometer (17 GHz)	16, 17, 45, 44
Microwave patrols	Berne University	67, 35
	Sagamore Hill	9, 73
	Ottawa/Penticton (2.8/2.7 GHz)	M. Bell, pers. comm.
	Toyakawa (1-9.4 GHz)	43, 44
Hα	Ottawa River Solar Observatory	9, 30
	Solar Optical Observing Network	9, 36
	Big Bear Solar Observatory	10, 38
	Udaipur	40, 74
	Meudon	34, 55, 80
White Light	Mauna Loa	41, 57, 86
Corona	SMM C/P	23, 54, 57, 77
	P78-1 Solwind	J. Karpen, pers. comm.
Ultraviolet	SMM UVSP	9, 20, 21, 29, 40
X-rays, Soft	SMM XRP/BCS	28, 40, 50, 83
	SMM XRP/FCS	50
	GOES	S.G.D.
	P78-1 Solex	20
Medium-Hard	HXIS	7, 24, 48, 51, 54, 60, 63, 68, 79
Hard	HXRBS	9, 37, 44, 50, 62, 67 69, 71, 78, 82, 83
Gamma Rays	GRE	73, 78

1.4.2.1 The Onset Phase

We defined two onset times for each of the primary flares in Table 1.4.2. One was the impulsive onset observed in hard X-rays (HXRBS). The other was the soft X-ray onset, commonly defined as the start of the rise of the flare flux profile. Webb determined the onset time in this manner, using the full sun 1-8Å X-ray flux recorded by the NOAA/GOES satellite for the events in this study. On average, the soft X-ray onset occurred ~ 2 minutes before the onset of the hard X-ray burst, in agreement with previous results (Svestka 1976). Schmahl, Strong and Waggett used background-corrected BCS light curves in a similar way to determine the

Table 1.4.1b
References for Tables 1.4.1, 1.4.2

1. Martin 1980
7. Machado *et al.*, 1982
8. Kundu *et al.*, 1982
9. Kundu *et al.*, 1985
10. Hurford and Zirin 1982
13. Lang 1979
14. Kundu 1981
15. Kundu and Shevgaonkar 1985
16. Kai *et al.*, 1983
17. Kosugi *et al.*, 1985
18. Willson 1983
20. Woodgate *et al.*, 1982
21. Woodgate *et al.*, 1981
22. Gary 1982
23. Sime *et al.*, 1980
24. Harrison *et al.*, 1985
27. Moore *et al.*, 1984
28. Wolfson *et al.*, 1983
29. Schmahl 1983
30. Gaizauskas 1984
34. Malherbe *et al.*, 1983
35. Simon *et al.*, 1984
36. Rust *et al.*, 1981
37. Rust *et al.*, 1980
38. Zirin 1983
39. de Jager *et al.*, 1983
40. Machado *et al.*, 1983
41. Rock *et al.*, 1983
42. Woodgate 1983
43. Enome *et al.*, 1981
44. Hoyng *et al.*, 1983
45. Kosugi and Shiomi 1983
46. Solar Geophys. Data *4418* 1981
47. Dwivedi *et al.*, 1984
48. Duijveman *et al.*, 1982
50. Strong *et al.*, 1984
51. Oord *et al.*, 1984
54. Simnett and Harrision 1984
55. Gosling *et al.*, 1976
57. Gary *et al.*, 1984
60. Schadee *et al.*, 1983
61. Martin *et al.*, 1983
62. Svestka *et al.*, 1982
63. Gaizauskas and Schadee 1984
67. Wiehl *et al.*, 1983
68. Hoyng *et al.*, 1981
69. Dulk and Dennis 1982
70. Tandberg-Hanssen *et al.*, 1983
71. Woodgate *et al.*, 1983
72. Gabriel *et al.*, 1981
73. Cliver *et al.*, 1984
74. Bhatnager 1981
75. Poland *et al.*, 1982
76. Neidig and Cliver 1983
77. Wagner 1982
78. Marsh *et al.*, 1981
79. Duijveman *et al.*, 1982b
80. Malherbe *et al.*, 1983
81. McKenna-Lawlor and Richter 1982
82. Wu *et al.*, 1983
83. Antonucci *et al.*, 1982a
84. Antonucci *et al.*, 1982b
86. MacQueen and Fisher 1983

soft X-ray onset times. These onset times and the hard-X-ray minus soft-X-ray time differences are shown in Table 1.4.2.

There was surprising agreement between the GOES and BCS (Ca XIX: ~ 3.2Å) timings, with onsets rarely differing by 1 minute in the two X-ray regimes. When differences arose the BCS data were used since its 6' x 6' (FWHM) field of view minimized confusion from flares in other regions. Since Ca XIX is formed at ~ 1.5 x 10^7 K, the onset profiles indicate the existence of pre-impulsive plasmas as hot as ~ 10^7 K.

Harrison, Schadee and Schrijver plotted onsets for a number of flares using the softest channel (3.5-5.5 keV) of the HXIS instrument. Several onset profiles are shown in Figure 1.4.1a. The profiles of Figure

1.4.1a were integrated over the full coarse field of view (6!' x 6!'2), and therefore may represent the sum of more than one onset source. Figure 1.4.1b illustrates the comparison of onsets for the full field of view and for four areas of a few pixels each at and near the flare site,

Table 1.4.2 Event Summary of SMY Precursors

Date (1980)	Flare[1] Peak Flux	Implsv. Onset (HXRBS)	SXR ΔT ΔT(2) ΔT(3)	Hα flare[4] or burst	Fil. Act. -Onset	Rising Loop or Trans.	Brightenings[5]	Microwave Patrol Intensity Change	Microwave Pol. Change	Radio Spectral Events	Event References
Mar 23	C7	1658.3	(4)/--	N	Y-1600; 1648		Hα	Inc:1648	Inc.+Rev.	NONE	10,47,78.
Mar 29	C31	2041.3	4/--	Y-2016			SXR,UV,Hα	?	Inc:2016	III, V	—
Apr 6	C7	0716.5	(7)/?							I, III	—
Apr 10	M4	0917.1	7/22				SXR,UV	Step:0903		NONE	7,27,40,48, 68,74,83,84.
Apr 30	M2	2022.1	4/28			Filling loop	SXR,UV,Hα	GRF:2020-2215		NONE	1,36,37,39,72.
May 15	M2	--		Y-2019				GRF:2030-2130 Inc:1935	Dec.+Rev.	III	10
June 19	M1	1838.2	(8)/(8)	N			UV,Hα	GRF:1742-1810	Dec.	I, III	10
June 22	M1	--	--/(~50)	N	Y-1250	Inferred	Hα	GRF:1300-1830		I, III	34,35,61.
June 25	M1	1551.3	11/31	Y-1522	Y-1530	Inferred	SXR,HXR, UV,Hα	Step:1520	Inc.+Rev.	I, III	8,9,14,20, 30,47.
June 26	M4	2339.8	1/8?	Y-2315	Y-2330		SXR,UV,Hα	GRF:2310-2330	NONE	I, II	29
June 28	C5	1845.3	3/--	N	Y?	HXIS trans. C/P chg.	SXR,UV	NONE		I, III	24,29,41.
June 29	C4	0233.0	0/24	N	Y?	HXIS trans. C/P trans.	SXR,UV,Hα	NONE		I: <0231	22,24,29,54, 57,70,77,81.
June 29	M4	1040.2	0/11	N	HeI Loop <1040	HXIS trans.	SXR,Hα?	NONE		III	24,29,54,67,77 81.
June 29	M4	1822.0	0/20	Y-1803		HXIS trans. MLO trans.	SXR,UV,Hα	GRF:1701-1741		III	23,24,29,54,75, 81,82,83.
July 1	M5	1626.8	5/8	Y-1626			Hα	Fall:1430 1600	Dec.:1619	NONE	10,47,67,72. 76,84.
Oct 11	X2	1304.3	1/--				SXR,UV	NONE		NONE	
Oct 11	C2	1740.2	0/(40)	Y-1700	N			SXR,Hα	GRF:1730-2045	III	10
Nov 2	C7	0207.8	2/--	N			SXR,UV	NONE		III, V	70
Nov 5	C6	2225.9 2232.2	0/--	Y for 2232 flare	Active Fibrils	SXR,HXR,Hα		GRF:2140-2220		III	28,42,44,45,46, 48,49,51,79,84.
Nov 8	M4	1440.3	1/8	Y-1354[6]			SXR,UV,Hα	PBI:1350-1600		III	71
Nov 11	C8	0626.3	1/>21	N			SXR,UV	GRF:0600-0650		III, V	—
Nov 11	M1	2051.4	10/23	?			SXR,UV,Hα	Rise:2030	Dec.	I, III: >2016	10
Nov 12	C5	2231.1	6/43	Y-2155	Y?		SXR,UV,Hα	GRF:2020-2030	Active	I, III	—
Nov 13	M1	--	--/?	?-1712			Hα	Rise:1718	Dec.	I	10
Nov 23	M2	1840.2	(7)/(50)	Y-1754, 1815			Hα	GRF:1833-1922	Dec.	NONE	10
May 1, 83								GRF:2051	Inc.		15

NOTES:
1. GOES-2 1-8Å flux: C1 = 10^{-6} w/m², M1 = 10-5 w/m², X1 = 10^{-4} w/m²
2. Time difference: HXRBS Impulsive onset minus SXR flare onset from BCS. GOES data used when BCS not available. GOES Data in ( ).
3. Time difference: HXRBS Impulsive onset minus earliest SXR precursor in BCS.
4. Hα subflare or flare, radio burst or X-ray peak in preflare interval. Y = Yes; N = No.
5. Preflare brightenings in active region; SXR = soft X-rays, HXR = hard X-rays (>15 keV), UV = UVSP, Hα = from Hα images.
6. No Hα flare patrol: 1415-1515 UT.

18:17-19:03, 28 June 1980. We shall return to this flare later, but for now we note that the full field-of-view integration shows an earlier onset than the flare pixels themselves, suggesting preheating away from the flare site. Pre-onset activity has been noted outside of the flare structure before, especially in the Skylab data (e.g., Van Hoven 1980, Webb 1983, Kahler and Buratti 1976) and SMM data (e.g., Machado et al., 1982). However, at this stage it is not clear how to compare the results from Skylab and SMM. For instance, the X-ray filters on the Skylab AS&E experiment defined plasmas of lower temperature (~ 1-5 x 10^6 K for the softest filter) than the SMM BCS (> 8 x 10^6 K) and HXIS (greater than or about 10^7 K) experiments. There was no hard X-ray detector aboard Skylab so it was not possible to compare directly the distribution of soft X-ray to impulsive onsets. Since the Solrad or GOES maximum almost always followed the impulsive maximum by a few minutes, the Skylab onset times would have to be modified for comparison with this study.

The fact that a gradual onset in soft X-rays or microwaves is always present suggests a thermal origin for the first phase of flares (e.g., Svestka 1976). Machado et al. (1982) have suggested that the preflare gradual phase is a manifestation of the same phenomenon as the post-flare gradual phase. However, while this is conceivably true of the soft X-ray emitting regions, the gradual phase in microwaves shows remarkable differences (in polarization or source-size changes) preflare and post-flare (e.g., Kundu et al., 1985, Hurford and Zirin 1982). More study is required to determine the nature of the gradual onset of flares. We show below, in examples reported by Team members, several physical interpretations in terms of heating, upheavals or reconfigurations of magnetic flux.

1.4.2.2 Flare Precursors

The SMM data base is much more continuous than that of Skylab, and it is therefore possible to make stronger distinctions about flare precursors. In the more sporadic Skylab coverage (Webb, 1983; Kahler and Buratti, 1976; Kahler, 1979), it was more difficult to distinguish precursors from the onset phase. When such a precursor was observed, we defined the onset of the flare "conservatively" by the last pre-impulsive phase minimum of the light curve.

Since it is difficult to distinguish a true precursor signal from the flare onset or rise phase itself, when it occurs within a few minutes of the impulsive phase (Kahler 1979), we emphasized analysis of observations from about 60 to 5 minutes before impulsive onset. SMM images sometimes revealed precursors which were physically distinct from the flare, during what would otherwise be defined as the gradual onset phase.

X-rays. Precursors in the high resolution X-ray photographs from Skylab appeared as loops or kernels close to, but not necessarily, at the flare site. Often these sources were multiple and small (several arc-sec). In many cases, the precursors were closely associated with activated filaments (Van Hoven 1980, Webb et al., 1976, Webb 1983).

Examples of all of these effects are present in our SMM data set. In terms of the SMM X-ray light curves, gradual-rise-and-fall (GRF) precursor signatures were frequently detected in X-rays (Figure 1.4.1) and microwaves. Such a signature is considered indicative of coronal heating and is strongly associated with filament activations (Martin 1980, Van Hoven 1980, Webb et al., 1976, Rust et al., 1985, Sheeley et al., 1975, Webb and Kundu 1978) and mass ejections (Harrison et al., 1985, Simnett and Harrison 1984, McKenna-Lawlor and Richter 1982, Kahler 1977).

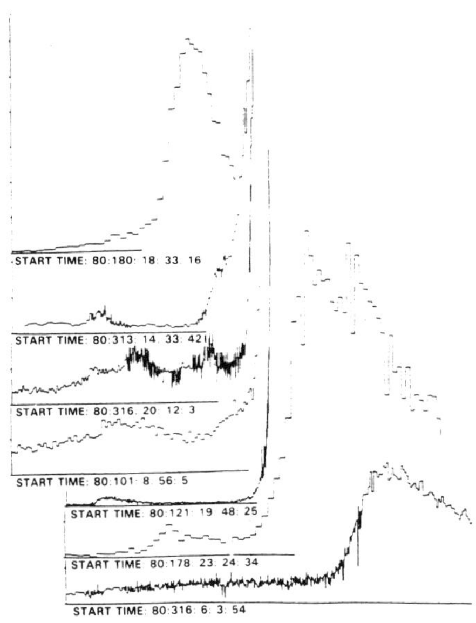

Figure 1.4.1a Onset profiles for a number of flares observed by HXIS in the softest (3.5-5.5 keV) band.

Microwaves. In the cm-λ regime as observed with interferometers, precursors may appear in the preflare hour as changes in circular polarization (Lang 1974, 1979), enhancements in polarization and intensity (Kundu 1981, Kundu et al., 1982, 1985, Kundu and Shevgaonkar 1985) small impulsive prebursts (Kai et al., 1983, Kosugi et al., 1985), or gradual rises in intensity accompanied by a decrease in fractional polarization (Hurford and Zirin 1982). Simultaneous observations with the VLA and Westerbork Synthesis Radio Telescope (WSRT) (Willson 1983) have shown examples of preburst 6 cm heating near the footpoint of one of the loops which flared. However, other observations have shown that preburst changes are not usually detected, since only 1 out of 8 bursts observed at 2, 6, and 20 cm showed pre-burst activity at the burst site (Willson and Lang 1984), and only 8 of 27 10.6 GHz bursts showed preburst activity (Hurford and Zirin 1982). X ray precursors were found a majority of the time in the flare active region (Webb 1983), namely, 17 out of 23 cases. So the question arises whether the lower rate of occurrence of microwave precursors is a selection effect, a threshold effect or a function of the pre-flare energetics or production mechanisms.

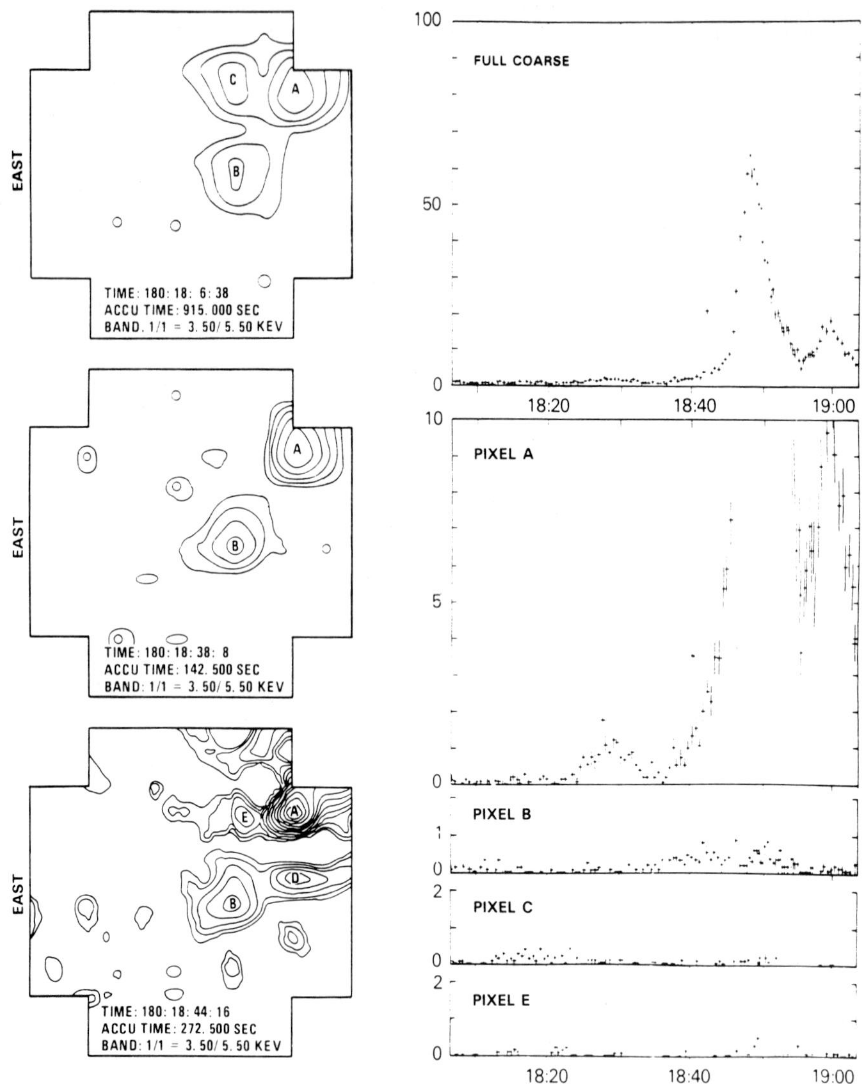

Figure 1.4.1b Comparison of time profiles integrated over the coarse field of view and sets of 1 to 4 individual pixels of the HXIS instrument.

Ultraviolet. Some examples of UV precursors include preflare surging motions in C IV (Kundu et al., 1985, Woodgate et al., 1982), and rising loops (Woodgate et al., 1981); these enhancements will be discussed for individual events below. In Section 1.4.4 Waggett and Bentley report on the correspondence of precursors in ultraviolet and X-rays.

Coronal White light. Various observers (Gary 1982, Sime et al., 1980, Harrison et al., 1985, Gary et al., 1984, Wagner 1982) have discussed the early appearance of coronal mass-ejection transients. SMY observations of transients in the low (HXIS) and mid-corona (Mauna Loa) have given a manifestation of pre-onset activity, which is clearer than the Skylab "forerunners" (Jackson and Hildner 1978). The physics of the relation of coronal transients to flares and their respective precursors remains unclear but some preliminary concepts will be presented in Section 1.4.7.

Common Precursor Factors. The variety of precursors seen during the SMM period is surprisingly large but as we shall show, there appear to be common factors that connect them. Emergence of flux at the photospheric level is one such factor, but does not appear to be a necessary condition for the precursor, as shown by the discussion of the 25 June 1980 event in Section 1.4.3.1

A particularly important question for flares in which a filament eruption occurs is whether the uplift of the filament signifies a reconfiguration of the magnetic field that causes the main phase of the flare to begin. There is no question that, typically, a significant amount of energy is released before the impulsive phase begins and before the most violent part of the filament eruption (Webb et al., 1976, Martin and Ramsey 1972, Moore et al., 1984). But precisely where the preflare heating occurs, relative to the observed filament motions and the flare site, is a more central question. Several important flares with well-observed preflare activity are described below in an attempt to access the important parameters and common features of such activity. These features will be summarized in the last section.

1.4.3 Specific Illustrative Events

We have selected 12 well-observed events from our preflare study to illustrate the diverse physical phenomena observed in the corona before flares. These include data with the best imaging and spectral coverage. This selection rules out spurious instrumental effects or unwarranted interpretations that might arise from the data from a single instrument. In a few cases the preflare and flaring periods have been thoroughly analyzed, and the interpretations are not likely to change significantly. But for most of these cases, the analyses are still very preliminary. The reader is cautioned, therefore, that this discussion is only meant to provide some initial summaries and interpretations of these preflare coronal manifestations.

1.4.3.1 A Filament Eruption Without Emerging Flux (25 June 1980)

The initial source of interest in this event was the set of preflare microwave maps made at 6 cm using the VLA. In the hour before the flare onset (1548 UT), according to Kundu (1981), the region around the flare site showed intensification in several compact (less than or about 20") sources whose polarization icreased up until flare onset. In the 15 minutes before onset, the polarization of one bipolar source reversed sign, and the subsequent 6 cm burst occurred at the site of that reversal (Kundu et al., 1982). It was deduced that magnetic changes were taking place during the pre-flare period (Kundu 1981). When he realized that the preflare period contained a well-observed filament activation, and that its subsequent eruption had been well observed in both on- and off-band Hα at the Ottawa River Solar Observatory (ORSO), Gaizauskas undertook an exhaustive analysis of the kinematics of the event (Kundu et al., 1985, Gaizauskas 1984). Woodgate (Woodgate et al., 1982), recognizing the significance of simultaneous C IV upflows before the same flare, also undertook a complete analysis of the UVSP dopplergrams. We briefly summarize the details of this event, which have been described at greater length elsewhere (Kundu et al., 1985, Schmahl 1983).

Figure 1.4.2 shows the time line of the preflare period as observed in hard X-rays, soft X-rays, Ultraviolet, Hα and microwaves. The main (1B) flare began at 1548 UT (HXRBS), with an earlier minor burst at 1522, which corresponded to an Hα subflare seen in the 1' x 1' UVSP field of view and recorded in the BCS Ca XIX and Fe XXV channels. (There were no imaging X-ray observations of this flare from either SMM or P78-1). Although the subflare had kernels within 20"-30" of the flare-associated filament, the filament motion was not affected. The filament showed steady transverse motion (see Figure 1.4.2) as early as ~ 3 hours before onset, with upward doppler shifts near its midpoint and axial flows and twisting motions along its length. Brightenings at 6 cm were seen in a 5 minute VLA map at the time of the subflare, but the source of emission was ~ 1' from the Hα and ultraviolet brightenings. It is likely that this prior radio brightening was related to the magnetic field changes taking place before the onset of the main flare. Just after the subflare, (15:33-15:38), a brightening and upflow occurred in the C IV dopplergram image, coincident with the rising portion of the Hα filament which also showed enhanced axial flows. The brightening and upflow reappeared more strongly from 15:42-15:46 in approximately the same location. Finally, a third brightening and upflow reappeared even more strongly at impulsive onset (15:49). By this time the transverse motion of the Hα filament had carried it further southward, so that the blue-shifted, impulsive, C IV brightening seen in the last UVSP image at 15:52 UT was clearly north of (and presumably below) the rising filament.

In the last VLA preflare map (15:30-15:40) there was a polarity reversal in one of the bright active region components ("B" in Kundu et al., 1982). The location was cospatial with the subsequent ultraviolet and microwave impulsive onset at ~ 15:50. This was interpreted as possibly due to the interaction of new magnetic flux with pre-existing flux, creating increased 6 cm opacity through the gyroresonance process.

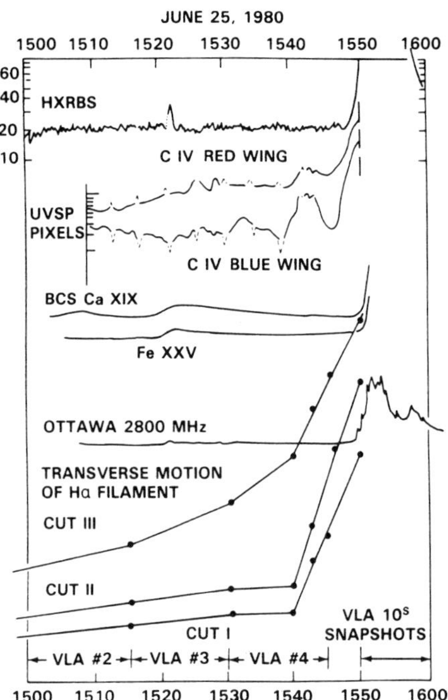

Figure 1.4.2 Time profile of hard, soft X-rays, UV, Hα and microwaves for the 25 June 1980 flare. (Courtesy B. Dennis, M. Bell). Note the two BCS maxima at 15:07 and 15:34 UT. Only the 2nd of these was recorded by UVSP.

However, careful examination of the ORSO Hα films (Gaizauskas 1984, Kundu et al., 1985) and magnetographic data revealed no signature of emerging flux at the chromospheric or photospheric level (Section 1.4.3b). The main conclusion was that changes in magnetic field strength were mainly coronal, with photospheric changes being gradual or evolutionary in this event.

Gaizauskas (1984) has argued that the instability of the filament developed out of a major disturbance in a magnetically connected structure in the same active region. The slow upheaval of the entire structure exhibited enhanced axial flows leading to a rapid twist, consistent with the weak kink instability (Sakurai, 1976; Hood and Priest, 1979; Sung and Cao 1983).

The three preflare rising motions, seen in the ultraviolet, occurred beneath the Hα filament (see Figure 1.4.3), which responded with enhancements of axial flows from its midpoint towards the eastern footpoint. The first two episodes of rising motions in C IV were of lower velocity and density than the third. Kundu et al. (1985) have suggested that the coronal conditions were such that the first two uplifts were not sufficient to trigger the final disruption of the filament, but that the

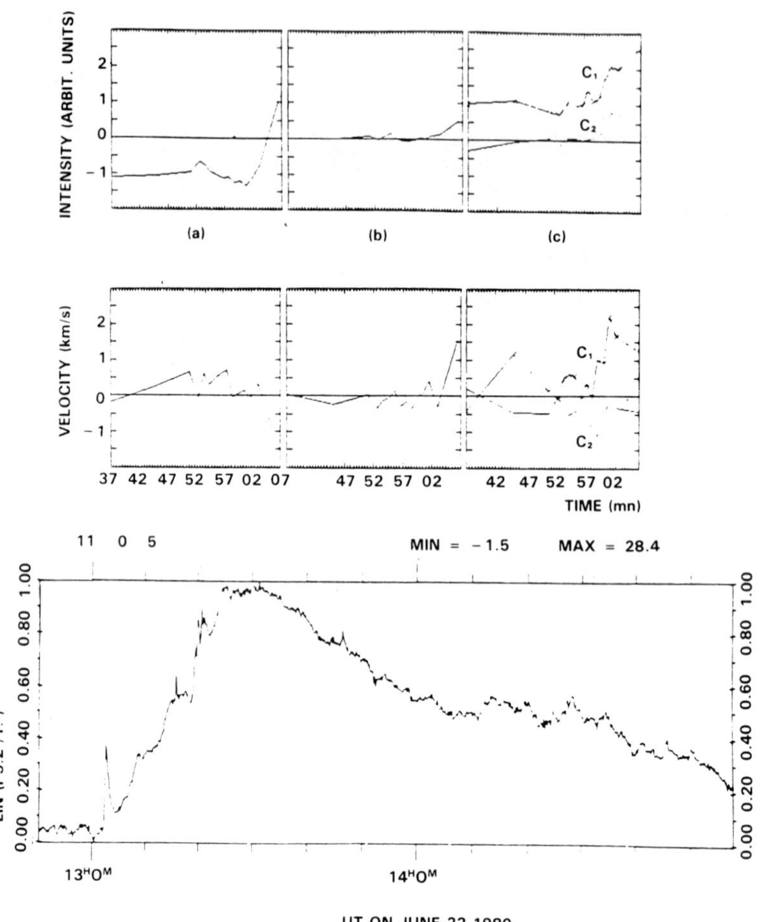

Figure 1.4.4 Time profiles of Hα and microwaves for the preflare period, 22 June 1980 (Simon *et al.*, 1984). Upper: Hα intensity. Middle: Hα velocity. Lower: Microwave flux.

third one was. The restructuring of magnetic fields before 15:45 UT, suggested by the VLA map must be related to the rising, twisting motions of the filament and its supporting field lines.

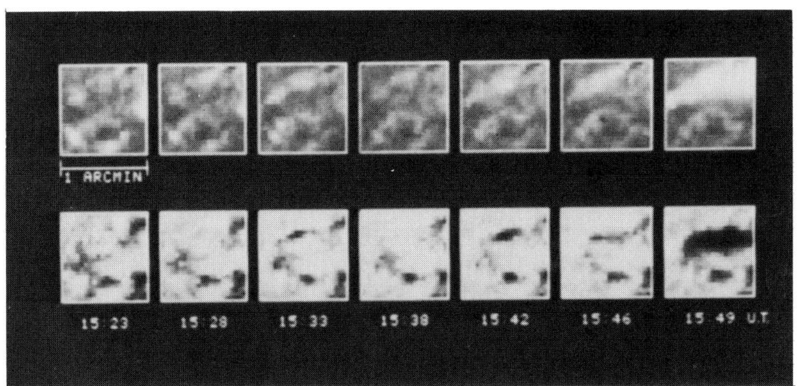

Figure 1.4.3 Rising motions and intensity fluctuations as shown in C IV dopplergrams and spectroheliograms for the preflare period, 25 June 1980. (Kundu *et al.*, 1985).

1.4.3.2 Filament Eruption with Colliding Poles (22 June 1980)

The preflare activation of a filament on June 22, 1980 serves as an interesting counterpoint to the event of June 25. The region in which the activity occurred, Hale 16918, was studied extensively as part of an SMY FBS interval (Martin et al., 1983), and the filament activity was initially described by Malherbe et al. (1983). At the workshop, Simon presented results of a more complete study of the event with an interesting interpretation in terms of current sheets (Simon 1984).

The hard X-ray burst associated with this event was not recorded by SMM because of orbital night. (P78-1 Monex data may exist, but has not been reduced). The microwave impulsive phase of the event occurred between 13:03 and 13:06 UT. At 5.2, 8.4, and 11.8 GHz, an impulsive burst was recorded at Berne from 13:02:40 to 13:03:40. At 3.2 GHz (Berne) and 2.8 GHz (Ottawa) the strongest impulsive burst occurred about two minutes later (13:05:20). Almost simultaneously (± 1 minute) another flare occurred in a neighboring region making the full-disk data difficult to interpret. This second flare occurred along the same neutral line and the two flares may have been related.

The Hα flare began in close coincidence with the impulsive bursts, with the central Hα intensity in the brightest kernel rising most rapidly from 13:03 to 13:06 UT (Malherbe et al., 1983). The Hα intensity and velocity profiles are shown in Figure 1.4.4. The filament associated with this flare showed activity as early as 12:36 UT in the form of red and blue shifts at various locations along its length (Figure 1.4.5). Figure 1.4.5 shows the Hα intensity and velocity maps at 13:00 U.T. Systematic blue shifts began at about 12:45 in the filament, where it passed through region designated "O" by Martin (1983). At the western end of the field of view, where the filament was darkest, the velocities were generally

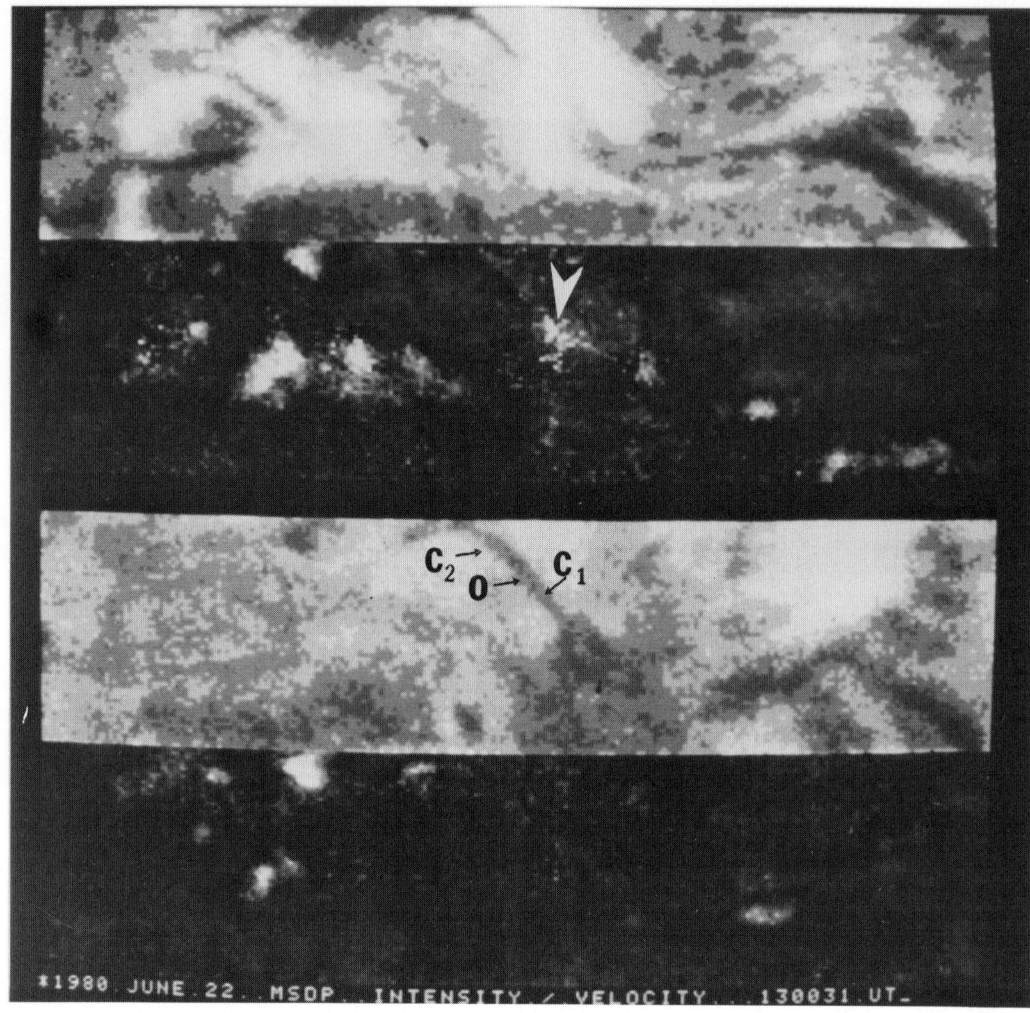

Figure 1.4.5 Intensity and Velocity in Hα at 13:00 UT, 22 June 1980. The arrow in the second panel shows the blue-shifted material at "O". C_1 and C_2 in the third planel show the locations of the features of Figure 1.4.4.

small but became redshifted during the main phase of the flare. At the opposite extremity of the filament (C_2) on the other side of "O", blue shifts also changed to red shifts during the main phase. Near the midpoint (C_1) of the filament and "O", the Doppler shifts became large toward the blue side. This behavior is qualitatively similar to that of the June 25 filament. The Hα profiles made north of the neutral line showed evidence of absorbing material moving transversely (northward) away from the neutral line between 13:00 and 13:03. The transverse velocity was ~ 35 km s^{-1} and the vertical velocity was at least 50 km s^{-1}. The ejected material remained connected to the filament at a point near "O" until 13:05, at the time of the impulsive phase.

During the rise of the filament, 12:59-13:04 UT, the brightest Hα knot near "O" moved systematically toward the neutral line with a transverse velocity of 20 km s^{-1}. On the other side of the neutral line, the knots did not show significant motion. At the time of the Hα explosive phase (~ 13:06) the absorbing material north of the neutral line separated. The filament reformed soon after the flare, as it did on June 25. The moving knot was interpreted (Simon et al., 1984) as the foot of a current sheet separating emerging and pre-existing fluxes.

Common Features of the June 22 and 25 Preflare Activity

Although the instrumentation observing the events of June 22 and 25 was different and the published descriptions emphasize different physical phenomena, it is clear that there were several common features in these events.

o Filament Kinematics - The upward motions of the filaments occurred near their midpoints, with downward motions near the ends. The driving force which lifted the magnetic arch apparently did not constrain the material from falling down the ends of the structures. In the analysis of both events it was concluded that not all of the filamentary material was ejected.

o Multiple Motions - Several kinds of motions were observed. In the June 25 case, there were axial and twisting motions as well as upflows and downflows. The upflows occurred as three successive events in the 20 minutes before onset. On June 22, upflows occurred during the preflare 25 minutes as three or four upheavals at the brightest point of the filament, but axial and twisting motions were not reported.

o Exciting Agent - The trajectory of the ejected material appeared to be directed away from a bright "exciting agent" at lower levels. On June 25, a bright surging feature was observed in the C IV line which could be inferred as triggering the eruption (Kundu et al., 1985, Woodgate et al., 1982). On June 22, the bright feature was an Hα knot which moved toward the neutral line as the dark absorbing structure on the other side moved away from it. The bright moving knot was interpreted as the footpoint of the current sheet between colliding lines of force (Simon et al., 1984). Although the two "exciting agents", seen at lower levels, were morphologically different, a common interpretation in terms of moving magnetic fields is possible for both.

o Filament Reformation - In both events, the filament reformed within a few minutes of the impulsive phase. This implies that the boundary conditions of the configuration, especially in the photosphere, remained sufficiently similar that the preflare magnetic field structure was restored postflare (Kundu et al., 1985).

o Dissimilarities: Is Emerging Flux Necessary? For the June 25 flare, a clear case has been made that no emerging flux existed below the filament. For the June 22 flare, the evidence is not so clear, but the colliding poles seen below the filament certainly are not characteristic of the classic emerging regions, which overlie diverging bipoles. However, such cancelling Magnetic Features (Martin, 1984) may have a similar effect to emerging flux in triggering flares (Priest, 1985). In both flares, the triggers for the eruptions may have been slow changes of magnetic field in the neighborhood of the filament. Similarly, one can ask whether the apparently different "exciting agents' (the surging C IV emission and the moving Hα knot) be interpreted by similar mechanisms. We return to this question after summarizing another event which shows both common and different features of preflare activity.

1.4.3.3 Rising Loop at the Limb (April 30, 1980)

Hα observers (e.g., Martin 1980, Rust et al., 1981, Rust et al., 1980, Webb 1983) have shown that Hα emission can precede the impulsive phase by a few to tens of minutes. The April 30, 1980 flare illustrates the case where the initial burst is preceded by a bright Hα mound at the site of developing new magnetic fields. (The flare occurred close to the limb where chromospheric footpoints typically are hard to detect.) The above-the-limb Hα emission was cospatial with a C IV loop (Woodgate et al., 1981) and was interpreted as an arch-filament system (Rust et al., 1981). Foreshortening at the limb, however, makes it impossible to determine the extent of the loop along the line of sight. For example, the loop might have been an elongated helical structure like a filament seen end-on, as suggested by line-of-sight flows seen in the UV, and by the fact that its southern footpoint was further onto the disk than the northern footpoint (Woodgate et al., 1981).

The preflare period for this event was well observed by all the SMM instruments and the details of the preflare activity were summarized by de Jager et al. (1983). Figure 1.4.6 shows the schematic UVSP loops (Woodgate et al., 1981) along with the HXIS light curves (de Jager et al., 1983). The two major HXIS structures in the flare, the "kernel" and the "tongue", both appeared spatially coincident (± 8") with the UVSP structures AB and DE (respectively). The brightenings in HXIS X-rays and in UVSP C IV were also in temporal coincidence as were the softer X-rays seen in the BCS Ca XIX and GOES 0.5-4Å channels.

According to the HXIS observers (de Jager et al., 1983), after the maximum of the "kernel" precursor at 19:55 until the onset of the flare at 20:22, the intensity decline was consistent with conductive losses (no

radiative losses) of a gas at $T_e \approx 1.5\text{-}2 \times 10^7$ K and $n_e \approx 3.2 \times 10^9$ cm^{-3}. In the "tongue", particle energization occurred continuously from the onset of the precursor to the impulsive rise.

During the rise phase (20:17-20:22) of the burst, the Hα and soft X-ray emission increased, and material rose in the C IV loop leg, with Doppler shifts (≈ 20 km/s) toward the observer. At the time of the main

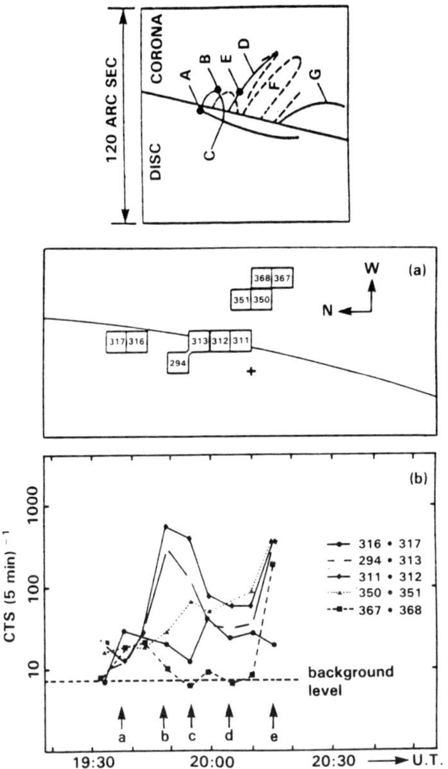

Figure 1.4.6 HXIS light curves (adapted from de Jager *et al.*, 1983) for the April 30, 1980 flare. Schematic loops (Woodgate *et al.*, 1981) are shown above.

burst, "breakout" occurred at the top of the loop where a new feature appeared in Hα.

Two explanations have been suggested (Woodgate et al., 1981) for the observed developments, the first being that a smaller loop filled with heated gas, pushing it upward into the larger loop and creating the flare at the junction. The second explanation was that the small loop became unstable at the top after the injection of gas and released the gas into the surrounding medium.

The first of these explanations is similar to the scenario developed for the June 25 flare (Kundu et al., 1985), where the destabilized Hα filament was accompanied by rising (C IV) loops. The question of the

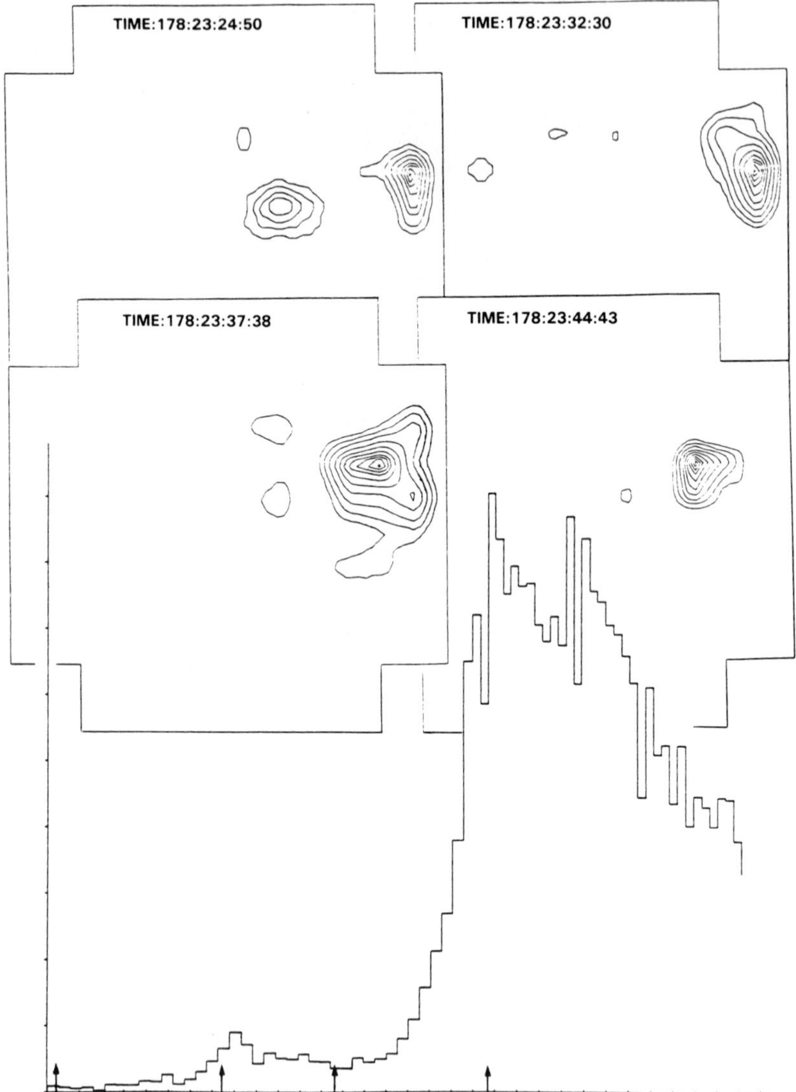

Figure 1.4.7 HXIS images showing the 23:32 precursor before the 23:40 flare on 26 June, 1980. The arrows on the time axis of the burst profile show the times of the individual maps.

existence of emerging flux in these events may not be as significant as the similarities in the triggering of the impulsive phase by a rising loop. Whether there was continuous high-energy preflare energization in the June events (as there appeared to be in the April 30 flare) is not known because HXIS did not observe the later pair.

The time profile of soft X-rays [BCS, GOES, HXIS] for April 30, the time profile of C IV intensity and Hα velocity for June 25, and the Hα velocity profile on June 22 all suggest that the flare "tries to start" and fails until the final agent (rising loop?) triggers the explosive phase.

1.4.3.4 X-ray Precursor Not at Flare Site (April 10, 1980)

The BCS Ca XIX and GOES flux started to rise at ~ 09:00, or about 20 minutes before the onset of the HXRBS burst (09:16). During this rise there was a Ca XIX burst (~ 09:05) which is considered a flare precursor. This peak was also recorded by HXIS (Machado et al., 1983) in the two softest channels. Preflare emission in N V was observed in three regions which were postulated (Machado et al., 1983) to be the footpoints of the subsequent flare loops. The 09:05 HXIS precursor appeared mainly at the southeastern and northern footpoints. The UVSP time profile in a 21" x 21" raster centered on the western footpoint showed impulsive brightening at that footpoint. Woodgate, Waggett and Bentley found that the small UVSP raster precluded an analysis of correlations between preflare ultraviolet bright points and FCS activity.

The combined HXIS and UVSP data imply that the precursor activity occurred in loops displaced ~ 8" x 16" away from the main hard X-ray brightening. Machado et al. (1983) estimated the emission measure (~ 5×10^{47} cm^{-3}) and temperature (~ 1.3×10^{7} K) for the precursor. The data permitted a multithermal interpretation, but counting statistics did not warrant the computation of differential emission measures.

The authors conjecture that the preflare gradual phase was a part of the overall gradual phase upon which the impulsive phase was superposed.

1.4.3.5 X-ray Preflare Emission From Filament Disruptions

On June 26 1980, Boulder Region 2522/30 produced what Martin classified as a "predictive filament" activation starting at approximately 23:30 UT. The BCS Ca XIX intensity showed a small precursor superposed on the rise at ~ 23:33. According to Harrison there were two precursors seen by HXIS at 23:35 (Figure 1.4.7), one located west (limbward) of the subsequent flare site and the other to the east. The BBSO Hα film showed that the precursor appeared as a subflare/surge in the penumbra of the leader spots to the west.

During liftoff (23:30-23:40) the activated Hα filament went from absorption into emission at 23:39:29 UT, close to the 23:39:35 impulsive onset. For the approximately two minutes of the Hα explosive phase, the filament rose rapidly, much like one leg of an expanding loop, after which ribbons formed on either side of the neutral line. Immediately following the flare, the filament reformed.

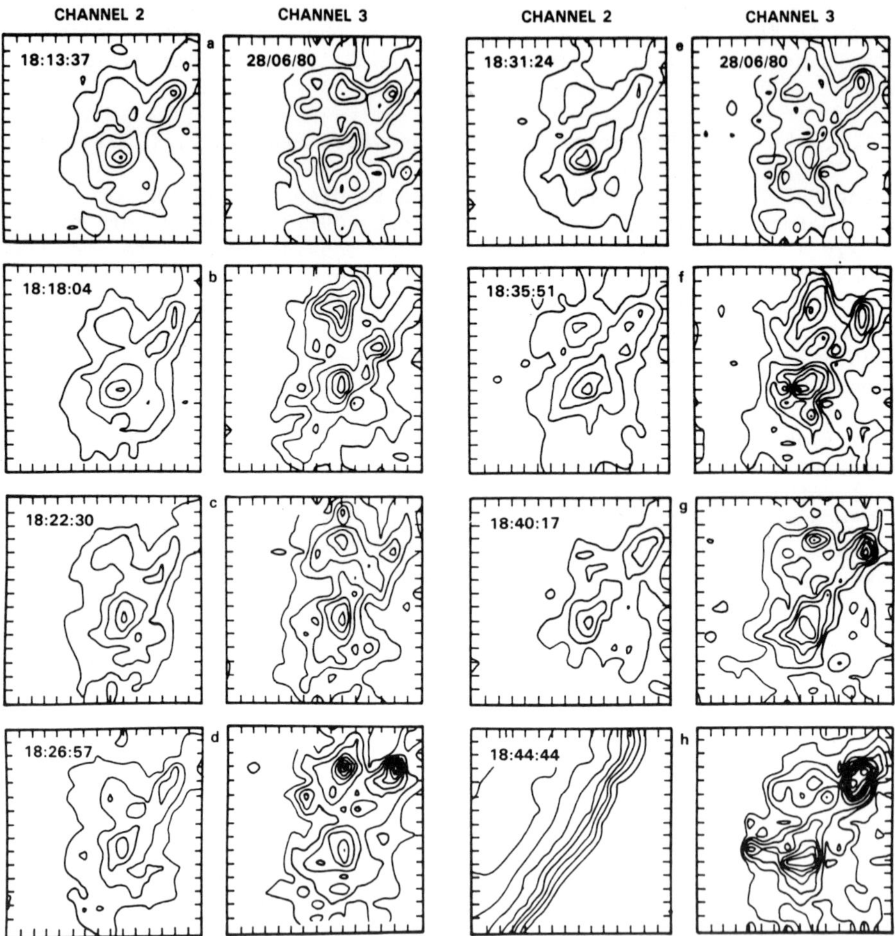

Figure 1.4.8 Ne IX and Mg XI (XRP) images showing a precursor, onset and the main phase of the 28 June, 1980 flare. The last Ne IX image is replaced by a white light image. The main flare can be seen in the north-west quadrant of the Mg XI image at 18:44 UT (h). The onset can be seen at the same position at 18:40 UT (g).

Preliminary coalignment suggests that the western HXIS precursor at ~ 23:32 coincided with the subflare/surge event. The eastern precursor was not obviously associated with any Hα activity. Several similarities in the preflare activity of the June 22nd, 25th and 26th events are discussed below.

On June 28, 1980 the leading portion of region 2522/30 was near the west limb and produced prominence and flare activity that was well observed at Mauna Loa Solar Observatory (MLSO) and by SMM. The first sign of activity in the hour before the flare was an eruptive prominence observed by MLSO from 17:12 - 18:44 UT (Rock et al., 1983). Subflares in the region occurred from 18:19 - 18:24. Associated activity was observed by UVSP in the Si IV line from 18:23 - 18:27. The flare itself appeared as two bright knots on the limb, seen in both Hα and Si IV. The preflare brightening occurred between the knots, then at the main flare site (18:23 - 18:27). Waggett and Bentley reported what may be an X-ray precursor in Mg XI inside the limb at ~ 18:18 18:26 (Figure 1.4.8b-d) and then subsequently at the flare site at 18:27 (Figure 1.4.8). The same precursor at the flare site was observed by HXIS in maps prepared by Schadee and Schrijver.

Hα flare onset was at ~ 18:24 before the onset of hard X-rays and continued as an upflow until at least 18:34 UT. The onset of the flare in X-rays appeared to start at 18:37 in the soft HXIS channel. The Mg XI images (Figure 1.4.8) showed onset at later than or about 18:42 and a double flare. In Si IV a brightening occurred near the northern flare knot at ~ 18:41, maximizing at ~ 18:45, the time of the hard X-ray burst. The Hα upflows apparently continued through 18:54, with motion paralleling the previous prominence eruption. The outflowing prominence material moved above the MLSO occulting disk at ~ 18:59 at the same position angle as the previous eruption.

Common Activity in the June Events

In the June 26-28 events, preheating of the filament was inferred from X-ray brightening during its liftoff. Both events had Hα "double-ribbons"; the June 28th flare was also double in ultraviolet and X-rays. Preflare brightenings were observed displaced from the flare site. On the 26th, two HXIS preflare brightenings occurred west and east of the flare site. On the 28th, the early Mg XI brightenings were displaced laterally and possibly above the limb flare. Both of these flares require considerably more analysis before firm conclusions can be drawn. Nevertheless, one striking similarity to the June 22 and 25 events stands out. In all four events preflare brightenings (in Hα on the 26th and in ultraviolet on the 28th) occurred beneath the filament or between the flare knots, suggesting that the flare was triggered from below. But the subflares preceding the June 25 and 26 flare, and the June 26th and 28th X-ray precursors were all displaced from the flare site and had no obvious relationship to the subsequent flare.

1.4.3.6 Homologous Flaring - November 5, 1980, 22:26 and 22:32 UT

Woodgate (1983) suggested that a majority of flares might be homologous in the sense of having footpoints reappearing very near the same places. The importance of homologous flares is that differences in initial conditions between flares can be minimized in order to isolate which factors are significant in terms of the site of flaring, timing, field strengths and energy release.

The November 5, 1980 flares were a good example of well-observed, homologous events. The hard X-ray profiles were similar (see Figure 1.4.9), although the second burst was an order of magnitude stronger. The microwave burst profiles were also similar, but the second burst at 17 GHz (Enome et al., 1981) was ~ 40 times larger than the first. The first burst was observed by the VLA at 15 GHz (Hoyng et al., 1983) and by various patrols from 1-17 GHz. The microwave spectrum went as $\nu^{2.9\pm0.1}$ up to 17 GHz, and the maximum of the spectrum was therefore above 17 GHz. The patrol data (Nobeyama, Nagoya, and Toyakawa: (Enome et al., 1981, Kosugi and Shiomi 1983, S.G.D. 1981) show that the second burst had a very similar spectrum, also with a maximum ≧ 17 GHz (but less than 35 GHz). The amplitudes of the two events were in rough proportions of ~ 40:1 at all frequencies from 1-17 GHz. Helium D3 films from BBSO showed that the two bursts were optically similar (see Chapter 5, Section 5A.5, Figure 5A.13). The bright D3 kernels of the impulsive phase appeared and disappeared with close simultaneity to the hard X-ray bursts, and the two "kernels" of the second event were cospatial with those of the first. One important difference between the two events was a weak "outlier" seen in D3 far from the main kernels.

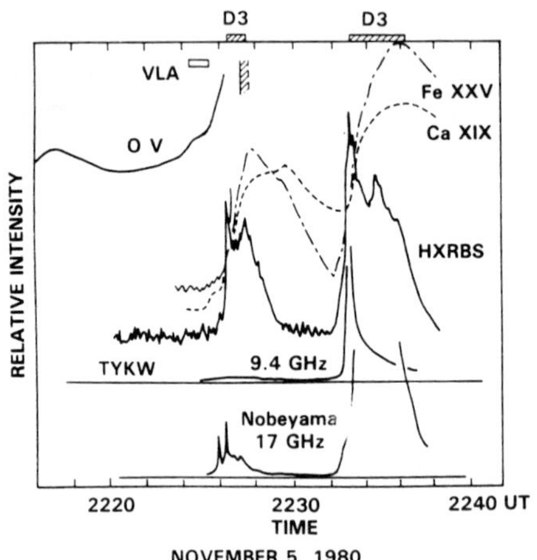

Figure 1.4.9 HXRBS, microwave, and BCS time profiles of the homologous pair, 22:36 and 22:32, 5 November 1980.

According to Martin (see Chapter 5) the outlier appeared to correspond to the weak impulsive source reported by HXIS (Duijveman 1982). Both flares showed strong Ca XIX blue shifts (~ 300 km/s) during the impulsive phase (Antonucci et al., 1984). This bears on the question: did the first flare trigger the second? X-ray observations have shown (Strong et al., 1984) that a flare closely following a previous one can be affected by the presence of thermal electrons exceeding a certain critical density which are "quenched" in the flux tube where the impulsive acceleration of the second flare takes place. If the critical density (~ 3×10^{-2} cm^{-3}) is exceeded, the beam electrons will lose their energy at high altitudes, and no chromospheric evaporation will occur (as was the case in the double flare of August 31, 1980). The measurements of strong Ca XIX blue shifts in both flares on November 5 indicate that "quenching" of the second flare did not occur. Since the observations all suggest that the main components of the two flares were co-spatial (not including the outlier), it is likely that both flares occurred in one flux tube and that the critical density was not reached.

Longevity of the X-ray Loops

Martens et al. (1985) conjectured that two fairly stable loops in region 2776 dominated the HXIS emission from November 5, 12:30 to November 6, 03:50. These loops were labelled AB and BC (Figure 1.4.10). During this time loop AB flared twice at 22:26 and 22:34 UT with several other flare-like brightenings at 15:04, 17:20, 20:47 and 23:50 on November 5. In the second flare the footpoint C of the impulsive phase was connected to the common footpoint B of the two loops. Apparently, loop BC was quite long-lived (Martens et al., 1985) with small variations correlated with the brightenings in loop AB.

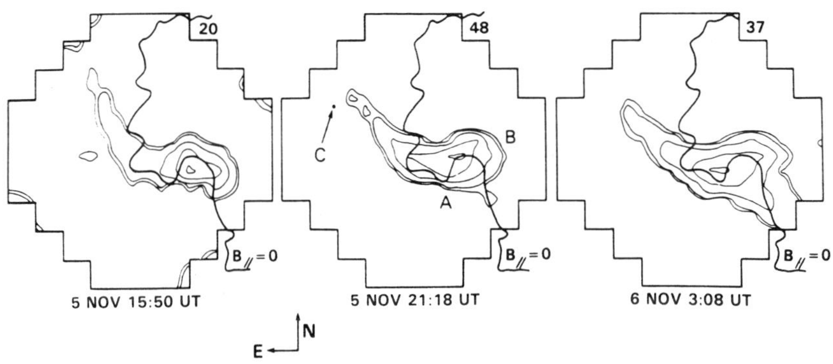

Figure 1.4.10 HXIS images of loops BC and AB preflare and postflare (Martens *et al.*, 1985).

Because loop BC was stable in emission we can assume that it was in static thermal equilibrium. We can therefore use scaling laws (Rosner et al., 1978) to derive the electron density (n_e) and the heating rate in the loop (Eh) from the observed loop length (L) and temperature (T_{obs}) from HXIS. (We note, however, that the usefulness of the static loop scaling law has been questioned by Roberts and Frankenthal (1980)). From the observed emission measure (Y) and the derived density an estimate of the emitting volume $V_{em} = Y/n_e^2$ of the loop could be made. This emitting volume was, surprisingly, much smaller than the observed volume $V_{obs} = L\pi r^2$ of the loop. The loop filling factor: $\phi = V_{em}/V_{obs}$ had an almost constant value of 10^{-3}. These results are summarized in Table 1.4.3.

Table 1.4.3
General Data on Loop BC, November 5/6, 1980

Mean Length	$(93.4 \pm 1.5) \times 10^8$ cm
Mean Diameter	$(11.4 \pm 0.3) \times 10^8$ cm
Mean Temperature	$(10.4 \pm 0.2) \times 10^6$ K
Mean Electron Density	$(9.8 \pm 0.4) \times 10^{10}$ cm^{-3}
Mean Filling Factor	$(1.4 \pm 0.3) \times 10^{-3}$
Mean Heating Rate	(0.64 ± 0.04) erg cm^{-3} sec^{-1}

Similar data on loop AB could not be derived, since it was unresolved by HXIS and its emission was highly variable. During quiescent periods loop AB had a temperature of 10^7 K and an emission measure per (8"x 8") pixel of about 1.5×10^{46} cm^{-3}. Duijveman et al. (1982) derived an electron density of 2.4×10^{10} cm^{-3} from the observed temperature and an emission measure of loop AB by assuming a filling factor of unity, while FCS observations of Ne IX were used to derive $n_e = 1.5 \times 10^{12}$ cm^{-3} (Wolfson 1983). Agreement between these observations is obtained by using a filling factor of 2.6×10^{-4}, which is of the same order of magnitude as that of loop BC.

Together these observations suggest the continuous dissipation of a current which is present in loops AB and BC, with the actual loss taking place in a very thin region. Clearly during flares the dissipation mechanism changes qualitatively in character.

It is still an open question whether the first and smaller burst late on November 5 triggered the second burst, or whether both were triggered by earlier events. The preflare manifestations of these flares have not yet been reported in sufficient detail to assess their significance. It appears, from the BBSO Hα film, that the first flare started in a fibril crossing the neutral line. This fibril went into emission at 22:23:29, approximately simultaneously with a rise in OV seen by UVSP, a small rise at 15 GHz (Hoyng et al., 1983) and a rise at 9400 MHz (Enome et al., 1981). Earlier subflares occurred on the same neutral line.

This double flare is important not only for the questions associated with homology, but as a possible test case for double flares in general. It has been estimated (Strong et al., 1984) that ~ 43% of the ~ 200 largest flares seen by the XRP in 1980 were "multiple" (on the basis that the Ca XIX flux did not fall to background between maxima). The importance of such "multiple" flaring for preflare studies lies in their possible relation to precursors, triggering and the repeated "attempts" of a flare to start.

1.4.4 Comparison of Preflare X-rays and Ultraviolet

Waggett, Bentley and Woodgate compared BCS and UVSP images for those of the 26 events in the Table 1.4.2 that showed pre-impulsive activity in the UVSP images. As both the FCS and UVSP are capable of performing a variety of different size rasters within large fields of view it is possible for them to be looking at different areas within the same active region. Coalignment of the instruments' fields of view left 10 events with good overlapping spatial and temporal coverage in both UVSP and FCS images.

Two time intervals were considered: the preflare period prior to the HXRBS onset and the impulsive phase prior to the HXRBS peak. The spatial separation of the UVSP preflare bright points and the FCS flare site were divided into three distance categories: less than 20" (adjacent to or at the flare site), between 20" and 40" (close to the flare site), and greater than 40" (far from the flare site). The results are given in Table 1.4.4 were the separation of the nearest preflare pixel is indicated by a "Y" in the appropriate distance column. The BCS column

Table 1.4.4 Details of the Nearest UVSP Preflare Brightening Pixels for the Ten Events Used in the XRP Comparison

Date	Distance of Pixel <20″	20-40″	>40″	BCS	BDIP	UVSP Line	Pixel Size (arc sec)	Preflare Intensity (UVSP C/S)	Impulsive Phase Intensity (UVSP C/S)
29/3/80 (2014 UT)			Y	N	N	CIV	3	1.42×10^4	3.42×10^4
27/4/80 (0106 UT)		Y(*)		N	?	CIV	3	3.42×10^4	NOT SEEN
30/4/80 (2023 UT)	Y			N	N	CIV	3	6.13×10^3	2.76×10^4
20/6/80 (0488 UT)			Y	N	N	CIV	3	2.92×10^3	2.34×10^4
29/6/80 (2022 UT)	Y			Y	N	CIV	3	1.08×10^4	1.66×10^4
10/7/80 (2126 UT)	Y			N	Y	SiIV	3	4.07×10^2	7.00×10^4
11/10/80 (1741 UT)			Y	N	N	OV	10	3.80×10^2	4.28×10^3
2/11/80 (0211 UT)	Y			N	?	OV	10	2.99×10^2	NOT SEEN
11/11/80 (2054 UT)	Y			Y	Y	OV	10	9.06×10^2	1.63×10^3
12/11/80 (2231 UT)	Y			Y	N	OV	10	1.46×10^2	2.59×10^2

indicates whether there was significant activity in the BCS Channel 1 light curve during the preflare period and the BDIP column indicates whether the preflare pixel brightened during the impulsive phase in the FCS data.

With only 10 events it is difficult to form meaningful conclusions. It is hoped that coordinated observations with special observing sequences during the SMM 2 mission will expand the sample. The inclusion of XRP data has improved the results of the UVSP analysis by confirming the position of the flare site and by removing events that were initially confusing. It is clear that the correlation of preflare UV bright point position and the flare site is good since 6 (possibly 7) of the nearest 10 preflare events are coincident with the subsequent flare site.

1.4.5 Preflare Microwave Intensity and Polarization Changes

We discussed in Section 1.4.3 some interferometer observations which showed preflare polarization changes. Although it may be true (Hurford and Zirin 1982, Willson and Lang 1984) that preflare polarization changes at centimeter wavelengths are not generally detected, there is evidence (Hurford and Zirin 1982) that certain microwave signatures are relatively reliable predictors of flares.

In a study of a sample of 81 major flares observed at 10.6 GHz with the Owens Valley Radio Observatory (OVRO) interferometer, Hurford and Zirin (1982) found a variety of preflare behavior. The most common preflare signature was a step-like increase in signal amplitude I, accompanied by a decrease or reversal in the polarization signal V during the last 10 to 60 minutes before the flare. This signature was found in the first half of the data base (Feb. 19 - Sept. 1, 1980), and when applied by a computer program to the second half (Sept. 1, 1980 - March 31, 1981) succeeded in "predicting" five out of 54 flares. This low success rate limits its practical value as a flare predictor but the microwave signatures do illustrate the coronal manifestation of some kind of magnetic activity. Figures 1.4.11a,b show two examples of flares observed at 10.6 GHz by OVRO on July 1 and October 11, 1980. The first shows a signature of increasing I and decreasing V, while the second shows only increasing V. Kundu (1981) suggested that the cause of microwave enhancements and polarization changes may be increasing magnetic field strength at the coronal level. The increasing magnetic field causes new loops to become optically thick at low harmonics (2nd, 3rd, and 4th) of the gyrofrequency. Depending on the relative orientation of the pre-existing and new loops, the polarization can increase (Lang 1974, 1979), flip (Kundu 1981, Kundu et al., 1982) or decrease (Hurford and Zirin 1982). Only two-dimensional inteferometry (VLA) can resolve the loop geometry, and there are still too few examples from OVRO to infer the loop geometry statistically. Hurford and Zirin (1982) showed in 3 cases that microwave changes were associated with subflares or filament changes. The July 1 event illustrates an example of a preflare brightening in Hα (at ~ 16:18), near the start of the increase in one of the polarization channels. The June 25 flare reported above in Section 1.4.3.1 illustrated another association between microwave and Hα changes.

The gradual rise in preflare microwave intensity can be observed with patrol instruments, and is presumably associated with "preheating" and gradual rises in soft X-rays (see Section 1.4.2). These gradual increases in microwaves are frequently associated with filament activations (Webb et al., 1976, Sheeley et al., 1975, Webb and Kundu 1978) and were observed by patrol instruments for the June 22, 25 and 26 flares discussed above. The OVRO amplitude for the July 1 X-ray flare (Figure 1.4.11a) showed a preflare increase that was not observed by patrol instruments or in the Ca XIX time profile. Thus the heating effects may have been too small to be observed in X-rays, but a microwave increase was observed possibly because of the extreme sensitivity to the magnetic field in the cyclotron emission mechanism. More recently the OVRO system has been made into a microwave spectrometer, examining typically 30 - 40 frequencies from 1 - 18 GHz with a time resolution of seconds. With this system Hurford (1983) found that some microwave precursors were characterized by narrow-band spectra with sharp high and low frequency cutoffs. He has interpreted these data in terms of gradual loop heating.

1.4.6 Non-Thermal Precursors

Long before SMM, it was argued that non-thermal processes occurred during the "buildup" phase of solar flares (Kane and Pick 1976) and in the absence of flares (Webb and Kundu 1978). Several preflare team members presented new evidence and theoretical arguments for non-thermal, non-flaring activity as seen in radio waves. Kosugi summarized a number of preflare activities observed at 17 GHz using the Nobeyama interferometer. He and his co-workers (Kai et al., 1983) examined 25 pairs of bursts which occurred within 10 to 50 minutes of each other. These pairs were cospatial (in one dimension) to $< 50''$ in 12 out of 15 cases. In most cases, the prebursts were impulsive, and therefore were not likely to be signatures of gradual preflare heating. However, they also noted that in more than half of the cases, Hα flaring started before the "preflare" burst, which argues in favor of preflare heating. They suggested three possible mechanisms for prebursts:

- A process related to the main energy release such as joule heating in current sheets.
- Pre-acceleration of electrons prior to the main acceleration.
- Manifestation of "leakage" of accelerated electrons.

They pointed out that the "leakage" mechanism is probably not consistent with the long time interval between bursts. Kosugi also showed evidence for statistical association in the number of type III (meter wave) bursts within minutes of prebursts at 17 GHz (Kosugi et al., 1985). No spatial locations were available for the type III bursts. Finally, Jackson reported that a study of spatially located type III's observed at Culgoora showed a statistically significant tendency to occur an hour or so prior to large Hα flares. If these associations between prebursts at 17 GHz, Hα flares and type III's are valid, then the mechanism of pre-acceleration appears to be favored.

Harrison presented a model (Simnett and Harrison 1984) of precursors in which 10^2-10^3 keV protons heat a high coronal loop, destabilize the pressure balance and heat the chromospheric plasma to produce the precursor X-rays. The acceleration mechanism is a small shock, which primarily heats protons within a large-scale magnetic loop. The model is directed primarily at the situation where precursors are widely separated preceding a coronal mass ejection with or without a flare.

There is evidence for electron acceleration in the absence of flares. Chiuderi-Drago pointed out that sometimes bright non-flaring

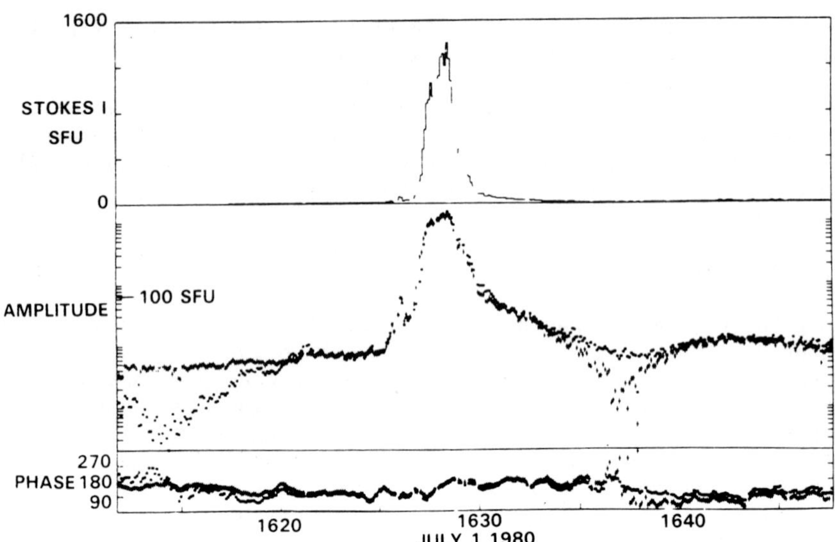

Figure 1.4.11a The July 1, 1980 flare time profiles at 10.6 GHz (Owens Valley Radio Observatory, (Hurford and Zirin 1982)) showing the "onset" signature of increasing I and decreasing V. The middle panels show the amplitudes of R and L (right and left circular polarizations); V = (R-L)/V.

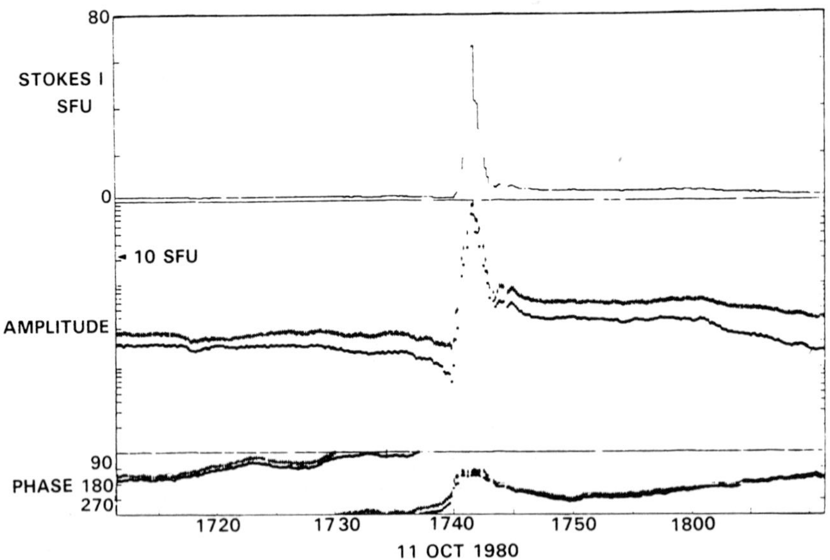

Figure 1.4.11b The Oct. 11, 1980 flare time profiles at 10.6 GHz (OVRO, Hurford and Zirin 1982).

sources ($T_b > 5 \times 10^6$ K) appear on microwave maps, and these sources may be explained (Chiuderi-Drago and Melozzi 1984) in terms of gyrosynchrotron emission from accelerated electrons trapped in coronal loops, where they may survive for ~ 10^2 sec. This is consistent with the lifetime of some precursors, and if continual acceleration occurs, the mechanism could explain some long-lived microwave sources. If the number of electrons is sufficiently high, then the gradual-rise-and-fall events commonly associated with filament disruptions (Webb et al., 1976, Sheeley et al., 1975), can be explained by a thermalization process (Webb and Kundu 1978) which follows the precursor acceleration.

1.4.7 Precursors of Coronal Mass Ejections

Although most of the SMM preflare data are for disk events, three coronal transient events on June 29 were well observed as region 2522/30 neared the limb.

All of these limb events were observed by HXIS and C/P (Harrison et al., 1985) and the 02:33 and 18:22 events were also well observed by XRP and UVSP. These were possibly homologous events because the flares occurred in the same location and had similar X-ray profiles with GRF precursors and long flare decay times (Stewart, 1984; Woodgate et al., 1984).

Figure 6.3.1a,b (Chapter 6) shows the HXIS flux profile for June 29 and the extrapolated C/P coronal transient onset times for two of the three events. The detailed HXIS data and their interpretation are presented in Chapter 6 and only summarized here.

A long-lived 160 MHz noise storm preceded the 02:33 flare but ended at 02:21 (Gary et al., 1974). The flare was associated with two moving white-light loops whose height-time profile extrapolated back to the surface about 6-8 minutes before flare onset. The HXIS precursor began at about 02:10. Shine prepared Figure 1.4.12 which shows the preflare period 02:17- 02:29 in Si IV and O IV rasters, as well as the flare itself (02:33 - 02:37). At the time of the first UVSP preflare rasters, the BCS, FCS and HXIS all recorded a brightening at the flare site at 02:19. Later, the point brightened again at ~ 02:28, but was not visible in X-rays. The onset of the Si IV/ O IV burst was at 02:33:56 and appeared as multi-pixel brightenings within ~ 10" of the preflare UV brightenings.

The 10:40 UT event was not well observed by the C/P or UVSP, but a preflare He I "jet" was observed (Schmahl 1983). The white-light loop transient associated with the 18:22 flare had a projected surface start before 18:15 (Sime et al., 1980, Harrison et al., 1985). A small Hα and X-ray flare occurred at 18:05 at a position slightly displaced from the later flare site. The UVSP observed a preflare OV loop and Fe XXI brightening (Poland et al., 1982), and BCS analysis revealed turbulent line broadening up to 4 minutes before onset (Antonucci et al., 1982).

The relationship between coronal mass ejection (CME's) transients and flares is far from clear, and it is farly well established that one may have CME's without flares and vice versa.

However, Harrison et al. (1985) argue that the flare precursor and the mass ejection precursor may be one and the same. In Section 3 of Chapter 6 Harrison gives examples in which the X-ray precursor of the

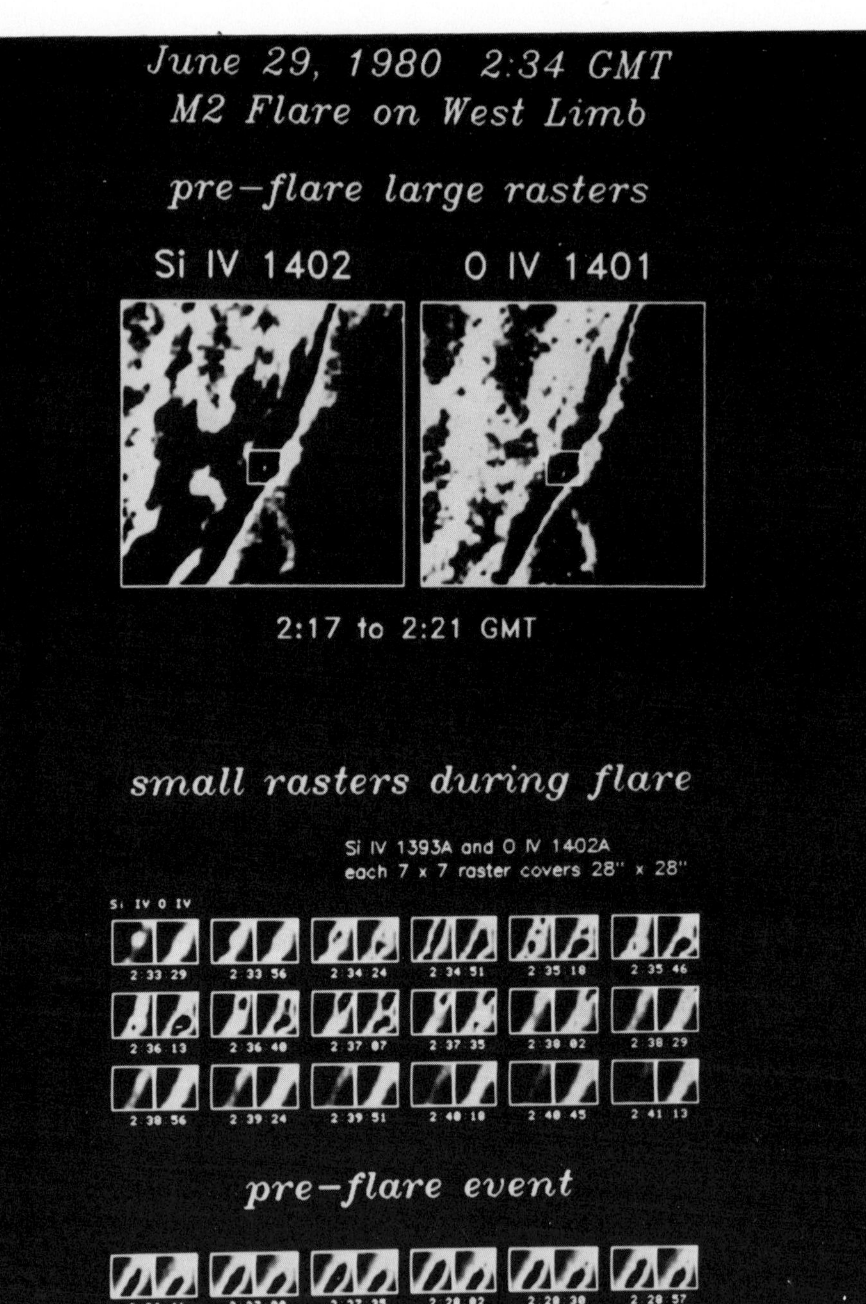

Figure 1.4.12 Preflare and flare images in Si IV and O IV (UVSP), 02:17-02:37, 29 June 1980.

mass ejection may be very small, as large as a flare, or "a lone precursor", without a folowing flare. At this stage of the analysis it is premature to assess the reality of the possible relationships among precursors of flares and CME's, but research along these lines may well provide a broader understanding of the role precursors play in the energy release process.

1.4.8 Short-Lived and Long-Lived HXIS Sources as Possible Precursors

Owing to its low background, particularly in its lowest energy band (3.5 - 5.5 keV), HXIS is capable of detecting very weak X-ray sources. The X-ray precursors reported by HXIS observers (Sections 1.4.3.3,.4 above) have been interpreted as thermal events, with temperatures 1 - 2 x 10^7 K. HXIS images often showed (Schadee et al., 1983) short-lived sources (SLS) and long-lived sources (LLS) in the 3.5 - 5.5 keV band. The short-lived sources (lifetime less than or about 15m) appeared indistinguishable _per se_ from HXIS precursors but did not always precede flares. The long-lived (hours-days) sources were of larger scale. Both had band-ratio temperatures of ~ 10^7 K.

In the context of our study, HXIS LLS's preceded two large flares with precursors, namely, May 21 20:55 and the June 22 flare discussed in Section 1.4.3.2. Although not part of this study, the May 21 X1 flare is one of the best analyzed SMM flares and is discussed elsewhere in this monograph. As in the June 22 case, it was preceded by a compression of pre-existing flux. See Section 3.5.4(iii), Harvey (1983). Discrete LLS's cospatial with the curving filament persisted for many hours on May 20 and 21. One source was located near the site of the EFR where the filament broadened then parted 10 minutes before impulsive onset.

Figures 1.4.13a (coarse FOV) and 1.4.13b (fine FOV) show accumulated HXIS images during 20 hours preceding a two-ribbon flare on June 21, 00:55. LLS's were frequently present, cospatial with the neutral line and filament curving from SE to WNW through the center of the FOV. LLS's occurred along the filament until its eruption before the flare of June 22, ~ 13:04. Even though the filament soon reformed, no further LLS's were observed after this event.

As evidenced by Figure 1.4.13, long-lived sources extend over a large area, often persisting for several hours. The characteristics of LLS's are: durations of tens of minutes to hours; temperature ~ 10^7 K; and emission measures of ~ 10^{46} cm^{-3}. They often show gradual intensity changes.

Most LLS's appear to result from activity along neutral lines. They may represent the high temperature tail of the thermalized plasma associated with filament activity as observed during Skylab (e.g., Webb et al., 1976, Webb and Kundu 1978, Kahler 1977). The Skylab soft X-ray filament enhancements had emission measures an order of magntitude higher and temperatures of an order of magnitude lower than the HXIS LLS's.

If the LLS's originate beneath the filament, then the model of Kopp and Pneuman (1976), Van Tend-Kuperus (1978) and Hood and Priest (1980) may be appicable. In that model the convergence of magnetic flux towards the neutral line in the photosphere results in energy dissipation by

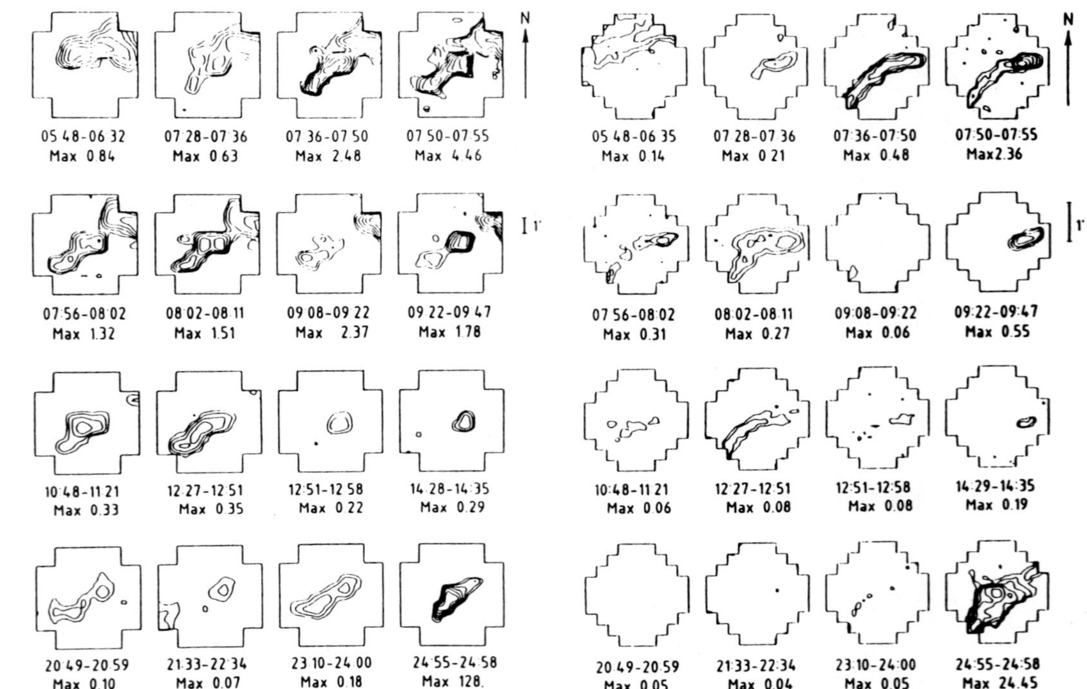

Figure 1.4.13 Long-Lived HXIS sources seen preflare June 22, 1980. Post-flare images do not show these sources. Coarse field-of-view maps are shown on the left and fine field-of-view maps are shown on the right.

reconnection below the filament in the corona. Interestingly, other observations (Athay et al., 1984) show evidence of magnetic flux converging at the neutral line in the Martin region. They suggest a process of continuing reconnection. At this stage it is not clear whether the X-ray emission results from filament activation, from continuing reconnections, or both.

1.4.9 Summary: Are all the Blind Men Looking at the Same Elephant?

We have reported a variety of coronal manifestations of precursors or preheating for flares and have found that almost everyone with a telescope sees something before flares. Whether an all-encompassing scenario will ever be developed is not at all clear at present. The clearest example of preflare activity appears to be activated filaments and their manifestations, which presumably are signatures of a changing magnetic field. But we have seen two similar eruptions, one without any evidence of emerging flux (Kundu et al., 1985) and the other with colliding poles (Simon et al., 1984). While the reconnection of flux is generally agreed to be required to energize a flare, the emergence of flux from below (at least on short timescales and in compact regions) does not appear to be a necessary condition. In some cases the cancelling of magnetic flux (Martin, 1984) by horizontal motions instead may provide the trigger (Priest, 1985).

We have found many similarities and some differences between these and previous observations. The similarities, besides the frequent involvement of filaments, include compact, multiple precursors which can occur both at and near (not at) the flare site, and the association between coronal sources and activity lower in the atmosphere (i.e., transition zone and chromosphere). Because of differences in instrumentation and improvement in multi-wavelength coverage with high time resolution, we have been able to identify several new aspects of preflare activity in the SMY data. These include the facts that: precursors were observed over a wide range of temperatures and heights; there were long-lived, hot ($> 10^7$ K) X-ray sources preceding some flares; and there was evidence for high energy phenomena, particularly electron and possibly proton acceleration before flares. The fairly rapid reformation of some filaments after their explosive eruption suggests that the photospheric boundary conditions remain unchanged, at least after some flares. Concerning filament-eruption flares, we saw suggestive examples in Section 1.4.3 of a flare exciting agent (at least as detected by its emission) first arising under the central portion of the filament.

Finally, our results leave us with several important questions. We have shown examples of preflare X-ray enhancements and small impulsive-like bursts. Are these signatures of a incremented instability, in which the flare "tries to start and fails" or are they signatures of a separate process that energizes the corona first with the flare following as a separate phenomenon?

Is the preflare gradual phase caused by the same mechanism as the postflare phase? Alternatively, does the gradual preheating occur through thermalization of an energetic population signified by nonthermal prebursts? Do microwave signatures signify changing coronal magnetic fields during the preflare hour, and (if so) what can we learn of the field strength and configurations? While the observational analysts continue to wrestle with the study of the vast SMM data store, the theoreticians must continue to synthesize and interpret these diverse phenomena in a consistent fashion.

1.5 REFERENCES

Antonucci, E., Gabriel, A.H., and Dennis, B.R.: The Energetics of Chromospheric Evaporation in Solar Flares: Astrophys. J. **287**, 1984, pp. 917-925.

Antonucci, E., Gabriel, A.H., Acton, L.W., Culhane, J.L., Doyle, J.G., Leibacher, J.W., Machado, M.E., Orwig, L.E. and Rapley, C.G.: Impulsive Phase of Flares in Soft X-Ray Emission, Solar Phys. **78**, 1982a, pp. 107-123.

Antonucci, E., Wolfson, C.J., Rapley, C.G., Acton, L.W., Culhane, J.L. and Gabriel, A.H.: Solar Observations Using the Soft X-Ray Polychromator Experiment on SMM, Proc. of the SMY International Workshop, Simferopol Moscow, IZMIRAN, 1, 1982b, pp. 62-76.

Athay, R. G., Jones, H. P., and Zirin, H.: Magnetic Shear I. Hale Region 16918. Astrophys. J. **288**, 1984, pp. 363.

Athay, R. G., White, O. R., Lites, B. W.; and Bruner, E. C. Jr.: Impulsive EUV Bursts Observed in C IV with OSO-8. Solar Phys. **66**, 1980, pp. 357-370.

Aydemir, A. and Barnes, D.: Sustained Self-Reversal in the Reversed Field Pinch, Phys. Rev. Lett. **52**, 1985, pp. 930-935.

Barnes, C.W. and Sturrock, P.A.: Astrophys. J. **174**, 1972, p. 659.

Bell, B. and Glazer, H.: Some Sunspot and Flare Statistics. Smithsonian Contr. Astrophys. **3**, 1959, pp. 25-38.

Bhatnagar, A.: Hα Solar Observations During SERF and FBS Intervals of April 6-12 and May 22-28, 1980, Proc. of the SMY International Workshop, Simferopol, **1**, 1981, pp. 202-214.

Bhattacharjee, A., Brunel, P. and Tajima, T.: Magnetic Reconnection Driven by the Coalescence Instability, Phys. Fluids **26**, 1983, pp. 3332-3337.

Birn, J. and Schindler K.: Two-Ribbon flares: Magnetostatic Equilibria, Ch. 6 of Solar Flare MHD (ed. E.R. Priest), Gordon and Breach (1981).

Birn, J., Goldstein, H., and Schindler, K.: A Theory of the Onset of Solar Eruptive Processes., Solar Phys. **57**, 1978, pp. 81-101.

Biskamp, D.: Effect of Secondary Tearing Instability on the Coalescence of Magnetic Islands, Phys. Letters **871**, 1982a, pp. 357-360.

Biskamp, D.: Dynamics of a Resistive Sheet Pinch, Z. Naturforsch **37a**, 1982b, pp. 840-847.

Biskamp, D.: Resistive MHD Processes, Physica Scripta, **T22**, 1982c, p. 405.

Born, R.: First Phase of Active Regions and their Relation to the Chromospheric Network, Solar Phys. **38**, 1974, pp. 379-388.

Brants, J.J., Cram. L.E., Zwaan, C.: An Emerging Active Region: Some Preliminary Results. The Physics of Sunspots, L.E. Cram and J.H. Thomas (eds.), Sacramento Peak Workshop, pp. 60-63.

Brants, J.J.: High-Resolution Spectroscopy of Active Regions II: Line Profile Interpretation Applied to an Emerging Flux Region. Solar Phys. **95**, 1985, pp. 15-36.
Bray, R.J., and Loughhead, R.E.: Sunspots. Chapman and Hall, London, 1964, pp. 226-236.
Browning, P.K. and Priest, E.R.: Magnetic Nonequilibrium of Buoyant Flux Tubes, Solar Phys. **92**, 1984, pp. 173-188.
Bruzek, A.: On Arch-Filament Systems in Spot groups, Solar Phys. **2**, 1967, pp. 451-461.
Bruzek, A.: On Small-Scale Mass Motion Associated with Flares. Mass Motions in Solar Flares and Related Phenomena, Y. Ohman(ed.), Nobel Symposium 9, John Wiley & Sons, 1968, pp. 67-70.
Bruzek, A.: Motions in Arch Filament Systems, Solar Phys. **8**, 1969, 29-36.
Bruzek, A., and DeMastus, H.L.: Flare-Associated Coronal Expansion Phenomena, Solar Phys. **12**, 1970, pp. 447-457.
Bruzek, A.: Some Observational Results on Moustaches, Solar Phys. **26**, 1972, pp. 94-107.
Bumba, V. and Obridko, V.N.: "Bartels" Active Longitudes, Sector Boundaries and Flare Activity, Solar Phys. **6**, 1969, pp. 104-110.
Bumba, V.: Radial Motions in Small and Young Sunspots. Bull. Astron. Inst. Czech. **18**, pp. 238-243.
Bumba, V. and Howard, R.: A Study of the Development of Active Regions on the Sun, Astrophys. J. **141**, 1981, pp. 413-470.
Canfield, R.C., Priest, E.R., and Rust, D.M.: A Model for the Solar Flare, Flare Related Magnetic Field Dynamics, Proceedings of a Conference, High Altitude Observatory, CO., 1974, pp. 361-371.
Canfield, R.C. and Fisher, R.R.: Magnetic Field Reconnection in the Flare of 18:28 UT 1975 August 10, Astrophys. J. Lett. **210**, 1976, pp. L149-L151.
Cargill, P.J. and Priest, E.R.: Slow-Shock Heating and the Kopp-Pneuman Model for 'Post'-Flare Loops, Solar Phys. **76**, 1982, pp. 357-375.
Cargill, P.J., Migliuolo, A. and Hood, A.W.: Activation of Solar Flares, Proc. Varenna Workshop on Plasma Astrophysics, 1984.
Cargill, P.J., Hood, A.W. and Migliuolo, S.: MHD Stability of Line-Tied Coronal Arcades, III Necessary and Sufficient Conditions, Astrophys. J., submitted.
Cheng. C. -C., Bruner, E. C., Tandberg-Hanssen, E., Woodgate, B. E., Shine, R. A., Kenny, P. J., Henze, W., and Poletto, G.: Observations of Solar Flare Transition Zone Plasmas from the Solar Maximum Mission. Astrophys. J. **253**, 1982, pp. 353-366.
Chiuderi Drago, F. and Melozzi, M.: Non-Thermal Radio Sources in Solar Active Regions Astron. Astrophys. **131**, 1984, pp. 103-110.
Cliver, E.W., Forrest, D.J., McGuire, R.E. and von Rosenvinge, T.T.: Nuclear Gamma Rays and Interplanetary Proton Events, in the 18th International Cosmic Ray conference, Late Papers vol., eds., Durgaprased et. al., 1984, Tata Institute, Coloba, Bombay, India.
Culhane, J.L. and Phillips, K.J.H.: Solar X-Ray Bursts at Energies Less than 10 keV Observed with OSO-4, Solar Phys., **11**, 1970, pp. 117-144.
de Jager, C., Machado, M.E., Schadee, A., Strong, K.T., Svestka, Z., Woodgate, B.E. and van Tend, W.: The Queen's Flare: Its Structure and

Development; Precursors, Pre-Flare Brightenings, and Aftermaths, Solar Phys. **84**, 1983, pp. 205-235.
deLoach, A. C., Hagyard, M. J., Rabin, D., Moore, R. L., Smith, J. B., Jr.; West, E. A.; and Tandberg-Hanssen, E.: Photospheric Electric Current and Transition Region Brightness within an Active Region. Solar Phys. **91**, 1984, pp. 235-242.
Duijveman, A., Hoyng, P. and Machado. M.E.: X-Ray Imaging of Three Flares During the Impulsive Phase, Solar Phys. **81**, 1982a, pp. 137-157.
Duijveman, A., Somov, B.V. and Spektor, A.R.: Evolution of a Flaring Loop after Injection of Energetic Electrons Solar Phys. **58**, 1982b, pp. 257-273.
Dulk, G.A. and Dennis, B.R.: Microwaves and Hard X-Rays from Solar Flares: Multithermal and Nonthermal Interpretations, Astrophys. J. **260**, 1982, pp. 844-875.
Dunn, J.M. and Martin, S.F.: An Attempt to Identify Flare Precursor Mass Motions in Real Time, Bull. Am. Astr. Soc. **12**, 1980, p. 904.
Dwivedi, B.N., Hudson, H.S., Kane, S.R., and Svestka, Z: Hα and Hard X-Ray Development in Two-Ribbon Flares, Solar Phys. **90**, 1984, pp. 331-342.
Einaudi, G. and Van Hoven G.: Stability of Diffuse Linear Pinch with Axial Boundaries, Phys. Fluids **24**, 1981, pp. 1092-1096.
Enome, S., Shibasaki, K., Takayanagi, T., and Takata, S., Atlas of Solar Bursts for 1980, Toyokawa Obs., WDC-C2, 1981.
Finn, J.M. and Kaw, P.K.: Coalescence Instability of Magnetic Islands, Phys. Fluids **20**, 1977, pp. 72-79.
Forbes, T.G. and Priest, E.R.: Numerical Study of Line-Tied Magnetic Reconnection, Solar Phys. **81**, 1982, pp. 303-324.
Forbes, T.G. and Priest, E.R.: A Numerical Experiment Relevant to LineTied Reconnection in Two-Ribbon Flares, Solar Phys. **34**, 1983a, pp. 169-188.
Forbes, T.G. and Priest, E.R.: Mass Upflows in "Post"-Flare Loops, Solar Phys. **88**, 1983b, pp. 211-218.
Forbes, T.G. and Priest, E.R.: Numerical Simulation of Reconnection in an Emerging Magnetic Flux Region, Solar Phys. **94**, 1984, pp. 315-340.
Forbes, T.G. and Priest, E.R.: Reconnection in Solar Flares, Proc. Solar Terrestrial Theory Workshop, Coolfont (ed. B.U.O. Sonnerup).
Furth, H.P., Killeen, J. and Rosenbluth, M.N.: Finite-resistivity Instabilities of a Sheet Pinch, Phys. Fluids **6**, 1963, pp. 459-484.
Gabriel, A.H. and 14 co-authors: Observations of the Limb Solar Flare on 1980 April 30 with the SMM X-Ray Polychromator, Astrophys. J. **244**, 1981, pp. L147-L151.
Gaizauskas, V.: Preflare Activations of Filaments Located Along Inversion Lines of Magnetic Polarity, Proc. Kunming Workshop, 1984 (in press).
Gaizauskas, V.: The Relation of Solar Flares to the Evolution and Proper Motions of Magnetic Fields, Adv. Space Res. (Eds. Svestka, Rust and Dryer) **2**, 1983, pp. 11-30.
Gaizauskas, V., Harvey, K.L., Harvey, J.W., Zwaan, C.: Large-Scale Patterns Formed by Solar Active Regions During the Ascending Phase of Cycle 21, Astrophys. J., **265**, 1983, pp. 1056-1065.
Gaizauskas, V. and McIntosh, P.S: On the Flare-Effectiveness of

Recurrent Patterns of Magnetic Fields, Proc. Workshop on Solar Terrestrial Predictions, Meudon, 1984, submitted.
Gary, D.: Radio Emission from Solar and Stellar Coronae, Ph.D. Thesis, 1982, Univ. of Colorado (University Microfilms).
Gary, D., Dulk, G.A., Illing, R., Sawyer, C., Wagner, W.J., McLean, D.J., and Hildner, E.: Type II Bursts, Shock Waves and Coronal Transients: The Event of 1980 June 29, 0233 UT, Astron. Astrophys. **134**, 1984, pp. 222-233.
Giovanelli, R.G.: The Relations between Eruptions and Sunspots, Astrophys. J. **89**, 1939, pp. 555-567.
Glackin, D.L.: Emerging Flux Region, Solar Phys. **43**, 1975, pp. 317-326.
Gosling, J.T., Hildner, E., MacQueen, R.M., Munro, R.H., Poland, A.I., and Ross, C.L.: The Speeds of Coronal Mass Ejection Events, Solar Phys. **48**, 1976, pp. 389-397.
Hagyard, M. J., Cumings, N. P., West, E. A., and Smith, J. E.: The MSFC Vector Magnetograph. Solar Phys. **80**, 1982, pp. 33-51.
Hagyard, M.J., Smith, J.B., Jr., Teuber,D., and West, E.A.: A Quantative Study Relating Observed Shear in Photospheric Magnetic Fields to Repeated Flaring, Solar Phys. **91**, 1984, pp. 115-126.
Hagyard, M. J.; West, E. A.; and Smith, J. B., Jr.,: Electric Currents in Active Regions. Proceedings of the Kunming Workshop on Solar Physics and InterPlanetary travelling Phenomena, Kuming, People's Republic of China, 1984, in press.
Harrison, R., Waggett, P., Bentley, R., Phillips, K.J.H., Bruner, M., Dryer, M. and Simnett, G.M: Solar Phys. 1985, in press.
Harvey, J., Gillespie, B., Miedaner, P., Slaughter, C.: Synoptic Solar Magnetic Field Maps for the Interval Including Carrington Rotations 1601-1680, May 5, 1973 - April 26, 1979. World Data Center A, Report UAG-77, 1980, (Boulder, CO).
Harvey, J.W.: Flare Build-up: 21 May 1980, Adv. Space Res. **2**, 1983, pp. 31-37.
Harvey, K.L. and Martin, S.F.: Ephemeral Active Regions, Solar Phys. **32**, 1973, pp. 389-402.
Harvey, K.L. and Harvey, J.W.: Observations of Moving Magnetic Features near Sunspots, Solar Phys. **28**, 1973, pp. 61-71.
Harvey, K. L., and Harvey, J. W.: A Study of the Magnetic and Velocity Fields in an Active Region, Solar Phys. **47**, 1976, pp. 233-246.
Heyvaerts, J., Priest, E.R. and Rust, D.M.: An Emerging Flux Model for the Solar Flare Phenomenon, Astrophys. J. **216**, 1977, pp. 123-137.
Heyvaerts, J., Lasry, J.M., Schatsman, M. and Witomsky, P., Blowing up of Two-Dimensional Magnetohydrostatic Equilibria by an Increase of Electrical Current or Pressure, Astron. Astrophys. **111**, 1982, pp. 104-112.
Heyvaerts, J. and Priest, E.R.: Coronal Heating by Reconnection in DC Current Systems: a Theory Based on Taylor's Hypothesis, Astron. Astrophys. **137**, 1984, pp. 63-68.
Hong Qin-Fang; Ding You-Ji; and Luan Di: The Rapidly Emerging Sunspot Group in SESC 2372 Region in April 1980. Acta Astronomica Sinica **23**, 1982, pp. 327-334.
Hood, A.W. and Priest, E.R.: Kink Instability of Coronal Loops as the Cause of Solar Flares, Solar Phys. **64**, 1979, pp. 303-321.

Hood, A.W. and Priest, E.R.: The Equilibrium of Solar Coronal Magnetic Loops, Astron. Astrophys. **77**, 1979, pp. 233-251.
Hood, A.W. and Priest, E.R.: Magnetic Instability of Coronal Arcades as the Origin of Two-Ribbon Flares, Solar Phys. **66**, 1980, pp. 113-134.
Hood, A.W. and Priest, E.L: Critical Condition for Magnetic Instabilities in Force-Free Coronal Loops, Geophys. Astrophys. Fluid Dynamics **17**, 1981, pp. 297-318.
Hood, A.W.: Magnetic Stability of Coronal Arcades Relevant to Two-Ribbon Flares, Solar Phys. **87**, 1983a, pp. 279-299.
Hood, A.W.: Stability of Magnetohydrostatic Atmospheres, Solar Phys. **89**, 1983b, pp. 235-242.
Hood, A.W.: An Energy Method for the Stability of Solar Magnetohydrostatic Atmospheres, Geophys. Astrophys. Fluid Dynamics **28**, 1984a, pp. 223-241.
Hood, A.W.: Stability of Magnetic Fields Relevant to Two-Ribbon Flares, Adv. in Space Research, 1984b, in press.
Hoyng, P., Marsh, K., Zirin, H., and Dennis, B.R.: Microwave and Hard X-ray Imaging of a Solar Flare on 1980 November 5, Astrophys. J. **268**, 1983, pp. 865-879.
Hoyng, P. and 23 co-authors: Hard X-Ray Imaging of Two Flares in Active Region 2372, Astrophys. J. **244**, 1981, pp. L153-L156.
Hoyng, P., Duijveman, A., Machado, M.E., Rust, D.M., Svestka, Z., Boelee, A., de Jager, C., Frost, K.J., Lafleur, H., Simnett, G.M., van Beek, H.F., and Woodgate, B.E.: Origin and Location of the Hard X-ray Emission in a Two-Ribbon Flare, Astrophys. J. **246**, 1981, pp. L155-L159.
Hurford, G.J. and Zirin, H: Interferometric Observations of Solar Flare Precursors at 10.6 GHz, Technical Report AFGL 3 TR-82-0117, March 1982.
Hurford, G.J.: The Owens Valley Frequency-Agile Interferometer, Technical Report AFGL-TR-83-108, March 1983.
Jackson, R. and Hildner, E.: Forerunners: Outer Rims of Solar Coronal Transients, Solar Phys. **60**, 1978, pp. 155-170.
Jackson, B.V.: Forerunners: Early Coronal Manifestations of Solar Mass Ejection Events, Solar Phys. **73**, pp. 133-144.
Jackson, B.V. and Sheridan, K.V.: Evidence for a Peak in the Number of Isolated Type III Bursts prior to Large Solar Flares, Proc. Astron. Soc. Aust. **3**, 1979, pp. 383-386.
Kahler, J.N. and Buratti, B.D.: Preflare X-Ray Morphology of Active Regions Observed with the AS&E Telescope on Skylab, Solar Phys. **47**, 1976, pp. 157-165.
Kahler, S.W.: Preflare Characteristics of Active Regions Observed in Soft X-rays, Solar Phys. **62**, 1979, pp. 347-357.
Kahler, S.: The Morphological and Statistical Properties of Solar X-Ray Events with Long Decay Times, Astrophys. J., **214**, 1977, pp. 891-897.
Kai, K., Nakajima, H. and Kosugi, T.: Radio Observations of Small Activity Prior to Main Energy Release, P.A.S.J. **35**, 1983, pp. 285-297.
Kane, S.R. and Pick, M.: Non-thermal Processes During the 'Build-up' Phase of Solar Flares and in the Absence of Flares, Solar Phys. **47**, 1976, pp. 293-304.
Karpen, J. and Boris;1985, submitted
Kawaguchi, I. and Kitai, R.: The Velocity Field Associated with the Birth

of Sunspots, Solar Phys. **45**, 1976, pp. 125-135.
Kiepenheuer, K.O.: Solar Activity. The Sun, G.P. Kuiper (ed.), University of Chicago Press, 1953, pp. 348-352.
Kopp., R.A. and Pneuman, G.W.: Magnetic Reconnection in the Corona and the Loop Prominence Phenomenon, Solar Phys. **50**, 1976, pp. 85-98.
Kosugi, T., Kai, K. and Nakajima, H.: Statistical Study of Type III Burst Activity Before Flares, to be submitted, 1985.
Kosugi, T. and Shiomi, Y., Solar Radio Activities, Nobeyama Solar Radio Observatory of the Tokyo Astron. Observatory, February 1983.
Krall, K. R., Smith, J. B., Jr., Hagyard, M. J., West, E. A. and Cumings, N.P.: Vector Magnetic Field Evolution, Energy Storage, and Associated Photospheric Velocity Shear within a Flare-Productive Active Region. Solar Phys. **79**, pp. 59-75.
Kundu, M.R., Schmahl, E.J., Velusamy, T. and Vlahos, L.: Radio Imaging of Solar Flares Using the Very Large Array: New Insight into Flare Process, Astron. Astrophys. **108**, 1982, pp. 188-194.
Kundu, M., Gaizauskas, V., Woodgate, B., Schmahl, E.J., Jones. H., and Shine, R. A Study of Flare Buildup from Simultaneous Observations in Microwave, Hα and UV Wavelengths, Astrophys. J. Suppl **57**, 1985, pp. 621-530.
Kundu, M.R., Solar Flare Observations at Centimeter Wavelengths Using the VLA, Proc. of the SMY International Workshop, Simferopol, Moskow, IZMIRAN, **1**, 1981, pp. 124.
Kundu, M.R. and Shevgaonkar, R.K.: VLA Observations of a Pre-Flare Solar Active Region and a Flare at 2, 6 and 20 Centimeter Wavelengths, Astrophys. J. **291**, 1985, pp. 860-864.
Kuperus, M. and van Tend, W.: The Eruption of Active Region Filaments and its Relation to the Triggering of a Solar Flare, Solar Phys. **71**, 1981, pp. 125-139.
Lang, K.R.,: High Resolution Interferometry of the Sun at 3.7 cm Wavelength, Solar Phys. **36**, 1974, pp. 351-367.
Lang, K.R,: in Solar Terrestrial Prediction Proceedings III, Solar Activity Predictions, 1979 (ed. R.F. Donnelly).
Levine, R.H.: EUV Structure of a Small Flare, Solar Phys. **56**, 1978, pp. 185-203.
Liggett, M. and Zirin, H.: Emerging Flux in Active Regions. Solar Phys. 1984, in press.
Lites, B. W.; and Hansen, E. R.: Ultraviolet Brightenings in Active Regions as Observed from OSO-8. Solar Phys. **55**, 1977, pp. 347-358.
Low, B. C.: Evolving Force-Free Magnetic Fields. I. The Development of the Preflare Stage. Astrophys. J., **212**, 1977, pp. 234-242.
Low, B. C.: Evolving Force-Free magnetic Fields. II. Stability of Field Configurations and the Accompanying Motion of the Medium. Astrophys. J. **217**, 1977, pp. 988-998.
MacQueen, R.M. and Fisher, R.R.: The Kinematics of Solar Inner Coronal Transients, Solar Phys., **89**, 1983, pp. 89-102.
Machado, M. E., Somov, B. V., Rovira, M. G. and De Jager, C.: The Flares of April 1980. Solar Phys. **85**, 1983, pp.157-184.
Machado, M.E., Duijveman, A. and Dennis, B.R.: Spatial and Temporal Evolution of Soft and Hard X-Ray Emission in a Solar flare. Solar Phys. **79**, 1982, pp. 85-106.

Machado, M.E., Somov, B.V., Rovira, M.R., and de Jager, C: The Flares of April 1980: A Case for Flares Caused by Interacting Field Structures, Solar Phys. **85**, 1983, pp. 157-184.

Malherbe, J.M., Simon, G., Mein. P., Mein, N., Schmieder, B., and Vial, J.C.: Preflare Heating of Filaments, Adv. Space Res., (eds. Svestka, Rust and Dryer) **2**, 1983, pp. 53-56.

Malherbe, J.M., Mein, P. and Schmieder, B: Mass Motions in a Quiescent Filament, in Advances in Space Research, (eds., Z. Svestka, D.M. Rust and M. Dryer) **2**, 1983, pp. 57-60.

Malherbe, J.M. and Priest, E.R.: Current Sheet Models for Solar Prominences, I, Magnetohydrostatics of Support and Evolution through Quasi-Static Models, Astron. Astrophys. **123**, 1983, pp. 80-88.

Malherbe, J.M., Priest, E.R., Forbes, T.G. and Heyvaerts, J.: Current Sheet Models for Solar Prominences, II, Energetics and Condensation Process, Astrophys. J. **127**, 1983, pp. 153-160.

Manheimer, W. and Boris, J.P.: Comments Plasma Phys. Conf. Fusion **3**, 15, 1977.

Mariska, J.T., Boris, J.P., Oran, E.S., Young, T.R., Jr., and Doschek, G.A., Astrophys. J. **255**, 1982, p. 783.

Marsh, K.A.: Ephemeral Region Flares and the Diffusion of the Network. Solar Phys. **59**, 197, pp. 105-113.

Marsh, K.A., Hurford, G.J., Zirin, H., Dulk, G.A., Dennis, B.R., Frost, K.J. and Orwig, L.E.: Properties of Solar Flare Electrons Deduced from Hard X-Ray and Spatially Resolved Microwave Observations, Astrophys. J. **251**, pp. 797-804.

Martens, P.C.H. and Kuperus, M.: Resonant Electrodynamic Heating and the Thermal Stability of Coronal Loops, Astron. Astrophys. **114**, 1982, pp. 324-327.

Martens, P.C.H., Van den Oord, G.H.J. and Hoyng, P.: Observations of Steady Anomalous Magnetic Heating in Thin Current Sheets, Solar Physics, **96**, 1985, pp 253-274.

Martin, S.F. and Ramsey, H.E.: 1972, in Solar Activity Observations and Predictions (eds. P. McIntosh and M. Dryer), MIT Press, Cambridge, MA, pp. 371-388.

Martin, S.F.: Preflare Conditions, Changes and Events, Solar Phys. **68**, 1980, pp. 217-236.

Martin, S.F., Dezso, L., Antalova, A., Jucera, A., and Harvey, K.L.: Emerging Magnetic Flux, Flares and Filaments - FBS Interval 16-23 June 1980 (eds. Z. Svestka, D. Rust and M. Dryer) Adv. Space Res. **2**, 1983, pp. 39-51.

Martin, S.F.: Early Signs of New Active Regions, Bull. American Astron. Soc. **15**, 1983, p. 971.

Martin, S.F.: Dynamic Signatures of Quiet Sun Magnetic Fields. Small-Scale Dynamical Processes, Sacramento Peak Workshop, 1984, in press.

Martres, M. -J., Michard, R., Soru-Iscovici, I. and Tsap, T. T.: Etude de la Localisation des Eruptions dans la Structure Magnetique Evolutive des Regions Actives Solaires. Solar Phys. **5**, 1968, pp. 187-206.

Martres, M. J., Soru-Escaut, I., and Rayrole, J.: An Attempt to Associate Observed Photospheric Motions with the Magnetic Field

Structure and Flare Occurrence in an Active Region. Solar Magnetic Fields, IAU Symposium No. 43, R. Howard (ed.), D. Reidel Publishing Co., Holland, 1971, pp. 435-442.
Martres, M. -J., Soru-Escaut, I., and Rayrole, J.: Relationship between Some Photospheric Motions and the Evolution of Active Centres. Solar Phys. **32**, 1973, pp. 365-378.
Martres, M. -J., Rayrole, J., Ribes, E., Semel, M. and Soru-Escaut, I.: On the Importance of Photospheric Velocities in Relation to Flares. Flare Related Magnetic Field Dynamics, Proceedings of a Conference, High Altitude Observatory, Boulder, Co., 1974, pp. 333-352.
Martres, M. -J. and Soru-Escaut, I.: The Relation of Flares to "Newly Emerging Flux' and "Evolving Magnetic Features". Solar Phys., **53**, 1977, pp. 225-231.
McConnell, D. and Kundu, M.R.: VLA Observations of Fine Structures in a Solar Active Region at 6 Centimeter Wavelength, Astrophys. J. **279**, 1984, pp. 421-426.
McIntosh, P.S.: The Birth and Evolution of Sunspots: Observations, The Physics of Sunspots, L.E. Cram and J.H. Thomas (eds.), Sacramento Peak Workshop, 1981, pp. 7-57.
McKenna-Lawlor, S.M.P. and Richter A.K.L: Physical Interpretation of Interdisciplinary Solar/Interplanetary Observations Relevant to the 27-29 June 1980 SMY/STIP Event No. 5, in Advances Space Res. (eds., Z. Svestka, D.M. Rust and M. Dryer), **2**, 1982, pp. 239-251.
Melville, J.P., Hood, A.W. and Priest, E.R.: Magnetic Equilibrium in Coronal Arcades, Solar Phys. **87**, 1983, pp. 301-307.
Melville, J.P., Hood, A.W. and Priest, E.R.: Magnetohydrostatic Structures in the Corona, Solar Phys. **92**, 1984, pp. 15- .
Migiuolo, S. and Cargill, P.J.:MHD Stability of Line-Tied Coronal Arcades, I, Astrophys. J. **271**, 1983, pp. 820-831.
Magiuolo, S., Cargill, PJ. and Hood, A.W.: MHD Stability of Line-Tied Coronal Arcades, II, Shearless Magnetic Fields, Astrophys. J. **281**, 1984, pp. 413-418.
Milne, A. and Priest, E.R.: Internal Structure of Reconnecting Current Sheets and the Emerging Flux Model for Solar Flares, Solar Phys. **72**, 1981, pp. 157-182.
Mok, Y. and Van Hoven, G: Resistive Magnetic Tearing in a Finite Length Pinch, Phys. Fluids **25**, 1982, pp. 636-642.
Molodensky, M. M.: Equilibrium and Stability of Force-Free Magnetic Field. Solar Phys. **39**, 1974, pp. 393-404.
Moore, R.L.,Hurford, G.J.,Jones, H. and Kane, S.R.: Magnetic Changes Observed in a Flare, Astrophys. J. **276**, 1984, pp. 379-390.
Moreton, G. E. and Severny, A. B.: Magnetic Fields and Flares in the Region CMP 20 September 1963. Solar Phys. **3**, 1968, pp. 282-297.
Mosher, J.M. and Acton, L.W.: X-rays, Filament Activity and Flare Prediction, Solar Phys. **66**, 1980, pp. 105-111.
Mouradian, Z., Martres, M. J., and Soru-Escaut, I.: The Emerging Magnetic Flux and the Elementary Eruptive Phenomenon, Solar Phys. **87**, 1983, pp. 309-328.
Nagy, I.: Sunspot Proper Motions in the Western Part of Hale Region 16864 (May 25029, 1980), Publ. Debrecen Obs., **5**, 1983, pp. 107-116.
Nakagawa, Y. and Raadu, M. A.: On Practical Representation of Magnetic

Field, Solar Phys. **25**, 1972, pp. 127-135.
Neidig, D.F. and Cliver, E.W.: A Catalog of Solar White-Light Flares (1859-1982), Including their Statistical Properties and Associated Emissions, AFGL-TR-83-0257, 1983, Hanscom AFB, MA.
van den Oord, B.V.D., Martens P.C.H. and Hoyng, P.: HXIS Observations of the Thermal Evolution of the Coronal Loop of November 5/6, 1980, Solar Phys., 1984, (to be submitted).
Parker, E.N.: Cosmical Magnetic Fields, Oxford University Press, 1979.
Parker, E.N.: Magnetic Reconnection and Magnetic Activity, Magnetic Reconnection in Space and Laboratory Plasmas (ed. E. Hones) A.G.U. Geomonograph Series, 1984, pp. 32-38.
Patty, S. R. and Hagyard, M. J.: Delta-Configurations: Flare Activity and Magnetic Field Structure, Solar Phys., 1984, submitted.
Petschek, H.E.: Magnetic Field Annihilation, AAS-NASA Symp. on Phys. of Solar Flares, NASA SP-1954, pp. 425-539.
Petrasso, R.D., Kahler, S.W., Krieger, A.S., Silk, J.K. and Vaiana, G.S.: The Location of the Site of Energy Release in a Solar X-Ray Subflare, Astrophys. J. Letters **L199**, 1975, pp. L127-L130.
Poland, A.I., Machado, M.E., Wolfson, C.J., Frost, K.J., Woodgate, B.E., Shine, R.A., Kenny, P.J., Cheng, C.-C., Tandberg-Hanssen, E.A., Bruner, E.C., Henze, W.: The Impulsive and Gradual Phases of a Solar Limb Flare as Observed from the Solar Maximum Mission Satellite, Solar Phys. **78**, 1982, pp. 201-213.
Porter, J. G., Toomre, J. and Gebbie, K. B.: Frequent UV Brightenings Observed in a Solar Active Region with SMM, Astrophys. J. **283**, 1984, pp.. 879-886.
Priest, E.R. and Milne, A.M.: Force-Free Magnetic Arcades Relevant to Two-Ribbon Flares, Solar Phys. **65**, 1980, pp. 315-346.
Priest, E.R.: Solar Flare Magnetohydrodynamics, Gordon and Breach, 1981a.
Priest, E.R: Flare Theories, Proc. 3rd European Solar Meeting (ed. C. Jordan) 1981b, pp. 203-232.
Priest, E.R.: Magnetic Reconnection at the Sun, Magnetic Reconnection in Space and Laboratory Plasmas (ed. E. Hones) A.G.U. Geomonograph, 1984a, pp.. 63-79.
Priest, E.R.:The MHD of Current Sheets, Rep. Prog. Phys. 1985, **48**, No. 7.
Pries, E.R.: Small-Scale Reconnetion, Proc. ESA Meeting on SOHO and CLUSTER, Garmisch, 1985a.
Priest, E.F.: Magnetohydrodynamic Theories of Solar Flares, Solar Phys., in press, 1985b.
Pritchett, P.L. and Wu, C.C.: Coalescence of Magnetic Islands, Phys. Fluids **22**, 1979, pp. 2140-2146.
Raadu, M.A.: Supp.ression of Kink Instability for Magnetic Flux Ropes in the Chromosphere, Solar Phys. **22**, 1972, pp. 425-433.
Rabin, D. M.; Moore, R. L.; and Hagyard, M. J.: A Case for Submergence of Magnetic Flux in a Solar Active Region. Astrophys. J. **287**, 1984, pp.. 404-411.
Rabin, D. and Moore, R. L.: Heating the Sun's Lower Transition Region with Fine-Scale Currents. Astrophys. J. **285**, 1984, pp. 359-367.
Ray, A. and Van Hoven, G.: Hydromagnetic Stability of Coronal Arcade Structures: the Effects of Photospheric Line Tying, Solar Phys. **79**,

1982, pp. 353-364.
Roberts, P.H.: Velocity Fields in Magnetically Disturbed Regions of the Hα Chromosphere, Ph.D. Thesis, California Institute of Technology, 1970.
Roberts, R. and Frankenthal, S.: The Thermal Statics of Coronal Loops, Solar Phys. **68**, 1980, pp. 103-109.
Rock, K., Fisher, R., Garcia, H., and Hasukawa: A Summary of Solar Activity Observed at the MLSO, NCAR/TN-221 STR, Nov. 1983, 1980-1983.
Rosner, R., Tucker, W. and Vaiana, G.L.: The Dynamics of the Quiescent Solar Corona, Astrophys. J. **220**, 1978, 643-665.
Roy, J.R. and Tang, F.: Slow X-Ray Bursts and Flares with Filament Disruption, Solar Phys. **42**, 1975, pp. 425-439.
Roy, J.-R. and Michalitsanos, A.F.: Chromospheric Activity Associated with Moving Magnetic Fields, Solar Phys. **35**, 1974, pp. 47-54.
Rust, D.M.: Chromospheric Explosions and Satellite Sunspots, Structures and Development of Solar Active Regions, IAU Symposium 35, K.O. Kiepenheuer (ed.), D. Reidel Publishing Co., Holland, 1968, pp. 77-84.
Rust, D.M.: Flares and Changing Magnetic Fields, Solar Phys. **25**, 1972, pp. 141-157.
Rust, D.M. and Roy, J.-R.: The Late June 1972 "CINOF" Flares, AFCRL-TR-75-0437, 1975, pp. 61-93.
Rust, D.M. and Bridges, C.A.: The Work of the Diode Array: He 10830 Observations of Spicules and Subflares, Solar Phys. 43, 1975, pp. 129-145.
Rust, D.M., Nakagawa, Y. and Neupert, W.M.: EUV Emission, Filament Activation and Magnetic Fields in a Slow-Rise Flare, Solar Phys. **41**, 1975, pp. 397-414.
Rust, D.M. and Hildner, E.: Expansion of an X-ray Coronal Arch into the Outer Corona, Solar Phys. **48**, 1976, pp. 381-387.
Rust, D.M.: Observations of Flare-Associated Magnetic Field Changes, Phil. Trans. R. Soc. London, Ser. A. **281**, 1976, pp. 427-433.
Rust, D. and Webb, D.F.: Soft X-Ray Observations of Large-Scale Coronal Active Region Brightenings, Solar Phys. **54**, 1977, pp. 403-417.
Rust, D.M., Benz, A.O., Hurford, G.J., Nelson, G., Pick, M., and Ruzdjak, V.: Optical and Radio Observations of the 29 March, 30 April and 7 June 1980 Flares, Astrophys. J. **244**, 1981, pp. L179-L183.
Rust, D.M., Buhmann, R.W., Dennis, B.R., Robinson, R.D., Willson, RR., Simon, M. and the SMM XRP team: Spatial and Temporal Corelation of High and Low Temperature Solar Flare Emissions, Bull. Am. Astr. Soc. **12**, 1980, p. 752.
Rutherford, P.H.: Nonlinear Growth of the Tearing Mode, Phys. Fluids **16**, 1973, pp. 1903-1908.
Sakurai, T.: Magnetohydrodynamic Interpretation of the Motion of Prominences, Publ. Astr. Soc. Japan **28**, 1976, pp. 177-198.
Scadee, A., deJager, C. and Svestka, Z.,: Enhanced X-Ray Emission Above 3.5 keV in Active Regions in the Absence of Flares, Solar Phys. **89**, 1983, 287-306.
Schadee, A. and Gaizauskas, V.: Identification of Two X-Ray Miniflares with Hα Subflares, Adv. Space Res. **4**, 1984, pp. 117-120 (XXV COSPAR, Graz).
Schindler, K., Birn, J. and Janicke, L.: Stability of Two-Dimensional

Preflare Structures, Solar Phys. **87**, 1983, pp. 103-134.
Schmahl, E.J.: Flare Buildup in X-rays, UV, Microwaves, and White Light, Adv. Space Res. (Eds. Svestka, Rust and Dryer) **2**, 1983, pp. 73-90.
Schmieder, B., Vial, J.C., Mein, P. and Tandberg-Hanssen, E.: Dynamics of a Surge Obsrved in the C IV and Hα Lines, Astron. Astrophys. **127**, 1983, pp. 337-344.
Schmieder, B., Mein, P., Martres, M.-J. and Tandberg-Hanssen, E.: Dynamic Evolution of Recurrent Mass Ejection Obsrved in Hα and C IV Lines, Solar Phys. **94**, 1984, pp. 133-154.
Schoolman, S.: Videomagnetograph Studies of Solar Magnetic Fields II: Field Changes in an Active Region, Solar Phys. **32**, 1973, pp. 379-388.
Sheeley, N.R.: The Evolution of the Photospheric Network, Solar Phys. **9**, 1969, pp. 347-357.
Sheeley, N.R., Jr., Bohlin, J.D., Brueckner, G.E., Purcell, J.D., Scherrer, V.E., Tousey, R., Smith, J.B.,Jr., Speich, D.M., Tandberg Hanssen, E., Wilson, R.M., deLoach, A.C., Hoover, R.B. and McGuire, J.P., Solar Phys. **45**, 1975, pp. 377-392.
Sheeley, N.R. and Golub, L.: Rapid Changes in the Fine Structure of a Coronal "Bright Point" and a Small Coronal "Active Region", Solar Phys. **63**, 1979, pp. 119-126.
Sheeley, N.R.: Temporal Variations of Loop Structures in the Solar Atmosphere, Solar Phys. **66**, 1980, pp. 79-87.
Sheeley, N.R.: The Overall Structure and Evolution of Active Regions. Solar Active Regions, F.Q. Orrall (ed.), Colorado Associated University Press, 1981, pp. 17-42.
Sime, D., Fisher R., and Munro, R.: Ground-Based Observations of the Corona Following the 29 June 1821 Flare, Bull. Am. Astr. Soc. **12**, 1980, p. 903.
Simnett, G.M. and Harrison, R.: The Relationship Between Coronal Mass Ejections and Solar Flares, Solar Phys., Advances in Space Research, COSPAR, 1984, **4**, pp. 279-282.
Simon, G., Mein, N., Mein, P. and Gesztely, L.: Preflare Activity of Solar Prominences, Solar Phys. **93**, 1984, pp. 325-336.
Smith, S.F. and Howard, R.: Magnetic Classification of Active Regions, Structure and Development of Solar Active Regions, IAU Symposium 35, K.O. Kiepenheuer (ed.), D. Reidel Publishing Co., Holland, 1968, pp. 33-42.
Smith, J. B., Jr., Krall, K. R., Hagyard, M. J., Cumings, N. P., West, E., Reichmann, E., and Smith, J. E.: Vector Magnetic Measurements of an Active Region. Bulletin Am. Astron. Soc. **11**, 1979, p. 440.
Smith, S.F. and Ramsey, H.E.: The Flare-Associated Filament Disappearance, Zeit. F. Phys. **60**, 1964, pp. 371-387.
Solar Geophysical Data, World Data Center A, Solar Radio Emission, Outstanding Occurrences, **4418**, 1981, pp. 5-45.
Sonnerup, B.U.O.:Magnetic Field Reconnection in a Highly Conducting Incompressible Fluid, J. Plasma Phys. **4**, 1970, pp. 161-174.
Steinolfson, R.S.: Energtics and the Resistive Tearing Mode: Effects of Joule Heating and Radiation, Phys. Fluids **26**, 1983, pp. 2590-2602.
Steinolfson, R.S.: Thermal Ripples in a Resistive and Radiative Instability, Astrophys. J. **281**, 1983, pp. 854-861.
Steinolfson, R.S. and Van Hven, G.: The Growth of the Tearing Mode;

Boundary and Scaling Effects, Phys. Fluids **26**, 1983, pp. 117-123.
Steinolfson, R.S. and Van Hoven, G.: Radiative Tearing: Magnetic Reconnection on a Fast Thermal-Instability Time-Scale, Astrophys. J. **276**, 1984a, pp. 391-398.
Steinolfson, R.S. and Van Hoven, G.: Nonlinear Evolution of the Resistive Tearing Mode, Phys. Fluids **27**, 1984b, pp. 1207-1214.
Stewart, R.T.: Homologous Type II Radio Bursts and Coronal Transients, submitted to Solar Physics, 1984.
Strong, K.T., Benz, A.O., Dennis, B.R., Leibacher, J.W., Mewe, R., Poland, A., Schrijver, H., Simnett, G.M., Smith, J.B., Jr. and Sylwester, J.: A Multi Wavelength Study of a Double Impulsive Flare, Solar Phys., **90**, 1984, pp. 325-344.
Sturrock, P: Flare Models, in Solar Flares (ed. P. Sturrock) Colo. Assoc. Univ. Press, Boulder, 1980, pp. 441-449.
Sung, Mu-Tao and Cao, Tian-Jun: The Instability of a Non-Static Plasma Column, Chin. Astron. Astrophys. **7**, 1983, pp. 159-164.
Svestka, Z., Dennis, B.R., Pick, M., Raoult, A., Rapley, C.G., Stewart, R.T. and Woodgate, B.E.: Unusual Coronal Activity Following the Flare of 6 November 1980, Solar Phys. **80**, 1982, pp. 143-159.
Svestka, Z., Stewart, R.T., Hoyng, P., van Tend, W., Acton, L.W., Gabriel, A.H., Rapley, C.G., Boelee, A., Bruner E.C.; de Jager, C., LaFleur, H., Nelson, G., Simnett, G.J., van Beek, H.F. and Wagner, W.: Observations of a Post-Flare Radio Burst in X-Rays, Solar Phys. **75**, 1982, pp. 305-329.
Svestka, Z: Solar Flares, D. Reidel, Dordrecht, Holland, 1976.
Tanaka, K. and Nakagawa, Y.: Force-Free Magnetic Fields and Flares of August 1972, Solar Phys. **33**, 1973, pp. 187-204.
Tandberg-Hanssen, E., Reichmann, E. and Woodgate, B.: Behavior of Transition-Region Lines During Impulsive Solar Flares Solar Phys. **86**, 1983, pp. 159-171.
Tang, F.: On the Origin of δ Spots, Solar Phys. 89, 1983, pp. 43-50.
Tang, F., Harvey, K., Bruner, M., Kent, B. and Antonucci, E.: Bright Point Study, Adv. Space Res. **2**, 1983, pp. 65-72.
Taylor, J.B.V.: Relaxation of Toroidal Plasma and Generation of Reverse Magnetic Fields, Phys. Rev. Lett., **33**, 1974, pp. 1139-1141.
Thomas, R.J. and Teske, R.G.: Solar Soft X-Rays and Solar Activity, Solar Phys. **16**, 1971, pp. 431-453.
Van Hoven, G.: The Preflare State, in Solar Flares (ed. P. Sturrock) Colo. Assoc. Univ. Press, Boulder, 1980, pp. 17-81.
Van Hoven, G., Steinolfson, R.S. and Tachi, T.: Energy Dynamics in Stressed Magnetic Fields: the Filamentation and Flare Instabilities, Astrophys. J. **268**, 1983, pp. 860-864.
Van Tend, W. and Kuperus, M.: The Development of Coronal Electric Current Systems in Active Regions and Their Relation to Filaments and Flares, Solar Phys. **59**, 1978, pp. 115-127.
Vasyliunas, V.H.: Theoretical Models of Magnetic Field Line Merging, 1, Rev. Geophys. Space Phys. **13**, 1975, pp. 303-336.
Vorpahl, J.R.: Flares Associated with EFR's (Emerging Flux Regions), Solar Phys. **28**, 1973, pp. 115-122.
Vorpahl, J.A., Gibson, E.G., Landecker, P.B., McKenzie, D.L. and Underwood, J.H.: Observations of the Structure and Evolution of Solar

Flares with a Soft X-Ray Telescope, Solar Phys. **45**, 1975, pp. 199-216.
Vrabec, D.: Streaming Magnetic Features Near Sunspots. Chromospheric Fine Structure, IAU Symposium 56, A.G. Athay (ed.), D. Reidel Publishing Co., Holland, 1974, pp. 201-231.
Waddell, B.V., Carreras B., Hicks, H.R., Holmes, R.A. and Lee D.K.: Mechanism for Major Disruption in Tokamaks, Phys. Rev. Letts. **41**, 1978, pp. 1386-1389.
Wagner, W.J.: SERF Studies of Mass Motions Arising in Flares, in Advances Space Res. (eds. Z. Svestka, D.M. Rust and M. Dryer), **2**, 1982, pp. 203-219.
Weart, S.R. and Zirin, H.: The Birth of Active Regions, Publ. Astron. Soc. Pacific **81**, 1969, pp. 270-273.
Weart, S.R.: The Birth and Growth of Sunspot Regions, Astrophys. J. **162**, 1980, pp. 987-992.
Webb, D.F., Krieger, A.S. and Rust., D.M.: Coronal X-Ray Enhancements with Hα Filament Disappearances, Solar Phys. **48**, 1976, pp. 159-186.
Webb, DF.: A Study of Coronal Precursors of Solar Flares, Technical Report AFGL-TR-83-0126, 1983.
Webb, D.F. and Kundu, M.R.: The Association of Nonthermal Electrons with Non-Flaring Coronal Transients, Solar Phys. **57**, 1978, pp. 155-173.
Wiehl, H., Batchelor, D.A., Crannell, C.J., Dennis, B.R. and Price, P.N.: Great Microwave Bursts and Hard X-Rays from Solar Flares, 1983, NASA TM85052.
Willson, R.: High Resolution Observations of Solar Radio Bursts at 2, 6 and 20 cm Wavelength, Solar Phys. **83**, 1983, pp. 285-303.
Willson, R. and Lang, K.R.: Very Large Array Observations of Solar Active Regions: IV Structure and Evolution of Radio Bursts from 20 cm Loops, Astrophys. J. **279**, 1984, pp. 427-437.
Wolfson, C.J., Doyle, J.G., Leibacher, J.W. and Phillips, K.J.H.: X-Ray Line Ratios from Helium-Like Ions: Updated Theory and SMM Flare Observations, Astrophys. J. **269**, 1983, pp. 319-328.
Wolfson, C.J.: Soft X-Ray Emission from Active Regions Shortly Before Solar Flares, Solar Phys. **76**, 1982, pp. 377-386.
Woodgate, B.E., Shine, R.A., Schmahl, E.J., Kundu, M.R., and Gaizauskas, V.: Upflows Immediately Prior to the Impulsive Phase of Solar Flares, Bull. Am. Astr. Soc. **14**, 1982, p. 898.
Woodgate, B.E., Shine, R.A., Brandt, J.C., Chapman,.R.D., Michalitsanos, A.E., Kenny, P.J., Bruner, E.C., Rehse, R.A., Schoolman, S.A., Cheng., C.-C., Tandberg-Hanssen, E., Athay, R.G., Beckers, J.M., Gurman, J.B., Henze, W. and Hyder C.L.: Observations of the 1980 April 30 Limb Flare by the Ultraviolet Spectrometer and Polarimeter on the Solar Maximum Mission, Astrophys. J.(Letters) **244**, 1981, L133-L135.
Woodgate, B.E.: Flare Buildup Studies - Homologous Flares Group Interim Report, Adv. in Space Research (eds. Z. Svestka, D. Rust and M. Dryer) **2**, 1983, pp. 61-64.
Woodgate, B.E., Shine, R.A., Poland, A.I. and Orwig, L.E.: Simultaneous Ultraviolet Line and Hard X-Ray Bursts in the Impulsive Phase of Solar Flares, Astrophys. J. **265**, 1983, pp. 530-534.
Woodgate, B.E., Martres, M.J., Smith, J.B.Jr., Strong, K.T., McCabe, M.K., Machado, M.E., Gaizauskas, V., Stewart, R.T., and Sturrock, P.A.: Progress in the Study of Homologous Flares on the Sun - Part II,

Advances in Space Research **4**, No. 7 (1984) pp 11-17.
Wu, S.T., Wang, S., Dryer, M., Poland, A.I., Sime, D.G., Wolfson, C.J., Orwig, L.E. and Maxwell, A.: Magnetohydrodynamic Simulation of the Coronal Transient Associated with the Solar Limb Flare of 1980, June 29, 18:21 UT, Solar Phys. **85**, 1983, 351-373.
Wu, S. T., Hu, Y. Q., Krall, K. R., Hagyard, M. J., and Smith, J. B., Jr.: Modeling of Energy Buildup for a Flare-Productive Region. Solar Phys. **90**, 1984, pp. 117-131.
Yang, C.K. and Sonnerup, B.U.O.: Compressible Magnetic Field Reconnection, A Slow Wave Model, Astrophys. J. **206**, 1976, pp. 570-582.
Zirin, H.: Active Regions I: The Occurrence of Solar Flares and the Development of Active Regions, Solar Phys. **14**, 1970, 328-341.
Zirin, H.: Fine Structure of Solar Magnetic Fields, Solar Phys. 22, 1972, pp. 34-38.
Zirin, H. and Tanaka, K.: The Flares of August 1972. Solar Phys. **32**, 1973, pp. 173-207.
Zirin, H.: The Magnetic Structure of Plages. Chromospheric Fine Structure, IAU Symposium 56, R.G. Athay (ed.), D. Reidel Publishing Co., Holland, 1974, pp. 161-175.
Zirin, H.: The Optical Flare, Proc. U.S. Japan Seminar, Solar Phys. **86**, 1983, pp. 173-184.
Zirin, H.: The 1981 July 26-27 Flares: Magnetic Developments Leading to and Following Flares, Astrophys. J. **274**, 1983, pp. 900-909.
Zwaan, C.: On the Appearance of Magnetic Flux in the Solar Photosphere, Solar Phys. **60**, 1978, pp. 213-240.
Zwaan, C.: Solar Magnetic Structure and the Solar Activity Cycle, Review of Observational Data - The Sun as a Star, S. Jordan (ed.), NASA SP-450, 1981, pp.. 163-179.
Zwaan, C., Brants, J.J. and Cram, L.E.: High-Resolution Spectroscopy of Active Regions I: Observing Procedures. Solar Phys. **95**, 1985, pp. 3-14.
Zweibel, E.: MHD Instabilities of Atmospheres with Magnetic Fields, Astrophys. J. **249**, 1981, pp. 731-745.
Zweibel, E. and Hundhausen, A.: Magnetostatic Atmospheres: a Family of Isothermal Solutions, Solar Phys. **76**, 1982, pp. 261-299.

CHAPTER 2: PARTICLE ACCELERATION

CHAPTER 2: PARTICLE ACCELERATION

TABLE OF CONTENTS

L. Vlahos, M. E. Machado, R. Ramaty, R. J. Murphy, C. Alissandrakis, T. Bai, D. Batchelor, A. O. Benz, E. Chupp, D. Ellison, P. Evenson, D. J. Forrest, G. Holman, S. R. Kane, P. Kaufmann, M. R. Kundu, R. P. Lin, A. Mackinnon, H. Nakajima, M. Pesses, M. Pick, J. Ryan, R. A. Schwartz, D. F. Smith, G. Trottet, S. Tsuneta, and G. Van Hoven.

Page

TABLE OF CONTENTS (Continued)

CHAPTER 2: PARTICLE ACCELERATION

L. Vlahos[1], M. E. Machado[2] R. Ramaty[3], R. J. Murphy[1,3], C. Alissandrakis[4], T. Bai[5], D. Batchelor[3], A. O. Benz[6], E. Chupp[7], D. Ellison[1], P. Evenson[8], D. J. Forrest[7], G. Holman[3], S. R. Kane[9], P. Kaufmann[10], M. R. Kundu[1], R. P. Lin[9], A. Mackinnon[11], H. Nakajima[3], M. Pesses[3], M. Pick[12], J. Ryan[7], R. A. Schwartz[9], D. F. Smith[13], G. Trottet[12], S. Tsuneta[14], and G. Van Hoven[15].

[1] University of Maryland, College Park, Maryland
[2] Observatorio de Fisica Cosmica -- CNIE, Argentina
[3] NASA Goddard Space Flight Center, Greenbelt, Maryland
[4] University of Athens, Athens, Greece
[5] Stanford University, Stanford, California
[6] Institute of Astronomy, Zurich, Switzerland
[7] University of New Hampshire, Durham, New Hampshire
[8] University of Delaware, Newark, Delaware
[9] University of California, Berkeley, California
[10] Instituto de Pesquisas Espaciasi, Brazil
[11] University of Glasgow, United Kingdom
[12] Meudon Observatory, France
[13] Berkeley Research Associates, Berkeley, California
[14] University of Tokyo, Tokyo, Japan
[15] University of California, Irvine, California

> Rest is a special form of motion
> G. Kirchhoff

2.1. INTRODUCTION

Electrons and ions are accelerated to high energies before, during and after the impulsive phase of flares. The presence of high energy particles at the Sun during a solar flare is inferred from the observed electromagnetic radiation resulting from the interaction of the energized particles with the ambient plasma and/or the magnetic field as well as from direct particle observations in interplanetary space. In this chapter we compile data from the SMM and HINOTORI satellites, particle detectors in several satellites and from ground based instruments and balloon flights and attempt to answer a number of fundamental questions that are stated below. We have also reviewed the progress made on the theory of mechanisms for particle acceleration in flares.

We define the term acceleration here as the preferential gain of energy by a small population of electrons and ions. Heating, on the other hand, is defined as the bulk energization of the ambient plasma. In other words, the development of a long nonthermal tail in the ambient distribution will be the result of "acceleration" but the increase of the random mean-square velocity of the ambient particles the result of "heating". The critical velocity, above which "acceleration" dominates heating, varies from flare to flare. The variability of the critical velocity has in the past created much discussion and divisions of flares into "thermal" or "nonthermal" classes. Another important "distinction" between "heating" and "acceleration" is the time that is required for the accelerated particles to reach the chromosphere and thermalize vs the time resolution of our instruments. For example, if the acceleration of the tail lasts only a few seconds, the propagation and thermalization of high energy particles can be faster than 10 secs, which is below the resolution of several current instruments. In this case, the division between thermal and nonthermal flares will be a time dependent phenomenon. Thus, one may argue (paraphrasing Kirchhoff's words) that "heating" is a special kind of acceleration. In the rest of this chapter we will show that heating and acceleration are always present in flares and we will discuss mechanisms that can achieve bulk heating and tail acceleration of the ambient plasma. We will adopt the more general term "energization" for bulk heating and acceleration.

As a primary goal for our study, we attempted to answer the following questions:

(1) What are the requirements for the coronal magnetic field structure in the vicinity of the energization source?

(2) What is the height (above the photosphere) of the energization source?

(3) Does the energization start before and continue after the impulsive phase?

(4) Is there a transition between coronal heating and flares? What are the microflares?

(5) Is there evidence for purely thermal, purely nonthermal or a hybrid type flare?

(6) What are the time characteristics of the energization source?

(7) Does every flare accelerate protons?

(8) What is the location of the interaction site of the ions and relativistic electrons?

(9) What are the energy spectra for ions and relativistic electrons? Does the spectrum vary from flare to flare?

(10) What is the relationship between particles at the Sun and in interplanetary space?

(11) Is there any evidence for more than one acceleration mechanism?

(12) Is there a single mechanism that will accelerate particles to all energies and also heat the plasma?

(13) How fast will the existing mechanisms accelerate electrons up to several MeV and ions to 1 GeV?

(14) If shocks are formed in a few seconds, can they be responsible for the prompt acceleration of ions and electrons? How are these shocks related to large-scale shocks which are responsible for the Type II bursts?

(15) Can the electron-cyclotron maser spread the acceleration region?

(16) Which of the acceleration mechanisms discussed above can explain the observed energy spectra?

We concentrate on these questions in Sections 2.2, 2.3 and 2.4. In Section 2.4 we also review the progress made during the last few years on mechanisms for particle acceleration in flares and in the last Section we summarize the still open observational and theoretical questions. We will attempt to answer questions (1)-(16) in Sections 2.2.7, 2.3.6 and 2.4.8. Hence, for a quick review of the status of our understanding of the problem of particle acceleration in flares the reader may go directly to these Sections and Section 2.5.

Section 2.2 was prepared by M. Machado and L. Vlahos from inputs from C. Alissandrakis, T. Bai, D. Batchelor, A. O. Benz, G. Holman, S. R. Kane, P. Kaufmann, M. R. Kundu, R. P. Lin, A. Mackinnon, H. Nakajima, M. Pick, J. Ryan, D. F. Smith, G. Trottet, S. Tsuneta. Section 2.3 was prepared by R. Ramaty and R. J. Murphy from contributions from T. Bai, E. Chupp, D. Ellison, P. Evenson, D. J. Forrest and M. Pesses and Section 2.4 was prepared by L. Vlahos from inputs from G. Holman, R. P. Lin, D. F. Smith and G. Van Hoven.

Finally, it is important to stress that this is a report of the discussions carried out during the Workshops and strongly reflects the opinions (and in many Sections even the biases) of the authors.

2.2 PHENOMENA ASSOCIATED WITH MILDLY-RELATIVISTIC ELECTRONS

In this Section we focus our discussion on phenomena associated with mildly relativistic electrons (10-400 keV) while in the next we concentrate on phenomena related to energetic ions and relativistic electrons ($\geq$ 500 keV). This division is in many ways artificial, since particles of all energies are produced during a flare. Thus, our discussion in this Section overlaps with Section 2.3 and vice-versa. In fact, our effort in this chapter will be to unify aspects related to subjects of Sections 2.2 and 2.3.

Hard X-ray imaging from the SMM and HINOTORI satellites and the stereoscopic hard X-ray observations made with the International Sun Earth Explorer 3 (ISEE-3) and Pioneer Venus Orbiter (PVO) spacecraft are reviewed in this Section. Imaging of microwave bursts is also one of our main new sources of information about particle acceleration. The results from the Very Large Array (VLA) telescope have made a large impact on our understanding of flare models. The spatial maps from the Nancay (France) Radioheliograph obtained with a high time resolution (0.04 secs) provide several new features of the topology of field lines near the acceleration site. The high time and spectral resolution of the Zurich radio spectrometer and 45 ft. radome-enclosed antenna at Itapetinga (Brazil) have opened a new window on the microinstabilities in flares. Balloon measurements with sensitive hard X-ray detectors have also been carried out with remarkable success.

2.2.1 Soft and Hard X-ray Source Structure, Location and Development

2.2.1.1 X-ray Imaging

Before the launch of the SMM and HINOTORI spacecraft, only isolated observations were available on the spatial structure of hard X-ray emission from flares. These were mainly provided by stereoscopic observations from two spacecraft (PVOs and ISEE-3, see Kane, 1983 and 2.2.1.2 below). Real imaging was first provided by the Hard X-ray Imaging Spectrometer (HXIS) aboard the SMM, and subsequently by the Hinotori hard X-ray telescopes (SXT).

The HXIS imaged simultaneously in six energy bands within 3.5-30.0 keV, with temporal resolution between 1.5 and 7 seconds and a spatial resolution of 8" x 8" (van Beek et al., 1980). The SXT's spatial resolution was 15" x 15" and the temporal resolution 7 seconds (Oda, 1983; Makishima, 1982; Tsuneta, 1984).

A heated controversy on the interpretation of impulsive phase hard X-ray emission motivated the early studies of hard X-ray images. Two competing models were, and still are, considered. The nonthermal model (Brown, 1971; Lin and Hudson, 1976; Hoyng et al., 1976) postulates that most of the flare energy is carried by a beam of fast electrons which is created within an active region loop and precipitates at its chromospheric footpoints, where it produces hard X-rays by thick target emission. On the other hand a qualitative model was developed, postulating that a large fraction of the hard X-ray emission at low energies (tens of keV) could be due to thermal bremsstrahlung (Brown et al., 1979; Smith and Lilliequist, 1979; Vlahos and Papadopoulos, 1979; Emslie and Vlahos, 1980). This model relies on the possibility of creating a hot source ($T \approx 5 \times 10^8$ K), confined by plasma instabilities which lead to ion acoustic turbulence at the expanding conduction fronts which move at the ion sound speed (see discussion in 2.2.6.2).

In the imaging data, for the range of energies covered by the HXIS and SXT, the distinction between the two models is, ideally, quite clear (see, e.g., Emslie, 1981b, and 2.2.1.3 below for the complications). The beam model predicts strong emission at the footpoints of loops, while the dissipative thermal alternate should show a bright, expanding source within the coronal loop, and minor contribution from the footpoints, due to the escaping tail of electrons which traverse the turbulent fronts.

Figure 2.2.1 (from Duijveman et al., 1982) shows that, at least in some cases, the HXIS observations seem to favor the nonthermal model. Widely separated footpoints are seen in three flares shown in the Figure, even as far as 70,000 km away from each other (November 5, 1980 footpoint C in Figure 2.2.1c). These footpoints overlay regions of enhanced chromospheric and transition zone emission, which brighten in temporal coincidence, in ultraviolet radiation, with the hard X-ray peaks (see Canfield et al. in this volume and references therein). Duijveman et al. (1982) analysed the events and concluded that the observations were consistent only with thick target emission in which the beam power implied a 20% acceleration efficiency during the early impulsive phase.

This result is not general however, and the HINOTORI investigators (Tanaka, 1983; Ohki et al., 1983; and Tsuneta, 1983b) have been able to identify at least three types of hard X-ray flares from the characteristics of the hard X-ray image, spectrum and impulsiveness of the time profile. The general characteristics of the three types (A, B and C) are listed in Table 2.2.1.

Table 2.2.1 Main Characteristics of Solar Hard X-Ray Flares

Type	Time Profile ($E \geq 20$ keV)	Hard X-ray spectrum ($E > 15$ keV)	Hard X-ray image ($E \sim 20$ keV)	Electron density (cm^{-3})	Magnetic field strength (Gauss)
A	$E \leq 40$ keV intense smooth time profile $E > 50$ keV no substantial emission with small spikes	very soft $\gamma \sim 7$–9 hot plasma ($T = 3 \times 10^7$ K) ($EM = 10^{49}$ cm^{-3}) dominantly contributed $E \leq 40$ keV. FeXXVI emission	small point-like hard X-ray source (~ 15 arcsec) low altitude (~ 5000 km)	$\sim 10^{11}$	≥ 330
B	*Impulsive phase* spiky with time scale of sec.	power-law (10 - 70 keV)	footpoint double source	$\leq 10^{10}$	
	Gradual phase smooth with time scale of min.	thermal sp. ($T =$ 3×10^7 K below 40 keV + power-law	coronal loop-like hard X-ray source	$\geq 10^{11}$	550
C	smooth time profile with time scale of min. even above 100 keV	power-law $\gamma \sim 3 - 5$ (30 - 150 keV) systematic hardening even in the decay phase	high altitude (~ 40000 km) coronal hard X-ray and micro- wave sources	3×10^{10}	50

The three flares shown in Figure 2.2.1 correspond to the type B, which are typical impulsive burst events. Their duration ranges from tens of seconds to minutes, and the time profile consists of an impulsive phase with spiky structure and effective power law index ranging from 3 to 5, and a gradual phase, generally softer, with smoother structure. During the gradual phase the hard X-ray morphology changes drastically, the footpoints disappear and a single elongated source is seen at high altitude. This behavior of type B flares is shown in Figure 2.2.2 (from Machado, 1983a; see Machado et al., 1982 for a complete discussion) for the April 10, 1980 event shown in Figure 2.2.1. We see a transition from footpoint to single source morphology of the 16-30 keV sources; similar behavior has been observed in the other two HXIS flares of Figure 2.2.1 (Hoyng et al., 1981; Duijveman et al., 1982 Machado et al., 1984b).

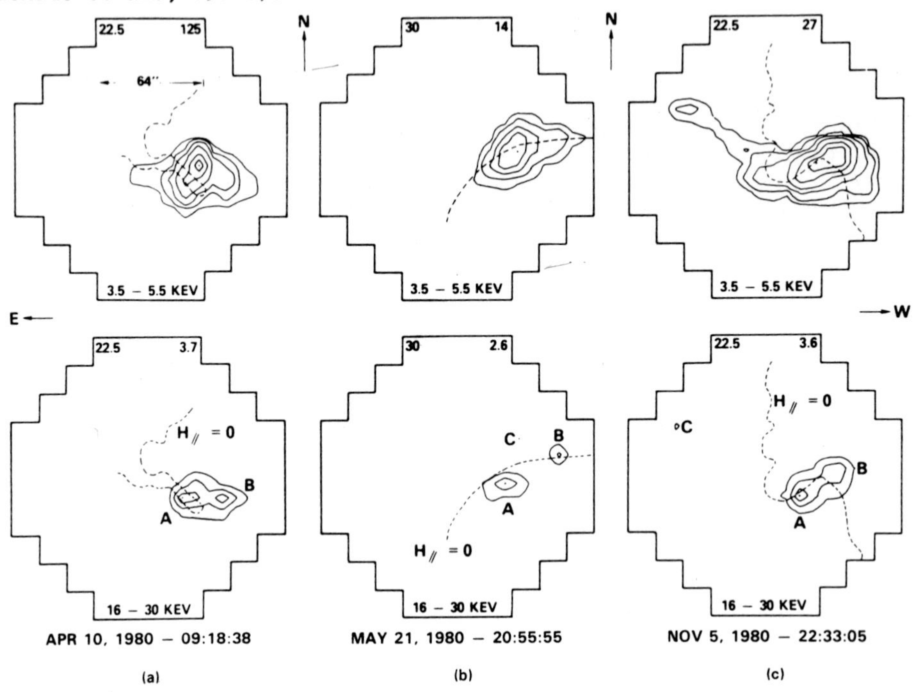

Figure 2.2.1 HXIS contour plots in soft (top) and hard (bottom) X-rays, for three flares discussed by Duijveman *et al.* (1982). The integration is over the impulsive spikes and the dashed lines show the magnetic neutral lines. The hard X-ray footpoints (16 - 30 keV) are labelled as A, B and C.

The transition from footpoint to single hard X-ray structure may reflect, as proposed by Machado et al. (1982) and Tsuneta (1983b), a change in the mode of the energy release from strong particle acceleration to plasma heating. A possible scenario on the way this may happen is described by Smith (1985, see Section 2.2.6.2) and Tsuneta (1985), but we should also be aware of the limitations of available imaging observations, discussed below. However, before reaching a definite conclusion on this subject, we must keep in mind that most of the SMM and HINOTORI type B flares (and all of type A) show single source structure (Duijveman and Hoyng, 1983; Takakura et al., 1985a).

There is, however, good evidence of high temperature plasma components in the gradual phase of some flares. These are given by the recent high resolution spectral observations of an impulsive flare, obtained with a solid state detector (Lin et al., 1981). In the impulsive

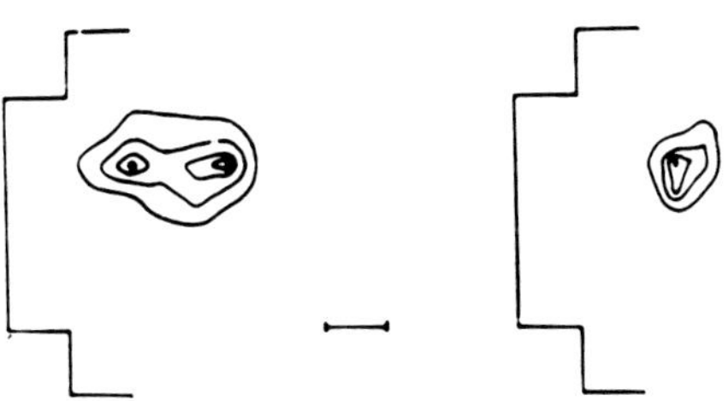

Figure 2.2.2 Hard X-ray (16 to 30 keV) observations of the April 10, 1980 flare. The doubled shaped structure (left) corresponds to the time of the impulsive burst (see Duijveman *et al.* (1982)), and the single structure to the gradual burst (see Machado *et al.* (1982)). The edge of the HXIS field of view is shown as reference. The soft X-ray emission (not shown) encompass the entire region with its maximum located at the position of the gradual component in the second image. The scale corresponds to 16 arc secs.

phase the spectrum is a power law, while in the gradual phase a hot thermal component with $T \approx 3 \times 10^7$ K appears as the power law gradually fades. Also, observations of a coronal source seen by the HXIS after the two ribbon flare of May 21, 1980 (cf. later phase of the flare shown in Figure 2.2.1.b), have been interpreted as evidence of a long lasting high temperature ($\geq 4 \times 10^7$ K) source (Hoyng et al., 1981; Duijveman, 1983). Duijveman (1983) discussed the heat balance of this source and found that its cooling rate by classical heat conduction would have been much larger than the saturated limit. He finds that the energy needed to maintain the hot source throughout its life time of several minutes is of the same order of magnitude as that needed to maintain the cooler (10^7 K) soft X-ray emitting component. These imaging and spectral observations show that high temperature plasma of about 3×10^7 K or more is generated during the development of at least some flares.

Further evidence of high temperature components in the hard X-ray emission is given from the analysis of the type A flares. Their integrated hard X-ray emission shows smooth time profiles, a steep power law index (7-9) and a duration $\geq$ 10 minutes. An example of type A flare is the July 17, 1981 flare observed by the HINOTORI (Tsuneta et al., 1984b). Line ratio analysis of the FeXXVI lines, detected throughout the flare development (Tanka et al., 1982; Moriyama et al., 1983) indicate the presence of 3 to 3.5×10^7 K plasma, with emission measure of the order of 10^{49} cm^{-3}. A possible interpretation of this type of flare is that intense heating occurs from the start of the flare, with a lesser amount of power being spent in particle acceleration (Tsuneta et al., 1984b). An example of this type of event as observed by the HXIS is the July 14, 1980 event described by Duijveman and Hoyng (1983).

Finally, the type C flares show long lasting time profiles with power law indices of 2 to 5 between 30 and 200 keV, which tend to decrease with time. An example of this type is the May 13, 1981 event (Tsuneta et al., 1984a), when a stationary hard X-ray source was observed at an altitude of $\sim 4 \times 10^4$ km, coincident with a gyrosynchrotron source at 35 GHz (Kawabata et al., 1983). These flares seem to belong to the microwave rich type (Kai and Kosugi, 1985) which are discussed in Section 2.2.4, and show relatively large energy-dependent delays in X-rays which we treat in Section 2.2.3.3 and 2.2.6.

A possible interpretation for the type C flares (Tsuneta et al., 1984a) invokes a coronal thick target trap model. As shown in Table 2.2.1 the target density of several type C flares was obtained by assuming that the delay is caused by complete trapping of nonthermal electrons (Bai and Ramaty, 1979; see 2.2.6 below). Also, Yoshimori et al. (1983) have found typical time delays of tens of seconds between MeV and lower energy hard X-ray emission while type B flares typically show delays of a few seconds. This may indicate differences in the particle acceleration timescales between type B and C flares. More details on the characteristics of these events can be found in the references we have listed.

It is worth pointing out that only a few events (less than ten total) from each spacecraft can be placed in one of the types mentioned above. The majority of the events observed do not fall in any of the above classes of flares. Thus, we believe that more complex magnetic structures and energization processes are at work during a flare (see discussion in Section 2.2.7).

From the data discussed above, it is clear that hard X-ray imaging has been achieved with SMM and HINOTORI. The imaging, however, is restricted to energies below 25-30 keV, with a spatial resolution of 8" (5800 km) at most. Let us now discuss some of the implications of these results, looking more closely at the data.

MacKinnon et al. (1985) emphasized that analyses of HXIS data to date have not adequately considered instrumental effects and data noise. The claim that three flares (April 10, May 21 and November 5, 1980) display "footpoint" emission, and therefore constitute evidence for the thick target beam interpretation of hard X-ray emission, has rested on morphological conclusions drawn from non-deconvolved images. Further, the count levels in these images are sometimes so low that consideration of photon shot noise must lead one to question the reality of morphological features. MacKinnon et al. (1985) developed a deconvolution routine, which takes into account all the instrumental effects, by use of the Maximum Entropy (ME) method. The advantages of this method, particularly the way it assesses reality of features are discussed in MacKinnon et al. (1985). MacKinnon et al. applied the above operation to images produced in the energy range 16-30 keV for the three HXIS flares which showed distinct bright points (Duijveman et al., 1982 and earlier references therein) and concluded that, in the 16-30 keV range, the presence of distinct bright points is stable to these procedures, and to the addition of noise (although other morphological features may be changed, as may such quantities as "contrast ratio"); it should also be stated that this is not always true in the 20-30 keV range due to poor counts statistics.

Further evidence for distinct bright points has taken the form of comparison of individual pixel time profiles, either to establish simultaneity of footpoint brightening or to distinguish the footpoint pixels from their neighbors (see Duijveman et al., 1982). MacKinnon et al. have investigated these conclusions quantitatively using cross-correlation coefficients. These findings, detailed in MacKinnon et al. (1985), vary slightly over the three flares, but in general they find that such comparisons do not serve to distinguish the "footpoints" either because the count statistics are not good enough, or because other, non-footpoint pixels also brighten simultaneously. Finally, they emphasize that all the above conclusions are based on band 5 data (16-30 keV), since the lower bands are not really "hard" X-rays. However, it has been pointed out that the correlation of points A and C in the November 5, 1980 (see Figure 2.2.1) flare is well borne out in the lower energy channels where the number of counts is much higher. MacKinnon et al. feel that this must be a question which requires careful consideration, in view of the undoubted role of hot (a few x 10^7 K) thermal plasmas in these energy bands.

2.2.1.2 Stereoscopic Observation

Simultaneous observations of solar hard X-ray bursts from two widely separated spacecraft has recently offered new possibilities for testing source models, in terms of both directivity and spatial distribution of the emission (Kane et al., 1979, 1982; Kane, 1981b). Such stereoscopic observations of the Sun, using the ISEE-3 and PVO spacecraft, have shown that most of the impulsive hard X-ray emission originates at altitudes $\leqq$ 2500 km above the photosphere (see Figure 2.2.3). The five events analyzed so far fall into two groups according to the occultation altitude involved. First, there is the series of three successive events occuring in a single active region on November 5, 1979, which were occulted from PVO at low chromospheric altitudes, increasing from about zero for the first event to about 2500 km for the third, due to the rotation of the Sun (Kane et al., 1982). For each of these events the ratio of occulted to unocculted flux was evaluated at photon energies of 150 and 350 keV, and for the third event the time evolution of this ratio was determined. Second, there are two events (October 5, 1978 and September 14, 1979) for which the occultation altitudes are coronal (25,000 km and 30,000 km respectively). Flux ratios are again available at two energies and their time evolution is known for the September 14 event. The main conclusions are: (a) about 90% of the impulsive X-ray emission and about 70% of gradual (extended) X-ray emission originate at altitudes $\lesssim$ 2500 km above the photosphere. In the 100-500 keV range, this altitude dependence is essentially independent of photon energy. (b) The brightness of the impulsive X-ray source decreases rapidly with increase in altitude, in a manner similar to that shown in Figure 2.2.3.

2.2.1.3 Implications of Hard X-ray Imaging and Stereoscopic Observations

Following the work of Brown and McClymont (1976) and Emslie (1981b), Machado et al. (1985) have computed the spatial distribution of hard X-rays in flare loops and the chromosphere by applying Brown and McClymont's method to the analysis of some well-observed SMM flares.

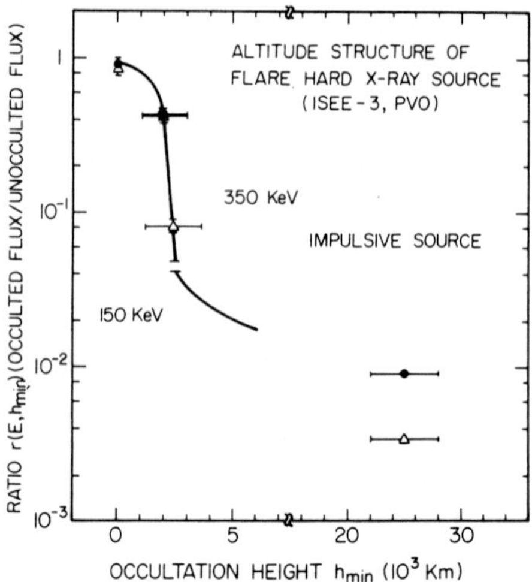

Figure 2.2.3 The ratio $r(E, h_{min})$ of occulted to unocculted X-ray flux plotted against the minimum altitude h_{min} observable from the occulted spacecraft (from Kane, 1983).

Their results show that, due to the combination of spatial resolution and rather low energy imaging, only under particular circumstances could chromospheric footpoints be seen in the images. This is readily seen from the fact that, under the best conditions,the flare loops have to cover three HXIS pixels (i.e. $\geq$ 15000 km) to be able to show separated footpoints. This implies that in order to have a strong chromosphere brightening at 20 keV, electrons with similar or higher energy must have a collisional mean free path equal to or larger than the above distance, or in other words the loop densities should be $\leq 4 \times 10^{10}$ cm^{-3}.

A transition from footpoints to single source hard X-ray structures was observed (cf. 2.2-2.1) in the November 5, 1980 flare studied by Duijveman et al. (1982). This transition occurred within the main flare region, where footpoints A and B were observed in the early flare phase. Figure 2.2.4 shows a light curve of the hard X-ray emission of the event, in which two hard X-ray peaks, P1 and P2, have been defined. P1 corresponds to the time when the footpoints were observed, while P2 (more gradual and softer) shows a single source in the hard X-ray (16-30 kev) images which is located between the two footpoints, coinciding with the locus of maximum emission in the soft X-ray images. An approximate estimate of the flare volume $V \approx 2.3 \times 10^{26}$ cm^3 can be obtained, leading to densities $n(P1) \approx 5 \times 10^{10}$ cm^{-3} and $n(P2) \approx 10^{11}$ cm^{-3} of the loop plasma during each peak (the density increase is presumably due to chromospheric evaporation). These densities are a lower limit, since a filling factor ≈ 1 is assumed (see Wolfson et al. (1983) for a critical discussion). The expected spatial distribution of hard X-rays can be calculated under simplified assumptions (cf. Brown and McClymont, 1976;

Machado et al., 1985) of the predominance of Coulomb losses and parallel injection of electrons along field lines. We should note that this is the most favorable case for footpoint prediction, since it neglects any effect (like e.g., pitch angle distribution, Leach and Petrosian, 1981) that could increase beam stopping.

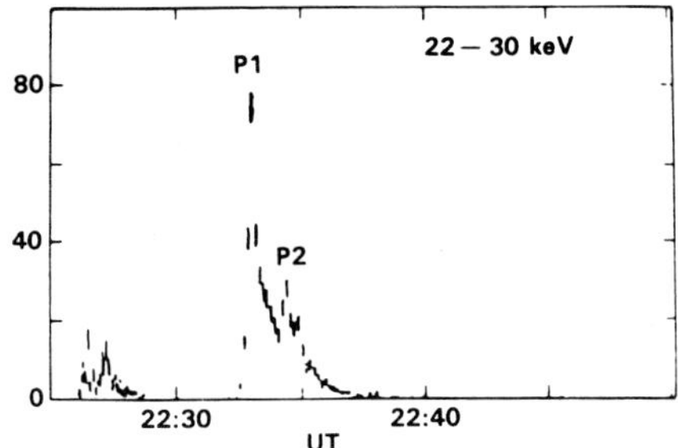

Figure 2.2.4 Hard X-ray emission light curves of the November 5, 1980 event. In the 22-30 keV lightcurve the two peaks are marked with P1 and P2 and correspond to the the points that the spatial distribution of hard X-rays has been computed.

Machado et al. (1985) calculated the intensity distribution of hard X-rays using idealized loop models. They analyzed three different models for the energy release (see Figure 2.2.5a). Case A represents a situation in which the acceleration site is located at the boundary between two pixels, presumably at the loop's apex, and the beam strength is symmetrical towards both sides. In case B it has been assumed that the acceleration region is at the middle of a pixel, and the beam is predominantly towards one side of the loop. Finally, in case C, the acceleration site is also located at the middle of a pixel, but beam strengths towards both sides are equal. The boxes shown in Figure 2.2.5a represent the pixels from the HXIS instrument, the total emission over half of a loop length assumed to cover two HXIS 80 pixels and the third pixel from the left is the "footpoint pixel". The "footpoint pixel" shows the emission of the chromospheric part of the hard X-ray distribution. In Figure 2.2.5b Machado et al. displayed the percentage of the total emission for all three cases using a photon energy ε x 19 keV. These idealized calculations clearly show a transition from footpoint to predominantly single structure in the hard X-ray distribution of P1 and P2. We also present in Figure 2.2.5b the results obtained from the convolution of the unconvolved distribution (c.f. earlier comments and Svestka et al., 1983). The general result here is that the convolution tends to decrease the footpoint/loop brightness ratio, a result consistent with the observations reported by Duijveman et al. (1982), Hoyng et al. (1981) and Machado et al. (1982). The Machado et al. results tend to reinforce the conclusions about the reality of hard X-ray footpoints, and provide a warning against the direct interpretation of single hard X-ray sources as indicative of regions heated by a mechanism different

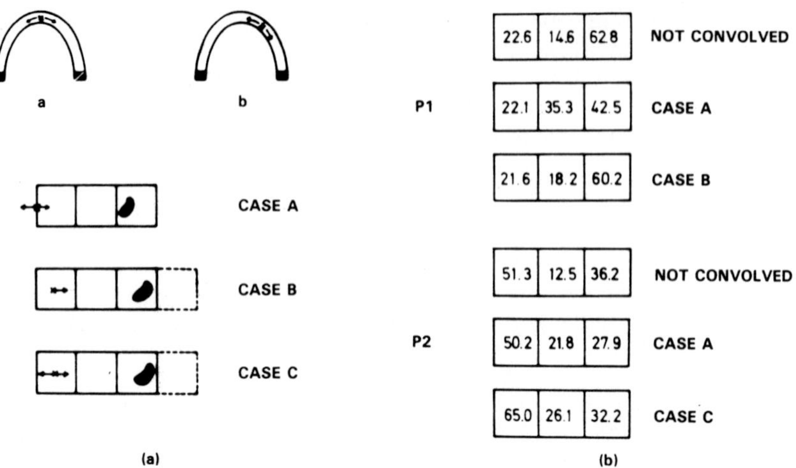

Figure 2.2.5 (a) The three cases of convolution discussed in the text. The cross marks the place where particle acceleration is presumed to occur and the arrows the predominant direction of beam injection. The shaded area in the third pixel is the chromospheric footpoint, which is assumed to be smaller than a HXIS 8" × 8" pixel and is shifted in location across the "footpoint pixel". Note that in cases B and C, due to its spatial overlap with its neighbor to the right, a fourth pixel should contain part of the footpoint emission. (b) Result of the intensity distribution of hard X-rays in percentage of the total emission. Cases A, B and C correspond to those shown in Figure 2.2.1. The "footpoint pixel" contains the total of the emission which should be spread in two (see cases B and C of Figure 2.2.1). Note the strong changes that can be expected by changing the location of the acceleration source and/or the footpoint location. In particular case A of P1 shows a large change in the brightness of the pixel located to the left of the footpoint, and case C of P2 the increase in brightness of the loop source.

from the one leading to acceleration (cf. implications of footpoint to single source transition in type B flares).

Another important aspect to take into account is the heating effect of beam particles along the loop, due to Coulomb collisions with the ambient plasma. Calculations of energy deposition rate as a function of column density, N(cm^{-2}), have been performed by many authors (Brown, 1972, 1973; Lin and Hudson, 1976; Emslie, 1978, 1980, 1983), generally in connection with chromospheric heating calculations. Machado et al. (1985) have been able to show that in the cases of high-density flare loops (like e.g., the July 14, 1980 event described by Duijveman and Hoyng, 1983) single sources are not only likely to appear because of particle stopping within the loop and high efficiency in the nonthermal bremsstrahlung production, but also because their localized heating causes an increase in the thermal contribution to the hard X-ray output below 25 keV (note also that if the heating is very large it invalidates the condition $E >> E_{th}$ of the thick target approximation, where E is the particle's energy and E_{th} the mean thermal energy of the particles in the target). It is also worth noting that these single source (type A or C) flares often show less "spiky" time profiles, which can result as a natural consequence of the fact that the temporal behavior is no longer exclusively related to time variations in the beam intensity but also to the conductive cooling timescale of the heated regions. A detailed analysis of this latter possibility has not yet been carried out. An alternative for the beam induced heating may also be related to the opposite case, i.e. low loop densities, which can lead to beam - plasma - return current instabilities and increase the beam losses due to non-collisional effects (Vlahos and Rowland (1984), Rowland and Vlahos (1984)). This is another field in which more work is needed before reaching definite conclusions.

Machado et al. concluded that, in spite of the instrumental limitations, the presence of footpoints in the hard X-ray images seems to give support to the thick target interpretation of the bursts. MacKinnon et al. (1985) on the other hand, feel that no aspect of the images demands such an interpretation uniquely, and find that some aspects of the data are difficult to accommodate in any conventional (thick target or dissipative thermal) model.

There are several pieces of evidence that indicate that a substantial fraction of the low energy ($E < 30$ keV) impulsive emission in flares is not purely due to thick target bremsstrahlung. Machado (1983b) reached this conclusion by the analysis of the energy and particle content of a compact flare loop, where a pure thick target analysis was shown to be incompatible with the parameters derived from the soft X-ray plasma.

Brown et al. (1983b), from the analysis of stereoscopic observations, find that the detailed quantitative dependence of occultation ratio on height, energy and time are not compatible with the basic thick target model as the sole source of the hard X-rays. Either emission from thermal sources or from magnetically trapped electrons has to be invoked to explain the observations.

Finally, Machado and Lerner (1984) re-analyzed the observations of a limb flare of April 13, 1980, which showed a bright X-ray (16-30 keV) source at the boundary between two distinct magnetic structures (see Machado et al., 1983). They find that the spatial distribution in intensity and spectral behavior of the hard X-rays is incompatible with a pure nonthermal interpretation. They conclude that a large fraction ($> 50\%$) of the emission in the 20 keV range is due to thermal bremsstrahlung of plasma with temperatures $> 5 \times 10^7$ K. The spatial distribution of the emission leads them to propose that the site of the maximum hard X-ray brightness is located where energy is released (at the region of interconnection between two field structures) both in the form of heating and particle acceleration.

2.2.2 Microwave Source Structure, Location and Development

Accelerated electrons produce microwave radiation through their interaction with the magnetic field. High resolution observations at cm-wavelengths have given important information about the magnetic structure of the flaring region. Observations at several frequencies can, in principle, provide valuable diagnostics of both the magnetic field and the distribution function of the energetic electrons as a function of time. However, so far there have been very few multi-frequency observations at high spatial resolution and consequently the discussion has been focused on the diagnosis of the magnetic field configuration.

Two dimensional images with the Very Large Array (VLA) radio telescope suggest that interacting magnetic loops and magnetic field reconnection have important roles to play in solar flares. This can occur as a result of emergence of new flux interacting with pre-existing flux, or as a consequence of rearrangement and/or reactivation (e.g., twisting) of two or more systems of loops. Kundu (1981) illustrated this phenomenon with a set of 6 cm observations made with the VLA (spatial resolution ~ 2") that pertains to changes in the coronal magnetic field configurations that took place before the onset of an impulsive burst observed on 14 May 1980 (Kundu, 1981; Kundu et al., 1982; Velusamy and Kundu, 1982). The burst appeared as a gradual component on which was superimposed a strong impulsive phase (duration ~ 2 minutes) in coincidence with a hard X-ray burst. Soft X-ray emission (1.6-25 keV) was associated with the gradual 6 cm burst (before the impulsive burst), as is to be expected. There was a delay of hard X-ray emission (> 28 keV) relative to 6 cm emission. The most remarkable feature of the 6 cm burst source evolution was that an intense emission extending along the north-south neutral line, possibly due to reconnections, appeared, just before the impulsive burst occurred, as opposed to the preflare and initial gradual emission being extended along an east-west neutral line. This north-south neutral line must be indicative of the appearance of a new system of loops. Ultimately the loop systems changed and developed into a quadrupole structure near the impulsive peak. This field configuration is reminiscent of flare models in which a current sheet develops at the interface between two closed loops. The impulsive energy release must have occurred due to magnetic reconnection of the field lines connecting the two oppositely polarised bipolar regions (Kundu et al., 1982).

A second burst observed by Kundu et al. (1984) on 24 June 1980, 19:57 UT, provides a good example of interacting loops being involved in triggering the onset of a 6 cm impulsive flare associated with a hard X-ray burst. It also provides evidence of preflare polarization changes on time scales of a minute or so, which may be related to coronal magnetic field configurations responsible for triggering the burst. The 6 cm burst source is complex, consisting initially of two oppositely polarized bipolar sources separated E-W by ~ 1.5' arc. The first brightening occurs in one component at 19:57:10 UT, the western component being much weaker at this time. It then brightens up at 19:58:05 UT, just at the onset of the impulsive rise of the burst and is accompanied by changes in its polarization structure. It then decays and splits into two weak sources separated E-W by ~ 12" arc. The eastern component brightens up at 19:58:41 UT, accompanied by significant polarization changes, including reversal of polarization. A third component appears approximately midway between the eastern and western component at 19:58:45 UT during the peak of the associated hard X-ray burst. The appearance of this source is again associated with polarization changes, in particular the clear appearance of several bipolar loops; its location overlaps two opposite polarities implying that it might be situated near the top of a loop. During the peak of the associated hard X-ray burst (1980 June 24, 19:57:00 event), a third (perhaps another bipolar) loop appears in between the previous two sources. Kundu believes that we are dealing with interaction between multiple loop structures, resultant formation of current sheets and magnetic field reconnection, which is responsible for the acceleration of electrons.

Lantos, Pick and Kundu (1984) combined observations of three solar radiobursts obtained with the VLA at 6 cm wavelengh and with Nancay Radioheliograph at 1.77m. A small change in the centimetric burst location by about 10" arc corresponds to a large change by about 0.5 R_0 in the related metric location. The metric bursts occur successively at two different locations separated by about 3.10^5 km. During the same period, an important change in the microwave burst source is observed. This may indicate the existence of discrete injection/acceleration regions and the presence of very divergent magnetic fields in agreement with the suggestions made by Kane et al. (1980).

The Westerbork Synthesis Radio Telescope (WSRT) was used by Alissandrakis and Kundu (1985) for solar observations at 6.16 cm with a spatial resolution as good as 30" and a time resolution of 10 sec. In spite of the limitations of one-dimensional fan-beam scans in total intensity (I) and circular polarization (V) of burst sources, several interesting features could be discovered in their structure.

Out of the 76 bursts observed, 57% consisted of two or more components in total intensity. An example of a burst with two components is shown in Figure 2.2.6a,b, where contours of 1-D brightness temperature as a function of position and time are plotted. In total intensity (I), the burst consists of two impulsive components, A and B, with their peaks separated by 260" and a total duration of about 4 minutes. The peaks are almost simultaneous with a possible delay of component B by no more than 5 sec with respect to component A. Component A is fairly symmetric with a width of 70" and a maximum 1-D brightness temperature of 6.5×10^7 K arc sec above the background; assuming a circular shape this value corresponds to a brightness temperature of about 10^7 K. The other component is asymmetric with a width of 110" and an estimated brightness temperature of about 4×10^6 K. Alissandrakis and Kundu pointed out that near the maximum the two components appear to be connected by a bridge of low intensity emission. Such interconnections between burst components are the rule rather than the exception in their sample of bursts. In the example shown there is a definite extension of component B in the direction of component A. The circular polarization map shows that both components as well as the bridge between them are polarized. Component A shows two peaks of opposite sense with the total intensity peak coinciding with the region of zero polarization; the degree of polarization at the V peaks is about 50%. The polarization of the other component is uniform with a 40% maximum near the I maximum. The sense of polarization of component B is the same as that of the nearest V peak of component A, as well as that of the bridge; the latter is almost 100% polarized. Such a polarization structure of 6 cm burst sources is quite common.

If we assume that the sense of circular polarization corresponds to the polarity of the magnetic field, we can interpret the observations in terms of a small flaring loop corresponding to component A and a larger loop connecting component A with component B. The large loop emits mainly at the footpoints with some emission from the rest of the loop which corresponds to the bridge; the emission from the top of the large loop is weak because it is located higher in the corona where the

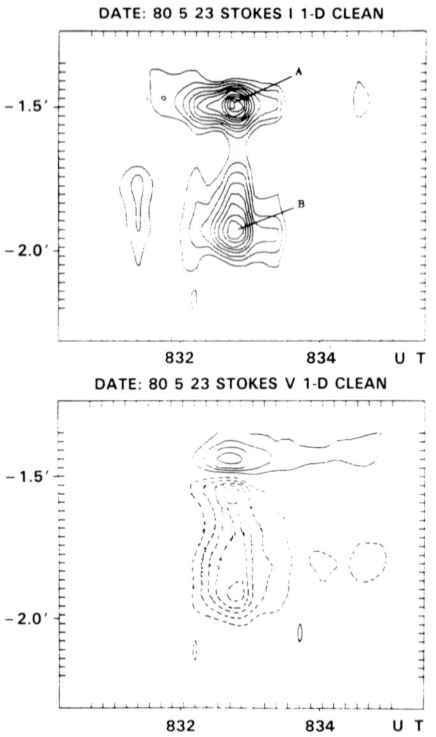

Figure 2.2.6 A two component burst observed with the WSRT at 6.16 cm. (a) The contours of equal brightness temperature (integrated in the direction perpendicular to the resolution) as a function of one-dimensional position and time in Stokes parameters I and V. (b) The I map is 10^7K arcsec with a contour interval of 0.5×10^7K arcsec, while the V map the lowest contour and the counter interval are 0.3×10^7K arcsec. Dashed lines show negative (left handed) circular polarization (from Alissandrakis and Kundu, 1985).

magnetic field is weak. This scenario is similar to the schematic model presented by Kundu and Shevgaonkar (1985) for the impulsive onset of the microwave burst radiation as a result of two interacting loops. However, as pointed out by Alissandrakis and Preka-Papadea (1984) the observed sense of circular polarization can be influenced by propagation effects in the corona outside of the flaring region, so that the polarization-inversion line does not necessarily coincide with the neutral line of the magnetic field. If polarization inversion does indeed take place, the observations can also be interpreted in terms of a single large loop connecting the two components and radiating predominantly at the footpoints.

Using the Nobeyama 17-GHz interferometer Nakajima et al. (1985) observed on November 8, 1980 a microwave burst occurring at a site (Hale region 17255) 8 x 10^5 km remote from the primary flare site (Hale region 17244). The time profiles of the secondary microwave bursts are similar in form to the primary bursts even in details. The overall time profiles of the secondary microwave bursts are delayed relative to those of the primary bursts by 11 or 25 secs. The velocity of a triggering agent inferred from this delay and the spatial separation is about 4 x 10^4 or 8 x 10^4 km s^{-1} and therefore is probably due to fast electrons which were transferred from the primary site to the secondary site along a huge coronal loop. The SMM-HXIS data showed that a new X-ray loop was excited in the region adjacent to the secondary microwave source. The X-ray loop was associated with a faint, compact Hα brightening at its footpoints. The event occurred twice with a similar behavior within a time interval of ~ 40 min and therefore the occurrence of the correlated events is not random. The observations suggest that a new flare (a sympathetic flare) was triggered at the secondary site by an energetic electron stream from the primary site. Similar observations were first reported by Kundu, Rust and Bobrowsky (1982) for a flare observed on May 14, 1980, with practically the same conclusions.

Heights and sizes of microwave burst sources at 17 GHz were obtained as shown in Figure 2.2.7. The events were selected from those which were observed with the 17 GHz one-dimensional interferometer between October 1978 and February 1981. An additional selection condition is that the longitude of the associated Hα flare is ≧ 70^0 and the peak flux density at 17 GHz is ≧ 50 sfu. The heights were estimated on the assumption that the microwave sources were above the corresponding Hα flares. Both the heights and sizes of the impulsive bursts (12 events) are roughly correlated and range from about 10 to 20 arc sec above the photosphere with an average value of 13 arc sec (10^4 km). The long-enduring bursts (2 events) are located higher (30 arc sec) and larger (35 arc sec) in size as compared to those of the impulsive bursts. Although SMM-HXIS and HINOTORI-SXT hard X-ray imaging observations show in several cases that the hard X-ray component of the impulsive burst is located in the chromosphere (e.g., Duijveman, Hoyng, and Machado, 1982; Tsuneta et al., 1983), the observations reported by Nakajima et al., 1985 show that the microwave emission from the impulsive burst comes from the corona. The VLA observations have often shown a compact (very small compared to the distance between Hα kernels) source of the impulsive bursts located spatially between Hα kernels (Marsh and Hurford, 1980; Velusamy and Kundu, 1982; Hoyng et al., 1983). On the other hand, the observation reported above shows that the source size and height are roughly the same. The height observations of the long-enduring bursts confirm the results reported by Kosugi et al. (1983) and Kawabata et al. (1983).

2.2.3 Time Structures and Time Delays in Radio and Hard X-rays

2.2.3.1 Centimeter-Decimeter Millisecond Pulses and Electron Cyclotron Masering

Spikes of durations less than 100 ms are well known in the 200 - 3000 MHz radio band. At meter wavelengths some have been reported near

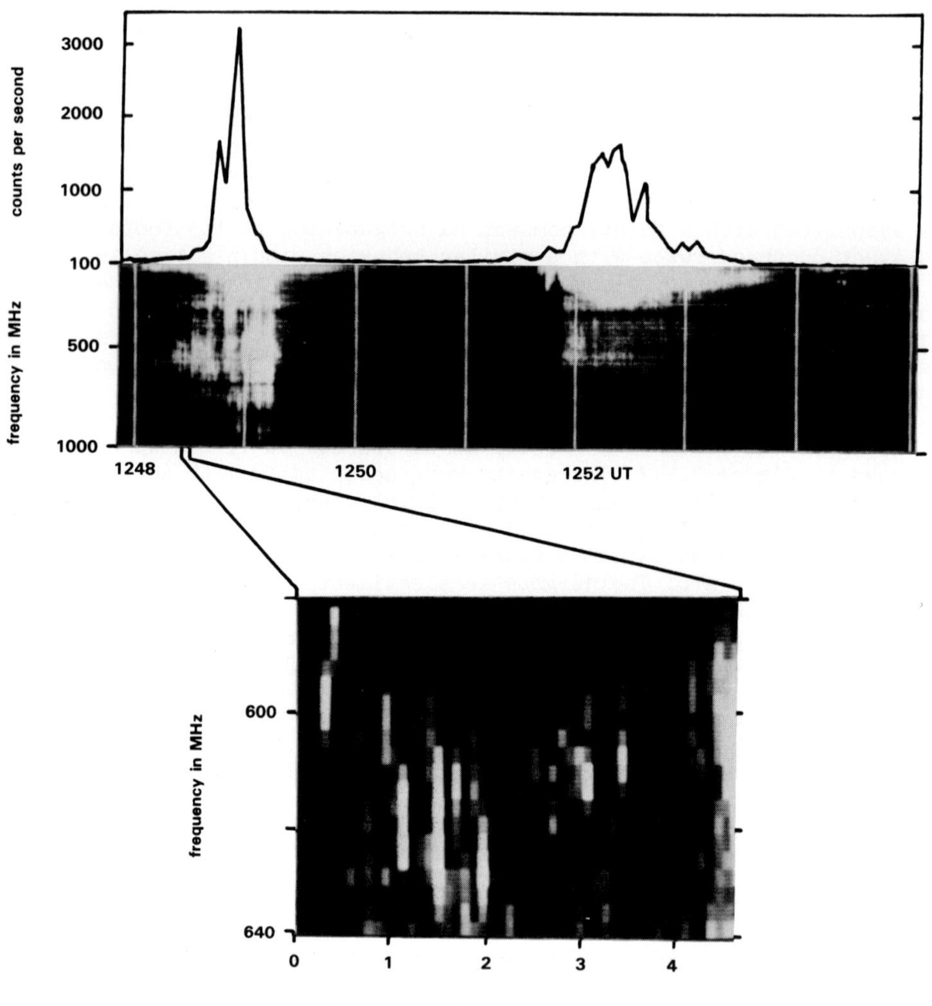

Figure 2.2.8 Top: Composed figure showing hard X-ray counts (> 30 keV, observed by HXRBS on board the Solar Maximum Mission) vs. time, of the double flare of August 31, 1980 and radio spectrogram registered by the analog spectrograph at Bleien (Zürich). The spectrogram shows type III bursts at low frequency having starting frequencies in correlation with the X-ray flux and spike activity above 300 MHz. Bottom: Blow-up of a small fraction of spectrogram produced from data of the digital spectrometer at Bleien (Zürich). The blow-up shows single spikes which are resolved in frequency (from Benz, 1984).

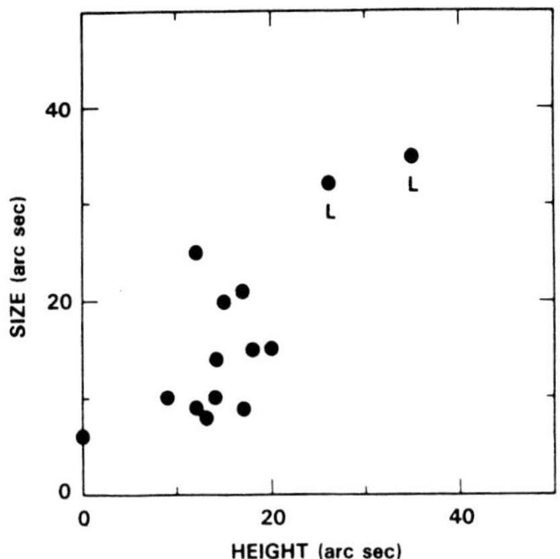

Figure 2.2.7 Heights and sizes of microwave burst sources at 17 GHz. L indicates long-enduring burst. The remaining events are impulsive (from Nakajima *et al.* (1984a)).

the starting frequency of type III bursts (Benz et al., 1982), at decimeter wavelengths as a part of type IV events (Droge, 1977) and at centimeter wavelengths superposed on a gradual event (Slottje, 1978). In an analysis of 600 short decimetric events (excluding type IV's), Benz, Aschwanden and Wiehl (1984) have found 36 events consisting only of spikes. An example of the data is presented in Figure 2.2.8 together with a hard X-ray time profile and a blow-up of some single spikes. A detailed analysis (Benz, 1984) shows that the groups of spikes are always associated with groups of metric type III bursts. The spikes tend to occur in the early phase of the type III groups and predominantly in the rising phase of hard X-rays. The half-power duration of the spikes is less than 100 ms, the time resolution of the instrument used. The spectrum of the spikes has been recorded and the typical half-power widths are 3-10 MHz at 500 MHz, i.e. about 1% of the center frequency. This puts a severe constraint on the spectral width of the radio emission and therefore on the generating mechanism. The most plausible interpretation is emission at the electron cyclotron frequency or harmonic (e.g., upper hybrid wave emission or cyclotron maser). Even then, the requirement on the homogeneity of the source is formidable: assuming a locally homogeneous corona with a magnetic field scale length of 10,000 km, the source size in the direction of the field gradient must be equal to or less than 100 km. This is less than the upper limit of the size imposed by time variation. Assuming this dimension for the lateral extent of the source, the lower limit of brightness temperature is up to 10^{15} K. Provided that the emission is radiated close to the plasma frequency, the source density amounts to about 3×10^9 cm^{-3}. The spikes have peak fluxes of up to 800 sfu and are circularly polarized. The polarization ranges from 25-100%. The sense of polarization is righthanded, opposite to most type III bursts occurring at lower frequencies at the same time.

The high brightness temperature of short duration (1-100 msec) spikes observed during the impulsive phase of some flares at microwave frequencies (~ 3 GHz) indicates that a coherent radiation mechanism is responsible. Coherent plasma radiation at the electron plasma frequency was originally suggested as the radiation mechanism (Slottje, 1978; Kuijpers, van der Post, and Slottje, 1981). Holman, Eichler and Kundu (1980) argued that electron cyclotron masering at frequencies just above the electron gyrofrequency or its second or third harmonic was a likely mechanism for the spike emission. As a third possibility, coherent emission at twice the upper hybrid frequency, has been suggested by Vlahos, Sharma and Papadopoulos (1983). Electron cyclotron masering has been the most highly studied of the three mechanisms. The mirroring of suprathermal electrons in a flaring loop naturally leds to a loss-cone particle distribution, which is unstable to electron cyclotron maser emission (Wu and Lee, 1979). The attractive features of this mechanism are that it is a linear process, not requiring wave-wave interactions, and the conditions for it to operate are essentially the same as those required for an incoherent microwave source: trapped, mildly relativistic electrons (roughly the same number as required for the incoherent emission) with moderately high pitch angles. The masering occurs as long as the loss-cone distribution of the mirrored electrons is maintained. As shown by Melrose and Dulk (1982) and Sharma, Vlahos, and Papadopoulos (1982), the saturated level of the emission is sufficient to provide the observed high brightness temperatures. The emission must escape thermal cyclotron absorption at the next higher harmonic, however, and this requirement favors second harmonic emission, since emission at the fundamental will generally not be able to escape the second harmonic absorption layer. Growth of the first harmonic poses a problem for second harmonic emission however, since the first harmonic growth can saturate the maser before the second harmonic is able to grow significantly. Sharma et al. (1982) and Sharma and Vlahos (1984) have shown that the first harmonic, extraordinary mode growth will be suppressed by the ambient thermal plasma if $\omega_e \gtrsim 0.4\ \Omega_c$, ($\omega_e$ is the plasma frequency; Ω_c is the gyrofrequency). The growth of the first harmonic ordinary mode is still large, however, so the conditions under which the second harmonic emission can grow and escape are still not entirely clear. Vlahos and Sharma (1984) analyzed the role of the filling of the loss-cone distribution and suggested that loss-cone driven electron cyclotron emission will be localized at the bottom of the corona and the emitted radiation will have a narrow bandwidth. This is in agreement with the observations reported above.

Finally, in a recent study Zaitsev, Stepanov and Sterlin (1985) suggested that the millisecond pulsations are due to a non-linear induced scattering of plasma waves by background plasma ions. They reduced the coupled non-linear system of equations, that describe the wave-particle interactions, to the well known Voltera equations which describe the "predator-prey" problem. The duration of the pulses (a few milliseconds) is used to determine the density of the energetic electrons that cause the radio emission.

2.2.3.2 Ultrafast Time Structure in Microwaves and Hard X-rays and their Time Delays

The use of antennas with large collecting areas has considerably improved the observation of solar bursts at centimeter and millimeter wavelengths with high sensitivity and time resolution (Kaufmann et al., 1975, 1982a; Butz et al., 1976; Tapping, 1983). The 45 ft. diameter radome-enclosed radio telescope, at Itapetinga, Brazil, operating at 22 GHz and 44 GHz, was extensively used during the period of SMM operation, providing high sensitivity (0.03 s.f.u. in single linear polarization) and high time resolution (1 ms) data; these data revealed new aspects of low level solar activity as well as fine time structures in larger bursts. In practically all the bursts studied with high sensitivity at mm-cm wavelengths, fine time structures ($<$ 1 sec) were identified superimposed on the slower time structures (seconds). The repetition rate of the ultrafast structures appear to be higher, for higher mean fluxes of 22 GHz bursts (see Figure 2.2.9). Kaufmann et al. (1980a, 1980b) suggested a possible interpretation of this behavior in terms of a quasi-quantization in energy of the burst response to the energetic injections. A similar suggestion was made earlier from the statistical properties of a collection of X-ray bursts (~ 10 keV) (Kaufmann et al., 1978). A trend similar to that shown in Figure 2.2.9 was found independently at 10.6 GHz (Wiehl and Matzler, 1980) but for bursts with larger flux and timescales. Kaufmann et al. (1980a) showed that for a given burst flux level S at 22 GHz there is a minimum repetition rate of ultrafast structures R, such as $S \leqq k.R$, where k is a constant. One of the faster repetition rates was found at the peak of an intense spike-like burst (Figure 2.2.10) which was also observed in hard X-rays by SMM-HXRBS (Kaufmann et al., 1984). A striking example obtained simultaneously in microwaves and hard X-rays is the burst of November 4, 1981 at 1928 UT (Takakura et al., 1983b).

High sensitivity 10.6 GHz data for the same burst was obtained with the 45-m antenna at Algonquin Radio Observatory, (Tapping, private communication). The presence of a "ripple" is evident at all microwave frequencies and is very significant at 30-40 keV range (HINOTORI-HXM). The ripple relative amplitude ($\Delta S/S$) is about 30% at 30-40 keV, 1% at 22 and 44 GHz and 0.4% at 10.6 GHz. The aparent lack of phase agreement for certain peaks might or might not be real. Confirmation of a nearly one-to-one correspondence of mm-cm vs hard X-ray association of superimposed ripples was obtained for the November 13, 1981, 1102 UT burst. The most important findings of such studies are: (a) the slow time structure (seconds) are often poorly correlated, or not correlated, between the four microwave frequencies (7, 10.6, 22 and 44 GHz) and 30-40 keV X-rays; (b) the superimposed "ripple" components are present and correlated (although phase differences might be present) in data obtained simultaneously by two radio observatories widely separated from each other (Brazil and Canada) and by the HINOTORI-HXM X-ray experiment.

The time structures in complex microwave bursts are frequently not correlated in time at various frequencies. Delays of peak emission at different microwave frequencies range from near coincidence to 3 sec, both toward higher and lower frequencies (Kaufmann et al., 1980a; 1982b).

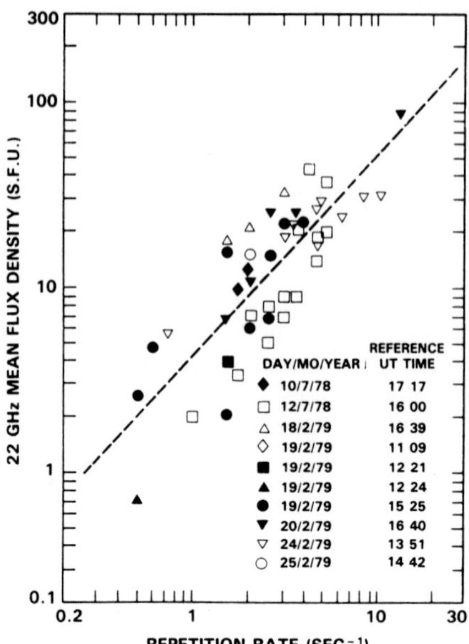

Figure 2.2.9 Scatter diagram of repetition rates $R(s^{-1})$ of fast time structures superimposed on solar bursts at 22 GHz against the mean flux value S (s.f.u.) for various bursts observed in 1978-1979 with the 13.7-m Itapetinga antenna (from Kaufmann *et al.*, 1980a).

Delays toward lower frequencies only have been reported by Uralov and Nefed'ev (1976) and Wiehl et al. (1980). One long-lasting pulsating burst (quasi-period 0.15 sec) has shown a systematic delay of 300 ms for 44 GHz pulses relative to 22 GHz pulses (Zodi et al., 1984). It might be meaningful, however, to stress that the faster time structures found seem to be well correlated (as the case of the "ripple" structures discussed above). In relation to hard X-rays, the microwave burst emission time structures often appear delayed in time. For relatively slower (and smoothed) time structures, the hard X-rays appear to occur 1-2 sec prior to microwave emission (Crannell et al., 1978).

There are several ways to interpret the time delays reported above, for example, convolution effects of multiple emitting kernels (Brown et al., 1980, 1983a; MacKinnon and Brown, 1984, see also discussion on Section 2.2.6.2) or the fact that the microwave emitting source may move in a varying magnetic field (Costa and Kaufmann, 1983) are among the suggested candidates. For the large delays between the microwave and hard X-ray peaks (several seconds), it has been suggested that microwave emission originates from another population than the one that produces the X-rays (Tandberg-Hanssen et al., 1984). Finally, the long-enduring persistent quasi-periodic pulsations in bursts, presenting pseudo-delays at different microwave frequencies, might be a phenomenon of a different

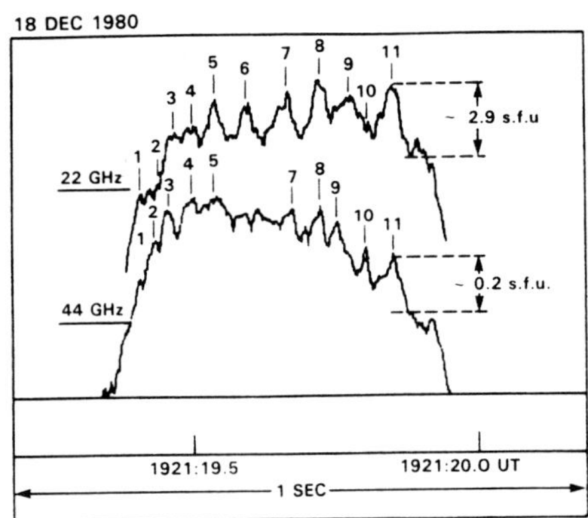

Figure 2.2.10 One-second section at the peak of an intense spike-like burst, displaying ultrafast time structures repeating every 30-60 ms at 22 GHz and 44 GHz (from Kaufmann *et al.*, 1984).

nature, and might be conceived as due to simple modulation of synchrotron emission by a varying magnetic field (Gaizauskas and Tapping, 1980; Zodi et al., 1984). Some bursts appear to be strictly coincident in time, at various microwave frequencies and X-ray energy ranges (to less than < 100 ms) (Kaufmann et al., 1984).

The impulsive phase X-ray and microwave emission examined with high sensitivity and high time resolution put several constraints on the models of the bursting region. Among the new observations that require theoretical interpretations are the "ripple" structures, the trend of flux vs. repetition rates, and the possible quasi-quantized energetic injections. Sturrock et al. (1985) suggest that "elementary flare bursts" may arise from the energy release of an array of "elementary flux tubes", which are nearly "quantized" in flux. As a stochastic process of reconnection sets in, by mode interaction, explosive reconnection of magnetic islands may develop in each tube, accounting for the ultrafast time structures (or "ripple") with subsecond timescales.

2.2.3.3 Time Delays in Hard X-ray Bursts

Before the launch of SMM, energy-dependent delay of hard X-rays had been observed only from a small number of flares (Bai and Ramaty, 1979; Vilmer, Kane and Trottet, 1982; Hudson et al., 1980). Hard X-ray delay was first observed from the two intense flares observed on August 4 and 7, 1972 (Hoyng, Brown and van Beek, 1976; Bai and Ramaty, 1979), which happen to be the first gamma-ray line flares (Chupp et al., 1973). Hudson et al. (1980) analyzed a very intense gamma-ray line flare observed with the first High Energy Astronomical Observatory (HEAO-1), and reported a delay of the continuum above 1 MeV with respect to the X-ray continuum about 40 keV. Vilmer, Kane and Trottet (1982) studied the hard X-ray delays exhibited in a flare observed with ISEE-3. The HXRBS experiment aboard SMM, which has a large area and good time resolution (71 cm^2 and 0.128 s in normal mode, respectively; cf. Orwig et al., 1980), is most suitable for studying energy-dependent delays of hard X-rays. In collaboration with the HXRBS group, Bai studied the delay of hard X-rays for many flares (Bai et al., 1983a; Bai and Dennis, 1985; Bai, Kiplinger and Dennis, 1985). A balloon-borne detector and the hard X-ray detector aboard HINOTORI also detected hard X-ray delays (Bai et al., 1983b; Ohki et al., 1983). The energy dependence of hard X-ray delays is not simple. In some flares the delay seems to increase smoothly with hard X-ray energy, but in others, it seems to show a sudden increase. For example, in the impulsive flares of June 27, 1980 (Bai et al., 1983b; Schwartz, 1984) and of February 26, 1981 (Bai and Dennis, 1985), the delay is negligibly small below a certain energy, and it suddenly increases above that energy. The energy at which a sudden increase occurs varies from burst to burst (Schwartz, 1984). In the August 4 and 7, 1972 flares, the delay increased gradually with increasing energy to about 5 s, and then for energies above ~ 150 keV it increased to ~ 15 s (Bai and Ramaty, 1979). In the flare of August 14, 1979 flare, the delay was about 10 ± 5 s for the energy channel 154-389 keV, but it increased to 32 ± 10 s for the next energy channel 389-874 keV (cf. Vilmer et al., 1982). (It is important to keep in mind that fast increases may also be the result of the fact that the energy channels are wider in higher energies). However, in other flares the delay seems to increase smoothly with hard X-ray energy (cf. Bai and Dennis, 1985). The energy-dependent delay of hard X-rays is equivalent to flattening of the hard X-ray spectrum. In flares with the delay increasing like a step function at a certain energy (such as the ones on June 27, 1980 and February 26, 1981), the spectral shape at low energies remains unchanged while the spectrum at high energies flattens as time progresses during the burst. If the delay is a smooth function of energy, the hard X-ray spectrum flattens with time both at low energies and high energies (Bai, Kiplinger and Dennis, 1985). Often single power law spectra give good fits to the data. The flares exhibiting hard X-ray delays form a small but significant fraction of the total number observed. Another important observational fact is that energy dependent hard X-ray delays have been mostly observed in flares which produced observable nuclear gamma-rays and/or energetic interplanetary protons (Bai and Ramaty, 1979; Hudson et al., 1980; Bai et al., 1983a, 1983b; Bai and Dennis, 1985; Ohki et al., 1983).

Figure 2.2.11a shows a smoothed plot of the 60-120 keV and 120-235 keV rates observed by the UC Berkeley balloon experiment during the impulsive phase of the 27 June 1980 flare (Schwartz, 1984). The smoothed rate during each 0.128 sec interval is computed by averaging the rates over the surrounding bins using a Gaussian weighting function with a 0.5 sec FWHM for the 60-120 keV rate and with a 1 sec FWHM for the two rates above 120 keV. This filters out the high frequency fluctuations, both real and statistical, but does not move the bursts centroids. The cross-correlation function has been computed between the rate pairs (22-33 keV, 60-120 keV), (60-120 keV, 120-235 keV), and (60-120 keV, > 235 keV) for the six bursts, A through F. The smoothed rates were used only for the rates above 120 keV. The delay times found by cross-correlating various energy channels are given in Table 2.2.2. For the

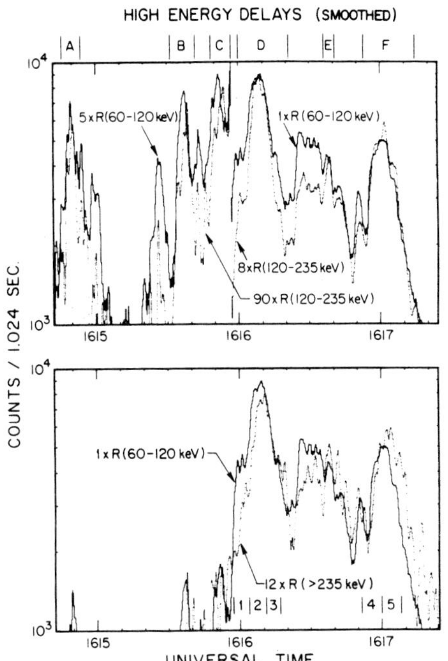

Figure 2.2.11 Delayed bursts of photon energies > 120 keV. (a) Delay of 120–135 keV profile with respect to 60–120 keV rate. (b) Delay of > 235 keV rate with respect to 60–120 keV rate (from Schwartz, 1984).

burst at 1616:38 UT, the > 235 keV rate is too low to accurately determine a centroid. The delay listed between the 22-33 keV and 60-120 keV rates is an upper limit based on the count rate statistics. Only burst C, at 1615:52 UT, has a real delay for the 60-120 keV rate of 0.128 seconds. Above 120 keV, all of these bursts show real delays. There is a delay of ~ 1-2 seconds in the 120-235 keV rate for the medium sized and shorter duration bursts A, B, C, and E. For the two most intense and longest duration bursts, D and F, the longest delay is for the > 235 keV rate with only a smaller delay for the 120-235 keV rate (Bai et al., 1983b). The lack of significant delays between the 22-33 keV and 60-120 keV channels make it unlikely that the large delays at higher energies can be explained purely by simultaneous injection at all energies followed by energy-dependent decay due to collisional energy loss (see bottom of Table 2.2.2). Figure 2.2.12 shows five spectra which were accumulated over the intervals marked in Figure 2.2.11. The evolution is similar over both bursts. The double power law becomes a single power law although the counting rate sensitivity is not enough to observe the hardening in detail. There are two important aspects of the spectral evolution which may provide important clues to the acceleration process. First, the power law exponent at low energies ($\lesssim$ 70 keV) does not change

Table 2.2.2 Cross Correlation Delays

Cross-Correlation Delays

Burst	A	B	C	D	E	F
Time after 1600 UT	14:48	15:37	15:52	16:08	16:38	17:02
(22–33, 60–120 keV)	<.05	<.05	<.20	<.05	<.05	<.05
(60–120, 120, 235 keV)	.8±.5	1.1±.5	1.9±5	.4±.25	.9–.5	.2±.25
(60–20 > 235 keV)	—	—	—	1.5±.5	—	1.88±.5
Delay Expected for Collisional Loss Process						
(22–33, 60–120 keV)	.13	.13	.60	.12	.26	.15

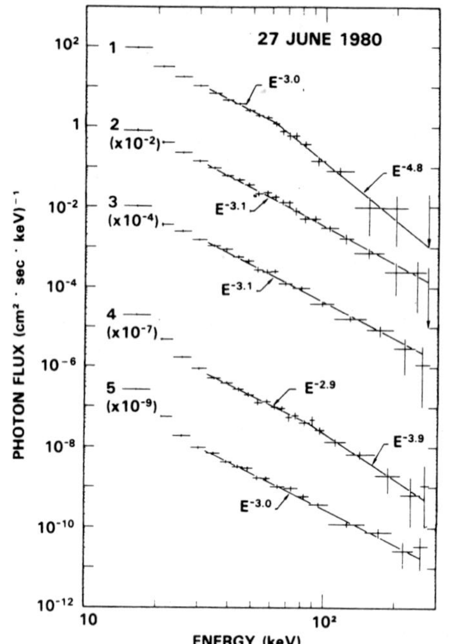

Figure 2.2.12 Five spectra accumulated over the intervals marked in Figure 2.2.11 (from Schwartz, 1984).

throughout the acceleration. Secondly, the spectrum at high energies hardens up to the point where the power law exponent is the same as at low energies, but not harder. It is not clear whether the spectral hardening occurs because the break in the spectrum has moved to very high ($\gtrsim$ 200 keV) energies or whether the entire high energy portion has hardened to form a single power law at all energies.

2.2.3.4 Hard X-ray Microflares

The U.C. Berkeley balloon flight of June 27, 1980 was the first to observe the Sun with high energy resolution ($\lesssim$ 1 keV) and sensitivity (50 cm^2 germanium plus 300 cm^2 scintillation detectors, both well collimated and actively shielded for low background) in the energy range $\gtrsim$ 20 keV (Lin et al., 1984). They discovered the phenomenon of solar hard X-ray microflares which have peak fluxes ~ 10-100 times less than in normal flares. These bursts occurred about once every five minutes through the 141 minutes of solar observations. Although they are associated with small increases in soft X-rays, their spectra are best fit by power laws which can extend up to $\gtrsim$ 70 keV. These microflares are thus probably nonthermal in origin. The integral number of events varies roughly inversely with the X-ray intensity (Figure 2.2.13), so that many more bursts may be occurring with peak fluxes below their sensitivity. The rate of energy released in these microflares may be significant compared to the rate of heating of the active corona (see also Athay, 1984 for similar conclusions and theoretical discussions by Parker, 1983a,b, and Heyvaerts and Priest, 1984). There is also some indication that these bursts may be made up of spikes of ~ 1 sec duration (Figure 2.2.14). Perhaps these are the real "elementary" bursts, a factor of 10^2-10^3 smaller than the elementary flare bursts reported by de Jager and de Jonge (1978). Kaufmann et al. (1985) reached a similar conclusion regarding such microbursts in the microwave domain (see Section 2.2.3.2). These hard X-ray microflares indicate that impulsive electron acceleration to above 20 keV energy is a very common phenomenon and may be the primary transient energy release mode in the solar corona.

2.2.3.5 Pre- and Post-Impulsive Phase Hard X-ray Pulses

Elliot (1969) proposed that flares could be the result of sudden precipitation of energetic ions stored high in the corona where their lifetime is long. Electrons, too, might be stored in a low density region where their collision loss rate would be low and then precipitate during the flare. This scenario allows the acceleration of electrons over a much longer timescale at much lower rate. The stored electrons, however, would radiate via bremsstrahlung. The high sensitivity of the UC Berkeley balloon hard X-ray measurements made on June 27, 1980 permit the study of the pre- and post-impulsive phase nonthermal emissions of a large flare (Figure 2.2.15) in great detail (Schwartz, 1984). Using the high sensitivity of the X-ray detectors upper limits have been set to the preflare flux during 1600 - 1610 UT. The three sigma upper limit to the flux at 20 keV is 8.3×10^{-4} (cm^{-2} sec keV)$^{-1}$. This gives an upper limit to the power law emission measure (Hudson, Canfield and Kane, 1978), $N_{20}n_i < 2.4 \times 10^{39}$ cm^{-3}, where N_{20} is the average number of electrons above 20 keV at any instant of time in a region with an ion density n_i. Conceivably, the electrons could be stored very high in the corona where the density could be as low as 1×10^5 cm^{-3}. This would give a 50 hour collisional lifetime for a 20 keV electron. Thus, up to 2×10^{35} electrons could have remained undetected. This is about the number of fast electrons in the small early burst at 1616:00 UT and it represents

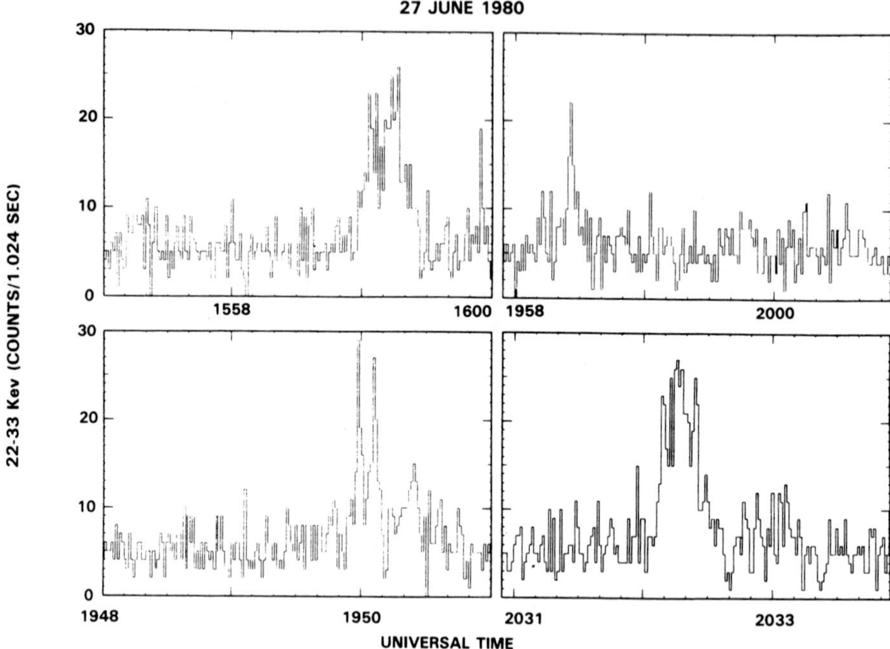

Figure 2.2.14 The four largest hard X-ray microflares are shown here at 1.024 sec resolution (from Lin *et al.,* 1983).

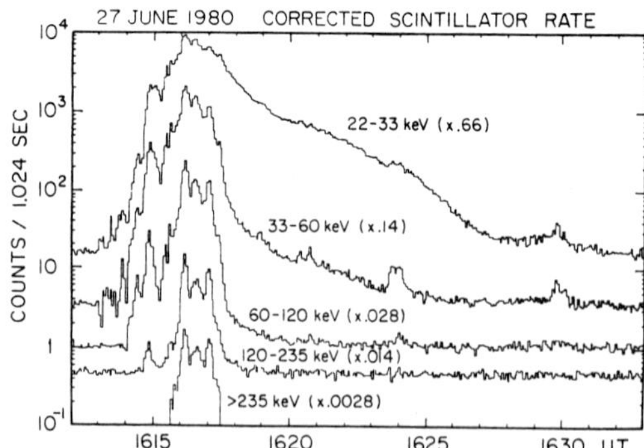

Figure 2.2.15 The hard X-ray burst observed by the scintillation detector. The low decay in the 22-23 keV channel lasts till ≥ 16:31 UT. This is due to the super-hot component. The small bursts of non-thermal emission occur till 16:31 UT (from Schwartz, 1984).

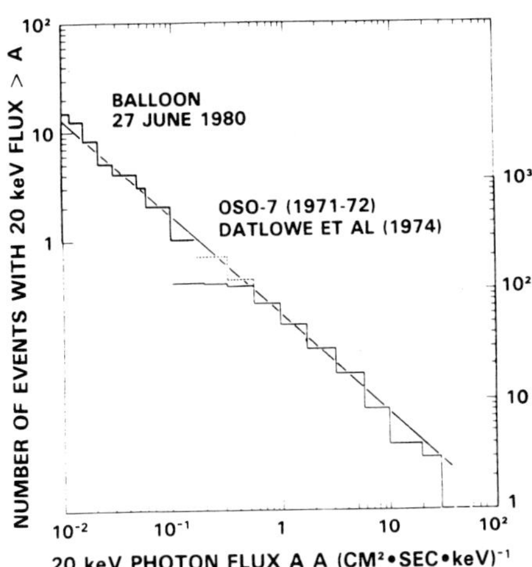

Figure 2.2.13 The distribution of the integral number of events versus peak 20 keV photon flux for the solar hard X-ray microflares observed in this balloon flight. Also shown for comparison is the distribution of solar flares hard X-ray bursts reported by Datlowe *et al.* (1974). The distributions have been arbitrarily moved vertically to show that their slopes are approximately the same (from Lin *et al.*, 1983).

less than 1% of the total accelerated electron population (see Figure 2.2.15). Schwartz (1984) concluded that while it is possible that a stored electron population could have triggered one of the early small bursts, the vast majority of the flare electrons could not have been stored in the corona but must be energized during the impulsive phase. The question is if there is any acceleration in the post-impulsive phase. In Figure 2.2.15 one can see that at ~ 1617:30 UT the > 60 keV X-ray flux falls to < 1% of peak intensity. Also, the 22-33 keV rate, mostly from the super-hot component, is falling more slowly. Of great interest for this discussion on electron acceleration is the series of impulsive bursts occurring during 1617:30 - 1630 UT and most clearly seen in the 30-60 keV rate. These post-impulsive phase bursts are similar to the impulsive phase bursts but have a peak intensity of about 0.5% of the largest impulsive phase peak. All the bursts contain fast spikes which rise and fall in 4-10 seconds. The spectral index γ, uncertain due to the large low energy continuum rates, is obtained by comparison with the count rates during the impulsive phase. These values are consistent with a nonthermal spectrum, similar to the bursts in the impulsive phase. Certainly, this continual bursting is evidence of electron acceleration throughout the post-impulsive phase, as proposed by Klein et al. (1983).

2.2.4 Microwave-Rich Flares

Figure 2.2.16 is a correlation diagram between HXRBS peak count rates and peak microwave fluxes; each point in this figure represents the peak HXRBS count rate and the peak flux density of 9 GHz microwaves for a particular flare. The frequency 9 GHz is chosen because for the majority of flares the microwave emission peaks near 9 GHz and because it is in a frequency range well observed world wide. As can be seen, there is a positive correlation between peak hard X-ray counts rates and peak microwave flux densities (cf. Kane, 1973), consistent with our understanding that both hard X-rays and microwaves are produced by energetic electrons of the same origin. Bai, Kiplinger, and Dennis (1985) defined the "microwave-richness index" (MRI) for each flare as follows:

$$\mathrm{MRI} = \frac{\text{peak flux density of 9 GHz microwaves (sfu)}}{\text{HXRBS peak count rate (counts/s)}} \times 10$$

Here the multiplication by 10 is to make the median value of MRI about 1. The diagonal straight lines in Figure 2.2.16 represent constant values of MRI. The line for MRI = 1 divides the population into roughly equal numbers. As can be seen from this figure, there is large scatter: MRI varies more than an order of magnitude (from less than 1/4 to more than 4). Bai, Kiplinger, Dennis (1985) studied the characteristics of the "microwave-rich flares" (with MRI > 4). They noticed that among the gamma-ray line flares studied by Bai and Dennis (1985) gradual gamma-ray line flares exhibit large delays of hard X-rays and large values of MRI. They studied 17 microwave-rich flares (12 flares in Figure 2.2.16 plus 5 microwave-rich flares observed in 1982), and found that these flares share many common characteristics. (1) Large values of MRI (> 4). This was the selection criterion. (2) Long durations of hard X-ray bursts. Microwave rich flares last several minutes, as opposed to the ordinary flares that usually last less than 1 minute. (3) Large H-alpha area. Except for one microwave-rich flare observed at the limb, all belong to H-alpha importance class 1 or higher, and 13 out of 17 belong to H-alpha class 2 or 3. (4) Long delay times (> 10s) of high-energy hard X-rays with respect to low-energy hard X-rays. In a given burst the delay time increases with hard X-ray energy. Such delays are equivalent to hardening of the X-ray spectrum with time during the burst (see Figure 2.2.17). (5) Long delay times (10 ~ 300s) of microwave time profiles with respect to low-energy hard X-ray time profiles. (6) Flat hard X-ray spectra. The average of the power-law spectral indices is 3.5. (Compare with 3.36, which is the value for the gamma-ray line flares studied by Bai and Dennis (1985)). (7) Association with type II and IV radio bursts. All of them produced type II or type IV bursts or both. Seven of the 13 microwave-rich flares have HXRBS peak count rates between 1000 and 5000 cts/s. Considering that the largest HXRBS count rates are of the order of 10^5 cts/s (Dennis et al., 1983), the above count rates are moderate. (8) Emission of nuclear gamma-rays. Only six of the 17 microwave-rich flares produced observable nuclear gamma-rays, but it is interesting to note that all the microwave-rich flares which did not produce observable nuclear gamma-rays have HXRBS peak rates < 4000 cts/s. None of the gamma-ray line flares observed during 1980 through 1981 have HXRBS peak rates < 4000 cts/s (Bai and Dennis, 1985). Therefore, the failure to observe nuclear gamma-rays from the microwave-rich flares with low HXRBS count rates is most likely to be due to the threshold effect of GRS.

Another interesting point is that the microwave-rich flares share all the characteristics of gamma-ray line flares. Detailed discussions on the correlation of microwave rich flares and gamma-ray line flares, as well as a possible scenario for their interpretation can be found in Bai and Dennis (1985).

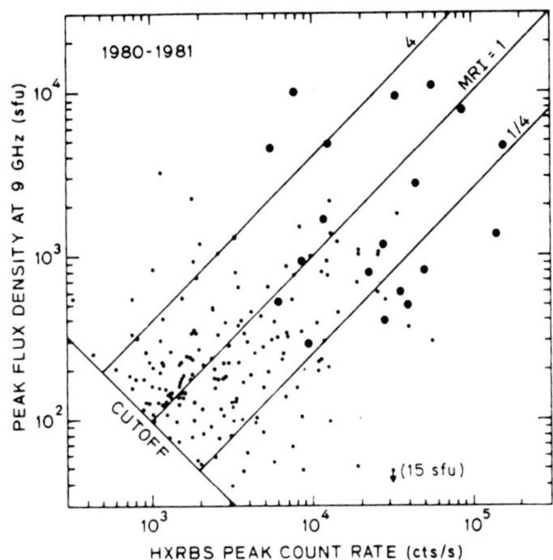

Figure 2.2.16 Correlation diagram between peak count rates measured by HXRBS and peak flux densities of 9 GHz microwaves for 1980 through 1981. Although there is quite a lot of scatter, there seems to be a positive correlation between these quantities. The median value of MRIs is about 1 (0.85 to be precise). The three straight diagonal lines indicate constant values of MRI, 1/4, I and 4. The large dots indicate GRL flares. Note that the HXRBS peak rates of the GRL flares are > 5000 cts/s (from Bai, Kiplinger, and Dennis, 1985).

2.2.5 Decimetric-Metric Observations and Comparison with X-ray Observations

Previous studies have already shown that type III bursts and soft X-ray increases are often observed several minutes prior to the occurrence of the flare itself (Kane et al., 1974). Evidence for hard X-rays observed before the flash phase was also reported by Kane and Pick (1976). A systematic study, using more sensitive spectrometers, was carried out by Benz et al. (1983b); they listed 45 major events observed by the HXRBS experiment aboard SMM. For most of these events, metric type III bursts and decimetric pulsation were observed preceding the hard X-ray emission. In 7 of the 45 flares, significant hard X-ray fluxes were observed before the rapid general exponential increase. This phase in the flare development was called "preflash phase". It usually lasts for about one minute. These observations give evidence for electron acceleration before the impulsive phase (see also Section 2.2.3.5).

It is commonly believed that electrons responsible for type III bursts and hard X-ray emission have a common origin (Kane, 1972, 1981a)

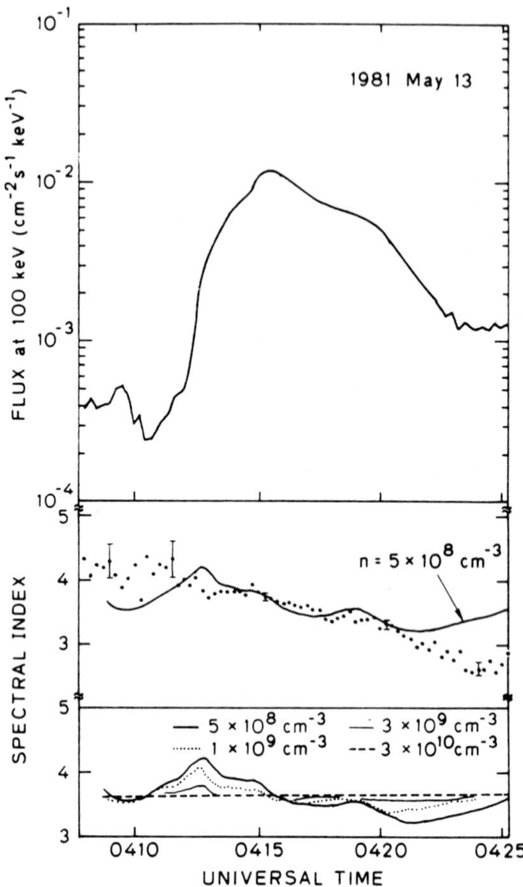

Figure 2.2.17 Spectral evolution of hard X-ray emission from the 1981 May 13 flare. The top panel shows the hard X-ray flux at 100 keV, and the middle panel shows the observed spectral evolution (with dots) together with a spectral evolution calculated using a perfect-trap model. In the last panel the spectral evolutions were obtained by using various values for the ambient density. In order to get a reasonable fit to the data, the ambient density should be as low as 5×10^{8} cm^{-3}. However, this density is incompatible with other observations as mentioned in the text, Tsuneta *et al.* (1984) found that the images of the hard X-ray source and the soft X-ray source are almost the same. These authors also deduced from the emission measure and the size of the soft X-ray source that the density of the flare loop is 3×10^{10} cm^{-3}. As the density increases, the spectral index change will be less and less, approaching the steady state case. For $n = 3 \times 10^{10}$ cm^{-3}, the resultant spectral index evolution is hardly different from a straight line, which is for the steady state case. Before the hard X-ray peak the spectral index is larger than the steady state case and after the peak it is smaller, but near the peak the spectral index is similar to that of the steady state case independent of the density (from Bai and Dennis, 1985).

since their temporal evolution is well correlated. Simultaneous observations of X-ray and radio emission with a time resolution of less than 1 sec have shed new light on our understanding of the electron acceleration process. The main conclusions can be summarized as follows: (a) Some hard X-rays peaks are well correlated with type III bursts and show delays of the order of or shorter than one second (Kane, Pick and Raoult, 1980; Benz et al., 1983b). The type III source may consist of several elementary components widely separated (by more than 100,000 kms) which radiate quasi-simultaneously or successively. This implies that the acceleration/injection region covers a wide range of magnetic fields (Mercier, 1975; Raoult and Pick, 1980). (b) Kane and Raoult (1981) reported an increase in the starting frequency of type III bursts during the development of the impulsive phase. This variation is correlated with an increase in the hard X-ray flux. As the type III bursts radiation is emitted at the local plasma frequency, the starting frequency corresponds to the density at the point where the electron beams become unstable. Thus this fast variation of the starting frequency may be explained either by a real variation of the electron density in the source (downward shift or compression of the injection/acceleration site) or to a variation in the distance from the acceleration site travelled by the electron beam before it becomes unstable (Kane et al., 1982). A systematic study was carried out by Raoult et al. (1985) to determine if the presence of an increase in the starting frequency of type III bursts influences the probability of their correlation with hard X-ray bursts. A total of 55 type III groups were selected which had been observed with the Nancay Radiospectrograph (Dumas et al., 1982) in the frequency range 450 - 150 MHz, and with the ISEE-3 X-ray spectrometer. Of the 55 events, 32 cases, (58%) were associated with X-ray emission. In this sample, 28 events (52%) showed an increase in the starting frequency. 75% of these events were associated with X-ray emission, resulting in significant improvement of the correlation. Conversely, 75% of the X-ray associated events show an increase in the starting frequency. Thus, an increase in starting frequency seems to be a significant factor that improves the association between type III burst groups and X-ray bursts. (c) However, Raoult et al. (1985) pointed out that among these 55 events, 15 events were associated with type V continuum visible in the frequency range 450 - 150 MHz with a typical duration of about one minute or less. All these events have an X-ray response. Thus, the presence of a type V continuum at frequencies > 150 MHz appears to be a decisive factor in increasing the correlation between X-rays and type III bursts.

Raoult et al. (1985) also suggested that type III/V events have a consistently large X-ray response. Stewart (1978) first reported that among a list of X-ray associated radio bursts, 80% contained a type V burst. They performed a detailed data analysis for meter events which have an X ray response. Their main findings are: (a) Pure type III burst groups are not associated with intense X-ray emission. The hard X ray bursts associated with these events have fluxes < 1 photon cm^{-2} sec^{-1} keV^{-1} at about 30 keV and are not detectable above 100 keV. The corresponding radio burst source is often multiple. The X-ray response around 30 keV closely follows the starting frequency evolution. (b) When a radio event is associated with strong X-ray emission (> 1 photon cm^{-2}

sec^{-1} keV^{-1} at 30 keV and detectable above 100 keV), a continuum emission (type V) in the range 450 - 150 MHz appears along with the type III groups. The typical evolution of these events is illustrated in Figures 2.2.18 and 2.2.19, and may be described as follows: The first part of the event, "preflash phase", contains only type III (or U) bursts coming from locations (A). Then a new source "B" (see Figure 2.2.19) appears at the

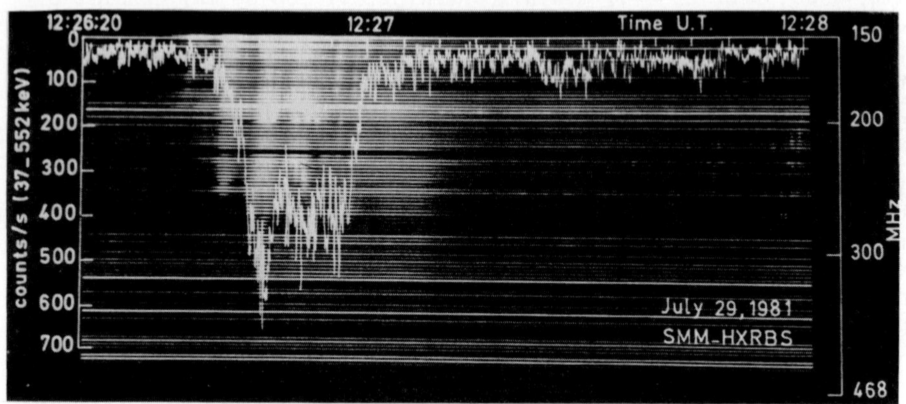

Figure 2.2.18 Type III-V bursts on 1981 July 29, observed with the Nancay Radiospectrograph (Dumas *et al.*, 1982) and the associated hard X-ray burst observed with the SMM-HXRBS experiment. Evolution of the X-ray emission compared to the evolution of the radio event. (From Raoult *et al.*, 1984).

time of the fast increase in the X-ray emission. This is also coincident with an increase in the radio starting frequency. At that time, one of the pre-existing type III burst sources (A9) becomes predominant. Sources B and A9 have similar sizes (29 - 39 arc at 169 MHz), and they fluctuate simultaneously within short time delays of less than one second. Thus both sources contribute to the spiky and smoothed parts of the radio emission identified as type III and type V bursts, respectively. There is an overall correlation between the radio flux and X-ray fluctuations, although there is no correlation on short timescales. When the rapid

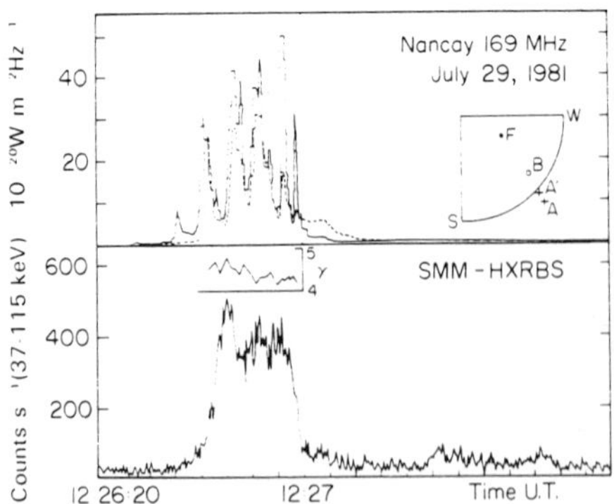

Figure 2.2.19 Top: Evolution of the radio flux from sources A, A′ (solid line) and B (broken line) see text. Bottom: Evolution of X-ray spectral index and hard X-ray emission. (From Raoult *et al.*, 1984).

radio fluctuations are no longer present, both X-ray and radio fluxes decrease sharply and the X-ray emission appears at energies below 30 keV. The total X-ray flux and radio flux decrease rather smoothly, the source B being usually the predominant one. The duration of the type V burst increases with increasing wavelength. Similarly the duration of the X-ray emission increases with decreasing energy.

The results described above have implications on the geometry of the magnetic field structure at the site of injection of electrons. The presence of sources A9 and B, which fluctuate quasi-simultaneously, implies that the electrons are quasi-simultaneously injected into two structures (or two unresolved groups of structures). According to Raoult et al. (1985) most flares that are associated with a small hard X-ray emission correspond to an electron injection/acceleration site that covers several diverging magnetic flux tubes. The fact that during the impulsive increase of the hard X-ray flux, the radio emission is reinforced in one pre-existing location and appears quasi-simultaneously in a new location, suggests that at the injection site the two magnetic structures interact. At that time the energy that is released in the interaction region increases sharply. Another possible interpretation was given by Sprangle and Vlahos (1983) and is discussed in Section 2.4.6.

Rust et al. (1980), Benz et al. (1983b), Aschwanden et al. (1985) and Dennis et al. (1984) have also studied the correlation of hard X-rays with decimetric radiation, which originates at lower altitudes. Their results on decimetric type III bursts reinforced many of the conclusions reported for metric bursts. Aschwanden et al. (1985) have found decimetric type III bursts to be associated with hard X-ray events in 45% of the cases. The association rate increases with the number of bursts per group, duration, bandwidth and maximum frequency of the group. Some single bursts (but not all) are correlated with hard X-ray spikes. In some cases the difference in time of maximum between type III and hard X-rays is a few tenths of a second, which may be significant. This may imply that ordinary cross-field drifts or diffusion from closed to open field lines are too slow. The acceleration of the electrons by intense electromagnetic waves, as proposed by Sprangle and Vlahos (1983) seems to be a likely interpretation (see Section 2.4.6 for details). These bursts occur at frequencies of 300 MHz to > 1 GHz, corresponding to densities $> 3 \times 10^9$ cm^{-3} (Benz et al., 1983b).

Strong et al. (1984) investigated a double impulsive flare in radio, soft and hard X-ray emissions. The decimetric radio emission of both events contains U bursts. In several cases they have harmonic structure. From the total duration and extent in frequency of the U bursts the geometry of the loop guiding the electron beam can be calculated. The average length of these loops is 94,000 km and 157,000 km, and the average height 24,000 km and 45,000 km in the two flares respectively. The U bursts are sometimes correlated in time with hard X-ray spikes (Figure 2.2.20). If the elongated soft X-ray source is interpreted as a loop, its projected size is only 30,000 km. Post-flare soft X-ray loops have been found in the second flare with footpoints separated by 115,000 km. The presence of loops of different sizes is also evident in the microwave spectrum which shows evidence for 3 peaks indicating sources with widely different magnetic field strengths.

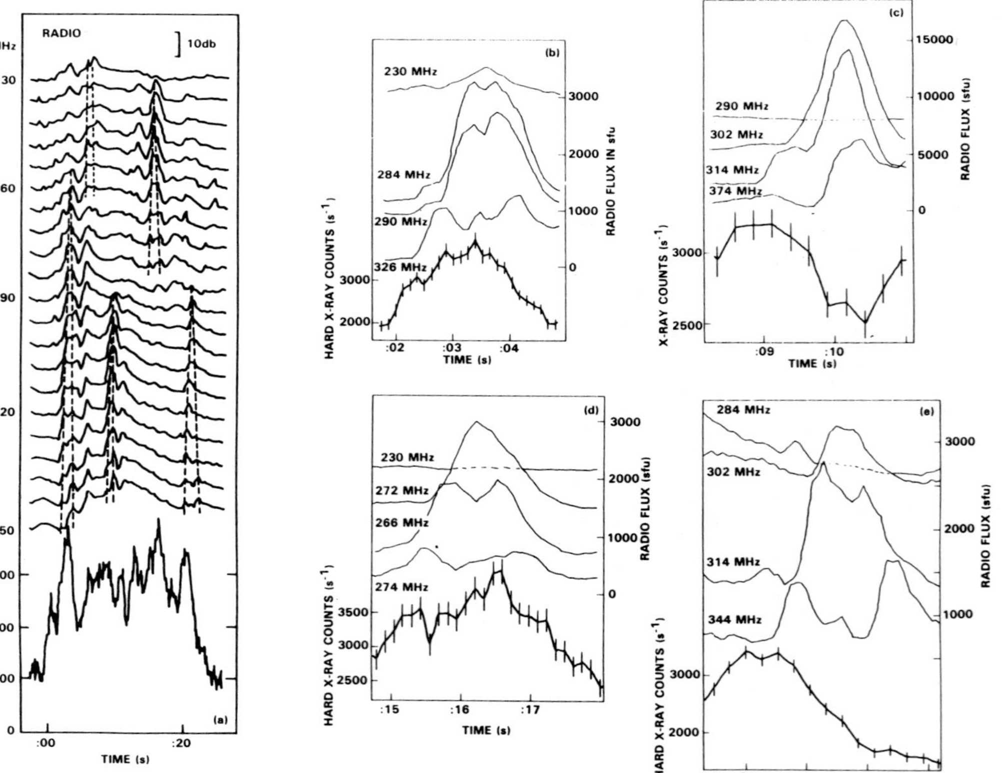

Figure 2.2.20 Correlation between U-bursts and hard X-ray spikes in the August 31, 1980, 1251 UT flare. Data from radio spectrometer at Bleien (Zürich) and HXRBS on SMM. Left: U-bursts are outlines in the upper part and compared with the X-ray count rate in the lower part. Right: Details of left figure showing correlation of X-ray spikes with rising part of U-bursts (from Strong *et al.*, 1984).

Apparently energetic particles have immediate access to small (soft X-ray) loops and large (U burst, post-flare) loops suggesting that the acceleration site is at the boundary or interface between the two loop systems.

The decimetric emission of flares can be divided into radiations which generally occur during the impulsive phase and the type IV emission generally observed after the impulsive phase. The impulsive phase bursts are found to vary considerably in shape (Wiehl et al., 1985). A large fraction can be interpreted as due to type III-like beam instabilities. The bursts may have some unexpected forms, however, such as narrow bandwidth ($\Delta\nu/\nu \leqq 0.2$), called blips by Benz et al. (1983a) or very high drift velocities (an example is shown in Figure 2.2.21). These deviations from the normal shape are probably caused by the disturbed properties of the ambient plasma.

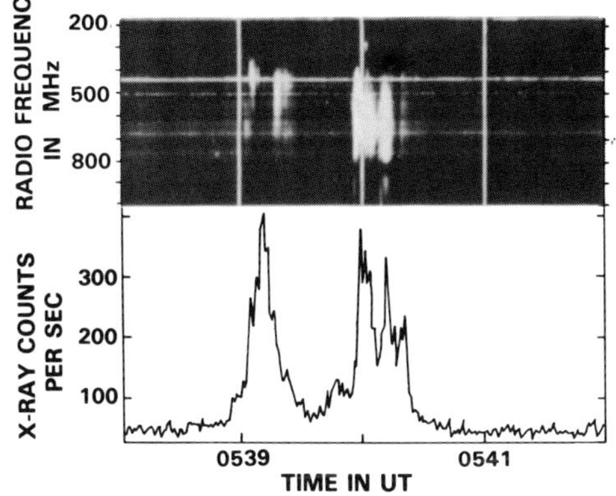

Figure 2.2.21 Top: Dynamic spectrogram of type III-like decimetric emission with very high drift rate, observed on May 19, 1980 with the analog spectrometer at Bleien (Zürich). Enhanced emission is bright, horizontal lines are terrestrial interference, and vertical lines are minute marks. Bottom: Hard X-ray count rate as observed at energies > 30 keV by HXRBS on SSM.

All decimetric bursts during the impulsive phase do not appear to be explained by particle beams. About 25% of all cases are in this category and they are strongly associated with hard X-rays (70%). These bursts have been divided into 4 classes by Wiehl et al. (1985).
1) Diffuse patches of emission probably originate from trapped particles either by synchrotron or loss-cone radiation. 2) Grass-like chains of small spikes resemble elements of metric type II bursts. They may be caused by shock waves. 3) Nonperiodic broadband pulsations with pre-flash hard X-ray emission. The decimetric emission in these cases precedes both hard X-ray and Hα emission, as shown by Benz et al. (1983b) from a study of 3 flares. However, a more general study of 45 such events (Aschwanden et al., 1985) has shown that pulsations usually start after the hard X-rays and end before them. Most important, pulsations and hard X-rays do not seem to correlate closely. Durations of single elements are between 20 and 100 ms. The elements are of similar bandwidth (several 100 MHz)

and have about the same low-frequency end. 4) Spikes of short (< 100 ms), narrowbanded (3-10 MHz) emission occur in large groups. They are associated with shorter and more impulsive hard X-ray bursts than the average. They tend to occur in the early impulsive phase (Benz, 1985). The single elements are scattered in a chaotic manner between ~ 400 and > 1000 MHz (corresponding to densities of 0.3 - 1 x 10^{10} cm^{-3}). Their circular polarization can be between 25 - 40%. They probably are similar to the microwave spikes observed at 2.6 GHz by Slottje (1978), probably also produced by electron cyclotron masering.

2.2.6. Discussion of Models for X-ray and Microwave Emission

Information about the accelerated electrons are obtained through models which depend on parameters such as local ambient density, temperature and magnetic field which are poorly known. Three major problems face us in our interpretation of the observations:

-- what is the relative role of thermal and nonthermal electrons in producing X-rays at different energies?
-- does nonthermal production of hard X-rays arise from beams of electrons (thick-target model) or from a trapped population of electrons or from a combination of both.
-- do the observations imply a single or a two step acceleration process?

We discuss below several attempts to model the energy release and answer some of the questions posed above.

2.2.6.1 Trap Plus Precipitation vs Two Step Acceleration Models

The two competing interpretations of the energy-dependent hard X-rays are trap plus precipitation models (Kane, 1974; Melrose and Brown, 1976; Bai and Ramaty, 1979; Vilmer et al., 1982; MacKinnon et al., 1983; Ryan, 1986) and second-step acceleration models (Bai and Ramaty, 1979; Bai, 1982; Bai et al., 1983a, 1983b; Bai and Dennis, 1985). Interestingly, the first paper that analyzed hard X-ray delays (Bai and Ramaty, 1979) invoked both interpretations, the trap model for small delays below 150 keV, and second-step acceleration for large delays (15 s) above 150 keV. Bai and Ramaty (1979) and Vilmer et al. (1982) used a pure trap model, and MacKinnon et al. (1983), Trottet and Vilmer (1983), and Ryan (1985) considered the effect of precipitation. MacKinnon et al. (1983) reported that, in the weak diffusion limit, precipitation does not change the essential nature of the trap model. A detailed discussion of trap models is given later in this Section.

We emphasize first that the second-step acceleration is different from the conventional "second-phase" acceleration proposed by Wild, Smerd and Weiss (1963), since the delay between the two steps is tens of seconds and not tens of minutes.

Bai and Dennis (1985), who have studied many flares exhibiting hard X-ray delays, note the following points favoring the second-step acceleration interpretation. (1) In impulsive flares which exhibit hard X-ray delays, the delay time as a function of hard X-ray energy is quite different from what is expected from the collisional trap plus precipitation models. Instead of increasing gradually with energy, the delay time exhibits a sudden increase at high energies. (2) In very gradual flares such as the ones observed on April 26 and May 13, 1981, the ambient density deduced with the trap model is of the order of 10^9 cm^{-3} (see Figure 2.2.17). This is too low to explain the observed emission measure of soft X-rays (assuming of course, that the hard and soft X-ray emitting regions are coincident). Actually, the density deduced for the May 13 flare from the observed emission measure and volume is 3×10^{10} cm^{-3} (Tsuneta et al., 1984a). The above argument does not exclude the possibility of trapping of energetic electrons in a huge loop, (this was proposed by Tsuneta et al., 1984a), but it proves that trapping is not the primary cause of large hard X-ray delays (or spectral flattening with time) observed in these gradual flares. (3) The association between hard X-ray delay and proton acceleration (see Section 2.3), is naturally explained by the second-step acceleration model. In Fermi type acceleration, stochastic acceleration by a fluctuating magnetic field, or shock acceleration, there exist threshold energies (or injection energies) for both electrons and protons above which the acceleration can overcome the Coulomb energy loss (e.g., Ginzburg and Syrovatskii, 1964; Sturrock, 1974; Ramaty, 1979). Therefore, when these kinds of acceleration mechanism accelerate protons to gamma-ray producing energies, they will also accelerate electrons with energies greater than the injection energy to higher energies. On the other hand, in trap models it is hard to see the connection between proton acceleration and trap electrons. (4) The total bremsstrahlung fluence above 270 keV is roughly proportional to the 4-8 MeV fluence (see Chupp 1982 and Figure 2.3.4). On the other hand, when the 4-8 MeV fluence is compared with the hard X-ray fluence above 30 keV, the correlation is very poor (Bai and Dennis, 1985). Actually many flares with large fluences in > 30 keV hard X-rays did not produce observable nuclear gamma-rays. With the second-step acceleration, this is easily explained. In the second-step acceleration model, both high energy electrons and gamma-ray producing protons are accelerated by the second step, hence we expect a good correlation between hard X-rays > 270 keV and 4-8 MeV fluences. On the other hand, the fluence of low-energy hard X-rays (> 30 keV), which is due to electrons accelerated by the first-step mechanism, is not expected to correlate well with the gamma-ray fluence, which is due to the second-step mechanism. (5) In trap models the photon spectrum is somewhat steeper at the beginning of the burst than in a thick-target beam model, and it gradually flattens to be about the same as the thick-target model near the peak. On the other hand, if the second-step acceleration is operating, the photon spectrum at the peak of the burst is expected to be flatter because of additional acceleration at high energies. Consistent with the second-step model, the photon spectrum measured at the peak of the burst is flatter on the average for gamma-ray line flares than for non-gamma-ray line flares, which do not in general show hard X-ray delays. The site of the second-step acceleration is proposed to be the corona instead of the chromosphere (Bai and Ramaty, 1979; Bai et al., 1983b); therefore, in this model at least, high energy electrons are assumed to be trapped in the corona. Hence, it is possible that in many flares hard X-ray delay is partly due to trapping and partly due to the second-step acceleration, as proposed by Bai and Ramaty (1979). It is usually difficult to determine their relative importance unless we know the ambient density of the flare loop (trap region). For the May 13, 1981 flare the ambient density is deduced to be 3×10^{10} cm^{-3} (Tsuneta et al., 1984), and for this density the hard X-ray delay is much smaller than the observed one. For the gradual flare of April 26, 1981, the same is true (Bai, Kiplinger and Dennis, 1985).

Let us now summarize the recent progress made on models that invoke trap and precipitation. Vilmer et al. (1982) applied the trap model to explain observations of high-energy X-ray delays, and McKinnon et al. (1983) considered the effect of precipitation on the trap model. Trottet and Vilmer (1983) have also studied the case where the precipitation from the trap is in the strong diffusion limit (e.g., wave-particle interaction). The basic ingredients of the model are: (1) a trap of uniform density n_o, (2) a continuous injection of nonthermal electrons in the trap, with constant spectral index γ, during a finite time t_o, (3) a time dependent injection function having a maximum at $t_o/2$, (4) energy losses entirely due to electron-electron collisions, (5) precipitation from the trap gives rise to a thick target component, either in the weak diffusion limit (Coulomb collisions) or in the strong diffusion limit (wave-particle interaction). The computed X-ray time profiles depend then on t_o, n_o and the precipitation process considered. In the weak diffusion limit, although hard X-ray emission starts simultaneously at all energies, the higher energy channels reach their maxima later than the lower ones. For given t_o and γ such delays are a function of n_o. Figure 2.2.22a shows that $\Delta t(E) = t_{max}(E) - t_o/2$ increases with increasing energy and decreasing density n_o. For a given energy E, $\Delta t(E)$ also increases with t_o. The ratio between the X-ray flux produced in the trap I_{trap} and the flux due to precipitated electrons I_{prec} depends neither on the energy nor on the density n_o. The total hard X-ray spectrum (trap + precipitation) hardens with time. When precipitation is in the strong diffusion limit, its rate depends on the particle energy and α_o^2/L (the ratio between the loss-cone angle and the characteristic length of the trap). The X-ray time profile depends on two characteristic times, the energy loss time t(E), which increases with E, and the precipitation time t_p which decreases with E. When t_p is larger than t_E (large scale loops or small α_o), $\Delta t(E)$ increases with energy, but does not exceed a few seconds. When t_p is of the order of or smaller than t_E, Figure 2.2.22a shows that $\Delta t(E)$ is very small, approximately constant with E and weakly dependent on n_o. In this last situation no observable delays are expected. On the contrary I_{trap}/I_{prec} is strongly dependent on the energy E (I_{trap}/I_{prec} decreases when E increases) and decreases when α_o^2/L increases. The hardness of the hard X-ray spectrum remains approximately constant with time. Moreover for the same injection function and trap density, the X-ray spectrum is somewhat harder than in the weak diffusion regime before the maximum (Trottet and Vilmer, 1983).

According to Trottet and Vilmer (1983), the main considerations that favor trap and precipitation models are as follows: (1) Hard X-ray imaging sometimes shows high and large X-ray sources, with power law spectra, suggesting a coronal thick target trap with continuous injection/acceleration of electrons (type C flares discussed in Section 2.2.1.1). (2) Some events exhibiting large delays (up to 1 min) have been successfully interpreted through trap and precipitation models (see Vilmer et al., 1982). The diversity of observed delays is easily explained by the variability of the trap density, injection time and nature of the scattering process. Certainly more work has to be done to describe more realistic situations, namely one has to develop time-dependent models where the inhomogeneity of the ambient medium and the angular distribution of the energetic particles are taken into account. A first approach to this problem is to look for general time-dependent solutions of the continuity equation. Vilmer et al. (1986) and Craig et al. (1985) have developed the mathematical framework that can be used for

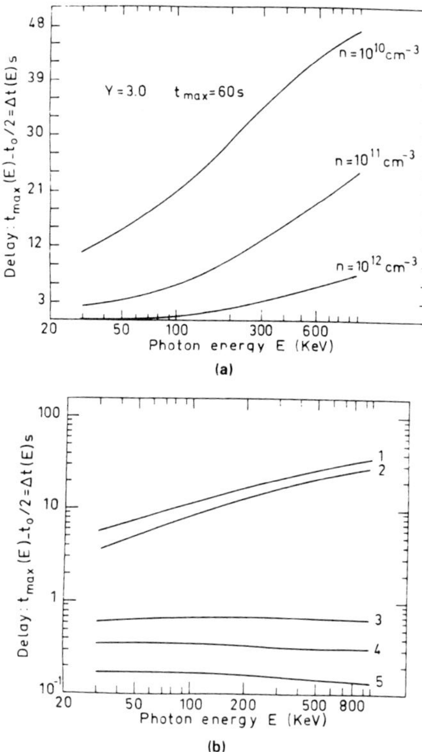

Figure 2.2.22 (a) Computed delays $\Delta t(E) = t_{max}(E) - t_0/2$ for $\gamma = 3.0$, $t_{max} = 60$ sec, for the weak diffusion case as a function of the photon energy E, for different values of the trap density. (b) Computed delays as a function of the photon energy E, using the same parameters as in (a), for different precipitating rates: perfect trap (1) weak diffusion (2) and strong diffusion (3), (4), (5) using $a_0^2/L \simeq 10^{-5}$, 2×10^{-5}, 5×10^{-5} respectively (from Trottet and Vilmer, 1983).

such a study. (3) The time lag between hard X-ray and γ-ray maxima is correlated with the γ-ray rise time.

Trottet and Vilmer (1983) have also argued that if a two-step acceleration is at work some difficulties arise. Indeed Chupp (1983) has shown that the ratio of the prompt γ-ray line fluence in the 4-7 MeV band to the 2.223 MeV line fluence is approximately constant from one flare to another. According to Ramaty (1985), this requires a constant spectral shape for the ions. Moreover the total electron bremsstrahlung fluence above 270 keV is roughly proportional to the 4-8 MeV excess fluence (Chupp et al., 1984b). This suggests that high energy electrons and ions are accelerated by the same process and that this process is common to all flares. Thus, if delays reflect a second step acceleration, they should be observed, without exceptions, for all flares producing γ-ray lines and X-rays above a few 100 keV. In fact some observations contradict such an interpretation. Let us illustrate this point by two examples reported by Rieger (1982). First, reverse delays between X-ray and γ-rays are clearly observed for the October 14, 1981 flare (4-7 MeV and 10-25 MeV channels peak before the 80-140 keV and 300 keV channels). Second, the June 21, 1980 flare exhibits variable delays from one peak to another, the first peaks occurring even simultaneously in all channels. In summary Trottet and Vilmer argued that even if a two-step acceleration process cannot be definitively ruled out, available observations of time delays may reflect the interaction between the accelerated particles and the ambient medium rather than the characteristics of the acceleration mechanism itself.

Ryan (1986) also considered independently the effects of particle trapping on the time profiles of hard X-rays and γ-rays. His results reinforce the work reported above. Ryan used three different models. The first is that of a closed trap with a finite density of matter within the trap providing the slowing down mechanism for the particles and the particle target for photon production. The two other models employ particle diffusion in a tenuous trap to allow particles to precipitate to denser regions of the solar atmosphere where they interact to produce the photons. The characteristics of all of these models are (1) to reduce the impulsiveness of the acceleration as it is seen in the high energy photons and (2) to produce delays in the maxima of the photon fluxes at various energies. These effects must be taken into account in searching for evidence of additional acceleration mechanisms. The constant density coronal trap which has been considered in the past for electrons below 200 keV can produce significant delays for electrons of energies > 0.5 MeV and larger effects still for γ-rays produced by ~ 20 MeV protons. Particle densities of 10^{10} cm^{-3} can produce delays in the γ-rays of several tens of seconds. If particles are injected impulsively at one point in the loop, they diffuse toward both ends of the trap precipitating to the loss regions of high density. With this process, there is an intrinsic delay in the precipitation rate and thus the photon flux due to the finite time required for the particles to diffuse to both ends of the loop. The rise and decay times of this process are also proportional to the size of the trap. It should also be noted that the particle propagation effect in the observed photon flux for the constant density trap is also a function of the size of the trap. The study by Rosner et al.

(1978) shows that the matter density in coronal non-flaring loops is inversely correlated with the length of the loop. Thus we have the situation where three mutually exclusive particle trap scenarios produce a reduced impulsiveness in the photon flux with respect to the particle acceleration or injection and the convolution of these effects with the acceleration profile produces a delay in the flux maxima with respect to the acceleration profile. In addition, the magnitude of these effects grows with the linear dimensions of the loop. The implications of this are that they complicate the search for and identification of a multi-step acceleration process and they limit the search for rapid fluctuations in photon flux, which is a signature or measure of the rapidity of the acceleration process.

2.2.6.2 Dissipative Thermal Model

We have emphasized in this section that heating and acceleration of the plasma tail occur nearly simultaneously in flares. This poses a fundamental problem: How does the flare-energized (hot + tail) plasma expand along the field lines? Since the plasma outside the energy release region is at coronal temperatures (several million degrees Kelvin), the energized plasma interfaces with a "cold" ambient plasma. The steep temperature and/or density gradients accompanying the rapid energization may give rise to D.C. and stochastic electric fields which contain most of the electrons, but allow the fastest electrons in the tail of the distribution to escape. Brown, Melrose and Spicer (1979) suggested (following similar work by Mannheimer (1977) in the pellet fusion plasma) that a return current, driven by the electrostatic potential at the interface, will set in and most probably will grow unstable, limiting the heat flux. This suggestion was followed by two extreme approaches: (1) Ignore the escaping electrons and use a fluid model to simulate the expansion of the hot plasma (see e.g., Smith and Harmony 1982 and references therein). (2) Describe qualitatively the hot plasma and concentrate on the escaping electrons (Vlahos and Papadopoulos, 1979 and Emslie and Vlahos, 1980). In the latter work it was also assumed that inside the energy release volume the tail was continuously replenished by sub-Dreicer electric fields. In reality both approaches were of a limited scope. The real problem is somewhere in between and we have to simulate the plasma below a critical velocity (which is not known) as a fluid and as particles above it. In other words, the need for a multifluid or Vlasov type simulation is obvious. Such a simulation is currently possible. It is worth mentioning that several qualitative suggestions, based on the dissipative thermal model, appeared in the last few years.

Brown et al. (1980) suggested that the energy release volume in a flaring loop may consist of many hot sources with lifetimes and sizes below the instrumental resolution. The overall hard X-ray burst emission is made up of a "convolution" of these "multiple kernels". They investigated the effective (time-integrated) spectrum of hard X-rays from one such kernel, and showed that the majority of observed spectra could be explained by invoking a spread in the parameters characterizing the kernels. The hardest spectra are not, however, amenable to such an intepretation.

Smith (1985) suggested the following scenario for solar hard X-ray bursts which may explain the evolution of type B flares (cf. 2.2.1.1). At the beginning of the impulsive phase, we often see brightening of footpoints which indicates that a significant fraction of the energy released is going into accelerated electrons. This could occur due to fast tearing modes in a loop leading to electron acceleration via the modified two-stream instability (see Section 2.4). After these electrons evaporate a sufficient amount of chromospheric plasma, which then travels back up the loop, the electron plasma beta, β_e, rises sufficiently to cut off the modified two-stream instability and the footpoint behavior ceases. The emission is then dominated by the primarily thermal single source near the top of the loop. There may still be some small regions in the loop where β_e is sufficiently small to allow acceleration of electrons required by the microwave emission.

Holman, Kundu, and Papadopoulos (1982) have shown that streaming suprathermal electrons will be isotropized by self-generated electrostatic waves (the "anomalous doppler resonance" instability) if the electron gyrofrequency (Ω_e) exceeds the plasma frequency (ω_e) somewhere along the loop, and if the minimum velocity in the suprathermal electron distribution is well above the mean thermal electron velocity in the ambient plasma. The first condition ($\Omega_e > \omega_e$) may hold in most flare loops, and the second condition will hold as long as the accelerated electron distribution does not extend down to the thermal distribution, or if the accelerated electrons escape into a cooler plasma. Holman, Kundu and Papadopoulos also show that if the suprathermal electrons are also responsible for the observed hard X-ray emission, the scattering of the particles can also lead to breaks in the hard X-ray spectrum. These breaks result from wave damping preventing all of the suprathermal electrons from being scattered. An important conclusion is that the microwave source structure does not necessarily indicate the location of the particle acceleration region. Similar conclusions can be reached from considerations of the loop geometry and the directivity of gyrosynchrotron emission (see Petrosian 1982).

Zaitsev and Stepanov (1983) showed that intense localized heating inside the energy release region may violate locally the condition that the plasma pressure is lower than the magnetic pressure, which in the past had permitted some numerical calculations of one-dimensional fluid models (e.g., Smith and Lilliequist, 1979). As a result, the magnetic field expands locally and sets up a local magnetic trap, and $B^2/8\pi \gtrsim nkT_e$ and the magnetic field compresses the plasma again. This cycle repeats and sets in an oscillatory motion. Zaitsev's and Stepanov's results may explain the periodic pulsations observed in hard X-ray and microwave bursts.

Batchelor et al. (1985) made a new analysis of the thermal flare model proposed by Brown, Melrose and Spicer (1979). They assumed that the model leads to the development of a quasi-Maxwellian electron distribution that explains both the impulsive hard X-rays and microwaves as opposed to our previous interpretation that allows a significant number of nonthermal electrons to escape from the thermal source. This implies that (a) the part of the microwave spectrum for which $f < f_{max}$ consists of optically thick emission, so the source area, A_o, can be calculated from the Rayleigh-Jeans law, and (b) the plasma temperature can be measured from the hard X-ray spectrum by determining the best fit to a single-temperature thermal bremsstrahlung function. Using (a) and (b), Batchelor et al. (1984) calculated A_o at the time of maximum hard X-ray flux. Assuming that the source was an arch, they estimated its half-length $L_o \simeq A_o^{1/2}$. The theoretical time scale of the burst would then be $\tau_o = L_o/c_s$, where $c_s = (kT_e/m_p)^{1/2}$ is the ion-sound speed, the speed of expansion of the source during the initial rise of impulsive emission. To test the prediction of the model, Batchelor et al. (1985) analyzed microwave observations made at the Bern Radio Observatory and hard X-ray observations obtained with the SMM-HXRBS experiment. The results are shown in Figure 2.2.23, which is a plot of log t_r vs log τ_o, where t_r is the measured rise time of the hard X-ray emission and τ_o is computed from independent spectral parameters only. For 17 disk flares, the best fit relationship is found to be $t_r \approx 0.51\ \tau_o^{1.5}$, which is within the statistical uncertainties of the predicted relationship, $t_r \approx \tau_o$. Three limb flares lie to the left of the disk flares on the diagram, consistent with the interpretation that they were partially occulted by the solar limb, which would result in reduced values of L_o and τ_o as observed. This result is in good agreement with the model, and is not explained by any other known flare models which have been considered. The main problem with the Batchelor et al. model, however, is that the behavior of the energetic electrons was not properly considered.

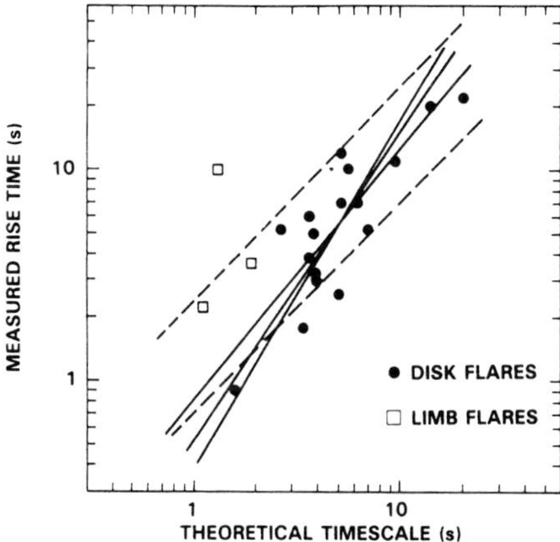

Figure 2.2.23 Correlation diagram of t_o and τ_o. Solid lines indicate best fits by linear least-squares fitting. Dashed lines are boundaries of the expected positions of disk points if the sources are arches from 2 to 4 times as long as they are thick (from Batchelor *et al.*, 1985).

2.2.7 Summary

We shall now return to the questions which we have posed in the introduction and which have guided our discussions during the workshops:

(1) What are the requirements for the coronal magnetic field structure in the vicinity of the energization source?

In the previous section we have shown a great deal of evidence suggesting that flares and strong particle acceleration do not generally occur in isolated magnetic structures (like an isolated flaring loop). Such evidence has been collected independently from soft and hard X-ray imaging observations, microwave imaging observations, and meter- wave one-dimensional imaging and decimetric observations. Simultaneous microwave/meter, microwave/X-ray and meter/X-ray observations have given support to the idea that during the impulsive phase several discrete injection/acceleration regions are present, connecting both open and closed field lines, the former associated in many cases with very divergent magnetic field lines. It is, of course, difficult to generalize the "small" sample of results presented in this section but we feel confident that in several cases (involving strong acceleration) the acceleration region must comprise a rather large volume encompassing regions of different topologies, as suggested schematically in Figure 2.2.24a and 2.2.24b. Such schematic models have been proposed earlier; however, the new wealth of space- and ground-based data obtained during the past solar maximum provide strong observational support to such models.

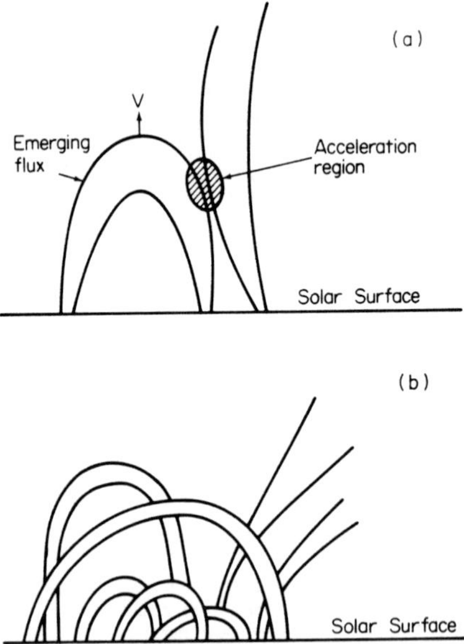

Figure 2.2.24 (a) The emerging flux model, (b) A catastrophic interaction of thousands of reconnecting layers.

(2) What is the height (above the photosphere) of the energization source?

A number of pieces of evidence in the past have placed the energization source in the low corona (microwave and decimetric burst observations). To this set we would like to add the observation on the starting frequency of type III burst and their correlation with hard X-ray bursts. We now believe that the acceleration source is in the low corona where the plasma density varies between 10^9 and 10^{10} cm^{-3}. The acceleration may start at lower densities and "drift" to higher densities with a variable speed or it is stationary in the low corona and the region where the beam becomes unstable to plasma waves "drifts" towards higher densities with time.

(3) Does the energization start before and continue after the impulsive phase?

We have presented evidence indicating that both heating and acceleration have signatures before and after the impulsive flare. This is contrary to the well-accepted scenario that slow heating starts before the impulsive phase, followed by intense acceleration during the flare and it ends up with a hot plasma that gradually cools off.

(4) Is there a transition between coronal heating and flares? What are the microflares?

High sensitivity hard X-ray detectors have dispelled the myth that the corona operates in two modes "heating" and "flaring". We have presented evidence suggesting that microflares may be occuring all the time in the corona. In other words, the transition from "flaring" to "heating" may be more gradual than commonly perceived and depends strongly on the sensitivity of available instruments. The presence of nonthermal tails at all times and microflares may be crucial requirements for the "coronal heating mechanisms".

(5) Is there evidence for a purely thermal, purely nonthermal or a hybrid type of flare?

This is an open question and may have an "energy-dependent" answer. Usually evidence for "purely thermal plasma" is provided by soft and lower energy hard X-ray bursts. However, gamma-rays and type III, IV and V bursts are not considered to be produced from a "purely thermal plasma". At the other extreme, a "purely nonthermal flare" is also a myth. We have presented much evidence indicating that a "hot component" is always present in flares. Indeed, we have emphasized that accelerated electrons can quickly "thermalize" and turn into a "hot plasma". In summary we feel that a hybrid model is the best resolution to this dilemma and as we shall see later theoretically it is the easiest to explain.

(6) What are the time characteristics of the energization source?

There is strong evidence that the time profiles of flares at different wavelengths sometimes show sub-second or even milli-second pulses. In several cases the pulses repeat at regular or quasi-regular intervals. The brightness temperature for each of these pulses is sometimes so high that a coherent emission mechanism must be invoked. Delays between microwave and hard X-ray pulses have also been reported. We believe that these fast pulses are evidence of "micro-injection" similar to the ones discussed earlier and a "flare" is composed of many micro-releases of energy. The understanding of such fast pulsations is still relatively poor.

(7) Is there any observational evidence for a two step acceleration mechanism?

A few key observations have guided our past thinking on particle acceleration in flares. One of them was the event analyzed by Frost and Dennis (1971). In this event, the impulsive phase was followed by a type II burst, which implies the presence of a shock, coinciding with the enhancement of relativistic particles. Thus the conclusion was drawn that during the impulsive phase (or first phase from the point of view of acceleration) mildly relativistic electrons were accelerated. This phase was followed several minutes later by a second phase which coincided with the formation of a shock that further accelerated ions and relativistic electrons. During the SMM workshops no evidence was presented for such delays (of the order of tens of minutes) between the acceleration of mildly relativistic and relativistic electrons and ions. The delays between pulses in different energy channels are of the order of seconds (10-50 secs). Thus, we must refer to the two phase acceleration rather as a "two step acceleration" (Bai and Ramaty, 1979) (two acceleration mechanisms operating in close proximity, with one being delayed from the other by 10-50 seconds). A novel suggestion was also made during the workshop, namely that we must search for one acceleration mechanism for particles of all energies and one possibly for heating. Such a mechanism must result in no delays for the acceleration of particles to higher energies. But then the question may be asked: How does one create delays out of a synchronous acceleration mechanism? The answer is by using a trapping and precipitation model. The debate between these two approaches was not resolved during the workshops and the arguments are presented in Section 2.2.6.

2.3 PHENOMENA ASSOCIATED WITH IONS AND RELATIVISTIC ELECTRONS IN SOLAR FLARES

Evidence for the acceleration of ions and relativistic electrons in solar flares is obtained primarily from gamma-ray line and continuum emissions and from neutron and charged-particle observations. Gamma-ray lines and neutrons result from nuclear interactions of accelerated protons and heavier ions with the ambient solar atmosphere, while gamma-ray continuum is due to electron bremsstrahlung and the superposition of broad and unresolved narrow gamma-ray lines.

In this section we present the gamma-ray and neutron observations and their implications and discuss the charged particle observations. We also examine the relationship between the acceleration of ions and other flare phenomena.

2.3.1 Gamma-Ray Observations

Gamma-ray lines and continuum have been observed from many flares. The first observations, carried out by detectors on OSO-7 (Chupp et al., 1973), were followed by observations on HEAO-1 (Hudson et al., 1980), HEAO-3 (Prince et al., 1982), SMM (Chupp et al., 1981) and HINOTORI (Yoshimori et al., 1983). The gamma-ray spectrometer (GRS) on SMM, in particular, has provided a broad base of data (e.g., Chupp 1984) which forms the basis of much of the discussion in this Section. In addition, the hard X-ray burst spectrometer (HXRBS) on SMM has provided important data regarding the temporal and spectral behavior of the X-ray continuum below ~ 0.3 MeV. We consider the spectra of the observed gamma rays, the timing of the fluxes in the various photon energy bands and the correlation of the gamma-ray data with other flare manifestations.

2.3.1.1 Gamma-ray Spectra

An example of a gamma-ray spectrum, observed by GRS from the April 27, 1981 limb flare, is given in Figure 2.3.1. Here the distribution of the net detector counts (the difference between source and background counts) is shown as a function of photon energy deposited in the detector, for energies > 0.27 MeV, the GRS detection threshold. As can be seen, this spectrum is a superposition of continuum emission, most likely due to electron bremsstrahlung, and narrow and broad lines resulting from ion interactions. The narrow lines are due to proton and alpha-particle

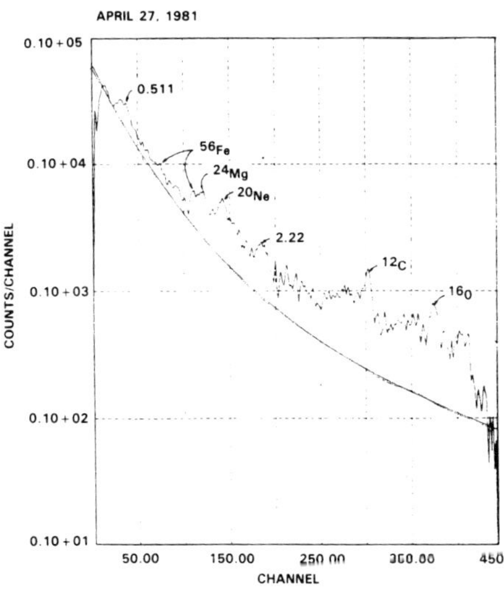

Figure 2.3.1 Observed GRS count spectrum of the April 27, 1981 flare. The solid curve is an estimate of the contribution of electron bremsstrahlung.

interactions with the ambient medium, while the broad lines are from the interactions of accelerated heavy particles with ambient H and He. As indicated, the strongest narrow lines are at 6.13 MeV from ^{16}O, at 4.44 MeV from ^{12}C, at 2.31 MeV from ^{14}N, at 2.223 MeV from neutron capture on hydrogen, at 1.634 MeV from ^{20}Ne, at 1.37 MeV from ^{24}Mg, at 0.85 MeV from ^{56}Fe and at 0.51 MeV from positron annihilation. The 2.223 MeV line, normally very strong for disk flares, is greatly suppressed in limb flares (Wang and Ramaty, 1974). Theoretical nuclear gamma-ray line spectra were calculated earlier by Ramaty, Kozlovsky and Lingenfelter (1979).

Because the contribution of the nuclear lines to the total emission below ~ 1 MeV is quite small, this component can be separated from the bremsstrahlung by fitting a power-law photon spectrum to the data below 1 MeV and then subtracting this power law from the data at higher energies. However, this technique can only approximate the nuclear contribution, since forms other than power laws could also be fit to the data < 1 MeV. Nevertheless, the fact that the excess radiation can indeed be attributed to nuclear lines is supported by the structure of the spectrum, which shows peaks at anticipated line energies, and by the vanishing of the excess above ~ 7.5 MeV, a feature characteristic of a spectrum dominated by nuclear lines. Also, the good correlation (Chupp 1982) of the excess fluence (time-integrated flux) in the 4-8 MeV band with the 2.223 MeV line fluence for several flares provides additional support for the nuclear origin of the 4-8 MeV excess. The 2.223 MeV line from solar flares is a signature of neutrons produced in nuclear reactions of flare-accelerated ions. That the gamma-ray emission from solar flares in the 4-8 MeV region is predominantly nuclear was first pointed out by Ramaty, Kozlovsky and Suri (1977) and Ibragimov and Kocharov (1977).

2.3.1.2 Time Dependences and Correlations with Other Flare Phenomena

Observations of the time dependences of the gamma-ray fluxes from solar flares provide a great deal of information on the acceleration and interaction of the energetic particles. These time dependences are determined by the temporal structure of the acceleration process, by the lag, due to propagation and trapping, between the acceleration and the interaction of the particles, and by the delay between the interaction of the particles and the emission of photons. Significant delays are caused by the finite capture time of the neutrons in the photosphere for the 2.223 MeV line (Wang and Ramaty, 1974) and both by the finite lifetimes of the various positron-emitting nuclei and the slowing-down times of the positrons for the 0.511 MeV line (e.g., Ramaty et al., 1983a). However, for the June 21, 1980 flare, the time profile of the 0.511 MeV line was analyzed in detail (Murphy and Ramaty, 1985), and it was found to depend predominantly on the delayed decay of the positron emitters. This implies a very short (< 10 sec) slowing-down and annihilation time.

On the other hand, bremsstrahlung and most nuclear line emissions are produced essentially instantaneously at the time of the interaction of the particles and therefore serve as the best tracers of the time dependences of the acceleration and interaction processes. Timing studies based on these radiations define the total duration of particle interaction in flares, as well as the overall temporal structure of the emission. But of particular interest is the temporal relationship between the fluxes in the various energy channels, as these data provide information on the relationship between ion and electron acceleration, and possibly on the existence of multiple acceleration steps.

The GRS gamma-ray observations > 0.3 MeV indicate a range of total flare durations from ~ 10 sec to over 1000 sec (e.g., Figure 2.3.2). The total emission in the majority of these events consists of at least a few emission pulses, each of which can be followed over a wide energy band. These separate emission pulses can be as short as ~ 10 sec and as long as ~ 100 sec and their duration within a given flare is roughly proportional to the total event duration. In a preliminary study (Gardner et al., 1981) of the separate emission pulses in several GRS events, it was found tha the time of flux maximum in the 4.1-6.4 MeV band occurred between $0^{\pm}1$ sec and 45 sec later than the corresponding maximum of the ~ 0.3 MeV flux, the delay being proportional to the emission pulse rise time. In addition to these, most gamma-ray line flares also show delays between the various gamma-ray and hard X-ray bands (Bai and Dennis, 1985). But it

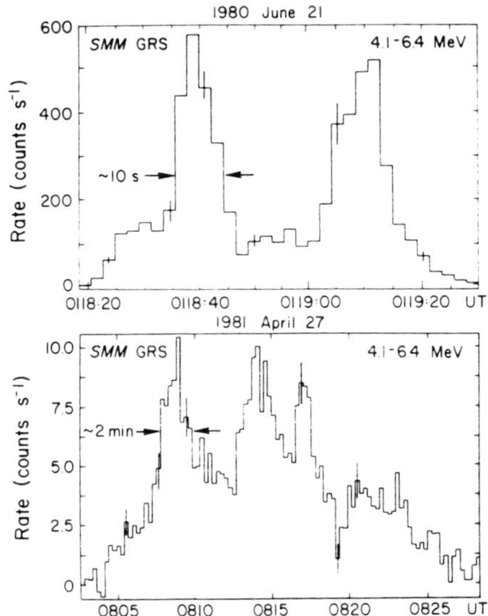

Figure 2.3.2 The observed time histories in an energy band dominated by gamma-ray lines from a fast gamma-ray flare (top panel) and a slow gamma-ray flare (bottom panel). From Forrest (1983).

is important to note that there are cases where no delays are detected. An example is shown in Figure 2.3.3, where the maxima in the various energy bands from 0.04 MeV to 25 MeV in several emission pulses are simultaneous to within the GRS instrumental resolution of ± 1 sec.

Another aspect of the timing studies is the relationship between the starting times of the fluxes in different energy channels. Forrest and Chupp (1983) have studied this relationship for the 40-65 keV flux and the 4.1-6.4 MeV flux in two impulsive flares. They found that the starting time, defined as the time when flux above background was first detected, was the same in each energy band within of ± 2 sec for the smaller flare and ± 0.8 sec for the larger flare, in spite of the fact that these two flares show evidence for a delay in the maxima of the fluxes of the same two energy bands.

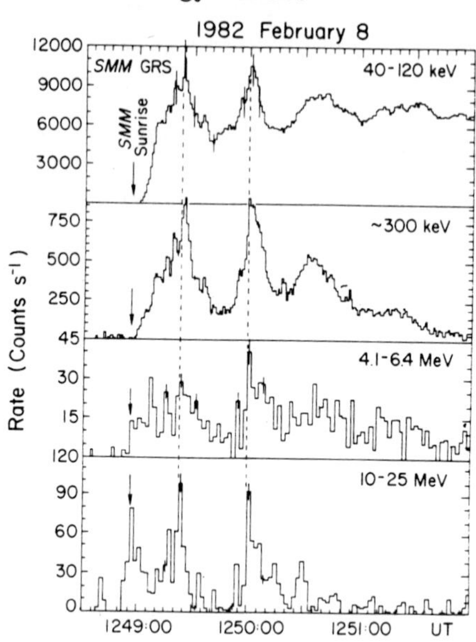

Figure 2.3.3 The observed time histories in 4 energy bands for a gamma-ray flare. From Chupp (1984).

In addition to the timing studies, information on the relationship between ion and relativistic electron acceleration can be obtained by comparing the bremsstrahlung with the nuclear gamma-ray emissions from many flares. In Figure 2.3.4, the ordinate gives the bremsstrahlung fluence > 27 MeV, found from a power-law fit for each solar flare event (see 2.3.1.1), and the abscissa gives the corresponding nuclear fluence above the power law for the energy range 4-8 MeV. As can be seen, for 4-8 MeV nuclear fluences greater than the GRS sensitivity threshold, relativistic electron bremsstrahlung is always accompanied by nuclear gamma-ray emission.

Gamma-ray emission is seen from flares of many different types, suggesting that ion acceleration could be a rather basic process. The first 2 ½ years of GRS data already show that (1) a gamma-ray event may be associated with any Hα class, (2) 20% (10 out of 50) of flares of class $\gtrsim$ 2B have associated 4-8 MeV excess (Cliver et al., 1982), (3) 75% of all GRS events have associated Hα class B (brilliant) emission, (4) 50% (13 out of 26) of GOES A X-ray events with peak intensity $\gtrsim$ X2 have significant 4-8 MeV excess, (5) GRS events are always associated with a solar microwave burst ($\gtrsim$ 1 GHz) and (6) 53% (19 out of 36) of 9 GHz bursts with peak flux density $\gtrsim$ 1200 solar flux units had significant 4-8 MeV excess (Cliver et al., 1983).

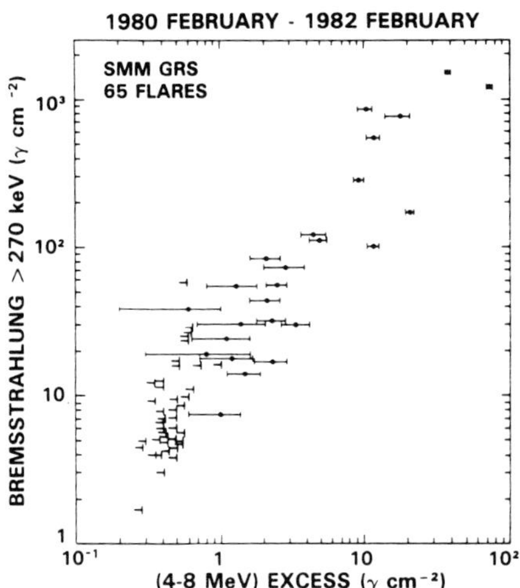

Figure 2.3.4 A correlation plot of the electron bremsstrahlung and nuclear line emission from 65 flares. From Forrest (1983).

2.3.2 Neutron Observations

Neutrons produced in solar flares have been observed directly by the GRS experiment (Chupp et al., 1982, 1983) and by neutron monitors on the ground (Debrunner et al., 1983). Neutron production in flares has also been inferred from observation of the 2.223 MeV line (e.g., Prince et al., 1983), and from the detection (Evenson et al., 1983) of prompt interplanetary protons resulting from the decay of neutrons produced in flares. The interplanetary proton flux (from the June 3, 1982 flare) due to neutron decay is shown in Figure 2.3.5. As we shall see, these data provide very important complementary information to that obtained from the direct neutron and the 2.223 MeV line observations. The prompt appearance of solar neutrons at Earth, indicated both by the ground based and SMM observations, require the prompt acceleration of high-energy protons in the flare. Protons should be accelerated to hundreds of MeV in less than 1 minute.

2.3.3 Implications of the Gamma-Ray and Neutron Observations

Gamma-ray and neutron observations of solar flares can provide information on the spectrum of the accelerated particles, on the total number and energy content in these particles, on the electron-to-proton ratio as a function of energy, on the anisotropy and interaction site of the accelerated particles, and, possibly, on the composition of the accelerated particles and the ambient medium (Ramaty et al., 1983a; Ramaty, 1985). To obtain this information, the observations must be compared (e.g., Murphy and Ramaty, 1985) with theoretical calculations which evaluate the expected gamma-ray and neutron emissions using the basic nuclear processes (Ramaty, Kozlovsky and Lingenfelter, 1979), particle interaction models (Ramaty, Kozlovsky and Lingenfelter, 1975; Zweibel and Haber, 1983) and various energetic particle spectra. The spectra that produce the best fits can then be compared to the predictions of particle acceleration models (e.g., Pesses, 1983; Lee and Ryan, 1985; Forman, Ramaty and Zweibel, 1985 and Section 2.4). The acceleration mechanisms that have been considered so far are the stochastic Fermi mechanism and shock acceleration. The former predicts (Ramaty, 1979) an ion spectrum, which, in the nonrelativistic range, is a Bessel function of the second kind, provided that the diffusion mean-free-path, λ, and the escape time from the acceleration region, T, are energy independent. The parameter that characterizes this Bessel function spectrum is $\alpha T = V^2T/\lambda c$, where V is the velocity of the scattering elements. Shock acceleration can predict many spectral forms; the simplest of these, however, is a power law in momentum, resulting from diffusive acceleration by an infinite planar shock with no losses.

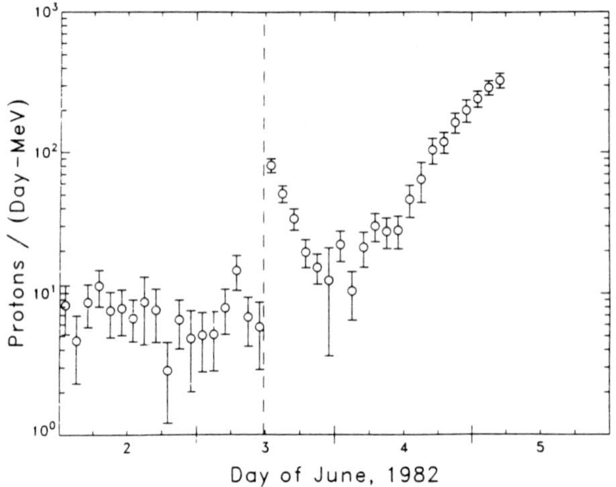

Figure 2.3.5 The flux of 25-45 MeV protons observed at ISEE-3. The gamma-ray arrival time is indicated by the dashed line. From Evenson *et al.* (1983).

The most appropriate interaction model for gamma-ray and neutron production in flares is the thick-target model in which the particles stop in the interaction region; the arguments that favor this model have been summarized recently (Murphy and Ramaty, 1985). The thick-target results presented here are based on the assumption that the angular distribution of the energetic particles in the interaction region is isotropic. The calculation of the yield of nuclear-line emission is essentially independent of this assumption because this emission is nearly isotropic with respect to the direction of the fast particles, but the calculation of the bremsstrahlung and high-energy neutron yields are quite sensitive to the angular distribution of the particles. Calculations that take into account anisotropy of the particles have not yet been published.

2.3.3.1 Energy Spectra of the Accelerated Particles

Gamma-ray and neutron observations can be used to test the validity of the functional forms of the energetic-particle spectra predicted by acceleration theories, as well as to set constraints on the values of the parameters that characterize these spectra. The relevant observations are (1) the ratio of the 4-8 MeV nuclear gamma-ray fluence to the 2.223 MeV line fluence, which provides a measure of the ion spectrum in the 10 to 100 MeV/nucleon range, and (2) the energy spectrum of neutrons released into interplanetary space, which provides information on the ion spectrum in the 100 to 1000 MeV/nucleon region. Technique (1) can only be used for disk flares, because for limb flares the 2.223 MeV line is strongly attenuated by Compton scattering in the photosphere (Wang and Ramaty, 1974). Values of αT, the parameter that characterizes the Bessel-function spectrum, and s, the spectral index of the power law, obtained (Murphy and Ramaty, 1985) by applying technique (1) to 8 disk flares are listed in Table 2.3.1 (events 1 through 8).

The value of αT for the June 21, 1980 limb flare (event 9 in Table 2.3.1) was derived from neutron observations. In Figure 2.3.6, the time-dependent neutron flux observed from the June 21, 1980 flare

Table 2.3.1 Energetic Particle Parameters (from Murphy and Ramaty 1985 except as noted)

	Bessel Function		Power Law		Interplanetary Observations	
Flare	αT	$N_p(>30MeV)$	S	$N_p(>30MeV)$	Spectral Index	$N_{p,esc}(>30MeV)$
1. Aug. 4, 1972	0.029 ± 0.004	1.0×10^{33}	3.3 ± 0.2	7.2×10^{32}	—	4.3×10^{34}
2. July 11, 1978	0.032	1.6×10^{33}	3.1	1.3×10^{33}	—	—
3. Nov. 9, 1979	0.018 ± 0.003	3.6×10^{32}	3.7 ± 0.2	2.6×10^{32}	—	—
4. June 7, 1980	0.021 ± 0.003	9.3×10^{31}	3.5 ± 0.2	6.6×10^{31}	$\alpha T \simeq 0.015$	8×10^{29}
5. July 1, 1980	0.025 ± 0.006	2.8×10^{31}	3.4 ± 0.2	1.9×10^{31}	—	$< 4 \times 10^{28}$
6. Nov. 6, 1980	0.025 ± 0.003	1.3×10^{32}	3.3 ± 0.2	1.0×10^{32}	—	3×10^{29}
7. April 10, 1981	0.019 ± 0.003	1.4×10^{32}	3.6 ± 0.2	1.0×10^{32}	—	—
8. June 3, 1982	0.034 ± 0.005	2.9×10^{33}	3.1 ± 0.1	2.2×10^{33}	$s \simeq 1.7$	3.6×10^{32}
9. June 21, 1980	0.025	7.2×10^{32}	—	—	$\alpha T \simeq 0.025$	1.5×10^{31}
*10. Dec. 9, 1981	—	$< 2 \times 10^{31}$	—	—	—	1.0×10^{32}

*from Cliver *et al.* (1983)

(Forrest, 1983) is compared with calculated neutron fluxes resulting from both Bessel-function spectra and power laws, normalized to the observed 4-7 MeV fluence. As can be seen, a Bessel-function spectrum, with αT = 0.025, can simultaneously fit both the shape and absolute normalization of the observed neutron flux; on the other hand, a power-law spectrum cannot provide such a simultaneous fit for any value of s. In Section 2.3.4.1 we compare the αT's derived from gamma-ray and neutron observations with the αT's obtained by fitting Bessel functions to the spectra observed in interplanetary space.

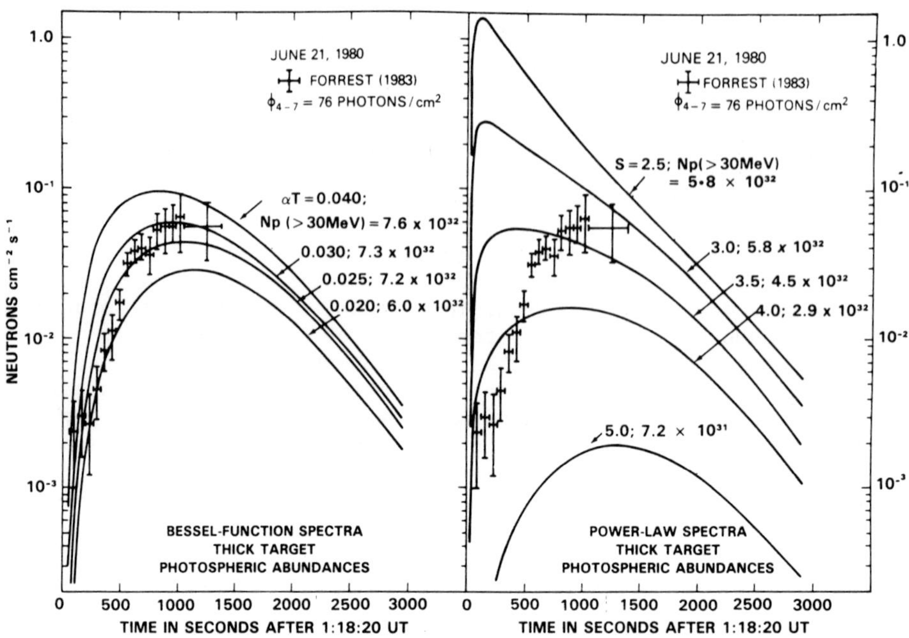

Figure 2.3.6 Calculated neutron time profiles for the June 21, 1980 flare and their comparison with observations. From Murphy and Ramaty (1985).

The only other flare for which published neutron time profiles are available is the June 3, 1982 flare (Chupp et al., 1983; Debrunner et al., 1983; Evenson et al., 1983). An analysis similar to that for the June 21, 1980 flare showed (Murphy and Ramaty, 1985) that a neutron time profile resulting from a Bessel-function proton spectrum again provided an acceptable fit to the data. The resultant value of $\alpha T \approx 0.04$ is in good agreement with that determined independently from the 4-7 MeV to 2.223 MeV flux ratio (see Table 2.3.1). This is the only flare for which the two techniques have so far been used simultaneously.

2.3.3.2 Total Particle Numbers and Energy Contents

In addition to setting constraints on energy spectra, the gamma-ray line and neutron observations also determine the total number of particles and the energy content in them, at least for particle energies above the gamma-ray production thresholds (generally a few MeV/nucleon). Total numbers of accelerated protons above 30 MeV, derived for the 9 flares discussed above, are also shown in Table 2.3.1. Also shown in this table is an upper limit on the number of protons that interact at the Sun for the December 9, 1981 flare, derived from the measured upper limit on the 2.223 MeV line fluence (Cliver et al., 1983) and assuming that αT = 0.025. The comparison of these numbers with the numbers of particles observed in interplanetary space is discussed in Section 2.3.4.2. Because of the chosen normalization, 30 MeV/nucleon, the proton numbers do not depend much on the assumed functional form of the spectrum, as can be seen in Table 2.3.1. On the other hand, the total energy contents do depend critically on the shape of the spectrum at low energies. Values of the energy content in ions of energies greater than 1 MeV/nucleon were derived for several flares using Bessel-function spectra (Ramaty, 1985): these energy contents range from ~ 5 $\times 10^{28}$ ergs for the July 1, 1980 flare to ~ 2 $\times$ 10^{30} ergs for the August 4, 1972 flare.

2.3.3.3 The Electron-to-Proton Ratio

Having described the derivation of the spectrum and normalization of the protons which produce the gamma-ray lines and neutrons, we proceed now to describe a similar derivation for the relativistic electrons which produce the gamma-ray continuum by bremsstrahlung. The relationship between an isotropic power-law electron distribution in the thick-target model and the resultant photon fluence was given previously (Ramaty and Murphy, 1984), but, as already mentioned, no calculations have yet been published for anisotropic distributions.

As discussed above, the gamma-ray continuum at energies below ~ 1 MeV is primarily electron bremsstrahlung, at energies between 1 and 8 MeV it is a superposition of bremsstrahlung and broad and unresolved narrow nuclear lines, while at higher energies, in addition to bremsstrahlung from primary electrons there could be a contribution from photons from π^0 decay and from bremsstrahlung by electrons and positrons from charged π decay. The gamma-ray continuum above 10 MeV from the June 21, 1980 flare should be mainly electron bremsstrahlung, since the proton spectrum derived from the neutron observations is too steep to yield many π mesons. The electron spectrum incident on the thick-target interaction region for this flare, deduced from the observed (Chupp, 1982) gamma-ray spectrum between 0.27 and 1 MeV and the integral fluence above 10 MeV (Rieger et al., 1983) can be approximated (Ramaty and Murphy, 1984) by a power law with spectral index ~ 3.5. This spectrum is shown in Figure 2.3.7, together with the proton spectrum obtained from the neutron and nuclear gamma-ray observations. In Section 2.3.4.3 we compare the e/p ratio implied by these results with that observed in interplanetary space.

The proton spectrum deduced for the June 3, 1982 flare ($\alpha T \approx 0.04$) implies a larger π-meson production relative to electron bremsstrahlung than for the June 21, 1980 flare. Consequently, the observed > 10 MeV fluence from the June 3 flare could contain an important contribution from π-meson decay. For more detail, see Murphy and Ramaty (1985).

2.3.3.4 Anisotropy and the Interaction Site of the Particles

Gamma-ray and neutron observations offer the opportunity to study the anisotropy of the accelerated particles. An important result concerns the neutrons that escape from the Sun. It was pointed out (Ramaty et al., 1983b) that the fact that such neutrons were seen from a limb flare (Chupp et al., 1982) indicates that the primary protons could not have penetrated to very large depths in the photosphere. The observed neutrons have energies of several hundred MeV and therefore

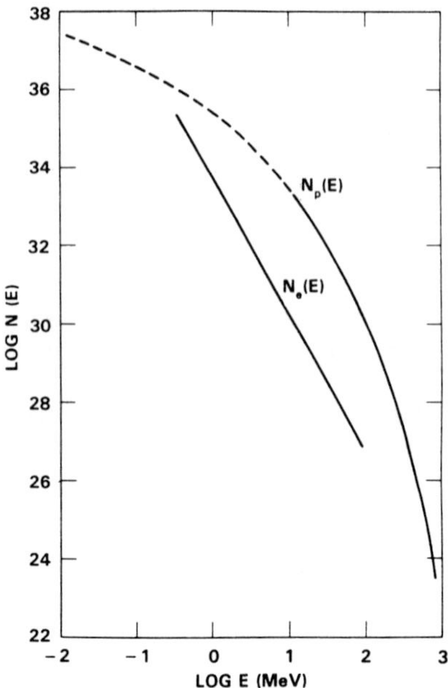

Figure 2.3.7 Calculated spectra and normalizations of the total numbers of protons and electrons accelerated in the June 21, 1980 flare. From Ramaty and Murphy (1984).

must be produced by protons of at least such energies. Such protons, should they travel in straight lines, would penetrate to great photospheric depths, on the order of a few tens of g/cm². Neutrons produced at such depths in limb flares could not escape from the photosphere in the direction of the Earth. The implication is that the primary protons must be stopped at much shallower depths, probably at the top of the photosphere. The stopping mechanism could be magnetic mirroring (Zweibel and Haber, 1983) or scattering off magnetic irregularities at the foot points of loops. These effects could also isotropize the particles. For the June 3, 1982 flare, Murphy and Ramaty (1985) find that the directional neutron flux at $50° \lesssim \theta \lesssim 90°$, obtained from interplanetary observations of protons from neutron decay, is essentially the same as that at $90° < \theta < 180°$, obtained from the 2.223 MeV line observations. Here θ is the angle between the normal to the photosphere and the direction of observation. This result seems to imply that the flux of protons up to energies of ~ 100 MeV in the interaction region cannot be very anisotropic.

On the other hand, Vestrand et al. (1985) found that flares with gamma-ray continuum > 0.27 MeV exhibit a center-to-limb variation, which would be consistent with the downward beaming of relativistic electrons. In addition, Rieger et al. (1983) have found that all GRS events with significant emission > 10 MeV are from flares near the solar limb. This observation implies the anisotropy of either the ultra-relativistic electrons, if the > 10 MeV emission is bremsstrahlung, or the very high energy protons, if this emission results from π mesons.

The gamma-ray line observations can, in principle, also provide information on the beaming of the ions at the interaction site through the widths of the lines and through Doppler shifts. The observational information on this issue, however, is not clear, primarily because detailed observations with high-resolution detectors have not yet been carried out. The basic theoretical ideas have been described by Ramaty and Crannell (1976) and by Kozlovsky and Ramaty (1977).

The gamma-ray observations suggest that the ambient density in the interaction site of the energetic particles should exceed a few times 10^{11} cm^{-3} and therefore this site should be the chromosphere. This result is based primarily on the observed (Share et al., 1983) time profiles of the 0.511 MeV line and their comparison (Ramaty and Murphy, 1984; Murphy and Ramaty, 1985) with theoretical time profiles which take into account the lifetimes of the various positron-emitting nuclei and the slowing-down and annihilation times of the positrons.

2.3.3.5 Compositions

Gamma-ray line spectra, such as shown in Figure 2.3.1, offer the opportunity for studying relative elemental compositions. The contribution of proton and alpha-particle interactions in the thick-target model is significantly larger than that of the heavier nuclei. Therefore, the structure of the total nuclear spectrum is much more sensitive to the relative abundances of the ambient medium, i.e., the chromosphere, than to those of the energetic particles. Studies in progress (Murphy 1985) indicate that the chromospheric abundances, deduced from the gamma-ray observations, could differ from those of the photosphere in a manner similar to that observed in the energetic particles (see 2.3.4.4).

2.3.4 Interplanetary Charged-Particle Observations

2.3.4.1 Energy Spectra

Interplanetary protons and heavier ions resulting from acceleration in solar flares are observed up to energies of several hundred MeV by instruments on spacecraft (e.g., McGuire and von Rosenvinge, 1985) and up to energies of about 10 GeV by ground based detectors (e.g., Debrunner et al., 1984). Flare-accelerated relativistic electrons are also observed by detectors on spacecrafts up to energies of tens of MeV (Lin et al., 1982; Evenson et al., 1984). The particle energy spectra deduced from these measurements are generally subject to uncertainties introduced by coronal and interplanetary propagation. These uncertainties, however, can be minimized by considering ony particle events that are well-connected magnetically to the detector and by constructing the particle energy spectra at times of maximum intensity at each energy (see McGuire and von Rosenvinge, 1985). This technique was used recently (McGuire and von Rosenvinge, 1985) to analyze the proton and alpha-particle energy spectra from a sample of particle events which show no evidence of interplanetary shock acceleration or multiple flare injection. It was found that many flares have proton spectra which are best fit by Bessel functions (see Section 2.3.3) over a broad energy range (from a few MeV to a few hundred MeV). An example is shown in Figure 2.3.8. The implied values for the whole sample, $\alpha T = 0.025 \pm 0.01$, are in good agreement with those derived for other flares from the gamma-ray and neutron observations (see Section 2.3.3.1). Moreover, for one flare, that of June 21, 1980, the Bessel-function spectrum with $\alpha T = 0.025$, derived from the neutron and gamma-ray observations, provides an acceptable fit to the interplanetary proton spectrum which was also observed (McDonald and Van Hollebeke, 1985) from this flare. These results provide support for the validity of the stochastic Fermi mechanism for ion acceleration in flares. But it should be noted that acceleration by planar shocks with losses (Forman, Ramaty and Zweibel, 1985) or by spherical shocks (Lee and Ryan, 1985, see also Section 2.4) can produce spectra that, in the energy range of interest, are indistinguishable from the Bessel-function spectra and therefore can also fit the observations. Furthermore, the observed alpha-particle spectra are generally steeper than the proton spectra, and this result is inconsistent with the assumptions of constant diffusion mean-free-path and escape time that are made in the simple treatment of stochastic Fermi acceleration (see Section 2.3.3). Stochastic Fermi acceleration with rigidity-dependent diffusion was treated by Barbosa (1979, see also Forman, Ramaty and Zweibel, 1985), but no detailed comparisons of the resultant spectra with data have yet been made.

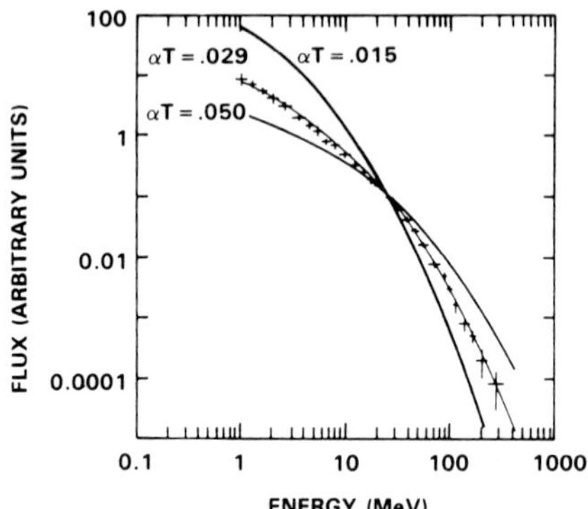

Figure 2.3.8 An example of fitting Bessel functions to the spectrum of the August 21, 1975 flare. From McGuire and von Rosenvinge (1985).

Another spectral form which fits the data of many flares is an exponential in rigidity, but no acceleration mechanism that predicts this form has yet been proposed. Occasionally, the proton spectra show (McGuire and von Rosenvinge, 1985) a complex structure that cannot be fit by any simple form. Such spectra could result from the combination of more than one acceleration process.

Particle acceleration at shock fronts is known to occur at many sites. As already mentioned (see Section 2.3.3) a prediction of acceleration by planar shocks with no losses is that the differential particle number per unit momentum should be a single power law. This implies (Ellison and Ramaty, 1985) that the differential particle flux per unit kinetic energy can be approximated by power laws in the nonrelativistic and ultrarelativistic limits, with the spectral index steepening by a factor of 2 above a kinetic energy equal to the particle rest-mass energy. The observed proton energy spectra for some flares can be fit (McGuire and von Rosenvinge, 1985) by single power laws up to kinetic energies of several hundred MeV, consistent with planar shock acceleration without losses. Moreover, the spectra predicted by such acceleration could also fit the proton spectra that are occasionally observed to extend up to 10 GeV. Recently, solar-flare proton spectra, obtained from the combination of spacecraft and ground-based data (Debrunner et al., 1984), show the characteristic steepening at ~ 1 GeV.

The energy spectra of relativistic interplanetary electrons from solar flares were studied by Lin, Mewaldt and Van Hollebeke (1982) and recently by Evenson et al. (1984). The observed spectra, if fit by power laws in kinetic energy over narrow energy intervals, show great variability from flare to flare. This is illustrated in Figure 2.3.9, where the filled symbols indicate gamma-ray flares. As can be seen, the relativistic electron spectra of such flares are among the hardest observed. For the June 21, 1980 flare, in particular, the electron spectral index, ~ 3.2, obtained from the interplanetary observations, is in good agreement with the index derived from the gamma-ray continuum (see Section 2.3.3.3). Power-law spectra at ultrarelativistic energies

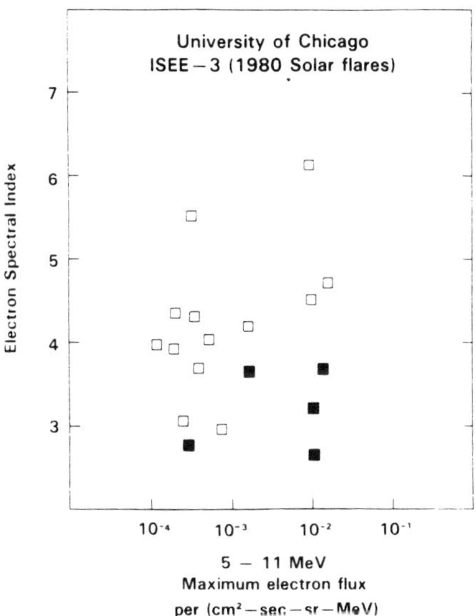

Figure 2.3.9 Spectral indices and fluxes from electron events observed in interplanetary space. The filled symbols indicate gamma-ray flares. From Evenson *et al.* (1984).

are consistent with both stochastic and shock acceleration with no losses.

2.3.4.2 Total Proton Numbers in Interplanetary Space

The number of protons that interact at the Sun and the number that escape into interplanetary space are listed for several flares in Table 2.3.1. The meager data base notwithstanding, these results appear to suggest at least 2 categories of events: (1) events in which the number of interacting ions greatly exceeds the number escaping (events 4, 5, 6 and 9), and (2) events in which the number of escaping ions exceeds the number interacting (events 1 and 10). It remains an unanswered question whether the variability of the ratio between the two particle numbers is caused by a variable escape probability from the Sun or by different acceleration processes. It is interesting to note that for the June 3, 1982 flare, the numbers of escaping and interacting ions differed only by less than an order of magnitude. The complex time dependences of the gamma-ray emissions from this flare (Chupp et al., 1983; Share et al., 1983) might require particle acceleration in at least two phases (Murphy and Ramaty, 1985).

2.3.4.3 The Electron-to-Proton Ratio

The ratio of interplanetary electron fluxes and proton fluxes at the same energy are shown in Figure 2.3.10. As can be seen, the largest values of this highly variable ratio correspond to gamma-ray flares. The time dependence of the fluxes of these electron-rich events are diffusive (see Figure 2.3.11), indicating that strong interplanetary shocks were not associated with these events.

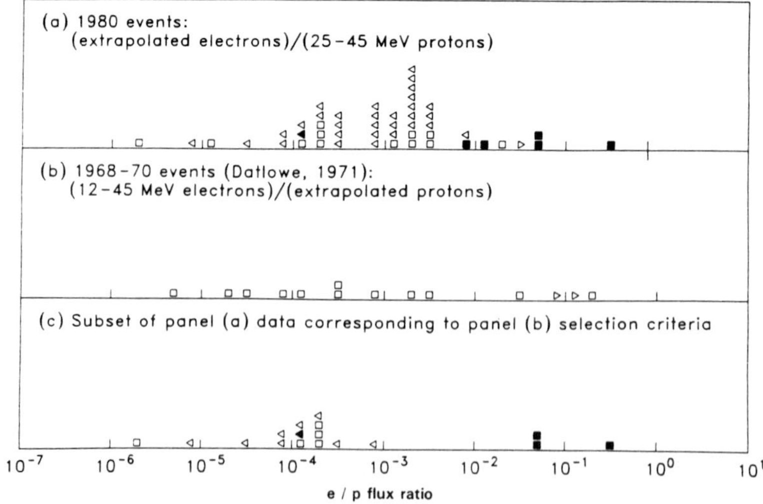

Figure 2.3.10 Electron-to-proton ratios. Square symbols indicate measurements, triangles indicate upper or lower limits and filled symbols indicate gamma-ray flares. From Evenson *et al.* (1984).

For the June 21, 1980 flare, which is an electron-rich as well as a gamma-ray event, the interplanetary electron-to-proton ratio can be compared with the corresponding ratio derived from the gamma-ray and neutron data. The interplanetary ratio at ~ 30 MeV is about 0.05 (Evenson et al., 1984), while the gamma-ray and neutron data (see Figure 2.3.7) imply that this ratio at 30 MeV is smaller by approximately a factor of 100. This difference could result from either the preferential escape of the relativistic electrons, or an anisotropy of the electrons in the gamma-ray production region. In the latter case, the isotropic bremsstrahlung calculation that was used to obtain the results of Figure 2.3.7 would underestimate the electron number.

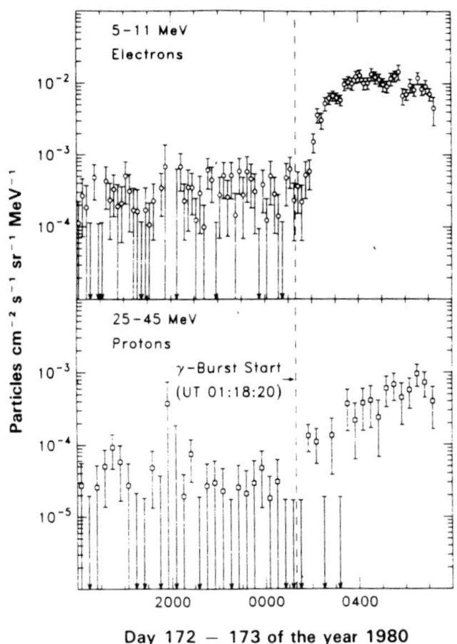

Figure 2.3.11 Interplanetary particle time profiles from the June 21, 1980 gamma-ray and electron-rich flare. From Evenson *et al.* (1984).

2.3.4.4 Compositions

The most recent results on the elemental composition of solar energetic particles were presented by Cook, Stone and Vogt (1984). It was found that, for flares for which the abundances are not a strong function of energy/nucleon, the observed compositions appear to vary about an average which could reflect the composition of the medium from which the particles are accelerated. This composition is different from that of the photosphere, but seems to be in agreement with the composition of the corona and the solar wind. As discussed in Section 2.3.3.5 the gamma-ray observations suggest that the composition of the chromosphere also differs from that of the photosphere. Thus, several independent methods of abundance determinations seem to imply that important compositional modifications occur during transport of matter from the photosphere to the rest of the solar atmosphere.

In addition to elemental composition, interplanetary particle observations also provide information on isotopic compositions (e.g., Mewaldt, Spalding and Stone, 1984) and charge (Gloeckler, 1985) states. We shall not discuss these here in detail, except to mention that the charge-state observations strongly suggest that the solar-flare particles seen in interplanetary space are probably accelerated in the corona. This conclusion is based on the observed charge-state distributions, which seem to agree with a plasma of coronal temperatures, and on the presence of singly-charged He, which implies that the amount of matter traversed during escape is very small. A similar conclusion follows also from the apparent absence of spallation products (^{2}H, ^{3}H, Li, Be, B) from the observed particles. We note, however, that at typical coronal temperatures singly-charged He is not the dominant charge state of He.

2.3.5 The Nature of the Ion and Relativistic Electron Acceleration Mechanisms

A variety of observational evidence (discussed in Section 2.2) has shown that energy release in flares is strongly related to the acceleration of nonrelativistic electrons. This acceleration has been commonly referred to as first-phase acceleration to distinguish it from such phenomena as Type II radio emission which were refered to as second-phase acceleration (Wild, Smerd and Weiss, 1963; De Jager, 1969; Ramaty et al., 1980). Prior to the SMM and other recent gamma-ray observations, it was believed that ion and relativistic electron acceleration occurs only in this second phase. However, the very impulsive nature of the gamma-ray emission and its close temporal association with the standard signatures of first-phase acceleration (e.g., hard X-ray emissions), demonstrate that ion acceleration can also take place in the first phase. This result, together with the fast onset and rapid rise of the gamma-ray fluxes, sets strong constraints on the nature of the acceleration mechanism. For example, it must be possible to accelerate ions and electrons to tens of MeV within a few seconds and ions to hundreds of MeV in less than a minute. Which of the various proposed flare acceleration mechanisms can satisfy these constraints remains an unanswered question.

We already discussed in Section 2.2.6 that first-phase acceleration might proceed in two steps, as proposed by Bai and Ramaty (1979), Bai (1982) and Bai et al. (1983b). The assumption of the two-step model that the ion and relativistic electron accelerations are closely coupled is supported by the data, in particular by the good correlation between the ~ 0.3 MeV bremsstrahlung and the 4-8 MeV nuclear emission (Figure 2.3.4). But there are conflicting points of view regarding the relationship between ion acceleration and nonrelativistic electron acceleration, which tests the basic postulate of the two-step model. The finding that there seems to be a set of characteristics that are common to only gamma-ray line flares (Bai et al., 1983a; Bai and Dennis, 1985) argues for a second acceleration step. These characteristics are: (1) For gamma-ray line flares, the time profiles of higher energy (> 100 keV) X-rays are delayed relative to those of lower energy X rays (~ 50 keV), or, equivalently, the hard X-ray spectrum flattens with time during the burst. In some instances, the time profiles of the 4-8 MeV nuclear gamma rays are also delayed relative to the X-rays (Bai, 1982). (2) The energy spectrum of the hard X-ray flux at the peak of emission is, on average, significantly flatter for gamma-ray line flares than for flares without detectable gamma-ray lines. (3) Gamma-ray line flares seem to be better correlated with Type II and Type IV radio bursts than are flares with no detectable gamma-ray lines.

On the other hand, it is not clear that gamma-ray flares always have these distinguishing characteristics and, furthermore, explanations other than second-step acceleration might exist for at least some of them. As mentioned in Section 2.3.1.2 the peaks of the fluxes in various energy channels can be simultaneous within the detector time resolution, as can be seen in Figure 2.3.3. In addition, the comparison of the starting times of the photon fluxes in the various channels could be more relevant in determining the reality of second-step acceleration than the comparison of the peak times because delays between the peak times could be due to energy-dependent Coulomb losses in a trap model (which was discussed extensively in Section 2.2.6). As noted earlier for two impulsive flares, the starting times of the 4.1-6.4 MeV and the 40-65 keV fluxes were essentially the same, even though these two fluxes show evidence for a delay in their times of maximum (Forrest and Chupp, 1983).

Given that there is substantial ion and relativistic electron acceleration in the first phase, it is natural to ask whether any further particle acceleration occurs in the second phase by shock acceleration in the corona, as originally proposed by Wild, Smerd and Weiss (1963). Support for such a concept comes from the poor correlation between the ion population that interacts at the Sun and that observed in interplanetary space which could be a manifestation of the two acceleration phases. Additional evidence comes from the observations of solar flare heavy ions in interplanetary space, wherein it appears that the source of the ions is a large volume in the corona (Mason et al., 1984). On the other hand, it is possible that this lack of correlation is simply due to the unequal and variable upward and downward escape conditions which are determined by unknown magnetic configurations. Furthermore, the good agreement between the proton spectra deduced from the gamma-ray observations and those observed in interplanetary space suggests that, for at least some flares, these two ion populations are commonly accelerated.

2.3.6 Summary

The opening of a new channel of information on the acceleration of particles in solar flares by routine gamma-ray and neutron observations is one of the major achievements of SMM. We summarized the most important available gamma-ray and neutron data and presented their interpretations in a manner as independent of specific models as possible. We then explored the relationships of these results with other flare phenomena.

As gamma-ray and neutron emissions are signatures of ion and relativistic electron acceleration, we have explored the relationship of these emissions to the other unambiguous manifestation of particle acceleration: the direct detection of accelerated particles in interplanetary space. We have also explored the relationships of the gamma-ray emission with other flare phenomena, in particular hard X-ray emission, in an attempt to resolve the question of multiple-step accelerations.

We return to the questions posed in the introduction:

1. What are the time characteristics of the energization source?

Gamma-ray emission from flares is prompt. Within detector sensitivity, the starting times of the nuclear gamma-ray emission and nonrelativistic bremsstrahlung X-ray emission are the same; but the peaks of the fluxes in the various energy channels can be delayed one relative to the other (Section 2.3.1.2). These results demonstrate that acceleration of ions and relativistic electrons are closely associated with the primary energy release in flares.

2. Does every flare accelerate protons?

The good correlation of the relativistic electron bremsstrahlung fluence > 0.27 MeV and the ion-associated nuclear excess fluence in the energy range 4-8 MeV suggest that, for 4-8 MeV fluences greater than the GRS sensitivity threshold, relativistic electron acceleration is always accompanied by ion acceleration. There also seems to be an overall association of gamma-ray emission with many diverse flare phenomena, suggesting that ion acceleration could be quite common.

3. What is the location of the interaction site of the energetic particles?

The gamma-ray line observations suggest that the interaction site of the ions and relativistic electrons is located at densities $> 10^{11}$ cm^{-3} (i.e., in the chromosphere) and the neutron observations imply that mirroring or scattering should stop the highest-energy ions from penetrating the photosphere (Section 2.3.3.4). The comparison of neutron release into interplanetary space and towards the photosphere is consistent with isotropic neutron production, at least up to 100 MeV (Section 2.3.4). But gamma-ray continuum observations suggest that relativistic electrons are anisotropic (Section 2.3.4).

4. What are the energy spectra for ions and relativistic electrons? Does the spectrum vary from flare to flare?

The energy spectra of the ions that interact at the Sun cannot be fit by single power laws in kinetic energy. Better fits are obtained with more curving spectra such as those resulting from stochastic Fermi acceleration or shock acceleration with losses (Section 2.2.3.1). The spectral index αT varies from flare to flare but over a fairly narrow range: $0.015 \lesssim \alpha T \lesssim 0.040$.

5. What is the relationship between particles at the Sun and in interplanetary space?

The spectra of flare protons observed in interplanetary space are in many cases similar to those of particles interacting at the Sun (Section 2.3.4.1). The range of spectral indices, $0.015 \lesssim \alpha T \lesssim 0.035$, is in reasonable agreement with the range deduced from the gamma-ray measurements, suggesting that, for most flares, a common mechanism could accelerate both particle populations. On the other hand, the numbers of ions inferred from the gamma-ray and interplanetary observations do not agree on a flare by flare basis. This could be due to unequal and variable upward and downward escape probabilities or to two acceleration phases (Section 2.3.5). The hardest relativistic electron spectra and the highest electron-to-proton ratios are observed from gamma-ray flares.

6. Is there any evidence for more that one acceleration mechanism?

The prompt nature of the gamma-ray emission from flares demonstrates that ion and relativistic electron acceleration can occur in the first phase of particle acceleration. A second phase of particle acceleration, due to the passage of a shock wave through the corona, probably also accelerates ions and relativistic electrons, but the gamma-ray signatures of these particles have not yet been identified. The possibility that first-phase acceleration could consist of multiple steps has also been considered, but the evidence for such steps remains equivocal.

2.4 THEORETICAL STUDIES OF PARTICLE ACCELERATION

In the last two Sections we specified the main requirements for the acceleration source. We now turn our attention to the recent development of ideas on bulk energization. It is well accepted today that the transformation of magnetic energy into plasma energy is the main process that provides the observed transient energization in solar flares. We present in this Section several mechanisms that can, under certain conditions, accelerate electrons and ions to high energies.

2.4.1 Particle Acceleration in Reconnecting Magnetic Fields

2.4.1.1 Resistive Tearing Instability

There have been a number of recent attempts (Van Hoven, 1979; Smith, 1980; Heyvaerts, 1981) to estimate the electric fields produced by the resistive tearing instability, but the results disagree. Most of this work concentrates on the dynamic energy-release mechanism (Furth et al., 1963; Coppi et al., 1976), with the aim of simulating the temporal development of the flare impulse. The principal reason for the lack of agreement on the acceleration performance of tearing reconnection is that the induced electric fields critically depend on the small-scale magnetic structure and on the energy-transport history as the instability nears the point of saturation, and such nonlinear behavior is poorly known. The situation (in the simplest case) is as follows: the resistive tearing instability (or reconnection) grows in a sheared magnetic field B(r) in which the field lines locally have the changing orientation of the steps of a spiral staircase. In the vicinity of that step which is perpendicular to the nearest wall (boundary), the magnetic field is decoupled from the plasma so that it is no longer frozen and can spontaneously tear and reconnect (Van Hoven, 1979). The reversal of the direction of the orthogonal components of the nearby field lines and their later merging are shown in Figure 2.4.1. As resistive tearing commences, the electric field changes from its pre-existing low equilibrium value of $\underset{\sim}{E}_o \approx \eta J_o \underset{\sim}{e}_z$, (where $\eta(T)$ is the resistivity and J_o is the current) in two essential ways.

The $\underset{\sim}{B}$ field-aligned z component ηJ_z grows at the reconnection point (Van Hoven, 1979), causing an increasing acceleration of electrons and protons along $\underset{\sim}{e}_z$. E_z can (according to some estimates) reach the Dreicer (1959) value E_D, at which the electric force becomes greater than the collisional drag, and the electrons will be freely accelerated. At the same time, the preexisting stationary plasma begins to move across the magnetic-field direction, as shown by the stippled arrows in Figure 2.4.1. The growing flow velocity v outside the resistive layer produces a rising $\underset{\sim}{E} = - \underset{\sim}{v} \times \underset{\sim}{B}$ electric field across $\underset{\sim}{B}$, which generates an increasing mass-independent drift velocity $(\underset{\sim}{E} \times B)/B^2$ of the plasma inside the resistive layer. Thus, there are two distinct mechanisms, available in a reconnecting field, for accelerating particles. The crucial (and yet incompletely answered) question is how large do these accelerating E fields become as the magnetic tearing proceeds through its energy supply. The answer depends sensitively on the time development of the field structure at the reconnection point, which (in turn) depends on the local resistivity and on the external boundary conditions. In addition, the most interesting limit for large electric fields is the "non-constant-ψ"

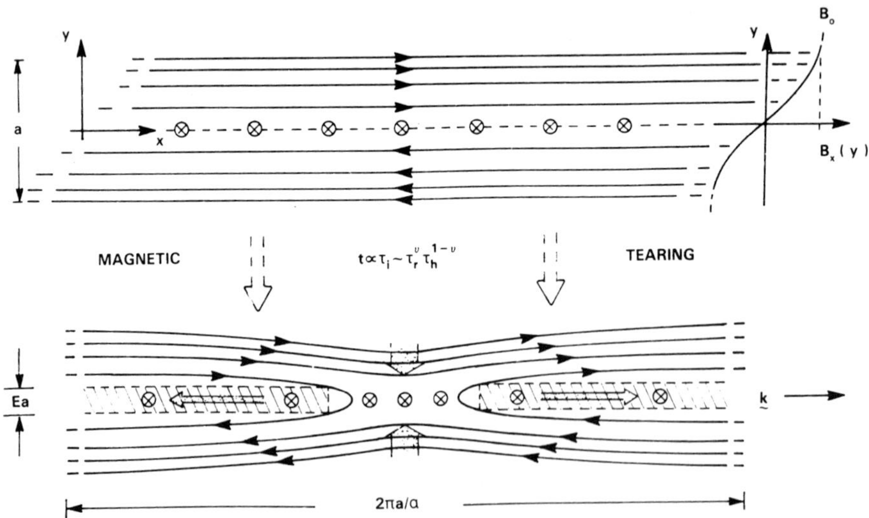

Figure 2.4.1 The development of magnetic tearing. The tail-of-the-arrow symbols show the direction of $\underset{\sim}{J}(x, 0)$ which is antiparallel to $\underset{\sim}{B}(x, 0)$. The stippled arrows show flow velocities.

(long-wavelength, roughly) case (Drake et al., 1978; Steinolfson and Van Hoven, 1983) for which the magnetic gradients are sharper, and the growth is faster and more prolonged. Most nonlinear computations presently available use boundary conditions chosen for mathematical convenience (not for solar applications), constant resistivity (set at an artificially high level to limit the expenditure of computer time) and short wavelengths (same reason). An entirely different scale of calculations, involving a full treatment of the energy transport, is needed to properly follow the variation of the local resistivity $\eta(T)$ as Joule heating, radiation losses and thermal conduction come into play (Van Hoven et al., 1984; Steinolfson and Van Hoven, 1984a). Relevant nonlinear computations of the resistive tearing instability have been performed by Van Hoven and Cross (1973), Schnack and Killeen (1978), Drake et al. (1978), Ugai (1982, 1983), and Steinolfson and Van Hoven (1984b).

The group at the University of California (Irvine) has investigated two aspects of reconnection growth and saturation. Nonlinear computations have been made which specifically evaluate electric fields, and linear calculations have been performed on a reasonably complete energy-transport model. The new nonlinear results (Steinolfson and Van Hoven, 1984b) apply to the canonical single-tearing layer, with periodic boundaries parallel and remote boundaries perpendicular to it, and constant resistivity. However, extreme care is given to the spatial resolution on which the electric fields depend, and on attaining the non-constant-ψ limit at relatively high levels of magnetic Reynolds number. The results have proved to be somewhat disappointing, as far as their potential for a flare model are concerned. Initially non-constant-ψ excitations do indeed grow faster (by an order of magnitude) and farther (by two orders in energy) than constant-ψ modes. However, the growth rate falls (by one-half to one order of magnitude) soon after the nonlinear threshold is crossed (island width ≈ resistive-layer thickness), thereby slowing the energy output. In addition, the outward Fermi-acceleration flow develops a stagnation point, and then reverses. Finally, a new magnetic island grows as the reconnection point bifurcates (see detailed discussion by Priest et al. in Chapter 1). This more complex and wider island structure holds down the parallel electric field to a value (weakly increasing with S up to 10^6) of order 10^{-3} of the Dreicer field (Dreicer, 1959). This study is continuing, with emphasis on the effects of boundary conditions at the ends of the tearing layer, which now cause a back pressure leading to the flow stagnation, and of energy transport which can accelerate the reconnection. Energy-transport effects are the prime focus of Van Hoven's and Steinolfson's second effort in this context. They have treated Joule heating, thermal conduction, and (especially) optically-thin radiation which proves to have the dominant influence in a reconnection layer. At coronal temperatures, the radiative loss function gives rise to thermal instabilities (Field 1965), so that a temperature perturbation spontaneously grows. Van Hoven et al. (1984) have discussed the linear mode structures and growth rates, and have demonstrated the existence of a tearing-like and a faster radiative mode at coronal temperatures. A crucial aspect is that the addition of radiation drives the temperature down at the reconnection point, in contrast to the expected behavior in which the local Ohmic-heating peak would raise the temperature (T_e). The radiative decrease of T_e has the effect of increasing the Coulomb resistivity $\eta \propto T_e^{-3/2}$, with the potential of accelerating the magnetic-tearing rate. Steinolfson and Van Hoven (1984a) have also examined the reconnection behavior of the faster radiative mode. It has a growth time which does not vary with temperature (resistivity) and is 30 times shorter (for coronal conditions) than that of the tearing mode. They have now shown that this mode, at a comparable level, exhibits 30% of the reconnected flux of the slower tearing mode. The radiative instability thus has the potential for providing fast magnetic-energy release. In addition the parallel electric field ηJ_z of the radiative mode is 200 times larger (at low levels) than that in the equivalent tearing mode. Nonlinear computations are needed to demonstrate the validity of these low-level indications.

2.4.1.2 Particle Dynamics Around the Neutral Point

The effects of turbulence in analytical and computational models have been excluded from most previous studies. It was assumed that the background magnetic field is initially smooth and that perturbations are symmetric and/or infinitesimal. There are indications (Matthaeus et al., 1984) that finite-amplitude fluctuations can lead to turbulence in X-point dynamics, thereby increasing reconnection rates. Consequently, turbulent reconnection might maintain strong X-point electric fields at high conductivity. The question of whether a sub-population of particles might be accelerated to high energies by the reconnection-zone electric field has been studied extensively (e.g., Speiser, 1967; Vasyliunas, 1980). A typical approach has been to calculate the trajectories of "test particles" in model MHD magnetic and electric fields. The electric field has usually been included parametrically. Sato et al. (1982) followed test particle orbits in the dynamic electric field of a symmetric, forced and non-turbulent MHD simulation. It was found that test particles do not spend much time near the X-type neutral points, but can easily be trapped near O-type neutral points. This is a crucial point in the assessment of the importance of reconnection as a particle accelerator. In non-steady incompressible simulations, strong accelerating electric fields do not appear near the O-points. A long residence time near the O-point is thus unlikely to produce significant particle acceleration. The short residence times of particles near the X-point region has been viewed as a limitation on the efficiency of the X-point acceleration mechanism. Matthaeus et al. (1984) followed orbits in the fields generated by an incompressible, MHD spectral-method simulation (see e.g., Matthaeus and Montgomery, 1981 and Matthaeus, 1982). The magnetic field configuration is a periodic sheet pinch which undergoes reconnection. Test particles are trapped in the reconnection region for a period of the order of an Alfven transit time in the large electric fields that characterize the turbulent reconnection process at the moderate magnetic Reynolds number ($S = \tau_R/\tau_A = 1000$, where τ_R is the resistive time and τ_A is the Alfven transit time) used in their simulation. They found that a small number of particles gained substantial energy at the end of their simulation.

2.4.1.3 The Coalescence Instability

Tajima (1982) used a fully self-consistent electromagnetic relativistic particle code (Langdon et al., 1976; Lin et al., 1974) to study the coalescence of two current filaments. In the solar atmosphere these filaments may represent two different loops. They showed that during the linear stage the energy release is small (comparable to the Sweet-Parker rate) but this stage is followed by the nonlinear phase of the coalescence instability (Wu et al., 1984; Leboeuf et al., 1982) which increases the reconnection rate by two or three orders of magnitude. An important result from their simulation is that the plasma compressibility leads to an explosive phase of loop coalescence and its overshoot results in amplitude oscillations in temperature (by adiabatic compression and decompression). These oscillations resemble in structure the double sub-peak amplitude profile of the 1980 June 7 and 1982 November 26 microwave emission Tajima et al. (1987) observed with the 17 GHz Nobeyama interferometer. Brunel et al. (1982) found that the ion distributions formed during the coalescence were characterized by intense heating and the presence of a long nonthermal tail. A word of caution about these results is in order here. Although the results seem very encouraging, the plasma parameters used in these simulations are far from those experienced in the solar environment. Thus, we can only use these results to stimulate new thinking and techniques for simulation, specifically oriented to solar physics in the near future.

2.4.1.4 Laboratory "Simulations" of Solar Flares

Another important tool for the exploration of the physical processes that take place during a flare is the use of laboratory experiments that "simulate" the solar environment and study the dynamics of the energy release process. Numerous such experiments have been performed in the past (e.g., Stenzel and Gekelman, 1985; Bratenhal and Baum, 1985). Recently, a group of scientists in U.S.S.R. have reported several major results obtained from laboratory experiments. (1) They detected the formation of current sheets, followed by increased plasma density and magnetic energy in the vicinity of the current sheet region (Somov et al., 1983; Altyntsev et al., 1984). (2) They have observed a sudden excitation of turbulence which was accompanied by fast energy release and particle acceleration, (Altyntsev et al., 1977; Altyntsev et al., 1981a). (3) Finally, during the non-linear stage the field structure is very complex and includes several magnetic islands (Altyntsev et al., 1983). The main result from the studies is that during a flare several fine, spatial and temporal, characteristics of the energy release can be studied in small laboratory experiment.

2.4.2 Electron Acceleration Along the Magnetic Field With Sub-Dreicer Electric Fields

In the simulations reported above the main question is: What is the strength of the electric field during the nonlinear phase of each of the instabilities? Let us now move one step forward and ask the question:

What will happen to the plasma in the presence of a pre-existing electric field? Holman (1985) studied qualitatively the acceleration of runaway electrons out of the thermal plasma, the simultaneous Joule heating of the plasma, and their implications for solar flares. He shows, in agreement with Spicer (1983), that the simple electric field acceleration of electrons is incapable of producing a large enough electron flux to explain the bulk of the observed hard X-ray emission from flares as non-thermal bremsstrahlung. For the bulk of the hard X-ray emission to be nonthermal, at least 10^4 oppositely directed current channels are required, or an acceleration mechanism that does not result in a net current in the acceleration region is required. He also finds, however, that if the bulk of the X-ray emission is thermal, a single current sheet can yield the required heating and acceleration time scales, and the required electron energies for the microwave emission. This is accomplished with an electric field that is much less than the Dreicer field ($E/E_D \sim 0.02$-0.1). To obtain the required heating time scale and electron energies, the resistivity in the current sheet must be much greater than the classical resistivity of a plasma with $n = 10^9$ cm^{-3}, $T_e = 10^7$ K. A plasma density of $\approx 10^{11}$ cm^{-3} is required in the flaring region or, if the density in the current sheet is less than 10^{11} cm^{-3}, the resistivity in the sheet must be anomalous. The identity of the microinstability that will enhance the resistivity is an open question.

Moghaddam-Taaheri et al. (1985) developed a quasi-linear code to study the time develpment of runaway tails in the presence of a D.C. electric field as well as Coulomb collisions. It is well known that particles with velocities larger than the critical velocity $v_{cr} = (E_D/E)^{1/2} v_e$ can overcome the drag force due to collisions and therefore can run away. As the bulk is depleted, the rate of particle flux in the tail decreases and causes the formation of a positive slope on the distribution of the runaway tail. This process sets up a spectrum of plasma waves with phase velocities on top of the runaway tail. The anomalous Doppler shift of these waves makes it possible for them to interact with fast electrons in front of the tail, when they acquire the appropriate parallel velocity to satisfy the resonance condition $k_\| v_\| = \omega_k + \Omega_e$. At this point, electrostatic turbulence can be driven by both Cherenkov and anomalous Doppler effects. Using different values of $E_\|$ Moghaddam-Taaheri et al. showed that for $E_\|/E_D \approx 0.2$ (the Dreicer field for a plasma with density $n_o = 10^{11}$ cm^{-3} and $T_e = 1$ keV is $E_D = 6 \times 10^{-2}$ V/m) or higher, electrons first stream along B and then are isotropized and thermalized. For $E_\| < 0.2\ E_D$ the anomalous Doppler-resonance scattering is weak and the tail is continuously accelerated to higher and higher velocities. They suggested that weak electric fields ($E_\| < 0.2\ E_D$) along a large fraction of the loop would be a better candidate for electron acceleration than the strong localized electric fields.

2.4.3 Lower Hybrid Waves

Tanaka and Papadopoulos (1983) have studied numerically the nonlinear development of cross-field current-driven lower hybrid waves (modified two-stream instability). They showed that for $\beta_e = (nkT_e/(B^2/8\pi)) < 0.3$ and ion drift velocity ≈ 2-$3\ V_i$ symmetric electron tails are formed, with average velocity ≈ 4-5 V_e. Smith (1985) has applied these results to the solar flare. He proposed that fast tearing modes occur in a current carrying loop (Spicer 1982). This leads to dissipation of the poloidal magnetic field and to bulk motion of the ions across the primarily toroidal field up to 0.3 v_A, (Drake, private communication), where the Alfven speed v_A is the speed in the poloidal field B_p. There are no calculations for the energetics of fast tearing, but calculations for slow tearing (Arion, 1984) indicate that 46% of the energy released goes into ion kinetic energy for the collisional tearing mode. It must be admitted that simulations (e.g., Terasawa, 1981) do not show such a large fraction of the energy going into ion kinetic energy; however, no simulations have approached the magnetic Reynolds number regime of 10^{10} - 10^{12} relevant for the Sun. For reasonable parameters (e.g., B_p = 200G, n = 1.5 x 10^{10} cm^{-3}, v_A = 3.5 x 10^8 cm s^{-1}, T_i = 3 x 10^7 K, $v_i^p = (kT_i/m_i)^{1/2}$ = 5 x 10^7 cm s^{-1}), the bulk velocity 0.3 v_A ~ 2-3 v_i, which is sufficient to excite the modified two stream instability. The modified two-stream instability is sensitive to the plasma beta (β) and saturates for levels of β ≥ 0.3. Thus once β increases due to additional heating and/or a density increase due to evaporated material traveling back up the loop, the instability turns off and efficient electron acceleration would no longer occur.

2.4.4 Fermi Acceleration and MHD Turbulence

Fermi or stochastic acceleration of particles in turbulent fields is defined as the process that causes particles to change their energy in a random manner with many increases and decreases that lead finally to stochastic acceleration. Stochastic acceleration can also result from resonant pitch-angle scattering from Alfven waves with wavelengths of the order of the particle gyroradius. A simple way of understanding this mechanism is to imagine the random walk of a particle colliding with randomly moving infinitely-heavy scatterers (see Ramaty, 1979 or Heyvaerts, 1981 for a detailed discussion). The solution of the diffusion equation for the particle distribution (Tverskoi, 1967) can be expressed in terms of modified Bessel functions for $E << mc^2$ (a case applicable to non-relativistic protons) and as an exponential for $E >> mc^2$ (a case relevant to electrons). There are no analytical solutions for energies in between. The model is defined only if the velocity of the particle is equal to or larger than the velocity of the randomly moving scatterers (e.g., Alfven velocity). In other words, this mechanism needs a pre-acceleration or heating mechanism to inject particles above the momentum $p_o = M_i V_A$. A similar conclusion can be reached from the analysis of MHD waves resonantly interacting with the particles. For example, a quasi-linear equation can be written that will connect the spectrum of the unstable waves with the diffusion coefficient and the minimum resonance velocity can be estimated from the resonance condition between the waves and particles. A particle can resonate with an Alfven wave only when $\omega = k_\parallel V_A = k_\parallel V_\parallel + n\Omega^*$ or $|V_\parallel - V_A| = V_A|n\Omega^*/\omega|$ (where Ω^* is the relativistic gyrofrequency). Since the highest turbulent frequency of the spectrum is less than the ion gyrofrequency Ω_i $|V_\parallel - V_A| > V_A|\Omega^*/\Omega_i|$ which implies that particle momentum must be larger than $M_i V_A$ to resonate. For a magnetic field strength $B \cong 500G$ and $n \sim 10^{10}$ cm^{-3} the Alfven speed is ~ 3000 km/s and the threshold velocity for ions is 0.1 MeV. Melrose (1983) has shown that magnetoacoustic turbulence with frequency $\omega \approx 30$ s^{-1} can accelerate ions from a threshold velocity of 0.1 MeV to 30 MeV in 2 secs if $(\delta B/B)^2 \approx 0.1$.

2.4.5 Shock Acceleration

Shock acceleration is currently one of the most intensely studied subjects in the space and astrophysical literature (Drury, 1983). During the workshop several new studies of shock acceleration were presented. We review here the main results.

2.4.5.1 Ion Acceleration

Lee and Ryan (1986) have calculated the temporally and spatially dependent distribution function of particles accelerated by a spherically symmetric coronal blast wave. The motivation here is to introduce an important geometrical effect into the calculation of an evolving acceleration to see if familiar features of accelerated particles can be reproduced. In order to carry out these calculations a few assumptions were made which limit the applicability and interpretation of the results. First, the particles are transported in a diffusive manner, thereby yielding less information about the earliest shock effects. Secondly, the density of the medium decreases as r^{-2} and the collisions are neglected making the calculations most relevant in dealing with high altitude and interplanetary acceleration as opposed to acceleration in the lower corona. Finally, a diffusion coefficient is chosen in a reasonable form but is independent of energy. This has an effect on the spectrum of the prompt arrival of particles at the earth. The three-dimensional geometry yields both the acceleration effects from the shock, the diffusive escape of particles from the finite-size acceleration region and the deceleration effects from the divergent flow behind the shock. The results of the calculations can be summarized as follows. The spectrum of particles accelerated by the shock is soft initially but hardens as time progresses, approaching a power law in the limit. The distribution of particles at a particular energy peaks in time as a function of that energy. The lower energy particles peak in number sooner and subsequently decay away. Higher energy particles take longer to accelerate as one would expect. The time to reach maximum is roughly proportional to the particle energy, although significant numbers of particles at high energies exist initially. The particle distribution peaks at the shock front and decays away in a roughly power law fashion ahead of and behind the shock.

Decker and Vlahos (1984) have studied the role of wave-particle interactions in the shock-drift acceleration mechanism. Between shock crossings the ions are permitted to interact with a pre-defined spectrum of MHD waves [e.g., Alfven (parallel and oblique)] assumed to exist upstream and downstream of the shock. The amplitude and spatial extent of wave activity in the upstream region is varied as a function of angle θ_{Bn} between the shock normal and upstream magnetic field to simulate a decrease in wave activity as θ_{Bn} approaches 90°. For each simulation run, they followed several thousand test ions which were sampled from a specified pre-acceleration distribution and ensemble-averaged results compared for the wave and non-wave situations. The results show that even a moderate level of wave activity can substantially change the results obtained in the absence of waves. This occurs because at a "single encounter" (during which the ion remains within a gyroradius of the shock), the waves perturb the ion's orbit, increase or decrease the number of shock crossings, and therefore increase or decrease the shock-drift energy gain relative to the no wave situation. In particular the presence of waves generally increases both the fraction and average energy of transmitted ions and produces a smaller, but much more energetic population of ions reflected by the shock.

Tsurutani and Lin (1984) have obtained some preliminary results on shock acceleration from a comprehensive study of 37 interplanetary shocks which were observed by the ISEE-3 spacecraft near 1 AU. In this study the normals to the shock surface and the shock speed were determined from plasma and magnetic field parameters, and the energetic particle response was categorized. The main finding was that particle acceleration to $\gtrsim 2$ keV for electrons and $\gtrsim 47$ keV for ions depended on the speed of the shock along the upstream magnetic field and on the ratio of the downstream to upstream magnetic field intensities. These results indicte that magnetostatic reflections of the particles off the shock itself play a very important role in the acceleration process. Figure 2.4.2 shows the 37 shock events plotted versus the shock speed in the upstream medium, and the angle θ_{Bn} between the shock normal and the upstream magnetic field (small θ_{Bn} implies quasi-parallel shock, θ_{Bn} near 90° implies quasi-perpendicular shock). The intensity of the particle acceleration due to each shock is indicated by the symbols: an open circle indicates little or no acceleration or a flux increase of < 20% above ambient pre-shock levels; a filled circle indicates an increase of ≥ 20% but < 200%; and a plus indicates increases of more than 200%. Typical uncertainties in θ_{Bn} are ~ 12°. We see that the open circles group toward the bottom left of the figure while pluses tend to the upper right, with filled circles in between. The lines are drawn for constant shock velocity along the upstream magnetic field, $v_{sB} = v_s/\cos\theta_{Bn}$. The line $v_{sB} = 250$ km/sec then separates those events with little or no particle acceleration from those with signficant particle acceleration. Figure 2.4.3 shows the same events with v_{sB} plotted against the ratio $|B_2|/|B_1|$ of downstream to upstream magnetic field strength. If the shock acts as a moving magnetic mirror,then this mirror will be least effective for small ratios of $|B_2|/|B_1|$ and vice versa. Figure 2.4.3 shows that this is the case: the open circles all have $|B_2|/|B_1| < 1.7$ and $v_{sB} < 250$ km/sec. Events with larger v_{sB} and small $|B_2|/|B_1|$ (< 1.7) only result in moderate events (filled circles), while events with both $|B_2|/|B_1| > 1.7$ and $v_{sB} > 250$ km/sec give rise to large events.

2.4.5.2 Electron Acceleration

Type II bursts, associated with flare-generated shock waves, require Langmuir waves for the production of radio emission at ω_e and at $2\omega_e$.

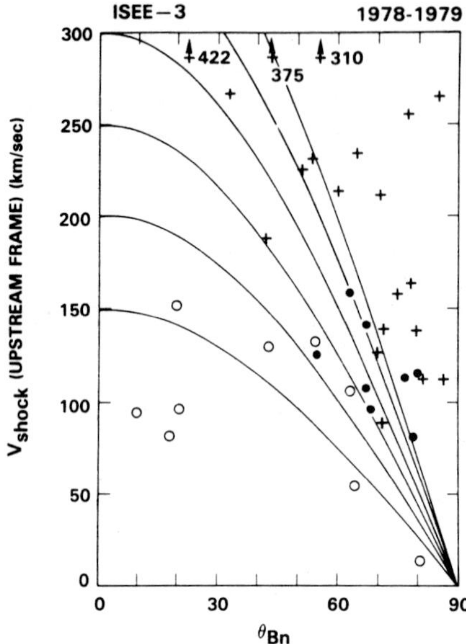

Figure 2.4.2 The shock speed (v_s) in the upstream medium is plotted versus the angle θ_{Bn} between the shock normal and the upstream magnetic field for 37 events observed with ISEE-3 (from Tsurutani and Lin, 1984).

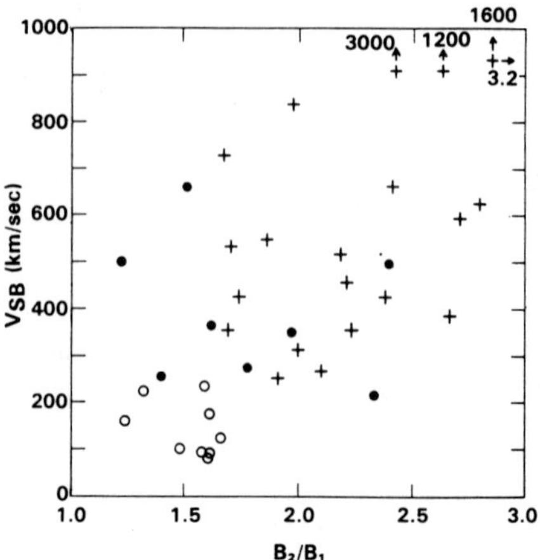

Figure 2.4.3 A scatter plot of the particle flux increases as function of the shock velocity (V_s) relative to the upstream medium (slow solar wind) and the shock normal angle θ_{Bn}. The particle flux increases (over the upstream ambient) are indicated by the type of event point used. Open circles represent events with little or no effects (less than 20% increases), solid circles are moderate events with 20-200% increases; and crosses are large events with > 200% increases. The contours are lines of constant velocity along the upstream magnetic field, $V_{SB} = V_s/\cos\theta_{Bn}$ (from Tsurutani and Lin, 1985).

These plasma waves are most easily generated by streaming suprathermal electrons, as for Type III bursts. Since shock-drift acceleration produces streaming suprathermal electrons upstream of the shock front, Holman and Pesses (1983) have studied the effectiveness of shock-drift acceleration for generating Type II bursts, and the observational consequences. They find that the required level of Langmuir turbulence can be generated with a relatively small number of accelerated electrons ($n_b/n \gtrsim 10^{-6}$) if the angle between the shock normal and the upstream magnetic field (ψ) is greater than 80° (for a 1000 km s^{-1} shock). Except for herringbone structure, which requires electron velocities ~ c/3, the electrons need only be accelerated to velocities that are a few times greater than the mean electron thermal velocity upstream of the shock. Band splitting is explained by the fact that when ψ is within a few degrees of 90°, no electrons are accelerated. The split bands are predicted to arise from different spatial locations (separated by ~ 1') upstream of the shock front. Observations of Type II bursts associated with coronal transients have shown that the emission sometimes arises from locations below and to the sides of the projected font of the white light transient. Since a shock front is expected both ahead of and at the sides of the transient, this indicates that special conditions are required for the generation of the Type II burst. Such a condition appears to be satisfied by shock drift acceleration, since the radio burst is only generated when ψ is large (when the shock is quasi-perpendicular).

Wu (1984) and Leroy and Mangeney (1984) have proposed independently that upstream electrons will be reflected and energized to high energies by the quasi-perpendicular shock. The theory is based on the adiabatic mirror reflection, in the appropriate reference frame, of incident electrons by the increase in magnetic field strength which takes place in the shock-transition region. The average energy per reflected electron scales for sufficiently large θ_{Bn} as $\approx [(4/\cos^2\theta_{Bn})^{1/2} m_e v_o^2]$, where v_o is the shock velocity. Since only particles with large perpendicular velocity will be reflected, this mechanism creates a ring electron distribution upstream of the shock. This distribution is unstable to plasma and maser instabilities and can be the source of Type II bursts. All the geometric effects discussed by Holman and Pesses (1983) are also present in this mechanism. It can be shown that for $\theta_{Bn} \approx 80\text{-}85°$ the flare accelerated electrons can be further energized to 5-10 times their initial energy. Tsurutani and Lin (1984) applied the same ideas to ions (see discussion above).

Electrons drift in the shock transition due to one of the following processes: (a) magnetic field gradient, (b) temperature gradient, or (c) ExB drift. Ions, on the other hand, do not drift if they are unmagnetized (e.g., when the ion gyroradius is larger than the shock transition thickness). When the electron drift velocity exceeds a certain threshold, cross-field current-driven instabilities are excited and are the source of energy dissipation in collisionless shocks. We discussed above that lower hybrid drift (or the modified two-stream) instability can become an efficient acceleration process. Vlahos et al. (1982) applied this mechanism to the acceleration of electrons during loop coronal transient events. Lower hybrid waves excited at the shock front propagate radially toward the center of the loop with phase velocity along the magnetic field which exceeds the thermal velocity. The lower hybrid waves stochastically accelerate the tail of the electron distribution inside the loop. Vlahos et al. discussed how the accelerated electrons are trapped in the moving loop and give a rough estimate of their radiation signature. They found that plasma radiation can explain the power observed in stationary and moving Type IV bursts.

2.4.6 Acceleration of Electrons by Intense Radio Waves

A number of observations have shown the nearly simultaneous release of secondary electrons streaming from the surface of the sun (Type III bursts) and the precipitating electrons associated with hard X-ray bursts (see Kane, Pick and Raoult, 1980; Kane, 1981b; Kane, Benz and Treumann, 1982; Dennis et al., 1984). One possible explanation for the accelerated electrons that are responsible for the Type III bursts is that they result from the primary precipitating electrons which drift out of a flaring loop and get into open field lines. However, this mechanism suffers from a number of difficulties; among them is the fact that the drift rate is exceedingly slow (Vlahos, 1979), contrary to observations. Achterberg and Kuijpers (1984) proposed cross-field diffusion from MHD turbulence, but the time scale for such diffusion is still much longer than the observed temporal correlation between X-rays and Type III bursts.

A recent parallel development has been the discovery of intense (10^{10} - 10^{11} Watts), narrow-band, highly polarized microwave bursts observed by Slottje (1978, 1979) and Zhao and Jin (1982). These observations have been interpreted as the signatures of unstable electron distributions, formed inside a flaring loop, by the flare-released precipitating electrons (Holman, Kundu and Eichler, 1980; Melrose and Dulk, 1982; Sharma, Vlahos and Papadopoulos, 1982; Vlahos, Sharma and Papadopoulos, 1983). These observations together with the lack of a viable explanation for the secondary accelerated electrons have motivated Sprangle and Vlahos (1983) to examine in detail the interaction of a circularly polarized electromagnetic (e.m.) wave propagating along a spatially varying, static, magnetic field as a possible acceleration mechanism responsible for the observed secondary electrons. In this process, the relativistic electron-cyclotron frequency and the wave phase change in such a way that the resonance between the particles and the wave is maintained in a uniform magnetic field (Kolomenskii and Lebedev, 1963; Roberts and Buchsbaum, 1964). For such a process, the intense, polarized, narrow-band e.m. waves observed by Slottje (1978, 1979) provide a link between the energetic electrons inside the flaring loop and those observed in the outer corona or interplanetary space (Type III bursts). An overall schematic of the acceleration process proposed by Sprangle and Vlahos (1983) is shown in Figure 2.4.4. Here the precipitating primary electrons are accelerated inside a flaring loop and stream toward the chromosphere. These precipitating electrons can then excite an intense, polarized, narrow-band e.m. wave. This e.m. wave is assumed to escape the flaring loop region and propagate along a open flux tube where it can accelerate the secondary electrons (see Figure 2.4.4). A question can be raised here about the possibility of reabsorption of the e.m. wave before it can escape from the flaring loop. This is especially important when the excited e.m. wave is near the fundamental electron-cyclotron frequency or upper-hybrid frequency. The main assumption here is that the excited wave approaches the second harmonic resonance with the particles inside the flaring loop where the total electron distribution is not Maxwellian. It is therefore amplified instead of damped (Sharma, Vlahos and Papadopoulos, 1982; Vlahos, Sharma and Papadopoulos, 1983). Sprangle and Vlahos (1983) concluded that the e.m. wave can accelerate approximately 10^{14} of the ambient electrons in the acceleration region to approximately 100 keV.

Kane, Benz and Treumann (1982) suggested that the close relation between type III and X-ray bursts is evidence for the fact that the acceleration region shares open and closed field lines. Vlahos believes that this mechanism will be competing with the one presented above, since the trapped electrons will become quickly the source for e.m. radiation and will accelerate eletrons together with the primary acceleration source. In other words primary and secondary accelerated electrons will be injected into open field lines during the flash phase of the flare.

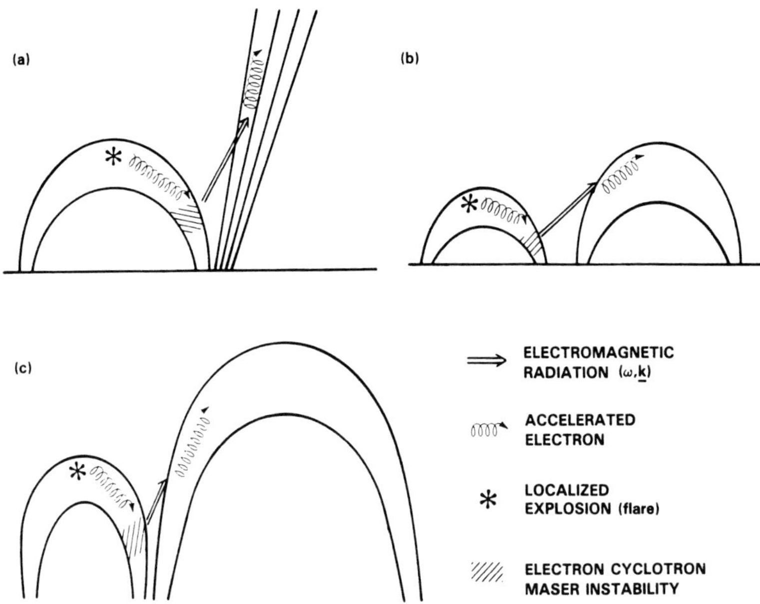

Figure 2.4.4 The development of the intense radio emission driven acceleration is shown for three different models. Flare released energetic electrons are precipitating towards an increasing magnetic field, reflected and form a loss-cone velocity distribution. Loss-cone driven electron cyclotron radiation can escape from the loop and accelerate electrons to high energies inside (a) open field lines, which will explain the hard X-ray/Type III correlation (b) nearby loop, which will explain the triggering of quadruple like structure observed by the VLA (c) large closed loop, which will explain the hard X-ray/Type U correlation (Strong *et al.* (1984), see Section 2.2.5 and Figure 2.2.21) or the hard X-ray/Type V correlation discussed in Section 2.2.

2.4.7 Preferential Acceleration of Heavy Ions

One of the more intriguing results of energetic solar particle flux measurements at ~ 1 AU is the discovery of anomalous enhancements in the abundance of some ionic species during occasional "unusual" events, often called ^{3}He-rich flares (Ramaty et al., 1980, and references therein). Three different types of enhancements can be distinguished in these events: (1) the enhancement of the isotope ^{3}He by as much as three and a half orders of magnitude, at energies ~ 1-10 MeV nucleon^{-1}; (2) enhancement of heavy (A > 4) ions by a factor of 5-10; (3) deviations of other charge states from solar abundances. These are often correlated with ^{3}He-rich flares. A review of the observations and theoretical interpretations is given by Kocharov and Kocharov (1984). Fisk (1978) proposed the most successful model so far since it seems to explain not only the enhancement (1), but also (2) and (3). It is based on selective preheating of the ^{3}He and certain heavy ions such as Fe by resonant interaction with ion cyclotron waves. Such waves can be driven unstable by electron currents or ion beams. In a pure hydrogen plasma, the ion cyclotron waves have a frequency above the proton cyclotron frequency Ω_H

(i.e., $\omega = (\Omega_H^2 + k^2c_s^2)^{1/2} \approx 1.2 - 1.4\ \Omega_H$ where c_s is the sound speed) (Kadomtsev, 1965). However, in the presence of a noticeable amount of doubly ionized $^4He^{++}$ and for some combinations of drift velocities and plasma parameters, waves near the $^4He^{++}$ cyclotron frequency can be excited. Since $\omega \approx 1.2\ \Omega_{^3He^{++}} \approx \Omega_{^3He^{++}}$, $^3He^{++}$ can be resonantly accelerated. In addition, ions with $A/Q \approx 3.3$ and $A/Q \approx 4.5$ can be accelerated by the harmonics (i.e. $\omega = n\Omega_i - k_z v_z$) although less efficiently. The subject of ion energization by ion cyclotron waves in multispecies plasmas has been recently examined by Papadopoulos, Gaffey and Palmadesso (1980) and Singh, Schunk, and Sojka (1981). It was noted that, for wave amplitudes exceeding a threshold, which depends on A/Q, ion acceleration does not require resonance but can be very strong even for significant frequency mismatch. The physical process responsible for this is resonance overlapping. The general theory was given by Chirikov (1979), while its application to lower hybrid heating of fusion plasmas can be found in several publications (Fukuyama et al., 1978; Karney and Bers, 1977; Hsu, 1982). It has been called non-resonant stochastic acceleration to be distinguished from the process requiring cyclotron resonance [i.e., $\omega = n\Omega_i - (k_z v_z)$], and was first shown by Papadopoulos et al. to exhibit a large selectivity in the ionic A/Q ratio. Varvoglis and Papadopoulos (1983) revised the model advanced by Fisk (1978) by including the proper nonlinear physics of particle energization by electrostatic ion cyclotron (EIC) waves. Their model retains two basic concepts of Fisk: the energization by EIC waves and the need for a second stage of acceleration. However, there is no need for exciting $^4He^{++}$ cyclotron waves,since the dominant process is non-resonant and can be accomplished by hydrogen cyclotron waves. The A/Q selectivity in the flux available for energization in the second stage process enters through the nonlinear saturation level $e\phi/T_H$, which in conventional theories (Dum and Dupree, 1970; Palmadesso et al., 1974) depends on the current that drives the instability.

2.4.8 Summary

In contrast to the dynamic and dramatic development of the observations, theoretical studies concerning particle acceleration in a solar environment during a flare are progressing at a steady rate. Part of the problem, of course, is our poor understanding of the energy-release process in flares, specifically our poor understanding of the reconnection theory in solar flares. Nevertheless, the workshop has stimulated several new ideas and provided many concrete questions for theoretical studies. Here, we pose several questions regarding theoretical studies of particle acceleration in solar flares.

(1) Is there a single mechanism that will accelerate particles to all energies and also heat the plasma?

We do not know of any such mechanism today. The mechanisms discussed by us either have a threshold for acceleration (like the stochastic Fermi acceleration and shock acceleration) or heat and accelerate the tail to mildly relativistic energies (Joule heating and sub-Dreicer E-field acceleration or Joule heating and lower hybrid waves). We then come to the conclusion that the existing theoretical understanding of particle acceleration favors the two step acceleration (with one important twist, the two mechanisms must operate simultaneously at the same place or in close proximity to each other).

(2) How fast will the existing mechanisms accelerate electrons up to several MeV and ions to 1 GeV?

Electric fields, lower hybrid waves and Joule heating can start, and then heat and accelerate the electrons to 10-100 keV in fractions of a second. If a shock or a turbulent spectrum of waves is already present, it is possible to accelerate the electrons further to relativistic energies in 2-50 secs. Melrose (1983) has already pointed out that magnetosonic turbulence can accelerate ions from 0.1 MeV to 30 MeV in 2 s but the question of pre-accelerating ions to 0.1 MeV and the driving force for the magnetosonic waves has remained open.

(3) If shocks are formed in a few seconds, can they be responsible for the prompt acceleration of ions and electrons? How are these shocks related to large-scale shocks which are responsible for the Type II bursts?

The shock acceleration as discussed here will not be appropriate for the prompt acceleration of particles, since almost all the above theories have assumed a plane quasi-parallel or quasi-perpendicular shock. In the vicinity of the energy release region the shock curvature and structure must be taken into account. The shocks developed inside the energy release volume during a flare are formed in the middle of an enhanced level of turbulence and laminar shock drift or purely diffusive acceleration calculations in a parallel shock are not valid. In summary, our current understanding of particle acceleration in shocks cannot give us an answer to the above question.

An important question regarding the formation of shocks inside the energy release region is the lack of evidence for Type II-like signatures in the decimeric radio frequencies. Such continuation for the Type III bursts is clearly present. This of course brings us to the next question of the relation of flares (and shocks developed inside the energy release volume) to large scale shocks, responsible for the Type II bursts and particle acceleration in space. This question remains unresolved.

(4) Can the electron-cyclotron maser spread the acceleration region?

We suggested that beams of electrons streaming towards converging magnetic field lines will be reflected and form loss-cone type velocity distributions in the low corona. A fraction ($\lesssim$ 1-5%) of the kinetic energy of the energetic electrons will drive electromagnetic waves that can easily spread the acceleration region to larger volume outside the flaring region.

(5) Which of the acceleration mechanisms discussed above can explain

the observed energy spectra (see discussion on Section 2.3)?

Most of our discussions on the acceleration mechanisms have been focussed on the energetics and not on the details of the energy spectrum of the accelerated particles. The only mechanisms that have successfully explained the observed spectra are stochastic Fermi acceleration and quasi-parallel diffusive shock acceleration. This of course does not exclude other mechanisms for flare acceleration but places important constraints on them. A detailed study of the energy spectrum of the accelerated particles must be the goal for any successful acceleration mechanism in solar flares.

Fermi acceleration and acceleration by quasi-parallel shocks although successful in explaining the energy spectrum, have not yet fully addressed other questions e.g., shock and turbulence formation time, formation of appropriate wave spectra from MHD waves, relation of MHD waves to energy release process, etc.

2.5 ACHIEVEMENTS -- OUTSTANDING QUESTIONS

As it is always the case in science, new and more precise measurements bring us higher in the helix of knowledge by replacing or revising older concepts with new ones. At the same time, however, new and more sophisticated questions are posed which demand even more sophisticated theoretical studies. Comparing the results reported in this chapter with the Skylab work which was reported in two articles by Kane et al. (1980) and Ramaty et al. (1980), one clearly sees this evolution on many concepts of particle acceleration in solar flares. The main achievements are highlighted below:

1. The X-ray imaging has, for the first time, provided evidence for discrete isolated footpoints during the impulsive phase. The footpoints gradually evolve and form a single source of hot plasma in the decay phase of some flares. Other flares never develop footpoints and the thermal plasma dominates their evolution. Finally, the majority of the existing X-ray imaging data are very complex and "refuse" to be placed in simple classes (e.g., A, B, C etc.). The rather simplistic division of flares to "thermal" and "nonthermal" seems to be a convenient abstraction for simple theoretical models. In reality thermal and nonthermal plasma is dialectically connected; "nonthermal plasma" is quickly thermalized and from a locally heated plasma energetic particles emerge. The X-ray imaging has provided evidence for the interconnection of thermal and nonthermal plasma.

2. The radio images have also put a mark on our understnding of evolution of the magnetic topologies that can lead to a flare. Interacting loops and bipolar structures commonly observed suggest that loop-like structures are the elementary components of the flare process.

The isolated loop structure has been assumed to be one of the possibilities for energy release in flares. There is now evidence that more complex magnetic topologies (e.g., interacting loops, emerging flux or even a catastrophic interaction of many loops) are at work during a flare. The loop, however, remains as the elementary structure that participates in these interactions.

3. The radio maps at meter wavelengths and detailed studies of the meter/X-ray correlation also point to the direction that the region of acceleration must encompass open and closed field lines, and that it is located at the low corona.

4. Analyzing the time evolution of the flare energy release process with high time resolution instruments has provided an unprecedented wealth of information that will take years to interpret physically. Pulsations arise in several wavelengths, some of them quasi-periodic, some of them chaotic. Fast pulses with durations that sometimes reach the instrumental resolution and delays between pulses in different wavelengths are among the new results. Some of the observed pulses are clearly the result of pulsations in the acceleration source, but in other cases the pulses are the result of the radiation mechanism.

5. For many years we believed that prior to a flare the Sun operates in a steady state norm that keeps the coronal plasma around 2×10^6 K. During a flare the energization mechanism will "process" this 2×10^6 K Maxwellian distribution to a new "heated" and "accelerated" plasma state. We have provided evidence that show that (1) the acceleration starts before the flash phase and continues after the impulsive phase of the flare, (2) "microflares" continuously occur in the corona and develop nonthermal tails. Thus, the acceleration may start from a distorted Maxwellian that is developed long before the impulsive phase. Furthermore, we presented evidence that suggests that the number of flares increases as the total energy per flare decreases, which indicates that there is no threshold for reconnection and "flaring" and "heating" smoothly join each other.

6. Intense, coherent, polarized microwave pulses sometimes occur during a flare. They may be the result of the conversion of kinetic energy of the precipitating electrons to electromagnetic radiation through the convergence of magnetic field lines.

7. Observations from gamma-ray detectors on SMM have dramatically changed our thinking of the way particles are accelerated in the Sun. The conventional two phase acceleration (a prompt first phase acceleration of mildly relativistic electrons followed by a slower second phase acceleration that energizes the ions and accelerates further the electrons to relativistic electrons) had to be abandoned, because in many flares fast synchronous (or nearly synchronous) pulses occur simultaneously in several energy channels. Hence, the concept of prompt acceleration of both electrons and ions to all energies in a few seconds is a new important discovery, which will change our thinking on particle acceleration. However, particles observed in interplanetary space may be accelerated by the second-phase acceleration.

8. We have evidence that the acceleration of relativistic electrons

and ions must be a common phenomenon during flares.

We have outlined several questions that need further study in Sections 2.2.7, 2.3.7 and 2.4.8. We believe that many detailed observational studies are still needed and we outline some of them here:

- Stereoscopic observations of hard X-rays which have been discussed briefly in this chapter, may provide information about the precipitating electrons and the validity of the thick target model.

- X-ray imaging at higher energies and hard X-ray polarization measurements will be valuable for knowing the critical energy that divides the thermal from the nonthermal component of flares.

- Detailed studies of the time structures of hard X-ray bursts may give us clues to the time development of the energy release mechanism.

- Detailed studies of hard X-ray and radio microflares and the pre- and post-impulsive phase activity are of fundamental importance.

- Spectral evolution during hard X-ray pulsations is an important source of information.

- Radio maps in different wavelengths with high time resolution accompanied by detailed models of the radiating source will play an important role in our understanding of the evolution of magnetic field topologies in the course of a flare.

- Detailed studies of the delays of the starting and peak time in different energy channels will restrict our choices on particle acceleration mechanisms. Spectral evolution is also a powerful tool to study the presence of more than one acceleration mechanism.

A number of outstanding theoretical problems remain open for future studies:

- Several new concepts have recently emerged concerning the energy release process during reconnection but usually these concepts address questions related to the energetics and not to particle acceleration. We believe that detailed studies on heating, energetic tail formation and energy spectra of electrons and ions energized during reconnection are of fundamental importance.

- The details of the magnetohydrodynamic waves that are excited during a flare are not well known. We believe that this is a major problem that deserves immediate attention. Of particular importance are questions related to the onset time, power spectrum of the excited waves, etc.

- Acceleration of ions, stability of propagating ion beams and the importance of energetic ions during a flare must be examined in detail.

o The mechanisms for electron and ion heating are not well understood and must be examined in detail.

o Prompt formation of shocks and shock acceleration of electrons and ions is a problem that must be reexamined in the light of the results presented in this chapter.

o The interaction of "hot" and "cold" plasma during a flare, the formation and propagation of the so called "conduction fronts" is a fundamental astrophysical problem. Detailed numerical, analytical and observational studies (using X-ray imaging) must continue.

o Magnetic models for the acceleration and radiation source using model velocity distributions must be developed for both electrons and ions in all energies. These models serve to restrict or eliminate acceleration mechanisms by comparing the resulting radiation signatures from the model with the data.

o The response of coronal plasma to sub-Dreicer and much stronger than the Dreicer electric fields is an area that we have not explored in detail. The formation of such potential drops in coronal conditions is also an open question.

o The time evolution of the acceleration mechanism is a higher order problem but the detailed observations, outlined above, demand such studies and must be placed in the agenda for careful study.

The achievements and outstanding questions posed in this chapter are of fundamental importance for the entire astrophysical community, since the impulsive energy release and particle acceleration observed during a flare is a phenomenon that must be occuring in many space and laboratory plasmas. We believe that our work will serve to generate new ideas and interpretations of the results presented during the workshop. We are aware that some of the questions posed by us are complex and will require detailed observational and theoretical studies for many years to come. We hope that our effort will be a guide for these studies.

2.6 ACKNOWLEDGEMENTS

The material presented in this chapter has been discussed by the members of team B in three separate one-week meetings of the Solar Maximum Mission Workshops at Greenbelt, Maryland in 1983-84. During the workshops, Marcos Machado led the discussion on "phenomena associated with mildly-relativistic electrons", Reuven Ramaty led the discussion on the "phenomena associated with ions and relativistic electrons" and Loukas Vlahos led the discussion on "theoretical studies of particle acceleration". The authors of this chapter would like to thank their colleagues, J. Brown, B. Dennis, M. Forman, G. Dulk, P. Cargill, A. Raoult, P. Sprangle, N. Vilmer, V. Tomozov and H. Wiehl, for many helpful discussions and suggestions.

2.7 REFERENCES

Achterberg, A. and Kuijpers, J. 1984, Astron. Astroph., **130**, 111.
Alissandrakis, C.E. and Kundu, M.R. 1985, in preparation.
Alissandrakis, C.E. and Preka-Papadema, P. 1984, Astron. Astroph., **139**, 507.
Altyntsev, A.T., Krasov, V.I. and Tomozov, V.M. 1977, Solar Phys., **55**, 69.
Altyntsev, A.T., Bardakov, V.M. and Krasov, V.I. 1981, ZhETF, **81**, 901.
Altyntsev, A.T., Krasov, V.I., Lebedev, N.V., Paperny, V.L. and Simonov, V.G. 1983, Preprint SibIZMIR No. 23-83, Irkutsk.
Altyntsev, A.T., Krasov, V.I. and Tomozov, V.M. 1984, Solar Flares and Plasma Experiments. Astronomiya, v. 25, Moscow, p. 99-191.
Arion, D. 1984, Ap. J., **277**, 841.
Aschwanden, M.J., Wiehl, H.J. Benz, A.O. and Kane, S.R. 1985, Solar Phys., **97**, 159.
Athay, R.G. 1984, Solar Phys., **93**, 123.
Bai, T. 1982, in Gamma Ray Transients and Related Astrophysical Phenomena, R. E. Lingenfelter, et al. (ed.), (New York: AIP), p. 409.
Bai, T. and Ramaty, R. 1979, Ap. J., **227**, 1072.
Bai, T. and Dennis, B.R. 1985, Ap. J., **292**, 699.
Bai, T., Kiplinger, A.L. and Dennis, B.R. 1985, Ap. J., submitted.
Bai, T., Dennis, B.R., Kiplinger, A.L., Orwig, L.E. and Frost, K.J. 1983a, Solar Phys., **86**, 409.
Bai, T., Hudson, H.S., Pelling, R.M., Lin, R.P., Schwartz and von Rosenvinge, T.T. 1983b, Astrophys. J., **267**, 433.
Barbosa, D.D. 1979, Ap. J., **233**, 383.
Batchelor, D.A., Crannell, C.J., Wiehl, H.J. and Magun, A. 1985, Ap. J., **295**, 258.
Benz, A.O. 1985, Solar Phys., in press.
Benz, A.O., Zlobec, P. and Jaeggi, M. 1982, Astr. Ap., **109**, 305.
Benz, A.O., Bernold, T.E.X. and Dennis, B.R. 1983a, Ap. J., **271**, 355.
Benz, A.O., Barrow, C.H., Dennis, B.R., Pick, M., Raoult, A. and Simnett, G. 1983b, Solar Phys., **83**, 267.
Benz, A.O., Achwanden, M.J. and Wiehl, J.J. 1984, Decimetric Radio Emission During Solar Flares, Kunming Workshop Proceedings.
Bratenahl, A. and Baum, P.J. 1985, in Unstable Current Systems and Plasma Instabilities in Astrophysics, ed. M. R. Kundu and G. D. Holman, D. Reidel, p. 147.
Brown, J.C. 1971, Solar Phys., **18**, 485.
Brown, J.C. 1972, Solar Phys., **26**, 441.
Brown, J.C. 1973, Solar Phys., **28**, 151.
Brown, J.C. and McClymont, A.N. 1976, Solar Phys., **41**, 135.
Brown, J.C., Melrose, D.B. and Spicer, D.S. 1979, Ap. J., **228**, 592.
Brown, J.C. and Smith, D.F. 1980, Rep. Prog. Phys., **43**, 125.
Brown, J.C., Craig, I.J.D. and Karpen, J.T. 1980, Solar Phys., **67**, 143.
Brown, J.C., Mackinnon, A.L., Zodi, A.M. and Kaufmann, P. 1983a, Astron. Astrophys., **123**, 10.
Brown, J.C. Carlaw, V.A., Cromwell, D. and Kane, S.R. 1983b, Solar Phys., **88**, 281.
Brunel, F., Tajima, T. and Dawson, J.M. 1982, Phys. Rev. Lett., **49**, 323.

Butz, M., Hirth, W., Furth, E. and Harth, W. 1976, Sond. Kelinheubacher Berichte, **19**, 345.
Chirikov, B.V. 1979, Phys. Rev., **52**, 264.
Chupp, E.L. 1982, in Gamma Ray Transients and Related Astrophysical Phenomena, R. E. Lingerfelter, et al. (ed.), (New York: AIP), p. 363.
Chupp, E.L. 1983, Solar Phys., **86**, 383.
Chupp, E.L. 1984, Ann. Rev. Astron. Astrophys., **22**, 359.
Chupp, E.L., et al., 1981a, Ap. J. (Letters), **244**, L171.
Chupp, E.L., Forrest, D.J., Higbie, P.R., Suri, A.N., Tsai, C. and Dunphy, P.P. 1973, Nature, **241**, 333.
Chupp, E.L., Forrest, D.J., Ryan, J.M., Heslin, J., Reppin, C., Pinkau, K., Kanbach, G., Rieger, E. and Share, G.H. 1982, Ap. J. (Letter), **263**, L95.
Chupp, E.L., Forrest, D.J., Kanbach, G., Share, G.H. 1983, 19th Int. Cosmic Ray Conf., (late papers Bangalore), p. 334.
Cliver, E., Share, G., Chupp, E., Matz, S., Howard, R., Koomen, M., McGuire, R., and Von Rosenvinge, T. 1982, Bull. AAS, 14, 4, 874.
Cliver, E.W. et al., 1983a.
Cliver, E.W., Forrest, D.J., McGuire, R.E. and Von Rosenvinge, T.T. 1983, in: 18th International Cosmic Ray Conference Papers (Late Papers), Bangalore, p. 342.
Cook, W.R., Stone, E.C. and Vogt, R.E. 1984, Ap. J., **279**, 827.
Coppi, B., Galvao, R., Pellat, R., Rosenbluth, M.N. and Rutherford, P.H. 1976, Sov. J. Plasma Phys., **2**, 533.
Costa, J.E.R. and Kaufmann, P. 1983, Astron. Astrophys. **119**, 131.
Crannell, C.J. Frost, K.J., Matzler, C., Ohki, K. and Saba, J.L. 1978, Ap. J., **223**, 620.
Craig, I.J.D., MacKinnon, A.L. and Vilmer, N. 1985, Astrophys. Space Sci., **116**, 377.
Datlowe, D.W., Elcan, M.J. and Hudson, H.S. 1974, Solar Phys., **39**, 127.
Debrunner, H., Fluckiger, E., Chupp, E.L. and Forrest, D.J. 1983,: in 18th International Cosmic Ray Conference Papers 4, Bangalore, p. 75.
Debrunner, H., Fluckiger, E., Lockwood, J.A. and McGuire, R.E. 1984, Journal Geoph. Res., **89**, 769.
Decker, R.B. and Vlahos, L. 1985, Journal Geoph. Res., **90**, 47.
DeJager, C. 1969: in COSPAR Symposium on Solar Flares and Space Research, eds. C. DeJager and Z. Svestka, p. 1.
DeJager, C. and De Jonge, G. 1978, Solar Phys., **58**, 127.
Dennis, B.R., et al., 1983, NASA TM 84998.
Dennis, B.R., Benz, A.O., Ranieri, M. and Simnett, G. 1984, Solar Phys., **90**, 383.
Drake, J.F., Pritchett, P.L. and Lee, Y.C. 1978, UCLA Phys Rept. PPG-341, unpublished.
Droge, F. 1977, Astron. Astroph., **57**, 285.
Dreicer, H. 1959, Phys. Rev., **115**, 238.
Drury, L.O.C. 1983, Rep. Progr. Phys., **46**, 973.
Duijveman, A. 1983, Solar Phys., **84**, 189.
Duijveman, A. and Hoyng, P. 1983, Solar Phys., **86**, 289.
Duijveman, A., Hoyng, P. and Machado, M.E. 1982, Solar Phys., **81**, 137.
Dum, C.T. and Dupree, T.D. 1970, Phys. Fluids, **13**, 2064.
Dumas, G., Caroubalos, C. and Bougeret, J.L. 1982, Solar Phys., **81**, 2.

Elliott, H. 1969, in Solar Flares and Space Research, (ed. C. de Jager and Z. Svestka), North Holland, Amsterdam, p. 356,
Ellison, D.C. and Ramaty, R. 1985, Advanced Space Res. (COSPAR), **4**, 7, 137.
Emslie, A.G. 1978, Ap. J., **224**, 241.
Emslie, A.G. 1980, Ap. J., **235**, 1055.
Emslie, A.G. 1981a, Astrophys. Letters, **22**, 171.
Emslie, A.G. 1981b, Ap. J., **245**, 711.
Emslie, A.G. 1983, Ap. J., **271**, 367.
Emslie, A.G. and Vlahos, L. 1980, Ap. J., **242**, 359.
Evenson, P., Meyer, P. and Pyle, K.R. 1983, Ap. J., **274**, 875.
Evenson, P., Meyer, P., Yanagita, S. and Forrest, D.J. 1984, Ap. J., **283**, 439.
Field, G.B. 1965, Astrophys. J., **142**, 531.
Fisk, L. A. 1978, Ap. J., **224**, 1048.
Forman, M.A., Ramaty, R. and Zweibel, E.G. 1985,: in The Physics of the Sun, eds. P. A. Sturrock et al., Reidel, Dordrecht.
Forrest, D.J. 1983: in Positron-Electron Pairs in Astrophysics, eds. M.L. Burns, A.K. Harding and R. Ramaty, A.I.P., N.Y., p. 3.
Forrest, D.J. and Chupp, E.L. 1983, Nature, **305**, 291.
Frost, K.J. and Dennis, B.R. 1971, Ap. J., **165**, 655.
Fukuyama, A., Momota, H. Itatani, R. and Takizuka, T. 1977, Phys. Rev. Lett., **38**, 101.
Furth, H.P., Killeen, J. and Rosenbluth, M.N. 1963, Phys. Fluids, **6**, 459.
Gaizauskas, V. and Tapping, K.F 1980, Ap. J., **241**, 804.
Gardner, B.M., Forrest, D.J., Chupp, E.L., et al., 1981, Bull. AAS, **13**, 903.
Ginzburg, V.L. and Syrovatskii, S.I. 1964, The Origin of Cosmic Rays, (New York: MacMillan).
Gloeckler, G. 1985, Advances Space Res. (COSPAR), 4, **2**, 127.
Heyvaerts, J. 1981, "Particle Acceleration in Solar Flares," in Solar Flare Magnetohydrodynamics, ed. E. R. Priest, Gordon and Breach, New York City, pp. 429-555.
Heyvaerts, J. and Priest, E.R. 1984, Astr. Astroph., **137**, 63.
Heyvaerts, J., Priest, E.R. and Rust, D.M. 1977, Ap. J., **216**, 123.
Holman, G.D. 1985, Ap. J., **293**, 584.
Holman, G.D. and Pesses, M.E. 1983, Ap. J., **267**, 837.
Holman, G.D., Eichler, D. and Kundu,M.R. 1980, in Proceedings of IAU Symposium 86, "Radio Physics of the Sun", Kundu, M.R. and Gergely, T.E., eds. (Dordrecht: Reidel), p. 457.
Holman, G.D., Kundu, M.R. and Papadopoulos, K. 1982, Ap. J., **257**, 354.
Hoyng, P., Brown, J.C. and van Beek, H.F. 1976, Solar Phys., **48**, 197.
Hoyng, P., Duijveman, A., Machado, M.E. et al., 1981, Ap. J. (Letters), **246**, L155.
Hoyng, P., Marsh, K.A., Zirin, H. and Dennis, B.R. 1983, Ap. J., **268**, 865.
Hsu, J.Y. 1982, Phys. Fluids, **25**, 179.
Hudson, H.S., Canfield, R.C. and Kane, S.R. 1978, Solar Phys., **60**, 137.
Hudson, H.S., Bai, T., Gruber, D.G., Matteson, J.L., Nolan, P.L. and Peterson, L.E. 1980, Ap. J. (Letters), **236**, L91.
Ibragimov, I.A. and Kocharov, G.E. 1977, Sov. Astron. Lett., **3(5)**, 221.

Kadomtsev, B.B. 1965, Plasma Turbulence (London: Academic), p. 73.
Kai, K. and Kosugi, T. 1985, Publications of Astr. Soc. Japan, **37**, 155.
Kane, S.R. 1972, Solar Phys., **27**, 174.
Kane, S.R. 1973, in High Energy Phenomena on the Sun, ed. R. Ramaty and R. G. Stone), (Washington: NASA), p. 55.
Kane, S.R. 1974, in Coronal Disturbances, G. Newkirk, Jr. (ed), IAU Symp. 57, Reidel, p. 105.
Kane, S.R. 1981a, Astrophys. Space Sci., **75**, 163.
Kane, S.R. 1981b, Ap. J., **247**, 1113.
Kane, S.R. 1983, Solar Phys., **86**, 355.
Kane, S.R. and Anderson, K.A. 1970, Ap. J., **162**, 1003.
Kane, S.R. and Pick, M. 1976, Solar Physics., **47**, 293.
Kane, S.R. and Raoult, A. 1981, Ap. J. (Letters), **248**, L77.
Kane, S.R., Pick, M. and Raoult, A. 1980, Astrophys. J., **241**, L113.
Kane, S,R., Benz, A.O. and Treumann, R.A. 1982, Ap. J., 263.
Kane, S.R., Kreplin, R.W., Martres, M.J., Pick, M. and Soru-Escaut, I. 1974, Solar Physics, **38**, 483.
Kane, S.R., Anderson, K.A., Evans, W.D., Klebesadel, R.W. and Laros, J. 1979, Ap. J. (Letters), **233**, L151.
Kane, S., Anderson, K., Evans, W., Klebesadel, R. and Laros, J. 1980, Ap. J., **239**, 85-88.
Kane, S.R., Fenimore, E.E., Klebesadel, R.W. and Laros, J.G. 1982, Astrophys. J., **254**, L53.
Karney, C.F.F. and Bers, A. 1977, Phys. Rev. Letters, **39**, 550.
Kaufmann, P., Iacomo, Jr. P., Koppe, E.H. Santos, P.M. Schaal, R. E. and Blakey, J.R. 1975, Solar Phys., **45**, 189.
Kaufmann, P., Piazza, L.R., Schaal, R.E. and Iacomo Jr., P. 1978, Ann. Geophys., **34**, 105.
Kaufmann, P., Strauss, F.M., Raffaelli, J.C. and Opher, R. 1980a: in Donnelly, R.F., ed. -- Solar Terrestrial Predictions Proceedings, V. 3, Solar Activity Predictions.
Kaufmann, P. Strauss, F.M., Opher, R. and Laporte, C. 1980b, Astron. Astrophys., **87**, 58.
Kaufmann, P. Strauss, F.M., Schaal, R.E. and Laporte, C. 1982a, Solar Phys., **78**, 389.
Kaufmann, P., Costa, J.E.R. and Strauss, F.M. 1982b, Solar Pys., **81**, 159.
Kaufmann, P., Strauss, F.M., Costa, J.E.R., Dennis, B.R., Kiplinger, A., Frost, K.J. and Orwig, L.E. 1983, Solar Phys., **84**, 311.
Kaufmann, P., Correia, E., Costa, J.E.R., Dennis, B.R. Hurford, G.J. and Brown, J.C. 1984, Solar Phys., **91**, 359.
Kaufmann, P., Correia, E., Costa, J.E.R., Sawant, H.S. and Zodi Vaz, A.M. 1985, Solar Phys., **95**, 155.
Kawabata, K., Ogawa, H. and Suzuki, I. 1983, Solar Phys., **86**, 247.
Klein, L., Anderson, K., Pick, K.M., Trottet, G., Vilmer, N. and Kane, S.R. 1983, Solar Phys., **84**, 295.
Kocharov, L.G. and Kocharov, G.E. 1984, Space Sci. Rev., **38**, 89.
Kolomenskii, A.A. and Lebedev, A.N. 1963, Soviet Phys. Dokl., **7**, 745.
Kosugi, T., Kai, K. and Suzuki, T. 1983, Solar Phys., **87**, 373.
Kozlovsky, B. and Ramaty, R. 1977, Astrophys. Lett., **19**, 19.
Kuijpers, J., van der Post, P. and Slottje, C. 1981, Astron. Astrophys., **103**, 331.

Kundu, M.R. 1981, Proc. SMY Workshop, Crimea, p. 26.
Kundu, M.R. and Shevgaonkar, R.K. 1985, Ap. J., **291**, 860.
Kundu, M.R., Rust, D.M. and Bobrowsky, M. 1983, Ap. J., **265**, 1084.
Kundu, M.R., Schmahl, E.J., Velusamy, T. and Vlahos, L. 1982, Astron. Astrophys., **108**, 188.
Kundu, M.R., Machado, M., Erskine, F.T., Rovira, M.G. and Schmahl. E.J. 1984, Astron. Astrophys., **132**, 241.
Langdon, A.B. and Lasinski, B.F. 1976, Methods in Computational Physics, Academic Press, N.Y., Vol. 16, p. 327.
Lantos, P., Pick, M. and Kundu, M. 1984, Ap. J., **283**, L71.
Leach, J. and Petrosian, V. 1981, Ap. J., **251**, 781.
Leboeuf, J.N., Tajima, T. and Dawson, J.M. 1982, Phys. Fluids, **25**, 784.
Lee, M.A. and Ryan, J.M. 1986, Ap. J., **303**, 829.
Leroy, M. and Mangeney, A. 1984, Ann. Geoph. Gauthier Villars, **2**, 449.
Lin, A.T., Dawson, J.M. and Okuda, H. 1974, Phys. Fluids, **17**, 1995.
Lin, R.P. and Hudson, H.S. 1976, Solar Phys., **50**, 153.
Lin, R.P. et al., 1981, Ap. J. (Letters), **251**, L109.
Lin, R.P., Mewaldt, R.A. and Van Hollebeke, M.A.I. 1982, Ap. J., **253**, 949.
Lin, R.P., Schwartz, R.A., Kane, S.R., Pelling, R.M., and Harley, K.C. 1984, Ap. J., **283**, 421.
Machado, M.E. 1983a, Adv. Space Res., **2**, No. 11, 115.
Machado, M.E. 1983b, Solar Phys., **89**, 133.
Machado, M.E. and Lerner, G. 1984, Adv. Space Res., **4**, 239.
Machado, M.E., Duijveman, A. and Dennis, B.R. 1982, Solar Phys., **79**, 85.
Machado, M.E., Rovira, M.G and Sneibrun, C. 1985, Solar Phys., **99**, 189.
Machado, M.E. Somov, B.V. Rovira, M.G.and de Jager, C. 1983, Solar Phys., **85**, 157.
MacKinnon, A.L. and Brown, J.C. 1983, Astron. Astrop., **132**, 229.
MacKinnon, A.L., Brown, J.C. and Hayward, J. 1985, Solar Phys. **99**, 231.
MacKinnon, A.L., Brown, J.C., Trottet, G. and Vilmer, N. 1983, Astron. Astrophys., **119**, 297.
Makishima, K. 1982, Proc. HINOTORI symposium on solar flares, ed. Y. Tanaka, et al., p. 120, published by the Institute of Space and Astronautical Science, Tokyo.
Mannheimer, W. 1977, Phys. Fluids, **20**, 165.
Marsh, K.A. and Hurford, G. 1980, Ap. J. (Letters), **240**, L111.
Mason, G.M., Gloeckler, G. and Hovestadt, D. 1984, Ap.J., **280**, 902.
Matzler, C., Bai, T., Crannell, C.J. and Frost, K.J. 1978, Ap. J., **223**, 1058.
Matthaeus, W.H. 1982, Geoph. Res. Lett., **9**, 660.
Matthaeus, W.H., Ambrosiano, J.J. and Goldstein, M.L. 1984, Phys. Rev. Lett, **L53**, 1449.
Matthaeus, W.H. and Montgomery, D.C. 1981, J. Plasma Phys., **25**, 11.
McDonald, F.B. and Van Hollebeke M.A.I. 1985, Ap. J. (Lett.), **290**, L67.
McGuire, R.E. and von Rosenvinge, T.T. 1985, Advances Space Res. (COSPAR), **4**, 2, 117.
Melrose, D.B. 1983, Solar Phys., **89**, 149.
Melrose, D.B. and Brown, J.C. 1976, Mon. Not. Roy. Astron. Soc., **176**, 15.
Melrose, D.B. and Dulk, G.A. 1982, Ap. J., **259**, 844.
Mercier, C. 1975, Solar Phys., **45**, 169.

Mewaldt, R.A., Spalding, J.D. and Stone, E.C. 1984, Ap. J., **280**, 892.
Moghaddam-Taaheri, E., Vlahos, L., Rowland, H.L., and Papadopoulos, K., 1985, Phys. Fluids, **28**, 3356.
Moriyama, F. et al., 1983, Annals of Tokyo Astron. Observatory, Vol. XIX, No. 2, 276.
Murphy, R.J. Ph.D. Dissertation, University of Maryland (in preparation 1985).
Murphy, R.J. and Ramaty, R. 1985, Advances Space Res. (COSPAR), **4**, 7, 127.
Nakajima, H., Dennis, B.R., Hoyng, P., Nelson, G., Kosugi, T. and Kai, K. 1985, Ap.J., **288**, 806.
Nakajima, H., Kosugi, T. and Enome, S. 1983, Nature, **305**, 292.
Oda, M. 1983, Adv. Space Res., **2**, p. 207.
Ohki, K., Takakura, T., Tsuneta, S. and Nitta, N. 1983, Solar Phys., **86**, 301.
Orwig, L.E., Frost, K.J. and Dennis, B.R. 1980, Solar Phys., **65**, 25.
Palmadesso, P.J., Coeffey, T.P., Ossakow, S.L. and Papadopoulos, K. 1974, Geophys. Res. Letters, **1**, 105.
Parker, E.N. 1983a, Ap. J., **264**, 635.
Parker, E.N. 1983b, Ap. J., **264**, 642.
Papadopoulos, K. Gaffey, J.D., Jr., and Palmadesso, P.J. 1980, Geophys. Res. Letters, **7**, 1014.
Pesses, M.E. 1983, Advance Space. Res., **2**, 11, 255.
Petrosian, V. 1982, Ap. J. (Letters), **255**, L85.
Prince, T.A., Ling, J.C., Mahoney, W.A., Riegler, G.R. and Jacobson, A.S. 1982, Ap. J. (Letters), **255**, L81.
Prince, T.A., Forrest, D.J., Chupp, E.L., Kanbach, G. and Share, G.H. 1983: in 18th International Cosmic Ray Conference Papers 4, Bangalore, p. 79.
Ramaty, R. 1979, Particle Acceleration Mechanisms in Astrophyics, American Institute of Physics, New York, p. 135.
Ramaty, R. 1985, in Holzer, T.E., Mahalas, D., Sturrock, P.A., Ulrich, R.K. (eds), Physics of the Sun, p. 1.
Ramaty, R. and Crannell, C.J. 1976, Ap. J., **203**, 766.
Ramaty, R., Kozlovsky, B. and Lingenfelter, R.E. 1975, Space Sci. Rev., **18**, 341.
Ramaty, R., Kozlovsky, B. and Lingenfelter, R.E. 1979, Ap. J. (Suppl.), **40**, 487.
Ramaty, R., Kozlovsky, B. and Suri, A.N. 1977, Ap. J., **214**, 617.
Ramaty, R. and Murphy, R.J. 1984, in High Energy Transients in Astrophysics, ed. S. Woosley, A.I.P., N.Y., p. 628.
Ramaty, R., Murphy, R.J., Kozlovsky, B. and Lingenfelter, R.E. 1983a, Solar Phys., **86**, 395.
Ramaty, R., Murphy, R.J., Kozlovsky, B. and Lingenfelter, R.E. 1983b, Ap. J. (Letters), **273**, L41.
Ramaty, R. et al., 1980, in Solar Flares, ed. P.A. Sturrock (Boulder: Colorado Associated University Press).
Raoult, A. and Pick, M. 1980, Astron. Astrop., **87**, 63.
Raoult, A., Pick, M. Dennis, B. and Kane, S.R. 1985, Ap. J., **299**, 1027.
Rieger, E., Reppin, C., Kanbach, G., Forrest, D.J., Chupp, E.L. and Share, G.H. 1983, in 18th International Cosmic Ray Conference Papers

(Late Papers), Bangalore, p. 238.
Rieger, E. 1982, in Hinotori Symp. on Solar Flares (Tokyo: Inst. Space Astronautical Sci.), p. 246.
Roberts, C.S. and Buchsbaum, S.J. 1964, Phys. Rev. A., 135, 381.
Rowland, H.L. and Vlahos, L. 1984, Astron. Astrop., **142**, 219.
Rosner, R., Tucker, W.H. and Vaiana, G.S. 1978, Ap. J., **220**, 643.
Rust D.M., Benz, A.O., Hurford, G.T., Nelson, G., Pick, M. and Ruzdjak, V. 1981, Ap. J. (Lett), **244**, L179.
Ryan, J. 1986, Solar Phys. **105**, 365.
Sato, T., Matsumoto, H. and Nagai, K. 1982, J. Geoph. Res., **87**, 6089.
Schnack, D.D. and Killeen, J. 1979, Nucl. Fusion, **19**, 877.
Schwartz, R.A. 1984, Ph.D. Thesis, University of California, Berkeley, CA.
Share, G.H., Chupp, E.L., Forrest, D.J. and Rieger, E. 1983, in Positron-Electron Pairs in Astrophys, eds. M.L. Burns, A.K. Harding and R. Ramaty, A.I.P., N.Y., p. 15.
Sharma, R.R. and Vlahos, L. 1984, Ap. J., **280**, 405.
Sharma, R.R., Vlahos, L. and Papadopoulos, K. 1982, Astron. Astrophys., **112**, 377.
Singh, N., Schunk, R.W. and Sojka, J.J. 1981, Geophys. Res. Letters, **8**, 1249.
Slottje, C. 1978, Nature, **275**, 520.
Slottje, C. 1979, in IAU Symposium 86, Radio Physics of the Sun, ed. M.R. Kundu and T.E. Gergely (Dordrecht: Reidel) p. 173.
Smith, D.F. 1980, Solar Phys., **66**, 135.
Smith, D.F. 1985, Ap. J., **288**, 801.
Smith, D.F. and Lilliequist, C.G. 1979, Ap. J., **232**, 582.
Smith, D.F. and Harmony, D.W. 1982, Ap. J., **252**, 800 (SH2).
Somov, B.V., Stepanov, V.E., Stepanyan, N.N and Tomozov, V.M. 1983, Phys. Solariterr., No. **20**, Potsdam, p. 5.
Speiser, T.W. 1967, J. Geoph. Res., **76**, 8211.
Spicer, D.S. 1982, Space Sci. Rev., **31**, 351.
Spicer, D.S. 1983, Advances in Space Research, **2**, 135.
Sprangle, P. and Vlahos, L. 1983, Ap. J., **273**, L95.
Steinolfson, R.S. and Van Hoven, G. 1983, Phys. Fluids, **26**, 117.
Steinolfson, R.S. and van Hoven, G. 1984a, Astrophys. J., **276**, 391.
Steinolfson, R.S. and Van Hoven, G. 1984b, Phys. Fluids, **27**, 1207.
Stewart, R.T. 1978, Solar Physics, **58**, 121.
Strong, K.T., Benz, A.O., Dennis, B.R., Leibacher, J.W., Mewe, R., Poland, A.I., Schrijver, J., Simnett, G., Smith, J.B. Jr. and Sylvester, J. 1984, Ap. J., in press.
Sturrock, P.A. 1974, in IAU Symp. 57, Coronal Disturbances, G. Newkirk (ed.), 437.
Sturrock, P.A., Kaufmann, P., Moore, R.L. and Smith, D. 1985, Solar Phys., **94**, 341.
Tajima, T. 1982, in Fusion Energy -- 1981, (International Energy Agency, ICTP, Trieste, 1982), p. 403.
Tajima, T., akai, J., Nakajima, H., Kosugi, T., Brunel, F., and Kundu, M.R., 1987, Ap.J., **321**, 1031.
Takakura, T. 1985, Proc. Kunming Workshop in Solar Physics and Interplanetary Traveling Phenomena.

Takakura, T., Ohki, K., Tsuneta, S. and Nitta, N. 1983a, Solar Phys., **86**, 313.
Takakura, T., Kaufmann, P. Costa, J.E.R., Degaonkar, S.S., Ohki, K. and Nitta, N. 1983b, Nature, **302**, 317.
Tandberg-Hanssen, E., Kaufmann, P., Reichmann, E.J., Teuber, D.L., Moore, R.L., Orwig, L.E. and Zirin, H. 1984, Solar Phys., **90**, 41.
Tanaka, K. 1983, IAU Col. 71, Activity in Red Dwarf Stars, eds. P. B. Byrne and M. Rodono, p. 307.
Tanaka, K., Watanabe, T., Nishi, K. and Akita, K. 1982, Ap. J., **254**, L59.
Tanaka, M. and Papadopoulos, K. 1983, Phys. Fluids, **26**, 1697.
Tapping, K.F. 1983, Solar Phys., **83**, 179.
Terasawa, T. 1981, J. Geophys. Res., **86**, 9007.
Trottet, G. and Vilmer, N. 1983, Proceedings of the 18th ICRC Conference, SP 3-15.
Tsuneta, S., Takakura, T., Nitta, N., Ohki, K., Makishima, K., Murakami, T., Oda, M., and Ogawara, Y. 1983, Solar Phys., **86**, 313.
Tsuneta, S., Takakura, T., Nitta, N., Ohki, K., Tanaka, K., Makishima, K., Murakami, T., Oda, M., and Ogawara, Y. 1984a, Ap. J., **280**, 887.
Tsuneta, S., Nitta, N., Ohki, K., Takakura, T., Tanaka, K., Makishima, K., Murakami, T., Oda, M., and Ogawara, Y. 1984b, Ap. J., **284**, 827.
Tsuneta, S. 1983a, Ph.D. Thesis.
Tsuneta, S. 1983b, in Proc. Japan-France Seminar on Active Phenomena in the Outer Atmospheres of Stars and the Sun, eds., J.-C. Pecker and Y.Uchida, p. 243, (Paris: C.N.R.S.).
Tsuneta, S. 1984, Annals of the Tokyo Astron. Obs., **20**, 1.
Tsuneta, S. 1985, Ap. J., **290**, 353.
Tsurutani, B.J. and Lin, R.P. 1985, J. Geoph. Res., **90**, 1.
Tverskoi, B.A. 1967, Soviet Phys. JETP, **25**, 317.
Ugai, M. 1982, Phys. Fluids, **25**, 1027.
Ugai, M. 1983, Phys. Fluids, **26**, 1569.
Uralov, A.M. and Nefed'ev, V.P. 1976, Astron. Zh., **83**, 1041.
Van Beek, H.F., Hoyng, P., Lafleur, B. and Simnett, G.M. 1980, Solar Phys., **65**, 39.
Van Hoven, G. 1976, Solar Phys., **49**, 95.
Van Hoven, G. 1979, Astrophys.J., **232**, 572.
Van Hoven, G. and Cross, M.A. 1973, Phys. Rev., **A7**, 1347.
Van Hoven, G., Tachi, T. and Steinolfson, R.S. 1984, Astrophys. J., **280**, 391.
Varvoglis, H.and Papadopoulos, K. 1983, Ap. J. (Lett.), **270**, L95.
Vasyliunas, V.M. 1980, J. Geoph. Res., **85**, 4616.
Velusamy, T. and Kundu, M.R. 1982, Ap. J., **258**, 388.
Vestrand, W.T., et al., 1985, paper presented at the 25th Meeting of the Committee on Space Research, Graz.
Vilmer, N., Kane, S.R. and Trottet, G. 1982, Astron. Astrophys., **108**, 306.
Vilmer, N., Trottet, G. and MacKinnon, A.L. 1986, Astron. and Astrophys., **156**, 64.
Vlahos, L. 1979, in IAU Symposium 86, Radio Physics of the Sun, ed. M.R. Kundu and T.E. Gergely (Dordrecht: Reidel), p. 173.
Vlahos, L. and Papadopoulos, K. 1979, Ap. J., **233**, 717.
Vlahos, L. and Rowland, H.L. 1984, Astr. Ap., **139**, 263.

Vlahos, L. and Sharma, R.R. 1984, Ap. J., **290**, 347.
Vlahos, L. and Gergely, T.E. and Papadopoulos, K. 1982, Ap. J., **258**, 812.
Vlahos, L., Sharma, R.R. and Papadopoulos, K. 1983, Ap. J., **275**.
Wang, H.T. and Ramaty, R. 1974, Solar Physics, **36**, 129.
Wiehl, H.J., Benz, A.O. and Aschwanden, M.J. 1985, Solar Phys., **95**, 167.
Wiehl, J.J. and Mätzler, C. 1980, Astron. Astrophys., **82**, 93.
Wiehl, H.J., Schochlin, W.A. and Magun, A. 1980, Astron. Astrophys., **92**, 260.
Wild, J.P., Smerd, S.F. and Weiss, A. A. 1963, Ann. Rev. Astron. Astrophys., **1**, 291.
Wolfson, J.D., Doyle, J.D., Leibacher, J.W. and Phillips, K.J.H. 1983, Ap. J., **269**, 1319.
Wu, C.S. 1984, J. Geoph. Res., **89**, 8857.
Wu, C.S. and Lee, L.C. 1979, Ap. J., **230**, 621.
Wu, C.C., Leboeuf, J.N., Tajima, T. and Dawson, J.N. 1984, Phys. Fluids, submitted.
Yoshimori, M., Okudaira, K., Hirasima, Y. and Kondo, I. 1983, Solar Phys., **86**, 375.
Zaitsev, V.V. and Stepanov, A.V. 1983, Sov. Astr. (Lett.), **8**, 132.
Zaitsev, V.V., Stepanov, A.V., and Sterlin, A.M. 1985, Sov. Astron. Lett., **11**, 192.
Zhao, R. and Jin, S. 1982, Scientia Sinica, **25**, 422.
Zodi, A.M., Kaufmann, P. and Zirin, H. 1984, Solar Phys., **92**, 283.
Zweibel, E.G. and Haber, D. 1983, Ap. J., **264**, 648.

CHAPTER 3: IMPULSIVE PHASE TRANSPORT

CHAPTER 3: IMPULSIVE PHASE TRANSPORT

TABLE OF CONTENTS

R.C. Canfield, E. Bely-Dubau, J.C. Brown, G.A. Dulk, A.G. Emslie, S. Enome, A.H. Gabriel, M.R. Kundu, D. Melrose, D.F. Neidig, K. Ohki, V. Petrosian, A. Poland, E. Rieger, K. Tanaka, H. Zirin

Page

TABLE OF CONTENTS

CHAPTER 3: IMPULSIVE PHASE TRANSPORT

Richard C. Canfield, University of California, San Diego
Francoise Bely-Dubau, Nice Observatory
John C. Brown, University of Glasgow
George A. Dulk, University of Colorado
A. Gordon Emslie, University of Alabama in Huntsville
Shinzo Enome, Nagoya University
Alan H. Gabriel, Rutherford Appleton Laboratory
Mukul R. Kundu, University of Maryland
Donald Melrose, University of Sydney
Donald F. Neidig, Air Force Geophysics Laboratory
K. Ohki, Tokyo Astronomical Observatory
Vahe Petrosian, Stanford University
Arthur Poland, Goddard Space Flight Center
Erich Rieger, Max-Planck-Institut fur Physik und Astrophysik
Katsuo Tanaka, Tokyo Astronomical Observatory
Harold Zirin, California Institute of Technology

3.1 INTRODUCTION

3.1.1 Motivation for Transport Studies

In the astrophysics community, "the solar flare problem" is generally considered to be how to accumulate sufficient magnetic energy in one active region and to subsequently release it on a sufficiently short time scale. Satisfactory solution of the solar flare problem will require at least two achievements by the solar physics community: first, convincing theoretical demonstration that one or more mechanisms of energy storage and release can occur; second, convincing observational demonstration that one (or more) of these theoretical processes actually does occur in the solar atmosphere. The contents of this Chapter essentially relate to the second problem, being largely concerned with how the energy released from magnetic form is transported through the solar atmosphere before escaping in the form of the radiant and mechanical energy signatures which we must interpret.

A central point of general agreement concerning all mechanisms suggested for dissipation of magnetic energy in flares is that the actual sites of reconnection must involve scale lengths well below currently, or forseeably, achievable spatial resolution. Consequently, observational evidence in support of a flare theory is necessarily indirect, in the sense of not involving measurement of plasma parameters in the primary dissipation regions. Such indirect evidence may be of several kinds. Firstly, circumstantial evidence may be obtained by spatial resolution of the geometry, on a larger scale, of the magnetic environment in which the mechanism operates. This may permit distinction between such options as emerging flux models and twisted arch models or between mechanisms driven by currents parallel to the magnetic field, as opposed to perpendicular. Spatial resolution also permits mapping of the paths of flare products. Secondly, temporal evidence can be obtained by use of high time resolution to set limits on instability growth rates, to imply the occurrence of repetitive or multiple dissipation, to indicate the production sequence of the various flare manifestations and, by causality arguments, to set an upper limit to the size of the primary dissipation site. Thirdly, thermodynamic evidence on the nature of primary dissipation is obtainable from the distribution, particularly in the impulsive phase, of the flare energy release over its various modes, i.e. fast particles, conduction etc. For example the Petschek mechanism releases a major fraction of the magnetic energy directly into bulk mass motion while the tearing mode initially results chiefly in plasma heating and particle acceleration.

Transport of energy away from the primary sites, and its ultimate thermalization, depend not only on the primary mechanism itself but also on the larger-scale structure of the active-region atmosphere in which the transport occurs. Consequently the study of energy transport as a diagnostic of flare mechanisms involves extensive theoretical modeling of the transport processes, as well as observational input, to provide the framework for interpreting the observations. Use of such transport studies to infer properties of the initiating disturbance is essentially an inverse problem, and so carries the danger of indeterminacy through mathematical ill-posedness. For example, the thermal structure of a conductively evolving atmosphere rapidly becomes only very weakly dependent on the heating function which initiated it, and so is a poor signature of this function. In such situations, the best strategy is to utilize jointly as many as possible independent signatures of the process to minimize indeterminacy. It is just such a combination of independent signatures of the flare process (much more compelling when taken together) that the coordinated observational approach of SMM has rendered possible.

In addition to these flare-oriented objectives, of course, the study of flare energy transport has contributions to make to the broader field of transport studies per se, such as in plasma and atomic physics.

3.1.2 Historical Perspective

Energy transport studies have become an increasingly prevalent means of investigating solar flares since around 1970 with the accompanying steady improvement in observational coverage and resolution in space, time, and spectrum, from the start of the Orbiting Solar Observatory (OSO) period onward. By the time of the Skylab Apollo Telescope Mount (ATM), considerable progress had been made toward obtaining the instrumentation needed for acquisition of high resolution data over as wide as possible a variety of wavelengths from optical to γ-rays. Thus the ATM package achieved spatial resolution of the order of arc seconds in the soft X-ray and ultraviolet ranges together with extensive UV line spectroscopy. Contemporarily though not simultaneously, other satellites (notably ESRO TD1A, OSO-7 and the Intercosmos series) were improving the quality of hard X-ray measurements and extending the spectral range upward in energy resulting in the detection of solar γ-ray lines. Though ATM itself was never designed as a flare mission it nevertheless made major contributions to progress in flare observations, in addition to its pioneering discoveries in relation to solar coronal structures and active regions (cf. Zirker 1977, Orrall 1981). In particular, ATM established the importance of loop structure in the magnetic configuration of many flares, and the compactness of bright XUV flare kernels. In addition, the use of ATM for flare studies (Sturrock 1980) delineated the limitations of such a package for answering some of the key questions concerning flares, and thereby provided important guidelines for the planning of subsequent missions, especially SMM.

The most important obstacles to progress before the launch of SMM were the lack of data of sufficiently high time resolution, the lack of data sufficiently early in (or prior to) the flare, and the lack of data coordination over a wide enough spectral range (Brown and Smith, 1980). Examining ATM in the light of these obstacles, we can see with hindsight that ATM did not respond sufficiently quickly to enable systematic studies of flare onset or impulsive phases. During ATM much information on flare spatial structure was recorded photographically in most characteristically "thermal" wavebands - from optical to soft X-rays - with resulting limitations on time resolution, on simultaneous spectral coverage, and on calibration. Typically "non-thermal" emissions such as hard X-rays, γ-rays, and microwaves were recorded with comparatively low sensitivity, and little or no coordination with the ATM experiments. A typical consequence of these problems was that testing of electron heated models of flare atmospheres involved construction of a spectroscopic model atmosphere from data obtained from a variety of places in a variety of flares, and use of hard X-ray data, devoid of spatial information, from yet different flares.

By contrast, SMM formed a coordinated package of instruments dedicated to study impulsive phase phenomena at high time resolution using pre-planned targets and automated response to flare onset triggers. In addition, the spacecraft carried short wavelength instruments of unprecedented sensitivity and time resolution (the Gamma-Ray Spectrometer, GRS, Forrest et al., 1980, and the Hard X-ray Burst Spectrometer, HXRBS, Orwig et al., 1980) and spatial resolution (the Hard X-ray Imaging Spectrometer, HXIS, van Beek et al., 1980) and the facility for high resolution atomic X-ray line Spectroscopy (the Soft X-ray Polychromator, XRP, Acton et al., 1980), with digital data recording, as well as the Ultraviolet Spectrometer and Polarimeter (UVSP, Woodgate et al., 1980). Furthermore, the package was supported by a wide range of other spaceborne instrumentation and an international network of ground based observations coordinated through the SMY (Svestka, Rust, and Dryer 1982). Ground support included rapid arc-second resolution in both microwaves, by the VLA, and in spectrally resolved optical lines by the Sac Peak Vacuum Tower. Hard X- and γ-rays were still observed without spatial resolution but with such sensitivity as to permit close temporal correlation with features in the longer wavelength images, as well as γ-ray nuclear abundance spectrometry with some simultaneous millimetric coverage at ultra high time resolution by Itapetinga. On the debit side, SMM was limited in soft X-ray spatial resolution compared to ATM, and in the restricted range of UV spectral coverage, of particular importance in modeling transition-region lines.

A significant contribution to our knowledge of impulsive-phase transport has come about as a result of the post-SMM launch of the Japanese Hinotori spacecraft, whose flare instruments (Solar Gamma-Ray Detector, SGR, Hard X-ray Monitors, HXM and FLM, Imaging X-ray Telescope, SXT, and Soft X-ray Crystal Spectrometer, SOX) are described by Kondo (1982). Early Hinotori results have been described in the Hinotori Symposium on Solar Flares (Tanaka et al., 1982) and Recent Advances in the Understanding of Solar Flares (Kane et al., 1983); later Hinotori results play an important role in this chapter.

In addition to the observational requirements already mentioned, there is obviously a need for improved theoretical modeling, particularly in the direction of making predictions which would be testable in terms of realizable data (Brown and Smith, 1980). The post-ATM period has indeed seen a major increase in the amount and sophistication of flare modeling work, particularly in respect of relaxation of earlier simplifying assumptions (such as hydrostatic equilibrium and optical thinness) in describing the atmospheric response, and of basic electrodynamic and plasma collective effects on the transport of charged beams.

The perceptive reader will find that many of the questions posed in the Skylab era have been answered in this chapter. Kane et al. (1980), in the impulsive phase chapter of Solar Flares: A Monograph from Skylab Solar Workshop II (Sturrock, 1980), posed three key questions:

1. Is the distribution of energetic electrons thermal or nonthermal?

2. Do the energetic particles (electrons), produced during the impulsive phase, provide the energy for the whole flare?

3. Among the models of the impulsive phase suggested so far, which ones are most consistent with observations?

The answers to these questions are fundamental to our understanding of space plasmas; if the models that are best supported by the observations require particle acceleration efficiencies $\gtrsim 0.1\%$ (Hudson, 1979), a substantial challenge is presented to the solar flare theorist.

Our work answers primarily the first and third questions, while the second motivated the Solar Flare Energetics group. Certainly there are still major gaps in our theoretical understanding of how energy is propagated by electron beams, as well as our observational understanding of spatial, temporal and spectral scales. However, one impulsive phase transport model now stands out above all others: the nonthermal-electron thick-target model, in which the dominant role in the transport of energy on impulsive-phase timescales (usually $\lesssim$ tens of seconds) is played by beams of electrons, mostly in the deka-keV range, whose velocity distribution function cannot be described by a single-temperature Maxwellian. These electrons are guided along a loop-like magnetic field structure, from an acceleration site in the corona; they heat and cause both thermal and nonthermal emission as they are fully thermalized in the loop plasma, a thick target. The reader should not get the impression that this model describes all flare energy transport, particularly on longer timescales, or that the nonthermal-electron thick-target model passes all the observational or theoretical tests we impose upon it below. Surely there is ample evidence for the impulsive-phase existence of nonthermal particles other than electrons, for example, and for thermal domination of later flare phases. However, our preoccupation with the nonthermal-electron thick-target model in this chapter is testimony to our finding that, of the models available, it does the best job of explaining the wide variety of impulsive-phase observations we have studied.

3.1.3 Overview of the Chapter

Some of the most striking recent results have come from observations in the highest energy ranges. Interpretation of SMM γ-ray line data has established the presence of protons ($E \gtrsim 10$ MeV) in regions of density $\gtrsim 10^{13}$ cm^{-3}, while the ISEE-3/PVO occultation data demonstrated that electrons of $E \geq 150$ keV are stopped deep in the chromosphere. Both these results suggest a thick target beam interpretation, in which the radiation is generated in the course of fully stopping the particles, in the lower solar atmosphere. On the other hand, ISEE-3/PVO data surprisingly show very little directivity at 350 keV, contrary to purely collisional transport models of the electron beam. Most notable of all high energy data is the striking demonstration, from limb brightening studies, that directivity is present in the continuum around 10 MeV. The first-ever images in the deka-keV range (10 - 100 keV) stimulate much of the work of this Chapter. While there remains considerable debate over their implications for theoretical models (in particular beams), the presence in some flares of impulsive hard X-ray footpoints, coincident in time and space with chromospheric emissions, has been clearly established by SMM and Hinotori. Hinotori results have also been used to suggest a provisional classification of hard X-ray flares, of which footpoint events are only one class. Simultaneous microwave data have permitted comparative morphology studies of hard X-rays and microwaves with resulting constraints on the electron and magnetic field distributions. Computations of hard X-ray polarization incorporating the effects of magnetic field curvature show that even the most recent low polarization results from Shuttle experiments are not incompatible with a beam model.

While the greatly improved spatial and temporal resolution and coordination of SMM data in the hard X-ray and EUV further support the view that these radiations have closely related origins, the relationship between the fluxes in these bands remains a theoretical enigma in terms of energy balance.

The inclusion of high resolution X-ray spectroscopy allowed detailed diagnostics of the hot plasma during the onset of the thermal phase, using recent developments in the atomic physics of dielectronic satellite spectra. Line profiles and shifts enabled the determination of turbulent energy content and upward velocities during the impulsive phase, resulting from the chromospheric evaporation process. Observations of X-ray and Kα line excitation, together with data on the soft X-ray plasma as a whole and on the hard X-ray source electrons, have demonstrated that electron beam excitation is not essential to explain the Kα observations, though not precluded by it.

Improved calculations of the production of the non-thermal red-shifted Lα line by capture processes on a descending proton beam have placed severe constraints on the flux of such beams and hence on flare models where protons play a central role energetically and in which there has been renewed interest recently.

At the time of ATM, white light flares were still regarded as a rare phenomenon suggestive of an exotic explanation. Studies with improved observational methods have shown that they are in fact of common occurrence. Coordinated temporal, spatial, and spectral information shows that the impulsive phase white light emission is spectrally compatible with a chromospheric origin, and temporally and energetically compatible with a thick-target electron beam as the power supply.

Interpretation of Hα profiles represents a particularly good example of the progress SMM achieved toward the ideals of rapid response, data coordination, and detailed modeling. Full radiative transfer models of the flaring chromosphere have permitted comparison on a pixel by pixel, instant by instant basis of Hα profiles observed at Sacramento Peak with those predicted according to chromospheric evaporation models driven respectively by thick target electron beams (observed by HXIS) and by thermal conduction. Results provide direct evidence in support of electron beam heating in some bright Hα kernels.

Finally, prior to the SMM era decimetric radio waves were considered to be a plasma diagnostic emission of little importance energetically. At this time serious consideration is being given to their role in heating both the corona (by reabsorption) and the chromosphere (by electron precipitation) in flares where conditions favor the onset of coherent amplification of these radio waves. The full ramifications of this process for flare energy transport are only now beginning to be appreciated.

3.2 IMPULSIVE PHASE OBSERVATIONS AND THEIR INTERPRETATION

3.2.1 Gamma-Ray Emission Above 10 MeV.

Photon emission above 10 MeV during solar flares was observed by the Gamma Ray Spectrometer (GRS) on board SMM. As of February 1984 the highest energy photons detected, energy ~ 80 MeV, came from an intense flare on June 3, 1982. In this contribution we discuss the timing between hard X-rays and gamma rays and present evidence for directivity of the highly energetic particles that give rise to the emission above 10 MeV.

3.2.1.1 Relative Hard X-ray and Gamma-Ray Timing

As a typical example we show in Figure 3.1 the time history of the flare of June 15, 1982 in different energy bands. The event has a simple time structure consisting mainly of one impulsive burst. The peak of the emission is simultaneous within ± 2 sec over more than three decades of energy from 30 keV - 50 MeV. The hard X-ray flux shown in the upper two

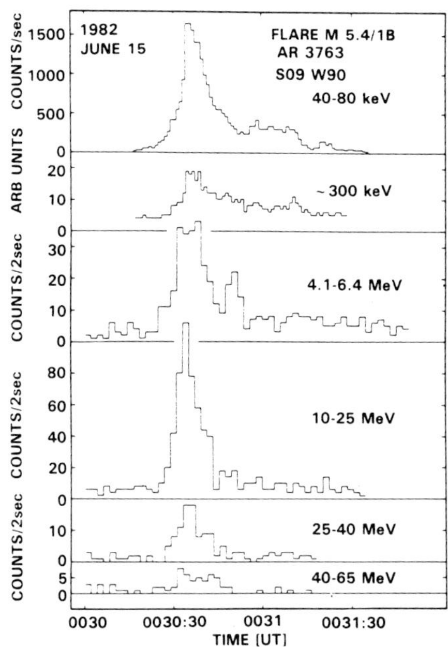

Figure 3.1 Time history of the flare of June 15, 1982 in various energy bands.

panels is assumed to originate from electron bremsstrahlung. The emission from 4.1 - 6.4 MeV (the nuclear energy band) is from nuclear lines and from bremsstrahlung of relativistic electrons (Forrest 1983). At energies greater than 10 MeV the gamma rays are expected to be produced by bremsstrahlung of very highly energetic electrons and by the decay of pions (π° and π^+), because the contribution of nuclear lines above 8 MeV is negligible (Crannell, Crannell, and Ramaty, 1979). The relative importance of the two processes for solar flares is discussed by Ramaty et al. (1983) and by Rieger et al. (1983).

Until February 1984, 14 flares were observed with emission above 10 MeV. Their time history has the following general characteristics: They are of short duration (~ 1 min) and very impulsive. Rise and fall times are on the order of seconds. They exhibit single or multiple peaks. The peak emissions of the hard X-ray and gamma rays are simultaneous within about ± 6 sec. There is no systematic delay of the gamma rays with respect to the X-rays.

Because of the simultaneity of the gamma- and X-rays, the energy loss of the highly energetic particles (electrons and/or ions) has to take place at ambient densities $> 10^{13}$ cm^{-3}, if we assume a thick target situation (Bai and Ramaty 1976).

3.2.1.2 Directivity of Highly Energetic Particles

The flares with photon emission above 10 MeV are at heliocentric angles of > 60° (Rieger et al., 1983). This is shown in Figure 3.2, where the location of the flares, known from Hα observations (NOAA: Solar Geophysical Data) is plotted. The arithmetic mean heliocentric angle of all 14 flares is 79°. If the radiation is isotropic the probability for a chance coincidence of such a distribution is ~ 2×10^{-7}. Therefore we conclude, that this "limb brightening" is the result of directivity of the radiation. This directivity must exist also in the primary highly energetic particles, because photons with energies above 10 MeV, if created by electrons (> 10 MeV) via bremsstrahlung or by protons (> 100 MeV) via pion decay, are emitted preferentially in the direction of the motion of these particles. At this time the most likely interpretation appears to be that the emitting particles are indeed travelling roughly normal to the sun-center direction (J. Cooper and V. Petrosian, private communication), as if they were near their mirroring point, for example.

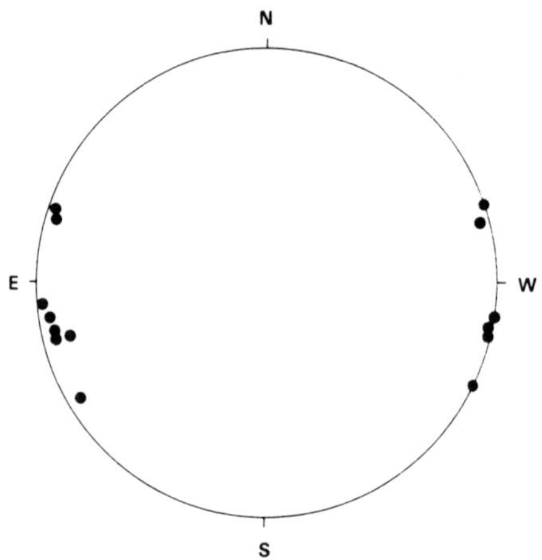

Figure 3.2 Solar disc position of flares with photon emission above 10 MeV.

To study this phenomenon in more detail it would be necessary to make stereoscopic observations with two detectors of the GRE type widely spaced in solar aspect angle, as was done by Kane et al. (1980, 1982) at lower energies.

3.2.2 Hard X-ray and Microwave Morphology

3.2.2.1 Observations

The strong correlation in the temporal evolution of hard X-rays and microwave radiation established over the years has led to the belief that the same or very closely related populations of electrons are responsible for both of these two different flare radiations. High spatial resolution observations at both wavelength ranges provide further information on the details of this correlation.

Since the first fanbeam observation of flares by Enome et al. (1969) with resolution 24" at 9.4 GHz, the new instruments such as the NRAO 3 element interferometer with resolution 10" at 2.7 GHz (Alissandrakis and Kundu 1975), the WSRT array with one dimensional resolution of 6" at 5 GHz (Alissandrakis and Kundu 1978, Kattenberg and Allaart 1981) and finally the VLA with two dimensional resolution of 1" at 15 GHz (Marsh et al., 1980, Lang et al., 1981 and Kundu et al., 1981) have steadily improved the resolution of microwave observations. VLA observations (Kundu et al., 1982, Dulk et al., 1983) and fanbeam observations at 35 GHz (Nagoya) and 17 GHz (Nobeyama) have continued, while the HXIS on SMM and the SXT on Hinotori have begun for the first time to provide spatially resolved images with resolution of about 10" at X-ray energies of less than 40 keV. Unfortunately, simultaneous microwave and hard X-ray high resolution images exist for only a few flares. Consequently, to some extent, we must still rely on the data of more numerous events with only either hard X-ray or microwave images.

We first summarize the results of observations where there exists only high resolution microwave data, then those with only X-ray data and, finally, the few cases with both X-ray and microwave observations.

a. Microwave Morphology. Studies of spatial structure of microwave radiation from flares have been summarized by Marsh and Hurford (1982) and Kundu (1983). It should be noted at the outset that, unlike the X-ray studies, the microwave observations (in particular the high resolution one-dimensional results from WSRT and the two-dimensional results from the VLA) have not been as systematic, continuous and extensive as the X-ray observations. Consequently, it is difficult to classify the microwave structure in well defined categories. In spite of this, these observations have shown some common features on which we will concentrate here, remembering that there may be more complex structures yet to be studied, analyzed, and classified.

The most general statement that can be made is that the brightest point of the microwave radiation during the impulsive phase occurs near a magnetic neutral line and not on an Hα kernel, and that in cases with simple field geometries there is a single dominant source. Whenever a secondary source is detected, that source lies near another neutral line (e.g., 26 May 1980 flare, Kundu 1983). It can be concluded that with few exceptions (see Kundu et al., 1982), the predominant microwave emission does not come from the footpoints of the flaring loop. In some flares the microwave source is approximately midway between the footpoints and is nearly equally but oppositely polarized (circular) on both sides of the maximum brightness spot (see Figure 3.3). The simplest interpretation of these observations is that the emission comes from the region around the top of a loop. Spectral variation of the structure and polarization then tell us something about the geometry of a loop and the pitch angle distribution of the radiating electrons (Petrosian 1982). However, there are flares where only one sense of polarization is observed (cf. Kundu 1983), and there are cases where the source is not located very high up in the corona (Kai et al., 1982). These features have been attributed to asymmetric magnetic loops (Kundu and Vlahos 1979) but, as we shall discuss below, such structures could also arise when a non-circular or sheared loop is viewed from an angle away from the line perpendicular to the field at the top of the loop.

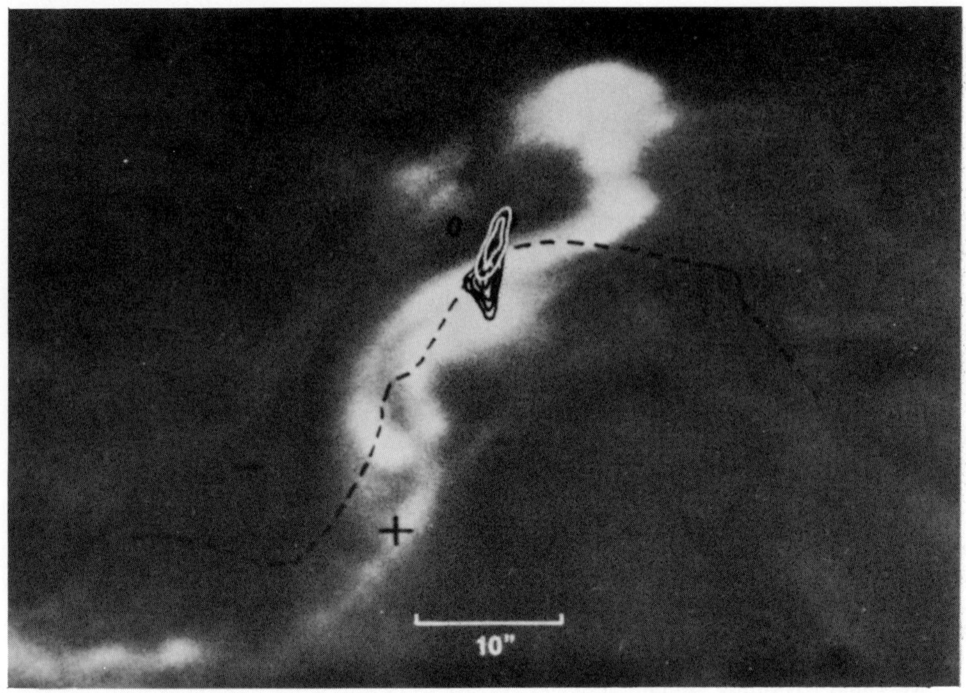

Figure 3.3 VLA map at 15 GHz in right circular (black) and left circular (white) polarization superimposed on Hα frames. The dashed line depicts the location of the magnetic neutral line (from Hoyng *et al.*, 1983).

b. X-ray Morphology. Prior to the era of SMM and Hinotori, the only information on hard X-ray spatial structure was obtained from stereoscopic observation by the PVO and ISEE-3 satellites (Kane et al., 1979, 1982). These observations do not provide images, but give easily-interpreted information on the height of the hard X-ray emitting regions. Furthermore, they extend to much higher X-ray energies than the imaging instruments and tell something about the variation with height of X-ray spectra, which turns out to be very useful in modeling (cf., Brown et. al., 1981, Leach and Petrosian 1983).

The first hard X-ray images of flares were obtained by the HXIS on SMM. The analysis of the few early flares emphasized X-ray structure consisting of two sources which were identified as the footpoints of flaring loops (Hoyng et. al., 1981). But as indicated more recently (Duijveman and Hoyng 1983), there are many flares where the bulk of the hard X-rays (> 16 keV) come from a single source.

Analysis of the data from SXT on Hinotori (Ohki et al., 1982, Tsuneta et al., 1982, Takakura et al., 1982, Ohki et al., 1983, Tsuneta et al., 1983, Takakura et al., 1983 and Kosugi et al., 1983) indicates the predominance of flare images consisting of a dominant single source which sometimes expands and become elongated, and sometimes shrinks. There are also rarer occurrences of double sources (not necessarily of equal intensity) which merge into a single source located between the original two sources as the flare progresses. Similar structures and evolution have now been shown to be present in the HXIS data (Machado 1983, and Machado et al., 1983).

Based on light curves and spectral images measured by instruments on board Hinotori, various classifications of the flares have been given by Ohki et al. (1983), Tanaka (1983) and Tsuneta (1983). For the following reasons, the classifications are preliminary at this time. They are based on only the strongest flares, which may not be representative of all flares. The majority are weaker and shorter-lived than the ones used for this classification. In addition, relatively few flares have been classified; the number of flares in each class is small. It should also be noted that they are based on only qualitative differences between spatial structures, spectra and the impulsiveness of the light curve. For a more thorough analysis the parameters describing these characteristics should be quantified and flares binned accordingly. The flares studied may be the extreme cases, so that further analysis may show a continuum of classes with no clear dividing line between the present categories.

The Hinotori data have been classified into three types, A, B, and C, as described in Table 3.1. Type A events are defined as those which have a smooth light curve below 40 keV (no discernable spiky features), a soft spectrum (spectral index > 6 for a power law fit, or an exponential spectrum within observational uncertainty) and consist of a single compact source. Two examples of such flares have been extensively studied

Table 3.1 Hinotori Flare Classification

OBSERVATIONAL CHARACTERISTICS

TYPE	TIME PROFILE	HARD X SPECTRUM	HARD X (E>15 keV) IMAGE	GAMMA-RAY TIMING
A	E<50 keV E>70 keV 10 min X-class only	Very soft $\gamma \sim 7-9$ hot plasma $T \sim 3-5 \times 10^7$ k $EM \sim 10^{49}/cm^3$ → Fe XXVI	Small point-like (~10") low-altitude (≤5000km) 10 arc sec	No γ-ray emission
B	≤50keV ≥50keV spike smooth	Impulsive phase →hard	Double sources footpoints Hα hard x	Delay ≤ several sec
		gradual phase →soft	Coronal loop-like 1min	No γ-ray emission
C	≤200 keV ← 1 hr → smooth spike X-class only	Power-law $\gamma \sim 3-5$ $\left\langle \frac{d\gamma}{dt} \right\rangle < 0$	High-altitude coronal source (stationary) Hα hard X 1min	Delay ~ tens of sec ~ min

(Tsuneta et al., 1984), the April 2nd and July 17th flares of 1981. The April 2nd flare (Figure 3.4) shows a single steady point source (perhaps barely resolved at 15"). Strong Fe XXVI lines are also observed throughout these events.

Type B flares are those with two distinct hard X-ray bright spots during the initial (impulsive) part of the flare, which evolve into a single source located somewhere between the initial bright regions.

HINOTORI SOLAR X-RAY IMAGE
SXT-2 30-50 KEV YR:MM:DD = 81:8:11
ONE PIXEL = 9.958 (ARCSEC)
CONTOUR: MIN = 1.0000 STEP = * 1.4142

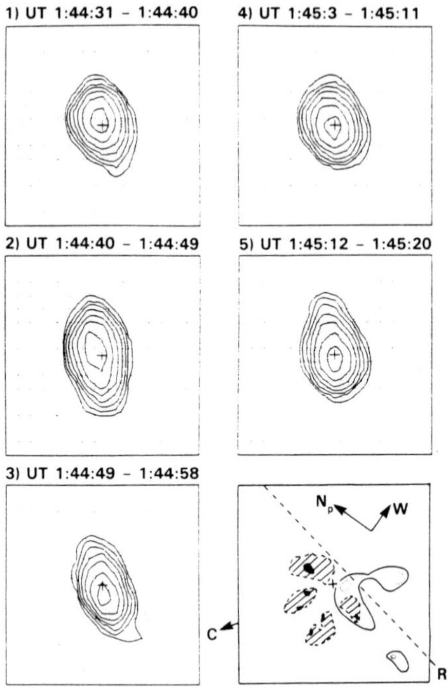

Figure 3.4 Example of a Type A flare, Aug. 11, 1981. The lower right-hand panel shows a sketch of the Hα flare at 01:55:45 UT and the direction of the peak of the one-dimensional brightness at 17 GHz at 01:45:10 UT. The other five panels show five different X-ray (25-50 keV) images (Takakura *et al.*, 1983).

During this evolution it is observed that spikiness of the light curve disappears and the spectrum softens. Hinotori events of July 20 (Figure 3.5) and October 15 of 1981 are two such events.

Finally, there are Type C flares which, like Type A flares, consist of a single source, but the source is more diffuse and is clearly displaced from the Hα kernel and, like the microwave sources described above, could be emanating from the top of a set of loops. The light curve is also smooth, even at energies greater than 49 keV, and the spectrum is hard. The May 13 (Figure 3.6) and April 27 flares of 1981 are two such events.

As mentioned above, the HXIS data also show flares with spatial structures similar to those described by Hinotori. Figure 3.7 shows some such examples.

It should be noted that some of the Type A flares may actually be of Type B, but unresolved, either because they are intrinsically compact or because they are elongated sources viewed along the major axis.

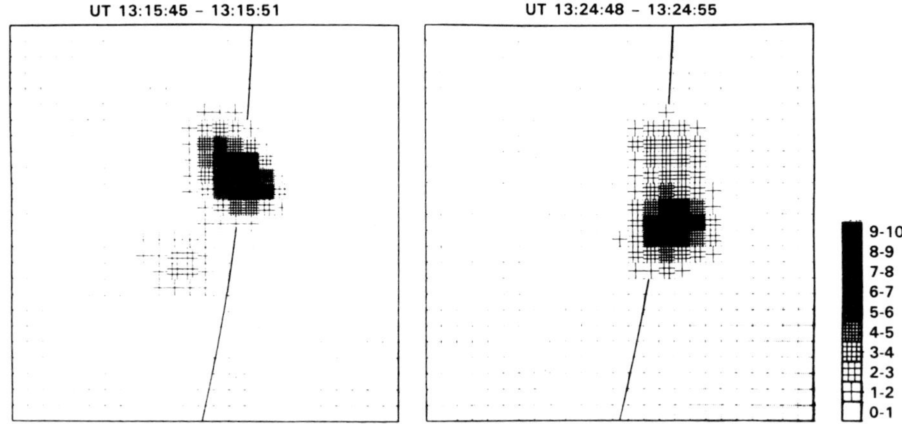

Figure 3.5 Example of a Type B flare, July 20, 1981: (a) impulsive phase, (b) gradual phase. Each image is 3′ in size and each pixel is 6″. The curved line bisecting the images shows the location of the solar limb. Note the evolution from a lower double to a higher single source (from Tsuneta *et al.*, 1983b).

c. Simultaneous Hard X-ray and Microwave Morphology. There are less than a dozen flares which are resolved (in one-dimension or two) at both hard X-ray and microwave energies. Here we describe the brightest of these and concentrate on the relation between the hard X-ray and microwave structure of the most important features. We begin by summarizing Hinotori observations of four events representing the three types described above, and then four SMM X-ray events that have also been observed in microwaves by the WSRT or the VLA, and fit them into categories A, B or C.

The August 11, 1981, Hinotori flare is classified as Type A, but as far as the X-ray image is concerned, it falls on the border line between Types A and C. Without a quantitative measure of the distinguishing parameters of the various types, one cannot resolve this ambiguity. In any event, the X-ray image appears to be a steady single elongated source (with a hint of a double source at the rising phase). The one-dimensional 17 GHz tracing shows a near coincidence of the peak microwave and X-ray positions (Figure 3.4). Both maxima seem to be located over a neutral magnetic line. (For details see Takakura et al., 1983.)

The April 1 and May 13, 1981, Hinotor flares are both classified as Type C. The May 13 flare has a very smooth time profile even above 100 keV, while the X-ray light curve for the April 1 flare is midway between the August 11 and May 13 flares. As shown in Figures 3.6 and 3.8, the 35 GHz microwave position agrees with the position of a single dominant X-ray source. Both emissions appear to come from the top of a loop or arcade of loops. However, there is some asymmetry in the one-dimensional microwave image. This asymmetry becomes stronger (indicating the existence of two sources) in the later phases of the flare, and in the case of the April 1 event (Figure 3.8) it may be interpreted as a double source. Unfortunately, there is no X-ray image for these periods.

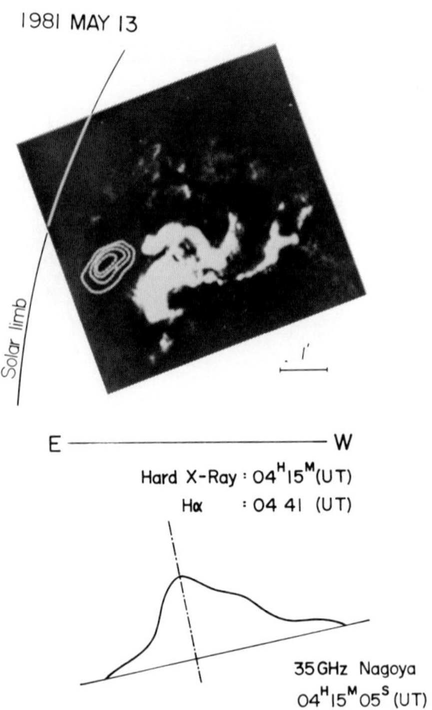

Figure 3.6 Example of a Type C flare, May 13, 1981. Comparison of the hard X-rays (17 to 40 keV), the Hα photograph and the one-dimensional radio images for the indicated times (from Tsuneta *et al.*, 1983b).

The October 15, 1981, Hinotori flare may be classified as Type B because initially it consists of two sources (one much brighter than the other). However it is not clear if these sources correspond to any Hα kernels. The 35 GHz image can be decomposed into a double source with the stronger microwave source corresponding with the weaker X-ray source (Figure 3.9). There is a secondary peak in the light curve during which the X-ray structure changes rapidly. The brighter source moves about 30 arcsec southward and the weak source becomes stronger. Finally, during the decaying phase these sources disappear and a third source appears in a new position. There are no microwave images for the second peak or the decay phase of this flare. Probably this is a Type B flare like the July 20, 1981, limb flare. However, one cannot rule out the less likely possibility that the four different bright regions correspond to the tops of four different loops.

The November 5, 1980 SMM event (Figure 3.10) described by Hoyng et al. (1983) has excellent overlapping HXIS and VLA coverage. As is typical, the 15 GHz radiation comes from above a neutral line, as shown in Figure 3.10. The hard X-ray and 9.4 GHz light curves show a Type B time profile and consist of three peaks, with the third peak being a soft, gradual one. The 16 to 30 keV HXIS counts in various pixels conform

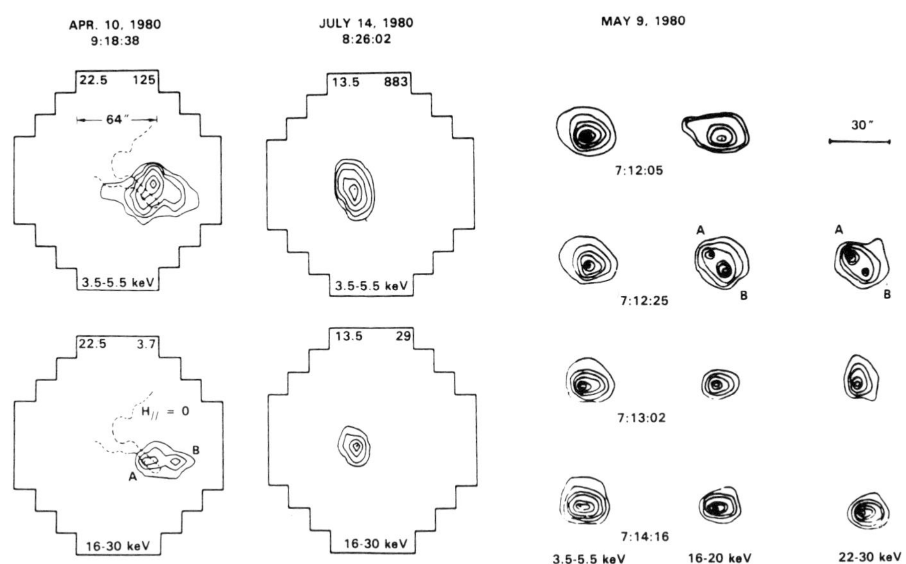

Figure 3.7 Examples from HXIS of double sources, single sources and evolving structures similar to those seen by Hinotori (from Duijveman and Hoyng 1983 and Machado 1983).

to the Type B character of this event in that, initially, there is less flux from the region over the neutral line than from the Hα bright spots, and during the second and third peaks most of the X-rays come from the approximate center of gravity of the bright region and coincide with the bright microwave source.

The July 13, 1980, SMM event described by Kattenberg et al. (1984) shows two sources in coincidence with the Hα bright patches. Unlike the Type B flares there is no gradual peak (or phase) in the light curve and no emergence of a single source in the middle. Unfortunately, the baseline of the WSRT one-dimensional microwave tracing is perpendicular to the line connecting the two X-ray sources so that we do not know if the microwave source is also double. However, the usual centrally-located microwave image is consistent with the data.

On June 24, 1980, two bursts at 15:20 and 19:57 were extensively mapped at 6 cm by the VLA with some coverage by the HXIS on SMM (Kundu et al., 1983). The 15:20 event shows a hard X-ray image displaced by 20 to 30 arcsec from the microwave source, the latter being typically bipolar (separated right and left circularly polarized structures, as in Figure 3.3). The softer X-ray and microwave emissions are co-spatial, presumably coming from the loop top, while the hard X-rays may come from a footpoint. This may therefore be a Type B event.

d. Summary. The sample of flares described above is obviously incomplete in many ways. In general, any sample is limited by selection effects associated with observing instruments. However, certain inferences about flare phenomena are affected less than others by such selection criteria. We hope that by concentrating on the gross features of the flare phenomenon the influence of the selection effects can be minimized.

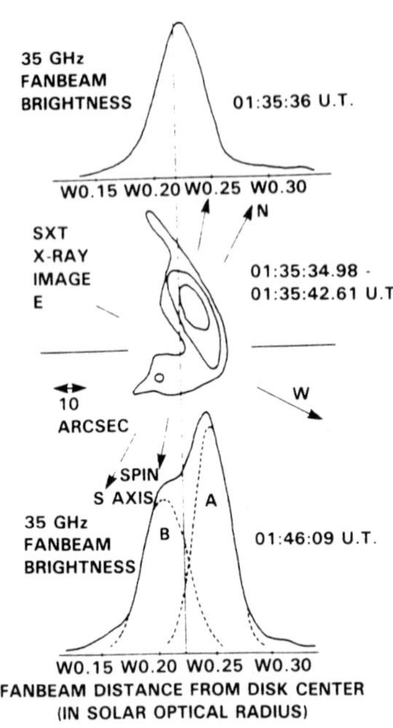

Figure 3.8 One-dimensional microwave tracing and the hard X-ray image from SXT of the April 1, 1981, flare (from Kawabata *et al.,* 1982).

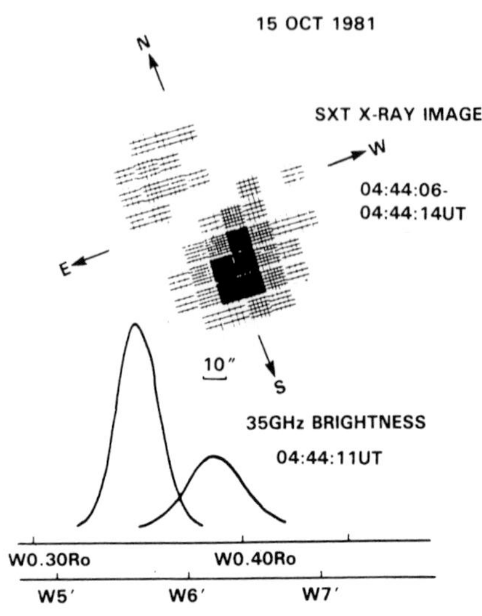

Figure 3.9 A comparison of the SXT hard X-ray image with the fanbeam radio brightness after restoration for the October 15, 1981, event (from Kawabata *et al.,* 1982). Whether the two images belong to footpoints of a single loop or emanate from top of the distinct loop is not known.

The prominent features of the X-ray and microwave morphology on the basis of the small number of events presently analyzed, can be summarized as follows:

1. For almost all flares where there exist simultaneous microwave and Hα images the microwave emission comes from a region above the neutral line (presumably around the top of a loop) between Hα bright patches (presumed to be the footpoints of loops). There are exceptions, when a minor microwave feature may be near an Hα bright patch (Dulk et al., 1983). Also, there are flares (e.g., Oct. 10, 1981) where because of coincidence of the microwave and hard X-ray images, it is assumed that all the emission is from footpoints. However, there is no direct evidence for microwave emission from the footpoints of loops.

2. In the majority of cases the X-ray emission comes from a single dominant source (Types A and C), which is found to be located above a neutral line and not on an Hα bright patch, except that for some compact Type A bursts higher resolution is needed for confirmation of this picture. The structure of Type B bursts

with two bright X-ray sources (normally coincident with Hα patches) evolves into a single source located between the Hα patches.

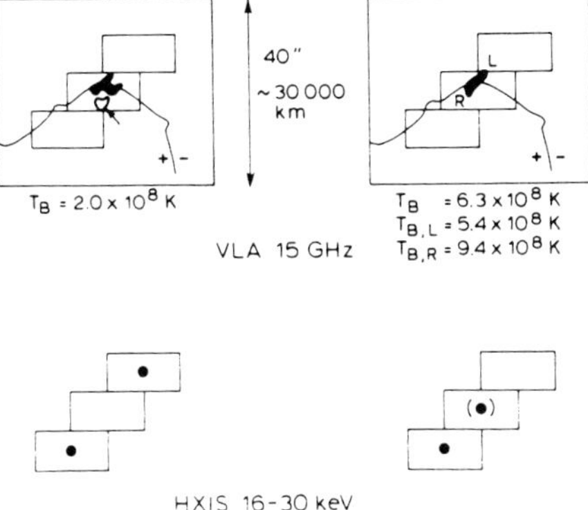

Figure 3.10 The microwave and hard X-ray images of November 5, 1980, flare. The lower panels are 8" × 16" in size and a dot indicates significant flux in excess of a single temperature fit (from Duijveman and Hoyng 1983). (*cf.* Figure 3.3)

3. The limited hard X-ray and microwave data obtained so far are consistent with, but are not direct evidence for, the conclusion that such large scale changes in the images occur on a hydrodynamic time scale, with velocity $v \approx 50$ to 200 km/sec.

4. The hard X-ray and microwave images, even though different in detail, have roughly the same location whenever both images are dominated by a single source. For bursts whose X-ray images are initially double, the single source that develops later most probably coincides with the microwave source, presumably located near the top of a loop or arcade of loops.

As we shall show in the next section, the above general features are consistent with the nonthermal model, whereby semi-relativistic electrons are injected in a closed loop somewhere in the corona. One aim of such observational and theoretical work is to determine what constraints the observations provide on model parameters such as field geometry, loop size and density, and the spectrum and pitch angle distribution of the accelerated electrons.

3.2.2.2 Interpretation

We now interpret the above data in the framework of the nonthermal models whereby electrons are accelerated to energies well beyond the thermal energy of the coronal plasma. These models are variously classified as thick target, thin target, trap, etc.

The traditional trap models require both a rapid convergence of the field lines and a low density. For electrons of energy E to mirror back and forth in the corona, one needs $\delta B/B >> \delta L/\lambda$, where δL is the length scale of the magnetic field variation δB, and λ is the mean free path of electrons of energy E. For the X-ray-producing electrons (energies $E \sim 20$ keV) to survive for time t one needs a value of density $n < 10^8$ (E/20 keV)$^{3/2}$(100/t sec) cm^{-3}.

The thin target model will be realized only in an open magnetic field configuration. Isotropic or outward streaming injections of electrons can be dismissed because, contrary to observations (Lin and Hudson 1971, 1976), they require as many electrons to reach the earth as are needed to produce the hard X-rays. If the electrons are highly beamed toward the chromosphere, the result will be similar to the thick-target closed-field configurations discussed below.

Consequently, we will consider only the general thick-target model where electrons are injected in a closed magnetic loop (Figure 3.11). We shall need to specify the site of injection of the electrons. If the site of injection is very deep, well below the transition zone (point 1 of Figure 3.11a), the electrons lose energy quickly, causing direct heating and evaporation of chromospheric plasma, and produce hardly any hard X-rays above the transition region. We shall not discuss this possibility here because it will fail to describe Type C bursts discussed above. On the other hand, if the injection is at lower densities (point 2 of Figure 3.11a), so that all of the outgoing particles eventually reach the top of the loop, then the situation is qualitatively similar to the injection at the top of the loop (point 3), which is what we shall assume.

a. Transport of Electrons. The transport of the nonthermal electrons is affected by many parameters. In particular, if the pitch angle distribution of the electrons is highly anisotropic (beamed electrons), they are subject to some instabilities which change the distribution on a time scale of the order of plasma oscillations. We shall not consider such cases here. Furthermore, if the beams constitute a high current, then a reverse current is set up. The electric field which drives the reverse current may then decelerate the original beam. These aspects of the transport are discussed in Section 3.3.1. Here we concentrate on the collisional effects and the role of large-scale static magnetic fields.

The solar flare plasma conditions allow for various simplifying assumptions, so that the variation with depth of the distribution of electrons $F(E,\mu,\tau)$, where $F\, dE\, d\mu\, d\tau$ is the electron flux in the energy range dE (in units of mc^2), pitch angle cosine range $d\mu$ and dimensionless column depth range $d\tau = dN/N_0$ ($N_0 = 5 \times 10^{22}$ cm^{-2} is the column depth required for stopping an electron with E = 1), is determined by a single parameter $d\ell nB/d\tau$, which describes the variatin of the magnetic field B along the loop, with respect to the distribution of plasma. Parameters of the injected electron spectrum $F(E,\mu,\tau) = AE^{-\delta}e^{(\mu^2-1)/\alpha_0^2}$ are the spectral index δ and the pitch angle dispersion α_0. Leach and Petrosian (1981) applied the Fokker-Planck method to electron transport in flare loops. Here we point out an aspect of their work that is relevant to the relationship between hard X-ray and microwave emission.

Let us initially consider loops with uniform magnetic field, i.e. $d\ell nB/d\tau = 0$. For nonrelativistic electrons, pitch angle diffusion and energy loss take place on comparable time scales. This means that models with widely different values of α_0^2 appear different only at the top of the loop. Such models are virtually indistinguishable lower down, where the electron distributions become nearly isotropic. Consequently, the radiation signature of spatially unresolved loops will tend to be closer to what one expects from an isotropic rather than a highly beamed pitch angle distribution. Note, however, that this is not true at relativistic energies (i.e. electrons producing the high-frequency microwave radiation and the continuum γ-rays). Relativistic electrons lose energy more quickly than they diffuse in pitch angle.

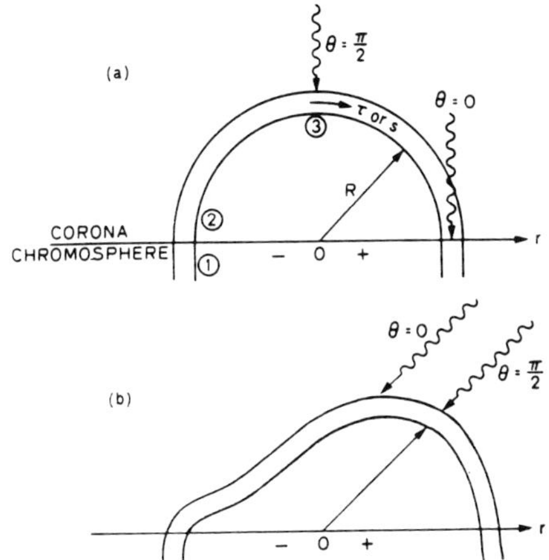

Figure 3.11 Description of the geometry of loops: (a) semi-circular loop; (b) asymmetric sheared loop. Circled points (1), (2) and (3) refer to the possible locations of injection of electrons. The wavy arrows show the line of sight and the value of θ used in equation (3). Note that ds and $d\tau = n\,ds/N_0$ are measured along the loop from the point of injection, assumed to be at (3). R is radius of the loop and r is a projected distance from the center of the loop.

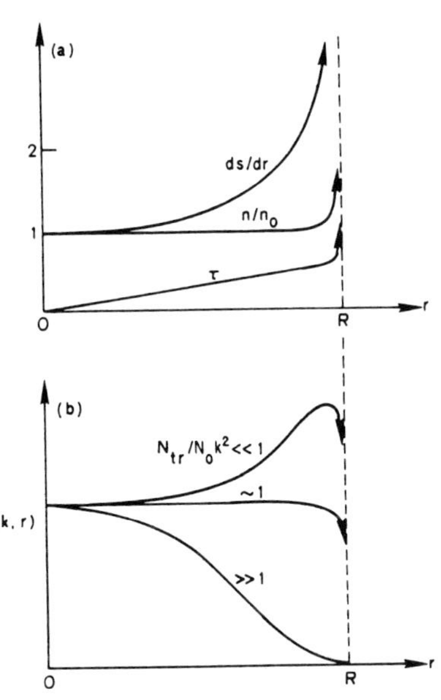

Figure 3.12 (a) Variation with projected distance (half of the loop) from the loop centers of various quantities. (b) Variation of expected hard X-ray brightness for the indicated values of N_{tr}/N_0k^2.

The effect of magnetic field convergence, i.e., of a non-zero $d\ell nB/d\tau$, is to mirror some of the electrons which do not reach the chromosphere and confine them to the top of the loop. However, such electrons also eventually scatter into smaller pitch angles, penetrate deeper, and thermalize. The overall effect of the non-zero $d\ell nB/d\tau$ is to isotropize the pitch angle distributions more quickly and enhance the effects described in the previous paragraph.

b. Non-thermal Emission. The most prominent aspects of the hard X-ray and microwave observations can readily be understood in the framework of a simple semi-circular loop model, as described above. Although this model is too simplified to use in interpreting the observations in detail, it shows that the geometry of the loop and the physical parameters of the plasma play a more significant role than emphasized in previous treatments. Following Leach and Petrosian (1983) and Petrosian (1982), we assume a semi-circular loop (radius R) in the corona with variable magnetic field which becomes vertical and uniform in the chromosphere (see Figure 3.11a). We also assume that the density no of the preflare plasma in the loop is constant in the corona and rapidly increasing in the transition region. The important parameters for our consideration are the column depth of the transition region, $N_{tr} = n_0 \pi R/2$, and $d\ell nB/d\tau$.

We begin with the spatial structure of X-rays in the range 16 to 50 keV. The X-ray emission is obtained from the electron flux F given by the transport analysis, and the bremsstrahlung radiation cross-section. At these energies the electrons are non-relativistic, and their radiation is nearly isotropic. The scaling between electron energy E and column depth τ directly translates into a similar scaling between τ and the photon energy k (in units of $m_e c^2$). Leach (1984) showed that the variation of the X-ray intensity I (normalized photon counts per energy inerval dk and depth interval $d\tau$) with k and τ can be described to within 20% by

$$I(k,\tau)d\tau \approx k^{-\delta+1}(1 + \tau k^2)^{\delta/2} d\tau/k^2 \tag{3.1}$$

This simple picture becomes complicated when we translate it into observables. Let us consider the circular loop of Figure 3.11a viewed from directly above. For comparison with observation we need the variation of the intensity projected on the solar disk I(k,r) with the projected distance r from the center of the loop:

$$I(k,r) = I(k,\tau)d\tau/dr = I(k,\tau)\ (n_o/N_o)\ (1-r^2/R^2)^{-1/2} \tag{3.2}$$

The last geometric term plays a significant role, as shown by ds/dr in Figure 3.12. Note, however, that even for non-circular geometry ds/dr (and $d\tau/dr$) can increase rapidly as the footpoints are approached. At a given photon energy k, if N_{tr} is small, or more precisely if $\tau_{tr} = N_{tr}/N_o << k^2$, then $I(k,\tau)$ will be nearly constant for $0 < r < R$. However, because of projection effects I(k,r) will rise with r, such that the footpoints at r = R will appear brighter than the top of the loop at r = 0 (cf., Figure 3.12b). For $N_{tr} >> k^2 N_o$ the brightness will decrease steadily from the top to the footpoints. The changeover of the location of the maximum brightness from r = 0 to r = R will occur quickly around $N_{tr} = N_o k^2 = 10^{20}\ cm^{-2}(k/22\ keV)^2$.

At photon energies less than 10 keV(k < .02) a low pre-flare value of $N_{tr} = 10^{19}\ cm^{-2}$ is sufficiently high to ensure that the maximum brightness occurs at the top of the loop. But such a low value of N_{tr} is not sufficient to confine higher-energy (say > 22 keV) electrons to the top of the loop. Such electrons will penetrate below the transition region so that their radiation at 22 keV will be concentrated on the footpoints. Only if $N_{tr} > 10^{20}\ cm^{-2}$ does the top becomes brighter than the footpoints at these energies also.

This general picture is nearly independent of the viewing angle and the field geometry because in general $d\tau/dr$ becomes very large near the footpoints. However, the details of the shapes of the curves drawn on Figure 3.12 do depend on the field geometry and the viewing angle.

This behavior therefore suggests that for the large diffuse Type C flares described in the first part of this section, $N_{tr} > 10^{20}$ cm^{-2} (perhaps $n \sim 10^{10}$ cm^{-3}, $L \sim 10^{10}$ cm), so that one sees the 10 - 40 keV photons from the tops of the loops. On the other hand, for the Type B events, initially $N_{tr} < 10^{20}$ cm^{-2} and the emission is concentrated in the footpoints. The asymmetry of the radiation from the two footpoints as commonly observed must then be attributed to an asymmetric field geometry (e.g., Figure 3.11b). As observed by Hinotori (Figure 3.5) and in some events by HXIS (Figure 3.7), the double images merge into a single coronal source as the flare progresses. This means that the column depth has increased from its initial low value to a value larger than 10^{20} cm^{-2}. As mentioned in connection with observations, such changes seem to occur on a hydrodynamic time scale, suggesting that the filling of the loop by the evaporated flare plasma may be the cause of the increase in the coronal column depth. It is straightforward to show that this is a distinct possibility (see the following chapter, Doschek et al., 1984).

For the majority of observed bursts the most likely location of the microwave source is around the top of the loop, as discussed in Section 3.2.2.1. However, the explanation for this observation is different than that presented above for hard X-rays. This is because the microwaves are the result of gyrosynchrotron emission by much higher energy electrons ($E \sim mc^2$) which, unlike the lower energy hard X-ray emitting electrons, cannot be confined easily. The confinement to the top of the loop of such high energy electrons by Coulomb collisions would require N_{tr} well above 10^{22} cm^{-2}. This is obviously unlikely and trapping in low density ($n < 10^{10}$ cm^{-3}) loops or some other explanation is needed to explain the observed origin of the microwave emission near the top of loops.

Microwave-producing electrons are clearly present throughout the whole loop but the efficiency of gyrosynchrotron emission varies throughout the loop and depends on the direction and strength of the magnetic field. The gyrosynchrotron radiation is strongest in the direction perpendicular to the field line. This factor alone, i.e., assuming everything else is equal throughout the loop, will favor a maximum brightness located between the footpoints. For example, the simple loop of Figure 3.11a, when viewed from above, will be brightest at $r = 0$. Note that this aspect is, to the first order, independent of the viewing angle. However, the gyrosynchrotron emission depends also on the number of electrons and the strength of the magnetic field. At a given frequency ν the emission from stronger field regions corresponds to lower harmonics of gyrofrequency ν_b. The lower harmonics are produced by lower energy electrons, $E \sim (\nu/\nu_b)^{1/2}$, which are more numerous. Since the field strength only decreases with the height above the photosphere, emission from higher regions could dominate only if this variation is slow. Finally, the strength of the gyrosynchrotron emission is a function of the pitch angle distribution of the electrons. Only if the pitch angle distribution of the accelerated electrons is broad will the emission from the coronal region be important. Otherwise, if electrons of kinetic energy $E \approx mc^2$ are beamed along the field line, they remain so throughout the corona and give little gyrosynchrotron radiation. The pitch angle distribution becomes broader and the radiation efficiency increases only below the transition region, i.e., at the footpoints. Another factor which also aids the relative enhancement of emission from the coronal part of the loop is the fact that gyrosynchrotron radiation is subject to various absorption and suppression effects (cf. Ramaty and Petrosian 1972), all of which are more important in the chromosphere than in the corona.

New observations of flare X-ray polarization by Tramiel et al. (1984) exhibit a low degree of linear polarization (< 5%). This result differs from earlier observation by Tindo and his collaborators (see e.g., Tindo et al., 1976). Such low polarization naturally will arise in the so-called thermal models or if the pitch angle distribution of the electrons is nearly isotropic. Earlier studies of nonthermal models (Haug 1972, Langer and Petrosian 1977, Bai and Ramaty 1978), which ignored the scattering of the nonthermal electrons, consequently found a high degree of linear polarization for highly-beamed electrons. Even Brown's (1972) calculation, which includes the scattering in an approximate fashion but does not include the curved loop geometry, has indicated a higher degree of polarization than observed. As a result it has been assumed that a low observed degree of polarization is a strong evidence against the non-thermal models.

Recent work by Leach, Emslie and Petrosian (1984) has examined the problem more fully, and does not confirm this expectation. The primary reason for a low degree of polarization is that no matter how strong the anisotropy of the injected particles, the average pitch angle distribution in the thick target is very broad (see Section 3.2.2.2). In addition, because the direction of the linear polarization varies along a loop, the integrated X-ray emission from the whole loop will have the characteristic of the emission from a highly broadened distribution and will show lower polarization than one would expect from the distribution of the injected electrons. Table 3.2 summarizes the effect of the spectral index δ and the pitch angle distribution dispersion parameter α_0^2 of the injected electrons, assuming the magnetic field to be uniform, and Figure 3.13 compares the observation of Tramiel et al. (1984) with three model calculations.

Table 3.2 Maximum Percentage Polarization at 16 keV

δ	α_0^2			
	∞*	0.4	0.04	0.01
3	≤ 5	≤ 5	≤ 5	6
4	8	8	11	13
5	10	11	20	21
6	10	16	26	26

*Electrons injected isotropically at the top of the loop.

A low degree of polarization can arise naturally from some nonthermal models. In particular, there is a strong effect of spectral index on polarization; compare tabulated polarizations for various δ values in Table 3.2. This is because the degree of polarization increases rapidly as the photon to electron energy ratio approaches unity ($k/E \rightarrow 1$). For steep spectra (δ large) the bulk of photons with energy k are produced by electrons of only slightly larger energies ($k/E \lesssim 1$). However, for harder spectra (δ small) the contribution of higher energy electrons (i.e., those with k/E considerably less than unity) increases, causing a reduction in the polarization.

Opposite polarization on both sides of the site of maximum brightness (cf. Figure 3.3) is a natural consequence of the simple loop geometry we have been discussing here. The circular polarization is (e.g., Petrosian and McTiernan 1983)

$$P_c = \cot\theta \left\{[3(1+\delta)\nu_b \sin\theta/\nu]/[1+(\nu_b \sin^2\theta/\nu\cos\theta^2]^{1/2}\right\}, \qquad (3.3)$$

where θ is the angle between the line of sight and the magnetic field lines. As evident $P_c = 0$ at $\theta = \pi/2$ and increases with opposite sense away from $\theta = \pi/2$ where the total microwave intensity is at its maximum. The simple circular loop, therefore, agrees with the qualitative features of the observations. Alteration of the field geometry changes the variation of P_c. In particular, for a highly asymmetric loop (see e.g., Figure 3.11b) it is possible that one sense of polarization becomes dominant. as often observed. Note that this is qualitatively similar to the field geometry proposed by Kundu and Vlahos (1979), though the details of their model and the one presented here are quite different.

3.2.3 Combined Soft X-ray and Hard X-ray Spectra

Analysis of the spatially integrated spectra from flares in the range 3 keV to 500 keV can yield much vital information regarding the separation of impulsive and thermal components of the flare. However, the interpretation is complex since several plasma regimes contribute to these spectra, and involve a range of atomic interaction processes. At the lower end, 3 keV to 7 keV, there are dominant line spectra of Fe XXVI, Fe XXV, Fe XXIV, Ca XIX, and in some cases Fe XXIII to Fe XVII. Although these can be excited by thermal or nonthermal electrons, the existence of these species requires that the excitation occur within hot thermal plasma.

Such hot thermal plasma also emits continuum radiation, which may extend up to photon energies of 10 keV, depending on the temperature. Non-thermal electrons, or streams, passing through the hot thermal plasma deposit energy, and contribute an additional component of the continuum spectrum. This process is normally represented by a thick target model. Power law electron distributions give rise to power law spectra. This deposition may take place within the hot plasma itself, or within cold dense plasma outside the hot region, as in the classic foot point emission models. Although this alternative makes no detectable difference to the continuum radiation emitted, it affects whether this mechanism contributes to the intensities of the above spectral lines. The range of mechanisms is thus large, and is further complicated since coverage of the region normally involves two or three spectrometers of quite

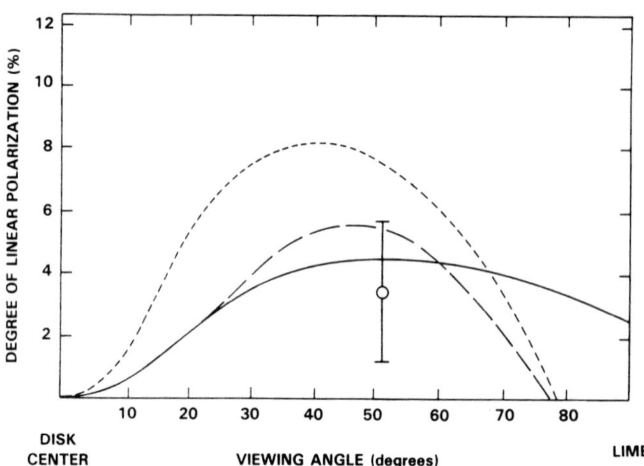

Figure 3.13 A comparison between the polarization measured for flare 2I of Tramiel *et al.* (1984) and the polarization calculated for three of the models of Leach and Petrosian (1983). For the solid line $\alpha_0^2 = 0.4$ and $d\ell nB/d\tau = 0.0$, the dashed line $\alpha_0^2 = 0.4$ and $d\ell nB/d\tau = 1.5 \times 10^{-9}$, and for the dotted line $\alpha_0^2 = 0.1$ and $d\ell nB/d\tau = 1.5 \times 10^{-9}$. A value of $d\ell nB/d\tau = 1.5 \times 10^{-9}$ corresponds to a twentyfold increase in the magnetic field strength from the top of the loop to the transition region at $N_{tr} = 10^{19}cm^2$. For a flare at disk center the viewing angle would be 0°; for one on the solar limb it would be 90° (from Leach *et al.,* 1984).

different characteristics. However, it is clear that the ambiguity in interpretation is greatly reduced by attempting a simultaneous analysis over a wide range of photon energies.

3.2.3.1. Line Spectra

The intensities of spectral lines, and their dependence on the plasma parameters, is determined by a range of atomic interaction quantities. For excitation of spectra from the ground state of the ion responsible, excitation and dielectronic capture rates are important. For helium and hydrogen-like ions, recent calculations by Bely-Dubau et al. (1982a,b) and Dubau et al. (1981) are widely used. Further work is in progress on the lower ions Fe XXIII to Fe XVII. These spectra lie in the narrow range 1.85 to 1.91 Å but cover a wide range of effective plasma temperatures. They are thus very valuable in extending the differential emission measure analysis downwards below 10^7 K, as well as in studying the transient ionization that occurs during rapid time variations of flare plasma. Moreover, where the ground configuration of the ion is complex (Fe XIX to Fe XXII), the intensity ratio of some of the lines can be used as an electron density diagnostic in the range 10^{12} to 10^{14} cm^{-3}. This could be especially valuable in the early part of the impulsive phase of hot flares.

For many purposes ionization balance calculations are also required for interpreting the line spectra. Many sets are available in the literature. However, some new data are becoming available from the solar flare spectra themselves. Antonucci et al. (1984, and further work in preparation) have used a study of the flare data from the BCS, Hinotori and P78-1 in order to derive ratios of Li-like to He-like ions in Ca and Fe as a function of temperaure. The ionization balance calculations of Jacobs et al. (1977, 1980) have been modified to take account of this work, and then used in some of the modeling described later.

3.2.3.2 Transient Ionization

It was earlier hoped that transient ionization effects would form an important diagnostic for the dynamic phases of flares. In this way, departures of the measured ionization ratios from their steady-state values would depend upon local electron density and the rate of change of temperature. It has since emerged that, due to the high electron densities, ionization is very close to steady-state over almost all phases of the flare for which statistically good data are obtainable. It has long been realized that the ratio of Li-like to He-like ions in flares is larger than theory would predict, by about a factor 2. This might be interpreted as a transient ionizing effect, were it not that it continues through all phases, including the flare decay. Koshelev and Kononov (1982) proposed a dynamic oscillating flare model in order to explain this effect by a continuous transient ionization condition. Doschek (1984) has attempted to examine such models on a more quantitative basis. He concluded that it is difficult to find an oscillating model which adequately predicts all the observed spectral intensities. A more likely explanation lies in a factor of 2 error in the ionization balance models derived from theory. This assumption is implicit in the rederivation from observations by Antonucci et al. (1984) mentioned above.

The existence of real departures from steady-state ionization are now being reported by some workers, who examine critically the earliest impulsive phase data (see later in this section). If substantiated, this would provide a valuable measure of the density at this time.

3.2.3.3 Hinotori Analysis

The Hinotori team has attempted to analyze jointly data from two of the instruments; the flare monitor (FLM) and the hard X-ray monitor (HXM). The FLM covers the range 5 to 20 keV with a resolution sufficient to separate line and continuum contributions. The HXM covers the range 18 keV to 400 keV. In an earlier analysis of FLM, Wantanabe (1984) found that while the continuum is best fitted by a power law in the early phase, it rapidly turns into a thermal spectrum. The time histories show that the thermal energy content is adequately represented by the time integral of the hard X-ray signal, indicating that the impulsive phase electrons may carry all the energy required for the flare. He concluded also that there is a strong tendency for all flares to produce a thermal component temperature of 2×10^7 K and density 10^{11} cm^{-3}.

Tanaka et al. (1984), in an attempt to understand the fluorescence origin of the Fe Kα line, have combined the spectra from FLM and HXM. The result, Figure 3.14, shows a smooth blend between the power law behavior above 10 keV with the thermal behavior below 10 keV. One is tempted to conclude that the plasma has a smooth distribution of electron energies merging from streams to thermal around 10 keV. This interpretation would

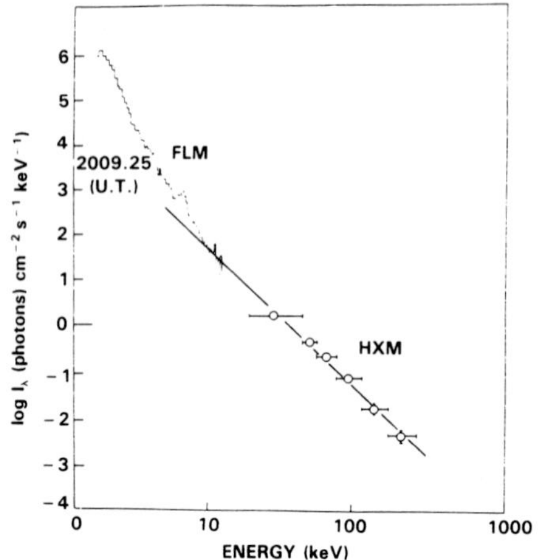

Figure 3.14 Combined spectra from the Hinotori FLM and HXM instruments, during the impulsive phase of a flare.

be valid only if both components represented thin target emission. If the streams dissipate in thick target emission, then the transformation from photon spectrum to electron spectrum would be different in the two cases.

3.2.3.4 SMM Anaysis

A method has been devised (Gabriel, Sherman, Bely-Dubau, Orwig and Schrijver, in preparation) to model the emitted spectrum based upon input in the form of a differential emission measure distribution plus electron beams. The model computes the predicted fluxes from 8 spectral lines in the Bent Crystal Spectrometer (BCS), the 6 channels of the Hard X-ray Imaging Spectrometer (HXIS) and the 15 channels of the Hard X-ray Burst Spectrometer (HXRBS). This then covers line emission from 10^7 K to 10^8 K and continuum energies from 3 keV to 260 keV. Beam energy distributions can be arbitrary and need not follow power laws. Thin target emission by the beam within the hot plasma is evaluated as well as the normal thick target emission, but double counting of this radiation is avoided. An inversion technique is not used; it is rather a question of trying various distributions to see which best fit the data. In this way it is possible to take proper account of the varying accuracy of the data. For example, a satellite to resonance line ratio in one BCS channel should be good to ≈ 10%, whereas absolute instrument calibration might be out by up to a factor of 2, although any such systematic errors must then always be present.

The method has been applied to the flare of 1822 UT 29 June 1980 at various times throughout the flare. The results, some of which are shown in Figure 3.15, are preliminary, but the following conclusions emerge:

1. A strong thermal component at 10^7 K initially, rising to $2x10^7$ K later, is always present.

2. This thermal component is not isothermal, but has a real width ≈ $0.5x10^7$ K.

3. At all times evaluated (even past the peak thermal emission) there is an additional high-energy component. If fitted with a thermal distribution, this is at T ≈ 10^8 K and has an emission measure 3 or more orders of magnitude lower than the primary component. It can also be fitted adequately with a power-law electron beam.

4. Early in the impulsive phase, the first thermal peak in the differential emission measure is smaller and at a lower temperature, while the higher energy component is relatively stronger. At the earliest time tested, it is very difficult to fit a thermal distribution to this feature. It is easier to fit a power law beam with a cut-off around 10 keV, with an electron spectral index of 5-7 on a thick target assumption. At this early stage, a good fit is obtained with the lines only if it is assumed that the ionization balance lags behind its steady-state value.

Figure 3.15 (top) shows ratios of calculated to observed intensities for two times during the flare, early in the impulsive phase and at the peak of the thermal phase. The anomalously high measured intensity of the Fe XXIV q line during the impulsive phase is an indication of the departure from ionization balance mentioned above. Figure 3.15 (bottom) shows the differential emission measures used in calculating these cases.

It could be interpreted that the lower thermal peak represents the main flare loop being filled with evaporated chromospheric plasma, having a temperature gradient, and temperature stabilization produced by conduction and radiation cooling. The higher energy component is produced by either or both of thick-target bremsstrahlung from the primary fast electrons and a high temperature region identifiable with the primary release. We thus have an indication that for this flare, the energy input continues at a reduced level until well after the peak of the thermal phase. Such an interpretation would be complicated, but not invalidated, if the observed emission arises from two independent flaring regions. In any case, the high energy component, at its peak of intensity, is precisely identical to the normally-assumed signature of the impulsive phase, i.e. a signature of nonthermal bremsstrahlung.

3.2.4 Iron Kα Emission

There has been great interest in the radiation mechanism of iron Kα, owing to its value as a diagnostic of nonthermal electron beams. The Kα lines at at 1.936 Å and 1.940 Å are radiated by inner-shell 1s-2p transitions following the removal of a K-shell (1s) electron from iron ions FeI-FeXI. The inner-shell ionization occurs either by photoion-

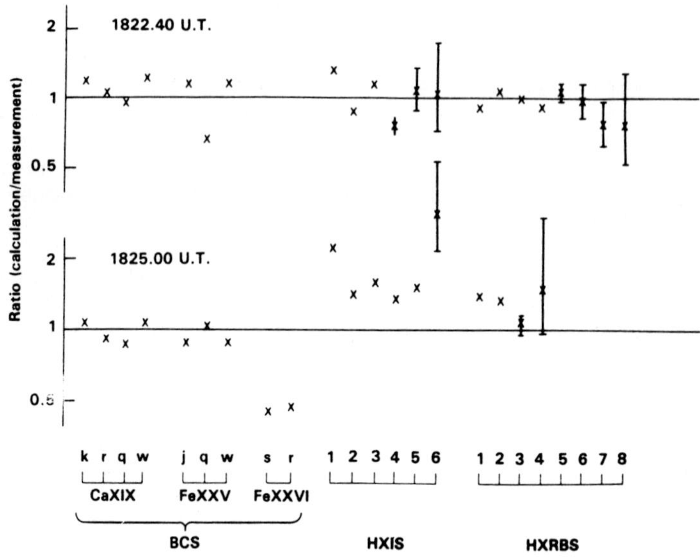

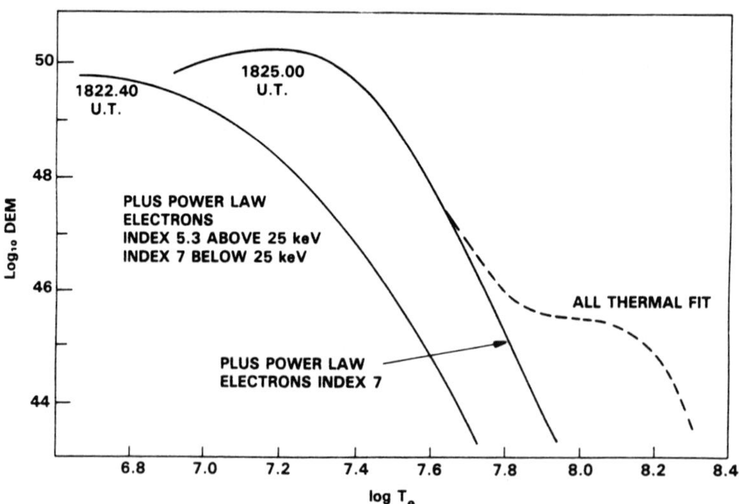

Figure 3.15 Top: Comparison of observed spectra and the model fit. Bottom: The differential emission measure used for the model fit shown above.

ization by X-rays of energy above the 7.11 keV ionization threshold (fluorescence) or by electron impact ionization. Since the chromosphere is transparent for 7 keV photons, irradiation of cool photospheric iron by X-rays of E > 7.11 keV could cause the emission (Neupert et al., 1967). Electron impact ionization occurs most effectively for electrons with an energy of about 25 keV. It has been suggested that nonthermal electrons associated with the hard X-ray burst excite Kα lines by collisionally ionizing the iron ions that are in lower ionization stages (Acton 1965, Phillips and Neupert 1973).

Observations of Kα lines in the previous cycle, though performed with relatively low spectral and temporal resolution, have suggested fluorescence as a plausible mechanism (Doschek et al., 1971; Tomblin 1972). During the present maximum more reliable observations of iron Kα have been made with much increased spectral and temporal resolution and increased sensitivity. The P78-1 SOLFLEX (Feldman et al., 1980) and the SMM XRP (Culhane et al., 1981) obtained clearly resolved Kα1 and Kα2 lines using flat Ge and bent Ge crystals respectively. The Hinotori SOX (Tanaka et al., 1982) obtained barely resolved Kα lines and the Kβ line at 1.755A by a SiO_2 crystal, and the Tansei IV satellite (Tanaka 1980) obtained Kα by a LiF crystal which possessed the highest sensitivity.

These observations have provided results which are in agreement. All flares show Kα emission which favors a fluorescence mechanism almost throughout the flare. The fluorescence origin has been quantitatively established by Parmar et al., (1984) from detailed comparison of the SMM observations of 40 large flares with Bai (1979)'s model of fluorescence. In Bai's model, which used a Monte-Carlo technique to follow iron Kα photons resulting from the photospheric absorption of X-rays, the Kα flux is evaluated as the product of three terms: integrated X-ray flux above 7.11 keV, fluorescent efficiency which depends on the X-ray source height and source temperature, and a term considering the absorption and scattering of emergent Kα photons, which depends on the heliocentric angle of the flare. For each of the flares Parmar et al., could show close similarity between the Kα light curve and that of the E > 7.11 keV X-ray flux. In fact, the Kα light curves were quite different from the hard X-ray burst profile of E > 50 keV. A remarkable center-to-limb variation was found for the Kα-to-continuum ratio for different flares at various heliocentric angles, and explained by Bai's model (Figure 3.16). Referring to Bai's model, the large dispersion in this ratio for flares at similar heliocentric angles is consistent with the variety of flare heights ranging from 0 to 0.05 solar radii for an iron abundance of 5.5 x 10^{-5} per H atom. Conversely, from the restriction for flare heights in short duration compact flares which were observed in the Skylab X-ray images (Pallavicini 1977), a plausible iron abundance has been estimated to be in the range from 5 x 10^{-5} to 6 x 10^{-5} per atom.

While the Kα light curve shows gradual variations similar to those of soft X-rays in most flares, transient Kα emissions in the very early phase of flare have also been reported in a limited number of flares. Culhane et al. (1981) reported a Kα light curve showing impulsive behavior. During the 15 seconds of the impulsive hard X-ray burst, the measured Kα flux was 2.5 ± 1.4 times that measured during the following 15 seconds (Parmar et al., 1984). However, this case has been considered

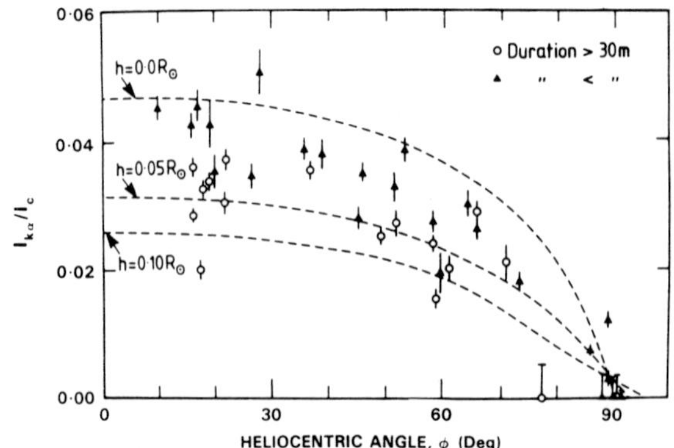

Figure 3.16 The ratio of Kα flux to the E > 7.11 keV continuum for various flares plotted against the heliocentric angle. The ratios predicted by Bai's model for an iron abundance of 5.5×10^{-5} per H atom, a temperature of 2.0 keV and three specific flare heights are shown as broken curves (Parmar *et al.*, 1984).

rather exceptional in the SMM data. The LiF crystal on the Tansei IV satellite detected a sudden enhancement of Kα emission that preceded an enhancement of the high temperature iron lines around 1.85A in three flares (Tanaka 1980). This enhancement occurred in the rising phase of the microwave burst at 17 GHz. The Hinotori observed similar, but less evident, cases in several (7-8) very impulsive flares. In these flares the Kα to Fe XXV resonance line ratio, which normally takes a fairly constant value of about 0.1, showed a very high value from 0.3 to 1.0 only in the initial phase, in close time coincidence with the hard X-ray burst.

These examples may provide candidates for the electron impact Kα emission. However, Tanaka et al. (1984) have suggested that the observed excess emission is mainly explained by fluorescence from the hard X-ray burst extending in energy to 7 keV. The high resolution continuum spectra from the Hinotori have revealed that a single power-law distribution prevails from 70 keV to below 10 keV in the early phase of some impulsive bursts (see Figure 3.14). Because in the very impulsive flares the thermal X-ray flux originating from 20 million degree plasma at 7 keV is less than the flux of the extended power-law distribution, fluorescence Kα shows impulsive behavior correlated with the hard X-ray burst. In the majority of flares, however, the thermal X-ray flux already dominates the 7 keV region when low-sensitivity crystal spectrometers can observe. One example of an impulsive event is shown in Figure 3.17. While the Kα light curve from the middle phase is explained by fluorescence due to thermal X-rays, the Kα flux in the early phase much exceeds the predicted values. The excess fluxes are explained by fluorescence from the power-law X-ray fluxes assuming that the source is located at 0 km. The transition from power-law X-ray fluorescence occurs rather smoothly around the peak of the hard X-ray burst, as expected from the fact that the thermal X-ray flux increases most rapidly at that time, with its flux increasing approximately in proportion to the time integral of the hard X-ray flux

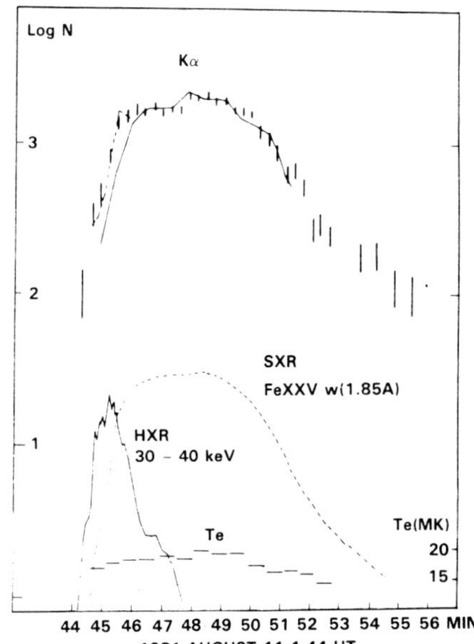

Figure 3.17 Kα light curve for an impulsive burst of 1981 August 11 (Observed by Hinotori), together with the Kα intensities predicted from fluorescence due to the thermal X-ray flux as derived from the iron line spectra (thick line) and those predicted from fluorescence due to the observed X-ray continuum which showed a power-law spectrum down to 7 keV (dashed line). Bars indicate ± σ range of the observed counts. Time profiles of the hard X-ray burst, Fe XXV resonance line intensity and electron temperature derived from the line spectrum are also shown.

(see, e.g., Tanaka et al., 1982). However, the conclusion that all Kα emission is due to fluorescence depends on assigning a height of zero to the power-law source. If this source is located high up in the corona, as has been observed in some gradual bursts observed by Hinotori, then an excess Kα flux remains to be explained, presumably by electron impact. In the impulsive bursts like this case, however, this seems unlikely.

This result may be compared with the thick target theory of the electron impact Kα developed by Emslie, Phillips, and Dennis (1985). The electron impact Kα yields predicted by this theory are always smaller than the Kα yields for fluorescence due to the power-law X-rays extending down to 7 keV, provided that the spectral index is larger than 3.5. Therefore, in most cases the electron impact Kα emission is veiled under the fluorescence Kα, even if the electron beam does exist. In conclusion, the electron impact Kα appears to be not necessary to explain the observations. On the other hand an evidence of low energy (below 10 keV) power-law photons may raise another problem related to the flare energetics because the energy included in the low energy electrons as derived from the thick target hard X-ray model is overwhelmingly large, probably exceeding the energy radiated in the UV and X-ray regions (Tanaka et al., 1984).

3.2.5 Ultraviolet and Hard X-ray Emission

The spatial, temporal and energetic relationships between impulsive ultraviolet emission and hard X-ray emission have attracted considerable interest as tests of flare particle transport models. Kane et al. (1980) reviewed the status of this field in the Skylab era. Two quite different UV observational approaches have been taken. Below, in Section 3.2.5.1, we review and interpret observations in the wavelength band from 10 to 1030Å. This band contains emission lines and continua that originate at a wide range of temperatures, from 10^4-10^7 K. We then turn to recent SMM observations, which are more straightforward to interpret-observations of the transition-zone line O V 1371Å, discussed in Section 3.2.5.2, and their intepretation in terms of both thermal and nonthermal energy-transport models.

3.2.5.1 10 - 1030Å and Hard X-ray Emission

An extensive set of calibrated concurrent measurements of impulsive hard X-ray and integrated 10-1030Å (henceforth EUV) emission has been obtained by Kane and Donnelly (1971), Donnelly and Kane (1978) and Kane, Frost and Donnelly (1979). The EUV observations have no spectral or spatial resolution, but good temporal resolution (~ 1 s). Although these measurements were made indirectly using the response of the ionosphere, subsequent direct meaurements from spacecraft by Horan, Kreplin and Fritz (1982) have fully validated the indirect method. The hard X-ray observations have no spatial resolution, but have sufficient spectral resolution to allow a power-law spectral fit, and also have good temporal resolution. All studies of the correlation between the EUV and hard X-ray bursts have found coincidence of peaks in the EUV flux and X-ray flux to within the time resolution of the instrument. This observation rules out certain types of thermal models (see Section 3.2.5.2b).

The observational datum of primary interest for each impulsive burst is the EUV/HXR ratio at the peak, where HXR is the flux in hard X-rays (> 10 keV), both measured at the top of the earth's atmosphere. Data for many impulsive bursts are shown in Figure 3.18. Donnelly and Kane (1978) and Kane, Frost, and Donnelly (1979) concluded that the peak EUV/HXR ratio is constant, rejecting nonlinearity in the peak EUV/HXR relationship seen in large flares as a selection effect. On the other hand, McClymont, Canfield and Fisher (1984) concluded that the peak EUV/HXR ratio is not constant, but rather the peak EUV $\propto$ HXR$^{1/2}$. The issue arises because the largest flares in Figure 3.18, for which the peak EUV/HXR ratio is smallest, tend to have been observed by a single spacecraft, TD1-A.

Until recently there has been general agreement that the observed EUV/HXR ratio is about 30 times smaller than the value predicted by typical electron beam fluxes in the thick-target nonthermal model, in which all the X-ray emission is produced by nonthermal bremsstrahlung and all the EUV emission by thermal emission from the collisionally-heated

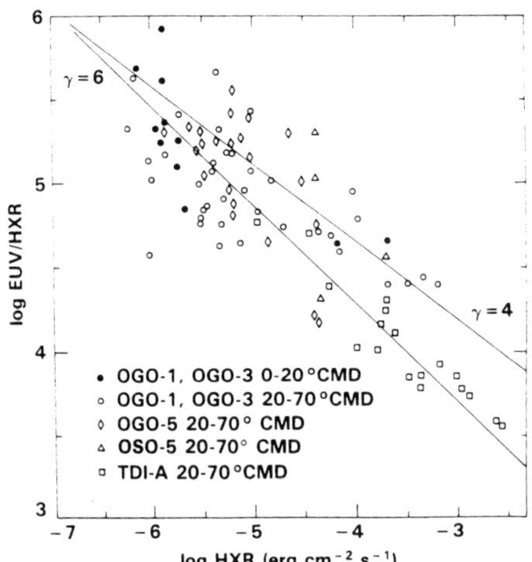

Figure 3.18 Observed relationship between the peak 10-1030 Å EUV/HXR ratio and the HXR energy flux above 10 keV measured at 1 AU, at peaks of impulsive bursts, from Kane and Donnelly (1971), Donnelly and Kane (1978), and Kane, Frost, and Donnelly (1979). McClymont, Canfield, and Fisher (1984) have multiplied the observations in the 20-70° central meridian distance (CMD) range by 2.5, as an approximate correction for the effect of limb darkening. Lines: Predictions of the thick-target nonthermal model, for two different values of the photon spectral index γ, and a universal beam cross-section of 3×10^{16} cm^2. From McClymont, Canfield, and Fisher (1984).

transition region and chromosphere (Donnelly and Kane (1978), Emslie, Brown and Donnelly (1978), Kane, Frost, and Donnelly (1979)). Donnelly and Kane (1978) suggested that the discrepancy could be explained by the "partial precipitation" model, in which the bulk of the X-ray emission is from electrons trapped in the corona and only a small fraction of the nonthermal electrons reach the chromosphere. However, Emslie, Brown and Donnelly (1978) pointed out that trapping of nonthermal electrons in the corona could not cut down the EUV flux sufficiently because electrons are scattered out of the trap onto the chromosphere after losing only roughly half of their energy (Melrose and Brown, 1976). Emslie, Brown, and Donnelly (1978) suggested that the thermal hard X-ray model (e.g., Brown 1973a) may be more appropriate.

The peak EUV/HXR ratio also shows a strong center-to-limb dependence. Donnelly and Kane (1978) showed that this dependence could be explained if the EUV emission comes from a source embedded in the chromosphere. They explained the dependence satisfactorily by a model in which the EUV emission comes from the bottoms of cylindrical wells of two different sizes. The idea that the EUV flare source is below the normal nonflaring chromosphere is compatible with observations of Hα spectra discussed in Section II.G below.

McClymont, Canfield and Fisher (1984) have looked at another aspect of the predictions of the thick-target model. Following previous authors (Brown 1973b, Lin and Hudson 1976, Emslie, Brown and Donnelly 1978), they calculated the depth of the flare transition region as a function of the electron beam flux per unit area. Considering only collisional energy losses, as in previous work, they find that the increase in depth of the transition region with increasing beam flux causes the EUV/HXR ratio to fall as the flux increases,

$$\mathrm{EUV/HXR} \propto (\mathrm{HXR/Area})^{(-\delta-2)/(\delta+2)} \quad , \qquad (3.4)$$

where the electron spectral index δ is related to the hard X-ray spectral index γ, $\delta = \gamma + 1$. The physical reason for the ratio to decrease with increasing flux is that the mass fraction of the beam-heated column that is filled with chromospheric material, which emits much of the EUV, decreases as flux increases. For the most commonly observed spectral indices, $\gamma \approx 4$, the predicted slope of the peak EUV/HXR ratio as a function of hard X-ray flux is in excellent agreement with the slope that they find in the observational data, if it is assumed that the beam cross-section is constant in all flares, i.e. only the electron energy input per unit area varies. Second, as shown in Figure 3.18, both the slope and the absolute value of the peak EUV/HXR ratio vs. HXR curve are correctly predicted if one assumes that the beam cross-section is of order 3×10^{16} cm^2, comparable to typically observed chromospheric flare kernel areas.

Were it not for the fact that extremely high beam fluxes are implied, one would conclude that the thick-target electron heating model, with a universal beam cross-section, explains the observations very well. With a 20 keV cutoff energy, the beam fluxes range from about 10^{10} ergs cm^{-2} s^{-1} for the smallest flares to 10^{14} ergs cm^{-2} s^{-1} for the largest. The majority of the observed flares have beam fluxes greater than 10^{11} ergs cm^{-2} s^{-1}, the commonly accepted upper limit for return current stability.

How, then, can the peak EUV/HXR ratio be understood, if beam fluxes greater than about 10^{11} erg cm^{-2} s^{-1} are disallowed? McClymont, Canfield and Fisher (1984) considered three processes which decrease the EUV/HXR ratio predicted by the thick target model. Firstly, if the preflare loop, in which electrons are acclerated, is hotter (and therefore denser) than a typical preflare loop--a state that might be inferred from the frequent observation of a soft X-ray precursor to the impulsive phase--the bulk of the electron beam energy may be deposited in the corona. They found that the drop in peak EUV/HXR ratio could be explained by this hypothesis, although rather high preflare temperatures ($\gtrsim 10^7$ K), and correspondingly high densities, were necessary to obtain the very low peak EUV/HXR ratio observed in large flares. Secondly, thermal hard X-rays may be important, as investigated previously. Thirdly, the effects of reverse (return) currents have not been considered fully in calculations of this type. Due to reverse currents, concentration of heating at the foot of the transition region may greatly increase the ability of electron beam heating to raise chromospheric plasma to coronal temperature, compared to pure collisional heating, thus reducing the emission measure in the temperature range where the EUV emission originates. No quantitative studies of this effect have yet been done.

3.2.5.2 OV and Hard X-ray Emission

a. Observations. Observations of individual EUV lines are potentially more powerful than measurements of the total 10-1030 Å EUV emission, discussed above. One such line, well observed by SMM, is the OV 1371 Å line. The SMM observations have demonstrated a close observational relationship between this line, formed in the transition region at a temperature of about 250,000 K, and hard X-rays. Progress has been made at defining temporal, spatial, energetic, and morphological relationships.

The temporal relationship between HXR and OV was studied in detail by Woodgate et al. (1983). In this work they studied the three SMM flares for which UV data were available on a time scale of 1.3 s, and HXR data (from 25--100 keV) were available every 0.128 s. Each flare displayed at least five impulsive bursts in both HXR and OV that were simultaneous to within the 1.3 s time resolution. Measurement of time lags for the peaks ($t_{OV} - t_{HXR}$) yielded an average of 0.3 ± 0.5 s. Thus, their results showed that peaks in emission that occurred in both OV and HXR were simultaneous to well within the time resolution.

The Woodgate et al. (1983) results also showed that there was a very good occurrence correlation between OV and HXR peaks. For the 16 HXR peaks there were 14 OV peaks (80%) and for 19 OV peaks there were 14 HXR peaks (74%). The correspondence is obviously quite good "with the primary difference being due to some small UV peaks late in the impulsive phase that do not appear in the hard X-rays". These small OV peaks may be associated with HXR bursts that are below the threshold of detectability, but may also indicate additional sources of heating.

Poland et al. (1984) have studied the relation between energy emitted in HXR vs. OV as a function of time for several flares. The results for the flare of 2 November 1980, whose light curves are shown in Figure 3.19, are presented in Figure 3.20. In the latter figure the boxes represent data points before flare maximum, the diamonds are within 5 s of flare maximum and the plusses are after maximum. It can be seen that there is a well-defined relation throughout the flare between the HXR and OV emission. Also, the premaximum and postmaximum emissions lie on almost the same line, with the postmaximum OV remaining slightly brighter than the premaximum emission for the same HXR emission. Ten flares were studied in this manner, with all showing similar results. Most of the flares show the rise phase and fall phase occurring on approximately the same line. However, some do show a significant deviation from this relationship in that OV emission is brighter during the decay phase than during the rise phase. There is also a significant variation in OV emission from flare to flare, but, this may in part be due to the variation in total background in OV from one flare to the next. When these data are plotted with the preflare background removed there is still a large variation in the OV intensity from one flare to the next.

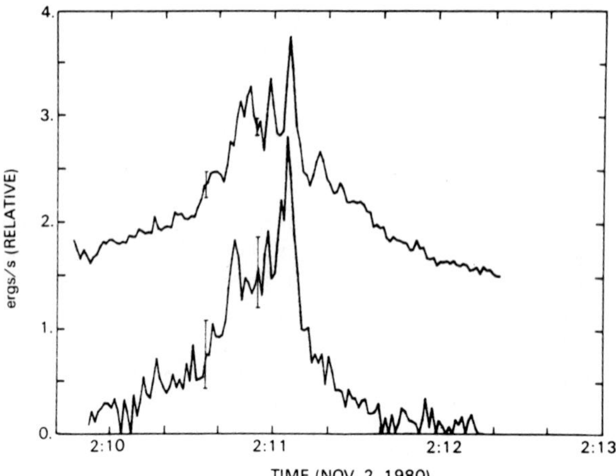

Figure 3.19 The upper curve shows emission in OV for the entire flare as a function of time (actual values are 10^{23} times larger than shown). The lower curve shows emission in HXR above 25 keV for the entire flare as a function of time (actual values are 3.84×10^{20} times larger than shown). Noise errors for a single measurement are shown at two different times for each plot.

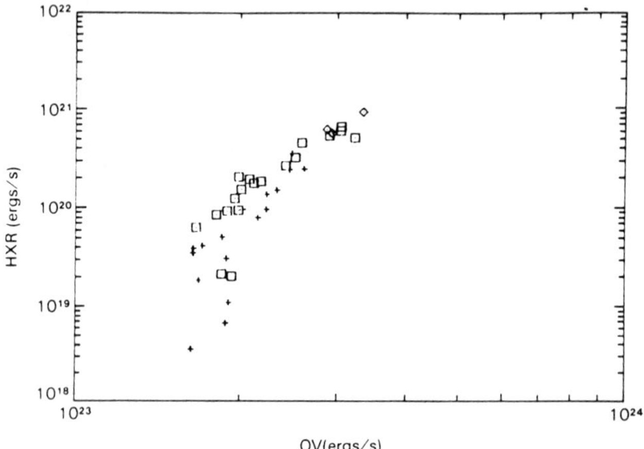

Figure 3.20 HXR emission as a function of OV emission in ergs s^{-1} for the flare shown in Figure 3.19. Squares represent data obtained before flare maximum, diamonds are within 5 s of maximum, and plusses are after flare maximum. Only every third point has been plotted for clarity of the figure.

The most important point to be made from these observations is that HXR and OV emission rise and fall together, with the rise and fall phases usually on approximately the same line, independent of time, during the entire event. How can this be understood? It suggests that the physical conditions in the transition region are directly related to the production of HXR, and independent of those flare quantities that are known to vary on gradual phase timescales, such as the temperature or emission measure of the thermal X-ray plasma. It is straightforward to show that on gradual phase timescales (~ 1 m) the emission measure at OV temperatures in a conductively dominated transition region is proportional to the transition region pressure, so the observation of relatively little intensity variation during a flare such as the one shown in Figure 3.20 implies that the transition region pressure in the emitting regions varies very little in the course of the observations (a period ~ 1 m). If the model of an electron beam in a single flare loop is to explain both the impulsive and gradual emissions, then apparently pressure equilibrium is either never reached or is substantially unaltered by the impulsive event. On the other hand, it may be that the impulsive phenomena arise in a different volume than the gradual phenomena; we explore the arguments on this issue below, pro and con.

To within SMM's ability to spatially resolve emitting regions in HXR and UV, the two emissions arise from the same location on the Sun. However, one must bear in mind that the UV observations are obtained at only 3" x 3" spatial resolution, at best, and more usually 10" x 10". X-rays between 3.5 and 30 keV are observed with a resolution of 7" x 7" (van Beek et al., 1980), at best. Harder X-rays, above 30 keV, are observed without spatial resolution. Hence, our knowledge of the cospatiality of these emissions is strongly constrained by observational limitations.

Spatially resolved observational information on the cospatiality of UV and HXR emission is limited to just a few flares. In a study of the June 29, 1980 flare Poland et al. (1982) showed that positions and changes in positions of UV transition region emission corresponded to positions and changes in position in 3.5 to 30 keV HXIS emission. These correspondences, together with the temporal correspondences between OV and higher-energy HXRBS hard X-ray light curves strongly suggest their cospatiality in this flare. Duijveman et al. (1982), in a study of the November 5, 1980 flare, reported cospatial HXR and UV bursts in one of the flare footpoints. Machado et al. (1982) found similar cospatiality in the April 10, 1980 flare. Although these two latter papers did not study the UV/HXR correlation in detail, the HXIS data quality is better than for the June 29 event.

In another SMM flare observation, Cheng et al. (1981) examined the SiIV and OIV UV emission with a resolution of 4" x 4" and found that each individual burst was seen in both UV and spatially unresolved hard X-rays, and that each component UV spike originated from a separate discrete flaring point. The density at these points was determined to be in the range of 5×10^{12} to 10^{13} cm^{-3}. There is an uncertainty in using this ratio since the two lines are formed at significantly different temperatures; the reliability issue is discussed in Cheng et al. (1982).

The close time-independent functional relationship between OV and HXR, shown in Figure 3.20, motivates us to ask what is the obervational evidence that indicates the relationship between the volumes in which impulsive phase emissions and gradual phase emissions arise? Two arguments suggest that impulsive phase emissions and gradual phase emissions arise in separate volumes. Firstly, it is known that in X-rays the emission is impulsive only for energies above approximately 5 keV; for lower X-ray energies the emission does not vary impulsively. For transition region emission, flare enhancements become increasingly impulsive above 10^4 K. This impulsiveness reaches a maximum near 2×10^5 K, decreases rapidly at 3×10^5 K, but remains significant to approximately 2×10^6 K, above which the emission is no longer impulsive (Donnelly and Hall 1973). Secondly, Widing and Hiei (1984) showed that impulsive emission in one flare arose from a single loop, while gradual phase emission arose from a separate location. The impulsive phase in this flare was seen in hard X-rays (20-30 keV) and in transition zone lines between approximately 80,000 K and 2×10^6 K. The gradual phase emission from this flare, seen in soft X-rays and UV lines above 2×10^6 K, was observed to arise from a separate region or loop near the impulsive loop.

On the other hand, several arguments can be made against the assertion that impulsive and gradual emissions occur in separate volumes. Duijveman, Somov, and Spektor (1983) showed that one can see impulsive variations in bands as low as the 3.5-8 keV band of HXIS. Also, there is a well-known correlation between the amount of hard X-ray emission and the slope of the soft X-ray light curve in a typical flare soft X-ray emitting line (see, e.g., Tanaka et al., 1982a, Machado (1983). Finally, in some cases impulsive-phase phenomena occur in loops that already produce considerable thermal emission (see, e.g., Duijveman, Somov, and Spektor 1983); it is quite possible that the impulsive soft X-ray component is masked within the overall emission, as a spectrum such as Figure 3.14 would suggest. Only with substantially better spatial resolution will this issue be decided in a totally satisfactory manner.

In summary, the impulsive phase observations clearly show that there is a strong physical link between the processes that form HXR and transition zone and chromospheric emission. This link was established for flares in general by the relation in energies of peak emission found by Kane and Donnelly (1971), see Section 3.2.5.1. The simultaneity of individual peaks in HXR and OV in several flares found by Woodgate et al. (1983), and the relation in energy emitted during several flares as described by Poland et al. (1984), places observational limits on the possible links. There cannot be a significant time delay (i.e. greater than 1 s) between the different excitations, and the energies exciting the emissions are functionally related. That the emissions occur from the same location to within 10" supports the concept of their being physically related, but this not a serious constraint since this large distance could allow quite separate locations in the same loop or even different loops.

b. Interpretation. Emslie and Nagai (1984) have studied the response of a coronal loop to both heating and nonthermal electron acceleration at the loop top, in order to determine the resulting structure of the transition region plasma ($T \approx 10^{5\pm1}$ K), and so the emission measure in, and intensity of, optically thin emissions formed in this region. The key quantity in the modeling of Emslie and Nagai (1984) is the logarithmic differential emission measure

$$\xi(T) = n^2 \, dz/d\ell nT \qquad (3.5)$$

where n is plasma density, z vertical distance, and T electron temperature. Under an optically thin approximation for the radiative losses from the gas, this quantity is directly proportional to the intensity in a given line, the other factors being atomic in nature and therefore independent of the structure of the atmosphere (e.g., Pottasch 1964). Thus determination of the evolution of $\xi(T)$ with time t gives a model prediction of the shape of the intensity versus time profile of the line emission being considered.

Figures 3.21 and 3.22 show the form of $\xi(T)$ for various times since the onset of energy input into the flare loop for the conductively heated and electron-heated models, respectively. It is readily apparent from

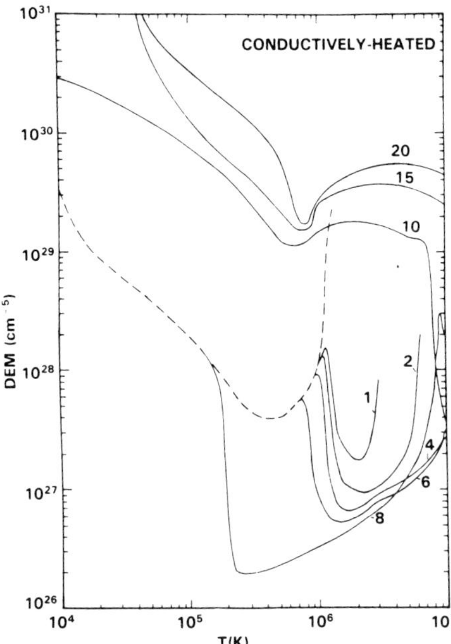

Figure 3.21 The logarithmic differential emission measure (DEM) ξ, equation (3.5), for various times (in seconds) in the impulsive phase, for the conductively heated model. The dashed line shows the initial (t = 0) profile. Note the behavior at transition region temperatures—constant for the first few seconds, followed by a sudden decrease and an equally abrupt increase. The peak at around T ≃ 10^7 K on the 8 s curve is real; it corresponds to the density enhancement that causes the large increase in ξ at transition region temperatures in the 10 s curve.

these figures that the evolution of ξ is markedly different in the two heating scenarios. Let us consider first the conductively-heated model, in which the rate of energy deposition varies linearly with time, up to a maximum of 2×10^{10} erg cm^{-2} s^{-1} after 30 seconds, and in which the energy deposition is spatially distributed over a Gaussian profile, symmetric about the loop apex, with dispersion $\sigma = 2 \times 10^8$ cm. In this model, there is virtually no change in ξ for the first few seconds, reflecting the fact that the thermal energy has not yet had time to propagate to the transition region layers. As soon as the heat front arrives (at t $\simeq$ 7 s for the parameters considered), a decrease in ξ results, due to the strong steepening of the temperature gradients produced by the large conductive flux $F_c(\propto T^{5/2}\ dT/dz)$, which evidently overcomes the increase in the n^2 factor (Equation (5)) in the determination of ξ. A few seconds later, however, the transition region plasma overheats to coronal temperatures, and the gas at the relevant

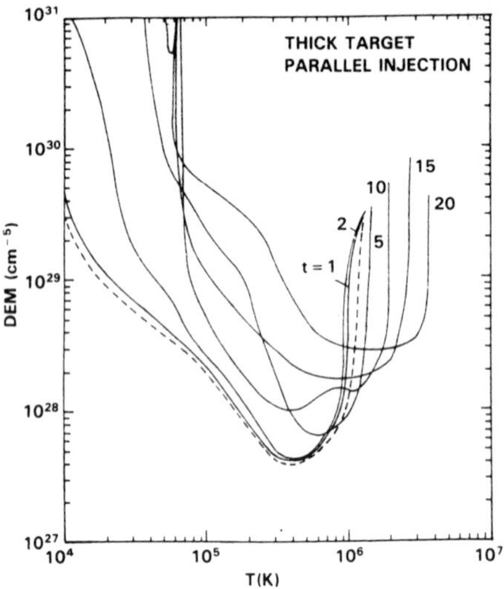

Figure 3.22 The logarithmic differential emission measure (DEM) ξ, equation (3.5), for various times in the impulsive phase, for the parallel-injection electron-heated model. Note the steady increase around transition-zone temperatures (T $\simeq 10^5$ K).

temperatures ($\simeq 10^5$ K) is now situated much deeper in the atmosphere, with a corresponding larger density. The density scale height is sufficiently small in the preflare chromospheric layers that the conduction front, which moves at speeds of order of the local ion sound speed $c_s = (kT/m_p)^{1/2}$, where k is Boltzmann's constant and m_p the proton mass, covers the distance to the "new" transition region in a very short time. Thus ξ abruptly rises by several orders of magnitude in a second or two. This predicted trend in $\xi(10^5$ K) (and so in the intensity of optically thin EUV emissions) is simply not reflected in the observations above, implying that such an energy mechanism is not a viable possibility.

On the other hand, for the thick target electron-heated model, in which energy is distributed throughout the flare loop with an identical time structure but with the broader vertical spatial structure resulting from consideration of the collisional dynamics of an electron beam in a relatively cool target (see Emslie 1978), the strong steepening of temperature gradients associated with a relatively local energy released region does not happen. Instead, the dominant effect controlling the behavior of ξ is the progressive overheating of denser and denser layers to coronal temperatures, resulting in a monotonic rise of the density at $T \approx 10^5$ K and so in ξ (10^5 K).

This correlation of EUV intensity with energy input [as measured by the instantaneous hard X-ray flux, which is roughly proportional to injected electron energy (Brown 1971)] is in accord with the above observations (see, e.g., Poland et al., 1982, 1984; Woodgate et al., 1983). Indeed, Woodgate et al. (1983) have used the close synchronism of EUV and hard X-ray emissions to rule out conductively heated models such as that discussed above. The results of Emslie and Nagai further support this rejection. However, they also demonstrate that the observed behavior is predicted by the thick target electron heated model. On reflection, it is by no means obvious that this should be so. The observed increase in intensity is only an order of magnitude or so even in large events (Poland et al., 1984), a factor well below that predicted by the n^2 factor in Equation (5), since n rises by over two orders of magnitude as a result of flare heating of the overlying layers (e.g., Machado et al., 1980, Emslie and Nagai 1984). Thus the increase in the temperature gradient factor must also play a role, and it is somewhat remarkable that these two large effects in opposite directions combine to produce a relatively small overall increase in ξ, as is observed. The underlying physical reasons for this behavior are not fully clear at present, and merit further investigation.

3.2.6 White Light Emission

We define white light flares as those detected in the optical continuum; thus their number is limited by observational considerations. In the last few years patrols at the National Solar observatory, Sacramento Peak (NSO/SP) and Big Bear Solar observatory (BBSO) have greatly increased the known number [see catalog of white light flares by Neidig and Cliver (1983)]. It is probable that all major flares have detectable continuum in the blue, although the number detected at longer wavelengths is still small. Thus the term "white light flares" simply means highly energetic flares. Our general motivation in studying white light flares is based on the great energy involved, the problem in transporting energy to the low atmospheric heights of significant continuum opacity, and a general fascination with intense flares. An important special interest here is in the nature of the continuum emission and how it may be produced.

3.2.6.1 Morphology

Our knowledge of the morphology of white light flares is limited by the small number of events for which good data are available. Nevertheless, four kinds of white light flare have been recorded: (1) Bright footpoints, usually at the edge of the sunspot umbra (Zirin and Neidig 1981); (2) Short-lived flashes (Zirin and Tanaka 1981) associated with impulsive X-ray spikes, up to a few arcsec diameter and 20 s lifetime, usually located along the neutral line of a delta configuration; (3) Moving fronts (Machado and Rust 1974; Zirin and Tanaka 1973) occurring after the impulsive phase and possibly associated with the hot thermal plasma; (4) Fixed bright points or ribbons (Zirin and Neidig 1981; Neidig and Cliver 1983), possibly associated with the hot thermal phase.

Except for the last, all the manifestations are moving, sometimes very fast, or jumping from point to point. The white light flare footpoints commonly appear at a number of locations and times throughout the whole event; presumably they are associated with the main bundle of field lines on which flare energy input occurs. Almost always at least one footpoint occurs in the penumbra, usually at the edge of the umbra, but other bright points may occur outside the sunspot, as do white light flare ribbons. The area of white light flare emission averages ~ 6 x 10^{17} cm^2 at flare maximum, with a duration ordinarily less than 10 minutes (Neidig and Cliver 1983).

The white-light kernels and fronts roughly match the emission observed in He D3 and the wings of Hα. The He D3 emission, which may cover a somewhat larger area than the associated white-light emission, becomes especially bright in the post-impulsive phase (for example, the July 1 flare (Zirin and Neidig 1981)). Probably all white light flares are accompanied by D3 emission.

Very little information is available regarding the white light flare relationship with spatially-resolved, high-energy emissions. In the June 6, 1982 flare (Ohki et al., in preparation) the Hinotori hard X-ray and soft X-ray images generally show spatial and temporal agreement with kernels of emission at λ3862Å. Videotapes show small kernels a few arcsec across, with changes occurring on time scales of a few seconds.

3.2.6.2 The Optical Spectrum

a. Observations. Spectral data on white light flares are usually obtained either from a series of broad band filtergams at several widely spaced points in the spectrum (e.g., Zirin and Neidig 1981) or from spectrograms with dispersions of several Å/mm (e.g., Neidig 1983). In filtergrams the variation of flare intensity with wavelength is affected by changes in seeing between the successive images, or even by temporal changes in the flare itself. Fast switching with video measurements removes some of these problems, and the seeing may be evaluated.

Another problem in interpreting the filtergram measurements is the possible effect of lines in the blue. Although the Balmer line emission can be minimized by using, for example, a narrow (15Å) filter at 3862Å (midway between H_8 and H_9), there still remains the possibility of emission by numerous photospheric lines (Hiei 1982).

The spectrograms, although able to record a large expanse of the spectrum in a single exposure, are typically obtained with lower spatial resolution, with the result that the flare emission may be highly diluted or the most intense kernels may not fall on the slit. Judging by the intensities observed, the spectra obtained so far have not recorded the most intense kernels. Intensities up to twice the continuum at visible wavelengths have been recorded with filters, while the spectral data have only revealed enhancements of 10-20% above the continuum. In addition, the variation of intensity with wavelength, as determined from spectrograms, may suffer from systematic errors due to chromatic aberration of atmospheric or instrumental origin. Thus while the gross features of the white light flare spectrum, the line intensities, and the Balmer jump can reliably be obtained from spectra, the slope of the continuum thus obtained, which might be used to derive the flare temperature or to identify the emission mechanism, should be treated with caution.

The filter measurements, as well as some of the spectrograms, show a fairly flat continuum in the visible with an increase below 4000 Å. Superposed on this there is often a relatively strong Balmer continuum. Figure 3.23 combines filter data on several flares; the data for July 1 and April 24 are from photographic measurements; that for May 9 is from high speed video measurements. Figure 3.24 shows spectrographic data in the vicinity of the Balmer jump for three representative flares.

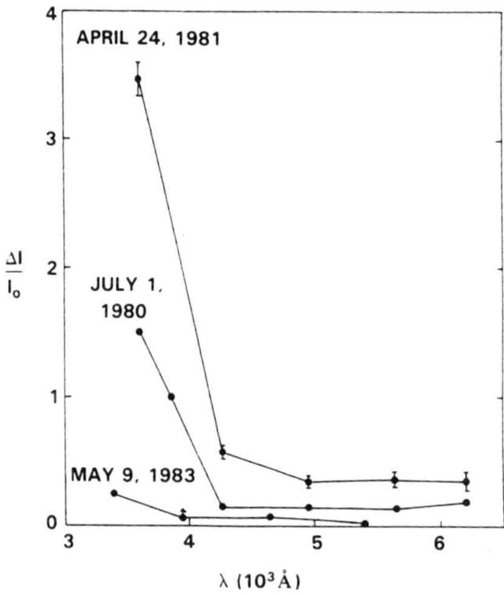

Figure 3.23 Broad-band filtergram measurements of the intensity enhancement [(flare − background)/background] at flare maximum for several white light flares. Measurements at 3610, 4275, 4957, 5650 and 6203 Å are from NSO/SP; 3400, 3862, 4642, and 5400 Å are from BBSO.

The Balmer continuum has been identified in spectrograms of six events (Hiei 1982; Hiei et al., 1982; Neidig and Wiborg 1984; Donati-Falchi et al., 1984), although several cases are known (e.g., Machado and Rust 1984; Boyer et al., 1985) in which a short wavelength continuum was present, but without a measurable Balmer jump. A Paschen jump was detected in the spectrum of the 24 April white light flare (Neidig and Wiborg 1984); unfortunately, the latter flare is the only known event with a infrared continuum sufficiently strong to permit a conclusive search for the Paschen jump.

In addition to the large intensities observed in the Balmer continuum, the increase in the flare contrast in the blue, especially below 4000Å, can be quite striking in some cases (Zirin 1980, Zirin an Neidig 1981). This "blue continuum" is probably related to the bluish color noted in a number of visual observations of white light flares, although the cause for it is not understood. An increase in the flare contrast could be expected simply as a result of the decline in intensity of the solar background at short wavelengths (this would be particularly evident in the broad band filtergrams where the solar background is blanketed by lines). On the other hand, intense brightenings ($\Delta I/I_0 \approx 1$)

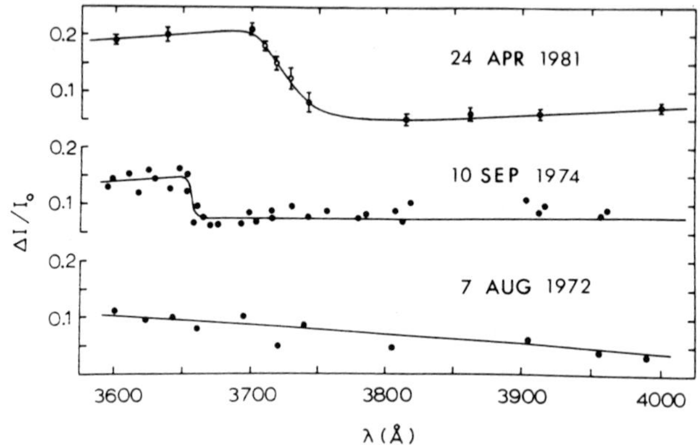

Figure 3.24 Spectrographic measurements for three white light flares: Apr. 24, 1981 (Neidig 1983); Sept. 10, 1974 (Hiei 1982); Aug. 7, 1972 (Machado and Rust 1974). I_0 refers to the intensity in the bright windows of the spectrum adjacent to the flare.

have been observed in filtergrams at 3835 or 3862Å when the corresponding emission at longer wavelengths was weak or absent (Zirin 1980). The latter observations indicate differences in contrast too large to be attributed entirely to differences in the intensity of the solar background. The question as to whether the blue continuum is truly anomalous, arising, perhaps, from some unidentified source of opacity or from photospheric lines, cannot be answered until more precise observations of the spectrum become available.

b. Interpretation. The principal task in interpreting the spectral observations is to identify the emission mechanism; this, in turn, determines the general atmospheric regime of the white light flare phenomena. The spectra in Figure 3.23 suggest a strong variation with wavelength in the opacity of the white light flare source, indicating an origin in layers that are not optically deep. The Balmer continuum (Figure 3.24), as well as the bright He D3 emission associated with white light flares, is characteristic of chromospheric densities (electron density 1-5 $\times 10^{13}$ cm^{-3}; see Machado and Rust 1974; Neidig and Wiborg 1984; Zirin 1983). Nevertheless, it is by no means certain that the emission underlying the Balmer continuum and extending to longer wavelengths is due to H_{fb} emission in general (the 24 April flare, which shows a Paschen jump, is a good candidate for an event dominated in the visible and infrared by H_{fb} emission). Hiei (1982) has pointed out that the continuum extending to longer wavelengths might be attributable to H^- emission, possibly originating in the photosphere; this hypothesis remains untested because information on the Paschen jump is not available for the flares studied by Hiei.

White light flares in which the Balmer jump is absent (e.g., August 7, in Figure 3.24) might suggest either an optically deep source with small temperature enhancement or an optically thin, very hot source with opacity dominated by free-free absorption. The latter possibility, however, might require a prohibitively large emission measure at high temperature ($n_e^2 \Delta z \geq 10^{35}$ cm^{-5} at 10^5 K), in order to produce a 10% enhancement at 4000Å. The spectrum of the August 7 flare was studied by Machado and Rust (1974) who concluded that the emission was Paschen continuum originating in the flare chromosphere. The latter interpretation, of course, would require Balmer continuum to be present in this flare, and therefore, unless the absence of the Balmer jump can be explained, a deeper, low-temperature source with H^- opacity might be indicated. In any case the height of the source in the August 7 flare would have to be greater than 300 km above $\tau_{5000} = 1$, as Machado and Rust found a probable absence of emission in lines formed below that level. With regard to a deep photospheric origin for white light flares in general, it is worth noting that strong reversals in the cores of photospheric absorption lines have not been observed in white light flare spectra.

Recent model calculations by Aboudaram et al. (1984) show that the white light flare does not arise from heating at $\tau_{5000} > 1$, because the predicted $\Delta I/I_0$ would then vary as $1/\lambda$, in contrast to the observations in Figure 3.23. Aboudaram et al., also show that the continuum cannot be explained by heating of the upper photosphere and temperature minimum region; such models produce $\Delta I/I_0$ increasing with wavelength in the visible and, in particular, they do not reproduce the observed increase in contrast below 4000Å.

Although the spectral data described here are fragmented and sometimes ambiguous, we note that the brightest white light flares observed with spectrographs (e.g., Donati-Falchi et al., 1984; Neidig and Wiborg 1984), as well as as the events observed with filters at several wavelengths, show strong Balmer continuum which must arise in a chromospheric-like regime.

In order to estimate the temperature in white light flares it is first required to know how the opacity varies with wavelength. But even if the latter is known, the interpretation will be limited by our incomplete knowledge of the optical thickness, departures from LTE, and by possible systematic errors in the data. If we assume H_{fb} emission, where the opacity varies as λ^3, the observed slope and intensity of the continuum in the visible region can satisfactorily be fitted to chromospheric temperatures and densities by making appropriate (although arbitrary) choices for geometric thickness and departures from LTE. In this way the white light flare continuum observed by Machado and Rust (1974) yielded temperatures of 8500-20,000 K. Similarly, the April 24 flare (well-observed with broad-band filters) can be fitted to

10,000-20,000 K. On the other hand, if H^- contributes to the white light flare continuum, then we might expect $T_e \lesssim 8000$ K for the layers in which this emission arises.

The temperature estimates discussed here do not allow for effects due to possible unknown sources of opacity in the blue. If such opacity is present, no further progress can be made until these sources are identified. Finally, we again remind the reader that the spectrograms obtained thus far may not accurately represent the brightest kernels of white light flares; hence a precise description of the brightest white light flare spectra, including possible contribution by line emission, remains in question.

3.2.6.3 Timing Relationships

In Figure 3.25 we compare the hard X-ray (HXR) burst profiles with the total power radiated in the white-light continuum for six white light flares. One event (Feb. 11, 1980) was obtained from Big Bear Solar Observatory (BBSO) at a single wavelength (4306Å); one event (July 1, 1980) is based on data from both BBSO (λ 3862Å) and NSO/SP (λλ 3610, 4275, 4957, 5650 and 6203Å); the remaining four events use NSO/SP data at five wavelengths. In computing the total power it is assumed that the

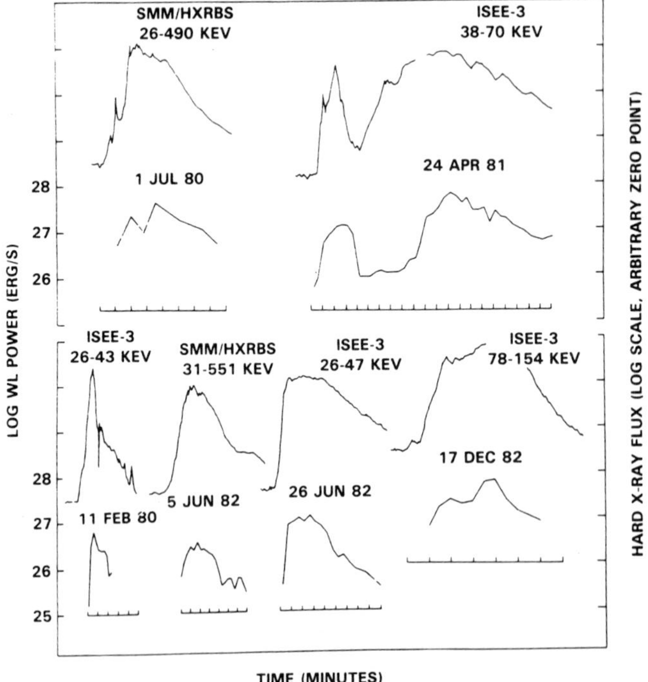

Figure 3.25 Comparison of total white-light power and hard X-ray profiles for six white light flares. Optical data are derived from broad-band filtergrams, as described in the text.

white light flare emission extends over the range 2500-10,000 Å. Time resolution varies from 15 s to 30 s except for the July 1 event (Zirin and Neidig 1981) where the resolution was 90 s in the five wavelengths of the NSO/SP data and 15 s in the BBSO data. In the case of the February 11 event we assume that the white light flare spectrum is qualitatively similar to the spectra in Figure 3.23.

The most striking feature of Figure 3.25 is the overall similarity between the white light and HXR. With the exception of the late phase of the July 1 flare, the correlations suggest that the HXR and white-light sources may be closely related, especially in the impulsive phase. This result confirms the conclusions of an earlier study (Rust and Hegwer 1975) which compared HXR and white-light intensity data at two wavelengths (4950 and 5900 Å).

The most unfavorable of the six comparisons in Figure 3.25 is the July 1 flare. In this case there is good agreement between hard X-rays and white light in the impulsive phase, but the brightest white light flare point occurred later and approximately coincided with the peak thermal energy content (Antonucci 1981) and Fe XXV emission (Phillips 1981). This suggests an association with the hot ($T \approx 20 \times 10^6$ K) thermal plasma in the flare. Similarly, Zirin and Tanaka (1973) observed a white light flare wave event in the hot thermal phase of the August 2, 1972 flare.

In summary, the HXR and white light are well-correlated in most cases, while in others the white light rises with the HXR and declines with the hot thermal plasma. In this sense the behavior of the white light in the impulsive and thermal phases bears some similarity to the rise and fall of Hα emission relative to hard and soft X-rays in non-white light flares (Zirin 1978).

3.2.6.4 Energetics

The principal datum that motivates our study of white light flares is the large observed continuum flux. For nine flares the mean peak flux is estimated to be 1.5×10^{10} ergs cm^{-2} s^{-1} between 2500 and 10,000 Å (Neidig and Cliver 1983), with the largest reported flux for a single event being 5×10^{10} erg s^{-1} cm^{-2} (Slonim and Korobova 1975). Allowing for the low spatial resolution of these observations, we suggest, for purposes of model calculations, that a reasonable upper limit on the flux is 10^{11} ergs cm^{-2} s^{-1}. The peak continuum flux far exceeds the combined flux of all the Balmer lines at the same location in the flare, although the line emission, being more widespread, may be comparable in total energy when integrated over time and space. The total peak continuum power, obtained by integrating over the white light flare area, is less sensitive to resolution, and is approximately 10^{28} erg s^{-1} for the largest white light flares.

These numbers are quite impressive, and they place severe constraints on the energy transport mechanisms in white light flares. A number of transport processes have been suggested in the literature, although none can be claimed to be satisfactory in general. Chromospheric heating by fast electrons has been a long-time favorite mechanism of modelers, and we shall demonstrate here that this mechanism might work in the impulsive phases of two well-observed white light flares, but fails in the post-impulsive (hot thermal) phase of one of these flares.

For both the impulsive and post-impulsive phases of the July 1 and April 24 flares, we list in Table 3.3: (1) the observed peak power in white light, (2) the electron energy E_o above which the power in the nonthermal electron spectrum (in a thick target approximation) is equal to the white-light power, (3) the stopping height $h(E_o)$ of an electron with energy E_o in the non-flaring atmosphere, (4) the column density $nz(E_o)$ of ionized gas that can be penetrated by an electron with energy E_o, and (5) the estimated column density nz(WL) required to produce the observed white-light intensity via H_{fb} emission at an assumed temperature of 10^4 K and a density equal to the density in the non-flaring atmosphere at the electron stopping height. The nonthermal electron data are derived from Dennis (1981), Batchelor (1984), Kane (1983), and Kane et al. (1985). We see that in the impulsive phase of both flares sufficient power in nonthermal electrons might, within limits of observational uncertainties, be transported through a column comparable in extent to that required to produce the amount of white-light emission. The only problem is producing the observed white light flare spectrum. Sufficient power is marginally available also in the post-impulsive phase of the 24 April flare, although for the post-impulsive phase of the July 1 flare the nonthermal electron model seems to fail completely.

Table 3.3 White Light Flare and Energetic Electron Transport Parameters

		Impulsive Phase	Post-Impulsive (Thermal Phase)
July 1, 1980	WL Power	2.3×10^{27} ergs s^{-1}	4.5×10^{27} erg s^{-1}
	E_0	53 keV	35 keV
	$h(E_0)$	880 km	1020 km
	$nz(E_0)$	1.2×10^{21} cm^{-2}	5.2×10^{20} cm^{-2}
	nz(WL)	3.9×10^{21} cm^{-2}	2.1×10^{22} cm^{-2}
Apr. 24, 1981	WL Power	1.4×10^{27} ergs s^{-1}	7.0×10^{27} s^{-1}
	E_0	119 keV	85 keV
	$h(E_0)$	650 km	730 km
	$nz(E_0)$	6.0×10^{21} cm^{-2}	3.1×10^{21} cm^{-2}
	nz(WL)	3.3×10^{20} cm^{-2}	3.3×10^{21} cm^{-2}

Heating by $E \gtrsim 1$ Mev protons in the impulsive phase white light flare was suggested by Najita and Orrall (1970) and Svestka (1970). The protons certainly have enough range, and the calculations of Lin and Hudson (1976) show that, at least in the August 4, 1972, flare, the proton flux alone was probably adequate to power a typical white light flare at column depths up to 10^{21} cm^{-2}. More recent work by Hudson and Dwivedi (1982) has shown that sufficient heating by protons may be possible even in the upper photosphere (column depth 5×10^{22} cm^{-2}), but that the heating at depths near $\tau_{5000} = 1$ (column depth 2×10^{24} cm^{-2}) is negligible. Lin and Hudson originally dismissed the proton heating mechanism because of an apparent delay in the 2.2 MeV γ-ray lines relative to the flare impulsive phase. But recent SMM data shows no delay in the energetic protons that produce (via collisions) the free neutrons responsible for the 2.2 MeV γ-ray line, so the delay is not a problem. This conclusion is corroborated by Ryan et al. (1983) who show, in addition, that while proton heating in the impulsive phase of the July 1 flare might have been possible, the proton event had ended before the post-impulsive white-light emission reached maximum. Thus, we conclude that heating by nonthermal electrons or protons cannot account for the post-impulsive phase white ight flare.

A more recently investigated mechanism for the white light flare is a two-stage transport process. We know that sufficient energy is available in E $\gtrsim$ keV electrons even in the post impulsive phase. These electrons are stopped near the 1000 km level, and the problem then is to efficiently transport the energy to regions of higher density to produce the optical continuum. Livshitz et al. (1981) have concluded that a shock wave may propagate downward from the heated chromospheric region with velocity 1000 km s^{-1}, and that this could produce heating at lower levels where the white light flare is seen. The principal difficulty here is the relatively short duration of the shock-induced effect (~ 10s), which is in contrast to observed white light flare durations as long as 10 minutes for a single bright kernel [see Figure 3.25, and Kane et al. (1985)].

The post-impulsive white light flare emission in the July 1 flare seems to be associated with the hot thermal phase of the event. We might be led, therefore, to consider heat conduction as a transport mechanism, were it not for the fact that sufficient conductive flux can be obtained only at temperatures of millions of degrees in the white light flare source. Clearly, this is incompatible with a chromospheric origin for the white light flare; moreover, it requires an emission measure several orders of magnitude larger than is observed from soft X-ray data. Thus we are unable to identify a satisfactory mechanism for the energy transport in post-impulsive phase white light flares.

3.2.7 Hα Emission

Study of the morphology, timing and spectrum of Hα ($T \approx 10^4$ K) emission during the impulsive phase, relative to hard X-rays and microwaves, bears on several basic questions of flare energy transport. Do flare heating effects occur nearly simultaneously (say within 1 s) throughout loops, down to Hα formation depths? Do fast electrons penetrate as far as the chromosphere? Is the observed depth dependence of flare heating in the chromosphere the dependence predicted by stopping of nonthermal electrons? Since the Skylab era much progress has been made at both observational and theoretical approaches to these questions.

3.2.7.1 Observations

During the recent solar maximum the usefulness of Hα observations was increased markedly by improvements in observing techniques that ensured, for the first time, during the impulsive phase, observations with high temporal, spatial and spectral resolution, coordinated with temporally, spatially, and spectrally resolved X-ray and microwave data.

a. Timing. The temporal resolution and precision of measurement in Hα and either hard X-rays or microwaves required to discriminate between conduction front and energetic particle models of the transport of energy from the corona to the chromosphere during the impulsive phase is approximately 1 s. It is also necessary to have spatial resolution, since not all parts of the chromospheric flare necessarily reflect the same transport process. In the past, the lack of adequate temporal resolution led to much confusion about the time delays between hard X-rays (or microwaves) and Hα (cf. Vorpahl 1972, Zirin 1978).

The first observations with sufficient resolution to be compelling have been obtained recently by Kaempfer and Schochlin (1982) and Kaempfer and Magun (1983). In a study of one flare with 1.4 s and 100 ms temporal resolution in Hα and microwaves respectively, Kaempfer and Magun found evidence for both fast electron transport, at one site of a flare, and hydrodynamic or nonclassical conductive transport, at other sites of the same flare. At the former site, they observed Hα and microwave synchronism to within two seconds; the lack of delays at different microwave frequencies also supported an energetic electron interpretation. At the latter sites, delays of about 10 s were observed. The authors show that these delays are consistent with the propagation of disturbances at about 2000 km s^{-1}, i.e. roughly the velocities expected for collisionless conduction fronts (Brown, Melrose and Spicer 1979).

b. Morphology. The spatial coincidence of Hα and hard X-ray emission during the impulsive phase of Type B flares supports the idea that both emissions often arise as a consequence of the transport of coronally accelerated nonthermal electrons to the footpoints of flare loops (see Section 3.2.2). Although this conclusion applies only to impulsive hard X-ray flares, not gradual ones, there is no controversy over this spatial relationship in the horizontal plane.

On the other hand, there has been a continuing controversy over the height of origin of impulsive chromospheric flare emission. On one side, Zirin (1978) has argued that impulsive Hα emission is produced by direct collisional ionization and recombination in an elevated source-prominence material-well above the quiet chromosphere. This belief is based on line center Hα filtergrams at and near the limb, during hard X-ray bursts. Zirin's observations show several events in which impulsive phase Hα brightening is seen in structures that typically extend 10^4 km above the limb. On the other side, theoretical models all show that the Hα emission of flares comes about primarily thermally, from deeper than the quiet chromosphere (see, e.g., Brown, Canfield and Robertson 1978 or Canfield, Gunkler and Ricchiazzi 1984). The latter picture is also what is inferred for the primarily chromospheric source of impulsive 10-1030Å EUV emission (Donnelly and Kane 1978), and is discussed in Section 3.2.5).

A recent observation of a limb flare, by Kurokawa (1983), bears on the controversy. It shows that flare emission sites observed at Hα line center ± 2.4Å in a limb flare show abrupt intensity changes directly correlated with microwave fluctuations. It also shows that these emission sites are brightest at Hα ± 1.2Å, where they appear about 1800 km above the photosphere, but are obscured by the quiet chromosphere at Hα line center. Kurokawa (1983) concluded that the main part of the flare chromosphere is confined to the low chromosphere, much lower than the surrounding undisturbed structures.

c. Spectra. Whereas before the recent solar maximum there were observations of spectral line profiles from flare kernels during the impulsive phase of perhaps one flare (Zirin and Tanaka 1973), there are now observations of hundreds. Many of these are accompanied by X-ray and microwave observations. Such observations of the Hα line (Acton et al., 1982, Gunkler et al., 1984, Canfield and Gunkler 1984) provide strong evidence for the propagation of nonthermal electrons down into the chromosphere during the impulsive phase.

Figure 3.26, from Canfield and Gunkler (1984), shows Hα line profile evidence for chromospheric heating by nonthermal electrons during the flare of 7 May 1980. Two of the times shown, 14:56:16 and 14:56:42 UT, are during the impulsive phase, and the hard X-ray emission at all energies above 30 keV was 3-4 times stronger at the second time. There were two distinct kernels in this flare, shown by the cross-hatching in Figure 3.26; 2-3 times more 16-30 keV emission was observed to come from the south (lower) kernel. During the time of impulsive hard X-ray emission, strong Hα wings develop. They disappear very quickly after the impulsive emission ends. The number of pixels that show such behavior rises and falls in direct proportion to the intensity of the hard X-ray emission.

3.2.7.2 Interpretation of Spectra

Since fast electrons lose several orders of magnitude more energy to heating than to bremsstrahlung X-ray photon production, one would expect to be able to see clear effects in thermally generated chromospheric emission. Ricchiazzi and Canfield (1983) recently developed static models of the effects of enhanced fluxes of nonthermal electrons, treated strictly collisionally, as well as thermal conduction and enhanced pressure from the flare corona, on the temperature, density, and ionization structure of the chromosphere. Their work assumed that the electrons last long enough to establish a new hydrostatic equilibrium in a closed loop (at least tens of seconds). Subsequent work, in an impulsive approximation, by Canfield, Gunkler and Ricchiazzi (1984), henceforth CGR, assumed that very little time (at most a few seconds) has passed since the start of electron heating, so no mass motions have had time to develop. These models superseded work carried out in the Skylab era by Brown, Canfield and Robertson (1978).

By using these model flare chromospheres to compute profiles of optically thick spectral lines, which are formed over a range of depths along the direction of propagation of the electrons, one can predict theoretical line profile signatures of the distribution of heating appropriate to fast electrons. CGR calculated Hα line profiles based on both hydrostatic and impulsive model chromospheres. They showed that both hydrostatic and impulsive model atmospheres heated by nonthermal (power-law) electrons characteristically produce wide and bright Hα profiles with a central reversal. Enhanced thermal conduction reduces the width and total intensity of the profiles. High thermal conduction alone cannot account for flare Hα enhancements. High coronal pressure dramatically increases the width and total intensity of the Hα profiles, while reducing the central reversal.

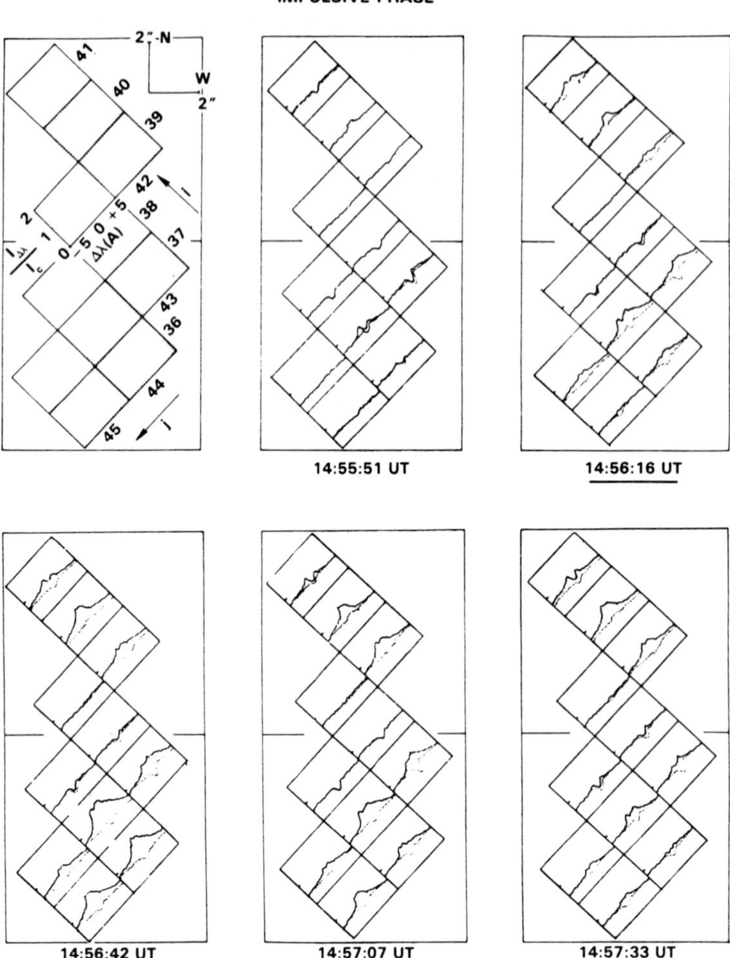

Figure 3.26 Impulsive-phase Hα profile observations. The two large contiguous squares framing each of the 6 panels indicate two 8" × 8" HXIS pixels; the smaller rectangles indicate 2" × 2.67" Hα pixels. The shaded Hα pixels indicate the north and south kernels. In each Hα pixel its spectrum at the indicated time $I(\Delta\lambda)$ is plotted in units of I_c, the observed quiet sun continuum intensity near the flare site. Each pixel's preimpulsive phase spectrum (145525 UT) is shown dotted, for comparison. From Canfield and Gunkler (1984).

CGR found that only high values of the flux of nonthermal electrons produced Hα profiles with obvious broad (Stark) wings. Their results are shown in Figure 3.27. In the upper panel, Hα profiles are shown for the impulsive model chromospheres; those in the lower panel are for hydrostatic models. In both panels it can be seen that pronounced wings develop only for values of electron energy flux (above 20 keV) above about 10^{10} ergs cm^{-2} s^{-1}. The calculations indicate that no combination of values of thermal conduction and coronal pressure can give the same strong Stark wings, accompanied by central reversal, as those associated with nonthermal electron heating.

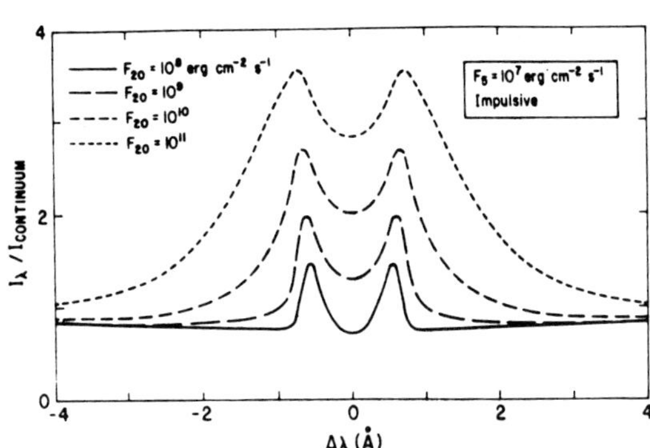

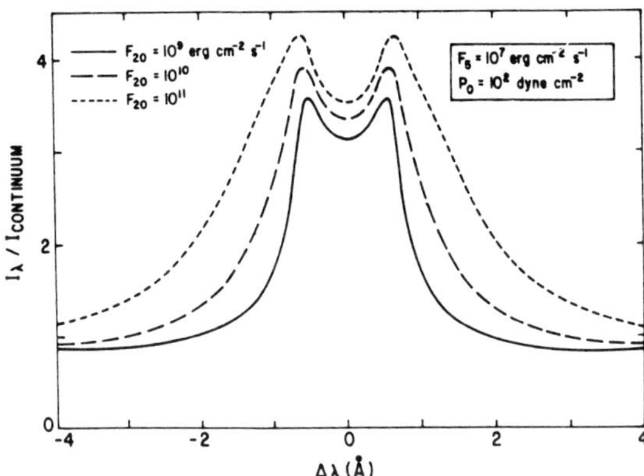

Figure 3.27 The effect on Hα line profiles of varying the energy flux of nonthermal electrons above 20 keV, F_{20}. The electron beam is assumed to be vertical, and the spectral index of the power-law electron energy distribution function is $\delta = 5$. Intensity is measured in units of the quiet sun preflare continuum near Hα. F_5 is the conductive flux from the corona, measured at $T = 10^5$ K. P_0 is the gas pressure at the top of the coronal loop, the assumed acceleration site. Top: impulsive model. Bottom: hydrostatic model. From Canfield, Gunkler and Ricchiazzi (1984).

The observations cited above show that Hα profiles develop obvious broad wings only in spatial and temporal coincidence with hard X-ray emission. Gunkler et al. (1984) and Canfield and Gunkler (1984) found, by quantitatively comparing the extent of observed Hα wings to the theoretical profiles of Figure 3.27, that values of $F_{20} \approx 10^{11}$ ergs cm^{-2} s^{-1} were implied. They showed that this same value was implied by combining the power of electrons above 20 keV inferred from the HXRBS observations with the observed Hα stark-wing area. From the point of view of transport theory, the implication of this result is that the electrons that produce impulsive hard X-rays indeed have a range that is in at least order-of-magnitude agreement with the predictions of a strictly collisional thick-target model of electron propagation. The fact that this inferred value of F_{20} is so large (approaching the return-current stability limit) is interesting. This is not the first time such large electron fluxes have been found from chromospheric spectra; using an inversion method, $F_{20} \approx 10^{12}$ ergs cm^{-2} s^{-1} was inferred from semiempirical flare chromospheric models (Machado et al., 1980) by Emslie, Brown and Machado (1981).

3.3 THEORETICAL STUDIES OF TRANSPORT PROCESSES

3.3.1 Electron Beams and Reverse Currents

The existence of large numbers of nonthermal electrons in solar flares has been supposed for some time now as a means of explaining, through the mechanism of electron-proton bremsstrahlung, the high flux of hard X-rays observed in large events (see e.g., Brown 1971). With observations (Tanaka, Nitta, and Watanabe 1984) now showing the hard X-ray continuum (i.e. line-subtracted) spectrum extending in an unbroken power-law down to energies $\lesssim$ 7 keV, it appears that, at least in some events, suprathermal bremsstrahlung-producing electrons exist at all energies above this value. This implies that an extremely large number of these electrons must be accelerated in the primary release process, whether through directed acceleration ("nonthermal" or "beamed" model) or bulk energization ("thermal" model). In particular, in the nonthermal model such hard X-ray observations lead to nonthermal electron injection rates $\gtrsim 10^{36}$ s^{-1} above 25 keV (Hoyng, Brown and van Beek 1976) and so, by extrapolation of typically observed power-laws, $\gtrsim 10^{38}$ above ~ 7 keV.

One of the major theoretical problems of the last decade has been to describe quantitatively the energy, momentum and charge transport effected by such strong beams. Early treatments (e.g., Brown 1972, 1973; Emslie 1978) neglected collective effects within the electrons comprising the beam, and instead concentrated on the mean behavior of individual electrons through Coulomb interactions with the ambient plasma particles. It was soon recognized, however, (Hoyng, Brown and van Beek 1976; Knight and Sturrock 1977; Hoyng, Knight and Spicer 1978) that the passage of such a large number of electrons through the stationary solar atmosphere constituted an extremely large current. This required that a reverse (or return) current be set up in the ambient plasma, in order to achieve both charge and current neutrality, and to provide fresh electrons to the acceleration site. We refer the reader to Knight and Sturrock (1977) for a discussion of the unacceptably large electrostatic and magnetic fields produced in the absence of a reverse current, and to Hoyng, Brown and van Beek (1976) for a discussion of the "electron number" problem associated with the depletion of electrons from the coronal acceleration site. This return current must be driven through a medium of finite resistivity, modifying (by an as yet unresolved amount - see below) the energetics of the primary electron beam. In addition, other collective effects, such as two-stream instabilities and collisionless pitch-angle dispersion, modify the dynamics of such beams, with implications for both their hard X-ray production and atmospheric heating. Finally, the beam may have yet another significant consequence beyond those of photon production and heating: the stopping of the beam may effect significant acceleration of the coronal and chromospheric plasma through which it passes. This section discusses the following aspects of nonthermal electron beams: the energetics of, and driving mechanism for, the return current, and collective instabilities driven by the beam itself. The motivation for such in-depth study of nonthermal electron beams is provided by recent observations from SMM and other spacecraft, which indicate a substantial nonthermal electron injection in some events (Brown, Hayward and Spicer 1981; Hoyng 1981; Hoyng et al., 1981).

3.3.1.1 Energetics of the Reverse Current

Emslie (1980) showed that the ohmic energy losses sustained in driving the reverse current through the finite resistivity of the ambient plasma could, in some cases, exceed the losses suffered through direct collisions of the beam electrons with ambient ones. For classical resistivity, the ratio of ohmic to collisional heating scales like $F\, n^{-1}\, T_e^{-3/2}$, where F is the injected electron flux, n the ambient density and T_e the electron temperature (Emslie 1980), and hence only for relatively large fluxes of electrons passing through relatively cool coronae could this ohmic heating dominate. This raises the question of the self-consistency of Emslie's (1980) treatment through considerations of energy balance and, perhaps more fundamentally, through the requirement that the electron drift velocity associated with the reverse current be below the threshold for plasma instabilities to develop, typically around the ion-sound speed $c_s = (k_B T_e/m_p)^{1/2}$, where k_B is Boltzmann's constant and m_p the proton mass (Fried and Gould 1961). These matters were addressed independently by Emslie (1981) and by Brown and Hayward (1982). Emslie (1981) considered the flux limitation imposed by the stability requirement on the reverse current drift velocity. He showed that this could reduce the importance of reverse current ohmic heating, except during the very early stages of the event, when low preflare coronal temperatures result in a very large plasma resistivity (see also Knight and Sturrock 1977), and also during the late phase of long events ($t \gtrsim 10$ s), by which time the ambient protons could be significantly heated by the hot electrons, resulting in an equilibration of ion and electron temperatures, an increased threshold drift velocity (Fried and Gould 1961) and so a larger possible ohmic to collisional energy loss ratio. Brown and Hayward (1982, see also Hayward 1984) considered the other aspect of the self-consistency problem, namely the determination of the ambient plasma electron temperature T_e through consideration of the energy equation. Including only dominant terms, this can be expressed as

$$\eta j^2 + \nabla \cdot F_c = 0, \tag{3.6}$$

viz. an energy balance between ohmic heating and thermal conductive cooling. Here η is the plasma resistivity, j the beam (or reverse) current density, and F_c the thermal conductive flux. Equation (3.6) assumes that direct collisional losses are unimportant, an assumption that can be tested a posteriori by comparing the ohmic losses in a plasma whose temperature follows from Equation (3.6) with the collisional losses in the same plasma. The collisional losses in turn depend on the density n and the average individual electron energy E. Thus the ratio of ohmic energy losses to collisional ones depends on F, E, T_e and n. In addition, from the values of n and T_e [found from solution of Equation (3.6)], we can test whether the beam flux exceeds the stability threshold nv_{crit}, where v_{crit} depends solely on T_e. From these considerations, Brown and Hayward (1982) deduced that when the reverse current is stable to the generation of plasma turbulence it is unimportant energetically; the effects ofan unstable reverse current are yet to be determined.

Emslie (1985) has recently contested the Brown and Hayward (1982) results, on the basis of the inadequacy of their treatment of the conduction term in (3.6) as $\kappa_o T_e^{7/2}/L_{es}^2$, where κ_o is the Spitzer (1962) coefficient of thermal conductivity and L_{es} the stopping distance of the beam electrons under the decelerating force produced by the charge separation electric field $E = \eta j$. He argued that the boundary condition determining the point at which the electrons stop is not in fact L_{es}, but instead the half-length L of the flare loop (beyond which Coulomb collisions in the high density chromosphere effectively stop the beam in a very short distance), and therefore recast Equation (3.6) in the form

$$\eta_o T_e^{-3/2} j^2 + d/dz\, (\kappa_o T_e^{5/2}\, dT_e/dz) = 0, \tag{3.7}$$

imposing a boundary condition on T_e at a fixed distance L corresponding to the half-length of the flare loop. Using this revised version of the energy equation, and also a more realistic power-law-type injected electron energy distribution, Emslie (1985) found, in contrast to the results of Brown and Hayward (1982), that a regime (albeit a small one in the available parameter space) where a stable reverse current dominates the energetics of the upper coronal plasma can exist. Physically, this is because the shorter and more physical boundary values of L imposed by Emslie result in a much lower apex temperature; since the resistive losses for a stable reverse current scale as $T_e^{-3/2}$ and the drift velocity stability threshold scales as $T_e^{1/2}$, a reduction in this peak temperature can have relatively large effect on the importance of ohmic heating, while having a relatively small effect on the maximum beam flux allowed to pass stably. For low densities, this stable reverse current heating is dominant, while for high densities collisional heating dominates. Thus there is a minimum energy input $\varepsilon = \max(\varepsilon_{collisions}, \varepsilon_{reverse\ current})$ into the coronal plasma, and, in turn, a minimum peak coronal temperature for a given hard X-ray flux, independent of coronal density. This relationship between hard X-ray burst intensity and peak coronal temperature (deduced from soft X-ray observations) in different events may afford observational tests of (a) the thick-target model and

(b) the relative role of reverse current and collisional heating in such a model.

3.3.1.2 Driving Mechanism

A fundamental issue for the physics of nonthermal electron beams is whether the reverse current is established electrostatically or inductively. Spicer (1982) and Spicer and Sudan (1984) have claimed that an assumption made by all previous authors, namely that the reverse current is in steady-state balance with the injected beam flux, and that it is established by the electrostatic field associated with the charge separation effected by the beam, is incorrect. Rather, they argued that the unneutralized current of the beam sets up, through Ampere's law, a self-magnetic field B which then establishes a return current inductively through Lenz's law. In addition, they argued that this return current decays resistively in the ambient plasma, so that eventually it can no longer effectively neutralize the beam; the resulting "bare beam" cannot propagate (for the reasons discussed by Knight and Sturrock 1977) and so turns off. They suggested that this may be the cause of the extremely rapid decays found in some hard X-ray events (Kiplinger et al., 1983). Brown and Bingham (1984) have criticized Spicer and Sudan's work on the grounds that they neglected the displacement current term in Ampere's law at the head of the beam. The inclusion of this term, according to Brown and Bingham, returns us to a scenario where the reverse current is indeed set up electrostatically, over a Debye length scale at the head of the beam; since the time required for the beam to traverse such a small distance is much smaller than timescales for hard X-ray bursts, they argued that a steady-state description (Knight and Sturrock 1977; Emslie 1980) is indeed valid. With regard to the resistive decay of the reverse current, they agreed with Spicer and Sudan that the finite curl of the E field (due to edge effects at the outside of the beam) leads to the establishment of an azimuthal magnetic field B . While this field would indeed choke off the beam over a short timescale (Alfven 1939; Lawson 1957) in the absence of a strong guide field, Brown and Bingham (1984) pointed out that the presence of such a guide field, as exists in flare loops, results only in increased helicity in the guiding field lines; as long as the azimuthal B field remains small or comparable to the guiding longitudinal field B , then little modification to the beam energetics results. The consensus reached is that, on the one hand, for beams spread out over large areas, edge effects are of minor importance, and so the predominant mechanism controlling the reverse current is the electrostatic field established by the displacement current at the head of the beam. On the other hand, for beams of small area, e.g., if the injection occurred over many sub-resolution elements (a "pepper-pot" scenario), the inductive edge effects would predominate. In such a case, "choking-off" of the beam (Spicer and Sudan 1984) may occur on timescales relevant to observations, but for this effect to have an appreciable hard X-ray signature, many elements of the "pepper-pot" would have to act coherently, and it is far from clear why this should occur. However, in both these cases it appears that the earlier steady-state electrostatic treatments of the reverse current energetics give valid results.

3.3.1.3 Collective Instabilities

Emslie and Smith (1984) showed how, due to the inverse square energy dependence of the Coulomb collision cross-section, the velocity distribution for a nonthermal electron beam injected into a cold plasma is preferentially depleted at low energies, resulting in the formation of a "hump" in the combined (background plus beam) distribution function. This type of distribution is well-known to be unstable to the growth of Langmuir waves; Emslie and Smith calculated the resulting wave level and discussed nonlinear interactions of these waves, through the process $\ell + \ell \rightarrow t$, (where ℓ is a Langmuir wave and t a transverse wave with angular frequency $\approx 2\omega_{pe}$, where ω_{pe} is the electron plasma frequency), to produce microwave radiation. They showed that the level of microwave radiation produced by this process is, for typical thick target beam parameters, much larger than observational upper limits on microwave fluxes from flares. This implies that either the nonthermal microwave radiation produced is strongly absorbed by gyroresonance absorption in the surrounding plasma, with corresponding strong implications for the energetics of the flare as a whole (Melrose and Dulk 1982b), or that the thick target electron heated model in its present form is incorrect.

Mok (1985) has examined another instability connected with the scattering of electron beams in the solar atmosphere - the anomalous Doppler resonance instability (Kadomstev and Pogutse 1968). This instability is driven by anisotropy in the electron velocity distribution $f(v_{\parallel}, v_{\perp})$ and operates through "beat" resonances of the form $\mathbf{k \cdot v} = \omega + n\Omega$, where k is the wavenumber, v the electron velocity, ω the electron plasma frequency, Ω the gyrofrequency and n an integer. Using a power-law pitch angle distribution centered in the direction of the guiding magnetic field lines, with a moderate degree of anisotropy ($\theta \approx 0.25$), Mok found that for low ω_{pe}/Ω (e.g., in the corona), no resonant amplification of waves through this instability results. However, with the increase in ω_{pe} caused by increasing density in the chromospheric regions of the loop, the ratio ω_{pe}/Ω becomes large enough so that the instability threshold is exceeded. This results in a growth of lower hybrid waves (Krall and Trivelpiece 1973), leading to fast collisionless scattering of the beam electrons into a more isotropic distribution. This has several effects. First, it causes the beam to suddenly slow down. Second, there is strong local heating produced near the region where the onset of instability occurs, possibly affecting the ratio of hard X-ray and EUV fluxes (cf. Section 3.2.5). Third, some microwaves are emitted in this region, although they may not represent a dominant contribution to the microwave emission during the impulsive phase.

3.3.1.4 Electron Beam Momentum

The strong observational evidence for the applicability of the thick target electron beam model to the transport of impulsively accelerated electrons has led many authors to carry out studies of its hydrodynamic and energetic consequences on the structure and dynamics of the target atmosphere (see, e.g., Sections 3.2.5 and 3.2.7). However, these studies have all focused their attention on the heating aspects of the electron beam; they have neglected the effect of direct momentum deposition in the equation of motion. As pointed out by Brown and Craig (1984), beam momentum is potentially of great interest to impulsive phase transport because it is transferred directly to the atmosphere, and hence produces an immediate acceleration a_B, whereas acceleration by nonhydrostatic thermal pressure gradients, a_P, takes time to build up.

Brown and Craig found that an electron beam, for high values of the energy flux, produces a beam acceleration that is much greater than solar gravity ($a_B \approx 100\ g_\odot$, for beam energy flux 10^{11} ergs cm^{-2} s^{-1}). They compared a_B to a_P on the assumption that beam heating is balanced by conduction. Their comparison implied that a_P was less that a_B for classical conduction and of the same order as a_B for saturated heat flux. Hence, they concluded that beam momentum deposition is important even after nonhydrostatic thermal pressure gradients have developed.

McClymont and Canfield (1984) came to quite a different conclusion. They estimated the time scale for a_P to exceed a_B, arguing that conduction and radiation can be neglected on the short time scales of interest, and so considered beam heating alone. They found that in the corona, a_B dominates a_P for at most the first few seconds; in the chromosphere it will do so even more briefly ($\lesssim 10^{-2}$ s). The basic reason for this is the large density gradient across the transition region in the preflare atmosphere. Since collisional heating gives rise to a temperature increase rate which is (locally) independent of density (Emslie 1978), the rapid heating of the preflare transition region layers to coronal temperatures gives rise to a large pressure gradient across this region. Hence they concluded that electron beam momentum deposition is not important to the global evolution of the flare atmosphere.

Both Brown and Craig (1984) and McClymont and Canfield (1984) also considered the hydrostatic case, i.e. an electron beam that has been on for long enough to reach a new pressure equilibrium. Because a_B can so greatly exceed gravity, Brown and Craig concluded that steady-state hydrostatic models require revision. However, McClymont and Canfield called attention to the fact that flare electrons are accelerated in closed loops. Due to containment of the heated plasma, the thermal coronal pressure is much greater than that due to gravity alone. As a result, the addition of beam pressure to gravitational and containment pressure has a negligible effect.

At this time no study has yet been carried out that examines the hydrodynamic importance of beam momentum deposition, taking fully consistent account of beam heating, conduction, radiation, and mass motion, in order to unambiguously demonstrate which of these conflicting views is correct. For hydrostatic models, on the other hand, it seems clear that the electron beam pressure can safely be neglected. However, under rather specialized conditions the momentum imparted to the atmosphere by proton beams may be significant (Brown and Craig 1984, Tamres, Canfield and McClymont 1985).

3.3.2 Proton Transport

The timing, spectrum and morphology of flare protons, when compared to electrons, provide important clues on the nature of flare acceleration and transport. Here we concentrate our attention on specific recent advances in understanding proton transport processes between the acceleration site and the lower solar atmosphere; the reader interested in interplanetary proton phenomena is referred to Forman, Ramaty and Zweibel (1985).

3.3.2.1 Lyman-α Charge-Exchange Emission

The broad-band X-ray and microwave spectrum of the impulsive phase of solar flares has been observed well enough so that we can confidently state that the most commonly accelerated nonthermal electrons are those in the deka-keV range. Are the most commonly accelerated protons also in this energy range? Or does the fact that gamma rays are the only observed emission due to protons mean that flares accelerate protons primarily to MeV energies? Suppose that flares typically accelerate protons, in numbers comparable to electrons, in the deka-keV range. What are the observable manifestations?

It is known that if energetic protons are injected into the largely un-ionized chromosphere, they will efficiently pick up electrons from ambient neutral hydrogen atoms (Orrall and Zirker 1976). The nonthermal protons thereby become nonthermal hydrogen atoms, and radiate the usual hydrogen emission spectrum, redshifted by the Doppler effect. Orrall and Zirker carried out an equilibrium calculation of charge exchange in an empirical model chromosphere that showed that the nonthermal Lα radiation would be readily observable above the quiet sun Lyman-α spectrum, for an input energy flux of 2×10^7 ergs cm^{-2} s^{-1}, assuming a power-law proton energy distribution with a 10 keV low-energy cutoff and exponent $\delta = 2.5$, 3, and 4. Recognizing that the ionization structure of the atmosphere would not long remain undisturbed by the energetic protons, they showed that for the beam parameters considered, the chromospheric ionization time was in the range $10^1 - 10^5$ s.

Recently Canfield and Chang (1984) examined the sensitivity of the Lyman-α photon emission spectrum to proton energy. Since the largest flares are thought to inject about 10^{11} ergs cm^{-2} s^{-1} into the lower atmosphere, they investigated the Lyman-α radiation that would be generated by this same flux, in the form of monoenergetic proton beams. Since the temperature and density structure of the atmosphere is of secondary relevance compared to its ionization structure, they assumed a uniform atmosphere. They adopted a 10% hydrogen ionized fraction, and assumed an equilibrium between the fast protons and the ambient chromosphere. Their result, shown in Figure 3.28, demonstrates a high level of emission over a wide range of proton energies.

It remains to be demonstrated that the radiation generated by charge exchange is produced when the atmosphere is heated (and perhaps largely ionized) by the proton beam. In particular, for what range of proton beam parameters will enough of the atmosphere remain sufficiently neutral to make charge exchange emission comparable to that calculated? Canfield and Chang (1984) showed that at high input beam fluxes the ionization time is much shorter than the charge exchange time. At high values of incident proton energy the ionization time tends to decrease more rapidly than the charge exchange time. At low values a proton accelerated at the top of a typical coronal loop does not reach the chromosphere. At low values of the incident energy flux F(0) the emission predicted by the steady-state calculation is too weak to be detectable above a typical preflare (active region) background. Finally, at high values the ionization time becomes short, making the emission difficult to detect. Notwithstanding, Canfield and Chang showed that some cases of potentially great interest (e.g., a beam of 300 keV protons and energy flux 10^8 ergs cm^{-2} s^{-1}) are well approximated by an equilibrium calculation for tens of seconds, and easily detected above the active-region Lyman-α background. Future calculations should examine time-dependence, a realistic target atmosphere, and spectra including Bessel functions (Forman, Ramaty, and Zweibel 1984).

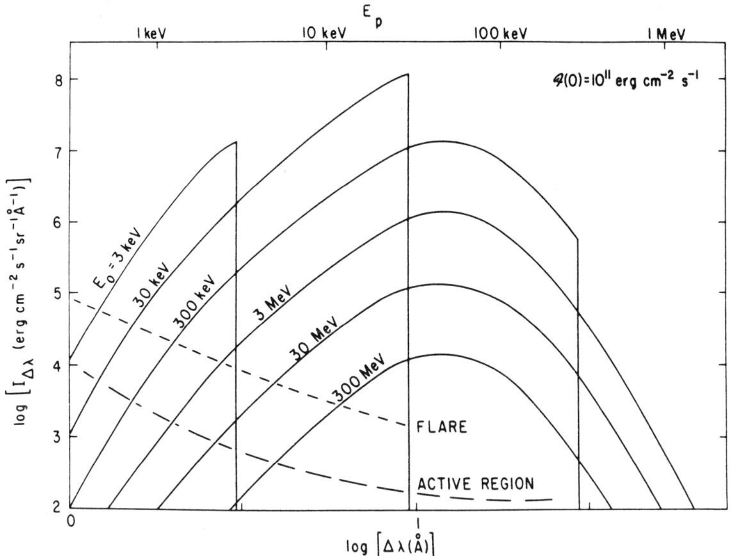

Figure 3.28 Solid curves: theoretical nonthermal Lyman-α spectra for monoenergetic vertical proton beams of energy flux 10^{11} ergs cm^{-2} s^{-1}. The short-dashed curve, marked F, is the average of two flares at Lyman-α flare maximum (Canfield and VanHoosier, 1980). The long-dashed curve is an active-region spectrum (Cohen 1982).

3.3.2.2 Heating of a Thick-Target Atmosphere

With the simultaneous operation of the Hard X-ray Burst Spectrometer (HXRBS) and Gamma Ray Experiment (GRE) on SMM, a few flares exist for which there are good data in both hard X-rays and γ-rays. These data provide information on the populations of energetic deka-keV electrons and Mev protons, respectively. Emslie (1983a) has used the available data from both experiments for the 1980 June 7 flare (see Kiplinger et al., 1983) to infer the energies in both electrons and protons for this event. Through thick target modeling of these emissions (Brown 1971, Ramaty 1984), Emslie has calculated the energy deposited per unit height by both electron and proton bombardment, as a function of depth in the atmosphere. Apart from the (relatively) unknown area factor, determination of the instantaneous energy input rate is possible only for electron heating, since hard X-rays are "prompt" data, providing information on the instantaneous target-averaged electron flux. On the other hand, some γ-rays (e.g., the deuterium formation 2.223 Mev line) are produced over a considerable period (compared to observational time scales) after the injection of the protons which lead ultimately to the γ-ray emission. We can calculate only the event-integrated energy deposition with reliability.

Proton spectra are in general harder (shallower) than electron spectra. Since Mev protons have a greater individual energy and stopping depth (Emslie 1978) than deka-keV electrons, it follows that in general proton heating dominates electron heating at large depths, while high in the atmosphere electron heating dominates. The "cross-over" point depends on the actual shape of the electron and proton spectra, and on the relative energy fluxes in the two sets of particles. Figure 3.29 shows, for the 1980 June 7 flare, the energy deposited per unit height interval by both electron and proton bombardment. It can be seen that for this event, electron heating dominates all the way down to a particle column density $N \approx 3 \times 10^{22}$ cm^{-2}. Since this level is below any reasonable location of the chromospheric flare (e.g., Machado et al., 1980; Ricchiazzi and Canfield 1983), it follows that electron heating dominates throughout most of the region that gives rise to observed flare emissions (with the exception of emissions from around the temperature minimum [see Machado, Emslie, and Brown 1978]). Since the 1980 June 7 event had a particularly high γ-ray to hard X-ray flux ratio (Kiplinger 1983), it follows that the latter conclusion is valid for all flare events (see

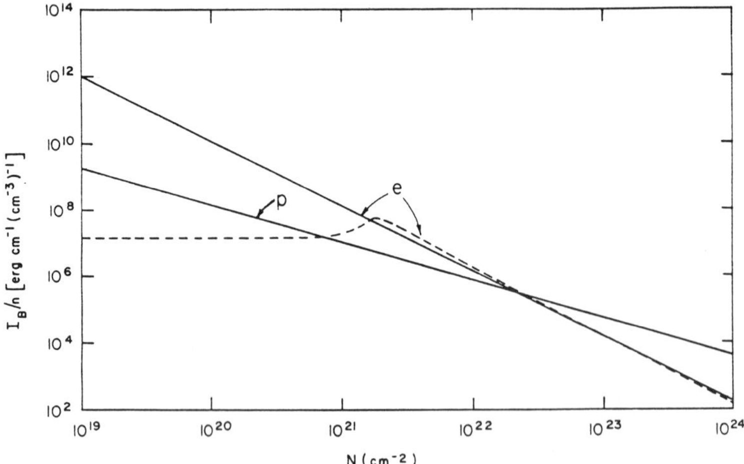

Figure 3.29 The quantity $I_B/n = \int\int\int [Q(n,y,z,t)/n(z)]\, dx\, dy\, dt$ (units ergs per cm of vertical height per ambient electron), representing the temporal and areal integral of the specific energy deposition rate Q as a function of column depth $N = \int n(z)dz$, for the 1980 June 7 event. From Emslie (1983a).

also Lin and Hudson 1976). However, Emslie (1983a) pointed out that this conclusion is critically dependent on the assumption of thick target emission for both the hard X-rays and the γ-rays. While Ramaty (1984) has conclusively demonstrated that γ-ray emission in solar flares is a thick target process, there is still considerable speculation (see, e.g., Brown and Smith 1980; Emslie 1983b) on whether hard X-rays are produced principally by thick target bremsstrahlung or by thermal bremsstrahlung from a confined ensemble of very hot ($\lesssim 10^8$ K) electrons. In the latter case, the number of precipitating electrons is reduced relative to the hard X-ray photon flux, and less chromospheric heating results (see Emslie and Vlahos 1980). Thus, if hard X-rays indeed turn out to be emitted principally by a hot thermal source, then the importance of proton heating in the chromospheric energy budget would merit closer examination.

3.3.3 Radiative Energy Transport by Amplified Decimetric Waves

Following the initial work on the application of maser theory to radio bursts from the Sun and stars (Holman et al., 1980; Melrose and Dulk 1982a) and a discussion of the energetic implications (Melrose and Dulk 1982b), Melrose and Dulk (1984) have carried out new work on the transport of energy by amplified radio radiation. These transport aspects are summarized below; the role of masers in particle acceleration is discussed in the previous chapter.

Strong theoretical arguments suggest that up to 50% of the energy in a flare should go into radiation at the cyclotron frequency Ω_e, the exact percentage depending on the degree of field line convergence toward the footpoints of the flux tubes, and the amount of fast electron energy lost through collisions. These arguments (based on an extension of then-existing theory by Wu and Lee in 1979) are reinforced by consideration of how the electrons accelerated in a flare lose the component of their energy perpendicular to the magnetic field in order to precipitate into the footpoints of magnetic flux tubes and hence heat the chromosphere and generate hard X-ray bursts. The perpendicular energy is radiated away through amplification of decimetric radio waves. The radiation travels from the flux tubes of energy release, across field lines, and is then reabsorbed at the second harmonic of Ω_e, i.e. at points where the field strength B is half its value in the source. From energetic considerations there seems to be enough energy in the radio-frequency (RF) radiation to heat the plasma of the absorbing region so that it produces the soft X-rays. In addition it is likely that this RF heating process can account for other properties of soft X-rays in the impulsive phases of flares: large source sizes, excess line widths, characteristic temperatures, and their temporal development.

3.3.3.1 Mechanism for Emission and Amplification

The generation of the high levels of RF radiation is thought to occur in an impulsive flare as follows.

(1) It is assumed that energy release occurs mainly near the tops of

one or more magnetic flux tubes of the kind illustrated in Figure 3.30. The accelerated (or heated) electrons are assumed to have energies E ~ 10 to 100 keV (or T ~ 10^8 to 10^9 K) and to have a nearly isotropic pitch angle distribution, i.e., $\langle E \quad \rangle \approx \langle E \rangle$.

(2) The fast electrons, with v $\gtrsim$ 0.1 c, travel down the legs of the flux tubes. those with moderate to large pitch angles α reflecting in the converging fields, and those with small α precipitating. In a typical travel time of ~ 1 s the electron pitch angle distribution $f(\alpha)$ in the legs is anisotropic, with no upgoing electrons with small α. The anisotropy represents a source of free energy because it creates a non-equilibrium distribution with an inverted population, i.e., there is a region of velocity space where $\partial f/\partial v_{\perp} > 0$.

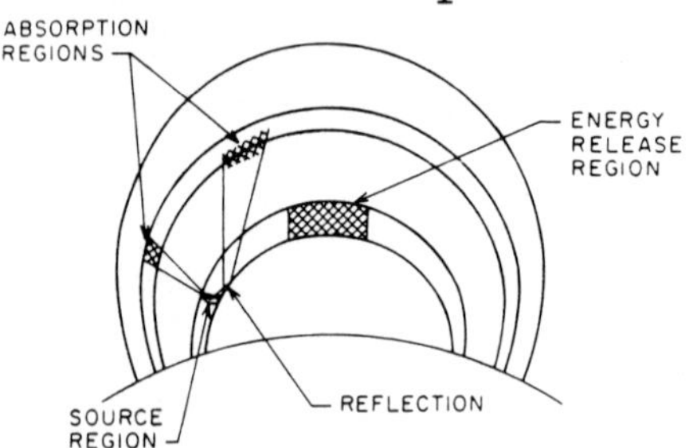

Figure 3.30 Schematic drawing of a sequence of loops showing an energy release region, the maser source region in one leg where the field strength is B, the cone of maser radiation with reflection of radiation directed toward higher field strengths, and the absorption regions where the field strength is B/2. From Melrose and Dulk (1984).

(3) Provided that the plasma frequency ω_p is $\leqq \Omega_e$, there can be a resonant transfer of energy from electrons to waves: in particular, electromagnetic waves at the cyclotron frequency can be amplified. The growth rate of this "cyclotron maser" is very large, ~ 10^7 s^{-1}, so in a few micoseconds the waves can be amplified by a factor of ~ e^{30} or so, at which time they saturate, having extracted most of the free energy. With B in the expected range of 100 to 1000 gauss, the amplified radiation is in the range 0.3 to 3 GHz, i.e., at decimetric wavelengths.

(4) Saturation of the maser is expected to occur by quasilinear relaxation: the maser converts the perpendicular energy of certain electrons into radiation and leaves them with mainly parallel energy, i.e. it diffuses them into the loss cone at the maximum possible rate. Provided that the density is low and/or the electron energy is large, so that collisions are infrequent, this is the only known process which can prevent most of the electrons from remaining trapped in converging, coronal flux tubes for many bounce periods, i.e. for several to many seconds. For example, 30 keV electrons trapped between mirror points 3 x 10^8 cm apart would bounce ~ 10^3 times (~ 10^2 s) in a plasma of $n_e = 10^9$ cm^{-3} before collisions would diffuse them into the loss cone (Section 3.2.2).

3.3.3.2 Energy Content

From studies of hard X-ray bursts it is known that the energy going into fast electrons in the impulsive phase is $E \sim 10^{27}$ to $\sim 10^{30}$ ergs s^{-1} (e.g., Brown and Smith 1980). Provided that the electrons are not highly collimated, about half of this is in the parallel component and, again assuming a low collision frequency in the coronal portion of the flux tubes, it is deposited in the low atmosphere at the footpoints of the loops of energy release. Most of the other half goes into the maser radiation, and then (as described below) into reabsorption and heating of the coronal plasma in a large volume, $V \sim 10^{26}$ to $\sim 10^{29}$ cm^3, surrounding the loops of energy release. Thus the average heating rate is $E/V \sim 1$ to ~ 10 ergs cm^{-3} s^{-1}, assuming that the larger values of E and V go together. On average this can increase the internal energy nkT of the preflare corona by a factor of 2 to 20 each second.

3.3.3.3 Reabsorption at Distant Locations

The RF radiation travels outward from the maser generation regions in the manner sketched in Figure 3.30. Only one maser emission region is shown, but there are expected to be $\gtrsim 10^7$ maser pulses per sec, each lasting about 1 ms, coming from various points in the legs of the flux tube, and from all flux tubes of energy release. Each pulse is narrowband and at a frequency ω slightly higher than Ω_e, but in total the pulses should cover a frequency range of an octave or so because B, hence Ω_e and ω, varies with position in the legs of the loops.

The maser pulses travel across field lines at nearly the speed of light until they encounter a region where the field strength is half its value at the source: this may be 10^3 to 10^4 km distant, depending on the gradient of B. Then gyroresonance absorption by the ambient electrons occurs at the second harmonic, very efficiently, with an optical depth $\tau \sim 10^4$. A series of pulses can heat a small volume of plasma, $\lesssim 10^{20}$ cm^3, to a few x 10^7 K in ~ 0.1 s. Temperatures above about 3 x 10^7 K energy are difficult to achieve by maser heating because of the temperature dependence of the collision frequency; since the mean free path is $\propto T_e^{1.5}$, further energy deposition leads to heating of a larger volume of plasma rather producing a higher temperature.

Turbulent plasma motions should occur because the maser pulses heat the ambient electrons in many, small, localized volumes, a result of the fine beaming and small absorption distance. The hot electrons tend to propagate out of the pockets of high pressure so created and thus provide an explanation for the highly broadened soft x-ray lines observed during impulsive flares.

The overall volume heated depends mainly on the field gradient in the vicinity of the loops of energy release. The model suggests that in a small flare a volume ~ 10^{26} cm^3, some 10 times larger than the volume of energy release, would be heated to $\gtrsim 10^7$ K in 10 seconds. In a large flare it is expected that the heated volume would be much greater than 10^{26} cm^3, as would the volume in which energy release occurs. Because the maser radiation is initially directed at a large angle ($\approx 70°$) to the field lines, most of the energy transport is across the field.

3.3.3.4 Secondary Effects

According to the RF heating model, the time profile of maser emission in simple impulsive flares should mimic the hard X-ray profile, and this should follow closely (within ~ 1 s) the profile of energy release. Much of the RF energy is deposited in the soft X-ray plasma and is radiated or conducted away, relatively slowly. Hence the slope of the soft X-ray intensity profile is expected to be approximately proportional to the hard X-ray intensity as observed. While this property is not unique to the RF heating model, it applies here to a soft X-ray emitting volume much larger than that of initial deka-keV energy release.

Densification of heated loops and the origin of the blue-shifted component of soft X-ray lines can occur in two ways. First, energy in the parallel component can "boil off" chromospheric material into the loops of energy release. Second, as other loops become heated by the maser radiation, enhanced conduction to the footpoints would boil off material there, though presumably not as fast.

The RF heating model has implications for enhancement of chromospheric evaporation, a major topic of the following chapter. As pointed out there, chromospheric evaporation can take place by both direct Coulomb heating from the parallel component of the nonthermal electron energy and by thermal conduction from energy deposited in the corona (possibly a slower process). Hence, if the RF heating mechanism heats loops other than those in which direct energy release has taken place, chromospheric evaporation may take place over a larger area of the chromosphere.

If the density in a flaring loop increases to the point where $\omega_p \gtrsim \Omega_e$ e.g. $n_e \geqq 10^{10}$ cm^{-3} if B = 300 gauss, the maser would stop. Or if the initial density in the flaring loop were such that $\omega_p \gtrsim \Omega_e$ there would be no maser, and probably only a gradual phase. However, in this high-density case, Coulomb collisions would scatter and degrade the energy of all but the highest energy electrons. Thus the conditions would be appropriate for a "high coronal flare" without an impulsive hard X-ray burst from footpoints.

3.3.3.5 Conclusions

As the RF heating radiation is entirely reabsorbed in the corona, it is not possible to observe it directly. However, there is strong evidence that this maser mechanism is the explanation for the very intense auroral kilometric radiation of Earth and for similar radiation from Jupiter and Saturn. From the Sun and stars, direct maser action (or a related process involving conversion of amplified waves) producing radiation just above the second harmonic is the favored explanation for "microwave spike bursts" of very high brightness and circular polarization; Enome (1983) has recently examined some of the properties of such bursts, and earlier studies have been made by Droge (1977), Slottje (1978) and Zhao (1982).

3.4 SUMMARY

In this chapter our main interest has been the transport of energy, momentum and charge during the dramatically nonthermal and energetic impulsive phase. Not surprisingly, the technological and operational advantages embodied in the SMM and Hinotori instruments, and their collective and coordinated use with forefront ground-based optical and radio instruments, have brought about many advances in our understanding of impulsive phase flare physics.

The topic most thoroughly explored in this chapter is that of the transport of nonthermal electrons. The thick-target electron beam model, in which electrons are presumed to be accelerated in the corona and typically thermalized primarily in the chromosphere and photosphere, is supported by observations throughout the electromagnetic spectrum. At the highest energies, the anisotropy of γ-ray emission above 10 MeV clearly indicates that these photons are emitted by anisotropically-directed particles. The timing of this high-energy γ-radiation with respect to lower-energy hard X-radiation implies that the energetic particles have short lifetimes. For collisional energy loss, this means that they are stopped in the chromosphere or below. Stereoscopic (two-spacecraft) observations at hard X-ray energies (up to 350 keV) imply that these lower-energy (but certainly nonthermal) electrons are also stopped deep in the chromosphere. Hard X-ray images show that, in spatially resolved flares whose radiation consists of impulsive bursts, the impulsive phase starts with X-radiation that comes mostly from the footpoints of coronal loops whose coronal component is outlined by microwaves. Recent X-ray polarization measurements at the low end of the deka-keV range show small values, but these are still large enough to be compatible with highly collimated particle injection somewhere in the coronal part of a loop. Preliminary analyses of combined hard and soft X-ray spectra appear to demand the presence of nonthermal electrons early in the impulsive phase. The thick-target electron-beam model accounts for both the close temporal coincidence of UV and hard X-ray bursts and the relative decrease of UV emission with increasing hard X-ray intensity. White-light emission from the largest flares is closely related temporally and energetically to the nonthermal electrons that produce deka-keV X-rays. Moreover, white-light spectra imply that commonly such emission originates in the chromosphere, where deka-keV electrons are stopped. Both timing and spectra of Hα emission imply that energetic nonthermal electrons penetratively heat the chromosphere during the impulsive phase. Finally, we now suspect that microwave spike bursts may require a level of microwave maser emission that dramatically enhances the precipitation of thick-target electrons.

Although the thick-target nonthermal-electron model meets with some success, it describes only a fraction of observed flares, and shows some puzzling and challenging failures. The limited amount of stereoscopic hard X-ray data at 350 keV shows little directivity, compared to the predictions of models of collisional electron transport in loops. If this trend is borne out by more comprehensive observations, it could provide a strong limitation on the thick-target model. The height dependence of the hard X-ray spectrum is systematically different from that predicted. The majority of images at the low end of the deka-keV energy range show single X-ray sources; in many flares, the X-ray producing electrons in this energy range are obviously confined to the corona. It now seems clear that direct collisions by nonthermal electrons are not necessary to understand the production of inner-shell Fe Kα emission; it appears that impulsive variability of these lines comes about because of irradiation of the chromosphere and photosphere by an impulsively varying spectrum that extends down to at least 7 keV. The thick-target model appears to predict much more EUV emission, in proportion to hard X-ray emission, than is observed. In some flares, white-light emission is observed well beyond the end of the impulsive phase, which appears utterly inexplicable in the thick-target model; such emission appears to correlate more strongly with the thermal X-ray emission of the hot ($T > 2 \times 10^7$ K) thermal plasma. Furthermore, some flares show spectra that imply that the white-light emission comes from deeper than the chromosphere, in regions that are collisionally inaccessible to deka-keV electrons. Finally, it is not well understood whether electron beams of the required flux can be maintained, in the presence of return-current instabilities known to exist in laboratory plasmas, and in view of the observational uncertainty in beam area.

Much of the work that we have described above leaves basic questions unanswered, requiring future observational and theoretical work. First, high spatial resolution, from microwaves through γ-rays, is the observational key to understanding impulsive-phase transport. What is the spatial scale of particle beams? Return currents? To what extent are they filamentary? What processes dominate the physics of return currents? What physical processes could disrupt beams? What role do they play in heating the ambient plasma? Does the relative size of hard and soft X-ray sources suggest an important energy transport role for microwave maser radiation? Second, what is the mass and charge composition of flare-accelerated particles? Do proton beams exist in the deka-keV range? Are energetic protons important to impulsive-phase momentum transport? Do an energetically significant number go undetected because of their inefficiency of bremsstrahlung production? Third, where are nonthermal particles accelerated within flare loops? What is their degree of anisotropy and their energy spectrum? How are they confined to the corona in those flares that do not show footpoint brightening in hard X-rays? How are they thermalized as they travel along flare loops? Is their transport dominated by particle collisions or by wave generation or by the emission or absorption of radiation? Are regions of particle precipitation identical to regions of explosive coronal and chromospheric hydrodynamic motion? Will the commonly accepted model of coronal loops as the basic building block of flare geometry withstand close scrutiny? Not until the answers to these questions are better understood can we claim to understand the basic physics of impulsive-phase transport.

3.5 REFERENCES

Aboudaram, J., Henoux, J. C., and Fang, C. 1984, private communication.

Acton, L. W. 1965, Nature, **207**, 737.

Acton, L. W., Canfield, R. C., Gunkler, T.A., Hudson, H.S., Kiplinger, A. L., and Leibacher, J. W. 1982, Ap. J., **263**, 409.

Acton, L. W., Culhane, J. L., Gabriel, A. H. et al., 1980, Solar Phys., **65**, 53.

Alfven, H. 1939, Phys. Rev., **55**, 425.
Alissandrakis, C. E. and Kundu, M. R. 1975, Solar Phys., **41**, 119.
Alissandrakis, C. E. and Kundu, M. R. 1978, Ap. J., **222**, 342.
Antonucci, E. 1981, July 1, 1980, Flare Study Group, unpublished.
Antonucci, E., Gabriel, A. H., Doyle, J. G., Dubau, J., Faucher, P., Jordan, C., and Veck, N. 1984, Astron. and Astrophys., **133**, 239.
Bai, T. 1979, Solar Phys., **62**, 113.
Bai, T., and Ramaty R. 1976, Solar Phys., **49**, 343.
Bai, T. and Ramaty, R. 1978, Ap. J., **219**, 705.
Batchelor, D. A. 1984, private communication.
Bely-Dubau, F., Dubau, J., Faucher, P., and Gabriel, A. H. 1982a, Mon. Not. R. Astron. Soc., **198**, 239.
Bely-Dubau, F. et al., 1982b, Mon. Not. R. Astron Soc., **201**, 1155.
Boyer, R. Machado, M. E., Rust, D. M., and Sotirovski, P. 1985, Solar Phys. **98**, 255.
Brown, J. C. 1971, Solar Phys., **18**, 489.
Brown, J. C. 1972, Solar Phys., **26**, 441.
Brown, J. C. 1973a, Solar Phys., **31**, 143.
Brown, J. C. 1973b, in G. Newkirk (ed.), Coronal Disturbances, Proc. I.A.U. Symp. 57, 395.
Brown, J. C. and Bingham, R. 1984, Astr. Ap., **131**, L11.
Brown, J. C., Canfield, R.C., and Robertson, M.H. 1978, Solar Phys., **57**, 399.
Brown, J. C. and Craig, I. J. D. 1984, Astr. Ap., **130**, L5.
Brown, J. C. and Hayward, J. 1982, Solar Phys., **80**, 129.
Brown, J. C., Hayward, J. and Spicer, D. S. 1981, Ap. J. (Letters), **145**, L91.
Brown, J. C., Melrose, D.B., and Spicer, D. S. 1979, Ap. J., **228**, 592.
Brown, J.C., and Smith, D. F. 1980, Rep. Prog. Phys., **43**, 125.
Candlestickmaker, S. 1972, Quart. J. Roy. Astron. Soc., **13**, 63.
Canfield, R.C., and Chang, C.-R. 1984, Ap. J., **295**, in press.
Canfield, R.C. and Gunkler, T.A. 1985, Ap. J., **288**, 353.
Canfield, R.C., Gunkler, T.A., and Ricchiazzi, P.J. 1984, Ap. J., **282**, 296.
Canfield, R.C., and VanHoosier, M.E. 1980, Solar Phys., **67**, 339.
Cheng, C.-C., Tandberg-Hanssen, E., and Orwig, L. E. 1984, Ap. J., **278**, 853.
Cheng, C.-C., Tandberg-Hanssen, E., Bruner, E. C., Orwig, L., Frost, K. J., Kenney, P. J., Woodgate, B. E., and Shine, R. A. 1981, Ap. J. (Letters), **248**, L39.
Cheng, C.-C., Bruner, E. C., Tandberg-Hanssen, E., Woodgate, B. E., Shine, R. A., Kenney, P. J., Henze, W., and Poletto, G. 1982, Ap. J., **253**, 353.
Cohen, L. 1982, An Atlas of Solar Spectra Between 1175 and 1959 Angstroms Recorded on Skylab with NRL's Apollo Telescope Mount Experiment, NASA Reference Publication 1069 (Washington: U.S. Government Printing Office).
Crannell, C. J., Crannell, H., and Ramaty, R. 1979, Astrophys. J., **229**, 762.
Culhane, J. L. et al., 1981, Ap. J. (Letters), **244**, L141.
Dennis, B. R. 1981, July 1, 1980, Flare Study Group, unpublished.

Donati-Falchi, A., Falciani, R. and Smaldone, L. A. 1984, Astron. Astrophys., **131**, 256
Donnelly, R. F. and Hall, L. A. 1974, Solar Phys., **31**, 411
Donnelly, R. F., and Kane, S. R. 1978, Ap. J., **222**, 1043.
Doschek, G. A. 1984, Ap. J., **283**, 404.
Doschek, G. A., Meekins, J. E., Kreplin, R. W., Chubb, T. A. and Friedman, H. 1971, Ap. J., **170**, 573.
Droge, F. 1977, Astr. Ap., **57**, 185.
Dubau, J., Gabriel, A. H., Loulergue, M.. Steenman-Clark, L., and Volonte, S. 1981, Mon. Not. R. Astron. Soc., **195**, 705.
Duijveman, A. and Hoyng, P. 1983, Solar Phys., **86**, 279.
Duijveman, A., Hoyng, P. and Machado, M. E. 1982, Solar Phys., **81**, 137.
Dulk, G. A., Bastian, T. S. and Hurford, G. J. 1983, Solar Phys., **86**, 219.
Emslie, A. G. 1978, Ap. J., **224**, 241.
Emslie, A. G. 1980, Ap. J., **235**, 1055.
Emslie, A. G. 1981, Ap. J., **249**, 817.
Emslie, A. G. 1983a, Solar Phys., **84**, 263.
Emslie, A. G. 1983b, Solar Phys., **86**, 133.
Emslie, A. G. 1985, Solar Phys., **98**, 281.
Emslie, A. G., Brown, J. C., and Donnelly, R. F. 1978, Solar Phys., **57**, 175.
Emslie. A. G., Brown, J. C., and Machado, M. E. 1981, Ap. J., **246**, 337.
Emslie. A. G. and Nagai, F. 1984, Ap. J., **279**, 896.
Emslie, A. G., Phillips, K. J. H., and Dennis, B. R. 1986, Solar Phys. **103** 89.
Emslie, A. G. and Smith, D. R. 1984, Ap. J., **279**, 882.
Emslie, A. G. and Vlahos, L. 1980, Ap. J., **242**, 359.
Enome, S. 1983, Solar Phys., **86**, 421.
Enome, S., Kakimura, T. and Tanaka, H. 1969, Solar Phys., **6**, 428.
Feldman, U., Doschek, G. A. and Kreplin, R. W. 1980, Ap. J., **238**, 265.
Forman, M.A., Ramaty, R., and Zweibel, E. 1985, in Physics of the Sun, ed. P.A. Sturrock, T.E. Holzer, D. Mihalas, and R.K. Ulrich.
Forrest, D. J. 1983, AIP Conf. Proc., 101, Positron-Electron Pairs in Astrophysics, M. L. Burns, A. K. Harding, R. Ramaty, eds, New York: American Inst. of Physics, pp. 3-14.
Forrest, D. J., Chupp, E. L. et al., 1980, Solar Phys., **65**, 15.
Fried, B. D. and Gould, R. N. 1961, Phys. Fluids, **4**, 139.
Gunkler, T.A., Canfield, R.C., Acton, L.W., and Kiplinger, A.L. 1984, Ap. J., **285**, 835.
Haug, E. 1972, Solar Phys., **25**, 425.
Hayward, J. 1984, unpublished.
Hiei, E. 1982, Solar Phys., **80**, 113.
Hiei, E., Tanaka, K., Watanabe, T., and Akita, K. 1982, Hinotori Symp. Solar Flares, p. 208.
Holman, G. D., Eichler. D., and Kundu, M. R. 1980, in IAU Symposium 86, Radio Physics of the Sun, ed. M. Kundu and T. Gergely (Dordrecht: Reidel). p. 341.
Horan, D.M., Kreplin, R.W., and Fritz, G.G. 1982, Ap. J., **225**, 797.
Hoyng, P., Brown, J. C. and van Beek, H. F. 1976, Solar Phys., **48**, 197.
Hoyng, P., Knight, J W. and Spicer, D. S. 1978, Solar Phys., **58**, 139.

Hoyng, P., Marsh, K. A., Zirin, H. and Dennis, B. R. 1983, Ap. J., **268**, 865.
Hoyng, P. et al., 1981, Ap. J. (Letters), **246**, L155.
Hudson, H. S. 1979, in Particle Accleration Mechanisms in Astrophysics, ed. J. Arons, C. Max, and C. McKee (New York: Amer. Inst. Phys), p. 115.
Hudson, H. S. and Dwivedi, B. N. 1982, Solar Phys., **76**, 45.
Jacobs, V. L., Davis, J., Kepple, P. C., and Blaha, M. 1977, Ap. J., **211**, 605.
Jacobs, V. L., Davis, J., Robertson, J. E., Blaha, M., Cain, J., and Davis, M. 1980, Ap. J., **239**, 1119.
Kadomtsev, B. B. and Pogutse, O. P. 1968, Soviet Phys. - JETP, **26**, 1146.
Kaempfer, N., and Schoechlin, W. 1982, Solar Phys., **78**, 215.
Kaempfer, N., and Magun, A. 1983, Ap. J., **274**, 910.
Kai, K., Kosugi, T. and Nakajima, H. 1982, Solar Phys., **75**, 331.
Kane, S. R. 1983, private communication.
Kane, S. R., Anderson, K. A., Evans, W. D., Klebesadel, R. W., and Laros, J. 1979, Ap. J. (Letters), **233**, L151.
Kane, S. R., Anderson, K. A., Evans, W. D., Klebesadel, R. W., and Laros, J. G. 1980, Ap. J., **239**, L85.
Kane, S. R., and Donnelly, R. F. 1971, Ap. J., **164**, 151.
Kane, S. R., Fenimore, E. E., Klebesadel, R. W., and Laros, J. G. 1982, Ap. J. (Letters), **254**, L53.
Kane, S. R., Frost, K. J., and Donnelly, R. F. 1979, Ap. J., **234**, 669.
Kane, S. R., Love, J., Neidig, D. F., and Cliver, E. W. 1984, Ap. J. (Letters), **290**, L45.
Kane, S. R. et al., 1980, in Solar Flares, ed. P.A. Sturrock (Boulder: Colorado Associated University Press).
Kane, S. R., Uchida, Y., Tanaka, K., and Hudson, H. S. 1983, Recent Advances in the Understanding of Solar Flares, Solar Phys., **86**.
Kattenberg, A. and Allaart, M. A. F. 1981, Thesis, Utrecht.
Kattenberg, A., Allaart, M., de Jager, C., Schadee, A., Schrijver, J., Shibasaki, K., Svestka, Z., and van Tend, W. 1983, Solar Phys., **88**, 315.
Kawabata, K. et al., 1982, Proc. Hinotori Symposium, eds. Y. Tanaka et al., ISAS, Tokyo, p. 168.
Kiplinger, A. L. 1983, private communitication.
Kiplinger, A. L., Dennis, B. R., Emslie, A. G., Frost, K. J. and Orwig, L. E. 1983, Ap. J. (Letters), **265**, L99.
Kiplinger, A. L., Dennis, B. R., Frost, K. J., and Orwig, L. E. 1983, Ap. J., **273**, 783.
Knight, J. W. and Sturrock, P. A. 1977, Ap. J., **218**, 306.
Kondo, I. 1982, in Hinotori Symposium on Solar Flares, ed. Tanaka et al. (Institute for Space and Aeronautical Science, Tokyo).
Koshelev, K. N., and Kononov, E. Ya. 1982, Solar Phys., **77**, 177.
Krall, N. A. and Trivelpiece, A. W. 1973, Principles of Plasma Physics, (New York: McGraw Hill).
Kundu, M. R. 1983, Solar Phys., **86**, 205.
Kundu, M. R. and Alissandrakis, C. E. 1975, Mon. Not. R. Astron. Soc., **173**, 65.
Kundu, M. R. and Alissandrakis, C. E. 1977, B.A.A.S., **9**, 328.

Kundu, M. R., Schmahl, E. J. and Rao, A. P. 1981, Astron. Astrophys., **94**, 72.
Kundu, M. R., Schmahl, E. J. and Velusamy, T. 1982, Ap. J., **253**, 963.
Kundu, M. R. and Vlahos, L. 1979, Ap. J., **232**, 595.
Kurokawa, H. 1983, Solar Phys., **86**, 195.
Lang, K. R., Willson, R. F. and Felli, M. 1981, Ap. J., **247**, 338.
Langer, S. H. and Petrosian, V. 1983, Ap. J., **215**, 666.
Lawson, J. D. 1957, J. Electron Contr., **3**, 587.
Leach, J. 1984, Ph.D. Thesis, Stanford University.
Leach, J., Emslie, A. G. and Petrosian, V. 1985, Solar Phys., **96**, 331.
Leach, J. and Petrosian, V. 1981, Ap. J., **251**, 781.
Leach, J. and Petrosian, V. 1983, Ap. J., **269**, 715.
Lin, R. P., and Hudson, H. S. 1971, Solar Phys., **17**, 412.
Lin, R. P., and Hudson, H. S. 1976, Solar Phys., **50**, 153.
Livshitz, M. A., Badalyan, O. G., Kosovichev, A. G., and Katsova, M. M. 1981, Solar Phys., **73**, 269.
Machado, M. E. 1982, Adv. Space Res., **2**, No. 11, 115.
Machado, M. E. 1983, Solar Phys., **89**, 133.
Machado, M. E., Avrett, E. H., Vernazza, R. W., and Noyes, R. W. 1980, Ap. J., **242**, 336.
Machado, M. E., Emslie, A. G. and Brown, J. C. 1978, Solar Phys., **58**, 363.
Machado, M. E., Duijveman, A. and Dennis, B. R. 182, Solar Phys., **79**, 85.
Machado, M. E. and Rust, D. M. 1974, Solar Phys., **38**, 499.
Machado, M. E. et al., 1983, Solar Phys., **85**, 157.
Marsh, K. A. and Hurford, G. J. 1980, Ap. J. (Letters), **240**, L111.
Marsh, K. A. and Hurford, G. J. 1982, Ann. Rev. Astron. Astrophys., **20**, 497.
Marsh, K. A., Hurford, G. J., Zirin, H., and Hjellming, R. M. 1980, Ap. J., **242**, 352.
McClymont, A. N. and Canfield, R. C. 1984, Astr. Ap., **136**, L1.
McClymont, A.N., Canfield, R.C., and Fisher, G.H. 1984, unpublished.
Melrose, D. B., and Brown, J. C. 1976, Mon. Not. R. Astron. Soc., **176**, 15.
Melrose, D. B., and Dulk, G. A. 1982a, Ap. J., **259**, 844.
Melrose, D. B., and Dulk, G. A. 1984b, Ap. J. (Letters), **259**, L41.
Melrose, D. B., and Dulk, G. A. 1984, Ap. J., **282**, 308.
Mewe, R., and Gronenchild, E. H. B. M. 1981, Astron. and Astrophys. Suppl., **45**, 11.
Mok, Y. 1985, Solar Phys., **96**, 181.
Najita, K. and Orrall, F. Q. 1970, Solar Phys., **15**, 176.
Neidig, D. F. 1983, Solar Phys., **85**, 285.
Neidig, D. F. and Cliver, E. W. 1983, AFGL-TR-83-0257, Hanscom AFB, MA.
Neidig, D. F. and Wiborg, P.H. 1984, Solar Phys., **92**, 217.
Neupert, W. M., Gates, W., Swartz, M. and Young, R. 1967, Ap. J. (Letters), **149**, L79.
Ohki, K. et al., 1982, Proc. Hinotori Symp., eds. Y. Tanaka et al., ISAS Tokyo, p. 102.
Ohki, K., Takakura, T., Tsuneta, S., and Nitta, N. 1983, Solar Phys., **86**, 301.

Orrall, F.Q. 1881, Solar Active Regions, (Boulder: Colorado Associated University Press).
Orrall, F.Q., and Zirker, J.B. 1976, Ap. J., **208**, 618.
Orwig, L. E.. Frost, K. J. and Dennis, B. R. 1980, Solar Phys., **65**, 25.
Parmar, A. N., Wolfson, C. J., Culhane, J. L., Phillips, K. J. H., Acton, L. W., Dennis, B. R. and Rapley, C. G. 1984, Ap. J., **279**, 866.
Petrosian, V. 1981, Ap. J., **251**, 727.
Petrosian, V. 1982, Ap. J. (Letters), **255**, L85.
Petrosian, V. and McTiernan, J. M. 1983, Phys. Fluids, **26**, 3023.
Phillips, K. J. H. 1981, July 1, 1980, Flare Study Group, unpublished.
Phillips, K. J. H. and Neupert, W. M. 1973, Solar Phys., **32**, 269.
Poland, A. et al. 1982, Solar Phys., **78**, 201.
Poland, A. I., Orwig, L. E., Mariska, J. T., Nakatsuka, R., and Auer, L. H. 1984, Ap. J., **280**, 467.
Pottasch, S. R. 1964, Space Science Rev., **3**, 816.
Ramaty, R. 1985, in Physics of the Sun, ed. P.A. Sturrock, T.E. Holzer, D. Mihalas, and R.K. Ulrich.
Ramaty, R., Murphy, R. J., Kozlovsky, B., and Lingenfelter, R. E. 1983, Solar Phys., **86**, 395.
Ramaty, R. and Petrosian, V. 1972, Ap. J., **178**, 241.
Ricchiazzi, P. J. and Canfield, R. C. 1983, Ap. J., **272**, 739.
Rieger, E., Reppin, C., Kanbach, G., Forrest, D. J., Chupp, E. L. and Share, G. H. 1983, Proc. 18th Int. Cosmic Ray Conf., Bangalore, 1983.
Rust, D. M. and Hegwer, F. 1975, Solar Phys., **40**, 141.
Ryan, J. M., Chupp, E. L.. Forrest, D. J., Matz, S. M., Rieger, E., Reppin, C., Kanbach, G., and Share, G. H. 1983, Ap. J. (Letters), **272**, L61.
Slonim, Yu. M. and Korobova, Z. B. 1975, Solar Phys., **40**, 397.
Slottje, C. 1978, Nature, **275**, 520.
Spicer, D. S. 1982, Adv. Space Res., **2**, No. 11, 135.
Spicer, D. S. and Sudan, R. N. 1984, Ap. J., **280**, 448.
Spitzer. L. W. 1962, Physics of Fully Ionized Gases, (2nd ed., New York: Interscience).
Sturrock, P. A. 1980, Solar Flares: A Monograph of Skylab Solar Workshop II, (Boulder Colorado Associated University Press).
Suemoto, Z. and Hiei, E. 1959, Publ. Astron. Soc. Japan, **11**, 185.
Svestka, Z. 1970, Solar Phys., **13**, 471.
Svestka, Z., Rust, D. M., and Dryer, M. 1982, Adv. Space Res., **2**, Number 11.
Takakura, T., Ohki, K., Tsuneta, S., Nitta, N., Makashima, K., Murakami, T., Ogawara, Y., and Oda, M. 1982, Proc. Hinotori Symposium, eds. Y. Tanaka et al., Tokyo, p. 142.
Takakura, T., Ohki, K., Tsuneta, S., and Nitta, N. 1983, Solar Phys., **86**, 323.
Tamres, D. H., Canfield, R. C., and McClymont, A. N. 1985, B.A.A.S., **17**, 634.
Tanaka, K. 1980, Proc. Japan-France Seminar on Solar Physics, ed. F. Moriyama and J. C. Henoux (JSPS), p. 219.
Tanaka, K. 1983, in Activity in Red Dwarf Stars, IAU Colloq. 71, eds. P. B. Byrne and M. Rodono, p. 307.
Tanaka, K., Akita, K, Watanabe, T., and Nishi, K. 1982, Proc. Hinotori

Symposium, eds. Y. Tanaka et al., Tokyo, p. 43.
Tanaka, K., Nitta, N. and Watanabe, T. 1982, Proc. Hinotori Symposium, eds. Y. Tanaka et al., Tokyo, p. 20.
Tanaka, K., Nitta, N. and Watanabe, T. 1984, Ap. J., **282**, 793.
Tanaka, K., Watanabe, T., Nishi, K. and Akita, K. 1982, Ap. J. (Letters), **254**, L59.
Tindo, I. P., Shuryghin, A. I. and Steffen, W. 1976, Solar Phys., **46**, 219.
Tomblin, F. F. 1972, Ap. J., **171**, 377.
Tramiel, L. J., Chanan, G. A. and Novick, R. 1984, Ap. J., **280**, 440.
Tsuneta, S. 1983, "Hard X-Ray Imaging of Solar Flares with the Imaging Hard X-Ray Telescope Aboard the HINOTORI Satellite," Ph.D. Thesis, Univ. of Tokyo, Japan.
Tsuneta, S. 1984, in Proceedings of Japan-France Joint Seminar, Active Phenomena in the Outer Atmosphere of the Sun and Stars, Paris.
Tsuneta, S. et al. 1982, Proc. Hinotori Symp. Solar Flares, Japan Inst. Space & Astronautical Sci., p. 130.
Tsuneta, S. et al. 1983a, Ap. J., **270**, L83.
Tsuneta, S., Takakura, T., Nitta, N., Ohki, K., Makashima, K., Murakami, T., Oda, M., and Ogawara, Y. 1983, Solar Phys., **86**, 313.
van Beek, H. F., Hoyng, P., La Fleur, B., and Simnett, G. M. 1980, Solar Phys., **65**, 39.
Vorpahl, J. 1972, Solar Phys., **26**, 397.
Watanabe, T. 1984, in Proceedings of Japan-France Joint Seminar, Active Phenomena in the Outer Atmosphere of the Sun and Stars, Paris.
Widing, K. G. and Hiei, E. 1984, Ap. J., **281**, 426.
Woodgate, B. E., Tandberg-Hanssen, E. A. et al., 1980, Solar Phys., **65**, 73.
Woodgate, B. E., Shine, R. A., Poland, A. I., and Orwig, L. E. 1983, Ap. J., **265**, 530.
Wu, C. S., and Lee, L. C. 1979, Ap. J., **230**, 621.
Zhao, Ren-yang 1982, Adv. Space Res., **2** No. 11, 177.
Zirin, H. 1978, Solar Phys., **58**, 95.
Zirin, H. 1980, Ap. J., **235**, 618.
Zirin, H. 1983, Ap. J., **274**, 900.
Zirin, H. and Neidig, D. F. 1981, Ap. J. (Letters), **248**, L45.
Zirin, H. and Tanaka, K. 1973, Solar Phys., **32**, 173.
Zirker, J.B. 1977, Coronal Holes and High Speed Wind Streams, (Boulder, Colorado Associated University Press).

CHAPTER 4: CHROMOSPHERIC EXPLOSIONS

CHAPTER 4: CHROMOSPHERIC EXPLOSIONS

TABLE OF CONTENTS

G. A. Doschek, S. K. Antiochos, E. Antonucci, C.-C. Cheng, J. L. Culhane, G. H. Fisher, C. Jordan, J. W. Leibacher, P. MacNiece, R. W. P. McWhirter, R. L. Moore, D. M. Rabin, D. M. Rust and R. A. Shine.

Page

TABLE OF CONTENTS (continued)

CHAPTER 4: CHROMOSPHERIC EXPLOSIONS

by

G.A. Doschek[a], S.K. Antiochos[b], E. Antonucci, C.-C.Cheng, J.L. Culhane, G.H. Fisher, C. Jordan, J.W.Leibacher, P. MacNeice, R.W.P. McWhirter, R.L. Moore, D.M. Rabin, D.M. Rust, and R.A. Shine

[a]Chapter editor and Team leader
[b]Underlined team members were responsible for substantial written contributions for the chapter.

4.1 INTRODUCTION

The subject of investigation for Team D is rather loosely called "Chromospheric Explosions". Chromospheric explosions include the response and relationship of the flare chromosphere and transition region to the hot coronal loops that reach temperatures of about 10^7K and higher. It also includes flare related phenomena such as surges and sprays. In particular, the team decided to focus on some of those aspects of coronal and chromospheric phenomena that are newly-observed by the complement of instruments flown on SMM and other recent spacecrafts such as P78-1 and Hinotori.

During the first two Workshops, it became apparent that the interpretation of some of the newly obtained X-ray spectra is quite controversial. In addition, there are differing opinions concerning theoretical aspects of flares, such as the role of electron beams versus thermal conduction in heating the chromosphere. The team therefore decided to focus on three controversial issues as Workshop projects, and to try to improve our understanding of these issues by presenting the work in terms of debates between two sub-teams. This was a reasonable approach because the team was not very large. An issue to be debated was written down precisely, and the team divided itself into two camps, pro and con. Each team member picked the position that at the time seemed most reasonable to that member. As the work progressed, some team members decided that the position of the other sub-team was stronger, but nevertheless remained on their initially chosen team and took the position of a devil's advocate. Thus, in this chapter individual members of the sub-teams are not identified, since their current feelings about the debated issues may not reflect their positions during the debates.

Each sub-team had a sub-team coordinator whose responsibility was to collect material from sub-team members, organize it into a coherent argument, and then present the argument as part of a debate during the final Workshop. The sub-team coordinators were also responsible for preparing position papers supporting their arguments. These position papers constitute this chapter.

After the debate on each of the three issues, a general discussion followed in which the entire team attempted to judge the relative strengths of the pro and con arguments. While these discussions mostly resulted in an impasse as far as completely endorsing or rejecting either the pro or con arguments, the weaknesses of the arguments were clearly identified.

The overall purpose of our debates was of course to resolve the issues chosen for debate. However, these issues are controversial either because of a lack of adequate experimental data or because of a lack of adequate theoretical developments. Therefore a diminished, but much more realizable goal of our debates, is to identify the loose ends in our observations, so that pertinent follow-on experiments to SMM can be devised, and to bring into sharp focus the limitations of the relevant theory and to identify what needs to be done to improve it.

The topics chosen for debate are given below in debate form:

Issue 1--Resolved: The blue-shifted components of X-ray spectral lines are signatures of "chromospheric evaporation". (Evaporated plasma is defined as cool chromospheric plasma that is heated to multimillion degree temperatures and therefore contributes to the X-ray emission of the soft X-ray flare. In the process of heating it moves upward from the chromosphere and fills coronal flux tubes.)
Sub-team coordinators for Issue 1--For: E. Antonucci, Against: G. A. Doschek

Issue 2--Resolved: The excess line broadening of UV and X-ray lines is accounted for by a "convective velocity distribution" in evaporation.
Sub-team coordinators for Issue 2--For: S. K. Antiochos, Against: D. M. Rabin

Issue 3--Resolved: Most chromospheric heating is driven by electron beams.
Sub-team coordinators for Issue 3--For: G. H. Fisher, Against: D. M. Rust

4.1.1 Rationale for Issue 1

One of the key experiments on SMM is the X-ray crystal spectrometer experiment (the XRP). The crystals in the spectrometer represent a quantum leap in spectral resolution compared to the crystals flown a decade ago by NASA on the Orbiting Solar Observatories (OSOs). Due to their high resolution and relatively high sensitivity, profiles of X-ray lines can be obtained with fairly good time resolution throughout the course of a soft X-ray (SXR) flare. Crystal spectrometers with essentially the same resolving power have also been flown on the DoD P78-1 spacecraft and on the Japanese Hinotori spacecraft.

These crystal spectrometer experiments have resulted in important extensions of observations concerning flares. One observation is that during the rise phase of flares the profiles of X-ray lines exhibit a blue-shifted component, characteristic of velocities between 100 and 500 km s^{-1} (Doschek et al., 1980, Antonucci et al., 1982). The intensity of the blue-shifted component is usually much less than the intensity of the non-Doppler shifted, or stationary component, throughout the rise phase of flares. However, for a few events the intensity ratio may be about 0.5 or greater at the onset of the events.

The observation of anisotropic bulk motions of high temperature plasma ($T > 2 \times 10^6$ K) is not new, and was reported in the analysis of previous X-ray (Korneev et al., 1980) and XUV data (Widing 1975). However, the time relationship of this emission to the emission from the bulk of the SXR emitting plasma could not be determined from the small number of observations of large flares. In addition, a number of other interesting characteristics of the blue-shifted emission have been found by Antonucci and Dennis (1983) and Antonucci, Gabriel, and Dennis (1984) from analysis of SMM data, assuming a two-component model composed of a stationary component, and a single blue-shifted component. Although the actual distribution of velocities in the flare plasma might involve more than two components, or in fact be a continuous distribution of velocities, the two-component approximation has been illuminating regarding the intensity and magnitude of the blue-shifted emission.

The two most important questions regarding the blue-shifted component are what is its origin, and what is the relationship of the blue-shifted emission to the stationary SXR emission. Doschek et al. (1980) suggested that the blue-shifted emission might be the signature of the "primary chromospheric evaporation" discussed by Antiochos and Sturrock (1978), but cautioned that the blue-shifted emission might also be a phenomenon similar to a surge, not necessarily physically associated with the magnetic loops that contain the bulk of the SXR plasma. The spectrometers on the P78-1 spacecraft lacked any spatial resolution, and therefore Doschek et al. (1980) could not distinguish between these two possibilities. The SXR spatial resolution of the SMM spectrometers is only ≈ 17", which is also not sufficient to distinguish between the two explanations.

Nevertheless, based on the presence of the blue-shifted emission during the rise phase of a large percentage of the flares observed by SMM, Antonucci et al. (1982) interpret the blue-shifted emission as chromospheric evaporation. That is, they propose that the primary flare energy release occurs in coronal flux tubes. Part of this energy is subsequently deposited by either conduction or transport by high energy particles into the chromosphere. This results in heating the chromosphere to multimillion degree temperatures. The heated gas moves upward into the flaring flux tubes, and the Doppler effect causes the X-ray emission from this upwelling plasma to be blueshifted relative to non-moving or stationary plasma already present in the flux tubes. The ablated or evaporated plasma is the main cause of the increase in emission measure of the SXR flare. This point of view was challenged by some members of the team. They felt that a close comparison of the characteristics of the blue-shifted emission to what is expected or predicted by numerical simulations of chromospheric evaporation revealed disturbing discrepancies. Resolving the origin of the blue-shifted emission is the primary reason for debating Issue 1.

4.1.2 Rationale for Issue 2

The reason for debating Issue 2 also revolves around the idea of chromospheric evaporation, and in addition is based on another important finding from the X-ray spectra. This concerns the considerable degree of turbulence found in flare plasma during the rise phase of SXR flares (e.g., Doschek, Kreplin, and Feldman 1979, Antonucci et al., 1982). The temperature of the plasma that we are referring to here is $> 6 \times 10^6$ K and $< 3 \times 10^7$ K. This is the temperature range of the so-called thermal flare plasma. The fact that considerable turbulence exists in flares is in itself not new and was discovered in the analysis of the high resolution X-ray spectra analyzed by Grineva et al. (1973) and was also found in the analysis of XUV Skylab data (e.g., Brueckner 1976, Cheng 1977). However, the new instruments have allowed the magnitude of the turbulence as a function of time during a flare to be determined, for a large number of flares.

The origin of this turbulence is unknown, and perhaps the term "random mass motion" is more appropriate. For the sake of brevity the term turbulence is adopted, with the caveat that the origin of the broadening may not be turbulence in the strict plasmaphysics definition of the word. It does not appear that the excess line widths are due to ion temperatures that are much higher than the electron temperature. The electron temperature is well-determined and therefore it is possible to compute the implied difference in ion and electron temperature accurately. At the known electron density of SXR flare plasmas, it is difficult to explain how this difference in temperature could be maintained over the observed intervals of flare rise times. Also, the quantity $\Delta\lambda/\lambda$, where $\Delta\lambda$ is the line width and λ is the wavelength, scales as $1/\sqrt{M_I}$ where M_I is the ion mass, if the broadening is predominately due to Doppler ion broadening. However, the observed values of $\Delta\lambda/\lambda$ are independent of ion mass, thus supporting the random mass motion interpretation. Therefore most investigators have assumed that the broadening is due to mass motions, and not instead due to an ion temperature that is much higher than the electron temperature.

It is a difficult task to answer the general question, what causes the non-thermal broadening of X-ray lines, since there are many possibilities, such as true plasma turbulence or simply spatially non--uniform heating in flux tubes which would produce hydrodynamic turbulence. Another possibility that the team thought interesting to investigate is that the turbulence is due to the distribution of velocities produced in chromospheric evaporation. More specifically, in the evaporation process plasma at a given temperature evaporates with a range of speeds. This is one of the results of numerical simulations, and would produce Doppler broadening of spectral lines. This occurs even in one dimensional (1D) numerical simulations. The reason for investigating this particular possibility is that it is possible to use numerical simulations to calculate the expected broadening. A similar calculation is very difficult for some of the other likely mechanisms, such as plasma turbulence. There are simply too many undetermined parameters to make meaningful calculations of all plausible broadening mechanisms. The team felt that it might be possible to determine whether or not convective evaporation is a viable broadening mechanism and therefore decided to debate the issue. Issue 1 and Issue 2 are intimately related.

4.1.3 Rationale for Issue 3

For some time now the relative importance of electron beams and conduction fronts in heating the flare chromosphere has been debated by the general solar physics community. A number of numerical simulations of chromospheric evaporation have been carried out (e.g., Nagai 1980, Cheng et al., 1983, MacNeice et al., 1984, Pallavicini et al., 1983, Fisher, Canfield and McClymont 1985b,c). These simulations contain ad hoc heating functions either in the form of coronal heating transported by conduction fronts or electron beams that deposit their energy in the chromosphere. At first sight it is not clear from these papers whether conductive heating or beam heating provides the closest correspondence to observational data. However, detailed comparisons of temperatures and upflow velocities from these model calculations do show significant differences between the two types of flare heating (Fisher, Canfield, and McClymont 1984). Furthermore, detaild modelling of the chromosphere and transition region has produced observational tests which can be used to study the spatial distribution of heating in flares (Canfield, Gunkler, and Ricchiazzi 1984, Fisher, Canfield, and McClymont 1985b,c).

The presence of an electron beam is inferred from the hard X-ray (HXR) bursts that frequently occur during the impulsive phase of flares (during the rise phase of SXRs). The SMM, P78-1, and Hinotori spacecraft carry HXR broadband spectrometers in addition to the crystal spectrometer experiments. A question of considerable importance to flare theory is whether the energy contained in a possible electron beam is sufficient to power the SXR event. Is the energy released in the chromosphere by the beam sufficient to drive chromospheric evaporation and produce the initial emission measure growth of the SXR flare? Feldman, Cheng, and Doschek (1982) argue that the HXR emission represents an energy release mode that is coincidental in time to, but not the cause of, the production of the SXR flare. They argue that the HXRs and SXRs are produced by a common energy release mechanism, but are otherwise unrelated, i.e., the energy in the nonthermal electrons does not power the SXR flare. Other investigators, while disagreeing with the position of Feldman, Cheng, and Doschek (1982), nevertheless also feel that the energy contained in electron beams is insufficient to power the SXR flare. They feel that chromospheric evaporation produces the SXR event and is caused by heating originating in coronal loops which propagates via conduction downward into the chromosphere. Because of the fundamental importance of these questions to the understanding of energy transport in flares, the team decided to debate Issue 3.

Finally, before commencing with the debate issues in Sections 4.3, 4.4, and 4.5, a brief description of the X-ray crystal spectrometer experiments is given in Section 4.2, since the data from these experiments are central to both Issues 1 and 2.

4.2 CRYSTAL SPECTROMETER EXPERIMENTS

The first high resolution solar spectra from hot plasmas were obtained by Grineva et al. (1973), from a few flares during 1970. The iron lines between 1.85 - 1.87 Å were observed with two quartz spectrometers on the Intercosmos-4 satellite, launched on October 14, 1970. The spectra were obtained with a resolution of 0.4m Å by scanning the solar disk with the satellite axis. The most important lines in the 1.85 - 1.87 Å range were identified using these spectra (Grineva et al., 1973, Gabriel 1972). However, emission from Fe II - Fe XXIII was not observed with high resolution.

The SOLFLEX (Solar Flare X-rays) experiment consists of four Bragg spectrometers operating in the interval 1.8 - 8.5 Å, where important SXR spectral lines are found. It was flown by the Naval Research Laboratory (NRL) on an Air Force spacecraft called P78-1, which was launched on February 24, 1979. A wavelength scan is completed by this instrument in about 56 seconds. The effective time resolution is variable and can be as short as about 10 s. Another crystal spectrometer experiment flown on P78-1 is the SOLEX (Solar X-rays) spectrometer package, built by the Aerospace Corporation. These crystals scan the longer wavelengths between 5 and 23 Å. The P78-1 instruments have been described recently by Doschek (1983).

The Soft X-ray Polychromator experiment (XRP) consists of two instruments: the Bent Crystal Spectrometer (BCS) and the Flat Crystal Spectrometer (FCS) (Acton et al., 1980). It was built by groups at Lockheed, Culham Laboratory, and the Mullard Space Sciences Laboratory and was flown on the Solar Maximum Mission satellite, which was launched on February 14, 1980. The Bent Crystal Spectrometer consists of eight curved crystals and position sensitive detectors to achieve high time resolution (effective time resolution of 1 second) as well as high spectral resolution. The spectrometer detects the SXR emission in the interval 1.77 - 3.23 Å. The Flat Crystal Spectrometer has seven Bragg crystals which simultaneously diffract X-rays in seven different wavelength regions in the interval 1.4 - 22.4 Å.

Two rotating crystal spectrometers, called SOX1 and SOX2, were flown by Tokyo University on the Hinotori (Astro-A) satellite launched February 21, 1981. These spectrometers observe the two wavelength ranges 1.72 - 1.95 and 1.83 - 1.89 Å (Tanaka et al., 1982a). They are flat spectrometers and the wavelength range is scanned by using the spin of the spacecraft itself. The time resolution is comparable to the effective time resolution of the BCS.

The spectrometers on the P78-1, SMM, and Hinotori satellites have operated continuously for many months, giving a large set of observations of intensities and profiles of lines, emitted by plasma at temperatures greater than 10^7 K, formed during solar flares. The spectral resolution of all these spectrometers is sufficient for measuring spectral line profiles. The intensities of lines are important for determining the electron temperature, electron density, emission measure distribution, and departure from ionization equilibrium. The spectral regions most studied are the wavelength bands near the resonance lines of helium-like iron (Fe XXV, 1.85 Å) and calcium (Ca XIX, 3.176 Å).

4.3 DEBATE OF ISSUE 1

The debate of Issue 1 began by defining precisely what is meant by chromospheric evaporation. The term was defined to mean the heating of chromospheric plasma to soft X-ray emitting temperatures by either thermal conduction and/or electron beams, and its consequent expansion upward into flux tubes. The evaporating plasma is rapidly heated to temperatures > 10^7 K, thereby producing the large SXR emission measure. The sub-team presenting the argument for the interpretation of the blue-shifts as indicators for chromospheric evaporation asserts that the evaporated plasma is the major cause of the increase of SXR flux and emission measure. The sub-team presenting the case against this interpretation did not attempt to claim that no upward moving plasma exists, but their contention is instead that the blue-shifts are not the major cause of the large SXR emission measure increase.

4.3.1 Argument for: The Blue-shifts Are Direct Evidence of Chromospheric Evaporation

The sub-team arguing for a chromospheric evaporation interpretation of the blue-shifts list several properties of the blue-shifts that either support or are consistent with this interpretation. Previously, the profile of the Fe XXV resonance line was reportd to be broadened, indicating turbulent motions of the order of 90 km s^{-1}, from observations obtained from the Intercosmos-4 satellite (Grineva et al., 1973). Most of the spectra were recorded during the peak or the decay phase of flares except for two obtained during the rise time. For the rise phase spectra a shift of the Fe XXV resonance line toward the blue was also observed, which implied motion of the SXR source toward the observer with velocities of 160 km s^{-1}. If line shifts are observed in peak phase or decay phase spectra, they are toward the red and the downward velocities deduced are of the order of 200 km s^{-1} (Korneev et al., 1980). An absolute wavelength calibration is available for these data.

From the line width analysis of several flares of class M and X observed with the SOLFLEX spectrometers, Doschek et al. (1980) and Feldman et al. (1980), conclude that line profiles are broader than the thermal width, computed from the electron temperature determined from the dielectronic satellite lines. The widths are greatest at flare onset, and decrease monotonically during the rise phase. Near maximum X-ray flux the broadening has decreased considerably. This effect is more evident in class X flares. In the decay phase of M and X flares the lines are narrower, with nonthermal velocities of about 60 km s^{-1}. The line broadening at flare onset is characterized by a nonthermal velocity parameter of about 150 km s^{-1}. Doschek et al. (1980) also reported the existence of a weak blue-shifted emission component of the Ca XIX resonance line, that was only apparent during the rise phase of X-class flares.

Two of the M flares reported by Feldman et al. (1980) show a relatively strong component of plasma (compared to non-moving plasma) moving outward from the sun at velocities of about 400 km s^{-1}, with a temperature probably near 15 x 10^6 K. This result is based on the observation of a blue-shifted component of both the Ca XIX and Fe XXV resonance lines, which has the form of an asymmetrical extension on the blue-wings of the lines. The blue-component is strongest in the earliest rise phase spectra. The X-ray flux in these spectra is about 50% of the flux in the stationary component. The flux of the dynamic component tends to slightly increase or remain constant during the flare rise phase. After flare maximum it is either very weak or absent. The dynamic and static components of the resonance line of Ca XIX are deconvolved under the assumption that both components produce Gaussian profiles. Wavelength shifts of the resonance line itself, as observed by Korneev et al. (1980), are not discussed by these authors.

The high time resolution observations of the flare SXR emission, obtained with the Bent Crystal Spectrometer on SMM, confirm that the X-ray line profiles of ions such as Ca XIX are broad and complex. These properties of the lines are found to be a systematic characteristic of the impulsive phase of flares (Antonucci et al., 1982). A more detailed discussion of the properties derived from BCS data is given in Subsection 1. The results obtained from the Hinotori observations in general confirm the existence of non-thermal profiles and blue extensions of spectral lines. Non-thermal velocities of about 200 km s^{-1} and upward velocities of about 400 km s^{-1} are derived at flare onset (Tanaka et al., 1982a,b).

4.3.1.1 General Properties of the Plasma Upflows

The general properties discussed above have been studied systematically using a large set of spectra observed during the rise phase of class M and X flares obtained with the BCS on SMM during 1980 and the results appear to be consistent with the hypothesis of chromospheric evaporation (Antonucci 1982, Antonucci and Dennis 1983, Antonucci, Gabriel, and Dennis 1984).

Evaporation occurs in response to local chromospheric heating as a consequence of the primary energy release during a flare, and as predicted by different models, it can be produced either by electron beam heating or conductive heating of the chromosphere. In the case of beam heating, the energy release is associated with emission in hard X-rays. Under the hypothesis that the blue-shifted components are due to chromospheric material heated during the energy release and driven up along the magnetic field lines, and considering the HXR impulsive emission as a temporal indicator of this energy deposition, upflows are expected as long as the impulsive HXR emission is observed. While evaporation takes place, the magnetically confined plasma accumulates in coronal flux tubes. The velocity of the evaporating plasma has to decrease with the increase in density of the plasma accumulating in the coronal region, because of the decrease in pressure difference between the chromosphere and the overlying coronal flare loop. In general, correlations between the parameters of the static and the dynamic components of spectral lines are expected, because in the theory of chromospheric evaporation they are causally related. Chromospheric evaporation is also expected to be related to the sites of energy deposition, that is at the footpoints of the flaring coronal loops.

The observational properties of the X-ray lines of class M and X flares confirm the above picture. In about 80% of the disk flares detected with the BCS, a blue-shifted component was observed in association with the spectra emitted during the impulsive phase. This component is absent in the emission of flares occurring beyond about 60 degrees in longitude. This indicates the presence of flows of a part of the SXR emitting plasma that are mainly upward. The observed upward velocities can be as high as 300-400 km s^{-1} at flare onset and they decrease during the rise of the soft X-ray flux. In the decay phase, the dynamic blue-shifted component of the SXR source is no longer observed. (Shifts between spectral components corresponding to velocities below 120 km s^{-1} cannot be measured reliably).

In both disk and limb flares, large non-thermal line broadenings are observed during the rise phase, and in general are insignificant during the decay phase. Upflows and non-thermal motions observed during the rise of the SXR emission are also observed simultaneously with the most intense HXR emission above 25 keV. This indicates a causal relation among the different phenomena, in particular, the HXRs, the SXRs and the plasma upflows, as expected if chromospheric evaporation is present. SXR and HXR emissions, from the region of the May 21, 1980 flare, are shown in Figure 4.1 as an example. Defining the parameter ΔT as the difference etween the Doppler and electron temperature, and defining v' as the velocity of the dynamic component of the SXR source, it can be seen that these quantities deviate considerably from zero during the impulsive SXR emission and the rise of the SXR flare. Both quantities moreover are largest at flare onset and decrease during the impulsive phase. They are negligible at flare maximum. Line-broadening motions in general precede line-shifting motions in a flare, as shown in the example of Figure 4.1. The quantity ΔT is taken here and subsequently as an indicator of nonthermal broadening, as opposed to line-shifting.

While the upflow velocity is decreasing, the density in the coronal region of the flare is increasing, as can be deduced from the evolution of the emission measure EM of the stationary SXR source for the event of May 21, 1980, derived assuming isothermal conditions at the electron temperature T_e. The quantities T_e and log (EM) are also shown in Figure 4.1. The temperature T_e is obtained from the intensity of the Ca XVIII dielectronic satellite lines. Since the area of the flare as observed in SXRs, in the energy band 3.5 - 8 keV of the Hard X-ray Imaging Spectrometer (HXIS), doe not change significantly during most of the impulsive phase, the increase in EM indicates a simultaneous increase in electron density of the flare plasma.

The emission of the material associated with the high speed upflows is in general weaker than that of the main SXR source, which produces unshifted spectral lines. Only in a few cases for a short period, on the order of 20 seconds at flare onset, is the intensity of the blue-shifted emission observed to be equal to that of the primary (unshifted) emission. Both of these emissions increase in absolute flux, although the flux of the dynamic component decreases with respect to the primary one during the impulsive phase (Antonucci, Gabriel, and Dennis 1984).

The upward velocities observed during the impulsive phase for the dynamic plasma component are correlated with the average rate of increase

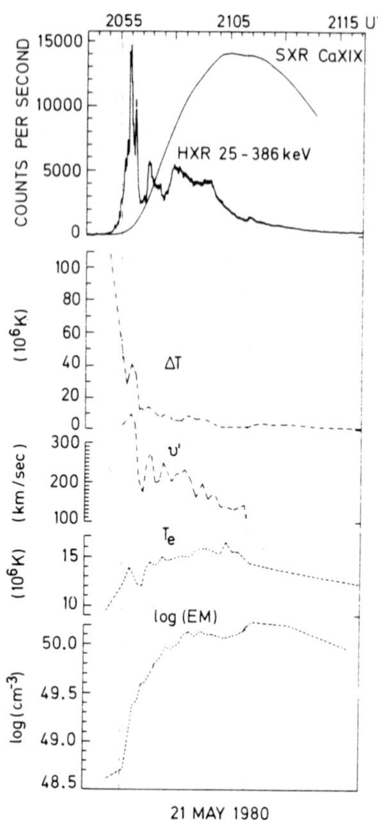

Figure 4.1 Temporal evolution of the following observed quantities for the flare on May 21, 1980: the SXR calcium line emission detected with the Bent Crystal Spectrometer integrated over the spectral range 3.165 to 3.231 Å; the HXR emission detected with the HXR Burst Spectrometer in the energy range from 25 to 386 keV; the parameter ΔT indicating the amount of nonthermal broadening in the line profiles, expressed in terms of equivalent temperature excess; the velocity v′ of the high speed upflows; the electron temperature T_e; and the logarithm of the emission measure EM of the principal SXR source. The last four physical parameters were derived from the analysis of BCS spectra in the Ca XIX channel, similar to the one shown in Figure 4.2. Upflows with $v' \geq 120$ km s^{-1} are observed during the time interval indicated by the dashed vertical lines.

of the emission measure of the coronal plasma, with a correlation coefficient of 0.8. In addition the peak emission measure of the upward moving plasma is also positively correlated with the peak emission of the principal coronal source emitting SXRs, with a correlation coefficient of

0.7 (Antonucci and Dennis 1983). The flare properties presented up to now are quite general.

In a few events it is also possible to observe, in addition to the blue-wing asymmetry or the blue-shifted component, a blue-shift of the resonance line of the primary component itself, as reported by Korneev et al. (1980). In the case of the May 21 event, the blue-shift of the dominant component is observed before the blue-shifted component appears (see discussion in Subsection 2).

4.3.1.2 SXR Line Spectra During the Impulsive Phase

The results presented in Subsection 1 are well illustrated by a sequence of calcium spectra detected during the May 21 flare, in Figure 4.2. The first significant SXR emission detected in the BCS spectral band from 3.165 to 3.231 A, where the Ca XIX resonance line and its associated satellites are observed, in Figure 4.2a, is an average over 120 seconds. The emission is integrated over the entire flare region, since the BCS field of view is 6 arc min x 6 arc min. The emission measure of the source at this time is approximately 5 x 10^{48} cm^{-3}. Although the counting

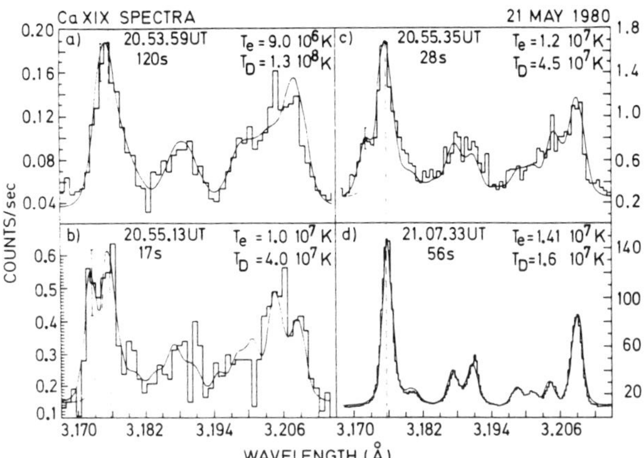

Figure 4.2 Sequence of SXR spectra obtained at four different times during the impulsive phase of the May 21, 1980 solar flare by the BCS in the Ca XIX spectral region. The last spectrum was recorded at the end of the impulsive phase. The smooth curve in each figure represents the synthesized spectrum computed for given values of the electron temperature T_e and the Doppler temperature T_D. The spectra shown in b and c are the sum of two synthetic spectra, one blue-shifted by 3.8 mÅ. The dashed lines represent the profile of the principal component expected for a given Doppler temperature. The time intervals represent the effective data accumulation periods. The times of the spectra are the mean times of the observation interval.

rate is low, approximate values can still be derived for the Doppler and electron temperatures as described later. The resonance line is significantly broadened with a Doppler temperature of about 1.3 x 10^8 K. This broadening must be nonthermal in origin since it would imply an ion temperature of one order of magnitude larger than the electron temperature, which is about 9 x 10^6 K. Hence nonthermal motions with a mean velocity of 220 km s^{-1} are expected in the flaring region.

The subsequent spectrum, in Figure 4.2b, shows the appearance of a blue-shifted component; the only observable unblended feature is on the blue side of the resonance line, i.e., the line at 3.177 Å. (This line is at the location of the vertical dashed lines in Figure 4.2.) The intensity of this feature exceeds by 7 standard deviations the intensity expected from the profile of the unshifted resonance line, i.e., the dashed profile in Figure 4.2b, computed for a Doppler temperature of 4 x 10^7 K. The two lines have comparable intensities and the relative blue-shift of the new feature is 3.8 mÅ, a value which implies high speed plasma flows with a line-of-sight velocity of about 360 km s^{-1} relative to the principal source. The line-of-sight is approximately radially outward since the flare occurred near disk center at S14, W15. In the spectrum of Figure 4.2c, the blue-shifted emission has decreased in intensity relative to the unshifted component with unchanged line-of-sight velocity. The line spectrum continues to consist of two components until the SXR emission reached its peak intensity at 21:07 UT. After this time it disappears, and the excess widths of the line profiles (over the thermal Doppler widths), assumed to be produced by nonthermal motions, disappear as well. Figure 4.2d shows an example of the one-component spectrum with thermal line profiles.

The principal component of the resonance line is itself blue-shifted during the early stages of the flare. The BCS has no absolute wavelength calibration and this effect is deduced by studying the relative wavelength shifts of the Ca XIX resonance line during a flare. The blue-shift can be determined by comparing the wavelength at the peak of the resonance line in Figure 4.2d, indicated with a vertical dashed line, with the corresponding peak wavelength in Figures 4.2a, b, c, and it is of the order of 0.8 mÅ. The corresponding speed of the plasma flow, about 80 km s^{-1}, decreases with time from this value and the effect is observed only in the early part of the impulsive phase. The high speed upflows, giving origin to the blue-shifted component, are observed later, in coincidence with a significant increase in the rise of the HXR emission. The initially observed 80 km s^{-1} upflow is interpreted as upward motion of the entire SXR source.

The theoretical spectra that best fit the data, shown as a continuous line superposed on the observed counting rates, are obtained from a synthesis of lines, computed for given electron and Doppler temperatures, following the technique described by Antonucci et al. (1982). Due to the large non-thermal broadenings and the low counting rates, the spectra detected early in a flare do not present well-resolved features. However, an estimate of the order of magnitude of the physical parameters is still possible. The two-component spectra, observed when the secondary spectrum is emitted by the high velocity plasma component, are fitted by the sum of two spectra, assuming the same source parameters for the two different sourcs. In general the impulsive phase spectra observed with the BCS are adequately fitted by a two component synthetic spectrum as in Figure 4.2c.

4.3.1.3 Spatial Location of the Upflows and the Site of Chromospheric Evaporation

In some flares, the simultaneous detection of the SXR spectral lines with the BCS and the SXR HXIS images in the 3.5 - 8 keV energy band enable the location of the SXR source to be determined at the time when the first upflows are observed. The HXIS, described in detail by van Beek et al. (1980), has 80 spatial resolution. In a few cases this instrument has resolved separate sources of 16 - 30 keV X-ray emission early in the impulsive phase. These sources have been identified with the footpoints of magnetic loops (Hoyng et al. (1981), Duijveman, Hoyng, and Machado 1982). The images observed during the May 21 event shown in Figure 4.3, indicate that at the time of the appearance of the high speed mass flows at 20.55.13 UT (second spectrum of Figure 4.2) the SXR emission is concentrated mainly at sites bright in SXRs, which presumably correspond to the footpoints of coronal loops. In Figure 4.3, the images of the SXR sources are compared with those of the HXR sources in the energy band 16 - 30 keV. These images were obtained during the development of the flare impulsive phase. The HXR bright points A and B are known to coincide with regions of opposite magnetic polarities and have been interpreted as the locations of incidence of beams of accelerated particles in the chromosphere early in the flare. Hence if this interpretation is correct, the SXR plasma and the high speed plasma upflows are spatially coincident with the sites of local chromospheric heating, when they are first observed. The HXR sources A and B presumably consist of a complex of unresolved sources, corresponding to a loop system connecting areas of different polarity. The coronal regions of the loops are still at low density, since the emission is weak between the footpoints.

At earlier times during the same event, in an interval of 120 seconds around 20.54.02 UT when the first spectrum is observed, the SXR emission is already predominantly concentrated at footpoint A, which remains the brightest until footpoints are observed. The presence at this time of a single spectral component, blue-shifted with respect to the rest wavelength, suggests that most of the emitting plasma is evaporating from A at low velocity in response to an initial heat input, simultaneous with HXR emission of moderate intensity.

The outward motion of the SXR source is observed for about three minutes. In approximately the same period, the centroid of the SXR emission moves from A to an intermediate position between the footpoints, in Figure 4.3. This change in configuration is consistent with the travel time to the coronal region of both the low and high speed upflows, which appear later on. The emission peak is interpreted as observed between A and B because of the density increase in the coronal region of the loop, since the evaporated material has reached the loop apex and accumulation has begun (Antonucci et al., 1984, Antonucci, Marocchi, and Simnett 1984).

The plasma injected at low velocity at flare onset is negligible in the total energy budget of the SXR flare, which is predominantly formed by the plasma moving at higher velocity. However, this phase is important for clarifying the initial conditions of the loops where chromospheric

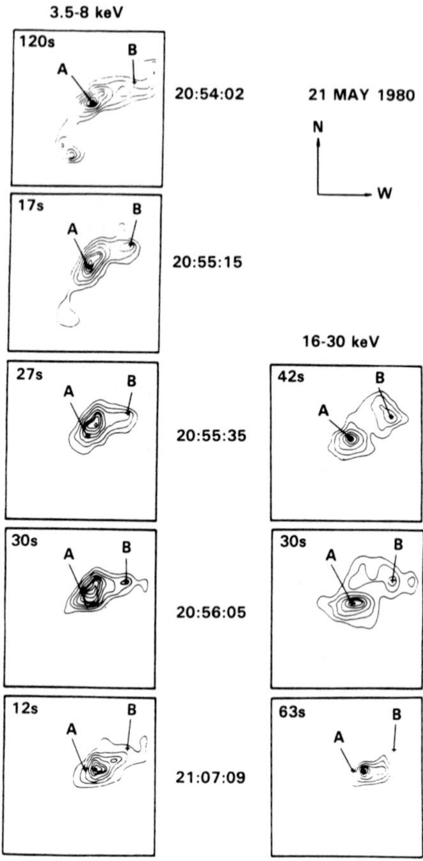

Figure 4.3 X-ray images of the flaring region, for the May 21, 1980 event, in the softer (3.5-8 keV) and harder (16-30 keV) energy bands of the HXIS. The contour levels correspond to 14, 26, 40, 55, 70, 80, 90, 98, 100% of the peak counting rate. For the hard X-ray images, the lowest contour level is 26%. The time intervals for data accumulation and the mean times of the images are shown. All images are deconvolved to allow for the triangular response of the HXIS collimator. The crosses represent the peak of the HXR emission at the two footpoints A and B, as measured in the first X-ray image in the energy band from 16 to 30 keV. This image is averaged over a period covering the two SXR images at 20.55.15 and 20.55.35 UT.

evaporation occurs. These loops are initially at low density except at the base. The precise initial density of the loops is unknown.

In addition, this phase corresponds to a low or moderately low value of the derivative of the observed HXR emission. Correspondingly, the energy flux, heating the chromosphere, increases slowly at first. This is probably the reason for an evaporation evidenced by a single component, indeed blue-shifted but by a small amount, which can be difficult to measure because of line-of-sight effects. In numerical simulations of flaring loops, the temporal profile of the energy input to the chromosphere should be chosen to simulate that inferred from the HXR emission, since it may be quite important in reproducing more accurately the observed flare profiles.

Under the interpretation given above, the dominant plasma source becomes stationary at the loop apex. However, it is still possible to infer from where evaporation occurs. This is done by comparing the ratio of the flux of the shifted to the unshifted spectral component, observed by the BCS, to the ratio of the SXR emission between 3.5 - 8 keV that arises from the two areas bright in HXRs, and identified with the foot-points, to the diffuse emission that arises between the footpoints, observed by HXIS. This test has been applied to the April 10, 1980 flare, a well-observed event discussed by Antonucci et al. (1982). The results suggest that during the impulsive phase the principal component in the BCS spectra is due to the emission integrated over the volume of the magnetic loop connecting the footpoints, while the blue-shifted component is localized in the proximity of the footpoints.

Referring now to earlier Skylab data, Hiei and Widing (1979) studied an event that occurred on January 21, 1974 that was optically classified as a-N subflare. This flare appears as a brightening of a single well-resolved loop. The position of the maximum electron density derived from the intensities of the Fe XV diagnostic lines, i.e., the maximum density at the temperature of formation of Fe XV, moves from one foot of the magnetic arch to the top of the loop with time. The concentration of the plasma near the footpoint at the beginning and its subsequent motion suggest that material is supplied from the lower atmosphere. The plasma in this case does not reach temperatures of 1×10^7 K. The transition region lines show blue-shifts corresponding to line-of-sight velocities of 120 km s^{-1}.

4.3.1.4 Energetics of Chromospheric Evaporation

For a definitive identification of upflows with evaporated chromospheric material, and to conclusively demonstrate that evaporation is the principal process in the formation of the SXR source in the coronal region, a detailed quantitative analysis is needed. With the data available, the amount of hot material supplied to the corona by the upflowing plasma cannot be established directly, since the amount of mass, enthalpy, and kinetic energy flowing into the coronal region cannot be derived from the parameters obtained from the SXR spectra alone.

However, the hypothesis of chromospheric evaporation can be indirectly explored by determining under what conditions the increase in emission in the principal spectral source, assumed to represent the

Table 4.1 Densities and Temperatures of the Coronal and Upflowing Plasmas

Flare	T_{max}	n_{e1}	n_{e2}	n_e'	T_e'	$T_{e,peak}$
		10^{11}	10^{11}	10^{11}	10^7	10^7
1980	UT	cm^{-3}	cm^{-3}	cm^{-3}	K	K
8 April	03.07	0.9-0.5	3.1-1.6	0.5-1.0	1.6-1.1	1.5
10 April	09.22	1.0-0.7	3.3-2.1	1.1-1.8	1.2-0.9	1.6
9 May	07.14	1.5-1.1	6.4-4.8	1.5-2.0	1.5-1.2	1.8
21 May	21.07	0.5-0.2	2.6-0.8	0.4-1.4	1.8-1.0	1.7
5 Nov	22.36	0.9-0.3	2.0-0.7	0.5-1.3	3.0-1.8	2.0

T_{max} — time of maximum emission.

n_{e1}, n_{e2} — electron densities of the coronal plasma at the onset and at the end, respectively, of the period of observed blue-shifts.

n_e', T_e' — electron density and temperature, respectively, of the upflowing plasma averaged over the period of observed blue-shifts.

$T_{e,peak}$ — peak value of the electron temperature of the coronal plasma.

The two values given for the densities and for T_e' correspond to the upper and lower limits on the volume (see explanation of Table 4.2).

Table 4.2 Energetics of Flares with Hard X-ray Footpoints

Flare Date	Soft X-Ray Coronal Plasma		Hard X-Rays
	ΔE_{SXR}	ΔE_{tot}	ΔE_{HXR}
1980	10^{30}	10^{30}	10^{30}
	ergs	ergs	ergs
8 April	2.9-3.9	5.3-4.5	9
10 April	1.9-2.4	3.6-3.1	5
9 May	1.6-1.7	4.5-3.3	5
21 May	8.2-16.	13.-14.	10
5 Nov	2.7-4.7	3.1-3.4	7

Explanation of column headings

ΔE_{SXR} — increase in the total energy of the coronal plasma including the radiative and conductive losses during the evaporation process.

ΔE_{tot} — total energy in radiative and conductive losses during the flare.

ΔE_{HXR} — total energy in fast electrons with energies above 25 keV derived from the HXR observations during the evaporation process assuming thick-target interactions.

The two values given for the parameters of the SXR coronal plasma are for two different geometries. The first value was calculated assuming a lower limit V_1 to the volume taken to be that of a semicircular loop with length $\pi\ell/2$ and cross-sectional area $A = d^2$. The second value was calculated using an upper limit V_2 to the volume taken to be ℓ^3.

coronal plasma, is consistent with the observed plasma upflows (Antonucci, Gabriel, and Dennis 1984). This is done under the assumption that the upflows are completely magnetically confined. It is also assumed that the plasma is flowing into the confinement region through the footpoint areas.

Only the average velocity v' and the average emission measure EM', of the upflowing plasma can be deduced with sufficient accuracy for the evaporating material. The temperature T_e' and the density n_e' of the upflow, however, can be derived indirectly. This is done by requiring that the material evaporating through the footpoints of the flaring loops supply sufficient mass and energy to account for the mass and thermal energy in the static coronal source at the end of the evaporation process. The physical conditions of the coronal source are well-defined except for those quantities that depend on the source volume. Upper limits to the volume can be inferred from the flare images obtained with the HXIS. Coronal loop footpoints, the most probable sites of chromospheric evaporation, can be identified from the impulsive phase HXR emission in the HXIS images, for a few flares reported in Tables 4.1 and 4.2. For these flares the upper limits to the footpoint cross-sections and distances between the footpoints are approximately known, and are selected for this quantitative test.

Following an approach described in Antonucci et al. (1982), the density of the upflowing plasma is derived by requiring that the integral of the electrons flowing in unit time through the two footpoints is equal to the increase of the electron number in the confining region during the evaporation time. Once n_e' is known, T_e' can be derived by requiring that the integrals over time of the enthalpy and kinetic energy transported by the upflowing plasma into the coronal region in unit time are equal to the simultaneous increase in the total thermal energy of the SXR emitting source. The total energy input corresponds to the sum of the thermal and turbulent energy increases during the evaporation process and of the energy losses, summed from flare onset. The energy losses of the coronal part of the loop are conduction to the chromosphere and in situ radiation from the coronal region.

The values of the electron density of the coronal plasma at the onset and at the end of the evaporation process, n_{e1} and n_{e2} respectively, and the peak values of the electron temperature T_e, are compared in Table 4.1 with the derived T_e' and n_e' values for the evaporating plasmas. The quantities T_e' and n_e' are average values over the evaporation period. For each quantity a range of values is given to account for the uncertainty of the inferred volume and to quantify the degree of dependence of the results on the flare volume. They correspond to a range of volumes within two limiting cases regarding the geometry of the source. In one case the source is chosen to be a semicircular loop connecting the two footpoints, and in the other case a cubic volume is assumed, where the linear dimension is taken to be the average linear extent of the SXR region.

Clearly, simple hydrodynamical considerations of the chromospheric evaporation model require that the pressure in the expanding evaporated plasma must exceed the pressure at the top of the loop. Since the temperature of the upflowing and stationary plasma are inferred to be approximately the same, this condition implies that n_e' must exceed n_e. The values tabulated in Table 4.1 for n_e' are believed to be consistent with this requirement, at least for volumes larger than a semi-circular loop. However, in this case also the results are presumably still valid. In fact, the values of n_e' are to be considered lower limits since the footpoint cross-sections are upper limits, due to the spatial resolution of the HXIS. In addition, it has been assumed that the magnetic fields completely confine the plasma flows in the corona, which is a restrictive condition. Higher densities would also result from overestimating the duration of the injection process. As shown in Table 4.1, for larger volumes, stil compatible with the observed dimensions of the flare region, even the lower limit of the evaporating plasma density n_e' becomes of the same order as the peak density in the coronal plasma n_{e2}, insuring that the density n_e' is larger than the coronal density while evaporation is observed.

The continuity equations used to derive the quantities in Table 4.1 are expressed in terms of two unknown parameters: $\alpha = v'/v$ and $\beta = T_e'/T_e$, where v and Te are respectively the volume and electron temperature for the plasma in the coronal region and v' and T_e' are the volume containing the blue-shifted component and its temperature (Antonucci et al., 1982). Since the emission measure of the coronal source is higher than that of the evaporating material, the pressure requirement on the chromospheric material in order that it is able to expand into the loop implies that the parameter α is less than unity. This corresponds to a situation where evaporated plasma is confined in the lower portions of the flaring loops.

Similarly, Tanaka, Ohki, and Zirin (1985) discuss a white light flare observed on June 6, 1982, which is the largest event detected by high resolution spectrometers during the recent solar maximum. Spectral information as well as SXR images in the energy band 5 - 10 keV have been obtained by the Hinotori satellite. The flare is preceded by a filament eruption. However, the authors conclude that the upflows observed at velocities of 400 km s^{-1} during the impulsive phase are not related to this phenomenon. Non-thermal motions are present throughout the impulsive phase. Bright points, assumed to be the footpoints of the magnetic arcade where the flare occurs, are observed in Hα and D3 emissions and it is assumed that semicircular loops connect the footpoints. By requiring mass balance during the impulsive phase, the density of the evaporating material should be twice as large as that of the coronal plasma to account for its formation. That is, the evaporation hypothesis is consistent with this result, if the rising plasma is confined to a small portion of the loop arcade, presumably near the footpoints, as expected. The energy balance presents some difficulty because of the large conduction losses. These losses integrated over the flare duration exceed 10^{32} ergs (one order of magnitude larger than the maximum thermal energy), since the flare plasma is estimated to be at the relatively high temperature of 4×10^7 K. This temperature is inferred from the Fe XXVI diagnostic lines. However, conduction losses may be reduced, e.g., by magnetic constriction, to a value lower than the thermal energy.

Hence, the observations of upflowing plasmas during the rise phase of SXR emission support the hypothesis that evaporation of heated chromospheric plasma is the main source of the SXR emitting plasma confined to coronal loops. This interpretation is quantitatively confirmed if the plasma upflows have densities equal to or greater than 10^{11} cm^{-3} and temperatures of the order of 1 - 2 x 10^7 K.

4.3.1.5 Direct Measurement of Material Evaporated During the Impulsive Phase

Evidence for the appearance of soft X-ray emitting plasma associated with a simultaneous disappearance of chromospheric material is reported by Acton et al. (1982). They have determined the amount of material that has been evaporated from the chromosphere during a compact flare, classified as C7, observed on May 7, 1980. The material evaporated appears to be more than enough to account for the emission measure of all the mass content at temperatures greater than 10^6 K, measured with the SMM instruments. These results are derived from a study of the profiles of the Hα line, obtained at the Sacramento Peak Observatory vacuum tower telescope. The amount of evaporated chromospheric material is estimated through a comparison of the observed profiles with the profiles predicted by a semi-empirical model of a flare chromosphere.

The same is found for another flare observed during SMM on June 24, 1980, as reported by Gunkler et al. (1984). In this flare, the maximum line-of-sight velocity is 300 km s^{-1} for the evaporated plasma emitting blue-shifted SXR, observed by the BCS. The amounts of chromospheric evaporation by conduction and non-thermal electrons are calculated and both mechanisms are found to be able to account for the observed increase in the coronal density. In this work the Hα profiles are analyzed by comparison with the theoretical profiles calculated by Canfield, Gunkler and Ricchiazzi (1984).

4.3.1.6 Early Phase of Chromospheric Evaporation

The blue-shift of the principal emission component observed in some events at flare onset has been interpreted in Section 4.3.1.3 as due to chromospheric material that was evaporated at low velocity in the confinement region. This interpretation is not corroborated by the same large set of data available for the high speed upflows. However, if this is the case, at this stage only moving, evaporated material is observed, since the coronal region is still at low density, permitting a direct estimate of the conditions of the moving SXR source.

Considering the May 21 event at the time of the first calcium emission (first spectrum in Figure 4.2), an upflow density of about 3 x 10^{10} cm^{-3} is derived by inferring the volume from the HXIS SXR images in Figure 4.3. Considering that the temperature of the moving evaporated material is 9 x 10^6 K, that its turbulent velocity is 220 km s^{-1}, and that its bulk velocity is 80 km s^{-1}, the rate of enthalpy, turbulent, and kinetic energy input into the loop through the footpoints is about 2 x 10^{27} ergs s^{-1}. The enthalpy and turbulent energy inputs are the most significant. The energy input rate to the coronal region due to the low-speed upflow is comparable with the energy input to the chromosphere associated with the electrons accelerated to energies of about 25 keV entering the thick target region.

The density and temperature derived for the slowly moving material do not differ much from the values indirectly derived for the high speed upflows, observed later on. The mass and energy input of this component to the corona at this stage is however negligible, as pointed out earlier.

4.3.1.7 Energy Input into the Chromosphere During Evaporation

The energy input into the chromosphere during the evaporation process can be derived easily under the hypothesis of thick target interactions by non-thermal electrons. This quantity can be computed from the HXR spectra measured with the Hard X-ray Burst Spectrometer (HXRBS), assuming a power-law electron spectrum with a cutoff at 25 keV. It is found that for the flares of Table 4.1, the rate of energy input by energetic electrons is sufficient to account for the increase in the thermal energy of the SXR emitting coronal plasma at all times during the impulsive phase. The total increase of coronal energy during the evaporation process ΔE_{SXR} is compared with the total energy input by non-thermal electrons ΔE_{HXR} in Table 4.2. The energy input cannot obviously be directly compared with the energy transferred by the upflows to the corona since this quantity has been derived under the requirement that it match the increase in coronal energy during the impulsive phase (Antonucci, Gabriel, and Dennis 1984). For the June 6, 1982 event, discussed by Tanaka, Ohki, and Zirin (1985), it is also found that the HXR emission implied a chromospheric energy input sufficient to account for the large conduction losses and the increase in thermal energy of the flare.

Hence, at least for the case of non-thermal electrons injected into the chromosphere, the HXR flux can be an indicator of an energy deposition indeed sufficient to account for the increase of coronal plasma energy content, while the upflows, under the conditions mentioned before, can provide the appropriate energy transfer mechanism.

The total energy ΔE_{tot} radiated and conducted away during these flares is also given in Table 4.2. From these values it appears that no energy input is in general required after the upflows are no longer observed.

4.3.1.8 Conclusions

The SMM data have allowed us to derive the general properties of the upflows of high temperature plasma observed during the rise of solar flares and to localize, in a few cases, their sources in the flare region. From these results we can conclude that the plasma rising into the corona is indeed the primary source of the thermal plasma observed in the corona during flares. This conclusion is not only supported by qualitative considerations, but also by quantitative tests on mass and energy balance during the impulsive phase. Within the present observational limitations, the conditions required for plasma upflows to account for the flare coronal plasma are consistent with the observations.

The most natural mechanism we can invoke to explain the origin of such plasma upflows is indeed chromospheric evaporation. This process is in fact predicted in any simulation of impulsively heated coronal loops. We note however that even in the alternative model proposed by the con team, for instance, upflows are still needed to supply most of the plasma contained in some of the magnetic configurations which brighten during a flare, although their origin in this case is of a different nature.

Since the quantitative tests presented here as supporting evidence for the main role of upflows in the formation of the flare coronal plasma are very important in our discussion, and since they are influenced more than other results by the present instrumental limitations (which do not allow, for instance, a direct measurement of the electron density of the blue-shifted component, or knowledge of the detailed topology of the magnetic configuration and its time evolution), it is worthwhile to stress a few points. For the events we have analyzed, a pressure difference sufficient to drive upflows into the coronal region where they are confined is easily obtained for values of the flare region volume compatible with the observations. Solving the continuity equations, the electron temperature of the upflows is found to be of the same order as that obtained for the coronal source. Its value, in fact, cannot be much lower otherwise the blue-shifted satellite lines due to dielectronic recombination would significantly increase and therefore they would be resolved and observed. Moreover, at least for a set of flare volumes considered in our discussion, the lower limit of the electron density of the upflows is of the same order as the density of the coronal loop at the end of the accumulation process, thereby exceeding the coronal density during evaporation. Therefore, the pressure of the upflows can be considered larger than the pressure in the loop at any time.

The emission measure of the stationary component varies according to the plasma input rate in the flaring loops, which is regulated by the electron density of the blue-shifted component and its velocity. The electron density is indirectly derived from the emission measure of the blue-shifted component which is also related to the volume where the plasma is propagating. Except at flare onset, the propagation region in a closed configuration represents only a part of the flux tube, since the rest of the loop is filled by stationary plasma whose emission is increasing as a flare progresses. For the flare volumes that satisfy the pressure requirement for plasma flows into the coronal loop, the effective volume occupied by upflows is found to be about 20% of the total volume. This value is averaged over the impulsive phase. The region where upflows are confined is presumably near the base of the flux tube. At flare onset, we can infer that the volume where upflows propagate is close to the total volume of the confining region.

There is evidence to suggest that the process of mass transfer to the corona often begins at low velocities. This is the phase when the blue-shifted component is observed to be dominant, while the stationary source is either weaker or absent. However, this phase is difficult to study since the blue-shift corresponds to low velocities and line-of-sight effects become important for its detection. In the few cases where this effect is well observed, the hard X-ray emission begins with low intensity, prior to the impulsive bursts. We can therefore deduce that coronal loops are first filled by low velocity plasma flows induced by a moderate chromospheric heating; the high velocity upflows only appear after the energy release becomes impulsive. In our opinion, a more accurate profile of the heating function accounting for a slow initial rise should be applied to coronal loops in the numerical simulations. This may be an important factor in determining the actual physical conditions of a coronal loop at the time of the major impulsive energy release.

4.3.2 Argument Against: The Blue-shifts Are Direct Evidence of Chromospheric Evaporation

The arguments against the blue-shift and chromospheric evaporation connection are based mainly on three observational properties of the X-ray and UV spectral lines. First, there are no instances in which a large blue-shifted component is the only component present during the rise phase. Second, some spectral lines should not show both blue-shifted and stationary components, but instead should show only a blue-shifted component. This is not found for one such well-observed line, the Fe XXI line at 1354.1 Å. Third, the electron density of the SXR flare is apparently high at flare onset, and therefore not as much chromospheric evaporation is necessary as has been previously assumed. However, this argument is not as strong as the first two mentioned. Below these arguments are discussed more fully.

4.3.2.1 Relative Intensities of Blue-shifted and Stationary Components of X-ray Lines

If the high density of SXR flares is primarily due to evaporating plasma from the chromosphere, then at some early time during the rise phase the large blue-shifted component should be the dominant spectral feature. The stationary component should be much weaker than the blue-shifted component or nonexistent. This is not observed in available flare spectra. The stationary component is always at least as strong as the blue-shifted component, implying an already existing high density loop, and contradicting predictions based on numerical simulation models.

The relationship between blue-shifted and stationary components has been made quantitative by Doschek et al. (1983) and Cheng, Karpen, and Doschek (1984), who discuss the shapes and Doppler shifts of spectral lines in terms of the gasdynamics of evaporation. The Doschek et al. (1983) paper discusses a flux tube that is heated in a localized region at the top of the loop. The conservation equations of mass, momentum, and energy are solved numerically for a two fluid plasma consisting of electrons and protons. The heating is symmetric, that is both sides of the loop receive the same amount of flare energy. A conduction front rushes down both sides of the loop and eventually drives chromospheric evaporation. (We note that actually only a half-loop was simulated in this calculation; see Cheng et al. (1983)). There are some possible differences between a half loop and a full loop simulation but these are of no consequence to our arguments.) The effects on spectral lines of evaporation for the special case of a symmetrically heated loop have been predicted by Doschek et al. (1983) and are summarized below according to temperature of line formation. The maximum temperature reached in the loop is about 20 x 10^6 K, at the top of the loop. Although the Doschek et al. (1983) results represent only a restricted range of possible input parameters, we believe that similar results should be obtained for a rather wide range of input parameters, assuming symmetric heating. Also, we note that chromospheric heating by electron beam simulations generally give evaporation velocities much larger than the observed blue-shifted velocities (Fisher, Canfield, and McClymont 1984). Thus, we restrict the simulation discussions to conduction heating models, because at least they give evaporation velocities that are comparable to the observed blue-shifted velocities.

Chromospheric and transition region lines ($T < 10^6$ K). These lines should show small red-shifts on the order of 20 km s^{-1}. The chromosphere and transition region are driven inward by evaporation. Because of their relatively high densities the inward velocities are small. The lower density evaporating plasma has been heated by conduction to temperatures of several million degrees and eaporates with much higher velocities.

Coronal lines (10^6 K $< T <$ 3 x 10^6 K). These lines are predicted to show essentially no Doppler shift (when viewed at Sun center). Gas at temperatures of one or two million degrees is at the dividing region between upward evaporating plasma and downward moving chromospheric and transition region plasma.

Low temperature flare lines (3 x 10^6 K $< T <$ 10 x 10^6 K). Spectral lines formed between about 3 and 10 x 10^6 K should show only a blue-shifted component, as long as evaporation persists. These lines are formed in the region that is evaporating with a large range of velocities. For a line formed near 10 x 10^6 K, the upward velocities range between 200 and 400 km s^{-1}. For a cooler line formed near 5 x 10^6 K, the upward velocities would be less, but still no stationary component should be present.

High temperature flare lines ($T >$ 10 x 10^6 K). These lines should show a blue-shifted as well as a stationary component. The stationary component arises from gas at the top of the loop where the temperature is highest. The abundances of high temperature ions are largest at the highest temperatures, and the contribution functions of the high temperature spectral lines are further increased at high temperature by the Boltzmann factor in their excitation rate coefficients. However, the density at the top of the loop is initially quite low compared to the density of the evaporating plasma, and the intensity of these lines is proportional to N_e^2. Thus the relative intensities of the blue-shifted and stationary components of high temperature lines at any given time during evaporation depend on the temperature, density, and velocity distribution of the gas within the loop. The results of the simulation show that during the initial stages of evaporation, the blue-shifted component should be dominant. As evaporation proceeds, the ratio of the blue-shifted to

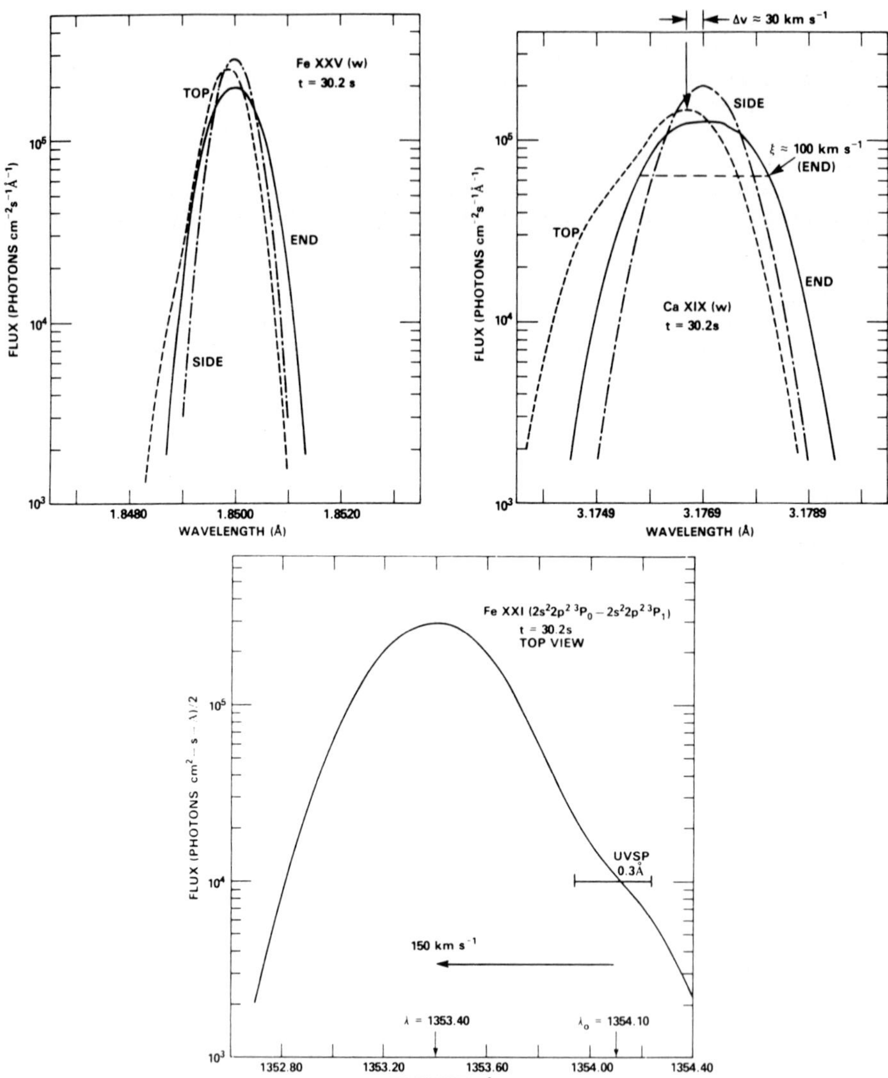

Figure 4.4 Computed line profiles of the resonance lines of Ca XIX, Fe XXV, and the forbidden line of Fe XXI at about 30 s after onset of flare heating in a loop heated at the top. Various line-of-sight viewing angles are shown. The half-length of the loop was 6700 km and the initial pre-flare loop density was 6×10^9 cm^{-3} at the loop top. See Cheng *et al.* (1983) and Doschek *et al.* (1983) for details.

stationary component decreases, and eventually the stationary component dominates.

The arguments concerning the high temperature lines given above cannot be rigorous because of the difficulty of including time-dependent ionization calculations in the simulations. That is, the Doschek et al. (1983) paper assumes ionization equilibrium in the calculation of line intensities and profiles. This assumption is not valid for times less than about 20 s in the Doschek et al. simulation. However, the effect of time-dependent ionization is strongest on the stationary component, because it has the lowest density. Qualitatively, the effect of including time-dependent ionization would therefore appear to be to reduce the ratio of the stationary to blue-shifted component, at least in some cases. Examples of computed profiles in this symmetric calculation are shown in Figure 4.4.

The description of line profiles and shifts given above might be modified if the loop size in the Doschek et al. (1983) simulations were greatly lengthened. However, the results are substantially modified if the heating is asymmetric, that is, if most of the flare energy is put into one side of the loop. In the Cheng, Karpen, and Doschek (1984) simulation of this situation, evaporation proceeds up both legs of a loop, but evaporation from the heated side of the loop is strongest. This results eventually in a net flow of plasma up the leg of the heated side of the loop and down the other leg.

The net flow of plasma within the loop has a dramatic effect on the line profiles. For cool flare lines formed between ≈ 3 and 10 x 10^6 K, the profile is split into both a blue-shifted and red-shifted profile. For the hottest flare lines there is blue-shifted and red-shifted emission, as well as emission at the rest wavelength. Examples of profiles from asymmetrically heated loops are shown in Figure 4.5.

A comparison of the numerical simulation results discussed above with observation reveals several disturbing discrepancies. For one, the observed blue-shifted component of the Ca XIX resonance line is much less intense relative to the stationary component, in contrast to what is predicted by the simulations for early rise times. The onset times of the flares in the P78-1 data base can be estimated to within seconds from noting when the proportional counter broadband X-ray data (also from P78-1) first show an increase in count rate. Another method of estimating onset time is to fit the early rise with an exponential (see Feldman, Doschek, and McKenzie 1984). The flux predicted from the exponential can then be extrapolated back in time and an approximate onset time can be determined. It is possible by measuring the angular steps at which the peak line emission occurs during the course of a flare to determine if the wavelength of peak emission changes with time. For the flares observed with P78-1, the wavelength of peak line emission appears to be nearly constant. It is clear that a strong, and in fact, usually totally dominant, stationary component is present within 10 s of the onset times of the flares.

Another disturbing aspect of the observed blue-shifted X-ray line components arises because they are seen not only in flares with relatively short rise times, i.e., < 2 minutes, but also in flares with very long rise times. An outstanding example is the June 5, 1979 flare

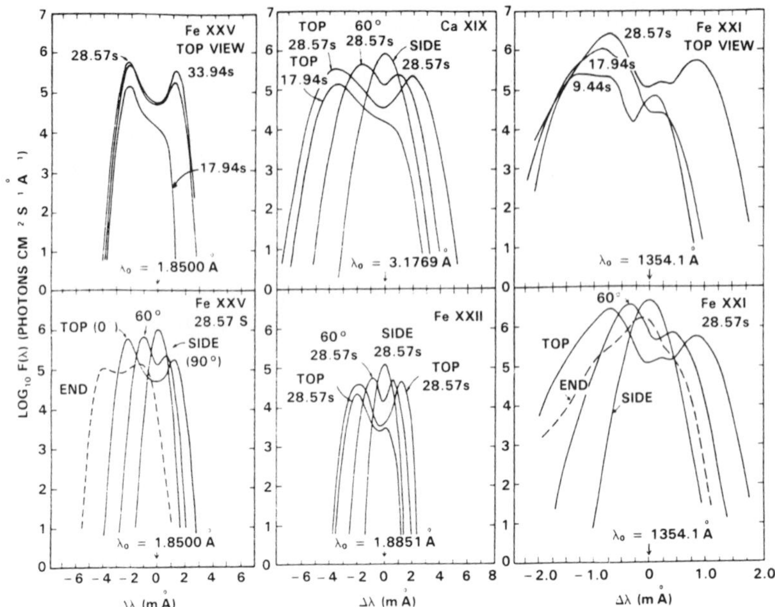

Figure 4.5 Computed line profiles of the lines shown for the asymmetric heating calculation described in the text. The profiles are shown at about 28 s after onset of flare heating. See Cheng, Karpen, and Doschek (1984) for details.

(see Doschek et al., 1980). In this flare, which had a rise time of about 30 minutes, the blue-shifted component is present during the rise phase indicating about the same velocity ($\approx$ 300 km s^{-1}) as found for more impulsive flares. In general for the X flares, the upward velocities indicated by the blue-shifts are much larger than necessary to account for the observed EM increase. Velocities less than 10 km s^{-1} are needed in some cases, so even if chromospheric evaporation occurs, there is a mismatch between the observed blue-shift velocities and the required chromospheric evaporation velocities. Note also that the intensity of the blue-shifted component for most flares is at all times much less than the intensity of the stationary component. (The ratio is about 0.2). The cases where the intensity ratio is larger, such as with the two M flares reported by Feldman et al. (1980), are exceptional.

To make the above remarks quantitative, consider a loop model that is qualitatively the same as adopted by the pro team in their attempts to support the evaporation theory. The model is modified by assuming for simplicity that the density of the evaporating plasma n_e' is essentially the same as the density in the loop. This is a rough prediction of the smulations discussed above and is also consistent with the observations and analysis used by the pro team. The rate of change of total number of electrons in the loop is,

$$\frac{d(n_e V)}{dt} = 2n_e' \, Av', \qquad (1)$$

where v' is the upflow velocity, V is the assumed constant loop volume (= LA, L = loop length), A is the cross-sectional loop area, and the factor of 2 accounts for both loop footpoints. The emission measure EM = n_e^2V, or

$$\frac{d(EM)}{dt} = 2n_e V \frac{dn_e}{dt} \qquad (2)$$

Assuming as mentioned for simplicity that $n_e' = n_e$,

$$\frac{d(EM)}{dt} = 4n_e^2 AL\ v'/L \quad , \qquad (3)$$

or,

$$\frac{d(EM)}{dt} = 4(EM)\ v'/L \quad . \qquad (4)$$

Finally,

$$(EM) = (EM)_o \exp(4\ v't/L) \quad . \qquad (5)$$

In order to explain the June 5, 1979 EM increase as due to evaporation at the speed of about 300 km s^{-1} indicated by the observed blue-shifts, equation (5), the rise time of about 30 minutes, and the ratio $(EM)/(EM)_o$ = 200 observed over the 30 minute interval, show that L ≈ 4.1 x 10^5 km or 9.4 arcmin. Such a length is improbably large. The problem can be circumvented by assuming the existence of several loops, such that V is not constant with time. However, in this case it is difficult to significantly increase the density in each loop, since multiple loops are needed to continuously increase V. Although in fact flares are usually composed of several loops, this discussion shows that the single loop chromospheric evaporation model does not account for all the observations.

We note that equations (1-5) also implicitly assume that n_e is constant in the loop, which is not a good assumption. However, in our simulations, n_e usually increases monotonically with decreasing temperature, and $n_e' > n_e$. The effect of a varying electron density can be approximated by assuming that $n_e' = \beta n_e$, where $\beta \gtrsim 1$. In this case the term 4v't/L in equation (5) becomes $4\ \beta^2 v't/L$. Since $\beta > 1$, the length of the 5 June flare loop inferred above would have to be even larger for a fixed value of EM/EM_o and a heating time of 30 minutes.

Finally, if flare loops are really heated asymmetrically, then the X-ray observations are at even greater variance with the predictions of the numerical simulations. There are only a few cases where red-shifted components of X-ray lines are seen, and these are confined to limb events (e.g., Kreplin, et al., 1985). On the other hand, there is ample evidence that real flare loops at least appear asymmetric in the brightness distribution of X-ray and EUV emission (e.g., Pallavicini, Serio, and Vaiana 1977).

4.3.2.2 Lack of Blue-shifts of Cool Flare Lines

Observations of cool flare lines ($6 \times 10^4 < T < 10^6$ K) are available from the NRL Skylab spectrograms and spectroheliograms, and from the UVSP SMM experiment. The Skylab spectoheliograph observed in the range from about 170 to 650 Å. It formed monochromatic images of the flare plasma. The predicted upflow velocities for chromospheric evaporation are between 200 and 400 km s^{-1}. As noted, the observed blue-shifted X-ray line component also implies velocities of about 300 km s^{-1}. At a wavelength of 200 Å these velocities correspond to Doppler shifts of between 0.2 and 0.4 Å. These shifts are easily measurable on the Skylab spectroheliograms for which the resolution is high enough to measure shifts of a few hundredths of an Angstrom. However, there are three difficulties with the spectroheliogram data. First, very few large flares were observed from Skylab since the experiment was performed near solar minimum. Second, the time resolution of the instrument was poor and only one or two events were actually observed during flare rise time.

Finally, since the slitless instrument forms images, a Doppler shift can be interpreted as a spatial shift, and vice-versa. Flares are frequently composed of more than one loop, and so it becomes very difficult to unambiguously separate Doppler and spatial shifts. The situation is well illustrated by the analyses of the 15 June 1973 flare presented by Cheng (1977) and Widing and Dere (1977). The images of the flare shown in those papers are quite complicated (even though the 15 June event had a typical SXR light curve), and at least two loop systems were involved. In spite of the caveats mentioned above, it was possible for Widing (1975) and Brueckner (1976) to identify blue-shifted emission features in the 15 June flare. In particular, the so-called "spike" (blue-shifted emission feature) is clearly a Doppler shifted feature (see Figure 4.6). However, this blue-shifted feature emanates from the top of the loop systems, rather than from the footpoints, and it shows up over a broad range of temperatures. The emission is observed in lines of He II, Fe XV, Fe XVI, and possibly Fe XXIV. The emission is observed in one rise phase spectrum (the only such spectrum available) and in two or three spectra obtained near maximum X-ray flux. This shows that at least some blue-shifted emission can arise from regions other than footpoints, and provides a counter example suggesting that the origin of the blue-shifts for at least some flares is located near the tops or sides of loops, rather than at their footpoints. In addition to the blue-shifted "spike" emission, the centroid of the Fe XXIV emission shifts about 5" from NW to SE in about four minutes around the time of peak X-ray flux. Widing and Dere (1977) suggest that another loop system may be activated. These observations simply point out the extreme geometric complexity of real flares. Very high spatial resolution is needed to distinguish small spatial shifts such as we are discussing. The spatial resolution of the Skylab spectroheliograph was 2".

The complexity of flares as seen with high spatial resolution implies that it is very difficult, if not impossible, to unambiguously identify individual footpoints of loops with the instruments on SMM and Hinotori. The spatial resolution of these instruments is at least a factor of 5 less than the Skylab instrument. Thus, the pro team arguments

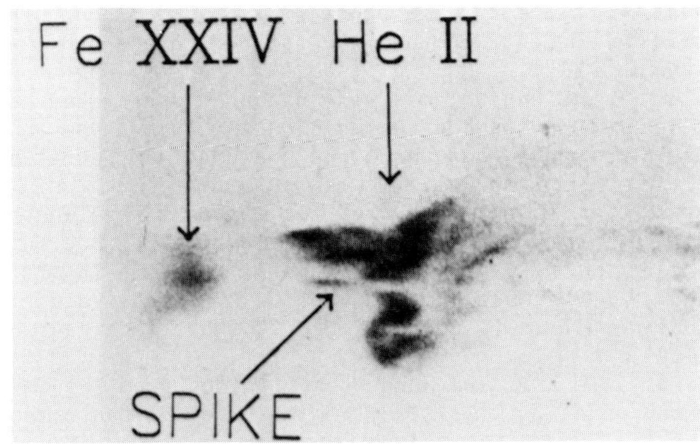

Figure 4.6 An NRL SO82-A Skylab spectroheliogram of the June 15, 1973 flare showing images of the flare in Fe XXIV (255.1 Å) and He II (256.3 Å). The feature marked "spike" is blue-shifted He II emission. The blue-shift corresponds to a speed of about 400 km s^{-1}. This feature is also apparent in lines of Fe XVI and Fe XV. Note that a superposition of the He II and Fe XXIV emission shows that the "spike" arises from the top of the loop system, rather than from the footpoints.

revolving around the locations of SXR and HXR emission, and analyses such as given by Antonucci et al. (1982) that depend significantly on knowledge of the SXR emitting volume, are inconclusive. They demonstrate only consistency with the evaporation model, and this consistency is made possible because the SXR volume is poorly known and the spatial resolution is coarse enough to be consistent with a proposed relationship even if the relationship does not in fact exist.

Finally, Acton et al. (1982) and the pro team argue that chromospheric evaporation was reported in the analysis of Skylab spectroheliograph data by Hiei and Widing (1979). This is true, but a close inspection of their paper reveals that the supposed evaporation does not occur in the loop or loops responsible for most of the Fe XXIV emission. It is these loops that would give rise to the high temperature X-ray lines presently under discussion. The loop discussed by Hiei and Widng (1979) was a cool loop with a temperature no more than 6 x 10^6 K. Furthermore, a further analysis of the same flare by Widing and Hiei (1984) shows that at least one location of HXR and XUV emission from this flare is physically completely separated from the loop or loops in which the SXR emission arises.

Another cool flare line observed from Skylab is the Fe XXI forbidden line, due to the magnetic dipole transition, $2p^2\ {}^3P_1 \rightarrow 2p^2\ {}^3P_0$, at 1354.1 Å. This line was identified independently by Jordan (1975) and Doschek et al. (1975). The line was observed with the NRL slit spectrograph, which had a resolution of 0.06 Å. Because the line is formed at about 8 x 10^6 K, and the flare plasma is turbulent, the line is quite broad and near X-ray maximum it can have a FWHM of about 0.5 Å.

Fe XXI data were obtained for a number of flares observed during the Skylab mission (e.g., Cheng, Feldman, and Doschek 1979). Recall from the previous discussion that for a line formed at 8×10^6 K, no stationary component should be present. Furthermore, at this temperature a spectral line such as the Fe XXI line should be shifted for as long as evaporation occurs. Even though not many flares were recorded during the rise phase, there were at least some spectra recorded during this phase and in any event the emission measure (and therefore presumably evaporation) frequently continues to increase somewhat after the peak X-ray flux is reached. Therefore, it is expected that the centroid wavelength of the Fe XXI line should vary considerably in different spectra. But this was not found to be the case. The wavelength of the Fe XXI line in all the spectra does not vary by more than about 0.02 Å, or equivalently, about 4.5 km s^{-1}. This result is in contradiction to what is expected from the evaporation model.

More recently, Fe XXI data have been obtained from the SMM UVSP instrument (Mason et al., 1985). They report blue-shifted components of the Fe XXI line in some of these data. However, a strong stationary component appears in all cases and the blue-shifted component velocity is much smaller than expected from the numerical simulations. Also, as Jordan (1985) has noted, there are several chromospheric lines in the region around the Fe XXI line, and care must therefore be exercised in interpreting possible blue or red wings to the Fe XXI line. Mason et al. (1985) have taken this into account in their analysis, however.

4.3.2.3 High Preflare Electron Densities

In the chromospheric evaporation picture, the high electron density of a flare is produced mainly by the large upward mass flux from the chromosphere. In this scenario the loop before flare onset is usually assumed to have a relatively low electron density, say considerably less than 10^{10} cm^{-3} (Nagai 1980). That is, if the initial loop density is much higher, there is no need to invoke evaporation to explain the large mass in the corona; it is there to begin with. Typical SXR flare densities at flare maximum have been estimated by different investigators to be between about 10^{11} and 10^{13} cm^{-3}. However, most values fall between 1 and 5×10^{11} cm^{-3}. If the initial loop density were also high, then in this case a large mass is already present at high altitudes, and chromospheric evaporation may not be the dominant or only mechanism that produces the high SXR flare density.

It is important to mention how the electron densities are measured. For temperatures less than or comparable to 6×10^6 K, densities can be derived from density sensitive spectral line ratios (e.g., Dere et al., 1979). Densities obtained from a number of different ratios all give values on the order of 10^{11} cm^{-3} or greater. For temperatures greater than 6×10^6 K, there are not adequate spectral line ratios. However from Skylab, P78-1, and SMM, there are images of flares in high temperature spectral lines, such as Fe XXIV 225 Å in the Skylab data. The Skylab data have the highest spatial resolution, about 2". From the incident flux in the image the emission measure n_e^2 V, can be derived. Then the volume V can be calculated from the image and plausible geometric assumptions about the shape of the flare. Thus, the electron density n_e can be calculated. This method also gives densities greater than 10^{11} cm^{-3} for the SXR flare. Finally, Feldman, Doschek, and Kreplin (1982) estimated electron densities greater than 10^{12} cm^{-3} for very impulsive flares that cool in a minute or so. This estimate results from the very short cooling times and large emission measures. The high density is obtained whether or not the cooling is due to conduction or to radiation. In summary, all evidence points to very high densities for SXR flares at times near peak SXR emission.

Measuring densities at flare onset times is much more difficult because X-ray fluxes are very low and high time resolution is needed. The best way at present to infer onset time densities involves neither a density sensitive line ratio nor an X-ray image. Instead, an attempt is made to use the fact that it takes a finite time after heating a plasma to strip the heavy elements to He-like and H-like ionization stages, i.e., to reach ionization equilibrium (IE). As explained in Gabriel (1972) and Doschek, Feldman, and Cowan (1981), it is possible to use combinations of line ratios to measure the departure from IE.

Figure 4.7 shows the IE line ratio diagnostic q/w and the abundances of He-like calcium and iron compared to the total element abundance as a function of the parameter $n_e t$, under the assumption of instantaneous heating of plasma from 1 - 2 x 10^6 K up to the values shown in the figure. That is, the time dependent rate equations for ionization and

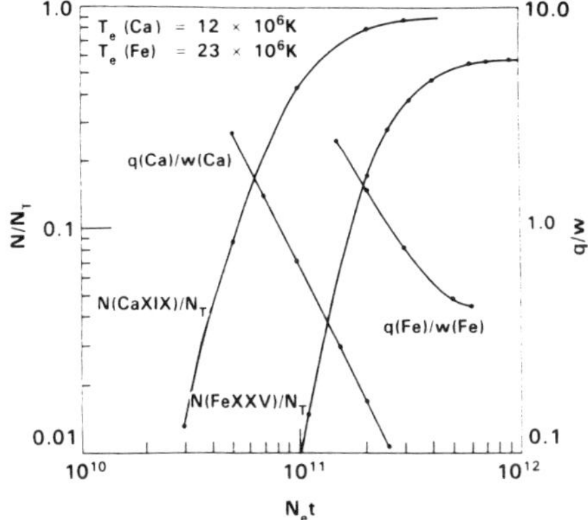

Figure 4.7 Transient ionization calculations for calcium and iron. The calculations assume calcium and iron ions are heated instantaneously from temperatures of 1-3 × 10^6 K to temperatures of 12 × 10^6 K and 23 × 10^6 K, respectively. The fractional He-like ion abundances and the key diagnostic ratio q/w are shown. The equilibrium ratio of q/w for calcium is about 0.09 in this calculation.

recombination were solved as a function of time assuming a step increase in temperature (and also a constant density). The procedure is described by Doschek (1984). The relevant parameter is $n_e t$, and not simply n_e or t.

These curves are to be understood as follows. The fractional abundance N(Ca XIX)/N_T and ratio q/w have values of 0.43 and 0.71 respectively, for $n_e t = 10^{11}$ $cm^{-3}s$. That is, if $n_e = 10^{11}$ cm^{-3}, then these values will be reached in a time of 1s. If $n_e = 10^{10}$ cm^{-3}, then these values are achieved in 10s, and so forth. From inspection of these curves, it would be difficult to measure a departure from IE using q/w ratios for $n_e t = 2 \times 10^{11}$ $cm^{-3}s$ for calcium and perhaps 3.5×10^{11} $cm^{-3}s$ for iron. Thus for $n_e = 10^{11}$ cm^{-3}, a time resolution better than about 3.5s is needed to detect departures from IE. However, for $n_e = 10^{10}$ cm^{-3}, departures could be observed in times of about 10 - 20s after flare onset.

Up to now, the earliest spectra observed during the rise phase either with P78-1 or SMM appear to be consistent with IE. Therefore it must be concluded that for most if not all flares the preflare loop density is at least 10^{10} cm^{-3}. It might be argued that in the evaporation theory this is not unexpected since plasma at transition region and chromospheric densities is being heated. Nevertheless, a strong stationary component of X-ray spectral lines is already present 10 - 20s after flare onset, and this must represent plasma already evaporated or initially present in loops. For impulsive flares, as opposed to the slow rise phase flare discussed earlier, the evaporation velocity needed to fill the loops in 10 - 20s is much larger than the velocity indicated by the blue-shifted component, and is not consistent with the evaporation scenario.

The conclusion is that the stationary component of X-ray spectral lines represents a high density plasma with $n_e > 10^{10}$ cm^{-3}, even at flare onset. It should be emphasized that this density is a lower limit. Observations within 10s of SXR flare onset are hampered by low counting rates and other instrumental problems.

In summary, a careful comparison of available X-ray data with predictions from numerical simulations of chromospheric evaporation reveal several striking discrepancies. The con team does not believe that these discrepancies can be resolved by either better data or a more adequate theory that for example would fully resolve the flare transition region. After all, the basic result of the simulations is the self-evident one that plasma flows upward into loops at about the sound speed, a result that can be qualitatively derived very simply. The upflowing plasma must be hot if it is heated in the conventional manner, i.e., either by conduction or beams. In this case some X-ray or EUV lines should show strong blue-shifts for the bulk of the emission, and this is not observed. Although the data base could certainly be improved upon, the con team takes the position that it is adequate to demonstrate the discrepancy. Finally, it would still be possible to have the mass of the SXR plasma supplied from the chromosphere without contradicting the X-ray observations if it is hypothesized that the chromospheric plasma is lifted into the corona by magnetic forces, similar perhaps to the eruption of a prominence (Pallavicini, Serio, and Vaiana 1977, Moore et al., 1980). In this case the plasma would be cold while lifted into the corona, and then heated after large upflow velocities cease. Such a view has been suggested by Hudson (1983). However, the con team remarks that the physics of this process is entirely different than the physics of the "popular" evaporation scenario under discussion. Hudson's (1983) suggestion represents a major departure from the prevailing view of how chromospheric evaporation occurs, and is beyond the scope of this debate.

4.3.2.4 An Alternative Multiple Loop Model

If chromospheric evaporation does not account for the high densities of soft X-ray flare plasma, then how are such high densiies achieved? Furthermore, since blue wings are observed on the wings of X-ray lines during the rise phase, another origin for them must be found if the chromospheric evaporation explanation is not accepted. The con team felt somewhat obligated to provide an alternative model that would account for all the observations, although strictly from the point of view of the debate this was not necessary.

The con team adopts the viewpoint that some of the blue-shifted material could be due to chromospheric evaporation, or ablation, but that this process is not the main cause of the high densities in soft X-ray flare plasmas. It is possible to envisage a rather different alternative model to that proposed by the pro team. In their model, the flare takes place within a single loop, although the energy to cause the flare is often considered to arise through magnetic field reconnection with another loop. In the model proposed below not only energy transfer, but mass transfer takes place during the reconnection process. This concept has, of course, been proposed previously in the context of high energy electrons leaking onto field lines above the original flaring region, causing for example radio emission. But the transfer of substantial amounts of mass during the impulsive phase has not been previously stressed. The concept is a natural consequence of the reconnection of two loops which have substantial helical components to their fields, in the same sense. The "reconnected" field would envelope both previous loops (see Figure 4.8).

Many authors have discussed flares observed both during the Skylab missions and SMM in terms of multiple loops. There is also a long history of flare models involving more than one loop and flare heating by magnetic energy dissipation or field line reconnection. Thus it seems natural to consider how the blue-shifted component could arise in these circumstances.

Consider a set of loops such that the large loops have a lower temperature and density. This is known to be the situation in loops associated with active regions (e.g., Gabriel and Jordan 1975). The emerging flux model as discussed by Heyvaerts, Priest, and Rust (1977) also gives a useful framework. With the emergence of new flux the pre-existing field configuration will be disturbed. This is observed through filament activation and disappearance. As part of this re-arrangement either the newly emerged flux or overlying flux loops may reach a configuration where reconnection can occur. The reconnection process will cause particle acceleration and plasma heating. It is unlikely that the pressure in the two loops (or more) will be identical. The reconnection will bring particles from the denser loop onto the field lines of the less dense loop(s) during the impulsive phase. The pressure difference could cause a net upflow of particles, unless the reconnection takes place very close to the loop apex. Thus, the observed blue-shifts could result from "injected" material rather than ablated plasma. In addition, particles accelerated during the reconnection phase can penetrate to the loop chromospheres giving rise to hard X-rays. Particles

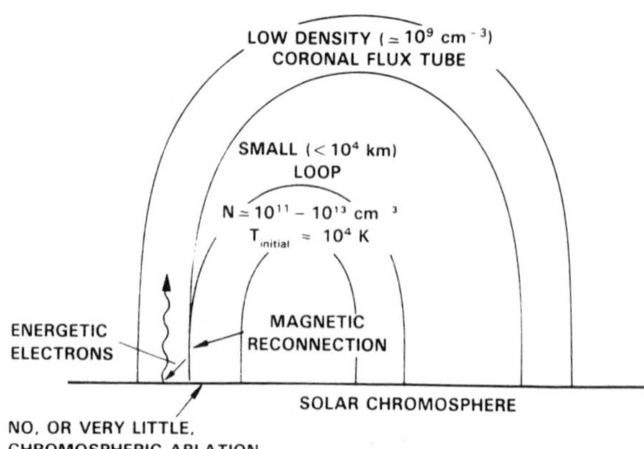

Figure 4.8 A possible alternate flare model that would produce blue-shifted components on X-ray emission lines while always maintaining a strong stationary component. Note that the proposed reconnection involves the *poloidal* components (not shown) of the magnetic fields defining each loop.

in the lower density loop will be able to reach the more distant footpoint in a non-symmetric situation. Heating of both the denser and less-dense loops will no doubt occur. In general the dramatic dynamic effects might be expected to occur in the less dense loops. The main soft X-ray emission could in principle arise from the accumulation of injected plasma, but clearly a range of flare behavior could result, depending on initial parameters and geometries. For example, the dense reconnecting loop when heated further through reconnection could also give rise to increased emission. This dense loop might have field lines that strongly and rapidly converge near the chromosphere, or conduction might be inhibited by anomalous effects. Therefore evaporation into the initially dense loop could be insignificant. Observations of this loop would reveal an initially dense loop with unshifted spectral lines, thus explaining the origin of high densities without invoking large Doppler shifts.

The sequence of events described by, for example, Feldman, Doschek, and Rosenberg (1977) and Doschek, Feldman, and Rosenberg (1977) for the June 15, 1973 flare, or by Dere and Cook (1979, 1983) and Dere et al. (1979) for the August 9, 1973 flare, finds a natural explanation in a multi-loop model, as indeed some of these authors discuss. Pressures that are higher in the hot regions than in the cool regions can be accounted for by the relative emission measures of different loops.

Whether or not the material is evaporated or injected, the time dependence of the blue wing flux can be examined to see whether or not it could "become" the static component - in this model arising from the "filled" portion of the loop. However, as pointed out by the pro team, there are no sufficient observations available to test this hypothesis. In particular, the temperature of the blue wing material is an important parameter which has not been measured. Studies of the blue wing asymmetry simultaneously over a range of coronal ions (O VII - Fe XXV) and in the transition region are urgently required.

Acton et al. (1982) computed total XRP band emission measures and compared these quantities with evaporation indicated by Hα analyses but not with any blue-shifted component in X-rays. Their pressure difference between the Hα and X-ray regions finds a natural explanation in a multi-loop model (see below). However, a re-examination of the Hα modelling removed this discrepancy (Canfield and Gunkler 1985). In a later paper Gunkler et al. (1984) summed the blue-wing emission measure and claimed agreement with the static component but they do not discuss the relative temperature or densities of the two components. Thus although there is evidence that the amounts of material emitting in X-rays is within an order of magnitude of that "evaporated", the hypothesis of identity must still be regarded as "not proven".

In summary, the con team regards the evaporation explanation of blue-shifted X-ray lines as inconclusively demonstrated. There are several disturbing discrepancies between observations and predictions based on the evaporation hypothesis. The quantitative arguments presented by the pro team are subject to large observational uncertainties and also depend on assumptions that at present cannot be verified. The blue-shifts that are observed may have an explanation in terms of the alternative model described above, although this is certainly not the only alternative explanation that could be proposed.

4.3.3 Recommendations for Further Research

The debate arguments given above have highlighted the strengths and weaknesses of present observations and theory. The purpose of this section is to summarize and recommend specific research that should clarify some of the problems concerning observation and theory that were discussed in the preceding sections.

Two specific experiments would shed considerable light on the role of evaporation in producing SXR flares. A weakness of the SMM experiment package is the low spatial resolution in SXRs. Similarly, weaknesses of the previous Skylab observations are the low time resolution and lack of hard X-ray spatial information. An experiment combining high spatial resolution (< 3") in both hard and soft X-rays is clearly desirable. Also, it is desirable to have monochromatic soft X-ray flare images, as obtained by the NRL S082-A slitless spectrograph on Skylab. However, because Doppler and spatial information are convolved in this instrument, a new experiment should consist of two such instruments, with the dispersions of their gratings at right angles to each other. Another way to achieve monochromatic spatial imaging in soft X-rays is to use the new technology of synthetic layer microstructures, i.e., multilayer coatings.

A second desirable experiment is to observe flares with X-ray crystal spectrometers that have considerably higher sensitivity than the present instruments on SMM, P78-1, and Hinotori. The reason is that when spectra are first observed with the present instrumentation, the electron temperature is already high (> 10×10^6 K). Thus, the initial early times in the heating of the preflare loop or loops are not accessible to these instruments. Higher sensitivity can be achieved most easily by increasing the effective collecting areas of the crystals, whether bent or flat. Spectra obtained at onset of flare heating might enable the unknown preflare loop densities to be determined, from a study of line ratios sensitive to transient ionization and recombination. Also, the blue-shifted plasma is strongest at flare onset, and therefore the characteristics of the upflowing plasma can be studied with minimum contamination from stationary sources. A new crystal that might be used for some of these studies is InSb (Deslattes 1985). Crystals covering the resonance lines of Mg XI, Mg XII, Si XIII, Si XIV, S XV, S XVI, Ca XIX, Fe XXV, and Fe XXVI are desirable, in order to determine the properties of the blue-shifted plasma as a function of electron temperature.

On the theoretical side, there are several problems that need clarification. First, it would be highly desirable to numerically simulate a real flare, i.e., the flare rise time, loop length, electron temperature, density, and emission measure indicated by the observations should, by adjusting the energy input, and initial loop conditions, be imitated by the numerical simulation. In this way a direct comparison could be made between computed X-ray line profiles and observed profiles. Other theoretical problems involving fluid simulations that need investigation are the one-dimensional assumption of current numerical simulations, and the effects of magnetic field divergence on upflowing plasma. In terms of a particle model of flares, anomalous conductivity and the predicted ultrathin transition region need investigation. The thickness of the predicted transition region based on fluid models is less than the mean free path of the electrons,and the effects of this on both plasma dynamics as well as on spectral line intensities is presently unclear.

4.4 DEBATE OF ISSUE 2

The pro team adopts the hypothesis that a suitable chosen distribution of convective velocities associated with chromospheric evaporation can account for the major properties of both the blue asymmetry and the nonthermal broadening of X-ray lines during the impulsive phase of flares. If the hypothesis can be confirmed, the case that chromosheric evaporation has been directly observed becomes strong enough to proceed to a detailed study of the properties of evaporation and their relationships to theoretical models. If the hypothesis cannot be confirmed, the interpretation of both the blue asymmetry and the nonthermal broadening is called into question. Might the blue asymmetry be produced by motions that are non-evaporative, or are the tail of the evaporative distribution rather than its primary signature? Is the broadening seen during flares just an intensification of the process that causes ubiquitous broadening elsewhere on the Sun? If so, what does this imply about the commonality between flares and the "steady heating" of active and quiet regions? Magnetically driven gas motions are also a plausible candidate for the source of broadening in flares and in non-flaring regions. If magnetic motions are involved, the line broadening in flares is another piece of evidence that time-variable magnetic geometries play a fundamental role in the flare process.

Before beginning the debate both pro and con teams set forth the observed characteristics of the broadening that need to be explained. The following general properties of nonthermal broadening emerge from the sample of flares studied thus far with the current X-ray spectrometers (Antonucci et al., 1982; Antonucci, Gabriel, and Dennis 1984, Doschek 1983):

a. The nonthermal broadening can be characterized approximately as isotropic Gaussian broadening with broadening parameters ξ (km s^{-1}).
b. ξ_{max} = 150 - 300 km s^{-1}, where ξ_{max} is the maximum nonthermal broadening measured during a given flare as observed in Ca XIX ($T_e \approx 15 \times 10^6$ K) or Fe XXV ($T_e \approx 30 \times 10^6$ K).
c. X-ray lines are broader than their thermal widths when first detected, before the peak of the hard X-rays.
d. The broadening diminishes markedly to $\xi \approx 0$ - 60 km s^{-1} by the time of the SXR maximum. (The broadening may increase again later in the decay phase; the present discussion is limited to the impulsive phase.)
e. There appears to be no correlation between ξ_{max} and the position of the flare on the disk.
f. Broadening is also seen in lower temperature lines such as Fe XXI at 1354.1 Å ($T_e \approx 10 \times 10^6$ K) and Mg XII at 8.42 Å ($T_e \approx 7 \times 10^6$ K).

Other properties, less securely known or more difficult to quantify, or found for only a single event, were also identified during the debate. For example:

a. The maximum broadening appears to coincide with the peak of the HXR burst.
b. The lower temperature transition region lines such as C IV at 1550 Å also show non-thermal widths during the rise phase; however, the turbulent velocities deduced from the symmetric broadening of this line are markedly lower, less than 50 km s^{-1} than for the 10^7 K lines. Shine (Workshop contribution) has studied the UVSP observations of line profiles. Unfortunately there are few wavelength scans due to the nature of the instrument; however, the cases that he has been able to find all show a relatively low turbulent width in the $\approx 10^5$ K lines. This is illustrated for three cases in Table 4.3. Note that higher velocities, greater than 100 km s^{-1}, are seen at times. However, these appear to be due to distinct and multiple emission features, probably from surge-like activity, and are not part of the symmetric broadening, which characteristically indicates a much lower velocity.
c. Leibacher has studied some SXR observations with high temporal and spatial resolution for two flares on November 5, 1980. Fortuitously, the lifetimes of the two flares overlapped such that the emission during the decay phase of the first event could be used as a wavelength reference for the emission during

the rise phase of the second. The data show that at the earliest phase of the second event, its Ca XIX emission actually consists of a relatively narrow component, turbulent velocities ≈ 80 km s^{-1}, but with a large blue-shift ≈ 350 km s^{-1}. The blue-shifted line evolves by broadening and moving toward the red, so that after ≈ 30 sec the emission from the second flare appears as a very broad line, turbulent velocity ≈ 170 km s^{-1}, and centered at the rest wavelength.

d. Although there seems to be little variation in non-thermal widths from disk center to limb, Antonucci finds an average velocity of approximately 100 km s^{-1} for disk flares and 120 km s^{-1} for seven flares of longitude exceeding 70°. A larger sample of limb events is required to confirm this result.

Some of the above properties of line profiles are illustrated in Figure 4.9, based on BCS data analyzed by Antonucci et al. (1982). The quantity T_i is the temperature deduced from the linewidths, while T_e is the electron temperature deduced from line ratios. Nonthermal broadening is present when $T_i > T_e$. The figure shows the temporal behavior of the nonthermal broadening for the flare of 10 April 1980. The temporal relationship of this broadening to the hard X-ray burst is also shown. The debate on Issue 2 somewhat surprisingly turned out to be more difficult than the Issue 1 debate. The Issue 2 debate overlaps considerably with the Issue 1 debate because the origin of the X-ray blue-shifts was also discussed along with the nonthermal broadening. If the nonthermal broadening can be explained as due to a distribution of velocities of evaporating plasma, then the blue-shifts would presumably have an origin in the evaporating plasma as well.

Table 4.3 Transition Region Line Widths for Impulsive Phase Flare Kernels

Line	Doppler Width	Event
O V 1371 A	37 - 46 km s^{-1}	Nov. 12, 1980 ≈ 17:00 UT
O IV 1401 A	20 - 30 km s^{-1}	June 29, 1980 ≈ 2:34 UT
Si IV 1402 A	20 - 30 km s^{-1}	June 29, 1980 ≈ 2:34 UT

4.4.1 Argument for: The Excess Line Broadening Is Due to Chromospheric Evaporation

The excess widths observed during the impulsive phase are almost certainly due to mass motions; however, the source of these motions is unknown. Arguments will be presented which strongly support the hypothesis that the observed widths are due to convective flows resulting from evaporation.

4.4.1.1 Amplitude of Evaporative Broadening Velocities

The numerical simulations to date have been concerned with the evaporation process in a single loop, rather than the distribution of

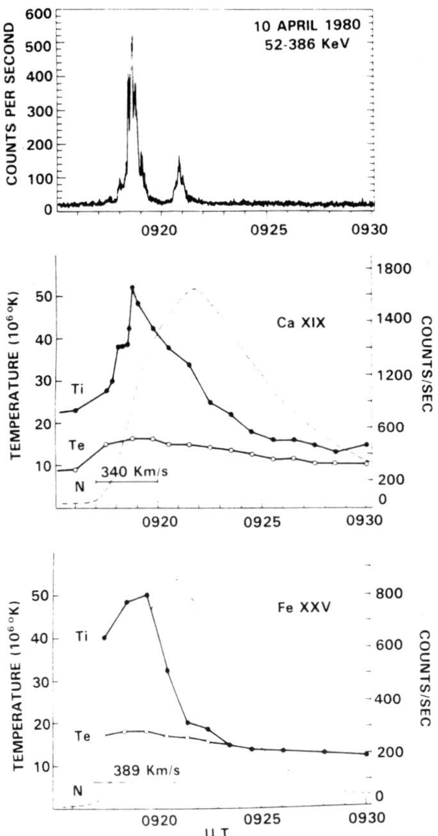

Figure 4.9 HXRBS and BCS results for the April 10, 1980 flare. T_i is a measure of the nonthermal broadening. See Antonucci *et al.* (1982) for details.

velocities in a multi-loop arcade, as would be the case in any real flare. Hence, it is not yet possible to compare directly observed line widths, or profiles in general, with a particular simulation. However, the simulations of single-loop evaporation have some key results from the viewpoint of non-thermal line widths:

The velocities produced by the evaporation process are large. Some of the electron beam simulations find peak velocities along the magnetic field as high as 1000 km s^{-1}. Fisher has studied analytically the question of maximum evaporation velocities. He finds that from general arguments, the absolute maximum velocity that can be produced is approximately twice the sound speed of the evaporated material, which implies velocities of up to 2000 km s^{-1} for temperatures of order 10^7 K. Of course, average velocities over the whole duration of the evaporation process will be significantly less than the above value; but, clearly bulk flow speeds of order several hundred km s^{-1} are to be expected from the evaporation process.

Another important result of the simulations is that, in general, the flows will not be all upward. If the flare heating and/or the loop geometry are not symmetric, as must be the usual case, then an evaporation flow from one foot point to the other will result. Cheng, Karpen, and Doschek (1984) have studied this process in detail by simulating a flare loop with an asymmetric heating function (see Issue 1 debate). As a result of this asymmetry the evaporation flow is stronger in the leg where the heat was deposited than in the other leg. The results of one of their simulations are shown in Figure 4.10 (see also Figure 4.5). The velocity profile along the loop is shown for various times. Note that initially upflows are present in both legs, but after approximately 30 s there are large downward flows in the unheated leg.

The final result pertains to the magnitude of the velocities at lower temperatures. Unfortunately, although the computed evaporation velocities at 10^7 K are fairly well known, those at 10^5 K are not. However, from quite general arguments evaporation velocities are expected to be of the order of the local plasma sound speed, so that the velocities at 10^5 K should be an order of magnitude or so less than those at 10^7 K. This result has been observed in the simulations that do resolve the transition region, albeit with lower impulsive heating than in a typical flare. Assuming that the flow speeds scale with the sound speed for larger heating rates as well, we expect evaporation velocities to be of order 50 km s^{-1} for material at 10^5 K.

4.4.1.2 Predicted Line Profiles

Qualitatively, the simulation results are in agreement with the obsevations. For a disk flare, both upflows and downflows should be observed, but with a stronger upward component since there is a net upward flow. Thus, 10^7 K spectral lines that are anomalously broad and with a blue excess are expected, and this is exactly what the observations show. The turbulent velocities should decrease with temperature, and again this is observed. At the limb we would expect to see instead of a blue excess a somewhat broader line, which also fits the observations.

Quantitatively, the agreement also appears to be good. First it must be emphasized that most solar flares do not consist of a single loop, i.e., a single activated flux tube. Flares generally have a complex magnetic geometry and a complex evolution. The observed emission at any particular time will be due to a distribution of loops, some of which may be in their evaporation phase, and others in their decay phase. Lacking accurate information as to the flare spatial and temporal structure, it is premature to attempt a detailed comparison of the evaporation model with the observed line profiles of any particular flare. Therefore the theoretical line profiles that have been calculated so far are only for idealized situations.

Leibacher has studied the effect on line broadening of the shear in velocity along a loop. He has taken one of the numerical simulations (Nagai 1980) and calculated the Ca XIX emission that would be observed by the XRP experiment from this loop model, including the instrument response. For the viewing angle he assumed the loop to be located halfway

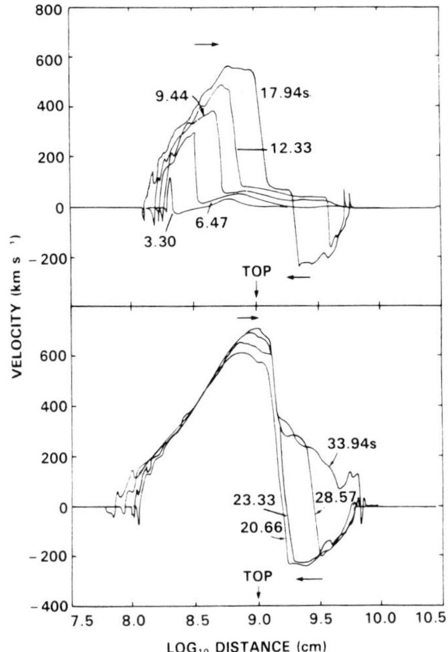

Figure 4.10 The fluid velocity as a function of position and time in the loop, from the numerical simulation of Cheng, Karpen, and Doschek (1984). Times are indicated in the figure. The top of the loop is also shown. Flare heating was deposited in a localized region about midway between the top of the loop and the left footpoint. Positive velocities indicate motion from left to right, i.e., upwards from the left footpoint. Negative velocities indicate upward motion from the right footpoint. The asymmetry between positive and negative velocities results from the asymmetric heat deposition.

between disk center and limb, and inclined at an angle of 60 degrees to the solar vertical. Figure 4.11 shows the evolution of the Ca XIX line intensity and width as predicted by this model. Note that in agreement with observations, the width is largest when the rate of increase of the calcium emission is largest (see Figure 4.11f). Leibacher obtains the very interesting result that even a single loop can produce a spectrum that matches the observations.

Of course a single loop will produce a different line profile when viewed from different orientations. Since flares appear similar either at disk center or at the limb, a distribution of loops is required to fit all the observations. It is well known that all flares consist of a combination of loops.

Antiochos has studied analytically the line profiles expected from an arcade of loops. He investigated the idealized case of a distribution of loops of identical size and shape and with a constant velocity flow throughout each loop. The orientations of the loops, i.e., inclinations

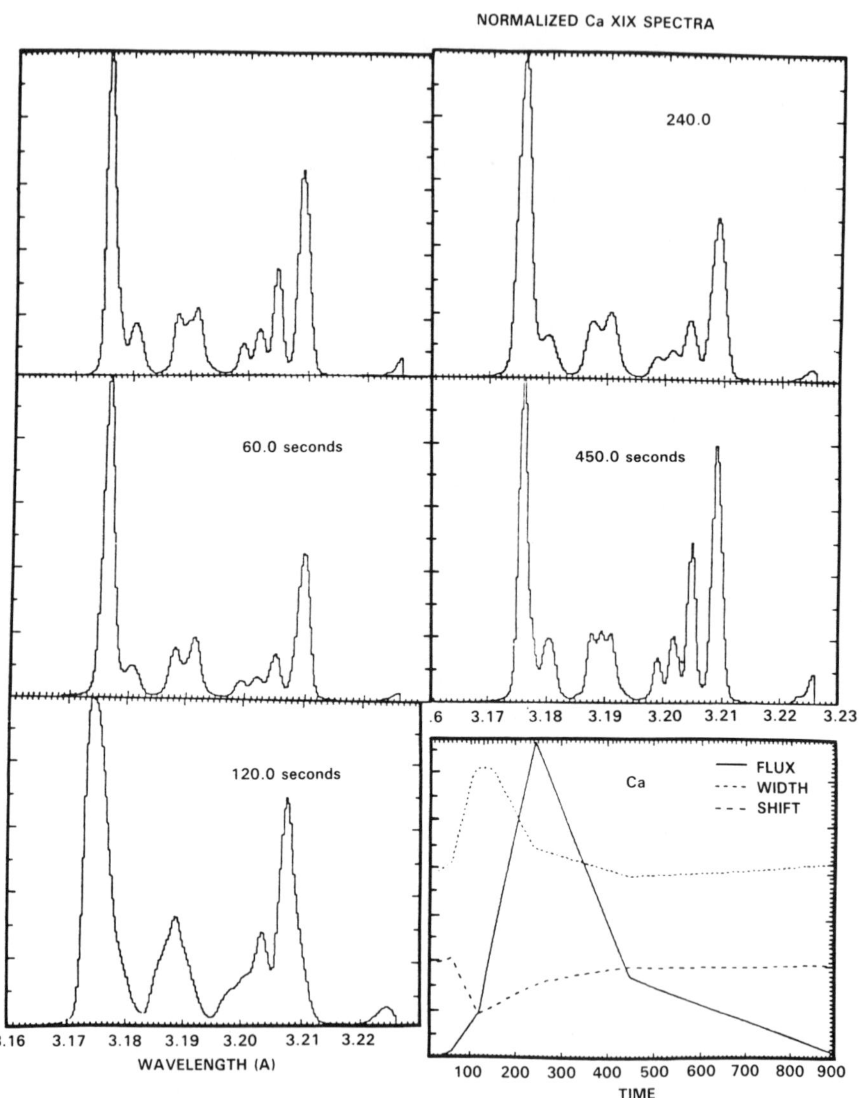

Figure 4.11 Profiles of the Ca line group computed from the simulation of Nagai (1980). See text for details.

of the loops and directions of the lines joining the footpoints on the solar surface, are assumed to be completely random. Enough loops are assumed to be present such that there are no preponderances of loops at any particular inclination angle or in any particular direction. We call this distribution a hemispherically symmetric distribution. The loop plasma was assumed to be at a constant temperature and density. The flow speed was assumed to be four times the ion sound speed, or a flow speed of approximately 300 km s^{-1} for Ca at 10^7 K. The line profiles produced by such a distribution, when viewed at disk center and at the limb, are shown in Figure 4.12. In terms of velocities, the disk width is 120 km s^{-1} and the limb width is 280 km s^{-1}.

The important points of this work are that typical velocities of evaporated plasma can produce some of the observed widths, and that the widths should be larger at the limb. In fact the limb profile shown in Figure 4.12 is much wider than any limb profile that has yet been observed. Also it is distinctly non-Gaussian. This is due to the artificial distribution of loops and velocities that have been assumed. A hemi-spherically symmetric distribution favors horizontal loops, so that a large fraction lie parallel to the solar surface. Coupled with a constant flow speed throughout the loops, this assumption will produce very broad lines when such a distribution is viewed from the sides, i.e, at the limb. In reality we expect the distribution of loops to be peaked more toward the vertical than the horizontal, and the flow speed to vary both from loop to loop and within each loop. The plasma temperature and density will also vary in the same manner. These effects will make the velocities appear much more random, and hence, will lessen the difference between the widths at disk center and at the limb. Additionally, a more random distribution will produce a more Gaussian line profile.

Cheng has also studied the effect of a distribution of loops on line. He uses only a small number of loops, but with a much more realistic plasma model than used to generate Figure 4.12. The distribution of velocity, temperature and density along the loop are

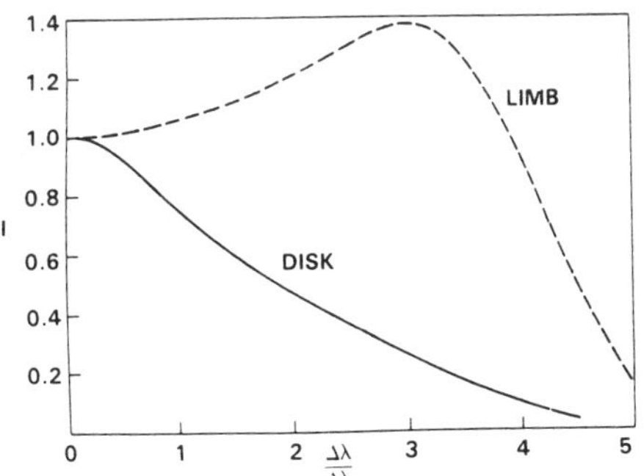

Figure 4.12 Computed Ca XIX resonance line profiles from the model by Antiochos (this Workshop). The profiles are computed for a disk flare and a limb flare. See text for discussion.

taken to be those at time = 28.57 s in his simulation of asymmetric flare heating (Figures 4.5, 4.10). Cheng has assumed three identical loops, all at disk center but inclined at angles of 0, 60, and 90 degrees to the solar vertical. The Fe XXV, Ca XIX, and Fe XXI line profiles produced by this combination are shown in Figure 4.13. Note than he finds not only a broadened line, but also a blue asymmetry due to the larger upward flow, as is observed. Again, the Ca XIX line profile in Figure 4.13 is wider than observed and is distinctly non-Gaussian. However, Cheng's assumptions are tantamount to having a flare in which all the loops are heated simultaneously, and then observing these loops at their time of peak evaporation. More physical assumptions will yield profiles closer to observed ones.

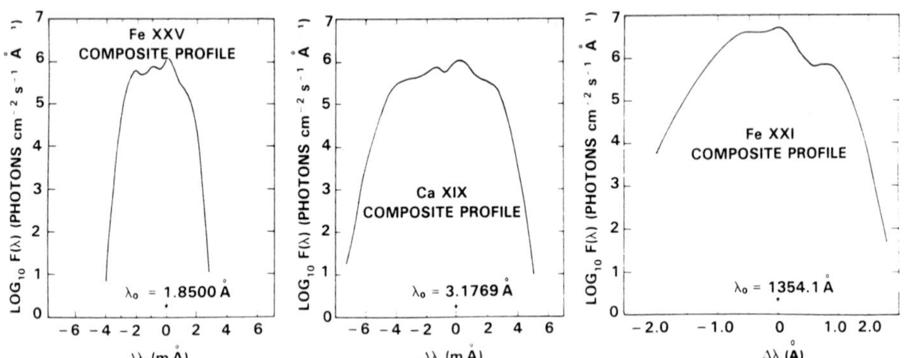

Figure 4.13 Line profiles predicted for an idealized 3-loop model based on the 1D simulations described in Cheng *et al.* (1983). The three loops are viewed at 0° (disk view), 60°, and 90° (limb view), respectively. The profiles are composites of all three loops. See text for discussion.

It is reemphasized that the calculations presented above are for unrealistically simple models, so that they cannot be expected to predict line profiles that agree in detail with the observations. The key point of this work is that it demonstrates how easily evaporation can produce line broadening. Even the grossly simplified models above predict line profiles that begin to resemble the observed ones.

4.4.1.3 Discussion of evaporation versus MHD mechanisms

So far the only well-defined model for the anomalous line broadening is that it is due to convective motions in a distribution of loops. The main alternative to evaporation-driven convection is that the turbulent velocities are due to some form of MHD motions, perhaps related to prominence eruptions. However no definite model using MHD motions has been proposed; hence this alternative is presently little more than speculation. Keeping this vagueness of the MHD hypothesis in mind, it is nevertheless possible to compare the two competing models.

The observational and theoretical results presented above constitute a very strong case in favor of evaporation. The major points supporting this model are the following:

Evaporation provides a natural explanation for the observed tight correlation between line broadening and rise in soft X-ray emission measure. If evaporation is the cause of the turbulent velocities, then the line should be broadest when most of the emission is originating from evaporating material at the beginning of the soft X-ray burst, and the line broadening should disappear when evaporation ceases. This will occur within a few minutes after the maximum in X-ray emission measure, since the time scale for convective motions to decay is of the order of the travel time across the length of the loop, $\approx$ 100 s for typical loop lengths and evaporative velocities.

There is no such clear explanation for the observations in terms of MHD motions. These are not well-correlated with the rise in X-ray emission measure. For large two-ribbon flares, such as the large flare on June 6, 1982 observed by Hinotori (Tanaka, Ohki, and Zirin 1985), the peak in MHD motions as indicated by the prominence ejection occurs well before the rapid rise in soft X-ray emission. Also, even if it is supposed that some form of MHD oscillations persists well after the prominence ejection, there is no good reason why such motions should end at the time of X-ray maximum. Alfven waves, for example should decay very slowly and would be present thoughout the flare duration. The magnitude of the turbulent velocities, and especially the strong temperature dependence of the velocities, are easily understood with the evaporative model. In evaporation the cool material always moves slower than the hot; hence, the line broadening will naturally decrease with decreasing temperature.

On the other hand, MHD flows often indicate the reverse situation, the most rapid velocities are in the coolest material. Sprays and prominence ejecta often show 10^4 K material accelerated to very high velocity, > 1000 km s^{-1}, far greater than any line shifts observed for the 10^7 K plasma. If MHD motions were the cause of the symmetric broadening, turbulent velocities at 10^5 K are expected to be as large or even larger than those at 10^7 K. To the knowledge of the pro team, this has never been observed.

Leibacher's observations of the November 5, 1980 flare provide further support for the evaporation model. The evolution of the Ca XIX line that he observed has no natural explanation in terms of MHD motions, but is exactly what is expected from evaporation. At the very earliest phase of a flare, the Ca XIX emission will originate from rapidly evaporating material in the first heated loop. For a flare at disk center a strongly blue-shifted line is expected, but without much broadening, just as is obtained from the single-loop simulations. Later in the event as the number of loops increases and the distribution of velocities becomes more complex, primarily a highly-broadened line should be observed.

From the theoretical work the major point supporting evaporation is that it can clearly produce broadening of the observed magnitude. The simulated line profiles all show at least as much width as is observed. If anything, the discrepancy is in the direction that evaporation is too efficient a broadening mechanism. MHD motions which are generally much faster than pressure driven ones would have an even greater difficulty in producing the relatively small (all quite subsonic) turbulent velocities inferred from the line widths. Note that any alternative model for the widths would first have to suppress somehow the strong evaporative broadening. Since evaporation is an almost inevitable consequence of flare heating, the pro team cannot see how evaporation can be suppressed.

In summary, the pro team concludes that evaporative motions are the cause of the anomalous line broadening. No doubt other models could be contrived that would also fit the observations, but these models would be just that -- contrived. With the evaporation model, on the other hand, line broadening follows naturally from flare heating.

4.4.2 Argument Against: The Excess Line Broadening Is Due to Chromospheric Evaporation

The pro team of Issue 1 has argued that the blue-wing excess observed in X-ray lines during the impulsive phase is the signature of chromospheric evaporation. If this is correct, there is no question that "about the right" velocities (100-400 km s^{-1}) are present in the evaporating flow to broaden the main component of the line by "about the right" amount. As yet, the case for convective broadening has not advanced much beyond this statement of plausibility; and if the blue-wing excess is not produced by evaporation, the plausibility evaporates as well.

Antonucci has shown that there is a moderate correlation (product moment correlation $r = 0.7$) between the maximum blue-shift of the blue component in a given flare and the maximum nonthermal broadening in that flare (see Issue 1 debate). If nonthermal broadening is caused by evaporative flow, there should be correlations between the parameters of the blue excess and the parameters of the broadening. However, as usual, the presence of correlations is a necessary but not sufficient condition for the hypothesis; the parameters could be correlated through a common underlying phenomenon, not necessarily chromospheric evaporation.

The temporal correlation between the broadening and the blue component is not clear, largely because of observational difficulties. Spectra from P78-1/SOLFLEX (Doschek 1983) are created by scanning. Neither the total scan time ($\approx$ 1 min) nor the time ($\approx$ 10 s) required to scan through a line such as Ca XIX is negligible during the impulsive phase of a flare; this complicates the determination of broadening and the comparison of broadening and blue excess. With SMM/ BCS data, it is often necessary to integrate over a comparable period of time (10-60 s) in order to build up enough signal to measure the blue component. Despite these difficulties, the observations do seem to indicate one aspect of the relationship between broadening and blue excess that is difficult to reconcile with the simplest picture in which the evaporative flow broadens the main component. Namely, the main component is sometimes measurably broadened for one or two minutes before there is any measurable blue component, and the degree of broadening may be as large ($\approx$ 100 km s^{-1}) as at any time during the impulsive phase (Antonucci and Dennis 1983; Antonucci, Gabriel, and Dennis 1984). To maintain a causal relationship between evaporative flow and broadening, it would seem necessary to suppose that a blue component was present but for some reason not detected. This emphasizes the importnce of thoroughly testing on noisy artificial data the algorithm used to fit the observed spectra (see Recommendations). We note that this result is the opposite of that found by Leibacher and discussed earlier. This difference may reflect a real difference among different flares, or perhaps it is partly due to the difficulty of deconvolving moving and stationary components of a line profile at very early flare rise times, when the line intensities are very weak.

4.4.2.1 Single-loop Simulations

Before constructing a multiloop model, it should be shown that a single-loop simulation can provide a good description of the evaporative flow field; otherwise little confidence can be placed on the results produced by a superposition of such models. It should be kept in mind that the goal of a simulation is not just to produce a blue excess, or nonthermal broadening, but to reproduce their observed relative evolution. In the first approximation, the evolution may be characterized by $\xi(t)$, the nonthermal broadening parameter, by $v_{bl}(t)$, the speed of the blue-shifted component, and by $R(t)$, the intensity ratio of the blue-shifted to the nearly unshifted ("main") component.

One good reason to stress relative behavior is that otherwise it is hard to know where in the time development of the model the observations begin. For example, in the centered-heating model of Cheng et al. (1983), for $t < 20$ s the blue-shifted component of Ca XIX dominates the unshifted component. Since this is never observed, one must either reject the model or assume that, because of the sensitivity limits of the detectors, the observations begin at a later stage.

The same model (Doschek et al., 1983; Cheng et al., 1983) can illustrate the danger of examining individual predictions in isolation. Figure 4.4 (see Section 4.3.2) shows the Ca XIX line profile 30.2 s after the onset of energy deposition. In a top view, the line shows a blue excess that would be fit by an evaporating component with $v_{bl} \approx 200$ km s^{-1}, $R \approx 0.2$. In an end view, the profile is convectively broadened by $\approx$ 100 km s^{-1}. These parameters are within the bounds of observation. However, at the same time ($t = 30.2$ s), the simulation predicts a gross shift of the Fe XXI line, about 150 km s^{-1} in the top view shown in Figure 4.4. As mentioned in Section 4.3, such shifts of the bulk of the emission have not been observed, either by Skylab (Doschek et al., 1983) or by the UVSP aboard SMM (Mason et al., 1985). We stress this disagreement to illustrate that the distribution of temperature and flow velocity in some single-loop models has points of serious disagreement with observation. It would therefore be premature to use such models as the basis for a multiloop simulation.

The comparison of Ca XIX and Fe XXI is an instance of a broader issue that needs more attention: the temperature dependence of the broadening, the blue excess, and the blue-shift of the main component. It does not appear that either existing observations or existing models have been probed sufficiently for the full extent of the information they might provide on this issue. The convective line broadening as a function of temperature can be considered in a general context, based on the simulations carried out to date. The following comments are probably independent of the simplifications of the 1D models, such as lack of transition region resolution and magnetic field tapering, etc. Assume for simplicity the case of a flare loop observed on the disk at zero longitude and heated symmetrically.

Consider first spectral lines formed over narrow temperature ranges with average temperatures of about $3 - 13 \times 10^6$ K. Ions such as Ca XVI, Ca XVII, Fe XXI, and Fe XXII are formed at temperatures within this range. After the loop or loops are filled with evaporated plasma these ions are formed near the footpoints of the loop and thus virtually the full spread in velocities will be observed. In the Cheng et al. (1983) simulation (Figure 4.4) $\xi \approx 50$ km s^{-1} at 30.2 s. However, since the ions are formed over a relatively narrow temperature range, zero velocities (loop top) are not observed and all of the long wavelength XUV lines of these ions should show a net Doppler shift as well as a nonthermal broadening. Invoking a multiloop simulation will not remove the Doppler shifts, since in the evaporation model a net flow of plasma upward into the loops must occur.

Now consider the highest temperature lines observed, i.e., the resonance lines of Fe XXV and Fe XXVI. They are emitted primarily at the top of the loop in a symmetric simulation, where $\xi = 0$. Since the temperature of most flares is $\approx 25 \times 10^6$ K, the line emissivities fall very rapidly towards the loop footpoints. Since the lines are formed near the loop tops, the Doppler shift and Doppler broadening tend to be much less than for the cooler lines, since motion is transverse to the line-of-sight. In the Cheng et al. (1983) simulation, $\xi = 35$ km s^{-1} for Fe XXV at 30.2 s.

Third, there is a group of lines such as the Ca XIX lines, that are formed at temperatures lower than Fe XXV, but higher than Fe XXI, and are formed over a broad temperature range. (Ca XIX is He-like). The Ca XIX lines are formed at the top of the loop as well as towards the footpoints. Thus these lines should show the maximum nonthermal broadening (≈ 86 km s^{-1} at 30.2 s in the Cheng et al. (1983) simulation) They show a larger blue-wing than the hotter ions, but they have a strong stationary component.

Fourth, quiet and active coronal lines from ions such as Fe XIII - Fe XVI should produce broadening that might depend critically on the accuracy of the models near the transition region, since their temperature of formation ($\approx 3 \times 10^6$ K) is near the turning point of evaporation and compression of the transition region. Similarly, transition region lines will be difficult to predict from the models for no other reason than that the transition regions have not been resolved in the numerical simulations.

In summary, symmetrically heated 1D loops produce non-thermal broadening that should be largest for lines such as Ca XIX and less for hotter (Fe XXV) and colder lines (Fe XVI). The innershell lines of Fe XXV, Fe XXIV - Fe XX should have different widths depending on ionization stage. On the contrary, the observations show large broadening for both Fe XXV and Ca XIX, with no significant decrease for the Fe XXV line.

Invoking an asymmetrically heated loop, rather than the symmetrical case, makes the situation even worse. Then downflows as well as upflows occur which broaden lines far beyond their observed widths (Cheng,

Karpen, and Doschek 1984). Furthermore, the line profiles in a single loop model would exhibit central reversals for many lines. These could be eliminated in a multiloop model, but the lines would probably be widened even more. Basically, the numerical simulations of single loop models carried out so far indicate substantial discrepancies with the observations that appear to be independent of the physical and numerical simplifications at present necessary to carry out the calculations.

Another feature of loop simulations that should be examined is the blue-shift of the main component as a function of time. (As is stressed in the Recommendations section, the division into "blue" and "main" or "unshifted" components is an approximation.) By the time the main component earns its name (by dominating the blue component), its blue-shift is small (< 30 km s^{-1}); but Antonucci, Gabriel, and Dennis (1984) have reported tentative indications of shifts of this order in the main component. If such shifts are confirmed, they bear on the broader question (Issue 1) of whether the observed blue excess is the primary signature of chromospheric evaporation. Antonucci et al. (1982) and Antonucci and Dennis (1983) have shown that reasonable densities ($n_e \approx 5 - 10 \times 10^{10}$ cm^{-3}) are inferred for the blue component if it is assumed to supply the mass necessary to produce the observed emission measure. However, densities an order of magnitude higher are not ruled out, and if the same argument is applied in reverse, small evaporative velocities ($v_{bl} \approx 10 - 40$ km s^{-1}) are inferred. The primary signature of evaporation would then be the small shift in the fitted wavelength of the main profile; what is here termed the blue component would be a high velocity tail or a secondary phenomenon, not the primary signature.

4.4.2.2 Multiloop Models

Two key observational points are that the broadening in each flare is approximately isotropic, and that the degree of broadening shows no strong trend with the position of the flare on the disk. Since, as might be expected, the line profiles produced by single-loop simulations of chromospheric evaporation show a marked dependence on the vantage point (Figure 4.5), a multiloop model is required to even approach a resemblance to the observations. Unfortunately for the pro side of this debate, a necessary characteristic of a multiloop model that is needed to satisfy the observations is generally a random orientation of the loops. Only in this way can the observational requirement of isotropy in the line widths be satisfied. However, although most flares are known to consist of several loops, the orientations of the loops are not usually random. Typically arcades of loops are involved, all oriented in the same sense. In this case it is expected that the line broadening would vary considerably among different flares, particularly between disk and limb flares. However, this is not observed. Furthermore, in order to explain the strong stationary components of lines such as Fe XXI, at least one loop is required that lies almost flat on the solar surface so that Doppler shifts are negligible. This can hardly be satisfied by all flares. And finally, the resultant line profiles (Figure 4.13), although greatly broadened, do not resemble Gaussian profiles, contrary to the observations.

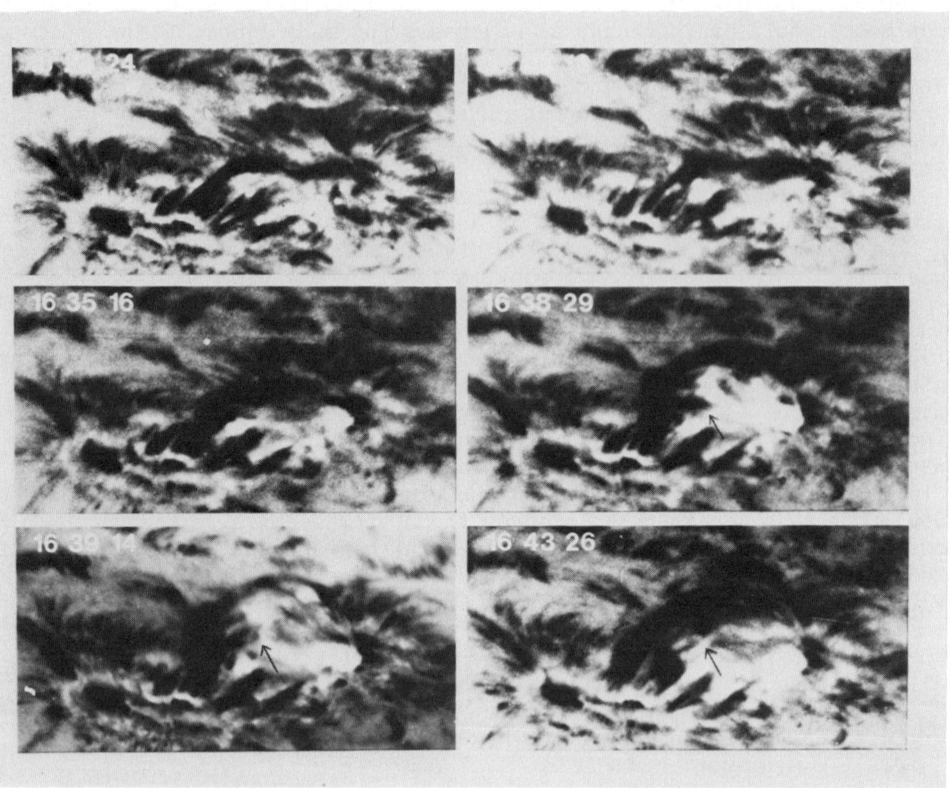

Figure 4.14 Complex motions of the magnetic field in a flare of 27 April 1972. Filtergrams 0.8 Å in the blue wing of Hα from the Big Bear Solar Observatory. The long dimension of the frames corresponds to approximately 10^5 km. The time sequence shows an overall expansion of the loop system at a peak velocity of about 70 km s^{-1}. The arrow in the frame at 1638 UT points to a dark feature that curves off to the left. The arrows in the frames at 1639 UT and 1643 UT are placed at the same spatial location as the first arrow. The dark feature shows motions of 75-100 km s^{-1} that have components in three directions and cannot be characterized as simple expansion. Note that at 1639 UT the dark feature appears to be braided with another dark feature, but by 1643 UT no braiding is apparent.

Unlike the nonthermal broadening, the blue excess is correlated with position on the disk: no flare located more than 60° from disk center has shown a detectable blue excess (Antonucci and Dennis 1983). In terms of a multiloop model, this implies that the blue excess is produced by X-ray hot gas that is still close to the footpoints of loops that emerge roughly perpendicular to the surface, while the broadening of the main component is produced by gas higher up in the loop. This is logical, but no multiloop models exist to demonstrate quantitative agreement with the observations.

4.4.2.3 Broadening Caused by Mechanisms Other than Evaporation

Nonthermal broadening of UV and X-ray lines is not confined to the impulsive phase of flares. On the contrary, nonthermal broadening is ubiquitous in non-flare spectra; only in quiescent prominences is there no evidence for $\xi > 0$ (Feldman and Doschek 1977). In quiet regions and coronal holes, $\xi \approx 20 - 25$ km s^{-1} in the lower transition region (Doschek, Mariska, and Feldman 1981) and 10 - 25 km s^{-1} in the corona (Cheng, Doschek, and Feldman 1979). In active regions, $\xi \approx 30 - 50$ km s^{-1} in the lower transition region (Brueckner 1975) and $\xi \approx 75$ km s^{-1} for preflare brightenings in Ca XIX (Antonucci et al., 1982). These broadenings appear to be isotropic: no differences have been established from center to limb.

Since isotropic broadening is observed nearly everywhere on the Sun, in both quiet and active regions, is it not perverse to invoke chromospheric evaporation -- a mechanism usually involved only during flares -- for flare-associated broadening? Would it not be more unifying to suggest instead that whatever processes cause the ubiquitous broadening are simply enhanced in flares, as are so many other processes? The cause of the ubiquitous broadening is unknown ("turbulent broadening" is at best a label, and perhaps a misnomer), but there is no reason to suppose that it is connected with chromospheric evaporation. Whatever the broadening mechanism, if it is the same in flares and non-flaring regions, it suggests a basic similarity between flares and "steady" atmospheric heating.

Magnetically driven motions should be considered as a specific alternative to evaporative motions as the source of nonthermal broadening. In cine Hα observations of flares, it is commonplace to observe motions with $v \approx 100$-200 km s^{-1} that appear to be motions of the magnetic field lines (together with their frozen-in plasma) rather than the flow of gas along stationary field lines. Moreover, these magnetic motions often appear to have an "unwinding" component, particularly evident in filament eruptions. Figure 4.14, a time-sequence of Hα observations, illustrates the complicated motions that accompany overall expansion during a flare. The broadening produced by such motions would have "about the right" value (a property shared by the evaporative hypothesis) and would also be approximately isotropic, as is observed. Unlike a postulated distribution of evaporative velocities, however, magnetic motions are observed, not only during fares but, to a lesser degree, all over the Sun. As to the relative timing of flare-associated broadening, it is relevant to cite the results of Martin and Ramsey

(1972), Moore et al. (1984) and Moore and Kahler (1984), that the fastest motions during a filament eruption usually occur at the peak of the impulsive phase.

There is one example that supports the importance of magnetic motions in the production of line broadening. Using data from the NRL slitless spectrograph aboard Skylab, Widing (1975) claimed that in the double ribbon flare of June 15, 1973, the area of greatest broadening in Fe XIV and Fe XXIV occurred over the magnetic neutral line, coincident with a filament eruption -- not at the footpoints of a loop. This possibility needs to be confirmed with future imaging instruments (see Recommendations).

Except in flare sprays, magnetic motions during flares with $v > 300$ km s^{-1} are uncommon, or at least uncommonly observed in Hα. Thus, it is not clear that magnetic levitation of heated material can account for the blue component of X-ray lines, for which $v_{bl} > 300$ km s^{-1} is typical. If the blue component is in fact due to evaporation, the magnetic motion alternative for broadening precludes the simplest kinematic picture in which both broadening and blue excess are produced by a common velocity field.

In summary, the hypothesis that convective evaporation produces the observed X-ray line widths in flares is no more than a hypothesis. It is not supported by any self-consistent physical theory. In fact, the 1D numerical simulations, although they indeed produce excess line broadening, predict relative widths and shifts that are not in agreement with the evaporation hypothesis. These disagreements are independent of the simplifications in the calculations.

4.4.3 Recommendations for Further Research

All but one of these recommendations can be pursued now, by theoretical work or with data from the repaired SMM, P78-1, or Hinotori. The last recommendation concerns the capabilities of a future spaceborne instrument designed to advance our understanding of the physics of the broadening and blue excess of X-ray lines.

a. Clarify the relative time evolution of the broadening and the blue excess of X-ray lines (both observations and models).

b. Clarify the temperature dependence of the broadening and the blue excess (observations and models).

c. Publish a detailed analysis of the performance of algorithms used to fit the observed spectra, including:

   i. Minimum detectable v_{bl} and accuracy of determination of ξ as a function of S/N (signal-to-noise ratio), v_{bl} and R.
   ii. Minimum detectable vbl and accuracy of determination of vbl as a function of S/N, ξ, and R.
   iii. Minimum detectable R as a function of S/N, v_{bl} and ξ.
   iv. Correlations between fitted values of ξ, v_{bl} and R.
   v. Minimum detectable velocity shift in the main component.

vi. Sensitivity of fitted parameters to the assumed electron temperature and broadening of the blue component (since the data are not sufficient to determine these parameters independently). It should be remembered that two-component fitting is always an approximation to what is undoubtedly a continuous distribution of velocities. When better loop simulations become available, it will be worth experimenting with velocity-amplitude distribution functions.

d. Using the results of Recommendation c, determine whether small blue-shifts (< 40 km s^{-1}) have been detected in the main component.
e. Determine for each instrument the absolute flux threshold at which a blue excess can be detected. It is important to understand why only one or two flares have been observed in which the blue-shifted component is nearly as strong as, or stronger than, the "stationary" component.
f. Examine Hα movies of flares also observed by X-ray spectrometers for evidence of magnetically driven motions.
g. Obtain spatially resolved SXR spectra. The relative locations of the plasma producing the blue excess, the broadened main component, and the HXRs will go a long way toward deciding the influence of chromospheric evaporation on the X-ray line profiles. Suggested characteristics of the detector are:

i. Angular resolution < 2".
ii. Temporal resolution < 10 s.
iii. Spectral resolution < 0.3 mÅ, comparable to that of existing spectrometers.
iv. Spectral ranges similar to those existing and previous instruments; it is important to include lines with characteristic temperatures in the range $1 - 10 \times 10^6$ K, as well as higher temperature lines.

4.5. DEBATE OF ISSUE 3

At the present time, there are basically two models concerned with how the energy could be transported to the chromosphere in order to drive the chromospheric evaporation process. One of these models, the non-thermal electron thick target model (Brown 1973), proposes that energy during the impulsive phase is supplied by a large flux of energetic electrons, whose source is generally supposed to be in the corona. It is also generally supposed that the electrons are accelerated to their high energies by rapid reconnection of coronal magnetic fields. In this model, the chromosphere is heated by Coulomb collisions from the flux of nonthermal electrons, and the chromospheric radiative output increases and evaporation can occur. Another model, the thermal conduction model, supposes that only the corona is heated during the impulsive phase, and that energy deposited in the corona drives evaporation from the chromosphere by thermal conduction. In the conduction model, the energy release can occur at various positions within coronal flux tubes, and could also be due to a reconnection process. The chromosphere is heated when a conduction front, generated at the site of energy release, propagates downward along the flux tube and deposits energy into the chromosphere. As with the beam model, the chromosphere reponds to this heat input both radiatively and dynamically.

Because thermal conduction and high energy electrons represent two different modes of energy transport and energy deposition into the chromosphere, the responses of the chromosphere to each of these modes of energy deposition should be different, at least to some degree. In the debate given below, one of the central issues is whether or not the physics of beam and conduction energy deposition is well enough understood such that the available observations can be used to decide which mode of energy deposition is more important in producing chromospheric heating during flares. Another central issue is, assuming that beam and conduction energy deposition physics is known very well, whether or not the observations are of sufficient quality to resolve differences between beam and conduction heating, and if not, what further observations are needed.

4.5.1 Argument for: Most Chromospheric Heating Is Driven by Electron Beams

In this position paper, the evidence in support of evaporation by the nonthermal electron thick target mode is examined. In Section 4.1, evidence is discussed which argues in favor of the thick target model from a comparison of spatial and temporal relationships between HXR emission, SXR bremsstrahlung, SXR line emission, EUV radiation from transition region temperatures, and Hα line profiles emitted from the chromosphere. In the subsequent two sections, some possible observational characteristics predicted by theoretical calculations of chromospheric evaporation by the thick target model are discussed. Section 4.2 begins with a review of what has been learned from numerical simulations of chromospheric evaporation. In Section 4.3, some possible evidence for hydrodynamic features predicted by the electron beam simulations is presented. Finally, the position of the electron beam team is summarized in Section 4.4.

4.5.1.1 Evidence of Impulsive Phase Emission from Observed Temporal and Spatial Relationships

a. Hard and soft X-ray bremsstrahlung. One of the reasons frequently invoked in favor of flare heating by nonthermal electrons is the relative timing of HXR and SXR emission. The peak in HXR emission during flares generally preceeds the peak in SXR emission. Qualitatively, this process is envisioned as follows. As bursts of nonthermal electrons generated by impulsive phenomena in the corona impinge on the chromosphere, nonthermal bremsstrahlung from the fast electrons is produced, giving rise to observed HXR bursts. As the fast electrons deposit their energy in the upper chromosphere, the chromospheric material is heated to coronal temperatures and subsequently expands to fill a coronal loop structure. As the loop fills with hot evaporated material, emission by thermal bremsstrahlung in the SXR spectral range increases (Nagai 1980). When the bursts of nonthermal electrons cease, the evaporation process allegedly stops, and the SXR emission measure reaches its peak value. As the coronal material cools by radiation and thermal conduction, the SXR emission begins to drop. Some aspects of these speculations are verified by detailed studies and others are not.

Although the HXR and SXR timing aspects of this picture hold true for a large number of flares, there are certainly examples of flares for which HXR and SXR emission do not appear to be causally related in the above manner (Svestka 1976).

For several particularly well observed flares, however, it is possible to go beyond the qualitative picture and compute the energy content of nonthermal electrons from HXR intensities and footpoint sizes observed by HXIS. This can then be compared to the thermal energy content in the heated coronal material, deducted by SXR spectra from Ca XIX lines and thermal bremsstrahlung. Antonucci, Gabriel, and Dennis (1984) (hereafter AGD) have done this for five large events observed by SMM. After correcting for radiation and conduction losses, they find hat the energy content in nonthermal electrons is quite compatible with the thermal energy observed in these flares. These results are shown in Table 4.2.

b. Hard X-rays and soft X-ray line emission. The most obvious interpretation of the blue-shifted components of X-ray lines is the upward expansion of "evaporated" material into the flare corona. Antonucci and Dennis (1983) compellingly argue for this interpretation by correlating the upward velocity inferred from Ca XIX blue-shifts for a number of different flares with the rate of change of SXR emission measure in each flare. AGD have also presented evidence that these blue-shifts are casually related to the presence of energetic electrons. The onset of the blue-wing asymmetry in Ca XIX line profiles coincides with the onset of HXR emission for the flare analyzed by these authors. This behavior is illustrated by Figure 4.1 (Section 4.3.1), which shows the temporal evolution of HXR emission, the velocity of the blue-shifted component, and the SXR emission measure for the May 21, 1980 event (AGD 1984). The fact that the blue-shifted material appears at the onset of HXR emission and disappears when HXR emission ceases argues strongly that upward coronal velocities associated with impulsive phase chromospheric evaporation are somehow powered by energetic electrons.

c. OV emission and hard X-rays. Evidence supporting the validity of the thick target description of chromospheric evaporation is the observation by SMM of a tight correlation in time between individual bursts of OV EUV emission and HXRs (Woodgate et al., 1983). The delay in individual OV bursts relative to HXR bursts in some flares is observed to be less than one second. These time delays are short enough to argue that the same energetic electrons responsible for HXR emission must also be responsible for the OV emission. This time correlation is difficult to explain if chromospheric evaporation is driven entirely by thermal conduction. Time delays of at least several seconds are typically seen in numerical simulations of conductive driven evaporation, where flare energy release is confined to the corona. This appears to be true even when the loop lengths are fairly short (Nagai 1980, Cheng et al., 1983,

Pallavicini et al., 1983). We note that OV observations should be made carefully, since the OV line contains some spectral blends. These blends are discussed by Jordan (1985). (Also, see the flare line list published by Cohen, Feldman, and Doschek 1978).

d. Hα emission and hard X-rays. Evidence that nonthermal electrons are involved in the chromospheric evaporation process comes from observations of chromospheric (Hα) line profiles during the impulsive phase of flares. The nonthermal electron thick target model predicts that the residual (unevaporated) flare chromosphere should be heated significantly by the imposed flux of nonthermal electrons (Brown 1973, Ricchiazzi and Canfield 1983). If, on the other hand, the only chromospheric heating mechanism invoked during the impulsive phase of flares is thermal conduction from the corona, it can be shown (Ricchiazzi and Canfield 1983) that the temperature structure of the unevaporated portion of the chromosphere is virtually unaffected. Since Hα profiles are formed in the flare chromosphere, observations of Hα profiles during flares yield useful information about the nature of the impulsive phase heating mechanism.

Calculations of Hα profiles by Canfield, Gunkler, and Ricchiazzi (1984) show distinct features of these profiles which can be linked to conditions in the flare chromosphere. One important result is that only the presence of nonthermal electrons can produce broad Stark wings, typically 1 to 5 Å away from line center. Recent observations of Hα profiles during flares were made at Sacramento Peak Observatory in conjunction with SMM observations (Acton et al., 1982, Gunkler et al., 1984, Canfield and Gunkler 1984). These observations show strong Stark-like wings in the Hα profiles coinciding spatially and temporally with hard X-ray emission observed by the HXIS instrument onboard SMM. These observations strongly suggest that nonthermal electrons heat the residual chromosphere during the impulsive phase of flares.

4.5.1.2 Some Results of Hydrodynamic Calculations

A number of theoretical hydrodynamic calculations of chromospheric evaporation in flares have been done, both for flare heating by nonthermal electrons, and for "thermal conduction" models where all or most of the energy is deposited in the corona. Since these calculations may shed some light on various aspects of the observations, the results of these calculations are now reviewed.

a. Thick target electron-driven evaporation. Numerical studies of chromospheric evaporation due to heating by collisions from an assumed flux of nonthermal electrons have been done by a number of authors (Kostyuk 1976, Kostyuk and Pikel'ner 1975, Somov, Syrovatskii, and Spektor 1981, Livshitz et al., 1981, Duijveman, Somov, and Spektor 1983, Nagai and Emslie 1983, MacNeice et al., 1984, Fisher, Canfield, and McClymont 1985a). These results show that evaporation driven by beams can be divided into two regimes, depending on the magnitude of the thick target energy input flux. Below a certain critical energy input flux, flare heating balances radiative losses everywhere in the flare chromosphere. The temperature in the flare chromosphere is raised a moderate amount by the thick target heating, but does not exceed 10^5 K. Hydrodynamically, the chromosphere responds to this moderate temperature increase by expanding upward. In the meantime, the flare heating in the corona raises the temperature there, increasing the conductive flux into the transition region. This slowly drives the transition region into the chromospheric material on coronal conductive time scales. This evaporation scenario is called "gentle" evaporation, because of the relatively gentle response of the atmosphere to the imposed heating function.

The threshold above which gentle evaporation no longer occurs can be found by setting the flare heating rate (per particle) at the column depth of the chromospheric material just below the preflare transition region equal to the product of the density there with the peak in the radiative loss function near T = 10^5 K. This procedure, suggested by Brown (1973) and Lin and Hudson (1976) has been verified with numerical calculations by MacNeice et al. (1984), and Fisher, Canfield, and McClymont (1984c), though an important modification has been suggested by McClymont, Canfield, and Fisher (1985), taking into account closed loop geometry and preflare coronal pressure.

When the flare energy input rate exceeds this threshold, the chromospheric material just below the preflare transition region is no longer able to radiate away the energy being supplied to it by the flare heating function. The material then rapidly heats up to coronal temperatures en masse. Since the time scale for this rapid temperature rise is usually short compared to hydrodynamic timescales in the heated region, this creates a large overpressure coinciding with the rapidly heating region. The heated material then expands explosively into the coronal material above it, at speeds near (roughly 60% to 100%) the $2.35c_s$ upper limit identified by Fisher, Canfield, and McClymont (1984). The quantity c_s here is just the sound speed in the explosively evaporated material. For this reason, we describe this scenario as "explosive" evaporation. Not only does the large overpressure drive the heated material upward, but it also drives mass motion downward into the flare chromosphere. Thus, velocities in the upper flare chromosphere are downward for explosive evaporation, in contrast to the upward velocities expected during gentle evaporation. This difference in sign of hydrodynamic velocities between explosive and gentle evaporation holds as well for velocities computed in the lower transition region, at temperatures around 10^5 K. In addition, it has been shown by Fisher, Canfield, and McClymont (1985a) that the downward moving region of the flare chromosphere during explosive evaporation must be cold and dense relative to its surroundings. This feature has been seen, though not properly explained, in a number of numerical simulations. Furthermore, this downward moving "chromospheric condensation" accretes matter very quickly, encompassing in a few seconds more material than is moving upward.

b. Conduction-driven evaporation. It has long been recognized that thermal conduction should be a very important mechanism for the evaporation of chromospheric material during flares. If we assume that thermal conduction is the only mechanism responsible for chromospheric evaporation, and that there is no other heating mechanism for the flare chromosphere other than quiescent heating, then it is possible to show that there are some differences between this and evaporation driven by nonthermal electron heating.

A number of numerical simulations have been done of chromospheric evaporation by thermal conduction (Nagai 1980, Somov, Sermulina, and Spektor 1982, Cheng et al., 1983, Pallavicini et al., 1983). These calculations typically show evaporation driven velocities of about half the sound speed, or at about 20% of the $2.35c_s$ upper limit. In recent simulations such as the revised SMM benchmark calculations, it can be seen that there is a very thin chromospheric condensation which stays just ahead of the moving conduction front. The conduction front plows its way through the chromosphere at a nearly constant mass flux rate. Just in front of the thin condensation, the hydrodynamic velocities, temperatures, and densities are unchanged from those before the flare. In the thin condensation itself, hydrodynamic velocities are downward, just as they were in the case of the chromospheric condensation formed in explosive evaporation by nonthermal electrons. One important difference, however, is that the thickness of the condensation for conduction driven evaporation is small at all times, at least in the revised SMM Benchmark caculation. The moving conduction front eats up the back end of the condensation nearly as quickly as the front of the condensation moves through the chromosphere. The velocities in the conduction front at transition region temperatures are downward. The temperature at which the velocity changes sign in this calculation is around 2×10^6 K. Importantly, there seems to be no threshold heating rate for which hydrodynamic velocities in the chromosphere or the transition region would change sign, as there is for evaporation by nonthermal electrons. Therefore, examination of velocities at transition region and chromospheric temperatures should yield important information about the nature of flare heating.

4.5.1.3 More Evidence in Favor of Evaporation by Nonthermal Electrons

From the theoretical hydrodynamic calculations, it is clear that one possible way to differentiate among flare heating mechanisms is to look for observational evidence of "explosive" evaporation. Such evidence could include the existence of upflow velocities near the $2.35c_s$ limit, or the sudden appearance of high densities at high coronal temperatures and their subsequent relaxation to lower values on coronal loop hydrodynamic time scales. Another possibility is to look for evidence of the threshold between "gentle" and "explosive" evaporation by looking for spatial changes of sign in observed velocities at chromospheric temperatures and comparing these with what is expected from observed fluxes of energy in nonthermal electrons. These effects should not be seen for conductive driven evaporation. Possible observations of explosive evaporation are not outlined.

a. Density near 2×10^6 K as a function of time. Doschek et al. (1981) derived the electron density near $T = 2 \times 10^6$ K as a function of time for the 8 April and 9 May 1980 flares, by analyzing emission from a forbidden transition of OVII seen from the P78-1 spacecraft during these flares. They find that the electron density increases by at least a factor of ten in a time no longer than a minute. The peak values of density, about 10^{12} cm^{-3}, are consistent with preflare chromospheric values. The electron densities then gradually decay in value over a timescale of several minutes.

Qualitatively, this is the scenario expected from explosive evaporation by nonthermal electrons, although the timescales are somewhat longer than predicted. As a portion of the chromosphere is rapidly heated en masse from chromospheric to coronal temperatures, the material is expected to pass through temperatures near 10^6 K while still at chromospheric densities. As the explosively heated material begins to expand, however, the density of this region should decrease. If evaporation is postulated as being due to a conduction front which impacts the chromosphere and then moves downward through it in a self similar fashion, the density near 10^6 K might be expected to suddenly increase, but to then maintain a more or less steady value as long as evaporation continues. Furthermore, unless the conduction front is moving exceedingly quickly, some hydrodynamic expansion would be expected to occur by the time that the material was heated up to 2 x 10^6 K, so that the densities would be noticeably less than chromospheric values.

b. Impulsive phase asymmetries in Hα profiles. Ichimoto and Kurokawa (1984) have recently reported that they see significant enhancements of red-wing asymmetry in Hα profiles observed during the impulsive phase of flares. They find that this red wing asymmetry corresponds to downward motion at Hα emitting regions of the chromosphere, with velocities between 40 and 100 km s^{-1}. These downward velocities are similar in magnitude to those seen in numerical flare simulations of explosive evaporation by nonthermal electrons. They claim downward motion in conduction model calculations, e.g., Cheng et al. (1983), does not produce sufficiently strong downward velocities to match those they measure from the asymmetries. Ichimoto and Kurakawa (1984) therefore conclude that they are seeing downward moving chromospheric condensations associated with explosive evaporation by nonthermal electrons.

4.5.1.4 Conclusions

The evidence in favor of driving chromospheric evaporation by fluxes of nonthermal electrons has been reviewed. It is found that there is enough energy in energetic electrons to power the SXR portion of the flares during the impulsive phase for five well observed flares. The observed blue-shifts in Ca XIX line profiles, which can be interpreted as signatures of chromospheric evaporation, begin with the onset of HXRs and terminate cotemporally with the HXR flux. Individual spikes of OV emission formed at transition region temperatures track corresponding spikes in HXR emission to better than one second. Hα profiles with Stark-like wings, indicating the presence of energetic electrons, are found to be cospatial and cotemporal with the HXIS images of HXR emission. Taken together, these findings strongly suggest that energetic fluxes of electrons penetrate deep into the solar atmosphere and are intimately connected to the chromospheric evaporation process.

4.5.2 Argument Against: Most Chromospheric Heating Is Driven by Electron Beams

The case for direct electron beam heating and subsequent evaporation of the chromosphere originally rested on the thick-target interpretation (Brown 1973) of HXR bursts and on the finding by Lin and Hudson (1976) that the electron beam energy, as calculated with the thick-target model, was greater than the coronal flare plasma energy in a number of large flares. Two important recent findings apparently have bolstered the case for direct electron beam effects in the chromosphere, namely, the close correlations found between hard X-ray and UV emissions (Woodgate et al., 1983), and the interpretation of Stark-Broadened Hα line profiles in terms of electron heating (Canfield, Gunkler, and Ricchiazzi 1984, Canfield and Gunkler 1985, and Gunkler et al., 1984). However, the thick-target model and arguments for chromospheric heating based only on energetics encounter severe difficulties when examined in detail. Furthermore, the recent UV and Hα results, while they argue for the existence of electron beams, do not prove that electron beams drive most chromospheric evaporation. A detailed discussion of these points is presented in the paragraphs that follow, starting with questions about the reality of the beams themselves. Then follows a discussion of UV and SXR observations that do not support the beam evaporation model. They tend, instead, to support a coronal heating/conduction-driven explanation of evaporation. Finally, computer simulations of flares are considered. The results are compared with observations of blue-shifts in SXR lines and of Hα line profiles. The results support conduction as the principal driver of evaporation, even when there is some evaporation of the upper chromosphere by a flux of nonthermal electrons.

4.5.2.1 Evidence Against Electron Beams

The reexaminations of the thick target model presented elsewhere in this book indicate that electron beams with the energy and dimensions required by observations may not be stable. The instabilities considered most likely to occur in intense beams will cause them to dissipate their energy as heat in the corona in a very small fraction of a second. But, as of now, there is no direct evidence that intense electron beams even exist in the corona. Attempts to measure polarization of X-rays have given either negative results, or the published positive results are suspect. Similarly, attempts to infer anisotropic X-ray emission indirectly by center-to-limb observations of flares have given negative results (e.g., Kane et al., 1982), although this result does not rule out some beam models. It simply means that highly anisotropic HXR emission is not observed. Thus, belief in electron beams must ultimately rely on belief in theoretical arguments.

If beams exist and if they reach the chromosphere, it is clear that they are short-lived phenomena and deposit all or most of their energy during the rise phase of the SXR event. Thus, any model that attempts to explain all chromospheric heating in flares as due to beams must necessarily adopt an impulsive energy input with a lifetime of only a minute or so. Such an energy input mechanism cannot explain the longevity of high temperature emission in SXR events (as evidenced by Fe XXV emission, see Doschek et al., 1980). All the theoretical models of SXR emission show that the temperature of the hottest part of the flare must immediately begin to decrease after cessation of the energy input (e.g., Cheng et al. 1983, Pallavicini et al., 1983).

Coronal heating, as evidenced by a rise in SXR emission, accompanies all flares. The same cannot be said for HXRs even in large two-ribbon flares (Dwivedi et al., 1984). If beams exist in the corona, then coronal heating, either by electron beam instabilities or by Coulomb collisions, provides a more consistent explanation of SXR flare phenomena than does the beam-in-the-chromosphere model. The range of phenomena that can be explained in a consistent manner is much greater.

Arguments for beams based on the close temporal correspondence between HXR and UV emissions are considerably weakened by the recent work of Cheng and Tandberg-Hanssen (1984), who observed UV bursts (Fig. 4.15) without concurrent HXR bursts. Evidently, something other than an electron beam is capable of causing flare-like bursts of transition region emission. Furthermore, the UV features that do flare simultaneously with HXR bursts begin to brighten at a time up to three minutes before HXRs are detected.

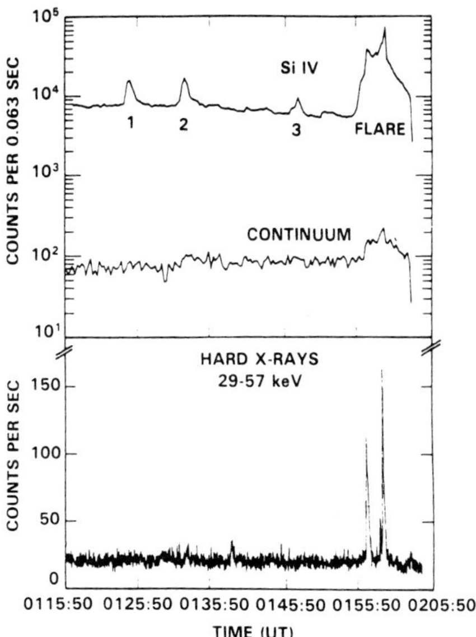

Figure 4.15 UV bursts without concurrent hard X-ray bursts (see Cheng and Tandberg-Hanssen 1984). There is also a flare in which both HXR and UV bursts occur.

Various thermal interpretations of HXR bursts have been proposed (e.g., Brown, Melrose, and Spicer 1979), and while a discussion of these models is outside the scope of this chapter, the impact of this alternative view of HXR bursts is important to this debate. If HXR bursts are thermal phenomena, there will be, of course, no beams in the chromosphere. Furthermore, the energy of the HXR-emitting plasma deposited in recent thermal models (Smith and Lilliquist 1979) is nearly as great as the energy deduced for electron beams from the thick-target model. Thus, the apparent agreement found for numerous flares by Antonucci (1982), and by others, between the calculated (thick-target model) electron beam energy and the SXR plasma energy shows only that a beam model offers one energetically consistent interpretation of the HXR-SXR sequence. More refined work -- measurements of energy release in HXRs and SXRs for example, with high spatial and temporal resolution -- will be required to distinguish between thermal and beam interpretations of coronal emission measure increases.

4.5.2.2. The Relative Importance of Conduction and Electron Beams

The temporal correspondence between UV and HXR bursts found by Woodgate et al. (1983) and the similar correspondence between chromo spheric emissions and HXR bursts reported by Feldman, Liggett, and Zirin (1983), Zirin (1978), Rust and Hegwer (1975) and others is a persuasive argument for the presence of electron beams. However, electron beam models predict much more UV emission than is observed, so in order to rescue the electron beam interpretation of the UV-HXR correspondence, it must be supposed that the beams involved are weak. Furthermore, the ubiquity of preflare UV brightenings implies that little energy is required to produce them. Therefore, the UV-HXR correspondence does not help establish that more chromospheric evaporation is driven directly by electron beams.

Tang and Moore (1982) showed that certain chromospheric brightenings removed by 1 - 4 x 10^5 km from flare cores brighten simultaneously with reverse drift type III bursts, which are certainly caused by electron beams. They concluded, however, that most of the energy reaching the chromosphere at those points was not carried by the beams, but rather by conduction or shock waves, in agreement with a similar study of remote brightenings by Rust and Webb (1977). The durations of the remote brightenings greatly exceeded those of the electron beams. Yet, if attention had been focussed on onset times only, it might have been concluded that electron beams were the only heating mechanism.

Rust, Simnett, and Smith (1985) studied the flare of November 5, 1980 and showed that arrival of the electron beam was followed by arrival of thermal energy originating at the main flare site. The thermal energy, detected in 3.5 to 11 keV X-rays, propagated with a velocity of $\cong$ 1700 km s^{-1} and was interpreted as an effect of thermal conduction. The energy in the thermal wave was enough to account for the chromospheric and coronal heating at the remote flare site.

What the above-mentioned study shows is that qualitative comparisons of electron beam signatures and chromospheric brightenings are not adequate to establish that most chromospheric heating is driven by beams. Thermal conduction can deliver the bulk of the required energy to evaporate the chromosphere even if it does not arrive there quite as quickly as the (weak) electron beams whose existence is generally accepted.

Gunkler et al. (1984) and Canfield and Gunkler (1985) have also helped clarify the relative importance of the roles played by beams and conduction in chromospheric evaporation. They studied two flares for which they have good spatial and temporal resolution in Hα and in X-rays between 3.5 and 30 keV. Some of their data are shown in Figure 4.16, where the HXR burst profile is compared with the Ca XIX (SXR) burst profile and the Hα light curve for the flare of 1522 UT June 24, 1980. Although there is a close correspondence between the Hα peak and the HXR peak, it can be seen that Hα emission continues for 3 min after the HXR

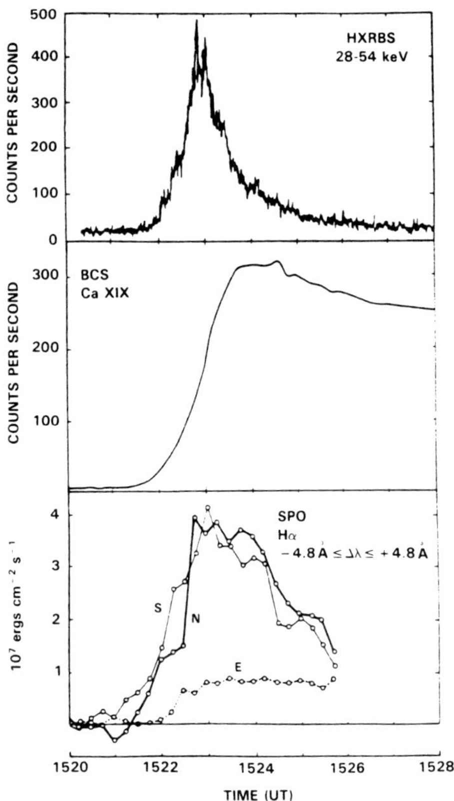

Figure 4.16 Hα data from Sacramento Peak Observatory compared with SMM BCS and HXRBS data for the June 24, 1980 flare. See text for discussion.

burst and that it began before it. Comparing observed and theoretical Hα profile (Canfield, Gunkler, and Ricchiazzi 1984), Gunkler et al. (1984) concluded that nonthermal electrons were at least 2-3 times less effective than conduction in evaporating the chromosphere.

In a reanalysis (after Acton et al., 1982) of the flare of 1455 UT 7 May 1980, Canfield and Gunkler (1985) find that the column depth of chromospheric plasma evaporated conductively reached 8×10^{20} cm^{-2}, whereas the inferred electron beam evaporated only 7×10^{19} cm^{-2}. Specifically, this means that the electron beam could not have raised the density in the coronal flare plasma to the observed level. They conclude that although the thick target model leads to an inferred beam energy of 3×10^{11} ergs cm^{-2} s^{-1} and although the broad wings seen in some of the Hα line profiles are consistent evidence for chromospheric heating by nonthermal electrons, it was conduction of heat to the chromosphere that produced the bulk of the SXR emitting plasma.

4.5.2.3 Simulations of Electron Beam Effects

Fisher, Canfield, and McClymont (1985a) have performed computer simulations that describe the hydrodynamic effects of chromospheric heating by electron beams. In addition, Fisher, Canfield, and McClymont (1984) found that there is an upper limit for upflows driven by chromospheric evaporation of roughly twice the sound speed of the evaporated plasma (see Section 4.5.1). For the electron energy fluxes inferred from observational limits, they found that the vertical velocities expected from his simulations ranged between 0.6 and ≈ 1 times the theoretical upper limit. However, most of the velocities during flares reported by Antonucci, Gabriel, and Dennis (1984) clustered about 0.2 and 0.25 times this limit. Thus their observations are not consistent with chromospheric evaporation by nonthermal electron beams. The calculations which best match both the observed Ca XIX temperatures and the upward velocities are those of the thermal model, with conductive fluxes not exceeding 10^{10} ergs cm^{-2} s^{-1}.

It might be supposed that the low observed velocities could be explained away with a model that takes into account the probable divergence of magnetic fields as they stretch into the corona. But, as Rabin and Moore (Workshop contribution) pointed out, the more divergent the fields are for evaporating material, the more convergent they are for precipitating electrons. An electron beam originating in the corona, therefore, will encounter magnetic mirrors higher and higher for electrons of larger and larger pitch angles. The proportion of energy lost to Coulomb collisions in the corona will increase at the expense of the proportion available to precipitating electrons.

A remark on the temperature of the material that is evaporated from the chromosphere is appropriate here. The temperature inferred from SXR observations is generally in the $10 - 20 \times 10^{6}$ K range, yet the only computer simulations of heating by electron beams which correspond to these temperatures (MacNeice et al., 1984) require very large > 10^{11} erg cm^{-2} s^{-1}) energy input rates. This severely restricts the conditions for direct chromospheric evaporation by electron beams. If, on the other hand, evaporation is driven by conduction from a very high temperature ($3 - 10 \times 10^{7}$ K) plasma, there is no difficulty in establishing the observed coronal flare plasma temperatures. The conclusion is that chromospheric evaporation by beam heating does not lead to a consistent interpretation of the observations of evaporated material.

For the sake of argument, the simulations of electron beam effects in the chromosphere have been accepted as plausible, at least. The UCSD flare simulations (Fisher, Canfield, and McClymont 1985a,b,c), in particular, indicate that chromospheric evaporation is driven by both electron beams and conduction. However, there are difficulties even in the assumption that electron beams reach the chromosphere. Analysis of density-sensitive X-ray line ratios (Doschek et al. 1981) shows that the preflare coronal density is at least 2 x 10^{10} cm^{-3}. A substantial portion of the electron beam energy would be lost in traversing such a corona (depending on the height of the beam acceleration site), so the question of whether electrons ever penetrate the chromosphere is an open one. If a substantial part of the beam energy is lost in the corona, it would end up being thermally conducted to the chromosphere. This view is consistent with the Hα flare profile studies of Gunkler et al. (1984), except that they feel that there is evidence for some energetic electrons in the chromosphere where the broadest Hα line profiles are seen.

4.5.2.4 Interpretation of Hα Line Profiles

In the static flare atmospheres that have been studied to date, broad Hα line wings are produced when the electron beam ionizes hydrogen at depths where the density is ≈ 10^{14} cm^{-3}. It is the combination of a high degree of ionization and high pressure that produces pronounced Stark broadening in Hα. However, to be completely consistent, one should compute the Hα line profile expected from the dynamic flare atmospheres simulated by, for example, Cheng et al. (1983), MacNeice et al., (1984), and Fisher, Canfield, and McClymont (1984). The curious feature of all these dynamic models is that a dense condensation appears in the chromosphere at a column depth of ≈ 3 x 10^{19} cm^{-2}, i.e., relatively high. Yet, the density may approach 10^{14} cm^{-3} and the pressure may be between 100 and 1000 dynes cm^{-2}. These condensations appear in model chromospheres heated by thermal conduction or by electron beams. It is left as an exercise for the reader to distinguish between the Hα line wings produced in such a condensation and those produced in the deeper layers of a chromosphere whose ionization state is enhanced by electron beam collisions.

Heating of the residual chromosphere (i.e., the chromosphere not removed by evaporation) may be accomplished either by a penetrating electron beam or by an equally penetrating flux of X-rays in the 0.1 - 10 keV range. Henoux and Nakagawa (1978) showed that X-radiation from a 10^7 K plasma can enhance the number density and temperature of electrons by factors of ten and two, respectively, just above the temperature minimum. Later, Henoux and Rust (1980) showed that a substantial portion of the Hα flux in a large two-ribbon flare could have resulted from chromospheric heating by SXR photons from the coronal flare plasma. The effect of X-ray phton heating on Hα line profiles has not yet been considered in detail, but the above-mentioned research indicates that such a study would be worthwhile in an effort to understand broad Hα line wings.

Canfield and Gunkler (1985) and Ichimoto and Kurokawa (1984) have attempted to distinguish between conduction-driven and beam-driven evaporation on the basis of asymmetries in Hα profiles. Hα profile data may eventually help to distinguish between evaporation mechanisms, but the blue asymmetries that were observed and attributed to gentle evaporation by weak beams of electrons could have been produced by the magnetically-driven motions that are common in flaring active regions. Flares are almost always associated with emerging magnetic flux and with erupting filaments. Emerging flux regions appear at flare kernels with upward velocities of $\approx$ 10 km s^{-1}. Filaments erupt with velocities as high as 600 km s^{-1}, but velocities of 50 to 200 km s^{-1} are more common.

Red asymmetries, like blue asymmetries, can result from mass motions. The conclusion is that care must be exercised in analyzing Hα profiles without high-resolution movies, which reveal most of the magnetically-driven phenomena. For example, line asymmetries can also be produced in moving optically thick plasmas, in which case an acceleration upwards produces a stronger red, not blue, wing.

4.5.2.5 Conclusions

It is very difficult to make the case that most evaporation is produced directly by beams, particularly when the "evaporation" is understood to be explosive, as defined by Fisher, Canfield, and McClymont (1984). The evidence for sufficiently intense beams, even in the corona, rests on theoretical arguments that become less tenable with each new study on instabilities. If the beams do exist and if they reach the chromosphere, computer simulations show that they should produce upward velocities about three times higher than observed, if the coronal temperature is to reach observed values. On the other hand, if the beam strength is lowered to give upflows consistent with observations, then the coronal plasma is far too cool to produce the observed soft X-ray flare.

Observations of hard X-ray emission from the feet of flare loops and of simultaneous UV emission show that weak fluxes of energetic electrons do impact the chromosphere. The few quantitative studies carried out to date show, however, that only a small fraction of the observed chromospheric evaporation is affected by the beams.

Although it is possible to cast doubt on many lines of evidence for electron beams in the chromosphere, a balanced view that debaters on both sides of the question might agree to is that electron beams probably heat the low corona and upper chromosphere, but their direct impact on evaporating the chromosphere is energetically unimportant when compared to conduction. This represents a major departure from the thick-target flare models that were popular before the Workshop.

4.5.3 Recommendations for Further Research

Studies of the spatially integrated emission spectra of flares in HXRs should be deemphasized in favor of high spatial and spectral resolution HXR and UV observations of preflare and flare onset behavior. Quantitative studies of UV and optical emissions at flare kernels should be conducted with the highest time resolution possible (e.g., $\approx$ 1s), in order to separate the effects of electron beams and conduction.

The numerical simulations of chromospheric evaporation suggest that one possible manifestation of "explosive" evaporation by nonthermal electrons is the sudden appearance of high densities at coronal temperatures, and their subsequent relaxation to lower values on coronal loop hydrodynamic timescales. Some possible examples of this behavior have been found in P78-1 observations and were discussed in Section 4.5.2.3. Such observations should be carried out in future experiments with much higher time resolution (a few seconds) and with concurrent HXR observations. The numerical calculations also suggest that it might be possible to detect the threshold between gentle and explosive evaporation due to nonthermal electrons by looking for a change of sign in chromospheric velocities as the energy flux in nonthermal electrons changes from low to high values. These velocities may be detectable by looking at asymmetries in Hα profiles.

The authors would like to thank Ms. Marlene Wedding and Ms. Judy Hardison for typing the manuscript.

4.6 REFERENCES

Acton, L.W., et al., 1980, Solar Phys., **65**, 53.
Acton, L.W., Canfield, R.C., Gunkler, T.A., Hudson, H.S., Kiplinger, A.L., Leibacher, J.W. 1982, Ap. J., **263**, 409.
Antiochos, S.K., and Sturrock, P.A. 1978, Ap. J., **220**, 1137.
Antonucci, E., et al., 1982, Solar Phys., **78**, 107.
Antonucci, E. 1982, Mem. Soc. Astr. Ital., **53**, 495.
Antonucci, E., and Dennis, B.R. 1983, Solar Phys., **86**, 67.
Antonucci, E., Gabriel, A.H., Dennis, B.R. 1984, Ap. J., **287**, 917.
Antonucci, E., Gabriel, A.H., Dennis, B.R., and Simnett, G.M. 1984, submitted to Solar Phys.
Antonucci, E., Marocchi, D., and Simnett, G.M. 1984, Adv. Space Res., in press.
Brown, J.C. 1973, Solar Phys., **31**, 143.
Brown, J.C., Melrose, D.B., and Spicer, D.S. 1979, Ap. J., **228**, 592.
Brueckner, G.E. 1975, in Solar Gamma-, X-, and EUV Radiation (IAU Symposium No. 68), ed. S.R. Kane (Dordrecht: Reidel), p. 135.
Brueckner, G.E. 1976, Phil. Trans. R. Soc. London A, **281**, 443.
Canfield, R.C., and Gunkler, T.A. 1985, **288**, 353.
Canfield, R.C., Gunkler, T.A., Ricchiazzi, P.J. 1984, Ap. J., **282**, 296.
Cheng, C.-C. 1977, Solar Phys., **55**, 413.
Cheng, C.-C., Doschek, G.A., and Feldman, U. 1979, Ap. J., **227**, 1037.
Cheng, C.-C., Feldman, U., and Doschek, G.A. 1979, Ap. J., **233**, 736.
Cheng, C.-C., Karpen, J.T., and Doschek, G.A. 1984, Ap. J., **286**, 787.
Cheng, C.-C., Oran, E.S., Doschek, G.A., Boris, J.P., and Mariska, J.T. 1983, Ap. J., **265**, 1090.
Cheng, C.-C., and Tandberg-Hanssen, E. 1984, preprint.
Cohen, L., Feldman, U., and Doschek, G.A. 1978, Ap. J. Suppl., **37**, 393.
Dere, K.P., and Cook, J.W. 1979, Ap. J., **229**, 772.
Dere, K.P., and Cook, J.W. 1983, Astr. Ap., **124**, 181.
Dere, K.P., Mason, H.E., Widing, K.G., and Bhatia, A.K. 1979, Ap. J. (Suppl.), **40**, 341.
Deslattes, R. 1985, private communication.
Doschek, G.A. 1983, Solar Phys., **86**, 9.
Doschek, G.A. 1984, Ap. J., **283**, 404.

Doschek, G.A., Cheng, C.-C., Oran, E.S., Boris, J.P., and Mariska, J.T. 1983, Ap. J., **265**, 1103.
Doschek, G.A., Feldman, U., and Cowan, R.D. 1981, Ap. J., **245**, 315.
Doschek, G.A., Feldman, U., Dere, K.P., Sandlin, G.D., VanHoosier, M.E., Brueckner, G.E., Purcell, J.D., and Tousey, R. 1975, Ap. J. (Letters), **196**, L83.
Doschek, G.A., Feldman, U., Kreplin, R.W., Cohen, L. 1980, Ap. J., **239**, 725.
Doschek, G.A., Feldman, U., Landecker, P.B., McKenzie, D.L. 1981, Ap. J., **249**, 372.
Doschek, G.A., Feldman, U., and Rosenberg, F.D. 1977, Ap. J., **215**, 329.
Doschek, G.A., Kreplin, R.W., Feldman, U. 1979, Ap. J. (Letters), **233**, L157.
Doschek, G.A., Mariska, J.T., and Feldman, U. 1981, M.N.R.A.S., **195**, 107.
Duijveman, A., Hoyng, P., Machado, M.E. 1982, Solar Phys., **81**, 137.
Duijveman, A., Somov, B.V., Spektor, A.R. 1983, Solar Phys., **88**, 257.
Dwivedi, B.N., Hudson, H.S., Kane, S.R., and Svestka, Z. 1984, Solar Phys., **90**, 331.
Feldman, U., Cheng, C.-C., and Doschek, G.A. 1982, Ap. J., **255**, 320.
Feldman, U., and Doschek, G.A. 1977, Ap. J. (Letters), **216**, L119.
Feldman, U., Doschek, G.A., and Kreplin, R.W. 1982, Ap. J., **260**, 885.
Feldman, U., Doschek, G.A., Kreplin, R.W., Mariska, J.T. 1980, Ap. J., **241**, 1175.
Feldman, U., Doschek, G.A., and McKenzie, D.L. 1984, Ap. J. (Letters), **276**, L53.
Feldman, U., Doschek, G.A., and Rosenberg, F.D. 1977, Ap. J., **215**, 652.
Feldman, U., Liggett, M., and Zirin, H. 1983, Ap. J., **271**, 832.
Fisher, G.H., Canfield, R.C., McClymont, A.N. 1984, Ap. J. (Letters), **281**, L79.
Fisher, G.H., Canfield, R.C., McClymont, A.N. 1985a, Ap. J., **289**, 414.
Fisher, G.H., Canfield, R.C., McClymont, A.N. 1985b, Ap. J., **289**, 425.
Fisher, G.H., Canfield, R.C., McClymont, A.N. 1985c, Ap. J., **289**, 434.
Gabriel, A.H. 1972, M.N.R.A.S., **160**, 99.
Gabriel, A.H., and Jordan, C. 1975, M.N.R.A.S., **173**, 397.
Grineva, U.I., et al., 1973, Solar Phys., **29**, 441.
Gunkler, T.A., Canfield, R.C., Acton, L.W., Kiplinger, A.L. 1984, Ap. J., **285**, 835.
Henoux, J.-C., and Nakagawa, Y. 1978, Astr. Ap., **66**, 385.
Henoux, J.-C., and Rust, D.M. 1980, Astr. Ap., **91**, 322.
Heyvaerts, J., Priest, E.R., and Rust, D.M. 1977, Ap. J., **216**, 123.
Hiei, E., and Widing, K.G. 1979, Solar Phys., **61**, 407.
Hoyng, P., et al., 1981, Ap. J., **246**, L155.
Hudson, H. 1983, private communication.
Ichimoto K., and Kurokawa, K. 1984, Solar Phys., submitted.
Jordan, C. 1975, private communication.
Jordan, C. 1985, submitted to Solar Phys.
Kane, S.R., Fenimore, E.E., Klebesadel, R.W., and Laros, J.G. 1982, Ap. J. (Letters), **254**, L53.
Korneev, V.V., Zhitnik, I.A., Mandelstam, S.L., Urnov, A.M. 1980, Solar Phys., **68**, 391.
Kostyuk, N.D., and Pikel'ner, S.R. 1975, Sov. Astron., **18**, 590.

Kostyuk, N.D. 1976, Sov. Astron., **20**, 206.
Kreplin, R.W., Doschek, G.A., Feldman, U., Seely, J.F., and Sheeley, N.R., Jr. 1985, Ap. J., in press.
Lin, R.P., and Hudson, H.S. 1976, Solar Phys., **50**, 153.
Livshits, M.A., Badalyan, O.G., Kosovichev, A.G., Katsova, M.M. 1981, Solar Phys., **73**, 269.
MacNeice, P., McWhirter, R.W.P., Spicer, D.S., and Burgess, A. 1984, Solar Phys., **90**, 357.
Martin, S.F., and Ramsey, H.E. 1972, in Solar Activity, Observations and Predictions, ed. P.S. McIntosh and M. Dryer (Cambridge: MIT Press), p. 371.
Mason, H.E., Shine, R.A., Gurman, J.B., and Harrison, R.A., 1985, preprint.
McClymont, A.N., Canfield, R.C., Fisher, G.H. 1985, in preparation.
Moore, R.L., et al., 1980, in Solar Flares, ed. P.A. Sturrock (Boulder: Colorado Associated University Press), p. 341.
Moore, R.L., Hurford, G.J., Jones, H.P., and Kane, S.R. 1984, Ap. J., **276**, 379.
Moore, R.L., and Kahler, S. 1984, in preparation.
Nagai, F. 1980, Solar Phys., **68**, 351.
Nagai, F., and Emslie, G. 1983, preprint.
Pallavicini, R., Serio, S., and Vaiana, G.S. 1977, Ap. J., **216**, 108.
Pallavicini, R., Peres, G., Serio, S., Vaiana, G., Acton, L., Leibacher, J., and Rosner, R. 1983, Ap. J., **270**, 270.
Ricchiazzi, P.J., and Canfield, R.C. 1983, Ap. J., **272**, 739.
Rust, D.M., and Hegwer, F. 1975, Solar Phys., **40**, 141.
Rust, D.M., Simnett, G.F., and Smith, D.F. 1985, Ap. J., **288**, 401.
Rust, D.M., and Webb, D.G. 1977, Solar Phys., **54**, 403.
Smith, D.F., and Lilliequist, C.G. 1979, Ap. J., **232**, 582.
Somov, B.V., Syrovatskii, S.I., Spektor, A.R. 1981, Solar Phys., **73**, 145.
Somov, B.V., Sermulina, B.J., Spektor, A.R. 1982, Solar Phys., **81**, 281.
Svestka, Z. 1976, Solar Flares, D. Reidel Publ. Co., Dordrecht, Holland.
Tanaka, K., Watanabe, T., Nishi, K., Akita, K. 1982a, Ap. J. (Letters), **254**, L59.
Tanaka, K., Akita, K., Watanabe, T., Nishi, K. 1982b, Hinotori Symposium on Solar Flares, Tokyo, 43.
Tanaka, K., Ohki, K., and Zirin, H. 1985, preprint.
Tanaka, K. 1984, presentation at the 3rd SMM Workshop.
Tang, F., and Moore, R.L. 1982, Solar Phys., **77**, 263.
Van Beek, H.F., Hoyng, P., Lafleur, B., Simnett, G.M. 1980, Solar Phys., **65**, 39.
Widing, K.G. 1975, in Solar Gamma-, X-, and EUV Radiation, page 153, proceedings of IAU Symposium No. 68, edited by S.R. Kane (D. Reidel Publ. Co., Dordrecht, Holland).
Widing, K.G., and Dere, K.P. 1977, Solar Phys., **55**, 431.
Widing, K.G., and Hiei, E. 1984, Ap. J., **281**, 426.
Woodgate, B.E., Shine, R.A., Poland, A.I., Orwig, L.E. 1983, Ap. J., **265**, 530.
Zirin, H. 1978, Solar Phys., **58**, 95.

CHAPTER 5: FLARE ENERGETICS

CHAPTER 5: FLARE ENERGETICS

TABLE OF CONTENTS

S. T. Wu, C. de Jager, B. R. Dennis, H. S. Hudson, G. M. Simnett, K. T. Strong, R. D. Bentley, P.L. Bornmann, M. E. Bruner, P. J. Cargill, C. J. Crannell, J. G. Doyle, C. L. Hyder, R. A. Kopp, J. R. Lemen, S. F. Martin, R. Pallavicini, G. Peres, S. Serio, J. Sylwester, and N. J. Veck

Page

TABLE OF CONTENTS (continued)

TABLE OF CONTENTS (continued)

CHAPTER 5

CHAPTER 5: FLARE ENERGETICS

S.T. Wu[1], C. de Jager[2], B. R. Dennis[3], H. S. Hudson[4], G. M. Simnett[5], K. T. Strong[6], R. D. Bentley[7], P. L. Bornmann[8], M. E. Bruner[6], P. J. Cargill[9], C. J. Crannell[3], J. G. Doyle[10], C. L. Hyder[11], R. A. Kopp[12], J. R. Lemen[7], S. F. Martin[13], R. Pallavicini[14], G. Peres[15], S. Serio[15], J. Sylwester[16], and N. J. Veck[17]

[1]University of Alabama, Huntsville, Alabama
[2]Space Research Laboratory, Utrecht, The Netherlands
[3]NASA Goddard Space Flight Center, Greenbelt, Maryland
[4]University of California at San Diego, California
[5]University of Birmingham, England
[6]Lockheed Palo Alto Research Laboratory, Palo Alto, California
[7]Mullard Space Science Laboratory, England
[8]Joint Institute for Laboratory Astrophysics, Boulder, Colorado
[9]National Center for Atmospheric Research, Boulder, Colorado
[10]Armagh Observatory, Northern Ireland
[11]High Altitude Observatory, Boulder, Colorado
[12]Los Alamos Scientific Laboratory, Los Alamos, New Mexico
[13]California Institute of Technology, Pasadena, California
[14]Observatorio Astrofisico di Arcetri, Firenze, Italy
[15]Observatorio Astronomico di Palermo, Palermo, Italy
[16]Instytut Astronomiczny, Wroclaw, Poland
[17]Marconi Research Center, Chelmsford, England

ABSTRACT

In this investigation of flare energetics, we have sought to establish a comprehensive and self-consistent picture of the sources and transport of energy within a flare. To achieve this goal, we chose five flares in 1980 that were well observed with instruments on the Solar Maximum Mission, and with other space-borne and ground-based instruments. The events were chosen to represent various types of flares. Details of the observations available for them and the corresponding physical parameters derived from these data are presented. The flares were studied from two perspectives, the impulsive and gradual phases, and then the results were compared to obtain the overall picture of the energetics of these flares. We also discuss the role that modeling can play in estimating the total energy of a flare when the observationally determined parameters are used as the input to a numerical model. Finally, a critique of our current understanding of flare energetics and the methods used to determine various energetics terms is outlined, and possible future directions of research in this area are suggested.

5.1 INTRODUCTION

S.T. Wu and C.J. Crannell

5.1.1 Objectives of Study

Understanding the flow of energy in a solar flare represents a good test of our knowledge of flare physics. Accordingly, this team has sought to identify the major sources and sinks of flare energy, together with the mechanisms by which the energy flows from one form to another. By doing this quantitatively we have sought to establish a comprehensive and self-consistent picture of the sources and transport of energy within a flare. To undertake this study, we use theoretical modeling and observations obtained with the Solar Maximum Mission (SMM), other satellites, and groundbased facilities. The perspective from which these objectives were pursued and the scope of the flare observations available for this effort are described in the following subsections.

5.1.2 Perspective-Skylab to SMM

Skylab Solar Workshop II (Sturrock, 1980) focused on observations of solar flares obtained with the first full-scale, manned, astronomical observatory in space. This observatory carried instruments covering the wavelength range from 2 to 7000A which encompasses the soft X-ray, UV, and visible light portions of the electromagnetic spectrum. Photographs of active regions and flares in progress provided unprecedented graphic documentation of the role of magnetic loops and multi-thermal plasmas in the storage and release of flare energy. The data from Skylab added to the results obtained during the preceding solar maximum from Orbiting Solar Observatories and other unmanned satellites, which had charted the dominant role of high-energy radiations in defining solar flare physics. SMM was designed to make coincident observations of as many aspects of solar flares as possible. The emphasis with SMM instrumentation was on comprehensive coverage, and despite unfortunate gaps, particularly in measurements of coronal lines, the observations available for this workshop reflect that emphasis. The observations obtained with SMM benefit, as well, from improvements in sensitivity and temporal resolution, particularly for the highest energy electromagnetic radiation. A major advance achieved with SMM is the capability of imaging X-rays with energies up to 30 keV. The availability of broad observational coverage guided the approach adopted by this team, which attempted to establish a global view of flare energetics. The key questions addressed from this perspective are presented in the following subsection.

5.1.3 Key Questions

A wide variety of electromagnetic and particle radiations are observed in solar flares. Although there are many questions about the mechanisms whereby these radiations are produced, there is a general consensus that conversion of magnetic energy provides the fundamental energy source. The location of the energy source itself is not known. To investigate the associated energy sources and their transport processes, we have classified the flare phases by their time structures as being either impulsive or gradual. The impulsive phase is usually characterized by hard X-ray and microwave spikes with timescales in the range from fractions of a second to a few tens of seconds. The gradual phase, on the other hand, has time-scales on the order of minutes or even hours. The work reported in this chapter addresses the following key questions about these two phases:

- How do we characterize the impulsive phase? Do all flares have an impulsive phase?
- What is the total energy content of the flare in the impulsive phase?
- What is the relative importance of the thermal and the non-thermal components of the impulsive phase of the flare?
- How does the energy in the sources associated with the gradual phase emissions compare with that attributable to impulsive phase energy sources?
- What are the dominant cooling mechanisms at different stages of the gradual phase?
- Do all the post-flare loops need continual energy input?
- Are there extended, late, flare-associated sources in the corona?

5.1.4 Approach Adopted for Study

The distribution of the total flare energy among various reservoirs and active phenomena is illustrated in Figure 5.1.1. In this diagram, we sketch, as a function of time throughout the pre-flare, impulsive, and gradual phases, the integral energy content in the following forms:

- radiant energy
- thermal plasma energy
- gravitational potential energy
- kinetic energy of accelerated particles
- kinetic energy in plasma flows and random motions.

We used this schematic representation to compare the magnitudes of different manifestations of flare energy. The underlying assumption is that the energy flows from the magnetic reservoir, which is continuously being fed by large-scale flows, so that Figure 5.1.1 represents a sudden decrease in the magnetic free energy. Figure 5.1.1 is highly nonlinear since it represents cumulative contributions of terms with very different magnitudes. It is mainly helpful for qualitatively visualizing the processes involved in the energetics analysis.

Similarly, Figure 5.1.2 has been used to illustrate the flow and transport of flare energy. In this scheme, we again show the five energy sources or sinks shown in Figure 5.1.1 but also include the following energetically significant interconnecting terms:

- direct heating
- conductive heating

- o radiative heating
- o heating by particles
- o radiative loss
- o evaporation
- o anti-evaporation ("coronal rain")
- o shock and turbulent heating
- o mechanical energy losses
- o shock and turbulent acceleration of particles
- o direct acceleration of particles
- o direct mass acceleration

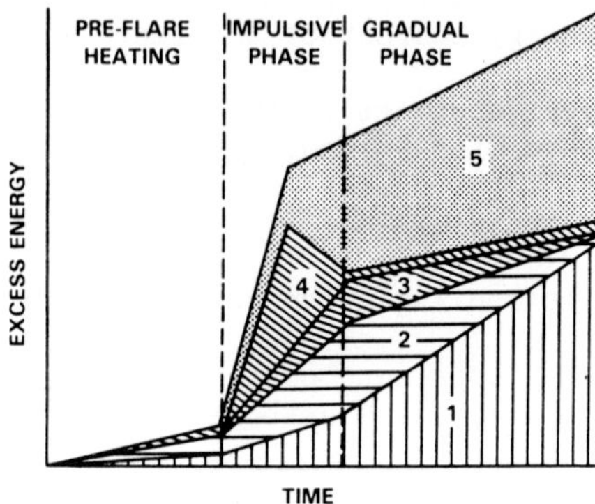

Figure 5.1.1 This diagram shows conceptually how the energy of a flare is partitioned into its various forms as a function of time. Plotted in the graph are the excess energies above the active region thermal energy, which is assumed to remain constant throughout the flare. The total energy of the system, including the magnetic free energy, is very large compared to the flare energy. During pre-flare heating, there is an increase in the radiant energy (1), the thermal energy of the plasma (2), and the gravitational potential energy (3) as material is carried into the coronal loops. During the impulsive phase, these quantities rise more rapidly with a significant increase in the energy in non-thermal particles (4) and mechanical energy (5). During the gradual phase, there may be some continued heating by non-thermal particles but eventually all the energy produced by the flare is left in the form of energy lost by radiation (1) or in the coronal transient (5).

Any of these terms could be broken down into finer detail: for example, the heating due to radiation represents, in general terms, the entire field of radiative transfer.

The energy content of various flare components can be estimated from the observable emissions associated with the interactions of these constituents with each other and with the ambient solar medium. In most cases, such estimates are model dependent, with the attendant disadvantage that the energy determinations are not unique. On the other hand, any systematic differences between the results obtained for the whole flare energetics with different models provide a means of discriminating among them.

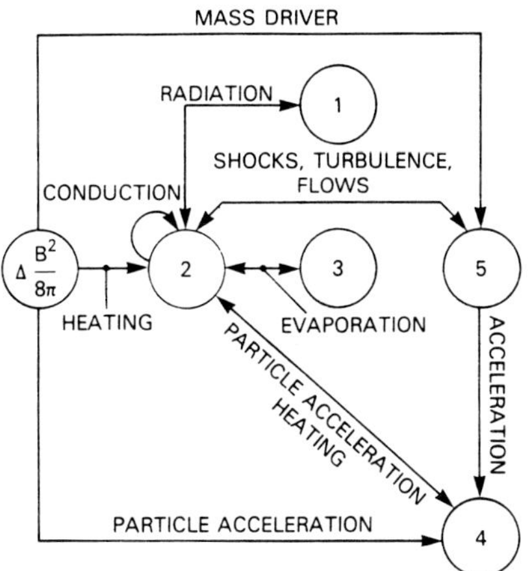

Figure 5.1.2 This diagram illustrates how energy is transferred between the various sinks or reservoirs (circles). The energy sinks are (1) radiation, (2) thermal plasma, (3) gravitational potential energy, (4) non-thermal particles, and (5) mechanical effects. The interconnecting lines indicate the possible mechanisms and directions of energy transfer between the sinks.

5.1.5 Data Characterization

The question of data presentation is an inevitable compromise between completeness and compactness. The approach that we follow in this chapter is that of presenting the data in parametric form, in which a theoretical relation involving a relatively small number of variables is used to represent the observations. Although the names of the parameters usually correspond to physically meaningful quantities such as temperature, density, or volume, the parameters themselves may not be accurate measures of these quantities. A temperature, for example, may accurately represent the slope of an observed spectrum over a limited energy range, even though the energy distribution of the source particles is non-Maxwellian.

For line profiles from the Ultraviolet Spectrometer and Polarimeter (UVSP) and for the Bent Crystal Spectrometer (BCS) and P78-1 emission line observations, we have used a Gaussian parameterization, stating the amplitude and central wavelength as functions of time throughout the period of the observations.

The Hard X-Ray Burst Spectrometer (HXRBS) and the Hard X-ray Imaging Spectrometer (HXIS), which are based on scintillation counters and proportional counters, respectively, are inherently broad-band instruments with spectral resolutions of a few points per decade of energy. Some HXRBS spectra are well represented by a thermal bremsstrahlung model, others by a power law, and some by both parametric forms. In either case, the practice followed has been to perform a convolution of the X-ray spectrum computed for an assumed electron energy distribution with the instrument response function and to adjust the parameters of the assumed distribution for best fit of the computed X-ray spectrum to the observations.

Ratios among the HXIS bands are a measure of the average slope of the spectrum, which, in turn, can be represented by a temperature parameter. Some measure of the consistency of this description can be found in the comparison of temperature parameters derived from different ratios.

The Flat Crystal Spectrometer (FCS) and BCS measure line spectra for which the thermal or power-law representations are inappropriate. Most FCS observations are made in a raster pattern in the seven home-position lines. These data are described for purposes of this chapter by a differential emission measure distribution computed by following the methods of Sylwester et al. (1980). This representation approximately reproduces the observed intensities and derives some compactness from the sum over the raster, but must still be presented as a curve for each observation. Further details are given in Section 5.3.

We alert our readers to the use of parameterization throughout the remainder of this chapter and urge them to draw careful distinctions between parameters used to characterize the data and the interpretation of the data in terms of physical quantities. The term "temperature" should be understood to mean temperature parameter when we are referring to the results of a thermal fit to an observed spectrum. In the power law representation of the spectrum, the parameters are the hardness index and the photon flux at a given energy. Electron densities derived from line ratios in the UVSP or P78-1 observations are referred to as "derived electron densities". Similarly, temperatures derived from either BCS or P78-1 line ratios should be understood to be "derived temperatures", even when this is not explicitly stated.

5.1.6 Observations and Their Interpretation

Details of the observations made with the different SMM instruments are given in the discussion of each of the flares studied. Here we summarize the types of observations that are available and their interpretations.

Observations of gamma-ray emission obtained with the Gamma Ray Spectrometer (GRS) were used to estimate the energy content of the energetic protons and ions (> 10 MeV) participating in selected flares. The observed fluxes and upper limits were used with the model of Ramaty (1982) to obtain scaled energy estimates and upper limits. Uncertainties in these values are dominated by the uncertainty in the choice of the appropriate spectral form. The observations constrain the numbers of protons and ions only for energies above 10 MeV.

Hard X-ray observations obtained with HXRBS and microwave observations obtained with the radio telescopes at Bern, Toyokawa, Nobeyama, and Nagoya, and with the Radio Solar Telescope Network (RSTN) were used to estimate the flare energy in the form of energetic electrons. Two models (a non-thermal, thick-target model and a thermal model) were used to obtain two different estimates of the electron energy associated with selected flares. In the non-thermal model, it is assumed that the hard X-rays are electron-ion bremsstrahlung produced as an electron beam arrives at the footpoints of a magnetic loop. Standard thick-target equations were used (Brown 1971, Hoyng et al., 1976, Lin and Hudson 1976) with a power-law spectral fit and a cut-off of 25 keV.

In the thermal model, it is assumed that the hard X-ray emission and the optically-thick microwave emission originate in a common hot plasma with the temperature, density, and source area consistent with the coincident observations. The effective temperature and emission measure are determined from a spectral fit to the hard X-ray observations, and the source area is determined from the microwave observations.

The energy components of the thermal plasma were estimated from combined data sets obtained with HXIS, BCS, and FCS. The HXIS images with 8 arcsec (FWHM) resolution were used to provide estimates of the flare area from which flare volumes were derived by assuming simple geometric models. The FCS also provides imaging but at a relatively lower resolution (14 arcsec FWHM). Thus, the FCS images were used mainly in a qualitative fashion to compare with the HXIS images, especially at lower temperatures (10^6 to 5×10^6 K).

All three instruments are capable of providing isothermal model estimates of the temperature and emission measure from which it is possible to derive the thermal energy. However, the preferred technique for deriving estimates of the thermal energy was to combine the observed fluxes from HXIS and BCS (and from FCS in at least one flare) in a multi-thermal model for the emission measure. The thermal energy of plasmas at temperatures above 10^7 K (2×10^6 K when FCS data were included) could then be obtained by using the volume derived from the HXIS measurement of the flare area and assuming a filling factor of unity.

The BCS is also capable of measuring two non-thermal energy components: the energy of plasma flows and the energy in the random motions of the turbulent plasma. During the rising portion of the soft X-ray flare, blue-ward broadening of the soft X-ray lines has been observed with BCS and interpreted as Doppler-shifted radiation from chromospherically evaporated plasma. The measured velocities of the upflowing plasma are typically 150 to 250 km s^{-1}, from which the kinetic energy was estimated. The energy in the random motion of the turbulent plasma is estimated from the symmetric line broadening of the soft X-ray spectra in excess of that expected from a thermal source with equal ionic and electron temperatures.

Information derivable from UVSP observations depends on the observing mode that was used. Most observations are relevant to the transition zone, although some chromospheric and coronal lines were also observed. Some modes are velocity sensitive, whereas others yield density-sensitive line ratios.

5.1.7 Flares Chosen for Study

We have chosen five flares from the SMM data set which we felt to be representative of different types of solar flares. Appendix 5A summarizes the properties of these flares, which were as follows:

- o a flare of moderate strength (GOES class M) from an extended active region producing frequent flares (1980 April 8 at 03:04 UT);
- o a large, two-ribbon flare which produced interplanetary protons at energies of > 40 MeV (1980 May 21 at 20:53 UT);
- o a moderate limb flare accompanied by a coronal transient (1980 June 29 at 18:03 and 18:22 UT);
- o a compact double flare (1980 August 31 at 12:48 and 12:52 UT);
- o a major double flare (1980 November 5 at 22:26 and 22:33 UT).

These flares will be familiar to readers of other chapters of this volume. We included double flares in an effort to study the effect of various initial conditions. All of the chosen flares had full SMM instrument coverage with the exception of that on June 29, for which the main flare site was not observed with UVSP; in addition there are no Coronograph/ Polarimeter (C/P) observations except for the June 29 event. Full details and bibliographies of published material on these flares appear in Appendix 5A; we summarize their properties in Table 5.1.1.

Table 5.1.1 Flares Chosen for Detailed Study

Date in 1980	Onset time (UT)	Location	Optical class	GOES class	Peak rate >28 keV (counts s^{-1})	Peak microwave flux at 8.8 GHz (s.f.u.)
Apr 8	03:04	N12 W13	1B	M4	3800	200
May 21	20:53	S14 W15	2B	X1	14300	1200
Jun 29	18:03	S29 W90	limb	M1	441	
	18:22	S29 W90	limb	M4	3177	74
Aug 31	12:48	N12 E28	−B	C6	7360	556
	12:52	N12 E28	−B	M2	3673	
Nov 5	22:26	N11 E07	1B	M1	3151	
	22:33	N11 E07	1B	M4	12730	3100

These all appear to be type B flares according to the classification scheme proposed originally by Tanaka (1983) and developed by Ohki et al. (1983) and Tsuneta (1983). None of the events we selected for detailed analysis is a type A or a type C flare. Type A flares are low-altitude, compact (< 10 arc sec) flares with very steep hard X-ray spectra and have been interpreted as purely thermal flares with temperatures of 3×10^7 to 5×10^7 K (Tsuneta et al., 1984a). Type C flares have virturally no impulsive phase, and the hard X-rays and microwaves appear to come, in at least one case, from altitudes of 5×10^4 km, near the top of a huge coronal loop (Tsuneta et al., 1984b). Cliver et al. (1984) have suggested that many flares that show similar characteristics to type C flares may be preceded some tens of minutes earlier by an apparently normal impulsive flare, presumably of type B. This classification scheme is described in more detail in Chapter 3 of this Workshop Report.

The concentration on type B flares was dictated by SMM observations, since the best observed flares were of this type. A further limitation of this work is the absence of a solar flare showing gamma-ray line emission. This again reflects the SMM data set in which there are no gamma-ray flares observed with the imaging instruments.

The growth of active-region magnetic energy has been estmated by using an MHD model given by Wu et al. (1983). This model is used to estimate the energy build-up during the pre-flare state on the basis of photospheric shearing motions. Wu et al. found that, in this model, the majority of the energy storage is magnetic; other forms (kinetic, potential, or thermal) are negligibly small in the pre-flare state. The physical basis of this MHD model is that observed shear motion at the photosphere is the result of motion of the magnetic loop footpoints. This movement causes the magnetic field to twist and hence magnetic energy to be stored. Because the movement is so slow (~ 100 m s^{-1}), the quasi-equilibrium state is always maintained so that the other energy modes do not increase as significantly as the magnetic energy.

5.1.8 Synopsis

In this investigation of flare energetics, observations characterizing the impulsive and gradual phases first were analyzed separately and then the results of the individual analyses were intercompared. Observational and analytic results related to the energetics of the impulsive phases of the events chosen for this study are presented in the following section. In Section 5.3, results obtained for the energetics of the gradual phases are presented. Investigation of the energetics relationships between the impulsive and gradual phases is described and their interpretation in terms of flare models is discussed in Section 5.4. In Section 5.5, an attempt is made to present a synthesis, based on the foregoing results, characterizing the total flare energy. Conclusions based on this work and observational requirements for addressing the remaining questions with instruments to be employed in the future are presented in Section 5.6. Appendix 5A describes the published results and observations of the five flares chosen for detailed study in this chapter. Appendix 5B contains a review of the observational characteristics of the impulsive phase of flares in general.

5.2 ENERGETICS OF THE IMPULSIVE PHASE

C. de Jager, M.E. Bruner, C.J. Crannell, B.R. Dennis, J.R. Lemen, and S.F. Martin

5.2.1 Introduction

In this section we concentrate on the impulsive phase, with particular emphasis on the energy content of the different components of the prime flares chosen for study (see Appendix 5A). From an energetics point of view, we are concerned primarily with high-energy electrons and protons and with plasma at temperatures above 10^4 K. The different forms of energy include the kinetic energy of fast particles, and the thermal, convective, and turbulent energies of hot plasma. Our approach is in contrast to that used by Canfield et al. (1980), who presented estimates of the total radiant energy in different wavelength ranges. We have also measured the radiant energy in soft and hard X-rays, and in Hα for one flare, and the results are discussed in Section 5.4. We find that the energy radiated during the impulsive phase is negligible compared with the other components of the total flare energy. A more detailed discussion is given of the radiant energy as determined from the new Hα observations because they give a better estimate of the Hα radiant energy from one flare than that obtained previously (see Section 5.2.3.5).

To estimate the energies in the different flare components, it is necessary to make various assumptions about the physical conditions in the flares. Unfortunately, these cannot be determined umambiguously from the available observations. In particular, the thermal or nonthermal nature of the hard X-ray source is still unresolved, in spite of considerable evidence favoring the existence of beams of fast electrons. The spectrum of electrons below 20 keV, where most of the total energy may reside, is subject to large uncertainties. Also, only upper limits can be placed on the energy in protons above a few MeV from the gamma ray observations of the flares considered here. There is no information at all on the possible existence of protons at 10 keV to 1 MeV, which have been postulated to be an important component during the impulsive phase (Simnett 1984).

For the thermal plasma emitting soft X-rays, the filling factor and density distribution are also largely unknown. In addition, the available observations for the prime flares studied here do not provide information on the energy of plasma with temperatures below 10^6 K.

As a result of these gaps in our knowledge, the energy of some potentially important components of flares cannot be estimated at all, and order-of-magnitude uncertainties exist in many of the estimates that can be made. Nevertheless, it is important in furthering our understanding of the flare processes to make these estimates as accurately as possible. Not only do they provide valuable limits on the total flare energy, but also, and possibly more importantly, they suggest how the energy is distributed among the components as a function of time during the various stages of the flare. The ability to determine these temporal variations in the distribution of the flare energy is improved markedly with the observations available for this study. Time resolutions as short as 0.1 s for the hard X-ray spectral observations and a few seconds for the X-ray imaging and soft X-ray spectral observations allow the flare energies to be determined on timescales an order of magnitude shorter than previously possible.

We begin Section 5.2.2 with a working definition of the impulsive phase and a brief review of the observations from which we obtained our energy estimates. We have concentrated on, but have not restricted ourselves to, the flares described in Appendix 5A that were selected by this team for detailed analysis. Estimates of the energies involved in energetic electrons, protons, and ions, chromospheric evaporation, and the thermal plasma at temperatures above 10^7 K are given in Section 5.2.3. A more detailed description of the analysis done for the 1980 April 8 flare is given in Section 5.2.4. Finally, in Section 5.2.5 the main results are summarized and the relation between soft and hard X-ray sources is discussed.

5.2.2 The Characteristics of the Impulsive Phase

A review of the observational characteristics of the impulsive phase of flares is given in Appendix 5B, and a more extensive discussion is contained in Chapter 2 of this report. The most important characteristic of this phase is the impulsive release of energy on timescales of less than or about 10 s. This is evidenced by the plots of the intensity of the hard X-ray flux versus time that are shown in Figures 5A.2, 4, 7, 10, and 12. The hard X-ray spectral observations allow us to estimate the energy in fast electrons as discussed in Section 5.2.3.1. The soft X-ray flux shown as a function of time in the same figures rises continuously during the time that the hard X-rays are observed. Measurement of the line and continuum emissions allows us to estimate the energy of the hot plasma during this period (Sections 5.2.3.3 and 5.2.3.4).

5.2.3 The Determination of Component Energies

In this section we describe the methods that we used to determine the energies of several of the energetically most important flare components during the impulsive phase. We considered the following flare components: (a) impulsive energetic electrons that produce hard X-rays, (b) energetic protons and ions that produce gamma rays, and (c) thermal plasma that produces soft X-rays.

Two different estimates of the energy in the fast electrons were obtained from the hard X-ray observations; the first was obtained under the assumption that the X-rays were produced in non-thermal, thick-target interactions, and the second was obtained under the assumption that the hard X-rays and microwaves were produced in the same thermal source at a temperature in excess of 10^8 K. The energy in the soft X-ray-emitting plasma includes the kinetic energies of the convective and turbulent plasma motions in addition to the thermal energy. The radiative and conductive energy losses from this thermal plasma were also considered but were generally negligible during the impulsive phase, as was the gravitational potential energy of the upflowing plasma.

5.2.3.1 The Energy in Impulsive Energetic Electrons

It is almost universally accepted that solar flare hard X-rays are electron-ion bremsstrahlung. Since the bremsstrahlung cross-section as a function of energy is well known, it is possible to determine the instantaneous energy spectrum of the emitting electrons from the measured X-ray spectrum. Unfortunately, the interpretation of this result depends on the temperature and density of the plasma in which the electrons are interacting. Three basic models have been proposed for the production of the hard X-rays: models based on thick- and thin-target interactions and thermal model (Brown 91971, Lin and Hundson 1976; Tucker 1975; Crannel et al., 1978). It is necessary to know, in any particular flare, which model (or combination of models) is correct, since the determination of the electron spectrum from the observed X-ray spectrum depends on the model assumed. More importantly, the correct model must be known to determine the role of the fast electrons in the overall flare energetics and ultimately to determine the fundamental energy release mechanism or mechanisms of the flare.

Recent observations have shown that the thick-target model is most likely to be the correct one, at least during the early part of the impulsive phase of some flares and for some fraction of the total hard X-ray flux. The arguments supporting this statement are presented in Appendix 5B. However, the subject is still controversial, and there is a strong possibility that a large fraction of the hard X-ray flux comes from a thermal source. This controversy is touched on in Appendix 5B, and a more detailed discussion can be found in Chapter 2 of this report.

From a spectral point of view, the hard X-ray observations used to make the energy estimates are consistent with either a thick-target or a thermal model. The X-ray spectrum can often be represented within the uncertainties of the data by either a power law or the exponential function expected for an isothermal source. Sometimes the thermal bremsstrahlung function fits the observations better on the rise and at the peak of impulsive bursts, whereas a power law fits better on the decline and in the valley between adjacent bursts (Crannell et al., 1978; Dennis, et al., 1981; Kiplinger et al., 1983a). However, a non-thermal source can produce an exponential spectrum, and a multi-temperature thermal source can generate a power-law spectrum (Brown 1974). In estimating the energy in fast electrons, therefore, we have chosen to make the calculations both for a thick-target model by assuming a power-law spectrum and for a single-temperature thermal model. The true model is probably some hybrid of the two, but we are unable to determine the proportions of the thermal and non-thermal components from the available observations.

(i) Non-thermal Model. The X-ray spectra of the impulsive bursts have been determined from HXIS and HXRBS observations by using techniques described by Batchelor (1984). By assuming thick-target interactions, we can determine the energy and the number of energetic electrons involved in the bombardment of the lower corona and the chromosphere from these X-ray spectra.

For calculating the energy in fast electrons, one usually assumes that the incident X-ray spectrum is a power law of the form

$$I_x(\varepsilon) = \alpha\ \varepsilon^{-\gamma} \text{ photons cm}^{-2}\text{ s}^{-1}\text{ keV}^{-1}, \qquad (5.2.1)$$

where $I_x(\varepsilon)$ is the differential flux of photons with energy ε in keV, and α and γ are parameters determined from least-squares fits to the spectral data. The standard thick-target-model equations are used to relate the measured X-ray spectrum to the spectrum of the electrons producing the X-rays (Brown 1971; Hoyng et al., 1976; Lin and Hudson 1976). The electron spectrum that results in a power-law X-ray spectrum is also a power law, of the form

$$I_e(E) = \beta\ E^{-\delta} \text{ electrons s}^{-1}\text{ keV}^{-1}, \qquad (5.2.2)$$

where $I_e(E)$ is the differential flux of electrons with energy E that enter the thick target. The parameters β and δ are related to α and γ by the following equations:

$$\beta = 3 \times 10^{33}\ a\ q(\gamma - 1)^2\ b(\gamma - 1/2,\ 1/2), \qquad (5.2.3)$$

$$\delta = \gamma + 1, \qquad (5.2.4)$$

where $b(\gamma - 1/2, 1/2)$ is the beta function.

The total energy in electrons that enter a thick target during a flare is obtained from the equation

$$W(> E_o) = \int_{E_o}^{\infty} \int \beta\ E^{-\delta+1}\ dE\ dt =$$

$$\int \beta E_o^{-\delta+2}\ (\delta - 2)^{-1}\ dt \qquad (5.2.5)$$

In this expression, the lower energy cut-off, E_o, must be imposed on the electron spectrum to ensure that W remains finite, but it is very difficult to determine the correct value of E_o from the observations. This difficulty introduces the largest uncertainty in the estimate of W. A cut-off in the electron spectrum at E_o would appear as a flattening of the X-ray spectrum at X-ray energies below E_o. The X-ray spectrum cannot get significantly flatter than ε^{-2}, however, as a result of the Bethe-Heitler formula for the bremsstrahlung cross section, regardless of the shape of the electron spectrum. Such a flattening at low energies is sometimes observed in the X-ray spectrum at the time of impulsive peaks in the flux, although it can be masked by the soft X-ray flux from the ~ 20×10^6 K plasma. As a suitable approximation, E_o = 25 keV has been taken for the calculation of the values of W given in Tables 5.2.1 and 5.2.2.

(ii) Thermal Model. An alternative model for the production of the hard X-rays is one in which the source is thermal and the electron spectrum can be represented with a single effective temperature. The X-ray continuum spectrum is then of the form expected from an electron population with a Maxwellian velocity distribution, i.e.,

$$I_x(\varepsilon) = 1.08 \times 10^{-24}\ Y\ \varepsilon^{-1}\ T^{-0.5}\ G(\varepsilon, T)\ \exp(-\varepsilon/kT)$$

$$\text{photons cm}^{-2}\ \text{s}^{-1}\ \text{keV}^{-1}, \qquad (5.2.6)$$

where Y is the thermal emission measure (in cm^{-3}), T is the effective temperature of the source (in degrees K), $G(\varepsilon, T)$ is the "total effective

Table 5.2.1 Flare Parameters Derived from HXIS and HXRBS Data

Date 1980	SXR Peak Time (UT)	Temperature (10^6K)	Emission Measure (10^{48} cm^{-3})	Volume (10^{26} cm^3)	Density (10^{10} cm^{-3})	Thermal Energy, U (10^{29} ergs)	Energy in Electrons W(>25 keV) (10^{29} ergs)
Apr 8	03:07	25.2	13.8	5.9	15	9.4	94
			5.2	62	2.9	19	
Apr 10	09:21:30	23.7	20.3	7.6	16	12	49
Apr 13	04:08	25.5	1.5	1.3	11	1.5	7
			2.3	36	2.5	9.6	
Apr 30	20:25	20.5	6.3	2.9	15	3.6	7
			1.9	28	2.6	6.2	
May 9	07:13:30	23.7	41.3	1.0	64	6.3	23
May 21	21:05	21.3	76	25	17	38	96/110
			27	40	8.2	29	
Jun 13	22:34:20	21.4	5.5	Edge of fine FOV			5
Jun 25	15:54			No HXIS Data			54
Jun 29	02:38	18.5	17	HXIS coarse FOV			49
Jun 29	10:43:15	18.5	28	HXIS coarse FOV			34
Jun 29	18:25:40	20.2	28	HXIS coarse FOV			13-93
Jul 1	16:28:50			Out of HXIS FOV			80
Jul 5	22:44:40	26.7	42	1.0-2.9	65-38	7-12	48
Jul 7	11:52:00	24.4	3.4	2.1	1.3	2.7	6
Jul 12	11:19:15			HXIS coarse FOV			
Jul 12	11:23:20			HXIS coarse FOV			142
Jul 14	08:27:00	18.4	76.5	4.3	42	14	12
Jul 21	03:00:40			No HXIS data			225
Aug 23	21:30	20.6	4.4	2.3	14	2.7	9
			2.8	14	4.5	5.3	
Aug 24	16:12:30	19.2	4.5	Edge of HXIS fine FOV			12
Aug 25	13:05:30	21.4	1.9	Edge of HXIS fine FOV			8
Aug 31	12:49:30	17.5	3.1	3.3	9.7	2.3	3.3
Aug 31	12:52:30	20.5	11.2	0.76	38	2.5	5
Sep 24	07:34:30	21.8	5.5	7	8.8	5.6	<7.5
Nov 5	22:28:17	20	3.1	3.3	9.7	2.6	1.6
Nov 5	22:35:30	25.9	14.2	14	10	15	28
			2.0	34	2.4	8.8	
Nov 6	17:27:30	29.1	2.4	0.32	28	1.1	Too steep
			4.2	62	2.6	19	
Nov 7	04:58:40	22.4	6.8	Edge of HXIS fine FOV			45
Nov 10	08:12	20.2	1.6	1.7	9.7	1.4	15
			3.7	11	5.8	5.3	
Nov 12	02:52	25.0	2.0	1.7	11	1.9	24
			2.9	23	3.6	8.5	
Nov 12	17:04	25.0	2.7	15	4.2	6.6	15
Nov 18	14:55	23.9	1.7	1.0	13	1.3	40
			7.0	16	6.6	10	

Note: In the cases above where there are two lines per flare, the first line corresponds to the "kernel" and the second to the "tongue" seen in the HXIS images. The temperatures of the kernel and tongue were assumed to be the same.

Table 5.2.2 Energetics of the Primary Flares

Date 1980 **HXR start time(UT)** **SXR peak time (UT)**	**Apr 8** **02:59:15** **03:07:00**	**May 21** **20:53:35** **21:05:00**	**Jun 29** **18:03:40** **18:04:00**	**Jun 29** **18:22:00** **18:25:40**	**Aug 31** **12:47:55** **12:49:30**	**Aug 31** **12:51:00** **12:52:30**	**Nov 5** **22:25:40** **22:28:17**	**Nov 5** **22:32:10** **22:35:30**
Area kernel ($10^{17}cm^2$)	8	22	4.8	HXIS coarse FOV	6	2	8	15
Area tongue ($10^{17}cm^2$)	53	40	—	—	—	—	—	35
Volume kernel ($10^{26}cm^3$)	6	25	1.6	(7)	3	0.8	3	14
Volume tongue ($10^{26}cm^3$)	62	40	—	—	—	—	—	34
HXIS Temp. (10^6K)	25	21	25	20	17	20	20	26
HXIS EM (total) ($10^{48}cm^{-3}$)	19	103	1.7	28	3	11	3	16
$C_{max,t}/C_{max,k}$	6%	20%	—	—	—	—	—	6%
$EM_k(10^{48}cm^{-3})$	14	76	—	—	3	11	—	14
$EM_t(10^{48}cm^{-3})$	5	27	—	—	—	—	—	2
Density(kernel) ($10^{10}cm^{-3}$)	15	17	10	(20)	10	38	10	10
Density(tongue) ($10^{10}cm^{-3}$)	3	8	—	—	—	—	—	2
Thermal energy (kernel)(10^{29} ergs)	9	38	1.7	(12)	2	2	3	15
Thermal energy (tongue)(10^{29} ergs)	19	29	—	—	—	—	—	9
Total thermal energy (10^{29} ergs)	28	67	1.7	(12)	2	2	3	24
Energy in electrons > 25 keV(10^{29} ergs)	94	96/110	4.8	13	3	5	1.6 - 12	28
Power law index γ	4.7/6.5	3.8/3.5	3.7-7	4.7	3.3	3.3	4	3.9

Note: Filling factor assumed to be unity.

Volume and thermal energy computed assuming a density of $2 \times 10^{11}cm^{-3}$ for the limb flare on 1980 June 29 at 18:25 UT.

Gaunt factor" as a function of ε and T as given by Mewe and Gronenschild (1981) and Matteson (1971), and k is the Boltzmann constant. The parameters Y and T are selected to give a least-squares fit to the HXRBS spectral data (Batchelor 1984).

The thermal energy in the isothermal plasma emitting the hard X-rays is given by the relation

$$E_{th} = 4.14 \times 10^{-16}\ Y\ T/n \text{ ergs} \tag{5.2.7}$$

where n is the electron density (in cm^{-3}). Since $Y = n^2 V_p$, where V_p is the volume of the plasma (in cm^3), the thermal energy can be rewritten as

$$E_{th} = 4.14 \times 10^{-16}\ Y^{1/2}\ V_p^{1/2}\ T \text{ ergs.} \tag{5.2.8}$$

Thus, either the density or the volume of the source must be known before E_{th} can be determined.

An estimate of the source volume can be obtained from the microwave data with the additional assumptions that the observed optically-thick microwave emission originates with the hard X-rays in a common, thermal source. In such a model, the frequency-dependent flux $I_\mu(f)$ in the optically-thick portion of the microwave spectrum is related to the projected source area A_μ (in cm^2) by the expression

$$I_\mu(f) = 1.38 \times 10^{-26}\ f^2\ A_\mu\ T, \tag{5.2.9}$$

where $I_\mu(f)$ is measured in solar flux units (1 sfu = 10^{-22} W m^{-2} Hz^{-1}) and f is the observing frequency in GHz (Crannell et al., 1978). This relation provides a means for determining the area of the source region through the use of the temperature determined from the hard X-ray observations and the microwave flux measured at fixed frequencies in the optically thick portion of the spectrum.

The volume of the source (V_p) can be estimated from A_μ by

$$V_p = A_\mu^{3/2}, \tag{5.2.10}$$

so that the expression for the thermal energy can be rewritten as

$$E_{th} = 4.14 \times 10^{-16}\ Y^{1/2}\ A_\mu^{3/4}\ T \text{ ergs.} \tag{5.2.11}$$

Similarly, the magnetic field associated with the source region can be inferred from the calculated temperature and the turnover frequency in the microwave spectrum, f_t (in GHz), according to the following simplified relation (Karpen, et al., 1979):

$$B = 1.86 \times 10^4 \, [f_t/(1.66 \times 10^8 + T)]^{1.14}. \qquad (5.2.12)$$

The values of the calculated parameters are given in Table 5.2.3.

Table 5.2.3 Parameters Calculated from Coincident Hard X-Ray and Microwave Analysis

Date 1980	Time (UT)	T (keV)	Y (10^{45} cm^{-3})	A_μ (10^{18} cm^2)	E_{th} (10^{29} erg)	Probability[1] (%)	f_t (GHz)	B (G)	Thick target (10^{27} erg s^{-1})	Probability[2] (%)
Apr 08	03:07:06	7	20	7.3	6.7	7	6	280	25	73
May 21	20:55:54	44	2.6	3.7	9.1	~0	7.4	114	57	30
	20:56:20	40	1.9	3.4	6.6	~0	7.6	125	41	46
	20:57:30	30	1.2	3.5	4.0	~0	7.9	166	33	75
	20:58:30	32	0.5	3.4	2.7	~0	6.4	124	16	4
	21:00:30	35	0.5	3.2	2.8	~0	6.4	116	17	~0
	21:03:20	39	0.2	2.6	1.7	~0	6.3	105	7	~0
Jun 29	18:23:20	16	3.3	33	19	~0	2.0	53	23	28
Aug 31	12:48:50	48	1.4	2.9	6.1	8	8.8	127	20	0.6
Nov 05	22:26:30	33	0.5	0.1	2.3	11	—	—	6.5	28
	22:33:03	37	3.3	4.9	11	~0	13.5	262	8.2	0.1

[1] The probability of obtaining a higher value of χ^2 than that obtained for the thermal fit to the HXRBS spectrum. A probability of ~0 indicates that a value of $\chi^2 >>$ the number of degrees of freedom was obtained.
[2] The probability of obtaining a higher value of χ^2 than that obtained for the power-law fit to the HXRBS spectrum.

5.2.3.2 Estimates of the Energy Content of the Energetic Protons and Ions

The three prime flares listed in Table 5.2.4 were observed with GRS at energies above 300 keV. However, none of them produced detectable nuclear line emission above the GRS sensitivity threshold. Therefore, although we have no direct measure of the gamma-ray flux resulting from the interactions of energetic protons and ions, we do have upper limits on this flux for each of the events which produced continuum emission above 300 keV. These upper limits can be used with the model described by Ramaty (1982) to deduce upper limits on the energy content of the energetic protons and ions above a few MeV. The values obtained in this fashion for each of the three flares are presented in Table 5.2.4.

In addition to these upper limits, an estimate of the energy content of the energetic protons and ions can be obtained from the observed continuum emission above 300 keV and an empirically determined scaling factor relating the continuum flux and the nuclear line emission. The values presented in the fourth and fifth columns of Table 5.2.4 are energy estimates obtained with this scaling procedure used in conjunction with the model described by Ramaty (1982), and the same spectral shape as was the upper limits. For two of the three flares, the estimates are significantly less than the corresponding upper limits. For one flare, the estimates are greater than the upper limits, but the difference is not significant.

Table 5.2.4 Estimates of the Energies in Protons and Ions

Event			Energy in Energetic Protons and Ions (ergs)			
Date 1980	Start time (UT)	Δt (s)	Scaled estimates		Upper limits	
			>10 MeV	>0 MeV	>10 MeV	>0 MeV
May 21	20:54:32	98.3	2.0×10^{27}	1.8×10^{28}	1.1×10^{28}	1.0×10^{29}
Aug 31	21:48:31	49.2	9.3×10^{26}	8.3×10^{27}	9.3×10^{27}	8.3×10^{28}
Nov 5	22:32:40	65.5	5.7×10^{27}	5.0×10^{28}	3.1×10^{27}	2.7×10^{28}

The values listed under the column heading ">10 MeV" are upper limits on the total energy content of protons and ions with kinetic energies greater than 10 MeV. The values listed under the column heading ">0 MeV" are upper limits on the total energy content of all protons and ions included in the energetic particle spectrum. Each event is designated by its date, start time, and Δt, which specifies the duration of the impulsive phase over which the flux values were integrated.

Realistic uncertainties in the values of the estimates and upper limits are dominated by the uncertainty in the choice of the appropriate spectral form. In Ramaty's model calculations, a Bessel function spectral form is used. Alternative choices, such as a power law with the low-energy cut-off as a free parameter, could change all of these values by several orders of magnitude, as the bulk of the energy is probably in protons and ions with energies < 1 MeV.

5.2.3.3 Thermal Energy of the Soft X-ray Emitting Plasma

As discussed in Appendix 5B, there is considerable observational evidence that the thermal plasma emitting soft X-rays, i.e., plasma with T greater than or about 3×10^7 K, is not isothermal, especially during the impulsive phase. Nevertheless, we have made estimates of the total thermal energy by assuming a single-temperature thermal source for this radiation. These estimates clearly are subject to large, systematic errors, but they should give reasonable results over restricted temperature ranges. In particular, they are expected to be more reliable at the end of the impulsive phase, when the thermal energy reaches its maximum value. At that time the plasma is closer to thermal equilibrium, and a single-temperature analysis should be more appropriate. Tables 5.2.1 and 5.2.2 show estimates of these peak thermal energies for direct comparison with the total energy in fast electrons entering a thick target during the impulsive phase.

A more comprehensive analysis technique that involves estimating the differential emission measure as a function of temperature was also used for the cases in which sufficient data were available (see Section 5.3.2). The results should give more accurate energy estimates, but again, because of the available observations and the uncertainty in the FCS calibration, they apply only in the restricted temperature range above 10^7 K.

For the single temperature analysis a technique was used that is similar to the one described in Section 5.2.3.1(ii) for the thermal analysis of the hard X-ray spectrum. The single temperature analysis was carried out using the six HXIS energy bands to derive a temperature and emission measure. The same equations used for the hard X-ray analysis were also used here with the addition of estimates of the line fluxes which contribute to the counting rate in the lower energy bands.

The estimates of the source volume, required to determine the thermal energy, were obtained from the HXIS fine-field-of-view images in bands 1, 2, and 3 (3.5 to 11 keV) after the smearing effects of the instrument collimators were removed. A simple iterative technique based on the method given by Svestka et al. (1983) was used to deconvolve the collimator response. Unlike Svestka et al., however, we continued the iterations until the χ^2 statistic stopped decreasing rather than when χ^2 became less than the number of degrees of freedom. In this way the most likely value of the source area was obtained. For the smaller sources, this was as much as a factor of 5 smaller than that which would have been obtained by the original method given by Svestka et al. Further details about the procedure are given by Dennis et al. (1984a).

In contrast to the microwave source area A_μ discussed in Section 5.2.3.1(ii), this measured source area is directly determined, but it is only the apparent area. The true area may be much smaller and may contain fine structure that was unresolved with the HXIS 8 arcsec FWHM spatial resolution. Again, we must assume some reasonable geometry to determine the source volume V_m from this measured area A_x. For most simple geometries

$$V_m \approx 0.5\ A_x^{3/2}. \qquad (5.2.13)$$

We include the filling factor ϕ such that

$$V_p = \phi\ V_m \qquad (5.2.14)$$

to allow for the possibility that the source was unresolved. Thus, we obtain the following relation for the energy in the plasma, assuming it to be isothermal:

$$E_{th}(< 10^8\ K) = 4.14 \times 10^{-16}\ (Y\ \phi\ V_m)^{1/2}\ T\ ergs. \qquad (5.2.15)$$

(c.f. Equation 5.2.8).

In many cases, the source at the time of the peak in soft X-rays was made up of at least two separate components: a small, compact source referred to as the "kernel" by de Jager et al. (1983) and a more extended source with lower counting rates per pixel, referred to as the "tongue". In the present analysis, it was possible to estimate the areas of these two sources separately. Further, by assuming that their temperatures were the same, it was possible to determine their relative contribution to the total emission measure. Thus, the thermal energy in each component could be determined separately (Dennis et al., 1984a). The different parameters computed for the kernel and the tongue for those flares in which they could be distinguished are listed in Tables 5.2.1 and 5.2.2.

5.2.3.4 Convective and Turbulent Motions

The BCS spectra in the Ca XIX and Fe XXV channels show a blue-shifted component and line broadening during the impulsive phase of many events. These have been interpreted as resulting from upward motions and turbulent motions, respectively (Antonucci et al., 1982). The energies associated with these bulk plasma motions can be significant and have been estimated, when possible, for the prime flares studied by this group.

The most extensive analysis of bulk motions has been carried out by Antonucci et al. (1982, 1984). In this analysis of 25 disk flares, the observed Ca XIX spectra were fitted with a synthesized spectrum consisting of the spectrum expected from a stationary, isothermal but turbulent source plus the spectrum from an upwardly moving source with the same electron and Doppler temperatures as the stationary source. In this way, the electron and Doppler temperatures and the emission measure of the source of the principal spectrum were derived together with the flow velocity and the emission measure of the moving source.

To compute the convective and turbulent energies from these parameters, it is necessary to know the mass of the plasma involved. Antonucci et al. (1984) were able to derive values for the mass of the upflowing plasma by requiring that the evaporating material supply sufficient mass and energy to account for the observed coronal plasma during the thermal phase of the flare. The volume of this plasma can be inferred from the HXIS images, as already discussed in Section 5.2.2.2. Furthermore, an estimate of the cross-sectional area and separation of the loop footpoints where the chromospheric evaporation is most likely to take place can be obtained from the high energy HXIS images in energy bands 5 and 6 between 16 and 30 keV.

From the values given by Antonucci et al. (1984), we have computed the convective (E_{con}) and turbulent (E_{tr}) energies for three of the prime flares. These energy values are given in Table 5.2.5, with the values given by the same authors for the increase (E_{SXR}) in the total energy of the coronal plasma plus the radiative and conductive losses (E_{loss}) during the evaporation process, i.e., $E_{SXR} = E_{th} + E_{tr} + E_{loss}$.

The two remaining prime flares not analysed by Antonucci et al. (1984), i.e., 1980 August 31 and June 29, showed very little blue shift or asymmetric line broadening. The 1980 June 29 flare occurred on the limb, and such flares typically do not have blue-shifted components or asymmetrically broadened X-ray line profiles during the impulsive phase. Presumably this is because the line of sight of the instrument is perpendicular to the direction of motion of the bulk of the moving plasma. However, symmetrical line broadening in excess of what would be expected from purely thermal motions is observed for limb flares, especially during the impulsive phase. The 1980 August 31 double flare was a compact event, which showed very little asymmetric broadening during either flare (Strong et al., 1984).

The procedure to determine the turbulent energy from the symmetrical line broadening was similar to that described above. Spectral fits were obtained over the full wavelength range in the BCS Ca XVII-XIX and Fe XXIV-XXV channels. A synthetic spectrum was computed by assuming an

Table 5.2.5 Parameters of the Chromospheric Evaporation

Date (1980)	Time (UT)	$\overline{n_e'}$ (10^{10} cm^{-3})	V′ (10^{26} cm^3)	$\overline{v'}$ (km s^{-1})	$v_{tr(peak)}$ (km s^{-1})	$\overline{E}_{con}$ (10^{29} ergs)	$E_{tr(peak)}$ (10^{29} ergs)	E_{SXR} (10^{29} ergs)
Apr 8	03:04	10	8	250	~90	0.4	0.05	29-39
May 21	20:53	14	24	200	~60	1.1	0.1	82-160
Nov 5	22.36	13	11	200	~60	0.5	0.04	27-47

isothermal model, and a best fit to the data was obtained which minimized the χ^2 parameter by varying the electron temperature, total line width, and emission measure. The excess line broadening was determined by subtracting the instrumental width and the contribution due to thermal motions (assuming that the ion and electron temperatures were equal). For the 1980 June 29 flare, the effect of the X-ray source size in the East-West direction (0.75 arc min) was included. The excess line broadening can be expressed as an ionic velocity. The values of this parameter obtained from the calcium and iron observations were found to be in good agreement with one another, giving confidence in the instrumental calibrations and justification for the isothermal assumption used in determining the excess line width.

The resultant turbulent energies are shown in Figure 5.3.8 and 5.3.9 for the two flares under consideration. The mass of the plasma was determined from a multi-thermal analysis for the emission measure, assuming that the volume remained constant throughout the flare. A filling factor of unity was used, and the volumes were assumed to remain constant equal to the values derived from the HXIS images given in Table 5.2.3. Since the mass was determined from a multi-thermal emission measure model for temperatures above 10^7 K, the calculated turbulent energies are appropriate for the high temperature component only. The results are summarized in Table 5.2.6. The turbulent energies are a factor of 20 to 100 less than the energies in fast electrons above 25 keV computed assuming thick-target interactions.

5.2.3.5 Radiant Energy in Hα

An estimate of the radiant energy in Hα during the impulsive phase of the second flare on 1980 November 5 was made on the basis of spectra taken with the multi-slit spectrograph operated at the San Fernando Observatory. No useful spectra were obtained until the instrument was triggered into the flare mode at 22:34:21 UT, the beginning of the second hard X-ray maximum of the second flare (see Figure 5A.12). After that time one spectrum was recorded on 35 mm film every 2 s with the slits stepping 2.6 arcsec between exposures. The extremities of the flare kernels at the flare maximum were approximately 31 arc sec apart in the direction of slit motion. Consequently, with the 51 arc sec slit spacing, the flare kernels were sampled in groups of 12 consecutive spectra with intervening intervals of 20 arc sec when the slits were not positioned on any bright part of the flare. The total radiant Hα intensity, ΔI, in excess of the quiet Sun was obtained from the photographically recorded spectra by using the relative continuum intensity given in the

Photometric Atlas of the Solar Spectrum (Minnert et al., 1940) and the Labs and Neckel (1968, 1970) absolute calibrations. The three series of spectra obtained before the end of the impulsive phase gave nearly the same value for ΔI: 1.5×10^{8} ergs s^{-1} sr^{-1}.

Table 5.2.6 Turbulent Energies

Date (1980)	Time (UT)	E_{tr}(peak) (erg)	Assumed volume (cm^3)
31 Aug	12:43:54	1.3×10^{28}	3×10^{26}
31 Aug	12:51:54	5.6×10^{27}	8×10^{25}
29 Jun	18:23:16	3.5×10^{28}	7×10^{25}

The total energy radiated in Hα can be determined from the measured intensities and the area $\underline{A}$ of the flare kernel of < 100 (arc sec)^2, determined rom the HeI D_3 filtergrams (see Appendix 5A). The average power $\underline{P}$ radiated in Hα during the second peak of the flare was computed from the relation (Canfield et al., 1980):

$$\underline{P} = 2\pi \Delta I \, \underline{A} = 4.9 \times 10^{26} \text{ ergs s}^{-1}.$$

Since the second peak lasted for ~ 120 s and contained only about a quarter of the total number of hard X-rays detected with HXRBS during the flare, our best estimate for the total Hα radiated energy during the impulsive phase of the flare is $4(\underline{P} \times 120 \text{ s}) = 2.3 \times 10^{29}$ ergs.

5.2.4 The 1980 April 8 Flare

The April 8 flare is unique among the set studied by us in that it was observed with the P78-1 satellite as well as with SMM. We have therefore chosen this event to illustrate the various approaches that we pursued in our investigation of flare energetics. Observations of hard X-ray spectra were obtained with HXRBS, images and X-ray spectra were obtained with HXIS, soft X-ray spectra were obtained with BCS, and images in the principal FCS set of lines were obtained throughout this event. UVSP observations produced a series of images in the density-sensitive lines of Si IV and O IV before the onset of the impulsive phase. The P78-1 observations include soft X-ray spectra, from which we selected the O VII, Ca XIX, and Fe XXV lines for analysis.

In Figure 5.2.1 we show the radiated energy in hard X-rays as a function of time for this flare as determined from power-law fits to the HXRBS spectral data. The power-law spectral index γ is plotted as a function of time in Figure 5.2.2, together with the power flux of electrons above 30 keV entering a thick target as derived from the power-law fits. Figure 5.2.3 shows the energy in electrons above 25 keV integrated from the start of the flare for comparison with the thermal energy in the soft X-ray emitting plasma. Two estimates of the thermal energy are plotted. The one marked "HXIS (single temp)" was obtained from isothermal fits to the HXIS data as described in Section 5.2.3.3. The source density was assumed to be 10^{10} cm^{-3} to obtain the plotted energy

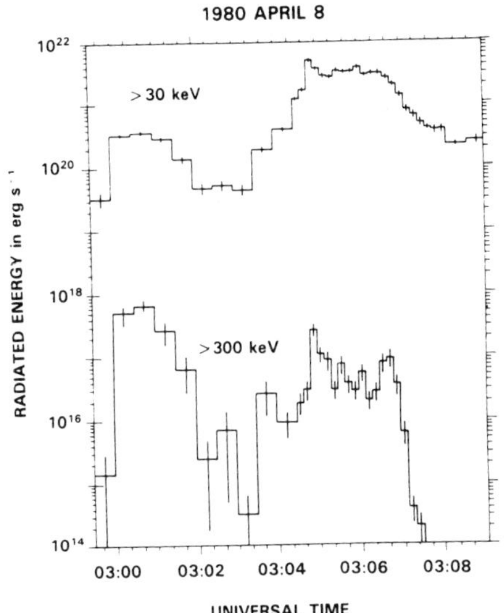

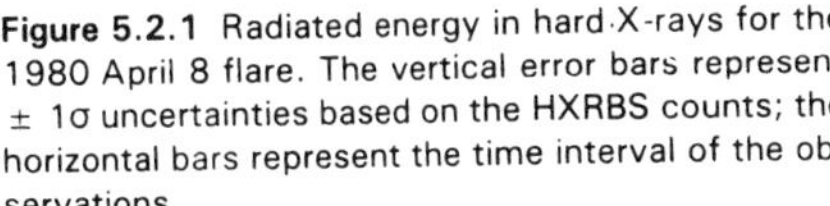
Figure 5.2.1 Radiated energy in hard X-rays for the 1980 April 8 flare. The vertical error bars represent ± 1σ uncertainties based on the HXRBS counts; the horizontal bars represent the time interval of the observations.

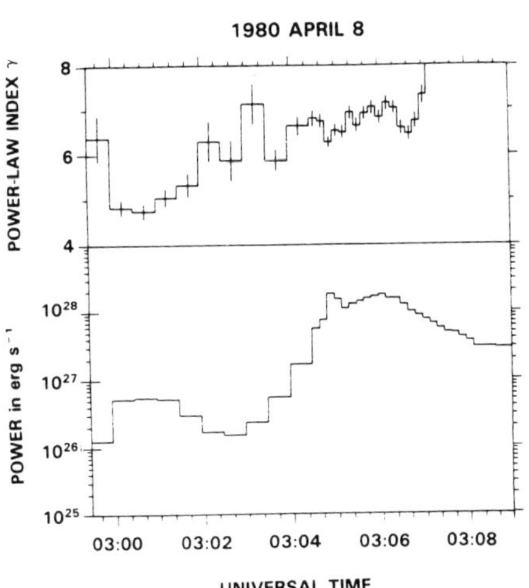

Figure 5.2.2 Power flux of electrons above 30 keV into a thick target computed from the HXRBS data for the 1980 April 8 flare. The top plot shows the HXRBS power-law spectral index on the same time scale.

from the temperature and emission measure resulting from the fitting procedure. The points marked "BCS and HXIS" were obtained from a full differential emission measure (DEM) analysis with both BCS and HXIS data. The DEM distribution was integrated over temperatures from 10^7 to 5 x 10^7 K, and the source volumes obtained from HXIS images were used to compute the total thermal energies plotted.

The effect of different assumptions about the temperature distribution and the densities or volumes can be seen by comparison of the two curves shown in Figure 5.2.3 for the thermal energy. If we force the single-temperature model to agree with the multi-thermal analysis at the end of the impulsive phase, say at 03:08 UT, by adjusting the assumed density, then the multi-thermal analysis gives about an order of magnitude more energy near the start of the impulsive phase, say at 03:04 UT. This is presumably because of a lower-temperature component with T near 10^7 K that does not give a significant contribution in the HXIS energy range, at least not so that the single temperature analysis can reveal its presence. This result suggests significant preheating before the appearance of hard X-rays for this particular flare.

Figure 5.2.4 shows the results of the analysis of the hard X-ray data, assuming that te source was thermal and that the electron spectrum could be represented with a single effective temperature. The effective temperature and emission measure are plotted as a function of time, as obtained from least-squares fits to the HXRBS data. The total energy of the plasma is also plotted as a function of time, assuming a source density of 10^9 cm^{-3}.

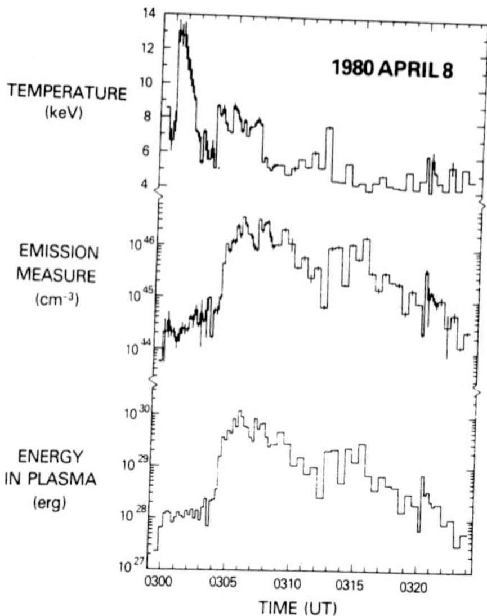

Figure 5.2.4 Temperature and emission measure as a function of time for the 1980 April 8 flare computed from HXRBS data between 27 and 467 keV using a single-temperature parameterization of the hard X-ray spectra. The total energy of the plasma is also plotted for a constant source density of 10^9 cm^{-3}.

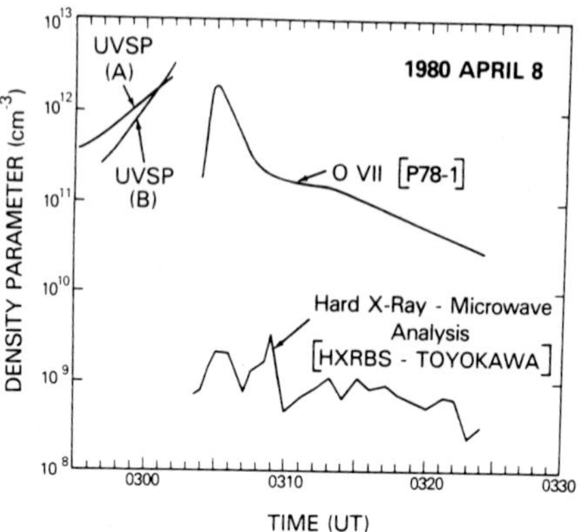

Figure 5.2.5 Source density as a function of time for the 1980 April 8 flare.

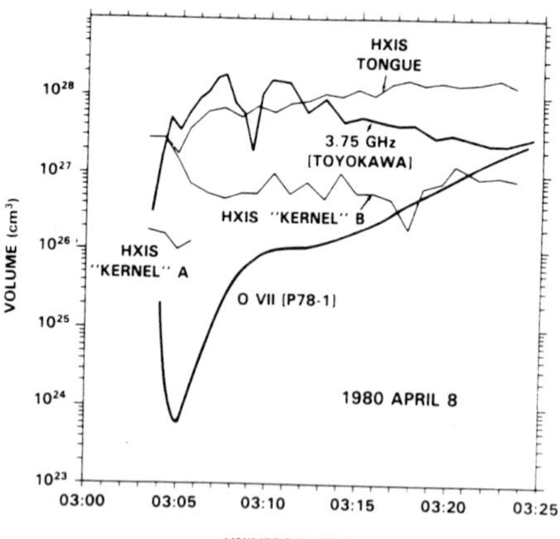

Figure 5.2.6 Source volumes for the 1980 April 8 flare determined from the HXIS soft X-ray images, the P78-1 density-sensitive line data, and the hard X-ray/microwave analysis discussed in the text.

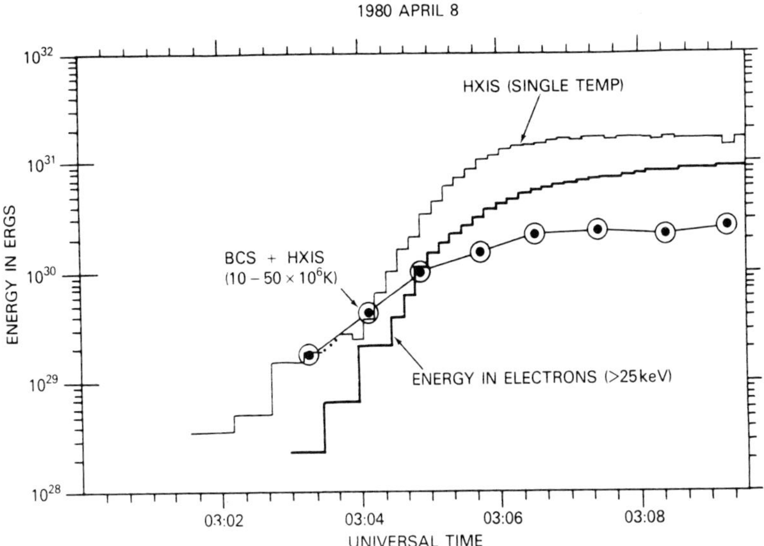

Figure 5.2.3 Energy in > 25 keV electrons and in the soft-X-ray-emitting thermal plasma as a function of time for the impulsive phase of the 1980 April 8 flare. The energy in electrons above 25 keV was computed from power-law fits to the HXRBS data assuming thick-target interactions. The energy in the soft-X-ray-emitting plasma was computed in two ways. In one case marked "HXIS (single temp)", a single temperature thermal spectrum was fitted to the HXIS spectral data by using the results of the HXIS count rate prediction program. In the other case, a full multi-temperature DEM analysis was carried out using BCS and HXIS data for temperatures from 10^7 to 5 × 10^7K.

The density determinations made possible with observations of the helium-like lines of O VII, obtained with the P78-1 satellite, enable us to obtain an estimate of the density and hence the volume of the soft X-ray emitting source at ~ 2 x 10^6 K (Doschek et al., 1981), and these are plotted as a function of time in Figures 5.2.5 and 5.2.6. During the impulsive phase of the flare, the density and hence the volume derived from the P78-1 observations are uncertain by as much as an order of magnitude, but after the peak in the density estimate at approximately 03:05 UT, the uncertainties are less than a factor of 3.

In addition to the satellite observations, records of microwave emissions from the the same flare are available from the Toyokawa Observatory (courtesy of S. Enome). Under the assumptions indicated in Section 5.2.3.1(ii), coincident analysis of the hard X-ray and microwave observations gives the area and hence an estimate of the apparent source volume associated with the most energetic electrons. These volume estimates, plotted as a function of time, together with those obtained from HXIS images at 3.5 to 11.5 keV are also shown in Figure 5.2.6. These observations show that the estimated volume of the 2 x 10^6 K plasma is 3 orders of magnitude less than the apparent volume of the hard X-ray source throughout the impulsive phase of the flare, but it increases monotonically until the two volumes are comparable at the end of the flare. Throughout the impulsive phase, the apparent volume estimated from coincident analysis of HXRBS and Toyokawa observations is comparable to the volume of the "tongue" source observed with HXIS.

The density of the hard X-ray and microwave source can be estimated from the apparent volume and the hard X-ray emission measure. This density parameter, shown in Figure 5.2.5, is seen to be relatively constant throughout the event and of the order of 10^9 cm^{-3}, justifying the use of this value in computing the thermal plasma energy plotted in Figure 5.2.4

The parameters characterizing the chromospheric evaporation for the 1980 April 8 flare are plotted versus time in Figure 5.2.7. The plotted parameters were determined from spectral fits to Ca XIX spectra determined with BCS as described in Section 5.2.3.4. The corresponding total energies are given in Table 5.2.5.

5.2.5 Discussion

We have computed the impulsive phase energies in various components of the prime flares selected by this group and the main results are summarized in Table 5.2.7.

The total energy in fast electrons during the impulsive phase and the thermal energy in the soft X-ray emitting plasma at the end of the impulsive phase are the two most important components of these flares from an energetics standpoint. If the hard X-rays are produced by thick-target interactions of the fast electrons, the total energy in electrons above 25 keV is almost embarassingly large compared with the energy of the soft-X-ray-emitting plasma. If the lower-energy cut-off is dropped to 10 keV, as appears reasonable, at least for the early part of the impulsive phase of some flares, then the calculated energy in fast electrons becomes even higher. A plot of the thermal energy of the soft-X-ray-emitting plasma against the energy in electrons above 25 keV, $\underline{W}$(> 25 keV), is shown in Figure 5.2.8. A strong correlation with a coefficient of 0.8 is found between these two quantities. Although this correlation is considerably better than that expected from the Big Flare Syndrome (Kahler 1982), correlations calculated for different models show similarly high correlation coefficients (see Section 5.4.2). There are other serious problems with this correlation, so that it cannot be used to argue in favor of thick-target interactions or of a causal relation between the fast electrons and the thermal plasma. These problems are discussed briefly here.

(1) The low-energy cut-off in the electron spectrum is unknown and may be different for different flares. This can cause order-of-magnitude errors in the estimates of the total energy in fast electrons.

(2) The filling factor for the thermal plasma is unknown and may be different for different flares. Introduction of a fractional filling factor will decrease the thermal energy by $\phi^{1/2}$. It has been argued that ϕ was as small as 10^{-2} during the flare on 1980 April 30 (de Jager et al., 1983), 3×10^{-3} to 3×10^{-2} during the flare on 1980 November 5 (Wolfson et al., 1983), and 10^{-4} for the same flare (Martens et al., 1984). Filling factors this small would decrease

the thermal energy by one or two orders of magnitude, thus destroying the similarity with the energy in fast electrons.

(3) The hard X-ray spectral fits may be contaminated by the high-energy tail of the thermal distribution, particularly toward the end of the impulsive phase when the soft X-ray flux is very

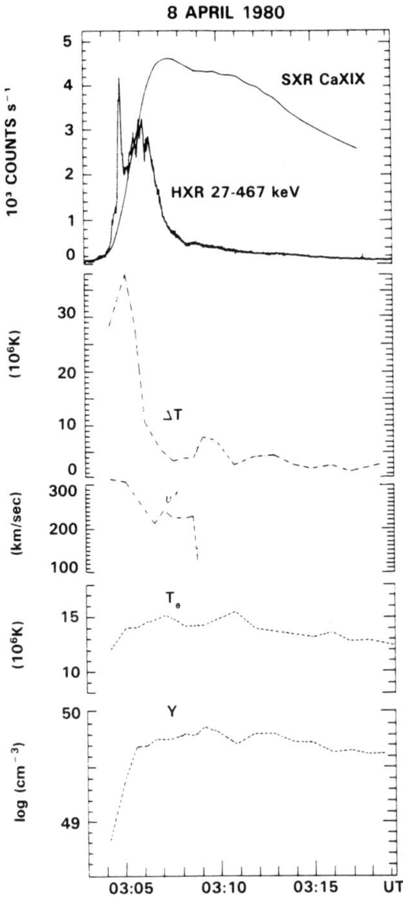

Figure 5.2.7 The time variation of the parameters characterizing the chromospheric evaporation for the 1980 April 8 flare. The following derived quantities are plotted as a function of time with the Ca XIX and hard X-ray rates shown for reference: the temperature difference ΔT between the ion and the electron temperatures, the upward component of the velocity v', the electron temperature T_e, and the Ca XIX emission measure Y (from Antonucci *et al.*, 1984.). Note that the plotted difference, ΔT, between the Doppler and electron temperature is a measure of the non-thermal line broadening attributed to plasma turbulence.

Table 5.2.7 Summary of Flare Energies (in units of 10^{29} ergs)

Date 1980	Time (UT)	W(>25 keV)	$E_{th}(>10^8K)$	$E_{th}(>10^7K)$	E_{SXR}	E_{tr}	E(Hα)
Apr 8	03:07	94	7	28	29-39	—	—
May 21	21:05	96/100	9	67	82-160	—	—
Jun 29	18:04	5	—	1.7	—	—	—
	18:26	13	19	12	—	3.5	—
Aug 31	12:49:30	3	6	2	—	1.3	—
	12:52:30	5	—	2	—	0.56	—
Nov 5	22:28:17	1.6-12	0.23	3	—	—	—
	22:35:30	28	11	24	27-47	—	2.3

W(>25 keV), total energy in electrons above 25 keV calculated from the hard X-ray spectra assuming thick-target interactions; $E_{th}(>10^8K)$, thermal energy at the time of the peak hard X-ray flux calculated from the hard X-ray spectrum assuming a source with a temperature of $>10^8K$; $E_{th}(>10^7K)$, thermal energy at the time of the peak soft X-ray flux computed from a multi-thermal analysis with $10^7 < T < 5 \times 10^7K$ and the HXIS source area; E_{SXR}, the increase in the thermal and turbulent energy of the plasma plus the total radiated and conducted energy losses up until the time of the peak in soft X-rays (see text); E_{tr}(peak), peak energy of the turbulent plasma motions; E(Hα), radiant energy in Hα integrated over the duration of the impulsive phase.

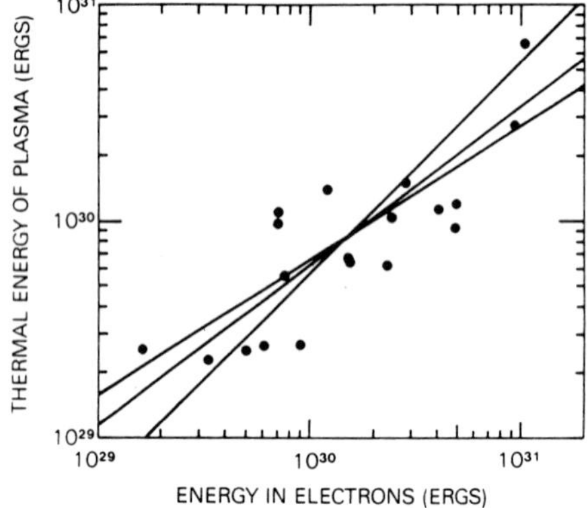

Figure 5.2.8 Plot of the thermal energy U at the time of the peak soft X-ray flux versus the energy in electrons W (> 25 KeV) entering a thick target integrated up to the same time. Each point corresponds to one of the 19 flares listed in Table 1, for which there is an estimate of U based on HXIS observations. Note that the plotted value for U is the sum of the energies in the ''tongue'' and ''kernel'' sources seen in the HXIS images.

large, and the high temperature component (3 x 10^7 K) detected by Lin et al. (1981) is expected to dominate. Also, when the X-ray spectrum becomes steeper than, say, ε^{-6} or ε^{-7}, the relatively poor energy resolution of scintillation counters makes it impossible to determine the true spectrum with any accuracy. As a consequence of these two factors, the energy in fast electrons may have been overestimated in some cases.

(4) Not all of the hard X-rays may have been produced in thick-target electron interactions. A considerable fraction of the

total X-ray flux may be produced from a thermal source, particularly in the later stages of the impulsive phase when footpoints can no longer be resolved in the hard X-ray images. If the X-rays are mainly from a thermal distribution of electrons, then a value of ~ 2 x 10^{10} cm^{-3} for the unknown density will make the energy in such a thermal source equal to the energy calculated assuming thick-target interactions (Smith and Lilliquist, 1979). With the thermal model described in Section 5.2.3.1 (ii) in which the hard X-rays and microwaves are assumed to come from a common source, the density and energy content of the energetic electrons are uniquely specified by the observed emissions. The total energy integrated over the whole flare, however, depends on the rate of energy input to the electrons and on unspecified loss mechanisms. It is, therefore, not known if the energetic electrons could supply the thermal energy of the plasma at temperatures greater than or about 10^7 K. Smith (1985) suggests that only 10% or less of the hard X-rays are produced in thick-target interactions, mostly in the first 1 or 2 min of the impulsive phase. He postulates that the energy release mechanism puts most of the energy into heating after that time. Rust (1984) also argues that only a small fraction of the observed chromospheric evaporation is produced by electron beams and that such beams are energetically unimportant when compared to conduction.

In addition to the total energies of the impulsive phase, we have determined the energies as a function of time for some flares. Figure 5.2.3 is a plot of the different components of the flare energy as a function of time for the impulsive phase of the 1980 April 8 flare; Figure 5.2.9 is a similar plot for the 1980 May 21 flare, from Antonucci et al. (1984). In both cases, the build-up in thermal energy of the soft X-ray emitting plasma is consistent with the depostion of energy into a thick target by fast electrons.

For the May 21 flare, when the impulsive phase was relatively long, lasting for about 10 min until ~ 21:05 UT, the radiation losses are significant. Depending on the assumed volume, the thermal energy is between 1 and 6 times larger than the accumulated radiation losses at 21:05 UT. For the other flares with shorter impulsive phases, the radiative losses tend to be less important, as can be seen from a comparison of E_{th} (> 10^7 K) and E_{SXR} in Table 5.2.7.

5.2.6 Conclusion

It is clear that with the present state of the observations, the study of flare energetics during the impulsive phase is a very poor way of discriminating between different flare models. Ideally, it would be possible to determine the fraction of the total energy that goes into accelerating electrons versus the fraction directly heating the plasma. We are clearly a long way from being able to make that differentiation, although the data, particularly the hard X-ray imaging observations, suggest that some significant fraction of the energy does go to accelerate electrons to energies above 20 keV.

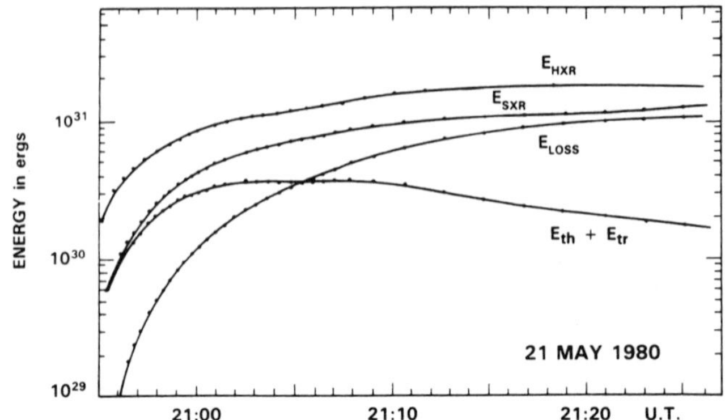

Figure 5.2.9 Temporal evolution of the following quantities during the 1980 May 21 flare from Antonucci *et al.* (1984): E_{HXR}, the total energy input to the chromosphere by non-thermal electrons above 25 keV from the flare onset time t_o up to the time t plotted; E_{SXR}, the total energy input to the soft X-ray emitting plasma; E_{th}, the thermal energy of the coronal plasma at time t; E_{tr}, the turbulent energy of the coronal plasma at time t; E_{loss}, the integral of the radiation and conductive loss rates from t_o to t. Note that $E_{SXR} = E_{th} + E_{tr} + E_{loss}$. The quantities derived from the soft X-ray emission were calculated assuming a lower limit volume of 2.4×10^{27} cm^3.

To improve this situation, future observations must address the fundamental constraints which have limited our analysis. The most important of these is the resolution of the thermal or non-thermal character of the hard X-ray source. It is not clear at present how this will be achieved, although hard X-ray imaging to higher energies and with better spatial and temporal resolution will be important. Also, measurements of the hard X-ray spectrum with higher spectral resolution than is possible with scintillation counters will allow spectra steeper than ε^{-7} to be determined and will also enable X-rays from plasma with temperatures up to 3 x 10^7 K to be separated more clearly from the higher-energy component. Moreover, by combining observations with improved spectral and spatial resolution, the spectrum of accelerated electrons below 20-30 keV can be determined, so that a better estimate of the total energy in these electrons can be made.

As far as improvements in the estimates of the energy in the soft-X-ray emitting plasma is concerned, it is clear that better estimates of the emission from plasma at temperatures below 10^7 K must be made and that the filling factor must be determined at all energies. A wide range of temperatures is involved during the impulsive phase, and single temperature analysis is inadequate, given the improvements in the observations. Multi-thermal analysis techniques, however, are limited in the detail with which the temperature distribution can be determined, even with more accurate observations.

5.3 THE ENERGETICS OF THE GRADUAL PHASE

K.T. Strong, R.D. Bentley, P.L. Bornmann, M.E. Bruner, P.J. Cargill, J.G. Doyle, J.R. Lemen, R. Pallavicini, G. Peres, S. Serio, G.M. Simnett, J. Sylwester, and N.J. Veck

5.3.1 Introduction

The gradual phase of a flare is best characterized by the smooth rise and decay of the soft X-rays (see Figure 5A.2). Just from inspection of the soft X-ray light curves of some typical flares, it is evident that the gradual phase comprises several different stages. During the rise of the X-ray emission, a completely different energy budget from that of the decay will be produced. Many flares show multiple peaks in their light curves, indicating that energy is released after the initial burst. The location of any such secondary energy release will also profoundly affect the relative importance of the terms in the energy budget of the flare.

The soft X-ray emission produced by the higher-temperature (> 10^7 K) plasma rises and decays more rapidly and peaks earlier than that produced by the cooler plasma (10^6 - 10^7 K). The impulsive phase usually occurs during the rise of the soft X-ray emission but there is no clear evidence of an impulsive component in the soft X-ray signal. Hence, in an energetics study the important question arises of what, if any, is the link between the impulsive and gradual phases.

We have avoided using the term "thermal phase" because it is misleading; a number of non-thermal energetics terms are associated with the gradual phase. The mass motions are an example; it is not clear whether they are peculiar to the early stages of a flare or whether they are also present later in the flare, which seems to be likely, as mass motions have been detected even in quiescent active regions (Acton et al., 1981). We therefore attempted to evaluate the non-thermal terms as well as the thermal energy throughout the gradual phase of the five prime flares.

As the flare observations made with SMM were not specifically designed to address the energetics problem, our choice of flares (see Appendix 5A) represents a compromise. We identified a number of different "types" of flares that, although sharing the same physical processes, had somewhat different physical characteristics.

The Skylab Solar Flares Workshop (Sturrock 1980a) laid the foundations for the methodology used in this section, and we compare our results with those in the chapter by Moore et al. (1980), who reached five main conclusions about the gradual phase:

- o the typical density of the soft X-ray emitting plasma is between 10^{11} and 10^{12} cm^{-3} for compact flares and between 10^{10} and 10^{11} cm^{-3} for a large-area flare;
- o cooling is by conduction and radiation in roughly equal proportions;
- o continual heating is needed in the decay phase of two-ribbon flares;
- o continual heating is probably not needed in compact events;

- most of the soft-X-ray-emitting plasma results from "chromospheric evaporation".

Our goal was to reexamine these problems with the data from SMM and other supporting instruments as well as to take advantage of recent theoretical advances. SMM is capable of measuring coronal temperatures more accurately and with a better cadence than has been possible before. The SMM data set is also unique in that the complete transit of an active region was observed, with soft X-ray and UV images being taken every few minute. We are therefore able to establish the pre-flare conditions of the region and see whether anything has changed as a result of the flare.

In the next subsection we describe the assumptions made in attempting to determine the required plasma parameters. The derived parameters for the five prime flares are presented, and the role of numerical simulations is discussed. Finally, we consider the overall implications of our results and discuss how both theory and observations have evolved since the Skylab workshop.

5.3.2 The Basic Physical Expressions

The quantities needed for this study are defined in terms of the fundamental plasma parameters of electron temperature (T_e), electron density (N_e), plasma volume (V), plasma velocity (v), and height (h). The four basic energies that we wish to obtain throughout the flares are the thermal energy (E_{th}) of the plasma,

$$E_{th} = 3\ N_e\ k\ T_e\ V \text{ ergs}; \tag{5.3.1}$$

the kinetic energy (E_k) of the plasma,

$$E_k = 1/2\ M_p\ N_e\ V\ v^2 \text{ ergs}; \tag{5.3.2}$$

the potential energy (E_p) of the coronal plasma,

$$E_p = M_p\ N_e\ V\ g_o\ h \text{ ergs} \tag{5.3.3}$$

(where M_p is the proton mass and g_o is the gravitational acceleration at the solar surface); and the stored ionization energy in the plasma (E_i), which is the sum of the ionization energy of every ion stage of all abundant elements existing up to the observed temperature. This latter component is dominated by the ionization energy of hydrogen and helium. Note that these energetics terms only apply in the above form for an isothermal plasma. However, the SMM data indicate that the coronal plasma during a flare is multi-thermal, and the integral forms of these equations must often be used (Section 5.3.2.1).

The energy flux terms and their corresponding timescales are as important as the individual energy terms for understanding the energy budget. There are three main terms: the conductive flux (F_c), which, in terms of the classical Spitzer-Harm theory of heat conduction, can be expressed as

$$F_c = - \kappa T_e^{5/2} dT_e/dh \text{ ergs cm}^{-2} \text{ s}^{-1}; \qquad (5.3.4)$$

the radiative flux (F_r),

$$F_r = - A^{-1} \int N_e^2 P_r \, dV \text{ ergs cm}^{-2} \text{ s}^{-1}; \qquad (5.3.5)$$

and the enthalpy flux (F_e),

$$F_e = (5/2A) K T_e d(N_e V)/dt \text{ ergs cm}^{-2} \text{ s}^{-1}. \qquad (5.3.6)$$

The corresponding decay timescales (in s) are

$$\tau_c = 2 \times 10^{-10} N_e L^2 T_e^{5/2} \qquad (5.3.7)$$

$$\tau_r = 10^7 T_e/N_e \qquad (5.3.8)$$

$$\tau_e = 1.2 N_e(dN_e/dt)^{-1}. \qquad (5.3.9)$$

It should be possible to derive all these quantities as a function of time throughout the flare from a knowledge of the plasma parameters and the geometry of the flaring loop. However, these quantities are not available directly from observations although the temperature and density can be inferred from soft X-ray line ratios. Generally, we derive the emission measure (N_e^2 V) from optically-thin line intensities. The SMM X-ray data can be used to derive the DEM. The temperature range available was 2 x 10^6 to 10^8 K, but no information was available in one of the most interesting temperature domains, namely 2 x 10^5 to 2 x 10^6 K. There is considerable uncertainty in the quantities needed to derive these energetics terms, especially the density. However, for the five prime flares, the values obtained for each of these energetics terms are discussed.

5.3.2.1 Multi-thermal Modeling

Several quantities are discussed above which express the energy content of the various flare components with the assumption of an isothermal plasma. However, there is strong evidence from SMM data that the coronal plasma is multithermal during flares (Figure 5.3.1). For this reason, an attempt was made to model the temperature distribution of the flare plasma from spectroscopic data obtained with various SMM

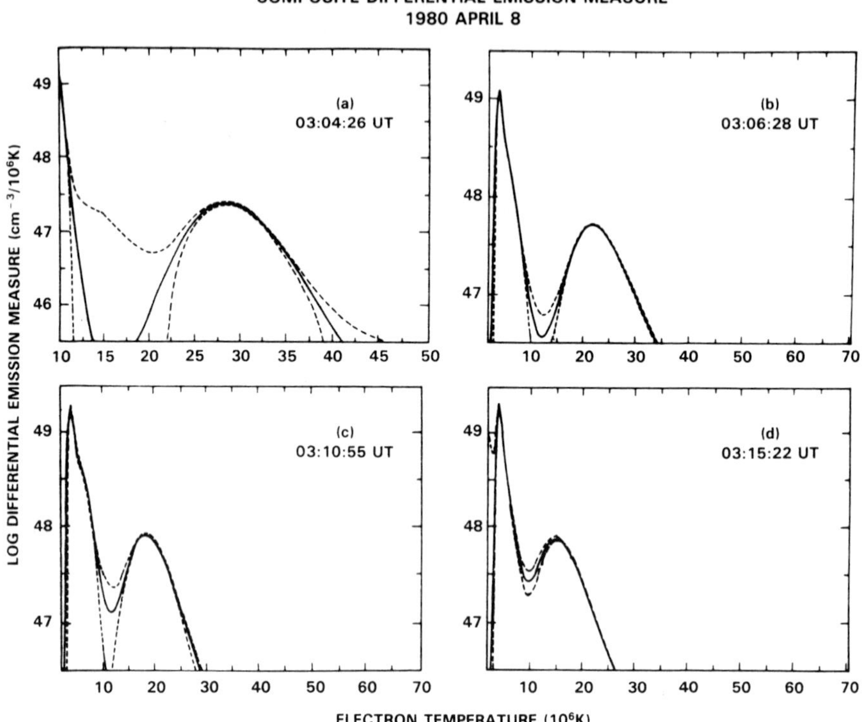

Figure 5.3.1 The DEM models calculated for four times during the 1980 April 8 flare. Note that, even during the rising portion of the gradual phase when the flare is at its hottest, there is a significant low temperature component. As the flare evolves the high temperature component of the DEM decays in temperature eventually merging with the low temperature component.

instruments. In this section we present the multithermal forms of the expressions for the thermal energy (E_{th}) and radiative flux (F_r) and discuss the techniques which were used to derive the temperature distribution of the plasma. The temperature distribution of the plasma may be characterized by the differential emission measure (DEM), q(T), which is defined as

$$q(T)\ dT = N_e^2\ dV. \tag{5.3.10}$$

With a multithermal assumption, several of the quantities given in Section 5.3.2 are modified: the expression for the radiative flux becomes

$$F_r = \int P_r(T)q(T)dT \text{ ergs s}^{-1}, \qquad (5.3.11)$$

where $P_r(T)$ is the radiative power loss function; the thermal energy, dE_{th}, of a volume element dV with temperature T and density N_e is given by

$$dE_{th} = 3\, N_e kTdV \text{ ergs.} \qquad (5.3.12)$$

If the density within the volume, V, is assumed to be uniform, Equation 5.3.12 may be integrated to yield

$$E_{th} = \frac{3kV^{1/2} \int Tq(T)dT}{(\int q(T)dT)^{1/2}} \text{ ergs.} \qquad (5.3.13)$$

Alternatively, if the pressure, p = nT, within the volume is assumed to be uniform, the thermal energy becomes,

$$E_{th} = 3k\, V^{1/2} \left[\int T^2\, q(T)\, dT \right]^{1/2} \text{ ergs.} \qquad (5.3.14)$$

The DEM was derived for the soft X-ray emitting thermal plasma by two different methods. The first method was an inversion technique initially developed by Withbroe (1976) and extended by Sylwester et al. (1980, 1984). This method has been applied to the five prime flares in this study by using data from four BCS lines and the six HXIS energy bands.

The four lines observed by BCS consist of the helium-like resonance line ($^1S_0-^1P_1$) in Ca XIX and Fe XXV and the lithium-like satellite line k in Ca XVIII ($^2P_{1/2}-^2D_{3/2}$) and j in Fe XXIV ($^2P_{3/2}-^2D_{5/2}$). These satellite lines are dielectronically populated and so their intensity ratios relative to the resonance lines are temperature sensitive and independent of ionization balance. The effects of blending with nearby lines were removed by fitting a fully synthesized spectrum to the data.

The HXIS coarse field-of-view data were analyzed by summing over the whole flare area rather than by studying individual pixels or groups of pixels, because it covers an area comparable with that of the BCS collimator (6 arc min FWHM). This approach probably does not affect estimates of the overall energy budget of the flare but it could possibly mask secondary energy releases at other sites away from the original flare location.

Any particular solution of q(T) is probably not unique. However, various tests were performed which showed that different but acceptable, solutions for q(T) generally conserved the total emission measure, $\int q(T)dT = \int N_e^2 dV$. A further limitation arises from the fact that BCS and HXIS are insensitive to emission from plasmas with temperatures below about 10^7 K, although a considerable amount of thermal energy is expected to be contained in these plasmas. Both of these problems could be lessened by including softer X-ray or UV data, which would extend the results to lower temperaturs and thus better constrain the range of acceptable solutions for the DEM. The lowest energy FCS channels are sensitive down to 2×10^6 K. However, since the cadence of the FCS observations was typically about 300 s per image, the FCS data were included for only one flare; 1980 August 31 double flare.

A second method for determining the DEM which combines data from BCS, HXIS, and HXRBS, has been developed by Bely-Dubau et al. (1984). In contrast to the approach just described, this method assumes a parametric description for the DEM. The parameters are adjusted until an acceptable agreement with the observations is achieved. So far, this method has only been applied to the 1980 June 29 flare. This technique is not subject to the problems which arise with an inversion technique when the emission functions cover a large proportion of the temperature range of the model. Eight BCS lines (the Ca XIX w, q, and k lines, Fe XXV w, q, and j lines, and Fe XXV r and s lines), the six HXIS energy bands, and the eight lowest energy HXRBS channels are included in the model. The model includes thermal and non-thermal components, and thus an important factor for energetics studies is the lower-energy cut-off for the nonthermal spectrum. Bely-Dubau et al. (1984) found that any cutoff value between 15 and 25 keV gave an acceptable fit to the data, but resulted in a considerable variation in the estimates for the energy budget of the flare.

Bornmann (1985a,b) has developed a new method for estimating the temperature and emission measure during the decay of the gradual phase from the shape of the light curves of soft X-ray lines. The rates at which the observed line fluxes decayed were not constant. For all but the highest temperature lines observed, the rate changed abruptly, causing the fluxes to fall at a more rapid rate later in the flare decay. These changes occurred at earlier times for lines formed at higher temperatures. Bornmann proposed that this behavior was due to the decreasing temperature of the flare plasma tracking the rise and subsequent fall of each line emissivity function. This explanation was used to empirically model the observed light curves and to estimate the temperature and the change in emission measure of the plasma as a function of time during the decay phase of the 1980 November 5 flare. This method provides a simple and independent measure of the temperature and cooling rate during the decay of the flare.

5.3.3 The Prime Flares

A more detailed description of the prime flares and a list of publications discussing their relevant aspects are given in Appendix 5A. In keeping with the impulsive phase section, we have emphasized the 1980 April 8 flare, but a brief discussion of the other prime events is given below.

5.3.3.1 The 1980 April 8 Flare

The M4 flare of 1980 April 8 originated in a small, isolated, bipolar region that was part of a delta configuration in NOAA active region 2372. The flare had an optical importance of 1B and showed multiple impulsive spikes in the HXRBS light curve, the brightest occurring at 03:05 UT. For details see Appendix 5A.

Both BCS and FCS showed that until 10 min before the flare, the level of coronal emission from the whole region was declining (25% in Ca XIX and about a factor of 2-3 in O VIII). The coronal temperature also appeared to be slowly dropping, changing from about 2.5 x 10^6 to less than 2 x 10^6 during the interval from 00:50 UT to 02:59 UT. At about 02:50 UT, the BCS Ca XIX lines started to increase in intensity. At 02:54 UT, the density measured with UVSP at the eastern footpoint began to rise exponentially, reaching a value of about 3 x 10^{12} cm^{-3} at 03:02 UT, when the measurement sequence was terminated. Both intensity and density were still rapidly increasing at this time, so that considerably larger values could have been reached at the time of the peak emission. Machado (1980) has identified the eastern footpoint with the isolated bipolar region discussed above. The electron density, as derived from the P78-1 O VII line ratios, reached its maximum of 10^{12} cm^{-3} at 03:05 UT, before any of the soft X-ray intensities had peaked. Since the characteristic temperatures of the two determinations differ by an order of magnitude, the fact that the derived densities are comparable suggests that the two regions were not in pressure equilibrium with each other and implies the existence of temperature gradients across the magnetic field. This result must be treated with caution, however, since the two determinations were not made simultaneously.

Subsequent development of the flare first involved the leader spot, via loops connecting it to the intermediate bipolar region, after which it spread to the trailer spot. Later in the gradual phase, the region of soft X-ray emission expanded southward. The HXIS images show that at least three discrete sources were involved during the rise phase, suggesting the existence of at least two flaring loops. A description of the magnetic field structure can be found in Chapter 1 of these proceedings.

The observations were analyzed along the lines discussed in Section 5.2.3. The HXRBS data can be represented by either a thermal or a power law spectral form. For analysis of the microwave data from Toyokawa, temperatures determined from fits to the thermal form of the HXRBS spectra were used to characterize the electron source of the hard X-ray emission for selected intervals throughout each burst. Observations obtained at 3.75 GHz were in the optically thick portion of the microwave spectrum throughout the impulsive and gradual phases of the flare and therefore enabled the computation of an effective area. The 3/2 power of an effective area was taken as an estimate of the size of the emitting volume. The maximum turbulent velocity derived from the BCS spectra was 120 km s^{-1}, assuming a single temperature determined from spectral analysis and neglecting the source size (Antonucci et al., 1982). The upflow was 200-300 km s^{-1}. Non-thermal line widths for the Fe XXV lines from P78-1 were derived by U. Feldman and S. Graham in work done specifically for this workshop. They assumed that the line profile was given by the convolution of three gaussians: one for the thermal width of the line, one for the turbulent velocity, and one for the angular size of the source. They assumed the angular size to be 30 arc sec and used the temperature value derived from P78-1 measurements of the j/w line ratio. Their results range from 100 to 300 km s^{-1}, bracketing the result of Antonucci et al. (1982).

We used the DEM distribution to represent the ensemble of soft X-ray observations of the gradual phase of the flare. We chose this approach because the DEM is proportional to both the observed intensities and the total radited power, even in the presence of inhomogenities in the radiating atmosphere. Four computations (Figure 5.3.1) of the DEM were made, corresponding to the four FCS rasters. BCS and HXIS data were selected for the times when the flaring plasma was being scanned with FCS. This technique is particularly powerful in placing constraints on the DEM distribution. The use of several isolated spectral lines provides intensity measurements for which the contribution function is well known. The inclusion of the four broad-band measurements from HXIS ensures that no important radiative energy sources have been omitted and provides a constraint on the total emission. Each DEM distribution shows two peaks, one near 2×10^7 K and the other near 4×10^6 K. The high-temperature peak is seen to move toward lower temperatures as the flare cools, as one would expect.

The temperature curves (see Figure 5.3.2) should be interpreted as indicators of the slope of the electron energy distribution in the energy range appropriate to each instrument; they do not necessarily represent physically meaningful quantities. With the exception of the temperature derived from HXRBS, the other temperatures agree to within a factor of 2. Assuming these estimates are a measure of the plasma temperature, they are not misleading, provided that one does not attempt to extrapolate outside the relevant electron energy range. The higher values of the HXRBS temperature parameter show that the electron spectrum producing the hard X-rays is relatively flatter than that producing the emission observed with the other instruments.

Density measurements for this flare are available from several observational sources (Figure 5.3.3). At the time of the maximum at 03:05 UT, the O VII density is probably dominated by material in the footpoints and does not reflect the average density in the loop system, since the evaporation phase has just begun.

Three volume estimates for the April 8 flare have been made: (1) the observed size in the HXIS fine field of view; (2) the apparent area derived from hard X-ray and microwave fluxes, and (3) using observed X-ray line intensities. The first two methods involve simple assumptions about the geometry of the emitting region to estimate the volume from the apparent area, whereas the third uses the derived electron density to compute the effective volume from the observed emission measure. The two estimates based on apparent areas agree fairly well during the impulsive phase, whereas the volume calculated from derived density and emission measure estimates is lower by a factor of 100 to 1000 (Figure 5.3.4). Although some of this discrepancy may be due to errors in the atomic data or in the underlying assumptions, the differences are important and have been cited as primary evidence for the existence of a small filling factor. The volumes derived from all these methods converge during the gradual phase. This probably indicates evolution of the geometry of the

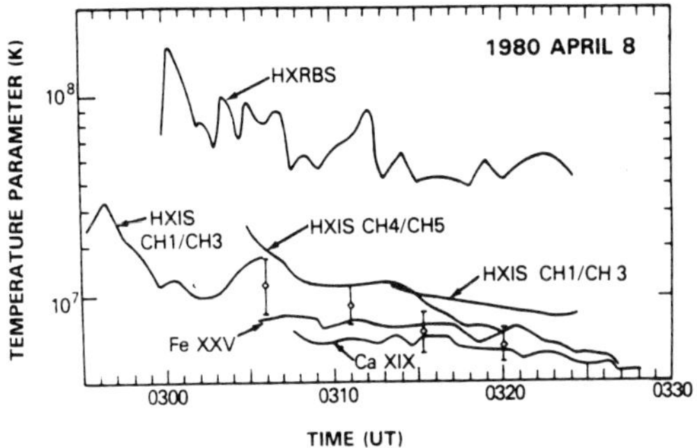

Figure 5.3.2 Evolution of the temperature parameter for 1980 April 8 derived from a number of different sources. The diamonds are derived from the high temperature component of the DEM calculations shown in Figure 5.3.1. The error bars represent the full width at half maximum of the distribution.

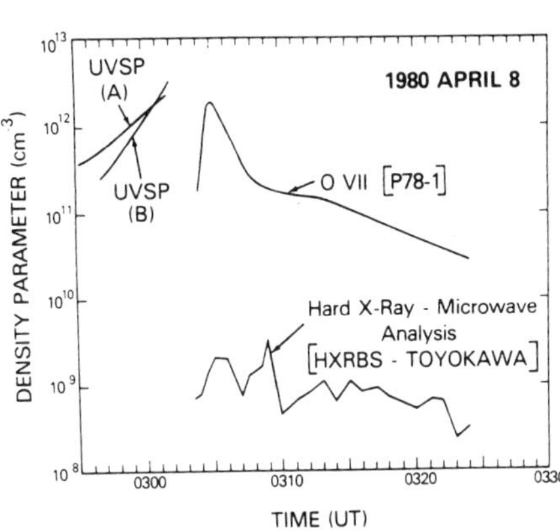

Figure 5.3.3 Evolution of the density parameter from UVSP Si IV/O IV line ratios and O VII measurements from P78-1, compared to the density derived from the hard X-ray—microwave analysis.

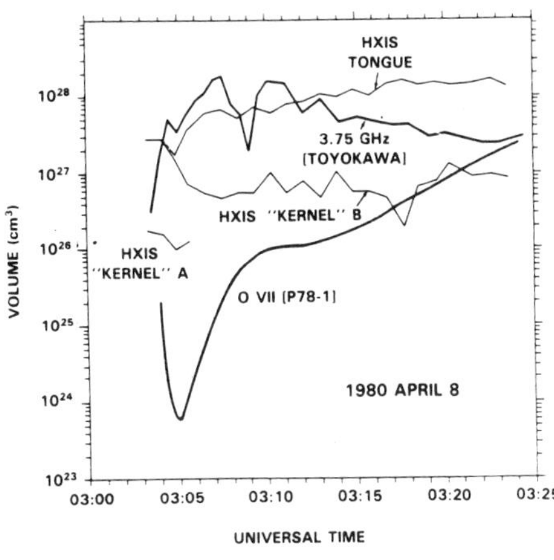

Figure 5.3.4 The history of the volume parameter from the 1980 April 8 flare from several different sources.

plasma during the gradual phase, with the filling factor tending toward unity. The evaluation of the filling factor is of obvious importance to the question of energetics, since the thermal energy content of the plasma is directly proportional to the total volume.

Figure 5.3.5 shows emission measures as functions of time determined from the BCS and HXRBS spectra. The P78-1 results of Doschek et al. (1981) are comparable. The four circles indicate the emission measures determined from the combined analysis of FCS, BCS, and HXIS data and correspond in time to the four FCS rasters.

Values of P_r and E_{th} (Figure 5.3.6) were determined for the high-temperature component (10^7 to 5×10^7 K) in a similar manner as for the 1980 June 29 flare (see below). The volume of 10^{27} cm^3 was estimated

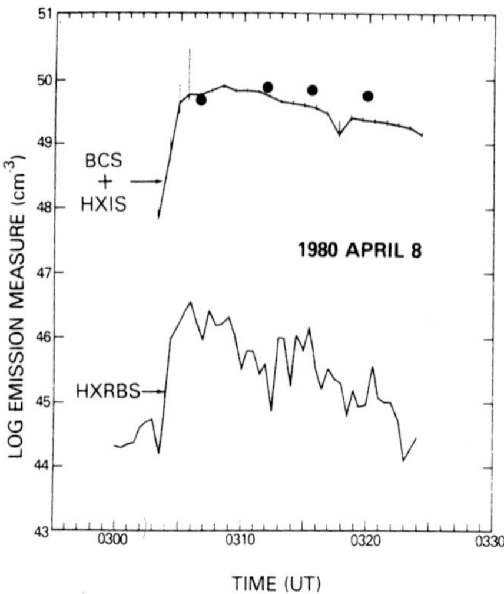

Figure 5.3.5 The evolution of the emission measure in the 1980 April 8 flare from the BCS, HXIS, and HXRBS data (thermal assumption). Note that the soft X-ray emission measure is over 3 orders of magnitude larger than that for the hard X-ray source. The BCS and HXIS value does not include information on the low temperature component ($< 10^7$K). The filled circles represent the emission measure calculated for temperature $> 2 \times 10^6$ K by including the FCS data with the BCS and HXIS observations. They show that the component between 2×10^6 K and 10^7 K is comparable to the component in the 10 to 20 $\times$ 10^6K range, at least later in the flare.

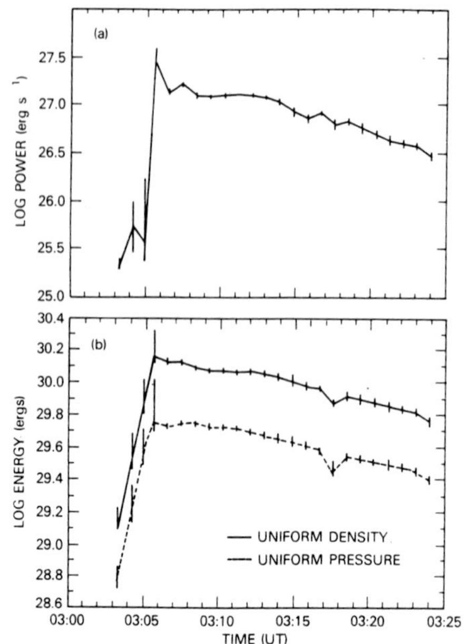

Figure 5.3.6 The evolution of (a) the radiated power (ergs s^{-1}) and (b) the thermal energy (ergs) for the 1980 April 8 flare from combined BCS and HXIS data. The error bars represent $\pm$ 1σ uncertainties. These quantities are calculated for the thermal plasma above 10^7 K only.

from HXIS fine-field-of-view images (Table 5.2.1). The peak value for E_{th} was found to be about 1.4 x 10^{30} ergs, which compares well with 9.4 x 10^{30} ergs for the total electron energy (Table 5.2.1). Note from Figure 5.3.1 that the low temperature component of the DEM is at least a factor of 10 larger than the high-temperature component which would mean that the thermal energy of the two components are approximately equivalent.

Figure 5.3.7 is the version of Figure 5.1.1 for this particular flare and shows the complete energy budget derived from the above physical quantities plotted. The kinetic energy in the upflows and in the turbulence was calculated using the P78-1 velocities and the BCS emission measures. The energy in the electrons determined from the thermal fit to the HXRBS spectra was used in Figure 5.3.7 rather than the thick-target non-thermal energy since the instantaneous energy in the electrons for that case is not significant. The gravitational potential energy and ionization energy are not plotted here, as they are too small to show on this scale. The peak gravitational energy is 3 x 10^{28} ergs.

Note that at 03:06 UT the total energy remains essentially constant. The slow drop in the total energy after this time (less than 30%) may be due to conductive losses which have not been included in the plot.

5.3.3.2 The 1980 May 21 Flare

This flare was chosen because it has exceptionally good data for the impulsive phase. However, the FCS and UVSP coverage during the gradual phase was very poor. They were in unsuitable modes during this event and so can contribute very little to the discussion of the energetics of the gradual phase, which is well described by Duijveman et al. (1983).

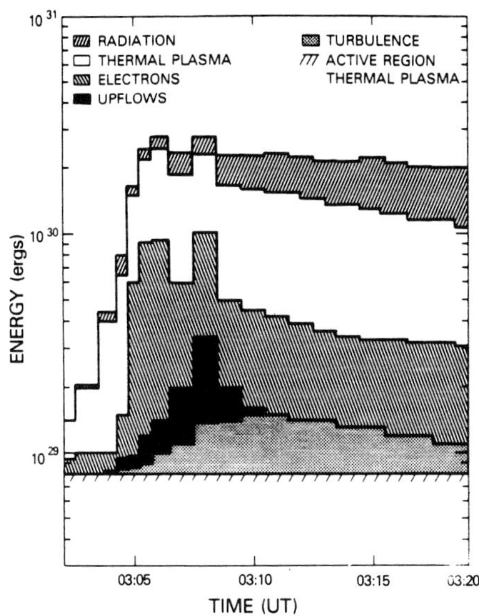

Figure 5.3.7 The distribution of the flare energy among various reservoirs and active phenomena as a function of time for the 1980 April 8 flare (c.f. Figure 5.1.1). The width of the histogram bins represents the time resolution for each of the parameters measured. Note that late in the flare (after 03:09 UT), the sum of the components remains effectively constant in spite of the fact that several of the components are varying.

5.3.3.3 The 1980 June 29 Flare

This event occurred on the limb, and so the determination of the volume of the flare is different than for a disc flare. Although we have the height structure and hence the area of the flare, the line-of-sight depth is still unknown. Further, it seems that part of the flare was probably behind the limb. The FCS images were used to determine the area of the flare, which gave a value of 2×10^{19} cm^2. The microwave data imply an area of 3.3×10^{19} cm2, although interferometric observations gave an area of an order of magnitude less.

Normally, one does not expect to find significant line shift in the BCS data for a limb event. However, this flare gave a velocity of 370 km

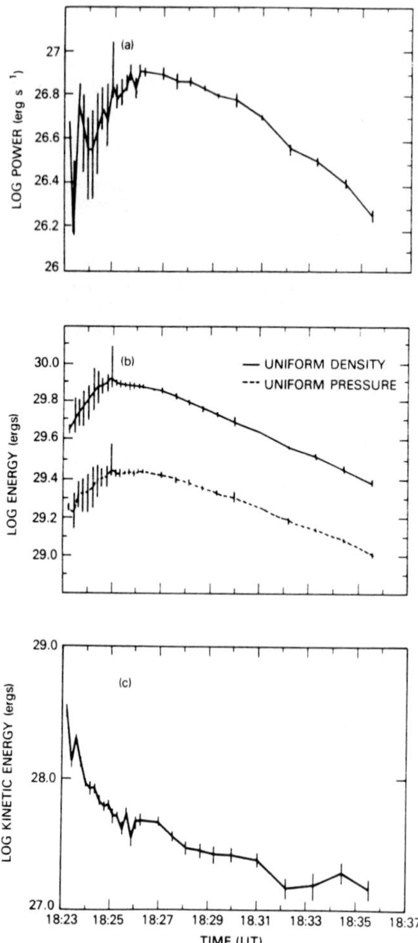

Figure 5.3.8 The radiated power (a), thermal energy (b), and kinetic energy (c) of the soft X-ray emitting plasma as functions of time for the 1980 June 29 flare. These quantities were derived from the BCS and HXIS data. The error bars represent $\pm$ 1σ uncertainties. The error bars on the kinetic energy estimates were calculated from the uncertainties of the line widths and electron temperatures but neglect the uncertainties of the estimated volume and emission measures.

s^{-1} at the flare onset. Turbulent velocities of about 200 km s^{-1} were estimated from the BCS Ca XIX line profiles. Vertical motions were measured by several instruments: 200 km s^{-1} in the lower corona, 500 km s^{-1} above 1.2 solar radii, and 1100 km s^{-1} from the type II burst.

A multi-thermal analysis of the high-temperature component of the flare plasma was made for this flare. For the determination of the DEM, the intensities of four lines observed by BCS and the integrated HXIS coarse-field-of-view fluxes were used. The DEM was then convolved with the radative power loss function, $P_r(T)$, of Summers and McWhirter (1979) to give the total radiative loss rate results shown in Figure 5.3.8(a). Since $P_r(T)$ is approximately constant with temperature near 2 x 10^7 K, the behavior of the radiative flux with time is similar to the behavior of the total emission measure with time. The value of E_{th} is shown in Figure 5.3.8(b), as calculated assuming either uniform pressure or uniform density throughout the volume of plasma with a given temperature. The volume was estimated to be 7 x 10^{26} cm^3 by assuming a density of 2 x 10^{11} cm^{-3}, which is typical of the estimated densities for the other principal flares we considered (see Table 5.2.4). The peak value of 7.8 x 10^{29} ergs for E_{th} calculated assuming uniform density is comparable to the total energy inferred (1.3 x 10^{30} ergs) for the non-thermal electrons above 25 keV. The turbulent energy derived from the BCS line widths was discussed in Section 5.2.3.4. The peak turbulent energy observed during the impulsive phase was approximately 3 x 10^{28} ergs.

There was a mass ejection during the 1980 June 29 event (Wu et al., 1983). The maximum temperature (2 x 10^6 K) and density (4 x 10^{11} cm^{-3}) derived from the XRP data, together with an assumed velocity, were used to simulate the mass ejection using an MHD model (Wu et al., 1983). The simulation produced spatially wide, large-amplitude, temporally steepened MHD waves but no shocks. This result seems to be supported by the fact that a type II radio burst was observed late in the event for only a few minutes. The density enhancements moved away from the Sun at the same velocity (1000 km s^{-1}) as determined with the Mauna Loa K-coronameter. However, the observation of the coronal transient included a rarefaction that does not appear in the simulation. The model enables us to derive the energy associated with the transient; the quantities are: peak temperature, 16.5 x 10^6 K; peak density, 3.0 x 10^{10} cm^{-3}; total volume, 1.0 x 10^{27} cm^3; and total energy, 4.0 x 10^{33} ergs. Comparison with the observed pB (polarization x brightness) signal, derived from C/P data to be 7 x 10^{-8}, gave a result that was a factor of 3 too high. This implies that the magnitude of the calculated density enhancement was high by at least 50% in the closed-field case and a factor of 3 higher in the open-field case. The peak values of the energetics terms of this flare are listed in Table 5.3.1, in which only part of the radiative energy could be estimated based on the available data. The computed total energy included all the energy sources (i.e. magnetic, potential, kinetic and thermal).

Using the alternative technique of Bely-Dubau et al., (1984) described above (Section 5.3.2.1), we find the energy content as a function of time given in Table 5.3.2.

Table 5.3.1 Energetics of the 1980 June 29 Flare

Energetics term	BCS + HXIS	Ca XIX Fe XXV	Fe XXI O V	HXIS	HXRBS
Emission measure (cm^{-3})	2×10^{49}	3.5×10^{49} 5.6×10^{49}	7.9×10^{43} 0.32×10^{43}	3.3×10^{45}	9×10^{45}
Radiative power (ergs s^{-1})	8×10^{26}	2×10^{27}		2.3×10^{28}	
Integrated thermal energy (ergs)	2.8×10^{26}				
Radiated energy (ergs)	3.7×10^{29}				

Table 5.3.2 Energy as a Function of Time for the 1980 June 29 Flare

Time (UT)	18:20:40	18:23:28	18:23:54	18:25:00	18:26:00	18:27:18
Thermal Energy (10^{30} ergs)	1.38	3.4	5.5	7.6	7.6	8.3
Energy Deposit rate (10^{27} ergs s^{-1})	1700	2000	26	30	4	2.5
Cutoff energy (keV)	10	12	26	26	26	26

5.3.3.4 The 1980 August 31 Flare

The properties of the impulsive double flare on 1980 August 31 were reported by Strong et al. (1984). By combining observations of the two events from several of the SMM instruments (XRP, HXIS, HXRBS, and UVSP) as well as radio and optical observatories, they concluded that the second energy release probably occurred in the corona plasma that was hotter and denser than that of the first flare. They inferred that the differences between the two flares resulted from changes in plasma properties brought about by the first energy release. Comparisons of the radio and X-ray data show that the primary energy release site was contiguous (for energetic electrons) with separate, closed magnetic structures. The loops which emitted the radio bursts were at least 10 times longer than those emitting the X-rays. There was no evidence for emerging magnetic flux in this region around the time of the flare; in fact, the field seemed to be decaying rapidly as the two leader spots approached each other. The primary energy release mechanism involved therefore seemed to be magnetic reconnection caused by the "collision" of these structures due to photospheric motions.

The density was determined from the emission measure and source size. The pre-flare temperature of the active region was found to be approximately 2.5×10^{6} K with an average emission measure of 3×10^{47} cm^{-3}. As there were no distinct structures visible, the volume was derived from the area of the FCS pixel and a typical soft X-ray scale height and was found to be 4×10^{28} cm^3. The mean electron density was therefore 3×10^9 cm^{-3} for the pre-flare active region. Although there were no pre-flare magnetograms, data taken just after this event showed that there was ample free magnetic energy available even after the flares, to supply all the energy observed. Over the previous days, observations of this active region had indicated that the two leading spots (see Section 5A.4) were moving towards one another. Although the velocities were relatively low (of the order of hundreds of meters per second), the energy involved in these photospheric motions must have been large and supplied the stored magnetic energy in the form of sheared magnetic fields.

The highest spatial resolution available on SMM (3 arcsec from UVSP), indicated that the X-ray emitting region in the first flare was 10^{26} cm^3 with an uncertainty of about a factor of 2. The same volume was assumed throughout both flares, as there was no direct measure of the volume for the second flare although the longer timescales of the second flare imply an increased volume.

The multi-thermal analysis covered the extended temperature range from 2×10^6 K to 5×10^7 K by using six FCS channels, four BCS lines, and the six HXIS fine-field-of-view channels. Figure 5.3.9 shows P_r, E_{th} and E_K as functions of time. The thermal analysis was not possible until

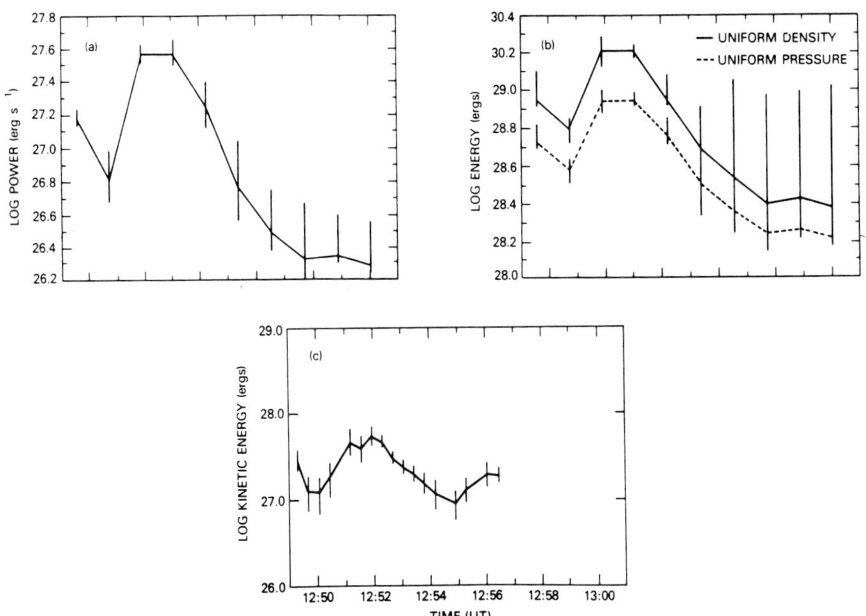

Figure 5.3.9 Same as Figure 5.3.8 for the two flares on 1980 August 31. Compare the peak kinetic energy from the second flare with that of the second flare on 29 June. There is nearly an order of magnitude difference showing that the role of mass motions in this second event are very much inhibited compared to both that of the first flare and the flare on 29 June.

the time of the peak of the first flare in soft X-rays as the count rates were low. In calculating E_{th}, a volume of 7.6×10^{25} cm^3 estimated from the HXIS fine-field-of-view images was used for both flares (see Table 5.2.2) even though the volume of the first flare was probably three times larger. The peak value of E_{th} for the second flare was approximately 1.6×10^{29} ergs and compares with a total electron energy of 5×10^{29} ergs.

The energy associated with the turbulent broadening was discussed in Section 5.2.3.4. The peak turbulent energy during the second flare was about 6×10^{27} ergs. It is interesting to note that the ratio of the peak value of Eth to the peak turbulent energy for this flare is approximately 27, which compares with 26 for the same ratio for the 1980 June 29 flare. This ratio is close to 10 for the first flare on 1980 August 31.

A comprehensive picture of the mass motions was derived from the BCS X-ray line profiles and from the source motions observed by HXIS. The uncertainties in the velocity measurements are minimal because the flare was compact and therefore not complicated by multiple flare sites. Using the peak integrated DEM and the above volume of 7.6×10^{25} cm^3, we estimated the mass of the "evaporated" plasma and hence the kinetic energy. The potential energy was determined at the peak of the soft X-ray emission from the source size and assuming a semicircular geometry for the loops. Again, by using the density and temperature, we estimated the ionization energy. The results of these calculations are listed in Table 5.3.3 for both flares.

This event illustrates multiple energy release, seemingly a common aspect of flares and a possible reason for so much diversity in flare characteristics. It establishes, for some cases at least, the corona as the locale for the primary energy release involving closed magnetic

Table 5.3.3 Non-Thermal Terms of the Energy Budget for the 1980 August 31 Flares

Energy term	Energy in first flare (ergs)	Energy in second flare (ergs)
Kinetic energy		
Turbulence	4×10^{29}	$< 3 \times 10^{27}$
Upflows	5×10^{28}	$< 2 \times 10^{27}$
Translational	7×10^{28}	1×10^{28}
Total	5×10^{29}	1×10^{28}
Potential energy	1×10^{27}	1.5×10^{27}
Ionization potential energy	3×10^{26}	3×10^{26}

structures. It also shows that structures of vastly different scale sizes can be involved in the same impulsive release. These are the sorts of morphological boundary conditions which are a prerequisite for continued progress in flare theory.

5.3.3.5 The 1980 November 5 Flare

This was one of the most comprehensively observed flares, with good groundbased data as well as good configurations for the SMM instruments.

It was another example of a double flare but with the components separated by more time than for the 1980 August 31 events. The geometry of the flare has been derived by Martens et al. (1984), who found the two principal loops observed during the flare to be 1.6×10^9 cm and 7×10^9 cm in length, with a diameter of 1.1×10^9 cm. The volume of the larger of the loops was therefore about 4×10^{28} cm^3 if it is assumed to be semicircular.

Using the higher density values from the He-like ion ratios (discussed below), we derived the energetics terms shown in Table 5.3.4. Figure 5.3.10 shows P_r and E_{th} for the two flares on November 5. Estimates of the temperature and emission measure were made from the following parameters: the shape of the FCS soft X-ray light curves (Bornmann 1985a,b), ratios of GOES count rates, DEM fits to the FCS data, satellite-to-resonance line ratios observed with BCS, ratios of HXIS count rates, and thermal fits to HXRBS data. The temperature rose to about 4×10^8 K in both flares and fell rapidly during the initial decay, relaxing to a nearly exponential decay late in the decay phase. The emission measure derived from HXRBS reached a maximum of 8×10^{45} cm^{-3} and 10^{47} cm^{-3} for the first and second flare, respectively. The soft X-ray instruments showed emission measures which began and ended near 10^{48} cm^{-3} and reached a maximum value of 3 to 6×10^{49} cm^{-3}.

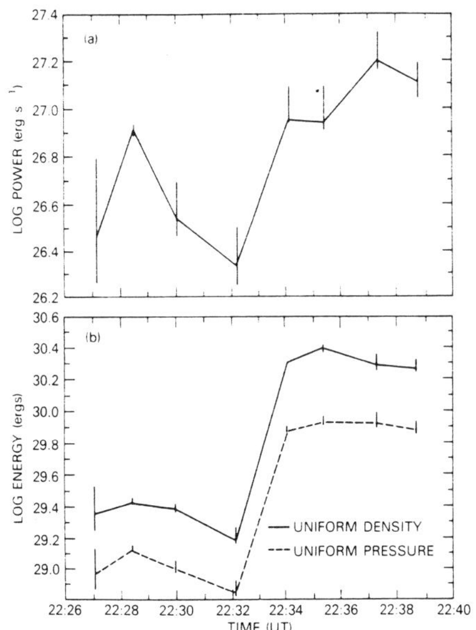

Figure 5.3.10 The evolution of (a) the radiated power (ergs s^{-1}) and (b) the thermal energy (ergs) for the 1980 November 5 flare from combined BCS and HXIS data. The error bars represent $\pm$ 1σ uncertainties. These quantities are calculated for the thermal plasma above 10^7 K only.

The electron density estimates during the second flare differ by roughly two orders of magnitude. Duijveman et al. (1982, 1983a) used the emission-measure volume technique, assuming a line-of-sight depth of the emitting region of 6000 km, to find electron densities in the two principal loops of 2.5×10^{10} and 9×10^{9} cm^{-3}. Density estimates made from FCS line ratios (Figure 5.3.11) indicate that the density fell from 1.5×10^{12} cm^{-3} at the time of peak soft X-ray flux to 3.0×10^{11} cm^{-3} near the end of the flare decay (Wolfson et al., 1982; Bornmann 1985a). For times when both density estimates are available, the emission-measure volume estimates are a factor of 30-50 lower than the line ratio

Table 5.3.4 Energetics Components in ergs for the 1980 November 5 Flare

Term	Peak	Decay	Total
Radiative losses	3×10^{28}	6.2×10^{26}	2.8×10^{29}
Thermal energy	7×10^{29}	3×10^{27}	
Mass motions (upflows)	4×10^{29}	2×10^{28}	
Mass motions (turbulence)	4×10^{28}		

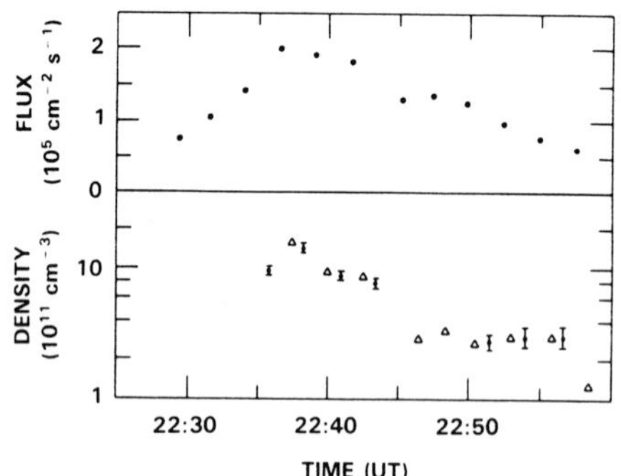

Figure 5.3.11 The evolution of the density and the Ne IX flux during the flare of 1980 November 5. The density is derived from the line ratios of He-like Ne IX (figure courtesy of C.J. Wolfson).

estimates. This discrepancy could be the result of emission from separate spatial regions, but this seems unlikely since both detectors sampled overlapping regions in space and energy. Alternatively, the line ratio estimate could be correct if the filling factor was 10^{-3} to 10^{-4}.

However, Wolfson et al. (1983) found that the emitting volume observed with FCS decreased from 3×10^{25} to 3×10^{24} cm^3 during the decay of the flare, assuming the temperature remained constant. The volume of an FCS pixel is about 10^{27} cm^3 ($A^{3/2}$), which gives a filling factor of between 3×10^{-3} and 3×10^{-2}. Thus, two separate estimates of the filling factor for this flare both indicate the need for filling factors of the order of 10^{-3}.

DEM analysis following the method of Withbroe et al. (1975) and Sylwester et al. (1980) using the FCS data indicates a high-temperature component which diminished in magnitude and temperature during the decay phase. A second, low-temperature component remained nearly constant with time. This DEM behavior has been observed in other flares examined by our group (cf. Figure 5.3.1).

Bornmann (1985a,b) used the shape of the FCS light curves to determine the temperature and emission measure during the decay of this flare (Figure 5.3.12). She assumed that the plasma was isothermal and that discrepencies between observed and calculated fluxes (which were less than a factor of 2) were due to incorrect assumptions for detector efficiencies and/or atomic abundances. Although the DEM and light curve

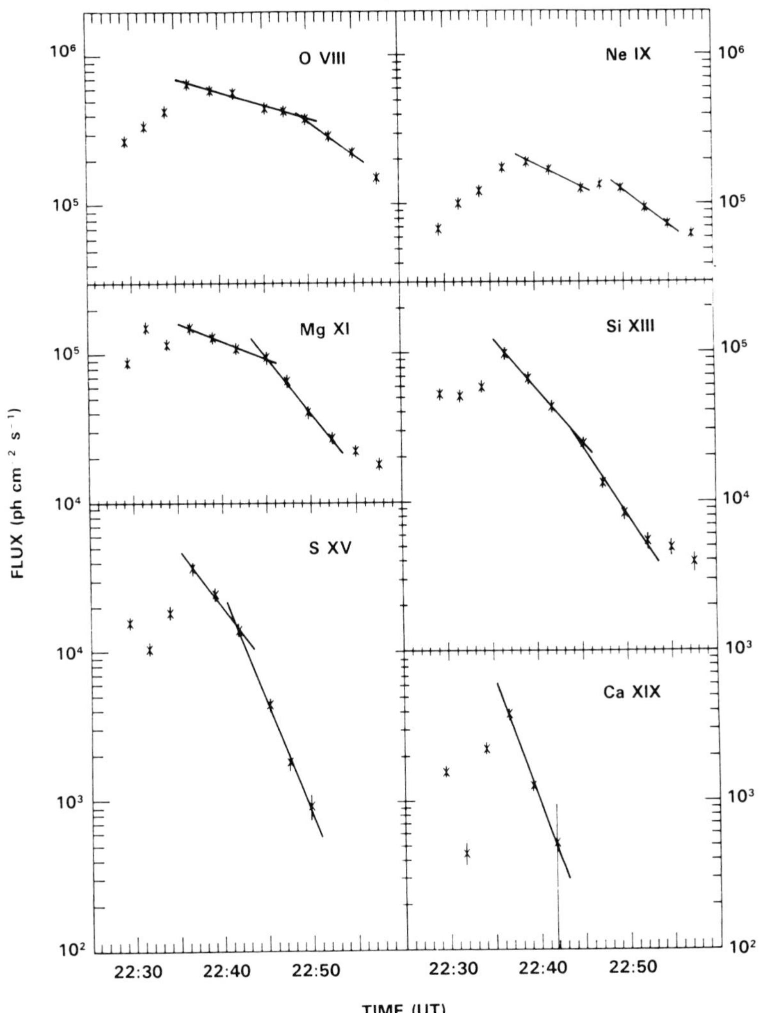

Figure 5.3.12 FCS fluxes for the 1980 November 5 flare illustrating the sharp changes in slope for each of the ions. Note how the ions formed at the higher temperatures have breaks earlier in time.

analyses both used the same data, they lead to different conclusions because of the manner in which the two techniques respond to possible problems in absolute flux calibrations; the DEM method assumes that the values given are correct, whereas the light curve method assumes that the plasma is isothermal and adjusts these values accordingly. The total emission measures in both cases agree to a factor of 2. The temperature derived from the light curve lies between the temperatures of the two peaks in the DEM. Further study of these discrepancies is in progress.

The volume derived from the soft X-ray emission measure and electron density (determined from line ratios) was found to increase from 7.2×10^{24} to 3.4×10^{25} cm^3 during the flare decay (Bornmann 1985a). The spatial extent of the flare, as recorded in the lowest-energy HXIS channel (3.5 to 5.5 keV) was not changing during this time, at least on the scale of the 8-arc-sec spatial resoluton. This indicates that the filling factor increased with time during the flare decay. The work done, if the increase in volume is due to expansion of the emitting material rather than to the addition of emitting material in neighboring regions, was $W = \int P\,dV > \int n_e\,k\,T\,dV$, or greater than 3.2×10^{28} ergs (Bornmann 1985a).

Bornmann (1985a) concluded that the mass of the soft X-ray emitting plasma remained constant during the decay phase, although the large errors on this measurement do not eliminate the possibility that the mass varied by a factor of two, reaching a maximum late in the decay phase.

The total radiative loss rate (Nagai 1980, Rosner et al., 1978) for the equivalent isothermal plasma derived from the temperature and emission measure was found to fall from 6.3×10^{26} to 1.9×10^{25} ergs s^{-1} during the decay of the second flare (Bornmann 1985a). The total radiative loss during this period was 2.8×10^{29} ergs.

The total thermal energy ($E_{tot} = n_e\,k\,T\,V$) of the plasma seen in the FCS pixel during the decay of the second flare was 4.2×10^{28} ergs at the time of peak soft X-ray flux, decreasing to 2.2×10^{28} ergs (Bornmann 1985a). Because FCS sampled only a small portion of the flare, the total thermal energy of the entire flare plasma was probably an order of magnitude larger. The thermal energy in the decay phase fell at a rate of 2.7×10^{25} erg s^{-1} (Bornmann 1985b), which is slower than the rate at which energy was lost by radiation. Therefore, additional heating during the decay phase is required.

The energy in mass motions fell off rapidly, at the rate of 1.8×10^{28} ergs s^{-1}. The conversion of the energy in mass motions into the observed radiative and thermal energies could be the source of additional heating, but the rate of decay of the mass motions was too rapid and would require an intermediate storage mechanism before appearing as thermal or radiant energy. The turbulent line broadening began at the same time that the blue shifts appeared and continued for several minutes. The BCS line widths indicate that the turbulent velocities peaked at 100 to 200 km s^{-1}, simultaneously with the hard X-ray peak, and fell essentially to zero during the decay phase. The peak turbulent energy, assuming an electron density of 10^{12} cm^{-3}, was 5.6×10^{27} ergs and fell at the slower rate of 5×10^{25} ergs s^{-1}. If this rate continued past the time at which turbulent broadening was observable, it could account for a significant fraction of the energy lost by radiation.

5.3.4 Flare Modeling

The above studies do not address the problem of the energy source function. Numerical simulations require an energy input function to reproduce the observed fluxes in the coronal lines. By adjusting the input energy profile to agree with the observed evolution of the X-ray line intensities, it should be possible to derive the heating function. The recent rise in interest in numerical simulations of solar flares is dealt with in more detail in Chapter 7. However, that section emphasizes the numerical techniques and differences between the codes. In contrast, the work done for our group by R. Pallavicini, G. Peres and S. Serio used a hydrocode to simulate two real flares.

The 1980 May 7 (15:00 UT) and November 12 (17:00 UT) flares were chosen for this study. In both cases the initial conditions for the models were determined from observations. The one-dimensional, time-dependent fluid equations were solved (see Peres et al., 1982; Pallavicini et al., 1983) subject to Gaussian energy input of the form

$$H = H_0 \, f(t) \, \exp[-(s-s_0)^2/(2\,\sigma^2)]. \qquad (5.3.15)$$

In the first stage of the analysis, s_0 was chosen so that the pulse was centered on the loop summit. In other work (Pallavicini et al., 1983; Peres et al., 1984), the fitting has been done for the pulse centered at the base of the loop and for energy deposition with electron beams. However, Pallavicini et al. (1983) showed that the exact location of the energy input did not make a significant difference to the global form of the computed light curves. The duration and amplitude of the heating during the rise phase of the flare are obtained by a best fit to the temperature and Fe XXV light curve. This approach is based on the results of Pallavicini et al., who showed that the Fe XXV light curves start to decay as soon as the heating level is decreased.

The May 7 flare was a compact, short-lived event. FCS observations show that it was confined within a single 15 arcsec pixel. An extensive analysis of this event, based on simultaneous SMM and Hα observations, has been given by Acton et al. (1982). The flare consisted of a single compact loop with a length (L) of 10^9 cm and cross-sectional area (A) of 9×10^{16} cm^2. The pre-flare temperature at the same location was determined to be $T_e = (2.5 \pm 0.5) \times 10^6$ K from the ratio of the intensites of the FCS Ne IX and Mg XI lines. The initial pressure and density have been inferred from the loop temperature and length by using the scaling law of Rosner et al. (1978). The derived initial pressure (P_0) was about 10 dynes cm^{-2}, implying a pre-flare coronal density of 10^{10} cm^{-3}.

The parameters and the time evolution function used to fit the observations were H_0 = 175 ergs cm^{-3} s^{-1}; s_0 = 5.7 x 10^8 cm; σ = 10^9 cm; $f(t) = t/150$ for $0 < t < 150$ s and $f(t) = \exp[(150 - t)/\tau]$ for $t > 150$ s.

In the initial model the decay time (τ) was assumed to be zero (i.e., an instantaneous switch-off of the heating function). This allows a good fit to the BCS Ca XIX and Fe XXV light curves during the rise phase of this flare (Figure 5.3.13). The lower-temperature lines from FCS cannot be used to further constrain the fit, as the temporal resolution available during this time was only 155 s. These parameters do not fit either of the BCS lines during the decay phase. The predicted fluxes decay faster than the observed ones. The decay time was increased to accommodate this problem; a value of about 60 s would give very likely the desired result. The large increase in density and temperature in the flare generally causes a decrease in all the characteristic timescales for a given loop model (e.g., the radiative cooling time, the conductive cooling time, and the sound travel time). This imposes a practical limitation on how far the evolution of the flare can be followed into the decay; in this work the model was limited to 20 s after the flare peak.

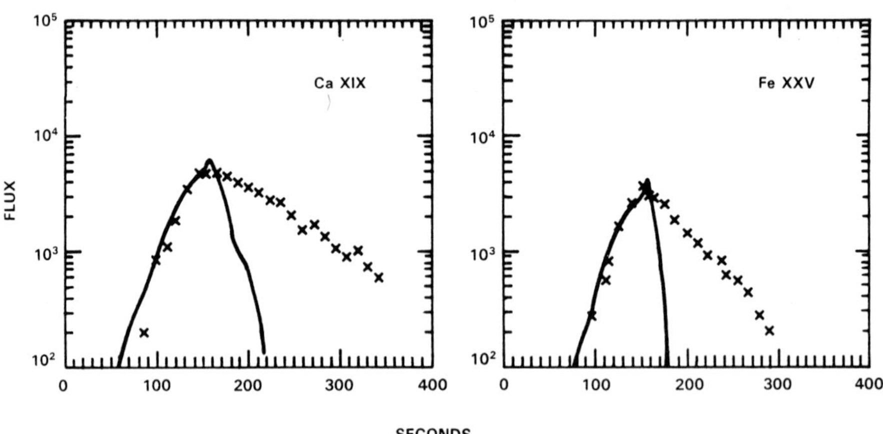

Figure 5.3.13 Predicted light curves (solid lines) from a loop model with an instantaneous switch off of the transient heating at t = 150s, compared to BCS data (crosses) for Ca XIX and Fe XXV (t = 0 corresponds to 14:56:36 UT). The disagreement between observed and predicted light curves in the decay phase indicates that additional heating is required after the peak in Fe XXV.

The thermal energy derived from these models was, typically, an order of magnitude larger than the gravitational and kinetic energies. From the modeling, the energy deposited during the rise phase of this flare was about 3×10^{29} ergs. Allowing for an additional energy deposition during the decay phase with $\tau = 60$ s, the total energy input was 5×10^{29} ergs. The energy deposition would increase if the energy were deposited at the loop footpoints. (For further details see Peres et al. (1985)).

The November 12 flare was substantially different from the May 7 flare in that it had a larger spatial extent (covering at least three 15-arc-sec pixels) and a somewhat lower density and pressure in the pre-flare state. For a more detailed description of this flare, see MacNiece et al. (1985). The soft X-ray light curves appear to be typical of most flares observed by FCS. Since the raster cadence was faster than for the May 7 flare, the simulations could be carried further.

The flare occurred in a loop with $L = 4 \times 10^9$ cm and $A = 2.5 \times 10^{17}$ cm^2. The pre-flare density and temperature were calculated as above to be 7×10^9 cm^{-3} and 3×10^6 K, respectively, corresponding to a P_0 of about

6 dynes cm^{-2}. The heat input was assumed to be at the loop summit (hence s_0 was 2 x 10^9 cm). The input was assumed to be 10 ergs cm^{-3} s^{-1} at a steady rate for a period of 180 s with a value for s of 5 x 10^8 cm during the rise phase of the flare. Several values of τ (0, 30, 60, and 140 s) were tried to fit the decay phase. The best fit to the observations was τ = 60 s, which was mainly determined from fitting the Fe XXV light curve, owing to the strong temperature dependence of the emissivity of this line.

Figure 5.3.14 shows a comparison of observations and numerical simulations for the FCS and BCS light curves. The predicted fluxes appear to be in agreement with observation for all lines except Ne IX. Not only is the general time evolution well reproduced, but the absolute flux values also agree with observations within the estimated calibration accuracy. The discrepancy found for Ne IX is not readily explained except in terms of instrumental effects, such as blending with nearby Fe XIX lines and/or crystal fluorescence. Note that the choice of τ = 60 s is made to fit only the Fe XXV light curve.

From the above results, it is possible to evaluate the energy budget of the November 12 flare. Figure 5.3.15 shows E_{th} as a function of time. Again, E_{th} dominated all the other sources of energy by an order of magnitude at any time during the flare. The total energy deposited into the loop for τ = 60 s was 7 x 10^{29} ergs.

The decay time we have adopted for the heating term has little direct relationship with the decay times of the light curves of the lines of lower ionization stages. In fact, the evolution of the light curves of the "cooler" ion is dominated by the evolution of emission measure and temperature (remember that during evolution, the loop's temperature is hardly uniform). Instead, as soon as the heating is switched off, the Fe XXV radiative efficiency is a very steep function of the temperature up to 70 x 10^6 K (see Figure 8 of Pallavicini et al., 1983). On the other hand, the emission measure of the cooler plasma (say at temperatures near the excitation temperature of O VIII) will keep increasing, essentially because the cooling of the hotter component will enrich the emission measure of the cooler one, if the maximum temperature in the loop is less than 70 x 10^6 K. Therefore, a decay of the Fe XXV line is a good indication of a decay in the heating term, due to steepness of the Fe XXV radiative efficiency.

Actually, in the simulations for the November 12 flare, if the heating is switched off abruptly, Fe XXV decays much faster than observed. For this reason we need some smoother decay for the heating. An exponential decay with τ = 60 s gives the best results for Fe. A posteriori, we also see that it gives agreement with the cooler lines, although for these last lines the value of τ, as explained, is less significant. For the May 7 flare light curves, we can infer that τ = 0 does not agree with the observations. The appropriateness of τ = 60 s is only guessed by the shape of the light curves and by our experience of many other simulations.

In summary, if one allows for heating in the decay phase of flares, good agreement with observations can be achieved. The success of this particular modeling effort lies in the fact that a reasonable fit to the form and, in most cases, the amplitude of the light curves of the remaining BCS and FCS lines was obtained, even though the parameters were selected to give a good fit only to the Fe XXV light curve. These results indicate that the bulk properties of the decay phase can be represented quite well by such a fluid model.

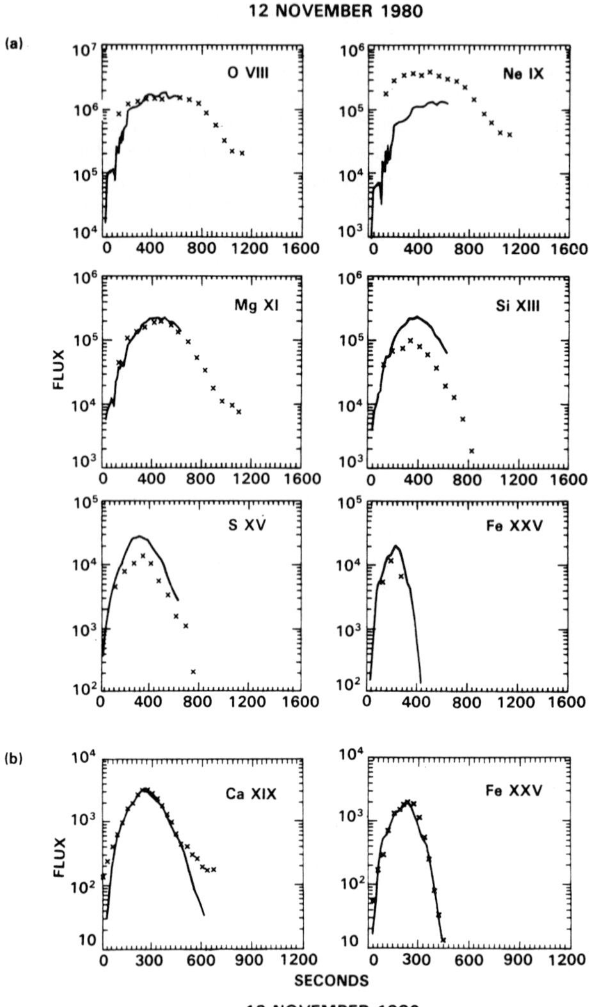

Figure 5.3.14 Comparison of the predicted (solid lines) and observed (crosses) fluxes for the XRP soft X-ray lines in the 1980 November 12 flare. The model has steady heating to t = 180s, followed by an exponential decay with an e-folding time of 60s (t = 0 corresponds to 17:00:00 UT). (a) Comparison for the six FCS channels. (b) Comparison with the BCS lines.

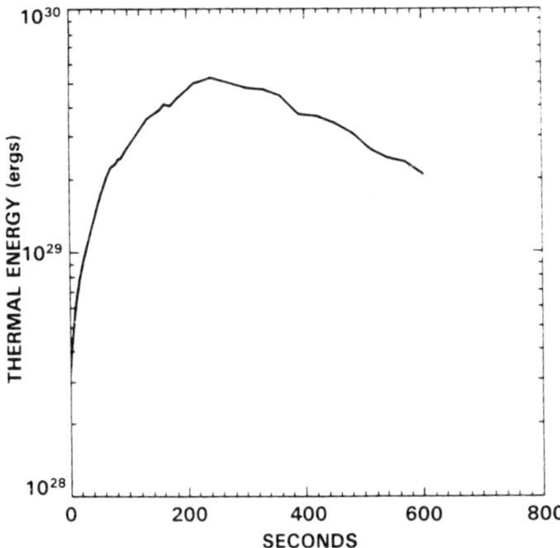

Figure 5.3.15 The evolution of the thermal energy in the 1980 November 5 flare as predicted by the dynamical loop model.

5.3.5 Discussion

So far in this section, we have outlined the methods of determining the physical parameters of our five prime flares and have discussed the energy budget at different times during those events. Having determined the temperature, density, and length of the flare loops, we could follow Moore et al. (1980) and construct a diagnostic diagram based on the order of magnitude cooling times (Equations 5.3.7-5.3.9) to determine whether our flares require long-term heating. We follow a similar procedure, but first it is important to record the progress that has been made in our theoretical understanding of flare cooling in recent years.

An important physical property of the flare plasma is how the DEM scales with temperature. We find that for $5.5 < \log(T_e) < 7$ (Dere and Cook, 1979; Widing and Spicer, 1980; Acton et al., 1983),

$$q = A\, N_e^2\, T\, (dT/ds)^{-1} \propto T^{\zeta}, \qquad (5.3.16)$$

where A is the emitting cross sectional area and ζ is typically 3.5. If we can find a cooling mechanism that gives ζ = 3.5, we may be able to find out how flares cool. Antiochos and Sturrock (1976) considered static conductive cooling in which the downward heat flux is dissipated in the chromosphere, presumably by radiation. They also included a non-uniform magnetic field structure in which the field is stronger at the footpoint of the loop. Underwood et. al. (1978) found that ζ = 0.5, and for the area change to have significant effect on the cooling time (a factor of about 3), the magnetic field had to change by a factor of 20 from the loop top to bottom. A refinement of this model was made by Dowdy et al. (1983), who considered a variety of flux-tube geometries, and concluded that the shape and amount of divergence were important in determining the magnitude of the inhibition of the heat flux. A slightly different approach to conductive cooling was taken by Van Hoven (1979), who considered the static cooling in a twisted flux tube with constant cross-sectional area and showed that the more the flux tube was twisted, the more the conduction was inhibited. Clearly, twisting gives rise to longer field lines, which makes conduction a less efficient mechanism. However, none of these variations in static conductive cooling makes any change in the form of q. Antiochos and Sturrock (1978) relaxed the static assumption and considered a model in which the chromosphere could respond to a downward heat flux by moving (or evaporating) upward. This model differs in an important way from that in their previous paper (Antiochos and Sturrock 1976), since the thermal energy of the flare plasma is conserved, so that all the energy in the downward heat flux at the base of the model goes into driving an upflow. They found that $\zeta = -0.5$ but noted that no energy is lost in radiation either implicitly or explicitly, so that the effect of evaporation will be overestimated. They did not provide a comparison between static and dynamic cooling. Since the conductive models did not appear to fit the slope of the DEM, Antiochos (1980) then tried radiative models. He found values of ζ between 1 and 2, depending on the form of the loss function chosen. Some of these models include a subsonic downflow, and as an extension to this, Antiochos and Sturrock (1982) included steady-state supersonic downflows driven by a pressure gradient between the loop summit and base. They found that ζ could then take on a value of 3.5 and concluded that the decay of the solar flares involved a large downward velocity at temperatures between 5×10^5 and 10^7 K. However, for the flare loop to drain at a reasonable rate (i.e., in a few minutes), they invoked an order-of-magnitude change in the flux tube width from top to bottom. In support of the idea of flare cooling via supersonic downflows, we note that Dere and Cook (1979) have observed large pressure differences between the top and base of a flare loop. However, BCS has not observed obvious red shifts of the order of magnitude required during flare decays. This could be due to a number of factors: the downflowing plasma could be at a lower temperature than BCS can see; and further, it is not an absolute spectrometer and so all the line shifts are relative.

One common feature of the above studies is that large changes in the cross-sectional area of the flux tubes with altitude are invoked. The magnetic field is concentrated in the lower atmosphere to strengths of 1-2 kG (Stenflo 1976). We have only a rough idea of the actual coronal magnetic field strength. Estimates are typically in the range of a few hundred gauss (Schmahl et al., 1982), giving a magnetic divergence of about 3 or 4. It also seems that most of the divergence occurs in the lower atmosphere (Stenflo 1976), i.e., in a region in which the physical validity of the above models is doubtful. It is therefore difficult to accept that flux divergence plays a significant role in the flare cooling process, and so the supersonic cooling model does not seem viable.

Finally, the decay phase of flares has been studied numerically (Antiochos and Krall, 1979; Antiochos, 1980; Nagai, 1980; Doschek et al., 1982; Pallavicini et al., 1983) with similar results. The first four of these papers all obtained large scale, supersonic downflows in the decay phase. The velocity of these flares was typically 50 - 70 km s^{-1}, implying that for them to be supersonic, the temperature had to be less than about 2 x 10^5 K. It is difficult to see how such flares can explain the slope of the emission curve up to temperatures of 10^7 K. The Pallavicini simulation, although run for long enough, does not discuss the dynamics of such flows. It therefore appears that the physical mechanism responsible for the $T^{3.5}$ slope of the DEM is still unclear. Supersonic downflows seem to come closest to explaining this phenomenon, but the problem of the extremely short draining timescales must be resolved. One potential solution is to relax the fluid approximation, as has been done in quiet-Sun studies by Shoub (1983).

Another problem in studying flare cooling is that of filling factors. As has been stated in Section 5.3.2, the possibility of filling factors of the order of 1% or less means that the electron density can be wrong by an order of magnitude if derived directly from the emission measure and estimated volume. The obvious immediate effect of this is to reduce the radiative cooling time and increase the conductive cooling time by the same factor. This effect can easily be allowed for in the cooling time formulae. However, such an adjustment may be too simplistic. It assumes that any fine structure giving rise to such a small filling factor is composed of long, thin strands which still communicate to the lower atmosphere and still lose energy by conduction, although at a reduced rate. An alternative picture is that the filamentation gives rise to small, isolated regions in the corona that do not communicate with the photosphere and so are unable to lose their energy by conduction. An example of this is the magnetic islands that arise in the tearing mode (Van Hoven, 1981). These small parcels of plasma can then only cool by radiation and so give cooling times inconsistent with the values derived from the equations simply adjusted for the filling factor. Clearly, our understanding of the coronal fine structure during flares will be important in determining the bulk properties of the decay phase of flares. It is worth noting that the three SMM flares (April 8, April 30, and November 5) in which we can compare the density derived from X-ray line ratios with that from the emission-measure volume approach are all large volume flares. An interesting goal for the renewed SMM operations should be to make equivalent observations of a compact flare.

A major source of error is that the behavior of a plasma in a temperature gradient is not well understood. The form of the conductive cooling time derived earlier relies on the Spitzer-Harm (1953, hereafter SH) description of heat conduction, which descibes a plasma only slightly displaced from a Maxwellian. In fact, under not very rigorous circumstances, the SH form of the heat conduction coefficient fails, and in flare plasmas it may fail badly. Rosner et al. (1984) have given a concise review of these failings, and we now paraphrase their arguments as applied to flare plasmas. The SH description of heat conduction is valid only if the mean free path (λ) is much smaller than the temperature scale height [L(T)], where the ratio of the two, R, is defined as

$$R = \lambda/L(T) = 10^{10} F_c/(N_e T^{3/2}) \ll 1. \qquad (5.3.17)$$

The inequality is strong in the sense that if R = 1/30, the SH calculation overestimates the heat flux by a factor of 2 (Bell et al., 1981). In the case where R = 1 (the free-streaming limit), the error is at least a factor 10, and laboratory results suggest that the error may be even larger. For a flare heat flux of 10^{10} ergs cm^{-2} s^{-1} and a density of 10^{12} cm^{-3} at a temperature of 10^6 K, i.e. numbers characteristic of the lower corona, we find R = 0.1. Care must be taken in numerical work to check how well the SH method models the heat flux; Craig and Davys (1984) outline a method that involves a gradual reduction of the heat flux as one moves away from the SH regime rather than the sudden jump used by many other authors.

Let us quantify these ideas by considering the five prime flares. Table 5.3.5 shows the principal parameters of each flare. The cooling times due to conduction and radiation with a filling factor of unity have, in some cases, large uncertainties as a result of lack of accurate information about the loop length. The other timescales given in Table 5.3.5 are as follows: τ_{motion}, the time constant of the temperature decay due to enthalpy flux as proposed by Veck et al. (1984) and defined by Equation 5.3.9; τ_{drain} (= L/C_s), where C_s is the sound speed, is the time taken for the material to drain from a loop assuming a large pressure gradient and supersonic flows.

First, note that τ_{drain} is much shorter than both τ_{obs} and τ_{motion}. The difference between these timescales is that the former allows for supersonic motions and the latter does not. Thus, the fact that $\tau_{drain} \ll \tau_{obs}$ suggests that the supersonic downflows postulated by Antiochos and Sturrock (1982) are not present. Consider the loop to be filled with a plasma for which the filling factor is unity and the thermal conduction

Table 5.3.5 The Principal Parameters and Timescales of the Prime Flares

Date 1980		Apr 8	May 21	Jun 29	Aug 31	Aug 31	Nov 5
Time (UT)		03:07	21:00	18:21	12:49	12:52	22:33
T_e (10^6K)		8.9	10.0	20.0	12.9	14.8	8.9
n_e ($10^{11}cm^{-3}$)		0.4	1.0	3.5	3.2	3.8	1.4
L (10^9cm)		1.6	3.2	6.3	0.5	0.5	1.6
τ_c	(s)	250	950	2400	50	46	1280
τ_r	(s)	4500	2000	1150	800	750	230
τ_{motion}	(s)	4600	3960	1163	720	600	
τ_{drain}	(s)	67	95	20 - 80	14	13	52
τ_{obs}	(s)	1200	12000	1300	25	45	600

is described by SH. Then the May 21 and November 5 flares both require long-term energy deposition, since the difference between their calculated and observed cooling times are so significant. The other flares, however, would not appear to require further energy deposition. This is consistent with the predictions of Moore et al. (1980).

If the filling factor is now set to 0.01 but still with SH, the radiative cooling times decrease by an order of magnitude. Then the May 21 and November 5 flares cool more quickly than before, but the June 29 flare would also require long-term heating, since it cools very effectively by radiation. Neither of the August 31 flares would require any heating, but their cooling is now dominated by radiation rather than by conduction. This is a function of our choice of filling factor, since a filling factor of 0.1 to 0.0001 would make this conclusion more doubtful. However, if thermal conduction is inhibited but the filling factor is 1.0, then the August 31 flares may need long-term heating, whereas the June 29 flare does not.

Even with these simple approximations, the overall picture of flare decay is confusing. All that can be concluded is that if the observed decay time is longer than the radiative cooling time for a filling factor of unity, then long-term energy deposition is required, since any smaller filling factor will simply make the plasma cool more quickly. However, if the flare appears to cool principally by conduction with the right timescale, care must be used since other factors can invalidate this conclusion. One must also be careful when invoking long-term heating, since the need for this can be removed by a combination of physical effects. Peres et al. (1984) found that the computed parameters decayed too rapidly when their model was compared with the observations, and so concluded that long-term heating was required. They found a very simple function that reproduced the observed light curves quite well. However, it is equally possible that the extended decay could be caused by a filling factor effect or by a conductive heat flux limitation. We can therefore only conclude that the question of how a flare plasma cools is less clear than it was before. In reality, our knowledge of the microphysics of plasmas is so poor that we must consider a large number of possible models before making any positive conclusions.

5.3.6 Summary of Conclusions

Let us address the five conclusions of Moore et al. (1980) and suggest how future theoretical efforts could clarify them further.

(1) We find that, under the assumption of filling factors equal to one, the densities obtained in these five flares are consistent with those of Moore et al. However, the SMM results suggest that the particles are contained within filamentary structures occupying no more than 1% of the observed volume (de Jager et al., 1983; Wolfson et al., 1983). Thus the densities quoted from the Skylab results are too small by at least an order of magnitude. The large uncertainties in the density and possibility of fine structure dictate our response to the remaining four questions.

(2) Depending upon the flare parameters, either conduction or radiation can dominate the cooling process. Mass motions may also play a role in energy transport, a point not considered by Moore et al. We therefore disagree with their conclusion that conduction and radiation

are generally equally important. For small filling factors, radiative cooling will dominate.

(3) In the May 21 flare, continued heating is needed, confirming the conclusions of Moore et al. The understanding of how this heating can occur has improved on the basis of the work by Forbes and Priest (1982, 1983a, b) and Cargill and Priest (1982, 1983), which is discussed in Chapter 1 of this report.

(4) The question of whether compact flares need long-term heating is now open again as a result of the filling factor problem. We do not know what type of filamentary structure exists in such flares. One possible way of heating such flares is via turbulence (Bornmann, 1985c).

(5) The chromospheric evaporation scenario proposed by Moore et al. seems to be confirmed by our work on these five flares.

Thus, two of the original conclusions of Moore et al. have been unambiguously confirmed by our work. Unfortunately, the SMM data have confused the other three; but they still could be true. A crucial issue for theorists and observers is to determine the nature of the fine structure in flares. This would appear to hold the key to our understanding of the decay phase.

5.4. RELATIONSHIPS AMONG THE PHASES

C.J. Crannell and H.S. Hudson

5.4.1 Introduction

The overall flare process involves phenomena we have characterized as the "impulsive" and "gradual" phases, following the X-ray signature first recognized by Kane (1969). In addition, evidence exists for a pre-flare phase in some flares, and recent SMM data have shown that a post-flare phase, in which extensive and energetically important coronal activity occurs, may also exist. The data to describe the pre-flare and post-flare phases are insufficient to place them properly into an overall picture of the energetics, aside from noting that these phases may indeed be significant from the energetics point of view. In this section, therefore, we review what is presently known and comment about the possible interactions among the flare structures involved.

5.4.2 Relationship Between Impulsive and Gradual Phases

The distinction between the impulsive and gradual phases of a flare was originally made by Kane (1969). The energetics relationship between these phases has been controversial from the beginning, when Kane and Donnelly (1971) showed that the large energy in 10-1030Å bursts correlated well with the energy inferred for 10 to 100 keV electrons, assuming non-thermal bremsstrahlung as an explanation of the hard X-ray bursts. This was the first real evidence that particle acceleration during the impulsive phase could have energetically significant consequences in the chromosphere, the source of the EUV flashes. We now have far better data with which to examine this question quantitatively, and this subsection deals with the investigation of the energetic relationship between the impulsive and gradual phases.

That the impulsive and gradual emissions are related can be seen in Figures 5.4.1 and 5.4.2, in which the peak counting rates of hard X-ray bursts (Figure 5.4.1) and the total hard X-ray counts (Figure 5.4.2) are plotted versus the peak Ca XIX soft X-ray counting rate. These scatter plots include all events observed with HXRBS and BCS which have both a peak hard X-ray counting rate greater than 100 counts s^{-1} and a peak soft

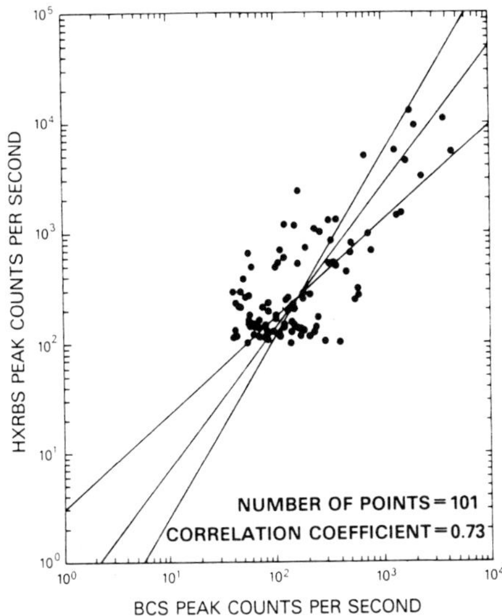

Figure 5.4.1 Scatter plot of the peak HXRBS counting rate versus the peak BCS counting rate in the Ca XIX channel for all flares recorded jointly by these two instruments in 1980 with >100 counts s^{-1} in HXRBS and >40 counts s^{-1} in BCS. The three lines were obtained from least-squares fits to the points minimizing the vertical, horizontal, or perpendicular distances of the points from the line.

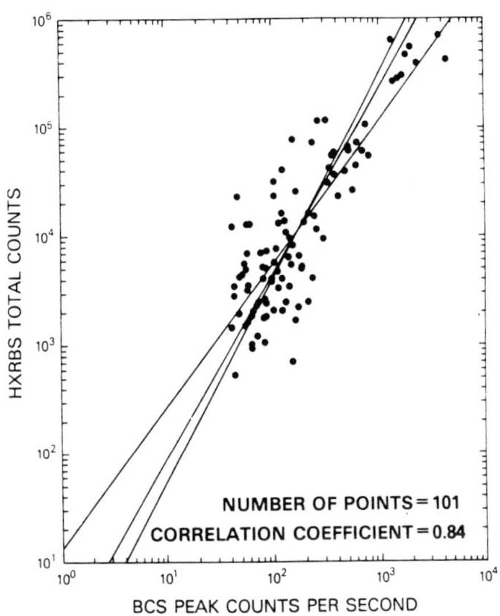

Figure 5.4.2 Same as Figure 5.4.1 but with the HXRBS total number of counts integrated over the duration of the flare (with a 40 counts s^{-1} background subtracted) substituted for the HXRBS peak rate.

X-ray counting rate greater than 40 counts s^{-1}. The total hard X-ray counts (TOTAL) are better correlated with the peak Ca XIX counting rate (BCS) than is the peak hard X-ray counting rate. This result supports the conclusion of Neupert (1968) that the gradual soft X-ray emission resembles an integral of the impulsive hard X-ray emission.

The significance of these results must be evaluated in view of the Big Flare Syndrome (BFS) identified by Kahler (1982). He found, quite simply, that bigger flares are bigger at all wavelengths. Quantitatively, the BFS is manifest as correlation coefficients of approximately 0.48, with a range of 0.3 to 0.65, between widely diverse parameters. The correlations shown in Figures 5.4.1 and 5.4.2 clearly indicate a closer relationship than would be expected from the BFS.

In an earlier investigation of the relationships between the energetics of the gradual and impulsive emissions, Crannell et al. (1982) found a good correlation between the energy in the energetic electrons, estimated from hard X-ray and microwave observations analyzed with a thermal model, and the energy in the soft X-ray emitting plasma estimated from SOLRAD observations. The value of the correlation coefficient dtermined in that work lies in the range 0.8 to 0.9, comparable to that found here for TOTAL versus BCS. In Section 5.2 a similar comparison based on a non-thermal, thick-target loop model of the impulsive emissions is presented. The correlation found there also is comparable to that for TOTAL versus BCS.

From these results, we conclude that the energetics of the gradual and impulsive emissions are more closely related than are parameters which characterize the BFS. On the other hand, we note with disappointment that the data we have considered are not sufficient to distinguish between the thermal and the non-thermal model of the impulsive emissions and, in fact, do not enable us to distinguish between the models and a non-model, i.e. TOTAL versus BCS.

5.4.3 The Pre-Flare Phase

The pre-flare conditions are the subject of an entire chapter of this workshop. As far as this chapter is concerned, it is important for a given flare theory that the energy source in the pre-flare state is adequate to supply the flare energy release. Most modern flare theories derive the energy input from the energy stored in stressed magnetic fields. The magnitude of the stresses can be estimated from the photospheric magnetic field distribution, with a model of the coronal field that permits field-aligned currents to maintain the stress (Tanaka and Nakagawa 1973). The amount of stored energy then follows from the model, which can derive the energy build-up rate from the photospheric boundary conditions. Such models of magnetic storage and free energy generation in general contain adequate energy for flares. This consistency of observations with models is mainly a result of inadequate data upon which to determine exact model parameters.

5.4.4 The Post-Flare Phase

A new phase of flare activity, namely a late, high-altitude, coronal phase, was first identified in SMM data for May 21/22 by Svestka et al. (1982a). Other similar events from SMM were reported by Lantos et al. (1982) and Svestka et al. (1982b and c), and in retrospect some of the large-scale structures observed by Webb and Kundu (1978) probably fall in this category.

Based upon HXIS observations, Svestka (private communication) has estimated that the total energy content of the post-flare arch structures in the 21 May event was between 1.5×10^{29} and 4.7×10^{30} ergs; the total energy in the analogous structure of the November 6 event was 1.2×10^{31} ergs. In both cases, the estimated energies are comparable to the energies of other major flare components. It is necessary in a complete picture of flare energetics to understand the relationship between these manifestations and those of the more well-known flare phases. It is premature, based upon the limited number of events -- not all of which may be of the same type -- to draw general conclusions yet.

5.4.5 Phenomena in the Distant Corona

Coronal transients, coronal mass ejections, interplanetary shock waves, and the like have an uncertain but important place in flare energetics. These phenomena can be observed by coronagraphs and by meter-wave radio telescopes, as well as by in situ techniques at larger distances from the Sun. The Skylab coronagraph observations provided the earliest comprehensive views of coronal transients (Rust et al., 1980), and observations have continued both in space (SMM and P78-1) and on the ground (notably with the Mauna Loa K-coronameter).

The relationship of these coronal phenomena with classical Hα "chromospheric flares," or with the high-energy flare events, remains problematical. There is no doubt that major flares often produce major coronal transients, but we have to guard against inferring a causal relationship: the BFS may confuse the picture (Kahler 1982). Indeed the suggested existence of "forerunner" coronal transients (Jackson and Hildner 1978) could imply that the coronal phenomena cause the flare rather than the other way round, and this is consistent with some theoretical views. The relationships are obscured by two major factors: there are only limited quantitative observations in the key inner corona, and in the outer corona there is confusion and uncertainty in the assignment of a given event to a given flare because of overlapping in time. Finally, it is known that coronal transients, especially with low speeds, may arise without the occurrence of a flare (Wagner, 1984). These events tend to fall in the "eruptive prominence" classification.

The energetics analysis of coronal phenomena has not advanced appreciably since the Skylab Workshop treatment (Webb et al., 1980). Among the prime flares studied by the energetics team in this chapter, only one (June 29) had C/P observations. However, even this limb flare was not satisfactory for quantitative energetics analysis because it could not properly be compared with the disk flares in the remainder of our list.

5.5 CHARACTERIZATION OF TOTAL FLARE ENERGY

H.S. Hudson

5.5.1 Statement of the Problem

5.5.1.1 Introduction

The total energy released by a solar flare has a certain distribution in form as well as a certain pattern of flow among the several forms, as described above. As data have grown more comprehensive, the definition of this distribution has improved; classical assessments are found in the works of Ellison (1963), Bruzek (1967), Smith and Smith (1963), and Smith and Gottlieb (1975). Most recently the Skylab flare workshop (Sturrock 1980) addressed this question in two surveys of a single well-observed flare on 1973 September 5. These surveys dealt with the radiant energy (Canfield et al., 1980) and the mechanical energy (Webb et al., 1980), and their results have become the definitive data on flare energetics despite acknowledged gaps in coverage and in theoretical understanding.

This section aims at updating our knowledge of this fundamental matter. Unfortunately, there are still limitations in data coverage, as described in detail below. In some areas, notably the X-ray and gamma-ray ranges, there have been striking improvements, as reported above. We summarize the improvements here and take the further step of attempting to fill in the gaps in coverage to estimate the total radiant energy. One purpose for doing this is to permit a comparison of the observed or estimated total with the upper limits derived from the precise total-irradiance monitor (the Active Cavity Radiometer Irradiance Monitor-ACRIM) on SMM.

5.5.1.2 Availability of Data

What are the key limitations in the data set available to us? The foregoing discussions have naturally emphasized the observed forms and have relied on theoretical considerations to bridge the gaps. Where are the largest gaps? We discuss these items briefly here and present recommendations for future observations in Section 5.6.

The most important omissions from the data set fall into two major areas: the radiant energy in optical and EUV wavelengths, and coronal observations of all types. The brightness of the quiet Sun makes the optical wavelengths particularly important, since an undetectably small optical continuum (for example) could rival the energy in wavelength ranges for which flares produce greater contrast. The Sacramento Peak observations from the white-light flare patrol show that continuum emission is not the rarity once thought (Neidig and Cliver 1984), and the line emission spectrum may contain even more energy. Even where spectrographic observations exist, these usually have covered only a fraction of the flare area or time profile, so that no definitive knowledge of the distributions even in a statistical sense now exists.

At EUV and XUV wavelengths a similar lack of coverage is striking; even though SMM carried an EUV instrument, its observations were extremely limited in coverage by a lack of imaging capability and appropriate telemetry bandwidth. SMM did not carry an XUV instrument, and so this vital wavelength range was totally omitted from consideration either from the diagnostic or the energetic point of view. Finally, considerable flare energy could appear at infrared wavelengths, but because of the simplicity of the emission mechanisms (H and H-free-free and free-bound continua should dominate), the omission of direct observations is less important than that of the optical and EUV-XUV ranges.

The optical and EUV-XUV data provide us with our best knowledge of the magnetic field distribution. Although vector-magnetograph observations began in a systematic way during the past solar maximum (Hagyard 1984), data that are extensive, precise, and sensitive enough to characterize the field adequately do not exist. As a result, we have only rough knowledge of the system of currents flowing in and around a flaring active region.

More serious omissions occurred in the area of coronal observations. The existing coronal observations have neither the sensitivity nor the diagnostic capability to contribute effectively to determining the key physical conditions in the corona. This is an enormous loss, because the coronal aspects of solar flares (at least in some cases, such as that of the well-documented Skylab flare of 1973 September 5) may dominate the total energy. In addition, the white-light coronagraph on board the SMM provided observations of only one of the well-observed prime flares chosen for detailed study here.

Because of these problems in availability of coronal data, this report does not address the non-radiant energy of a solar flare. This problem is reserved for a future solar maximum.

5.5.2 Techniques for Estimating Radiant Energy

5.5.2.1 Direct Observation

The radiant energy would ideally be determined from a full knowledge of the specific intensity as a function of wavelength, position, and time. Unfortunately, we must deal with integrals or small samples of this abstract function. Since our goal is to define the integral radiant energy in broad wavelength bands, fully detailed spectral information is important only for diagnostic purposes. At the highest energies, the coverage from broad-band soft and hard X-ray and gamma-ray detectors adequately describes the total flare radiation. The very large contrast of flare radiation at these wavelengths makes background subtraction, even for the disk-integrated radiation, relatively simple.

5.5.2.2 Differential Emission Measure (DEM)

Although observations are rarely available for wavelengths longer than about 20Å, there is generally sufficient diagnostic information to characterize a DEM, as described in Section 5.3.2.1. With such a tool one can turn to a tabulation of the characteristic radiations for a plasma of the proper temperatures and, with assumptions about plasma conditions (abundances and state of equilibrium), estimate the total theoretical luminosity. Cox and Tucker (1969) provided one of the first systematic tabulations of characteristic plasma radiations.

The DEM approach works well enough for the present application at temperatures above a few million Kelvin, and this technique provides a much better characterization of the X-ray emission from flares than was previously possible. Unfortunately, the missing wavelength ranges contain emission lines necessary to define the temperature domain from a few million Kelvin down through the transition region, and so this domain requires an extrapolation to obtain complete coverage.

5.5.2.3 Scaling

For the lowest temperatures, including those responsible for the dominant optical-EUV-XUV radiations, the DEM approach cannot work in principle because the fundamental assumption of this techinique, i.e., that the strong resonance lines (and possibly the continuum) are optically thin, is incorrect. For these wavelengths, the only possible route to obtaining energy estimates is through scaling from a well-observed representative wavelength such as Hα. Unfortunately, even the Hα line seldom is observed with simultaneous line profile and imaging data, so that it cannot be used as a reference; furthermore, no detailed studies of the errors induced by the scaling approach exist. Is the rest of the Balmer series inferrable from the Hα line alone? Is the line-to-continuum ratio approximately a constant from flare to flare, or across the space or time profiles of a given flare? Is the scaling approach adequate for regions of different morphological class -- for example "broad line" and "narrow line" regions -- separately? The answers to these questions do not exist at present and will require the accumulation of a more complete data base.

5.5.3 Determination of Radiant Energies

5.5.3.1 Measured Energies of Different Components

For each of the prime flares, Table 5.5.1 gives the total radiant energies derived from the data in different energy or wavelength ranges. These observations do not cover the entire spectrum, as discussed in detail above. The table entries cannot be simply summed because of overlap, for example between the 0.5-4Å and the 1-8Å GOES broad-band soft X-ray entries. Suitable optical data were only available for a single flare (1980 November 5). By "suitable" we mean observations of Hα with sufficient spatial, temporal, and spectral coverage to permit a rough estimate of the excess radiation. Such data come either from the Multi-Slit Spectrograph (see Section 5.2) or else from Sacramento Peak CCD observations (e.g., Gunkler et al., 1984), and unfortunately such observations are presently available only infrequently.

The purpose of Table 5.5.1 is to enable us to make intercomparisons among these observable components for these and other solar flares, and

Table 5.5.1 Component Radiant Energies (in ergs)

	Radiant Energy (ergs)				
Date in 1980 SXR Peak Time (UT)	Apr 8 03:07	May 21 21:05	Jun 29 18:26	Aug 31 12:52	Nov 5 22:35
> 1 MeV*	—	$<1\times10^{23}$	—	$<1\times10^{23}$	$<3\times10^{22}$
> 300 keV†	$<8\times10^{21}$	3×10^{22}	$<8\times10^{21}$	2×10^{22}	8×10^{22}
> 25 keV	1.5×10^{24}	1.1×10^{25}	1.1×10^{24}	1.2×10^{24}	2.8×10^{24}
0.5 - 4Å	1.7×10^{28}	9×10^{28}	—	—	1.3×10^{28}
1 - 8Å	1.9×10^{28}	6×10^{29}	1.0×10^{28}	—	8×10^{28}
Hα	—	—	—	—	2.3×10^{29}

*2σ upper limits based on nuclear gamma-ray line component, only.

†2σ upper limits and values from positive detections based on power-law fits to the observed continuum between 300 keV and 1 MeV, integrated to infinity.

to allow us to make comparisons with other, possibly related, phenomena such as stellar flares.

5.5.3.2 Estimates of Total Radiant Energy

With these component radiant energies and the methods described above, we have made estimates of the broad components of total flare energy. These are given in Table 5.5.2, as measured by the three techniques: the direct estimations give the energy above 25 keV; the DEM analysis gives the soft X-ray energy, essentially 1 - 25 keV; and the scaling method gives the optical-EUV-XUV component. The typical

Table 5.5.2 Total Energies (in ergs) of Component Radiations

Date 1980	SXR Peak Time (UT)	Soft X-ray $T > 2\times10^6$K	$T > 5\times10^6$K	$T > 10^7$K	Scaled $T < 2\times10^6$K	Hard X-rays > 25keV	Hα
8 Apr	03:07	—	—	1.1×10^{30}	—	1.5×10^{24}	—
21 May	21:05	—	5.8×10^{30}	2.6×10^{30}	—	1.1×10^{25}	—
29 Jun	18:04	—	—		—	5.4×10^{22}	—
	18:26	—	—	3.8×10^{30}	—	1.1×10^{24}	—
31 Aug	12:49	7.4×10^{28}	—	4.5×10^{27}	—	6×10^{23}	—
	12:52	7.6×10^{29}	—	5.6×10^{28}	—	6×10^{23}	—
5 Nov	22:28	—	1.3×10^{29}	5.1×10^{28}	—	3.3×10^{23}	
	22:35	—	4.0×10^{29}	2.9×10^{29}	1.8×10^{30}	2.5×10^{24}	2.3×10^{29}

uncertainties of these components are, respectively, 50%, 20%, and a factor of 10. For the direct estimates, the uncertainty comes from the photometric accuracy of the X-ray and gamma-ray detectors at higher energies; for the integrated energy above 25 keV this accuracy is dominated by the lack of spectral resolution in the determination of steep spectral distributions. In the soft X-ray region, we have a good check on the accuracy of the DEM approach from the broad-band soft X-ray photometry from GOES data. As noted in Section 5.3, the DEM in the 10^6 to 25 x 10^6 K range tends to show a bimodal distribution, with one "low-temperature" peak in the vicinity of normal active-region temperatures and one "high-temperature peak" above 10^7 K. Since these are distinctly resolved, we have made separate entries for them. In the rest of the spectrum, there are insufficient data to estimate the ucertainties, and the result has little significance.

5.5.4 Comparison with Total-Irradiance Upper Limits

The radiant energies estimated in Table 5.5.2 could in principle be observed by using ACRIM. A preliminary search for flare effects in the ACRIM data was carried out by Hudson and Willson (1983), with the result that only upper limits could be established. Similar limits appear in Table 5.5.3 for the prime flares discussed above. These limits consist of comparisons between the SMM orbit containing the soft X-ray flare maximum and adjacent orbits; the data themselves appear in Figure 5.5.1. In no case was a significant excess detected. The table expresses the results in terms of 5σ upper limits on total radiant power and energy over the one-orbit interval indicated. Hudson and Willson (1983) give further details on the treatment of data.

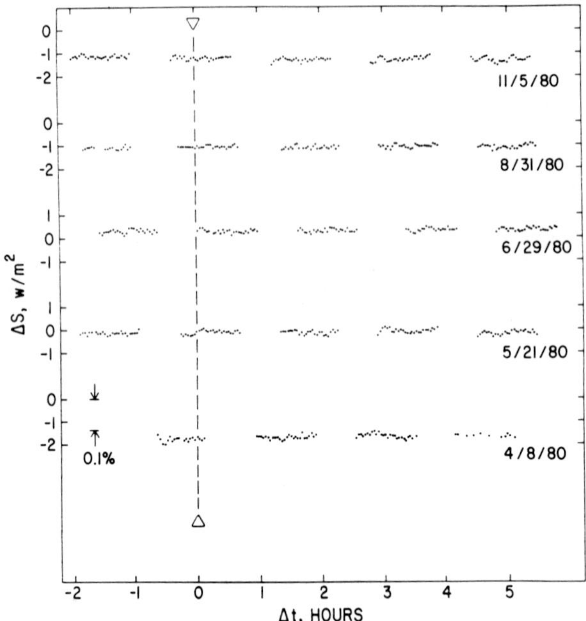

Figure 5.5.1 Radiant energy measured with ACRIM for 2 hours before and 6 hours after the prime flares listed in Table 5.5.3.

Table 5.5.3 Comparison of ACRIM Upper Limits with Total Radiant Energy

Estimates for Prime Flares

Date 1980	Time interval (UT)	Power limit (10^{29} ergs s^{-1})	Energy limit (10^{32} ergs)
Apr 8	02:29 - 03:24	3.2	5.0
May 21	20:48 - 21:47	2.7	4.4
Jun 29	18:04 - 19:03	2.1	3.4
Aug 31	12:34 - 13:31	1.8	3.0
Nov 11	22:06 -- 23:03	2.1	3.4

5.5.5 Conclusions

The estimates of total radiant energy in the prime flares lie well below the ACRIM upper limits. This is consistent with our knowledge of the energy distribution in solar flares. Insufficient data exist for us to be very firm about this conclusion, however, and major energetic components could exist undetected, especially in the EUV-XUV and optical bands. In addition, the radiant energy cannot quantitatively be compared at this time with non-radiant terms because of even larger uncertainties in the latter.

5.6 CONCLUSIONS

H.S. Hudson

5.6.1 Introduction

In this chapter we have tried to carry out a detailed accounting of the energy of a solar flare, intending that a clear view of the sources, sinks, and conditions for transport of energy would allow us to obtain a unique insight into the physics of flares. This tool was applied to several "key questions" stated in the introduction to this chapter, and in this section we give the responses to these questions as they appeared to follow from the energetics analysis. Unfortunately, the tool was generally not sharp enough - there are many key data missing from the observations - and the responses to the key questions are in some cases not definitive. As a result we have added a further section below, giving our suggestions for further observations needed to fill in the gaps and to give authoritative answers to the questions at some time in the future.

5.6.2 Responses to Key Questions

5.6.2.1 How Do We Characterize the Impulsive and Gradual Phases?

The classical definition of the impulsive phase depends on the presence of spiky (timescales of < 10 s) hard X-ray bursts (Kane 1969). The presence of this emission correlates well with the occurrence of microwave bursts whose spiky time profile, high polarization, and spectral maximum establish gyrosynchrotron radiation from energetic electrons as the emission mechanism. The microwaves thus serve as an effective substitute for the hard X-rays in defining the time of the impulsive burst.

The microwave spectrum also has an equivalent to the gradual phase, namely the "post-burst increase."The source of the post-burst increase is identified with the thermal soft X-ray source (Hudson and Ohki 1972). It is clear that this definitely "thermal" region - thermal in the sense of a Maxwellian, although optically thin, plasma - must coexist with the energetic electrons of the impulsive phase. A physically meaningful definition of the gradual phase therefore requires that the twophases overlap. We have therefore adopted this convention in describing flare energetics.

5.6.2.2 Do all Flares Have an Impulsive Phase?

The answer to this question is a qualified "yes," as judged from the correlation established between hard and soft X-ray occurrence shown in Figure 5.4.1 and 5.4.2. All of the flares considered fall near the correlation line with a scatter of $\sim$ 0.5 (rms) in the logarithm. At the faint end, flares for which we have only upper limits on hard X-rays may appear to be purely thermal or gradual, but the absence of hard X-rays appears to be no more than a threshold effect within the bounds of the observed correlation.

One should note that a full answer to this question depends on knowing what defines a solar flare. Many flare-like effects appear in different environments, ranging from the photosphere to the middle corona, and extreme cases may exist that have not been selected as data for the correlation plot. It is quite clear that the Hinotori "B" and "C" flare classifications (see Section 5.1.5) represent distinctly different physical phenomena, for which there are different values for the hard/soft ratio.

5.6.2.3 What is the Total Energy Content of the Flare in the Impulsive Phase?

This question is discussed in detail in Section 5.2 and 5B.

5.6.2.4 What is the Relative Importance of the Thermal and Non-thermal Components of the Impulsive Phase of the Flare?

The thermal components of a flare often display a distribution of temperatures. In the later gradual phase, this distribution usually has relaxed into the bimodal emission-measure distribution described above, with peaks at about 3×10^6 and 15×10^6 K. At earlier times, higher temperatures are best seen in data with high spectral resolution (Lin et al., 1981). During the impulsive phase itself, the hard X-ray spectrum in the 20 to 100 keV range is often well represented by an isothermal, free-free emission spectrum. The same data may be equally well represented by a power law or by a broken power law. The distinction between thermal and non-thermal radiations therefore becomes quite fuzzy if based on the spectrum alone. The inclusion of the microwave data does not help very much, since the key low-frequency emission (i.e., the microwaves produced by the same electron population responsible for the hard X-rays) is optically thick to self-absorption and cannot yield physical parameters inside the source in a model-independent manner. On this basis, the present data alone are not sufficient to provide an unambiguous answer to this question.

One may go beyond the hard X-ray and microwave spectra to clarify the distinction between thermal and non-thermal distributions. Additional evidence from observations include time correlations with other emission such as EUV or white light, correlation with definitely non-thermal processes such as ion acceleration as represented by gamma-radiation; to this we may add insight gained from theoretical knowledge. This need to resort to secondary characteristics has historically resulted in enormous confusion and controversy, and it would be fair to state that at present this controversy continues unabated. The SMM observations and data obtained from other sources do not permit us to present a clear consensus answer to this question. The most likely scenario, however, is given in Section 5.5 of this chapter. The chief result of this scenario is an answer to the key question under consideration, namely that non-thermal electrons accelerated during the impulsive phase have a dominant energetic role and that the thermal sources of soft X-rays are only one of several subordinate effects produced by this inherently non-thermal energy release.

5.6.2.5 How does the Gradual-Phase Energy Compare with the Impulsive-Phase Energy?

The answer to this question is model-dependent because of the ambiguity of the thermal/non-thermal question. We therefore give two answers in Table 5.6.1 based on the thermal [E_{th}(> 10^8 K)] and nonthermal [W(> 25 keV)] impulsive energies and the gradual thermal energies [E_{th}(> 10^7 K)] given in Table 5.2.7.

From Table 5.6.1 one can see that in the thermal case, the impulsive phase tends to be relatively unimportant and the main flare energy release occurs gradually throughout the duration of the flare. In the non-thermal case, it is possible that the impulsive phase contains a large fraction of the total flare energy.

5.6.2.6 What are the Dominant Cooling Mechanisms at Different Stages of the Gradual Phase?

The answer to this question depends crucially on the presence of unresolved fine structure in the soft X-ray sources. Decreasing the flux-tube diameter while holding the length and emission measure fixed requires higher densities, thus enhancing the radiative cooling rate. SMM has provided diagnostic evidence for small filling factors, but the observations are not comprehensive enough, nor at high enough spatial resolution, to permit this question to be answered in a model-independent fashion.

Table 5.6.1 Ratio of Impulsive to Gradual Energies

Date 1980	Time (UT)	Thermal Ratio	Non-thermal Ratio
Apr 8	03:07	0.25	3.4
May 21	21:05	0.13	1.5
Jun 19	18:04	—	2.9
	18:26	1.6	1.1
Aug 31	12:49	3.0	1.5
	12:52	—	2.5
Nov 5	22:28	0.080.5	-4
	22.35	0.46	1.2

The "Thermal Ratio" is given by $E_{th}(>10^8K)/E_{th}(>10^7K)$ from Table 5.2.7.

The "Non-Thermal Ratio" is given by $W(>25keV)/E_{th}$ $(>10^7K)$ from Table 5.2.7.

5.6.2.7 Do all the Post-Flare Loops Need Continual Energy Input?

In large, two-ribbon flares such as the 1980 May 21 flare, the reduction of the radiative cooling time due to filametary fine structure would worsen the discrepancy between the observed long cooling time and the predicted shorter time. Thus the SMM data confirm the need for continued energy input in the late phases of such flares, as reported in the Skylab workshop for the classical loop-prominence systems of the gradual phases of such flares. For compact flares, the observational situation is still vague.

The post-gradual phase phenomena found high in the corona after some major flares have an uncertain energetic link with the flare proper. At the other extreme, the "forerunner" of a coronal mass ejection sometimes appears to precede the associated flare. In both of these cases, there are simply insufficient data to understand the direction of energy transport or the forms of energy storage. These phenomena are energetically significant, however, and since they are remote from the flare it would be reasonable to assume that their energy sources are different.

5.6.2.8 Are There Extended, Late, Flare-Associated Sources in the Corona?

Yes, but there is no systematic knowledge of their relationship to the flares because of the lack of observations. They are energetically important.

5.6.3 Worthwhile Observations in the Future

The key questions addressed above are not sophisticated ones, and yet the present data have not proven capable of answering them all unambiguously. We can identify several reasons for this: first, the diagnostic capability of the available techniques is neither great enough nor fully executed in data of moderate resolution, such as those provided by the SMM observations. Second, the emphasis on diagnostic data has resulted in a lack of attention to basic data designed to define the energetics. Third, coronal observations have made insufficient technical progress and certainly have not provided an adequate data base. As a result, our knowledge of the physics of the lower corona above active regions is inadequate to define a correct conceptual model of flare evolution. Finally, the ground-based observatories have only begun to provide data that are sufficiently comprehensive to define morphology and energetics quantitatively. There is much room for improvement in ground-based observations at optical wavelengths.

As is evident from the ambivalent answers to some of the key questions, the information presently available does not distinguish between thermal and/or non-thermal models of solar flares. To address that fundamental issue, observations are required with sufficient spatial and temporal resolution to distinguish between large thermal plasma volumes, on the one-hand, and beams of energetic electrons which stream through comparably large volumes at speeds of one third or more times the speed of light, on the other. These observations must be obtained in the appropriate spectral range to characterize mildly relativistic electrons, from several tens to several hundreds of keV.

Part of the resolution of these problems will come from the high resolution facilities planned for space in the future: the Solar Optical

Telescope, the Pinhole/Occulter Facility, and other instruments leading up to the full-fledged Advanced Solar Observatory. However, it is a strong recommendation of this group that there be, in the next solar maximum, a renewed effort to obtain substantially improved ground-based observations at both optical and radio wavelengths.

5.6.4 Final Statement

This chapter has described our approach to understanding flare physics through analysis of the energy transport and storage in flares. Such an approach can only succeed if accurate, comprehensive data exist and can be understood in the context of a correct model for the flare phenomenon. Neither of these requirements has been met, and so the exercise has to be considered a failure. The introduction to this chapter describes the energy storage and flow in terms of the diagrams shown in Figure 5.1.1 and Figure 5.1.2. It would be safe to bet that not one single entry in these diagrams is presently known to better than an order of magnitude. Nevertheless, we feel that the attempt has been worth the effort, since by failing at this basic level to understand solar flares, we bring into question all more sophisticated channels of analysis.

Despite pessimism about our present state of knowledge, there is little doubt that remarkable progress has been made. We expect that a future exercise along these lines, hopefully during the forthcoming solar maximum, will be considerably more successful.

APPENDIX 5A. FLARES CHOSEN FOR THE ENERGETICS STUDY

G.M. Simnett, R.D. Bentley, P.L. Bornmann, M. Bruner, and B.R. Dennis

5A.0 Introduction

In this appendix we present brief descriptions of the flares which were chosen for the energetics study. A listing of the dates, times, locations, and classifications of these flares is given in Section 5.1 with the rationale for choosing this particular group of flares. More detailed discussions of these flares can be found in Sections 5.2 and 5.3.

5A.1 1980 April 8 at 03:04 UT

Bibliography: Antonucci, E. et al., 1984, Ap.J., 287, 917.
Cheng, C.C. et al., 1982, Ap.J., 253, 353.
Doschek G. et al., 1981, Ap.J., 249, 372.
Krall, K.R. et al., 1983, Solar Phys., 79, 59.
Machado, M.E. et al., 1983, Solar Phys., 85, 157.
Machado, M.E. and Somov, B.V., 1983, Adv. in Space Sci., 2, 101.
Strong, K.T. et al., 1984, Proc. of 25th COSPAR meeting, Graz, Austria.

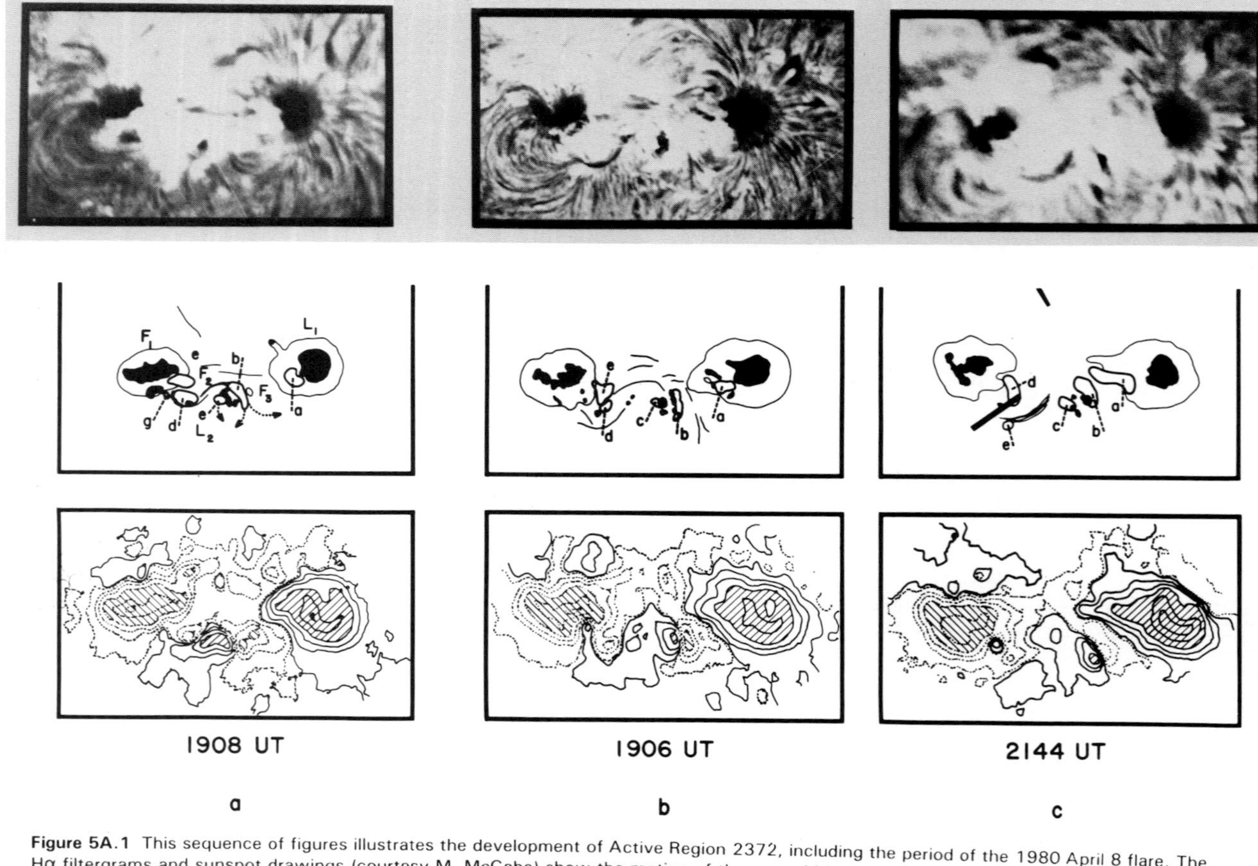

Figure 5A.1 This sequence of figures illustrates the development of Active Region 2372, including the period of the 1980 April 8 flare. The Hα filtergrams and sunspot drawings (courtesy M. McCabe) show the motion of the central bipole towards the leader spot. The equivalent MSFC magnetograms illustrate how the field changed and simplified as the region developed; the April 8 flare marked the end of this particular phase in the evolution of the region.

This flare occurred in Hale region 16747 (Boulder no. 2372), which had been very active in the preceding days. The region was characterized by two large sunspots of opposite polarity, separated by a small, isolated bipolar area in a delta configuration. The longitudinal magnetic field is shown in Figure 5A.1, which also illustrates how the region changed during the preceding days. The isolated pole moved quite rapidly toward the leading sunspot in the days before April 8, leaving the magnetic neutral line highly deformed, with the transverse component of the field showing high magnetic shear.

The light curve for this flare is shown in Figure 5A.2, which illustrates the behavior of X-rays at 26-53 keV (HXRBS) and at 3.9 keV (BCS Ca XIX). Note that the hard X-ray flux observed before 03:03 UT is

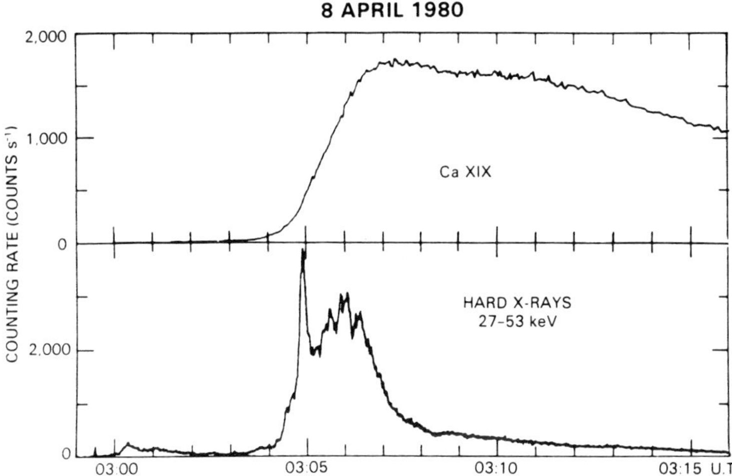

Figure 5A.2 Time profiles of the soft X-ray count rate in the BCS Ca XIX channel and the HXRBS channels 1 and 2 for the 1980 April 8 flare. Note that the hard X-ray flux peaking at 03:00:20 UT is not seen with BCS or HXIS and is believed to originate from a different active region on the west limb.

believed to originate from a different active region on the west limb, since no increase at that time was observed with the BCS or HXIS in their restricted fields of view. The relatively simple time profiles mask an extremely complex flare with a hierarchy of extended magnetic loop structures which became energized at different times and which had substantial temperature structure. Within the loops there appears to be a continuous evolution of the volumes confining the plasma, coupled with significant mass motions during the first few minutes of the impulsive phase. In fact, the maximum upflow of 310 km s^{-1} was observed around 03:04 UT, before the sharp increase in hard X-rays. The peak turbulent velocity was 120 km s^{-1}. One question we need to address is whether it is important to the flare energetics to consider the detailed morphology or whether it is adequate to analyze simply the full-flare light curves.

The spatial evolution of the flare as imaged in 11.5 to 16 keV X-rays is presented in Figure 5A.3 from the onset of the hard X-ray burst at 03:04:06 UT to late in the decay at 03:21:19 UT. The pre-flare

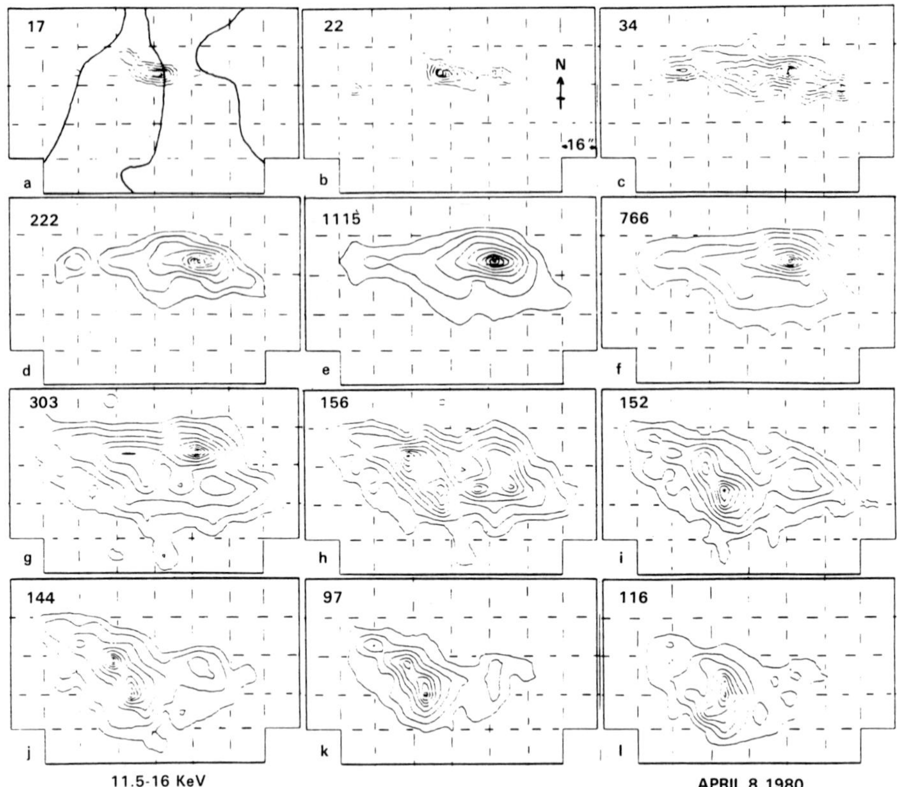

Figure 5A.3 Contour maps of the 11.5 to 16 keV X-ray flux for the 1980 April 8 flare at the following times: (a) 03:04:06 – 03:04:22, (b) 03:04:24 – 03:04:44, (c) 03:04:46 – 03:05:05, (d) 03:05:07 – 03:06:16, (e) 03:06:21 – 03:08:34, (f) 03:08:39 – 03:10:25, (g) 03:10:30 – 03:12:15, (h) 03:12:20 – 03:13:57, (i) 03:14:03 – 03:15:47, (j) 03:15:52 – 03:17:39, (k) 03:17:43 – 03:19:29, (l) 03:19:34 – 03:21:19 UT. These maps were obtained from the HXIS data by deconvolving the collimator response using the iterative technique described by Svestka *et al.* (1983).

brightening in fact starts as early as 02:45 UT and is followed by a second brightening from the same position at 02:59 UT, which leads into the impulsive phase. The position of the magnetic neutral line is drawn on Figure 5A.3(a); the pre-flare brightening is from the region of maximum shear in the transverse field. However, Figure 5A.3(c), which spans the peak in the hard X-ray burst, shows that the X-rays at that time came predominantly from two regions, with a secondary brightening to the east (left). This suggests a small, weak bipolar region that developed to the east of the neutral line near the large, trailing sunspot. The same two bright regions were resolved at 3.5–5.5 keV, before the peak in the hard X-ray burst. In the following minutes, the peak emission at all wavelengths shifted to an area between these two regions, and the hard X-ray emission entered a period of slow, monotonic decay. It was during this period that a region to the east brightened, and Figure

5A.3(h) shows the image accumulated between 03:12:20 and 03:13:57 UT, when this region had reached maximum intensity. There are three other resolved bright points in this image which are either persistent or visible at other wavelengths, or both.

5A.2 1980 May 21 at 20:53 UT

Bibliography: Antonucci, E. and Dennis, B.R. 1983, Solar Phys, 86, 67.
Antonucci, E., Dennis, B.R., Gabriel, A.H., and Simnett, G.M., 1985, Solar Phys., in press.
Antonucci, E., Gabriel, A.H., and Dennis, B.R., 1984, Ap.J., 287, 917.
Bely-Dubau et al., 1983, MNRAS, 201, 1135
Duijveman et al., 1982, Solar Phys., 81, 137.
Duijveman, A. 1983, Solar Phys., 84, 189.
Harvey, J.W., 1983, Adv. in Space Res., 2, 31.
Hoyng, P. et al., 1981, Ap.J., 246, L155
Svestka, Z. et al., 1982, Solar Phys., 75, 305.
Svestka, Z. et al., 1982, Solar Phys., 78, 271.

This event was a classical two-ribbon flare, starting with filament activity above the magnetic neutral line that led into the impulsive phase as the filament started to rise. It was from Hale region 16850, which had produced only subflares with very weak X-ray emission during the previous 100 h; similarly, it was followed by small, weak flares. This was therefore an example of an isolated large flare. The intensity time profile is shown in Figure 5A.4 for 27 to 54 keV X-rays (HXRBS) and the Ca XIX channel (BCS). Once the thermal phase of the flare began, the plasma had an extremely hot component, exceeding 30 x 10^6 K, until late in the decay phase.

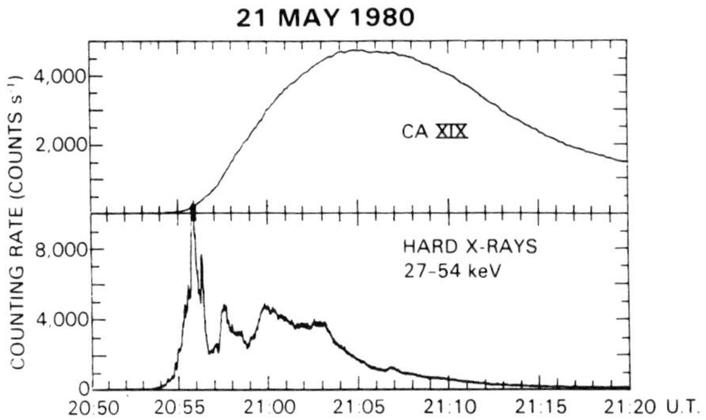

Figure 5A.4 Same as Figure 5A.2 for the 1980 May 21 flare.

Figures 5A.5(a) and (c) show the appearance of the flare region in Hα both before and during the main phase of the flare, and Figures 5A.5(b) and (d) show magnetograms for two times before the flare. The bulk of the 16 to 30 keV X-rays during the main impulsive hard X-ray spike appear at position A, with a significant second bright point at B about 27000 km away on the other side of the filament. The 16 to 30 keV X-ray features are superimposed on Figure 5A.5(a) to illustrate this. The X-ray images have been interpreted as providing the first observational evidence of footpoint brightening from the chromospheric base of a magnetic loop caused by deposition of non-thermal particles (Hoyng et al., 1981).

In addition to the hard X-ray features, HXIS studied the morphology of the soft X-ray emission above 3.5 keV, both along and across the filament, as the flare evolved. By 20:48 UT, the soft X-ray flux was enhanced along the filament, which had begun to separate at that time. In this pre-flare period there was already soft X-ray emission from A (Figure 5A.5), which appeared as a barely resolved magnetic loop at the site of the newly emerged magnetic flux. There were many resolved X-ray bright points early in the flare, suggesting a variety of magnetic loops.

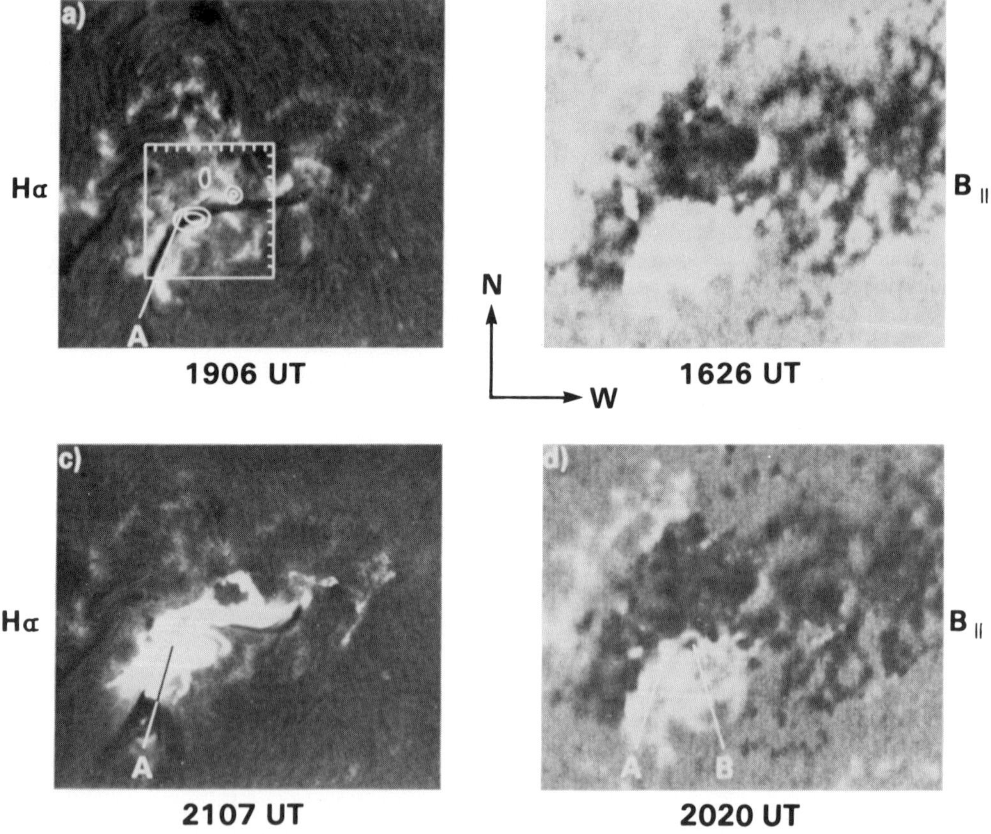

Figure 5A.5 Hα images and magnetograms for the 1980 May 21 flare from Hoyng *et al.* (1981). (a) The Hα image at 19:06 UT with the contours of the 16 to 30 keV X-ray bright patches superimposed. The X-ray bright patches were observed with HXIS during the two main hard X-ray peaks at 20:56 UT. The frame around the X-ray contours measures 96 × 96 arcsec. (b) Line-of-sight component of the underlying magnetic field at 16:26 UT. (c) The Hα flare at 21:07 UT. (d) The magnetic fields at 20:20 UT. 'A' marks the location of the new sunspot (positive flux), and B gives the location of new negative flux.

These disappeared at the onset of the impulsive phase just after the filament began to rise.

The Ca XIX line profile began to exhibit the signature of mass motion during this period, and a series of line scans is presented in Figure 5A.6. At 20:53 UT the entire soft X-ray source is observed to be moving upwards at 80 km s^{-1}. By 20:57 UT the motion of the bulk plasma flow diminished to the lowest detectable value of 30 km s^{-1}. However, at the start of the first hard X-ray burst, additional material was injected at a velocity of 370 km s^{-1}, and this is observed as a separate feature in the blue wing of the resonance line in Figure 5A.6(b). This high velocity upflow coincided with a spectral hardening of the X-ray emission to the power law index of -4, a value maintained until 21:04 UT. The detection of blue-shifted material continues to 21:05 UT. Therefore, the energy driving the upflow appears to be deposited in the chromosphere for as long as the spectral index remained around -4, but no longer.

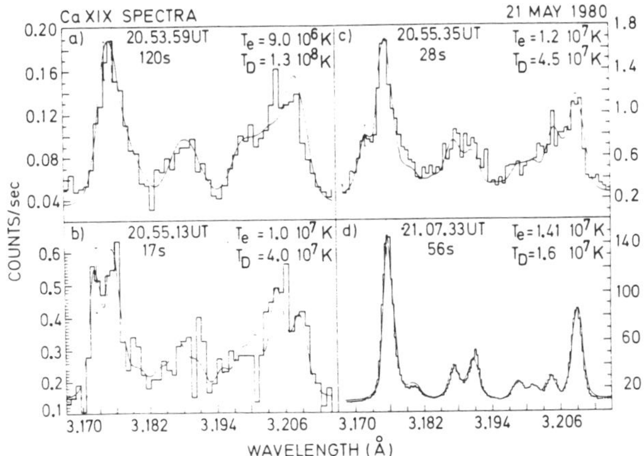

Figure 5A.6 Sequence of soft X-ray spectra obtained at four times during the impulsive phase of the 1980 May 21 flare in the BCS channel covering the Ca XIX spectral region from 3.165 to 3.231 Å. The smooth curve in each figure represents the synthesized spectrum computed for given values of the electron temperature T_e and the Doppler temperature T_D. In (b) and (c), two synthesized spectra, with one blue-shifted by 3.8 mÅ, are summed to form the spectrum represented by the smooth curves. The dashed lines in the same figures represent the profile of the principal component expected for the given Doppler temperature. The time given for each spectrum is the mean time of the observation interval, and the accumulation period is given below this time.

A type II radio burst starts at 20:57 UT, and the flare generated a prompt, energetic (> 40 MeV) proton event at 1 AU plus a traveling interplanetary shock which accelerated low energy protons as it moved outward. A type I radio noise storm was seen during the early hours of this motion, accompanied by weak soft X-ray emission from a large (> 105 km radius) coronal arch located above the flare site. The arch was visible in soft X-rays for over 10 h.

5A.3 1980 June 29 at 18:03 and 18:22 UT

Bibliography: Bentley, R.D. et al., 1984, Solar Phys., in preparation.
Harrison, R.A. et al., 1985, Solar Phys., submitted.
Poland, A.I. et al., 1982, Solar Phys., 78, 201.
Schmahl, E.J., 1983, Adv. in Space Sci., 2, 73.
Simnett, G.M. and Harrison, R.A., 1984, Proc. 25th COSPAR meeting, Graz, Austria.
Simnett, G.M. and Harrison, R.A., 1984, Solar Phys., (submitted).
Wu, S.T. et al., 1983, Solar Phys., 85, 351.

These two flares are discussed together, since they are related by the coronal disturbances seen both in white light and at X-ray and radio wavelengths. After the first flare at 18:03 UT, a coronal X-ray source was observed by HXIS to move slowly outwards. A coronal mass ejection was detected by C/P until 20:03 UT, and the onset of the event was observed with the Mark-3 K-Coronometer on Mauna Loa from around 18:18 UT, several minutes before the onset of the second (main) flare at 18:22 UT. However, a type II radio burst was only observed for a few minutes from 18:33 to 18:36 UT, and from the observed drift rates it appeared to be associated primarily with the second flare. The velocity measured by the K-Coronameter is 500 km s^{-1} and the velocity derived from the type II burst is 1100 km s^{-1}. The observations are summarized in Figure 5A.7, which shows the intensity-time profiles for 28 to 55 keV X-rays (HXRBS) and the Ca XIX channel (BCS).

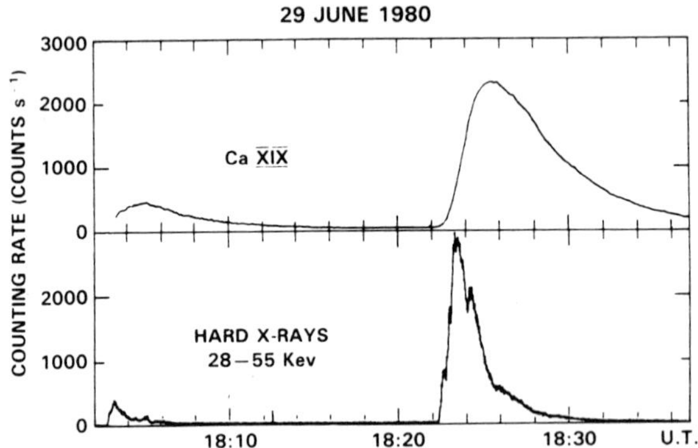

Figure 5A.7 Soft (top) and hard (bottom) X-ray time profiles for the 1980 June 29 flares similar to Figure 5A.2. Note that SMM emerged from night at 18:02:40 UT during the first of the two flares.

Although both flares occurred on the west limb, they were not from precisely the same location. The first flare was centered approximately 16 arc sec south of the main flare and occurred in coincidence with a subflare from S11 W36, which reached a maximum in Hα at 18:06 UT. The first flare was still visible in soft X-rays when the second flare started. For many hours previously, a large system of loops stretching high into the corona was visible with both HXIS and FCS.

Figure 5A.8 shows the development of the flares as observed in various wavelengths ranging from Hα to Mg XI. Figure 5A.8(a) is an FCS image in Mg XI taken from 18:08:08 to 18:13:07 UT; the dotted line marks the position of the limb as seen by the FCS white-light sensor. The actual limb would be the best-fit smooth curve through these points. The position of the UVSP field of view is outlined in the NW corner. Figure 5A.8(b) shows a smaller FCS raster, again in Mg XI taken from 18:26:27 to

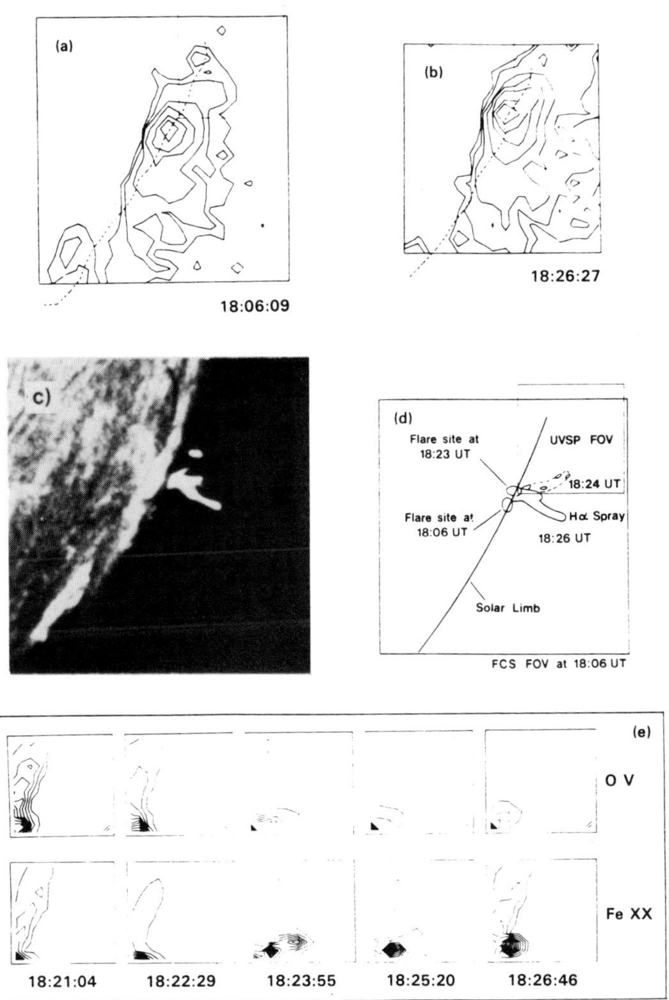

Figure 5A.8 Images of the 1980 June 29 flares in different wavelength ranges at the UT times indicated. (a) FCS image in Mg XI during the first flare. The dotted line marks the position of the limb as seen by the FCS white light sensor. The UVSP field of view is shown in the NW corner of the image. (b) Similar to (a) for the second flare. (c) Hα picture taken at the Ramey Observatory at 18:20 UT during the second flare. (d) Sketch of the solar limb and lower corona summarizing the imaging observations of the two flares. (e) UVSP images in OV and Fe XXI during the second flare.

18:30:54 UT, covering the initial decay of the second flare. The enhanced coronal X-ray emission is clearly visible in both images. Figure 5A.8(c) shows an Hα picture taken at the Ramey Observatory at 18:26 UT. This follows a small bright limb surge at 18:20 UT, after which a mass of ejected material, subtending 3 arc min at the Earth, became visible. The origin of the second flare is just beyond the west limb. Figure 5A.8(d) is a sketch of the solar limb and lower corona, summarizing these observations.

Figure 5A.8(e) shows UVSP images in O V and Fe XXI during the rise and decay of the second flare. There is clearly a path into the corona to the northwest during the initial stage of the flare, which became less prominent following flare maximum. This is consistent with the southerly swing seen in Hα.

The hard X-ray burst shown in Figure 5A.7 consists of a series of spikes, followed by a smooth, slow decay from 18:25:30 UT. The X-ray spectrum is hardest during the initial rise. The microwave emission shows similar multiple spike structure, and the burst was strongly right-circularly polarized with a peak frequency of 2.2 GHz. At the onset of the second flare, a series of type III and V radio bursts was observed in coincidence with a blue-shifted feature seen in the Fe XXV line profile. This is coincident with the southerly swing in the Hα spray, suggesting that the motion of the spray had a significant longitudinal component in the solar reference frame.

5A.4 1980 August 31 at 12:48 and 12:52 UT

Bibliography: Strong, K.T. et al., 1984, Solar Phys., 91, 325.

These flares involved two distinctive and separate energy releases in a compact flare region which produced quite different responses in the radiated emissions and mechanical mass motions. The longitudinal magnetic field structure of the region is shown in Figure 5A.9(a), in which a negative intrusion is visible between the two leading positive sunspots. Figure 5A.9(b) shows the magnetic neutral line and the positions of maximum shear in the longitudinal field. It is from one of these highly sheared regions that the bright flare points A, B, and C occurred, whereas a weak brightening at D shows that there was some interaction with more distant parts of the region.

Figure 5A.10 presents the time profiles for 28 to 500 keV X-rays, the Ca XIX resonance line (3.176Å), and the microwave intensity at 15.4 GHz. The gross differences between these curves have proved valuable in interpreting the various modes of energy release and energy transport. In the first flare, the impulsive emissions show two main spikes, one at 12:48:42 UT and the other at 12:48:51 UT. The Ca XIX emission was delayed significantly; the peak intensity occurred at 12:49:06 UT, by which time the impulsive emissions had decayed by an order of magnitude from their peak.

In contrast, the onset of the second flare was more noticeable in Ca XIX than in hard X-rays or microwaves, and there was a precursor rise, which by 12:51:25 UT had reached a level equal to the maximum of the

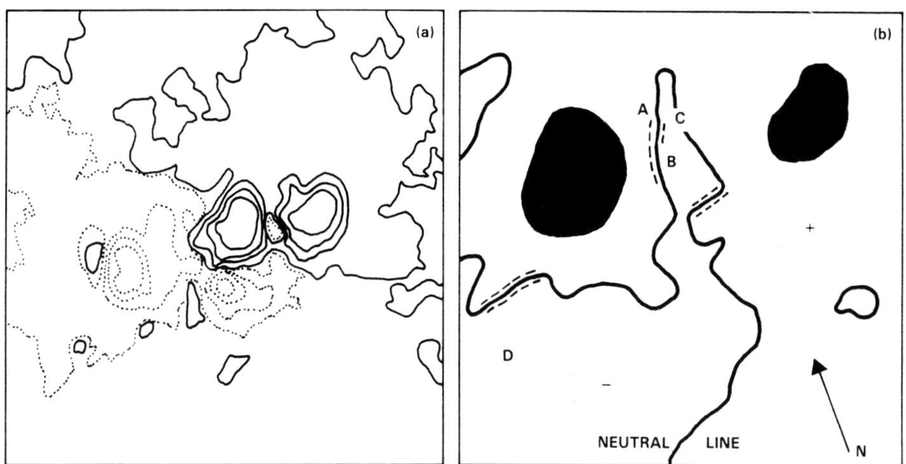

Figure 5A.9 The magnetic structure of NOAA Active Region 2646. (a) A magnetogram showing the longitudinal fields; note the location of the three sunspots, and the small negative intrusion between the two leading positive spots. Positive and negative polarities are indicated by solid and broken lines, respectively. The magnetogram is approximately 4 arc min square. (b) A schematic diagram of the flare site showing the location of the spot umbrae (shaded area), the neutral line (solid line), and the areas of maximum shear in the transverse field component (dashed lines). The main sites that are discussed in the text are labeled A-D.

first flare (see Figure 5A.10). It was accompanied by a weak increase in hard X-rays but no discernable change in the microwave flux. At 12:51:35 UT the main Ca XIX increase started, this time accompanied by hard X-ray emission but a relatively weak microwave event. The unusual features of the second flare were the close temporal correspondence of the peaks of the Ca XIX and hard X-ray emission and the rapid decay of the Ca XIX emission. There was also a large variation in the brightness of the second flare compared with that of the first at different wave-lengths. In hard X-rays (> 28 keV) and in 15.4 GHz microwaves, the second flare was weaker by a factor of 2 and 33 ± 6, respectively. In soft X-rays, the second flare was brighter by a factor of 5.8 in the Ca XIX resonance line, 3 in O VIII, and 8 in Fe XXV. The flares occurred at slightly different locations within the active region, as can be seen from the X-ray images in Figure 5A.11. The first flare, shown in an 8 to 16 keV image in Figure 5A.11(a), had two bright points aligned approximately E-W and on either side of the neutral line. The 16 to 30 keV image confirmed this. There was a blue shift in Ca XIX corresponding to a velocity of 60 ± 20 km s^{-1} and a line broadening corresponding to a turbulent velocity of 190 ± 40 km s^{-1}. A blue shift, corresponding to a velocity of around 180 km s^{-1} was also seen with UVSP in the Fe XXI line at 1354Å, whereas the line broadenings were equivalent to turbulent velocities in the range 66-150 km s^{-1}.

The hard X-rays at the onset of the second flare shown in Figure 5A.11(b) were from an area to the north near point A of Figure 5A.9(b). They subsequently appeared at a point about 6 arcsec to the south (Figure 5A.11(c)) very close to the eastern bright point shown in Figure

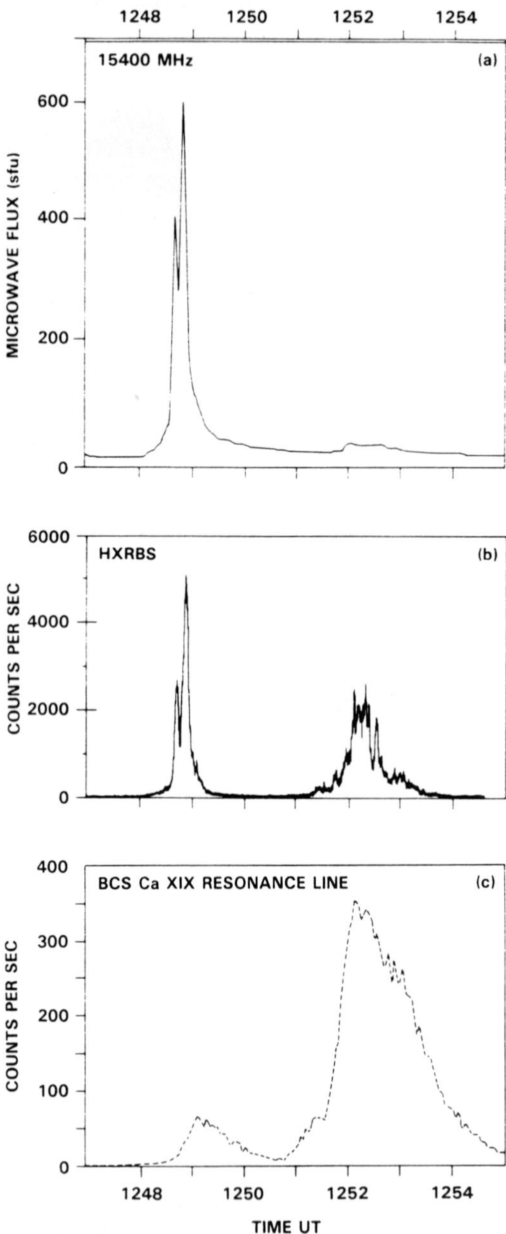

Figure 5A.10 Light curves of the two flares on 1980 August 31. (a) Microwave flux at 15.4 GHz. (b) Hard X-ray (> 28 keV) count rate. (c) Soft X-ray (Ca XIX resonance line at 3.176 Å) light curve. Note the variation in the relative brightness of the two flares at various wavelengths.

5A.11(a). There were no indications of significant blue shifts from the second flare. In fact, to the contrary, the blend of O I and C I at 1355.8Å observed with UVSP had the red wing enhanced, suggesting downflows of material at chromospheric temperatures. On account of the different locations and the very distinct characteristics of the two flares, it seems that heated plasma from the first flare broke out of the

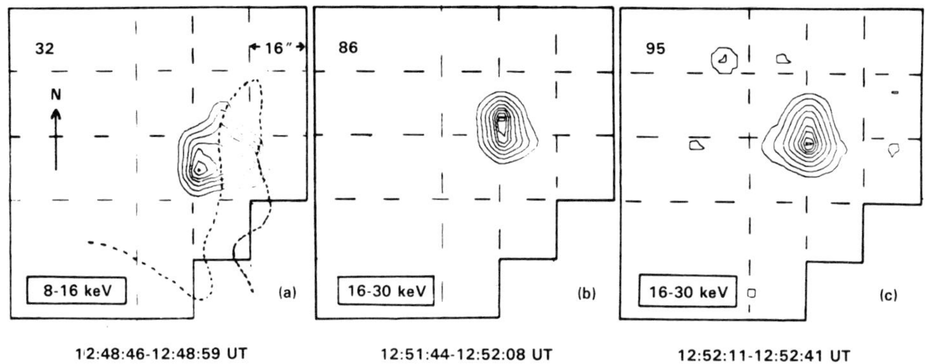

Figure 5A.11 HXIS contour plots showing the location of the hard X-rays at the times indicated during the two flares on 1980 August 31. The contours were obtained by deconvolving the collimator response with the iterative technique described by Svestka *et al.* (1983). The dotted line in (a) shows the location of the magnetic neutral line.

flare loop into an adjacent structure to the north, where it triggered the release of a larger amount of energy to produce the second flare. However, the evolution of the first flare resulted in a completely different set of starting parameters, such as densities and temperatures, governing the development of the second flare.

The X-ray spectrum is very similar in shape at the hard X-ray peaks in both flares. The principal difference lies in the region below 15 keV, which is considerably enhanced at the peak of the second flare. In this region the emission is dominated by thermal radiation at $(20 \pm 5) \times 10^6$ K.

5A.5 1980 November 5 at 22:26 and 22:33 UT

Bibliography: Bornmann, P.L., 1985a, Ap.J., in press.
Bornmann, P.L., 1985b, Solar Phys., submitted.
Dennis, B.R. et al., 1984, Proc. 2nd Indo-US Workshop on Solar Terrestrial Physics, New Delhi, India.
Duijveman, A. et al., 1982, Solar Phys. 81, 137.
Duivjeman, A. et al., 1983, Solar Phys. 88, 257.
Hoyng, P. et al., 1983, Ap.J. 268, 865.
Martens, P.C.H. et al., 1985, Solar Phys., 96, 253.
Phillips, K.J.H. et al., 1982, Ap.J. 256, 774.
Rust, D.M. et al., 1985, ApJ., 288, 401.
Rust, D.M. and Somov, 1984, Solar Phys., 93, 95.
Wolfson, C.J. et al., 1982, Ap.J. 269, 319.

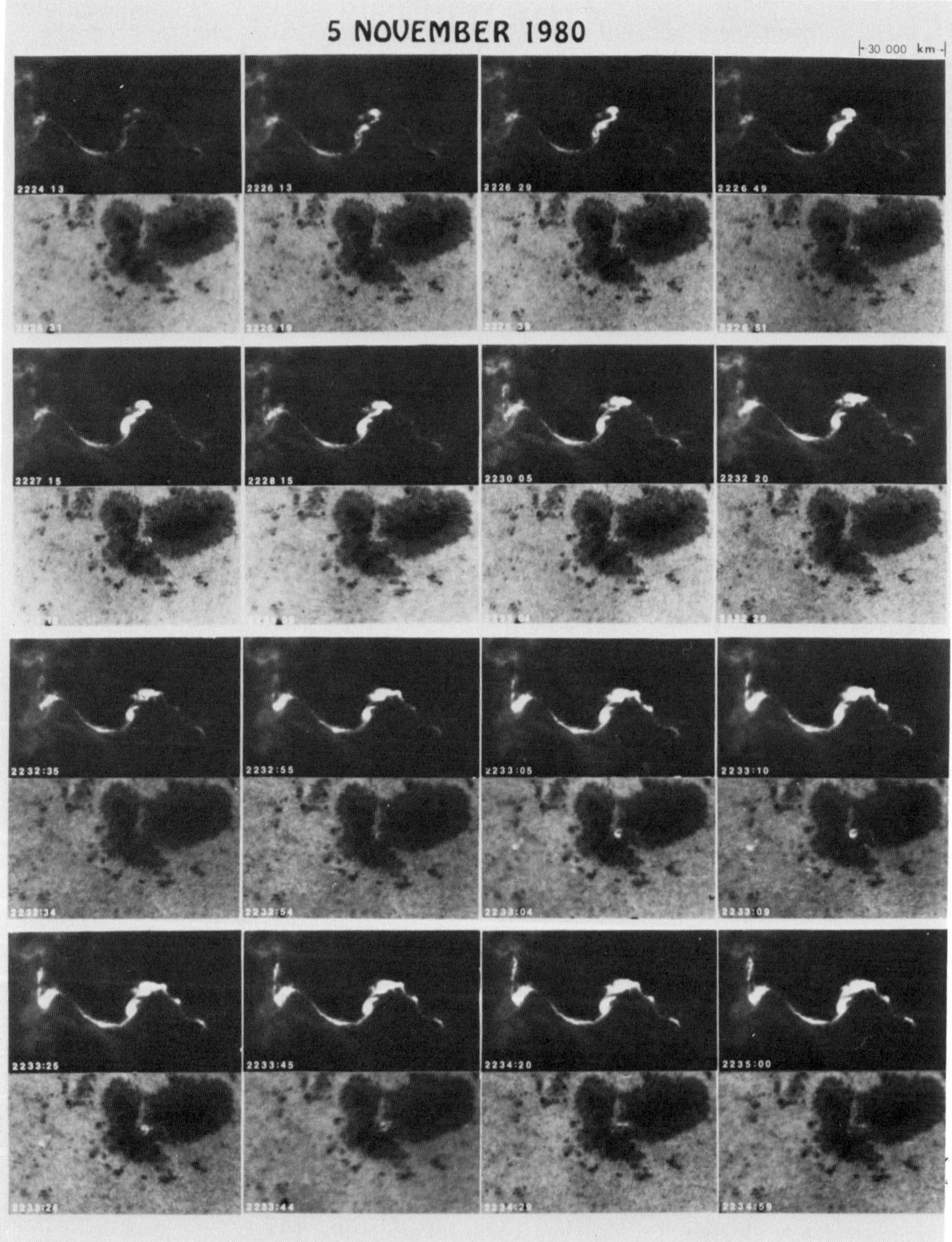

Figure 5A.13 (a) Images of first of the two 1980 November 5 flares in Hα and in HeI D_3 taken at the times indicated. The first and third rows of photographs are Hα center-line images, and the second and fourth rows are HeI D_3 images. (b) As for Figure 5A.13(a) for the second flare on 1980 November 5. Note that this flare began during the decay of the earlier flare and followed the same pattern of development.

These two flares occurred in Hale region 17244 (Boulder no. 2776), which included a group of sunspots in a delta configuration (opposite polarity umbra within a single penumbra). For at least 20 min before the onset of the first flare, there was 3.5 to 5.5 keV X-ray emission from the photospheric magnetic neutral line, with occasional bright points at places which subsequently featured prominently in the main flare. The intensity-time profiles are shown in Figure 5A.12 for 29 to 57 keV X-rays (HXRBS) and 3.5 to 5.5 keV X-rays (HXIS).

These two flares are included in our study because of the complete data coverage available. Ground-based optical, magnetogram, and radio data augment the satellite data from SMM and GOES. The entire flaring region was observed with HXIS, HXRBS, and ground-based stations. Although BCS was turned off during the decay of the second flare, data were available at all other times. This flare was also chosen because density diagnostics are available from the FCS data. The FCS was pointed at the brightest point in the flare, covering a 14 x 14 arcsec (FWHM) portion of the entire flaring region. The UVSP observed another portion of the flare, which did not overlap with the region seen by FCS. The VLA observed all of the first flare but only the decay of the second flare.

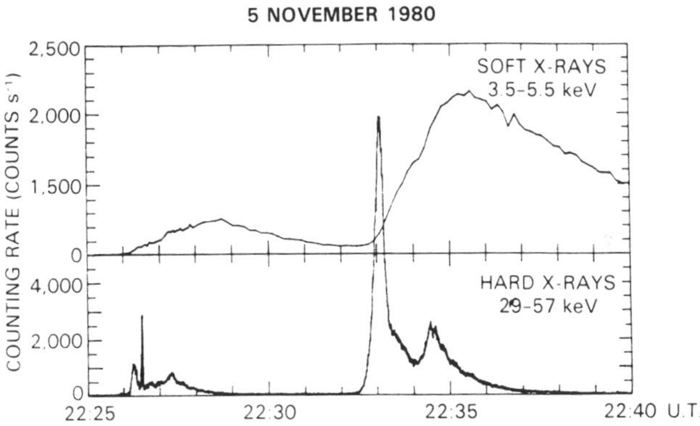

Figure 5A.12 Time profiles of the soft X-ray counting rate in HXIS band 1 (3.5 - 5.5 keV) and in HXRBS channels 1 and 2 (29 - 57 keV) for the two flares on 1980 November 5. The HXIS counts were from selected coarse-field pixels and the rate was corrected for the instrumental dead time.

The flares were well observed with the Hα Mulitslit Spectrograph at Big Bear Solar Observatory and consequently are illustrated by these observations in somewhat greater detail than is available for the other flares. Figures 5A.13(a) and (b) show a sequence of images of the two flares in both centerline Hα and HeI D_3 at 5876Å. Both flares were classified as two-ribbon flares, although the ribbons remained stationary and did not separate during either flare. The flares were nearly cospatial and can be regarded as homologous; the main difference between them lies in their overall brightness.

Several important features are apparent from the Hα and HeI D_3 observations:

o The HeI D_3 emission points began at the same time as the > 28 keV X-rays and continued for approximately the same duration in both flares (see Figure 5A.14).

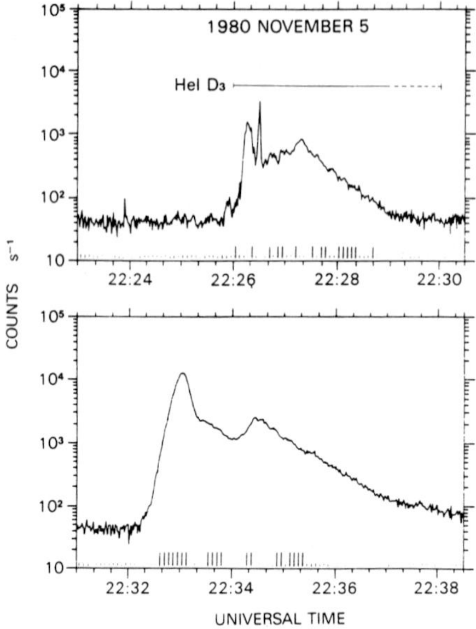

Figure 5A.14 The hard X-ray time profiles of the two flares on 1980 November 5 on a logarithmic scale with the times of the HeI D_3 observations also indicated. The short, evenly spaced, vertical lines just above the horizontal axis indicate when D_3 images were taken every 5 s on film at the Big Bear Solar Observatory. The longer vertical lines indicate the images which show obvious new or enhanced flare points in D_3. The entire impulsive phase is characterized by the presence of D_3 although not all new flare points are spatially resolved. The detection of these resolvable new points, however, is confirming evidence that energy is continuously released throughout the impulsive phase, not just during the rise.

o The temporal fine structure seen in hard X-rays is also a property of the HeI D_3 emission; each emission point was relatively short lived, usually lasting no more than 1 min.

o Both flares were a succession of many rapidly forming and decaying bright points extending on both sides of the flare kernels. The last of the bright points to develop were those most distant from the kernels, and they were the least bright. The exception was the bright point to the northeast, which peaked in intensity around 22:33:09 UT, slightly later than the hard X-ray peak from this location as discussed below.

o The flare sites which had bright HeI D_3 emission also had Hα line profiles typically more than 3Å wide (FWHM).

o After the initial brightening of the Hα kernels in the second flare, a rebrightening occurred coincident with the secondary hard X-ray maximum at about 22:34:20 UT (Figure 5A.12).

The first flare at 22:26 UT has been studied by Hoyng et al. (1983) with optical, hard X-ray (HXIS), and microwave (VLA) images. Figures 5A.15 (a) and (b) are sketches of the center of the active region showig the locations of the various flare components in Hα and HeI D_3 relative to the neutral line. The 15-GHz source and the boundaries of the prominent HXIS pixels are also indicated. The region appears to consist of a set of low-lying loops, probably highly sheared, whose footpoints lie in the Hα strands S1 and S2. Above these loops is an overlying loop whose footpoints appear to be in the Hα kernels labeled a and e in Figure 5A.15(b). The VLA images show a bright area that lies across the neutral line at the center of the flaring region. This 15-GHz source had a

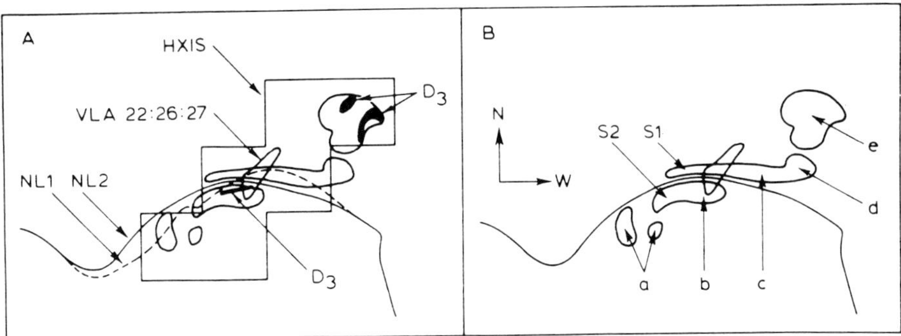

Figure 5A.15 Composite line drawings indicating the locations of the various flare components: the Hα flare kernels labelled 'a' through 'e' and the strands labelled S1 and S2; two possible positions for the neutral line, the one labelled NL1 being the measured location and the other labelled NL2 being a possible actual location that is everywhere within 2.5 arcsec of NL1; the VLA source at the time indicated; the outline of the six 8 × 8 arcsec HXIS pixels containing the points marked A, T, and B in Figure 5A.16(b); and the HeI D_3 emission patches (the one south of the neutral line being very weak).

circular polarization of up to 80% during the first flare with a brightness temperature of > 10^9 K. The sense of polarization reversed at the magnetic neutral line. The evolution of the hard X-ray images during the first flare is shown in Figures 5A.16(a), (b), and (c). At the onset of the first flare [Figure 5A.16(a)], the majority of the 16 to 30 keV photons came from the position labeled B in Figure 5A.16(b), which is coincident with the location of the bright He ID_3 kernels. By the time of the narrow intense spike at 22:26:30 UT (Figure 5A.12), the hard X-rays were coming from the position labeled T in Figure 5A.16(b), coincident with the location of the microwave source. The 22 to 30 keV image from 22:26:12 to 22:26:38 UT [Figure 5A.16(b)] shows emission from points A, B, and T. It is presumed that T is at the top of the magnetic loop with footpoints at A and B. This loop was situated over the smaller loops which presumably linked the bright Hα strands S1 and S2. Hoyng et al. (1983) estimated that the magnetic field at the neutral line was 700 ± 160 and 480 ± 110 G during the first two hard X-ray peaks, respectively. The loop AB is interpreted as a small, low-lying loop with a strong

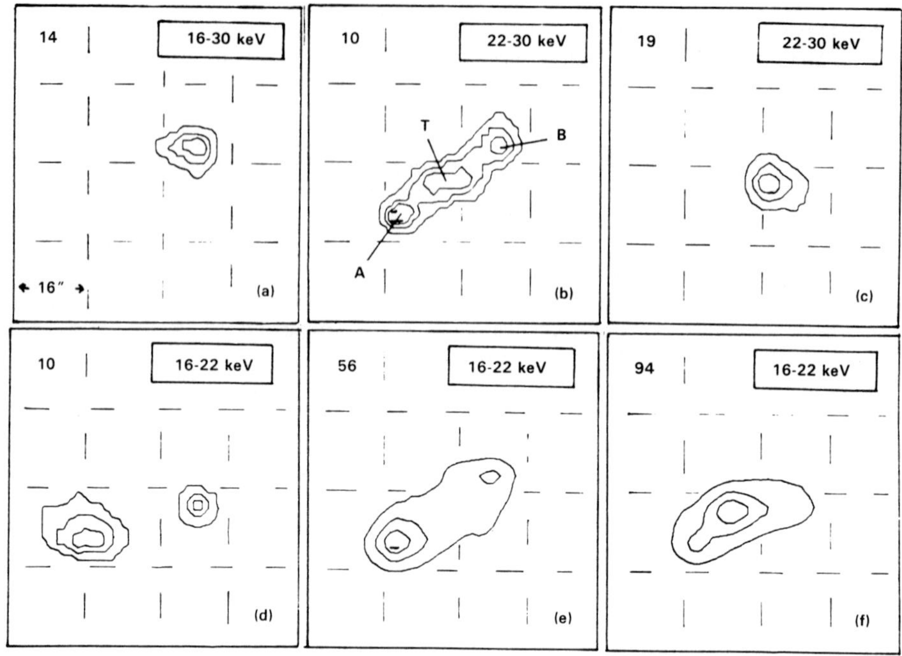

Figure 5A.16 HXIS contour maps of the hard X-ray flux at the following times during the flares on 1980 November 5: (a) 22:26:11 – 22:26:19 UT (start of first flare), (b) 22:26:11 – 22:26:37 UT (first two peaks of first flare), (c) 22:26:39 – 22:27:30 UT (third peak of first flare), (d) 22:32:34 – 22:32:47 UT (start of second flare), (e) 22:33:12 – 22:33:25 UT (decay of first peak of second flare), (f) 22:34:26 – 22:34:39 UT (second peak of second flare). The number in the top left hand corner of each map is the highest number of counts per pixel in the image. Points marked A, B, and T in (b) are believed to be the footpoints and the top, respectively, of a magnetic loop; they are discussed in the text.

horizontal field. From the optical data, Hoyng et al. concluded that the magnetic field rearranged itself over a small area in the center of the active region during the first flare and that a major rearrangement of the magnetic field took place during the second flare on a much larger spatial scale. During the final hard X-ray spike at 22:27:20 UT and on the decay of the first flare [Figure 5A.16(c)], the hard X-rays came almost exclusively from position T, although there is still a resolved bright point at lower energies from B.

At the onset of the second flare [Figure 5A.16(d)], the 16 to 22 keV X-rays came initially from region A and from a point to the south of B, at the position of the end of the bright Hα strand S1 seen in the first flare [Figure 5A.15(b)]. After the onset, the hard X-ray intensity continued the rapid rise and the bright points switched to position B and to a remote point at the eastern end of the filament [Figure 5A.17(a)]. Within 10 to 15 s, this remote emission had died away leaving a bright point at A [Figure 5A.16(e)]. During the final hard X-ray peak at 22:34:30 UT (Figure 5A.12), the hard X-rays were once more concentrated near T [Figure 5A.16(f)]. The appearance of the sequential hard X-ray brightenings from a number of distinct, and in one case widely separated, points suggests that there is a hierarchy of magnetic loops involved. The distant emission shown in Figure 5A.17(a), over 7 x 10^4 km away from the main flare site, is from the end of a structure which becomes completely filled with hot, X-ray emitting plasma during the decay of the flare [Figure 5A.17(b)]. Bright points corresponding to A, T, and the initial hard X-ray bright point south of B are clearly resolved.

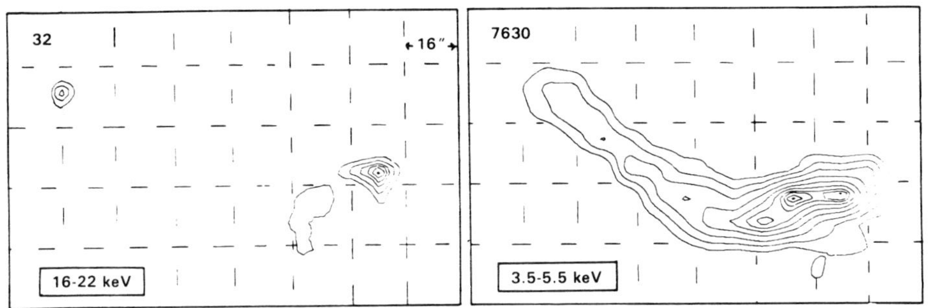

Figure 5A.17 HXIS contour maps of the second flare on 1980 November 5. (a) Map in hard X-rays at the time of the main impulsive spike, showing the two bright patches A and B on the right (west) and the third bright area in the top left (northeast) corner. (b) Map in soft X-rays during the decay of the second flare. The number in the top left-hand corner of each map is the highest number of counts per pixel in the image.

An estimate of the density of the soft-X-ray-emitting plasma is available during the decay of the second flare when FCS was operating in a spectral scanning mode. The measured Ne IX intercombination-to-forbidden-line ratio is density sensitive and indicates a maximum density of 1.5 x 10^{12} cm^{-3} at the time of the peak in the soft X-ray emission.

The Ca XIX and Fe XXV resonance line profiles observed with BCS during the impulsive phase of the second flare show broadening and extended blue wings visible for 30 s, beginning at the time of the peak hard X-ray flux. The blue shifts correspond to velocities between 200 and 500 km s^{-1}, and the line widths indicate that the turbulent velocities reached values of 100-200 km s^{-1}.

APPENDIX 5B. A REVIEW OF IMPULSIVE PHASE PHENOMENA

C. de Jager

5B.0 Introduction

In this appendix we present a brief review of impulsive phase phenomena in support of the models used in this chapter to compute the energies of the different components of the flares under study. A more complete review is given in Chapter 2 of this Workshop proceedings.

We begin with the observational characteristics of the impulsive phase, followed by the evidence for multi-thermal or non-thermal phenomena. The significance of time delays between hard X-rays and microwaves is discussed in terms of electron beams and Alfven waves, two-step acceleration, and secondary bursts at large distances from the primary source. Observations indicating the occurrence of chromospheric evaporation, coronal explosions, and thermal conduction fronts are reviewed briefly, followed by the gamma ray and neutron results. Finally, a preferred flare scenario and energy source are presented involving the interactions in a complex of magnetic loops with the consequent reconnection and electron acceleration.

5B.1 Observational Characteristics of the Impulsive Phase

5B.1.1 Main Characteristics

The main characteristic of the impulsive phase is the impulsive release of energy on timescales of less than or about 10 s. This is demonstrated best by intensity-time plots in hard X-rays and microwaves. These show a number of bursts (Figures 5A.2, 4, 7, 10, and 12), often with considerable time structure down to greater than or about 30 ms (Kiplinger et al., 1984, Dennis et al., 1984). The individual complex bursts are rarely shorter than 10 s or longer than 100 s. The more complex and long-lasting bursts are associated with complex Hα flares, particularly two-ribbon flares (Dwivedi et al., 1984). There is no strict energy range for impulsive bursts: sometimes flares show impulsive characteristics only at energies above > 40 keV (Tsuneta et al., 1984a), but usually impulsive X-ray bursts are also seen at lower energies (Tanaka et al., 1984). When the X-rays are imaged, impulsive bursts can sometimes be located in certain pixels at energies as low as 5 keV (de Jager and Boelee 1984; de Jager et al., 1984).

The impulsive nature of flares has been detected in other emissions besides hard X-rays and microwaves. Peaks in the emission of the transition zone lines, O V at 1371Å, C IV at 1548Å4, and Si IV at 1402Å4 have been observed with UVSP to coincide to within 1 s with peaks in the hard X-ray emission (Woodgate et al., 1983, Poland et al., 1984, Tandberg-Hanssen et al., 1983, Cheng et al., 1984, Dennis et al., 1984). Observations with UVSP in the EUV continuum have shown temporal structure coincident with similar structure in the hard X-ray emission to better than 0.25 s (Woodgate 1984). Coincident structure is also sometimes seen in Hα (Acton et al., 1982, Gunkler et al., 1984, Kampfer and Magun 1984). The coincidence of hard X-ray bursts with the appearance of emission kernels in HeI D_3 has been observed for the two flares of 1980 November 5 (see Figures 5A.14). At the other end of the spectrum, the hard X-ray bursts are closely correlated with gamma-ray bursts, the two kinds sometimes occurring simultaneously to within 2 s (Chupp and Forrest 1983).

5B.1.2 Spatial Investigations

The picture given above can be amplified by spatial investigations. At the time of the impulsive hard X-ray bursts, the source of radiation is restricted to small areas, in some cases situated in pairs with components on either side of the magnetic inversion line (Duijveman et al., 1982). These "footpoints' are clearly visible in medium/hard X-ray images (16-30 keV) but they can also be discerned in images taken at lower energies, particularly if care is taken in background subtraction. However, MacKinnon et al. (1984) have shown for the same flares analysed by Duijveman et al. that only ~ 10% of the hard X-rays within the HXIS fine field of view actually come from the footpoints. Thus, the observations of two bright areas in the 16 to 30 keV X-ray images does not necessarily mean that the major source is electron beams interacting in the thick target at the footpoints.

Similar bright compact sources or "kernels' are also observed in UV line emissions such as O IV (1401Å4) and Si IV (1402Å) (Cheng et al., 1984, Tandberg-Hanssen et al., 1983, 1984). Images taken in optical lines such as Hα and Mg I (Cheng et al., 1984) show that such kernels generally coincide with the location of the X-ray footpoints. These observations are inconsistent with a flare model in which hard X-rays are produced at the top of a loop, followed by the formation of a thermal conduction front propagating downward to the footpoints (Woodgate et al., 1983).

There is much evidence that the impulsive emissions are concentrated in the low coronal regions, the transition region, and even in the chromosphere. Kane et al. (1982) found that sources with photon energies of greater than or about 100 keV are located at altitudes of less than or about 2500 km. In addition, HXIS observations of the X-ray emission of the limb flares of 1980 April 30 (de Jager et al., 1983) and 1980 November 18 (Simnett and Strong 1984) show that the X-ray sources are located at low altitudes during the impulsive phase. For the flare of 1980 November 1 at 19:15 UT, images in the Fe XXI line show that at least during the first minute of the impulsive phase, the hot plasma component (10^7 K) was confined to the feet of the flare arch (Tandberg-Hanssen et al., 1984).

5B.1.3 Multi-thermal and/or Non-thermal Phenomena

During the first part of the impulsive phase, i.e., during a period of typically 1 min but sometimes longer (see Figure 5A.4), there are clear indications of the occurrence of multi-thermal or non-thermal phenomena. The temperature parameters derived from ratios of counts in different HXIS energy bands at that time have different values; this clearly suggests a multi-thermal or non-thermal situation. Thermalization is restored towards the end of the impulsive phase when all X-ray energy band ratios yield the same temperature values in the majority of flares (de Jager, 1985b). Bely-Dubau et al. (1984) found that they could not fit the combined BCS, HXIS, and HXRBS line and continuum spectral data with a purely thermal DEM model during the impulsive phase of the flare on 1980 June 29 at 18:22 UT. The deviations from such a model can be matched by assuming an additional component to the X-ray spectrum, as expected from a power-law distribution of electrons. MacNeice et al. (1985) also find for a flare on 1980 November 12 that the ratio of the flux in the Fe XXV line measured with the FCS to the continuum emission measured with HXIS is inconsisten with a thermal model. They conclude that there must be a significant non-thermal continuum contribution to the footpoint emission recorded by HXIS, even in the lowest channels from 3.5 to 8 keV.

The same situation is valid for the relative values of the electron temperature (derived from spectral fits to continuum observations) and the ion temperature (derived from Doppler widths of spectral lines). These also differ considerably in the first part of the impulsive phase and equalize later, usually also within 1 or 2 min. The Doppler temperatures are higher than the electron temperatures (Antonucci et al., 1984). Figure 5B.1 illustrates the similarity between these two different indicators of non-thermal and multi-thermal phenomena.

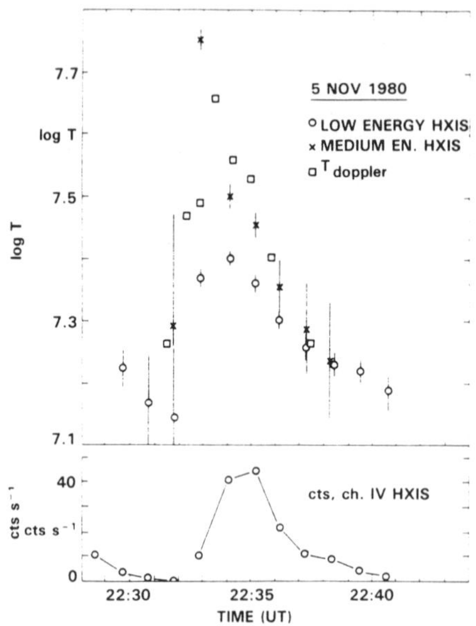

Figure 5B.1 Non-thermal or multi-thermal phenomena at the onset of the impulsive phase of the flare on 1980 April 8. 'Temperature' measurements from different HXIS energy band ratios show a similar behavior to the Doppler and electron temperatures measured at the same time with BCS (de Jager 1958b).

Additional evidence is provided by observations of the time variation of the Hα line profile. The energy output of a flare in that line ($|\Delta\lambda|$ < 5Å4) has a time dependence similar to that of hard X-rays (Figure 5B.2; Canfield and Gunkler 1985). The enhanced Hα emission of the flare of 1980 May 7 is due to increased emission in the wings of the line and agrees with theoretical line profiles calculated assuming bombardment of a thick target by energetic electron beams. These Hα observations therefore support the idea that the impulsive burst begins with a non-thermal event, i.e., the bombardment of the transition layer and/or the chromosphere by beams of energetic electrons.

This scenario is also supported by the X-ray imaging observations, as shown for the first time by Duijveman et al. (1982) for the flare of 1980 November 5 and later for that of 1980 May 9 at 07:10 UT by Machado (1983). Machado showed that the acceleration takes place during the early part of the impulsive hard X-ray event. One should note, however that a substantial amount of the emission at lower energies (less than or about

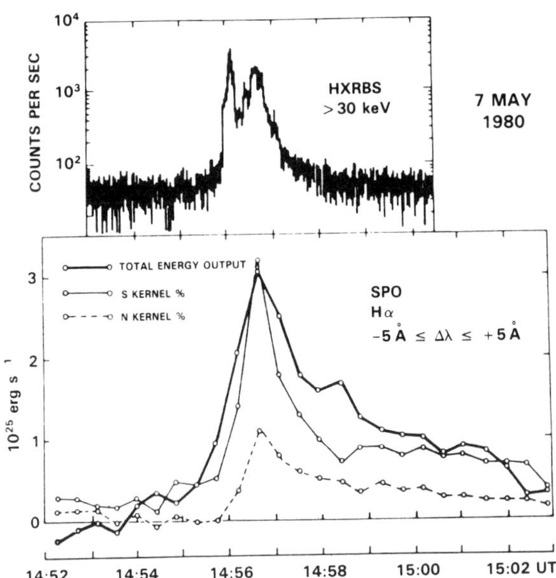

Figure 5B.2 Similarity between the hard X-ray response and the Hα output for the flare of 1980 May 7 (Canfield and Gunkler 1984). The hard X-ray data sum all counts at energies above 30 keV. The Hα data sum all power within 5 Å of pre-flare line center, subtracting the average value during a preflare reference period. For the Hα data, circles indicate values and times of observations. The heavy solid curve indicates the integrated Hα (± 5 Å) energy output (left scale). Percentage contributions of the south and north kernels are indicated by the light solid and dashed lines respectively (right scale).

10 keV) is of thermal origin at that time. Also, HXRBS and HXIS observations show that only 10% of the total hard X-ray emission originates in the footpoints; the rest comes from a larger area with a scale size of $\approx 10^4$ km (MacKinnon et al., 1984).

We therefore infer from the observed impulsive phenomena that an instability occurs "somewhere" in the solar atmosphere, followed by electron acceleration and footpoint bombardment by beams of energetic electrons. The location and nature of that instability is still unknown.

5B.1.4 Impulsive-Phase Microwave Bursts

Information on the origin of the non-thermal phenomena occurring during the onset of hard X-ray bursts can be obtained from the observation of impulsive microwave bursts. The peak fluxes of hard X-ray bursts F_x and of the corresponding microwave bursts F_μ are well correlated over three orders of magnitude in F_x and F_μ according to the relation $F_\mu \propto F_x^b$ (Kai et al., 1985). The scatter about this relation is much less than an order of magnitude. This correlation is remarkable at first sight because the microwave bursts are caused by gyrosynchrotron emission at high harmonics (10-100) of the gyrofrequency. Thus, the microwaves are mainly emitted by electrons with energies of the order of hundreds of keV, much higher than the tens of keV required to produce the hard X-ray bursts. However, the additional observation that the time variations of the microwave and hard X-bursts are very similar shows that both kinds ofelectrons must have the same origin, and the observed time variations must be caused by variations in the acceleration or injection of the electrons.

It is also very significant that the microwave sources are located at higher altitudes than the X-ray sources (Kane et al., 1983), in a medium with electron densities $n_e \approx 6 \times 10^8$ cm^{-3} (Enome 1983). This supports the suggestion that the source of the microwaves consists of high energy electrons trapped in elevated parts of the pre-flare loops (Hoyng et al., 1981, Marsh and Hurford 1982, Gary and Tang 1984), in contrast to the hard X-rays, which come from the feet of the loops. A possible confirmation of this model can be made from observations of the flare of 1980 May 28 at 17:47 UT, for which Gary and Tang (1985) found that the number of electrons responsible for the microwave emission is equal to that causing the hard X-ray bursts. The flare was a single spike in both hard X-rays and microwaves, and this facilitated the interpretation. Such equality was not found in previous comparisons because the thick-target model was adopted for the radio emission as well as for the X-ray emission. It now seems that the microwaves come from electrons near the top of the magnetic loops and so the thick-target model is not appropriate for the microwaves.

The model elaborated above is supplemented by data on size, magnetic field, and location of the microwave sources. The sizes are of the order of arc seconds. Spatially resolved images of microwave emission in the frequency range 5 - 15 MHz (Marsh and Hurford 1982) show that the sources are usually located near the magnetic inversion line and between Hα brightenings. These sources have diameters of 2 - 5 arcsec at 15 GHz and 10 - 15 arcsec at 5 GHz.

The emission properties demand a magnetic field B of the order 200-600 G. For the flare of 1980 June 29, B ≈ 220 G. A model was proposed that consisted of a hot kernel with an electron temperature of $\approx 10^9$ K surrounded by a "cooler" gas of 10^8 K (Dulk and Dennis 1982). The flare of 1980 March 29 at 09:18 UT emitted a hard X-ray burst of only 10 s duration and a simultaneous microwave burst, the latter observed in the frequency range from 0.9 to 10.4 GHz. The hard X-ray emission spectrum can be represented as thermal bremsstrahlung from a plasma with a temperature of $\approx 5 \times 10^8$ K, whereas the radio emission has been interpreted as gyrosynchrotron radiation from a loop with a magnetic field of 120 G (Batchelor et al., 1984).

Summarizing this section, we conclude that the impulsive microwave bursts are caused by greater than or about 100 keV electrons emitting gyrosynchrotron radiation in a source at higher altitude than those emitting hard X-rays. These microwave sources are small, with dimensions of arc seconds, and are located above the magnetic inversion line; they occur in magnetic fields of a few hundred gauss.

5B.1.5 Simultaneity of Burst Emissions; Fine Structures; Acceleration Problems

Even when the sources of the bursts are separated by large distances, of up to 2×10^5 km (Kattenberg et al. 1983), the bursts may still occur simultaneously, to within 1 s. Short time delays are sometimes observed, however, between the microwave and hard X-ray bursts. On 1980 November 1 an X-ray burst occurred 6 ± 3 s before a microwave burst (Tandberg-Hanssen et al., 1984). Nakajima et al. (1985) describe observations of "secondary microwave bursts" emitted by sources separated by 10^5 to 10^6 km from the primary source with time delays of 2-5 s. The evidence suggests that the secondary bursts were produced by electrons with energies of 10-100 keV, channeled from the primary source along huge coronal loops to the secondary location.

On 1980 June 7, flares were observed in which hard X-ray bursts preceded the microwave and gamma ray bursts by 1-2 s (Nakajima et al., 1983) (see Figure 5B.3). These observations are consistent with the two-step acceleration model proposed by Bai and Ramaty (1979) and Bai (1982). In this model, 10 to 100 keV electrons that produce the hard X-ray bursts are accelerated first, while another population of ~ 100 keV electrons causes the first microwave bursts. Ions and electrons are both accelerated to > 10^3 keV a few seconds later, giving rise to gamma rays and delayed microwave emission.

Large time differences, of up to 5 s, between hard X-ray bursts (< 100 keV) and microwave and harder X-ray bursts (> 100 keV) are ascribed to the time needed to build up a population of > 100 keV electrons (Takakura et al., 1983) and to path lengths. The latter aspect is dealt with easily on the basis of an Alfven wave travelling along a semicircular loop with a field B and an electron density n_e. An Alfven wave could travel between footpoints separated by the distances given in Table 5B.1 in a time of 4 s (chosen arbitrarily). Thus, most of the observed footpoint separations fit well with delay times of a few seconds, assuming that an Alfven wave is the connecting agent.

Table 5B.1 Distances in 10^3 km travelled by an Alfven wave in 4 s

$n_e(cm^{-3})$	B = 30 G	B = 300 G	B = 1000 G
10^8	16	160	530
10^9	5.5	55	180
10^{10}	1.6	16	50

High time resolution observations at mm-microwaves show fine structures with characteristic timescales as short as 30 to 60 ms. A few bursts inspected in detail show some correlation with similar structures in hard X-ray bursts (Kaufmann et al., 1980, Wiehl and Matzler 1980). Also, in these subsecond bursts, the microwave bursts are often delayed in time compared with the corresponding hard X-ray structures; delays may reach values up to ~ 0.4 s (Cornell et al., 1984), but there is a case of a 1.5 s delay. The most acceptable explanation seems to be that the delayed bursts are from another population of energetic electrons (≈ 100 keV) decoupled from, but accelerated nearly simultaneously with, the population that produced the hard X-rays (Tandberg-Hanssen et al., 1984). Gary and Tang (1985), however, note that the major part of the delay may be caused by the fact that the decay time for the microwave bursts can be considerably longer (because of the lower density) than that of the hard X-ray bursts.

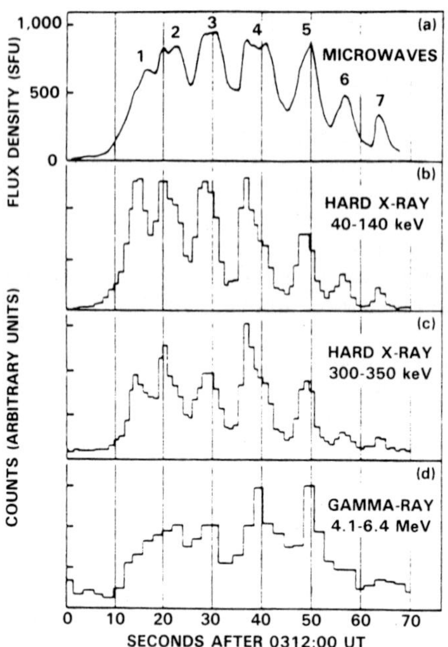

Figure 5B.3 Time profiles of the following emissions for the flare of 1980 June 7 at 03:12 UT from Nakajima *et al.* (1983): (a) 17 GHz microwaves, (b) 40 to 140 KeV X-rays, (c) 300 to 350 keV X-rays, and (d) 4.1 to 6.4 MeV gamma rays.

These subsecond bursts are apparently a new class of bursts with characteristic times shorter by a factor of ten than the elementary flare bursts defined by de Jager and de Jonge (1978). The rate of energy release in the subsecond bursts is of the order of 10^{27}-10^{29} ergs s^{-1} (Kaufmann et al., 1984), i.e., similar to that of longer bursts (Brown and Smith 1980). These subsecond bursts, as well as the elementary flare bursts, may be related to multiple energy injections. It remains to be investigated whether they are related to multi-thermal phenomena or to bursts in individual flare loops.

We conclude that information on the acceleration processes can be derived from a study of time differences between bursts in different energy ranges. Such studies show that acceleration of electrons to ~ 100 keV and of both electrons and ions to ~ 10^3 keV generally takes place in a time interval of less than a few seconds. It is not necessarily correct, though, to claim that all these particles are accelerated "simultaneously', because clear time differences have been observed (Bai and Dennis 1985). Melrose (1983) has given evidence that secondary acceleration, assumed to occur stochastically by hydromagnetic turbulence, is responsible for the occurrence of high-energy particles (greater than or about 20 MeV per nucleon). The first step, pre-acceleration, should yield particles with energies up to ~ 100 keV per nucleon. It is ascribed to local heating of ions to ~ 10^9 K or to acceleration by potential electric fields.

5B.1.6 Chromospheric Evaporation

The phenomenon of chromospheric evaporation (ablation) occurs during the early part of the impulsive phase. In high-resolution soft X-ray line spectra, the line profiles show evidence for high-speed plasma upflows with velocities up to 400 km s^{-1} and turbulent (non-thermal) mass motions of ~ 100 km s^{-1} (Antonucci et al., 1982, 1984). Simultaneously, the Hα line profile is broadened and modified at the footpoints, consistent with the bombardment of electrons and chromospheric heating. These phenomena are restricted to the Hα kernels in the impulsive phase (Badalyan and Livshits 1982, Ichimoto and Kurokawa, 1984, Canfield and Gunkler 1984, Gunkler et al., 1984).

5B.1.7 Convective Motions

Convective motions occur as a consequence of the heating of the evaporated gas. The observed velocities, ranging between 150 and 350 km s^{-1} are in agreement with theoretical views on convection in a coronal plasma (Fisher et al., 1984, MacNeice et al., 1984). An illustration of these phenomena is given in Figure 5B.4 for the flare of 1980 May 21 at 21:00 UT (Antonucci et al., 1984). The total number of electrons involved is 3 x 10^{37}, and the number of chromospheric atoms is 7 x 10^{37} (Acton et al., 1982). An energy flux of (3-10) x 10^{9} ergs cm^{-2} s^{-1} (MacNeice et al., 1984, Fisher et al., 1984) is needed, and the total energy that must be supplied to make the convection possible is (1-4) x 10^{30} ergs. The density of the upward moving plasma is > 4 x 10^{10} cm^{-3} (Antonucci et al., 1984). For the Queens' flare (1980 April 30), 10^{37} electrons were involved and the densities of the convected plasma and of the footpoint kernel were 10^{11} cm^{-3} and 4.5 x 10^{11} cm^{-3}, respectively (de Jager et al., 1983).

5B.1.8 Coronal Explosions

Coronal explosions have been detected in HXIS images of four flares (de Jager and Boelee 1984, de Jager et al., 1984, de Jager 1985a). They were discovered by noting for each pixel in the flaring area the time t_m at which maximum intensity is reached. The earliest t_m values are found in small areas (10^8 km^2) near the flare footpoints. From there, the t_m-isochrones, being the lines of equal time of intensity maximum in the flare, spread out laterlly over the flaring area with velocities that are initially of the order of a few hundred to more than 10^3 km s^{-1}, but that decrease rapidly in the course of a few minutes (Figure 5B.5). In the same way as there are generally two footpoints, there are also two "sources' (i.e. areas where t_m is earliest), although it is not yet certain that the "sources' are identical to the footpoints. From each "source' the isochrones spread out as a wave-like feature. The two waves, from the two sources, may meet each other and merge after a few minutes.

These waves are not thermal or conduction fronts as can easily be ascertained: the variation of temperature in the various flare-pixels is not correlated with that of the intensity in these pixels. Since the t_m-isochrones connect points where maximum intensity is reached for any

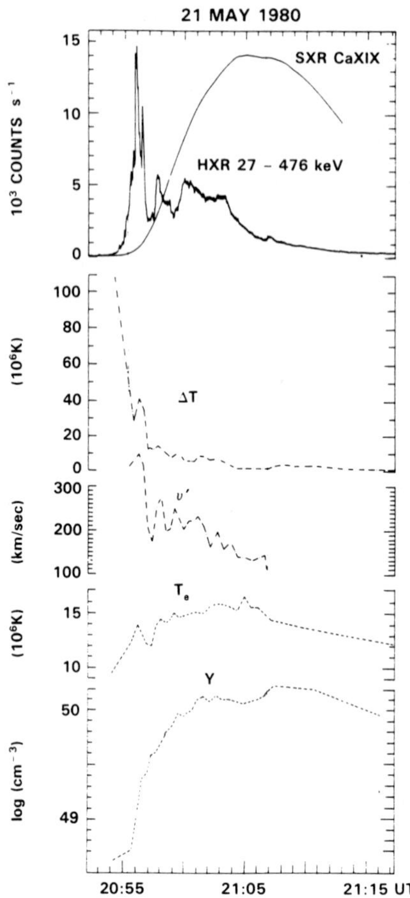

Figure 5B.4 Time variation of parameters characterizing the chromospheric evaporation for the 1980 May 21 flare from Antonucci *et al.* (1984). The following derived quantities are plotted as a function of time with the Ca XIX and hard X-ray rates shown for reference: the temperature difference ΔT between the ion and the electron temperatures, the upward velocity component v′, the electron temperature T_e, and the Ca XIX emission measure Y.

pixel, they connect points of maximum emission measure and, assuming the volume V not to vary too rapidly, the t_m-isochrones thus connect points where the density n_e is maximum in the various pixels. Therefore, a coronal explosion is a density wave, which may be identified with a shock wave, either hydrodynamical or magneto-hydrodnamical. The most likely explanation is that the explosion is the manifestation of gas that, after having emerged from the flare footpoint by ablation and consequent convective motions, spreads out over a large area, with a strong lateral velocity component.

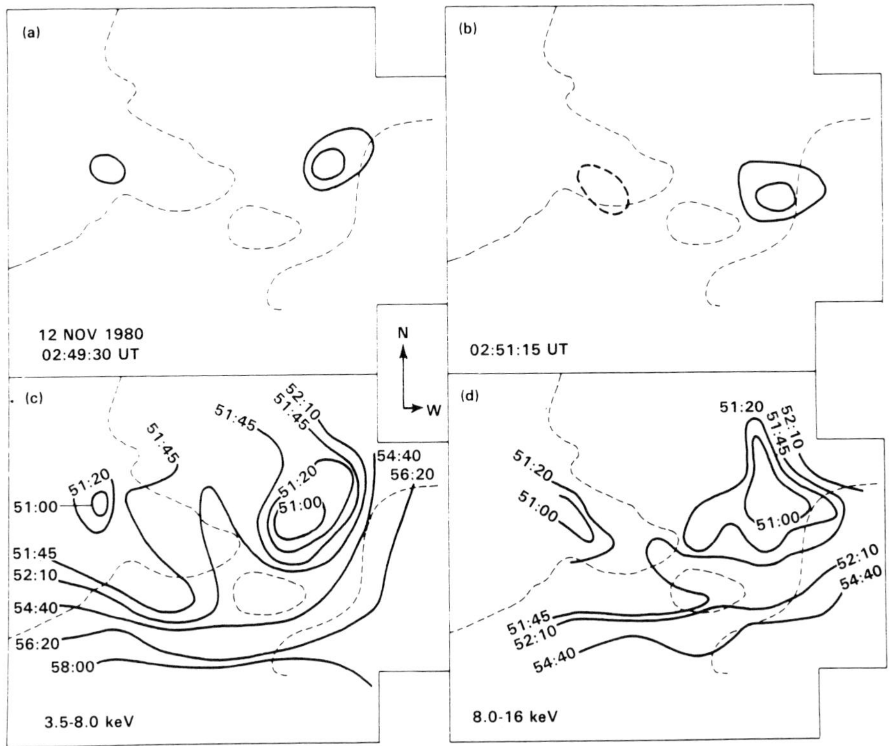

Figure 5B.5 Coronal explosion in the flare of 1980 November 12 at 02:50 UT from de Jager (1985a). (a) and (b) Location of hard X-ray footpoints at the times of the two main bursts of the impulsive phase. The dashed lines indicate the locations of magnetic inversion. (c) and (d) The explosion seen in soft (3.5 to 8.0 keV) and medium (8.0 to 16 keV) energy X-rays. The curves are labeled with the time in minutes and seconds after 02:00 UT.

The energy of the coronal explosions is at least an order of magnitude smaller than the total thermal energy content of the flare at the time of the explosion (de Jager 1985a).

5B.1.9 Thermal Conduction Fronts

Evidence for thermal conduction fronts has been obtained in a few flares (Rust et al., 1985). Their signature is the expansion of the X-ray emitting region along a loop with velocities between 800 and 1700 km s^{-1}. They differ from coronal explosions in that with conduction fronts the motions are directed along loops and the velocities remain high and virtually constant along the loop. The velocities are consistent with elementary considerations of thermal conduction with a heat flux of $\approx 10^{10}$ erg s^{-1} cm^{-2}. This corresponds to an energy release of 10^{28} erg s^{-1}.

5B.1.10 Gamma Rays and Neutrons from Flares

Observations of the gamma-ray and neutron flux and the analysis of their spectra offers a means of deriving the total kinetic energy W of the flare particles in the high-energy range (Ramaty 1982). Derived values for W(> 1 MeV) have an upper limit of 2.5×10^{30} ergs (for the flare of 1972 August 4), whereas the smallest reported value is 5×10^{28} ergs (1980 July 1).

Energetic solar neutrons with energies above 50 MeV have been detected with GRS after at least two flares on 1980 June 21 at 01:18 UT (Chupp et al., 1982) and 1982 June 3 at 11:42 UT (Chupp et al., 1983, Evenson et al., 1983). In the case of the first flare, impulsive photon emission from 10 keV to > 65 MeV lasting for ~ 66 s was followed by a transient flux of 50-600 MeV neutrons incident over a 17-min period. These observations indicate the emission of 3×10^{28} neutrons sr^{-1} with energies of > 50 MeV, requiring the rapid acceleration on a timescale of << 60 s of protons to GeV energies during the impulsive phase of the flare. Ramaty et al. (1983) have estimated for this flare that the total energy content in the protons and nuclei was 2.4×10^{30} ergs, with 75% of this energy residing in > 1 MeV particles. In contrast, the energy content in electrons of energies greater than 0.3 MeV was only 5×10^{28} ergs.

The neutron flux observed after the second flare was an order of magnitude more intense but with a similar spectral shape. Prince et al. (1983) have used the measured flux in the 2.223 MeV gamma ray line to determine an upper limit of 3.9×10^{28} neutrons MeV^{-1} sr^{-1} on the flux below 20 MeV, which is the major source of this line. This indicates that the neutron spectrum must flatten at energies below ~ 50 MeV.

5B.2 Summary and Scenario

The impulsive phase is a period during which hard X-ray bursts and microwave bursts occur nearly simultaneously, as well as bursts in other energy ranges (EUV, gamma). These emissions have a non-thermal character, but only for about 10^2 s, after which time thermalization is restored. During the impulsive phase, upward movements of plasma are also observed, indicating chromospheric evaporation (ablation) followed by convection. The thermal energy content of the flare increases steadily, reaching a maximum value at the end of the impulsive phase. Coronal explosions can also occur as a result of the large energy input in a small volume.

The flare model that results from these conclusions is that of a flare consisting of a system of loops. These loops may be interwoven and are often not resolved as loop complexes. Many observations show only a single or a few loops, but in other cases the observations can only be interpreted in terms of several or even many unresolved loops and therefore appearing as one. One such case is the flare of 1980 November 1, 19:15 UT (Tandberg-Hanssen et al., 1984). Also, the fact that the so-called "tongue' of a flare has a small filling factor (de Jager et al., 1983, Woodgate et al., 1983) suggests the existence of many unresolved magnetic structures or of many thin loops in a confined region.

Interaction between these loops on field lines that undergo violent motions and drastic changes of direction leads to magnetic reconnection (Moore et al., 1984) with the acceleration of electrons and ions. The literature describes a few cases in which the interaction of loop-systems leading to the flare was seen quite clearly. The Queens' Flare (1980 April 30 at 20:20 UT; de Jager et al., 1983) was a limb flare in which the interacting loops could be identified, and the site and time of the interaction was determined unambiguously. Another case is the disk flare of 1980 June 24 at 15:22 UT (Gunkler et al., 1984). These observations show that the point of reconnection (probably identical to the site of particle acceleration) is normally not at the "top" of the loop. Mouradian et al. (1983) suggest the existence of "elementary eruptive phenomena' consisting of two systems of loops, a cold ($T \approx 10^4$ K) "surging arch' and a hot "flaring arch'. This concept is interesting, but it remains to be seen whether it applies to most of the observed flares.

Magnetic field reconnection leads to electron acceleration and the consequent emission of microwave bursts by gyrosynchrotron radiation and footpoint bombardment by the electron beams. The latter effect heats the chromospheric/ transition layer part of the footpoint area, causes ablation (evaporation), convective ascent of heaed gas, and coronal explosions.

5B.3 The Energy Source of Flares

Observations show the close connection between magnetic field twisting, energy build-up, and energy release of the solar magnetic field. Consequently, it is currently assumed that the shearing motion of footpoints of flux tubes is the origin of the solar flare energy (Wu and Hu 1982, Akasofu 1984). On the basis of this idea and the linear force-free magnetic field theory, Tanaka and Nakagawa (1973) studied some active regions and a bipolar sunspot. Yang and Chang (1981), and Yang et al. (1983) solved the first order force-free differential equations. From the observational data of penumbral filament twisting and their force-free field solution, they computed the energy excess, ΔM, of the force-free field at different times, where $\Delta M = M_f - M_p$, and M_f and M_p are the magnetic energies of the force-free and the potential fields, respectively. Comparing their results with solar flare data, they found that when ΔM was between 10^{28} and 5×10^{30} ergs, a subflare may occur; if $\Delta M \approx 10^{30}$-10^{31} ergs, medium sized flares can appear, whereas $\Delta M > 10^{32}$ ergs is the condition for the appearance of a large flare. The first quoted values are of the same order as those derived from the observations of the impulsive phase of the prime flares discussed in Section 5.2.

These considerations, applied to AR 2372 (Boulder number) in the period 1980 April 5 to 7 show how a self-consistent MHD model can be used to study the magnetic energy build-up in that region. It is found that almost all of the energy for the flare is stored in the magnetic field, the magnitude of the shearing motions being the crucial parameter (Wu et al., 1984). It is therefore important that a reduction of the magnetic shear has been observed during or after several flares (Tanaka et al., 1980, Farnik et al., 1983).

5.7 ACKNOWLEDGEMENTS

We are grateful for contributions to this effort by E. Antonucci, H.M. Chang, U. Feldman, S. Graham, P. Hoyng, P. Kaufmann, M. Machado, D. Rust, and S. Walton. KTS and MEB acknowledge the support of NASA contract no. NAS5 23758 and the Lockheed Independent Research Program. The National Center for Atmospheric Research is sponsored by the National Science Foundation. RP, GP, and SS have been supported by the Italian Consiglio Nazionale delle Riceiche/Piamo Spaziale Nazionale and by the Italian Ministry of Education. STW would like to acknowledge support from NASA Grant (NAGW-9) and the SMM Guest Investigator Program. We are grateful to Drs. R.L. Moore and S.K. Antiochos for their careful reading of the manuscript and constructive criticisms as referees. We also thank Mrs. G. Wharen, Dr. Y. Strong and Dr. J.L.R. Saba for their help in preparing this chapter for publication. The VLA observations reported here were made at the National Radio Astronomy Observatory, which is operated by the Associated Universities Inc., under contract with the National Science Foundation

5.8 REFERENCES

Acton, L.W. et al., 1980, Solar Phys., **65**, 53.

Acton, L.W. et al., 1981, Ap.J. Lett., **244**, L137.

Acton, L.W., Canfield, R.C., Gunkler, T.A., Hudson, H.S., Kiplinger, A.L., Leibacher, J.W., 1982, Ap.J., **263**, 409.

Acton, L.W. et al., 1983, Solar Phys., **86**, 79.

Akasofu, S.-I. 1984, Plan. Space Sci., in press.

Antiochos, S.K., 1980, Ap.J., **241**, 385.

Antiochos, S.K. and Krall, K., 1979, Ap.J., **229**, 788.

Antiochos, S.K. and Sturrock, P.A., 1976, Solar Phys., **49**, 359.

Antiochos, S.K. and Sturrock, P.A., 1978, Ap.J., **220**, 1137.

Antiochos, S.K. and Sturrock, P.A., 1982, Ap.J., **254**, 343.

Antonucci, E. and Dennis, B.R., 1983, Solar Phys., **86**, 67.

Antonucci, E., Dennis, B.R., Gabriel, A.H., and Simnett G.M., 1985, Solar Phys., **96**, 129.

Antonucci, E., Gabriel, A.H., Dennis, B.R., 1984, Ap.J., **287**, 917.

Antonucci, E., Gabriel, A.H., Acton, L.W., Culhane, J.L., Doyle, J.G., Leibacher, J.W., Machado, M.E., Orwig, L.E., 1982, Solar Phys., **78**, 107.

Badalayan, O.G. and Livshits, M.A., 1982, Astron. Zhwin., **59**, 975.

Bai, T. and Dennis, B.R., 1985, Ap.J., **292**, 699.

Bai, T. and Ramaty, R., 1979, Ap.J., **227**, 1072.

Bai, T., 1982, Gamma Ray Transients and Related Astrophysical Phenomena (R.E. Lingenfelter, H.S. Hudson, and D.M. Worral, eds.), AIP Conference Proceeding, No. 77, p. 409.

Batchelor, D.A., 1984, Energetic Electrons in Impulsive Solar Flares, PhD. Thesis, Univ. of North Carolina at Chapel Hill; also NASA TM86102.

Batchelor, D.A., Benz, A.O., Wiehl, H.J., 1984, Ap.J., **280**, 879.

Bell, A.R., Evans, R.G., and Nicholas, D.J., 1981, Phys. Rev. Lett., **46**, 243.

Bely-Dubau, F. et al., 1982, Mon. Not. R. Astr. Soc., **201**, 1155.
Bely-Dubau, F., Gabriel, A.H., Sherman, J.C., Orwig, L.E., and Schrijver, J., 1984, Proc. COSPAR Symposium, Graz, Austria, to be published.
Bentley, R.D. et al., 1984, Astron. and Astrophys., in press.
Bornmann, P.L., 1985a, Ap.J., in press.
Bornmann, P.L., 1985b, Solar Phys., submitted.
Bornmann, P.L., 1985c, Ph.D. Thesis, Univ. of Colorado, Boulder,
Brown, J.C., 1971, Solar Phys., **18**, 489.
Brown, J.C., 1974, "Coronal Disturbances" (G.A. Newkirk, ed.), IAU Symp. 57, 395.
Brown, J.C. and Smith, D.F., 1980, Rep. Prog. Phys., **43**, 125.
Bruzek, A., 1967, in J. Xanthakis (ed.), Solar Physics, Interscience, New York, p. 414.
Canfield, R.C. et al., 1980, Solar Flares: A Monograph from the Skylab Solar Workshop II, (P.A. Sturrock, ed.), Colorado Assoc. Univ. Press, Boulder, p. 451.
Canfield, R.C. and Gunkler, T.A., 1985, Ap.J., **288**, 353.
Cargill, P.J. and Priest, E.R., 1982, Solar Phys., **76**, 357.
Cargill, P.J. and Priest, E.R., 1983, Ap.J., **266**, 383.
Cheng, C.-C. et al., 1982, Ap.J., **253**, 353.
Cheng, C.-C., Tandberg-Hanssen, E., Orwig, L.E., 1984, Ap.J., **278**, 853.
Chupp, E.L., Forrest, D.J., Ryan, J.M., Heslin, J., Reppin, C., Pinkau, K., Kanbach, G., Reiger, E., and Share, G.N., 1982, Ap.J. Letters, **263**, L95.
Chupp, E.L. and Forrest, D.J., 1983, Report to the SMM and STP Workshops on the Results from the SMM Gamma Ray Spectrometer Experiment.
Chupp, E.L., Forrest, D.J., Share, G.H., Kanbach, G., Debrunner, H., and Fluekiger, E., 1983, Proc. 18th Int. Cosmic Ray Conference, Bangalore, India.
Cliver, E.C., Dennis, B.R., Kiplinger, A.L., Kane, S.R., Neidig, D.F., Sheeley, N.R. (Jr.), and Koomen, M.J., 1985, Ap.J., submitted.
Cornell, M.E., Hurford, G.J., Kiplinger, A.L., and Dennis, B.R., 1984, Ap.J., **279**, 875.
Cox, D.P. and Tucker, W.A., 1969, Ap.J., **157**, 1157.
Craig, I.J.D., and Davys, J.W., 1984, Solar Phys., **90**, 343.
Crannell, C.J., Frost, K.J., Matzler, C., Ohki, K, and Saba, J.R., 1978, Ap.J., **223**, 620.
Crannell, C.J., Karpen, J.T. and Thomas, R.J., 1982, Ap.J., **253**, 975.
Culhane, J.L., Vessecky, J.F., and Phillips, K.J.H., 1970, Solar phys., **15**, 395.
de Jager, C. and de Jonge, G., 1978, Solar Phys., **58**, 127.
de Jager, C., Machado, M.E., Schadee, A., Strong, K.T., Svestka, Z., Woodgate, B.E., and van Tend, W., 1983, Solar Phys., **84**, 205.
de Jager, C., 1985a, Solar Phys., **96**, 143.
de Jager, C., 1985b, Kernel Heating and Ablation in the Impulsive Phase of Two Solar Flares, Solar Phys, submitted.
de Jager, C. and Boelee, A., 1984, Solar Phys., **92**, 227.
de Jager, C., Boelee, A., and Rust, D.M., 1984, Solar Phys, **92**, 245.
Dennis, B.R., Frost, K.J. and Orwig, L.E., 1981, Ap.J. Letters, **244**, L167.
Dennis, B.R., Kiplinger, A.L., Orwig, L.E., and Frost, K.J., 1984, Proc.

2nd Indo-US Workshop on Solar Terrestrial Physics, New Delhi, India.
Dennis, B.R., Lemen, J., and Simnett, G., 1985, The Relationship Between Hard and Soft X-rays in Flares Observed with the Solar Maximum Mission during 1980, preprint.
Dere, K.P. and Cook, J.W., 1979, Ap.J., **229**, 772.
Doschek, G.A., Feldman, U., Kreplin, R.W., and Mariska, J.T., 1980, Ap.J. Letters, **241**, L175.
Doschek, G.A., Feldman, U., Landecker, P.B., and McKenzie, D.L., 1981, Ap.J., **249**, 372.
Doschek, G.A., Boris, J.P., Cheng, C.C., Mariska, J.T., and Oran, E.S., 1982, Ap.J., **258**, 373.
Dowdy, J.F., Moore, R.L., and Wu, S.T., 1983, BAAS, **15**, 700.
Doyle, J.G., and Raymond, J., 1984, Solar Phys., **90**, 97.
Dulk, G.A., and Dennis, B.R., 1982, Ap.J., **260**, 875.
Duijveman, A., Hoyng, P., and Machado, M., 1982, Solar Phys., **81**, 137.
Duijveman, A., 1983, Solar Phys., **84**, 189.
Duijveman, A. and Hoyng, P., 1983, Solar Phys, **86**, 279.
Duijveman, A., Somov, B.V., and Spektor, A.R., 1983, Solar Phys., **88**, 257.
Dwivedi, B.N., Hudson, H.S., Kane, S.R., and Svestka, Z., 1984, Solar Phys., **90**, 331.
Ellison, M.A., 1963, Proc. Roy. Obs. Edinburgh, **4**, 62.
Enome, S.L, 1983, Solar Phys., **86**, 421.
Evenson, P., Meyer, P., and Pijl, K.R., 1983, Ap.J., **274**, 875.
Farnik, F., Kaastra, J., Kalman, B., Karlicky, M., Slottje, C., and Valnicek, B., 1983, Solar Phys., **89**, 355.
Feldman, U., Cheng, C.-C., Doschek, G.A., 1982, Ap.J., **255**, 320.
Fisher, G.H. Canfield, R.C., and McClymont, A.N., 1984, Flare Loop Radiative Hydrodynamics VI, Ap.J., submitted.
Forbes, T.G. and Priest, E.R., 1982, Solar Phys., **81**, 303.
Forbes, T.G. and Priest, E.R., 1983a, Solar Phys., **84**, 169.
Forbes, T.G. and Priest, E.R., 1983b, Solar Phys., **88**, 211.
Forrest, D.J. and Chupp, E.L., 1983, Nature, **305**, 291.
Gary, D.E. and Tang, F., 1985, Ap.J., **288**, 385.
Gunkler, T.A., Canfield, R.C., Acton, L.W., and Kiplinger, A.L., 1984, Ap.J., **285**, 835.
Hagyard, M., 1984, Proc. Kunming Workshop, Chen Biao and C. de Jager (eds.), to be published.
Harrison, R.A. et al., 1985, Solar Phys., **97**, 387.
Harvey, J.W. 1983, Adv. Space Res., **2**, 31.
Hoyng, P., Brown, J.C. and van Beek, H.F., 1976, Solar Phys., **48**, 197.
Hoyng, P. et al., 1981, Ap.J., **246**, L155.
Hoyng, P., Marsh, K.A., Zirin, H., Dennis, B.R., 1983, Ap.J., **268**, 865.
Hudson, H.S. and Ohki, K., 1972, Solar Phys., **23**, 155.
Hudson, H.S. and Willson, R.C., 1983, Solar Phys., **86**, 123.
Ichimoto, K. and Kurkawa, K., 1984, Publ. Astron. Soc. Japan, submitted.
Jackson, B.V. and Hildner, E., 1978, Solar Phys., **60**, 155.
Kahler, S.W., 1982, J.G.R., **87**, 3439.
Kai, K., Kosugi, T., and Nitta, N., 1985, Publ. Astron. Soc. Japan, **37**, 155.

Kaempfer, N. and Magun, A., 1984, Comparison of the Electron Flux Derived from the Temporal Fine Structures in Hα, Hard X-rays and Microwaves, Institute of Applied Physics, University of Bern, Switzerland, preprint.
Kane, S.R., 1969, Ap.J. Letters, **157**, L139.
Kane, S.R. and Donnelly, R.F., 1971, Ap.J., **164**, 151.
Kane, S.R., Fenimore, E.E., Klebesadel, R.W., and Laros, J.G., 1982, Ap.J. Letters, **254**, L53.
Kane, S.R., Kai, K., Kosugi, T., Enome, S., Landecker, P.B., and McKenzie, D.L., 1983, Ap.J., **271**, 376.
Karpen, J.T., Crannell, C.J., and Frost, K.J., 1979, Ap.J., **234**, 370.
Kattenberg, A., Allaart, M., de Jager, C., Schadee, A., Schrijver, J., Shibasaki, K., Svestka, Z., and van Tend, W., 1983, Solar Phys, **88**, 315.
Kaufmann, P., Strauss, F.M., Opha, R., and Laporte, C., 1980, Astron. Astrophys., **87**, 58.
Kaufmann, P., Correia, E., Costa, J.E.R., Sawant, H.S., and Zodi Vaz, A.M.. 1984, in preparation.
Kiplinger, A.L., Dennis, B.R., Emslie, A.G., Frost, K.J., and Orwig, L.E., 1983, Ap.J. Letters, **265**, L99.
Kiplinger, A.L. et al., 1984, Ap.J. Letters, in press.
Kopp, R.A. and Pneuman, G.W., 1976, Solar Phys., **50**, 85.
Krall, K.R. et al., 1983, Solar Phys., **79**, 59.
Labs, D. and Neckel, H., 1968, Z.Ap., **69**, 1.
Labs, D. and Neckel, H., 1970, Solar Phys., **15**, 79.
Lantos, P., Kerdraon, A., Rapley, C.G., and Bentley, R.D., 1981, Astron. Astrophys, **101**, 33.
Lin, R.P. and Hudson, H.S., 1976, Solar Phys., **50**, 153.
Lin, R.P., Schwartz, R.A., Pelling, R.M., and Hurley, K.C., 1981, Ap.J. Letters, **251**, L109.
MacKinnon, A.L., Brown, J.C., and Hayward, J., 1984, preprint.
MacNeice, P., McWhirter, R.W.P., Spicer, D.S., and Burgess, A., 1984, Solar Phys., **90**, 357.
MacNeice, P., Pallavicini, R., Mason, H.E., Simnett, G., Antonucci, E., Shine, R.A., Rust, D.M., Jordan, C., and Dennis, B.R., 1985, Solar Physics, in press.
Machado, M.E., 1983, Solar Phys., **89**, 133.
Machado, M.E. et al., 1983, Solar Phys., **85**, 157.
Machado, M.E. and Somov, B.V., 1983, Adv. Space Sci., **2**, 101.
Marsh, K.A. and Hurford, G.J., 1982, Ann. Rev. Astron. Astrophys., **20**, 497.
Martens, P.C.H., van den Oord, G.H.J., and Hoyng, P., 1985, Solar Phys., 253.
Matteson, J.L., 1971, Ph. D. thesis, UCSD, San Diego, California.
Melrose, D.B., 1983, Solar Phys., **89**, 149.
Melrose, D.B. and Dulk, G.A., 1982, Ap.J., **259**, 844.
Mewe, R. and Gronenschild, E.H.B.M., 1981, Astron. Astrophys. Suppl. Series, **45**, 11.
Mewe, R., Schrijver, J., and Sylwester, J., 19, Astron. and Astrophys. Suppl., **40**, 323.
Minnert, M., Mulders, G.F.W., Houtgast, J., 1940, Photometric Atlas of

the Solar Spectrum, Amsterdam, Kampert and Helm.
Moore, R. et al., 1980, Solar Flares: A Monograph from the Skylab Solar Workshop II. (P.A. Sturrock, ed.), Colorado Assoc. Univ. Press, Boulder, p. 341.
Moore, R.L., Hurford, G.J., Jones, J.P. and Kane, S.R., 1984, Ap.J., **236**, 379.
Mouradian, Z., Martres, M.J., and Soru-Escaut, I., 1983, Solar Phys. **87**, 309.
Nagai, F., 1980, Solar Phys., **68**, 351.
Nakajima, H., Kosugi, T., Kai, K., and Enome, S., 1983, Nature, **305**, 292.
Nakajima, H., Dennis, B.R., Hoyng, P., Nelson, G., Kosugi, T., and Kai, K., 1985, Ap.J., **288**, 806.
Neidig, D.F. and Cliver, E.W., 1984, A Catalog of Solar White-Light Flares (1859-1982), AFGL-TR-83-0257.
Neupert, W.M., 1968, Ap.J. Letters, **153**, L59.
Ohki, K., Takakura, T., Tsuneta, S., and Nitta, N., 1983, Solar Phys., **86**, 301.
Pallavicini, R., Peres, G., Serio, S., Vaiana, G.S., Acton, L.W., Leibacher, J.W., and Rosner, R., 1983, Ap.J., **270**.
Peres, G., Rosner, R., Serio, S., and Vaiana, G.S., 1982, Ap.J., **252**, 791.
Peres, G., Serio, S., and Pallavicini, R., 1985, in preparation.
Phillips, K.J.H. et al., 1982, Ap.J., **256**, 774.
Pneuman, G.W., 1982, Solar Phys., **78**, 229.
Poland, A.I., Orwig, L.E., Mariska, J.T., Nakatsuka, R.S., and Auer, L.H., 1984, Ap.J., **280**, 457.
Poland, A.I. et al., 1982, Solar Phys., **78**, 201.
Prince, T.A. et al., 1983, Proc. 18th Int. Cosmic Ray Conf., Bangalore, India, paper SP3-3.
Ramaty, R., 1982, Nuclear Processes in Solar Flares, NASA Techn. Mem. 83904 (Chapter 2 in The Physics of the Sun, eds. T.E. Holzer et al., Reidel, Dordrecht, in press).
Ramaty, R., Murphy, R.J., Kozlovsky, B., and Lingenfelter, R.E., 1983, Ap.J. Letters, **273**, L41.
Rosner, R., Low, B.C., and Holzer, T.E., 1985, The Physics of the Sun, eds. T.E. Holzer et al., Reidel, Dordrecht, in press.
Rosner, R., Tucker, W., and Viaina, G., 1978, Ap.J., **240**, 643.
Rust, D.M., 1984, Adv. Space Res., **4**, 191.
Rust, D.M. et al., 1980, Solar Flares: A Monograph from the Skylab Solar Workshop II, (P.A. Sturrock, ed.), Colorado Assoc. Univ. Press, p. 273.
Rust, D.M., Simnett, G.M., and Smith, D.F., 1985, Ap.J., **288**, 401.
Rust, D.M. and Somov, B., 1984, Solar Phys., **93**, 95.
Schmahl, E.J., Kundu, M.R., Strong, K.T., Bentley, R.D., Smith, J.B., Jr, and Krall, K.R., 1982, Solar Phys., **80**, 233.
Schmahl, E.J., 1983, Adv. Space Sci., **2**, 73.
Shoub, E.C., 1983, Ap.J., **266**, 339.
Simnett, G.M. and Harrison, R.A., 1984, Proc. 25th COSPAR Meeting, Graz, Austria.
Simnett, G.M. and Harrison, R.A., 1984, Solar Phys., submitted.
Simnett, G.M. and Strong, K.T., 1984, Ap.J., **284**, 839.

Simnett, G.M., 1985, Conference Papers, 19th International Cosmic Ray Conference, NASA CP2376, 4, 70.
Smith, D.F., 1985, Ap.J., **288**, 801.
Smith, E.v.P. and Smith, H.J., 1963, Solar Flares, (MacMillan).
Smith, E.v.P. and Gottlieb, D., 1975, Space Science Reviews, **16**, 771.
Sprangle, P. and Vlahos, L, 1983, Ap.J. Letters, **273**, L95.
Spitzer L.J. and Harm, R., 1953, Phys. Rev., **89**, 977.
Stenflo, J.O., 1976, IAU Coll., **36**, 143.
Strong, K.T. et al., 1984, Solar Phys., **91**, 325.
Strong, K.T. et al., 1984, Proc. 25th COSPAR Meeting, Graz, Austria.
Sturrock, P.A., 1980, "Solar Flares: A Monograph from the Skylab Solar Workshop II", Colorado Assoc. Univ. Press, Boulder.
Summers, H.P. and McWhirter, R.W.P., 1979, J. Phys., **B12**, 2387.
Svestka, Z., Kopecky, M., and Blaha, M., 1962, Bull. Astron. Obs. Czech. **13**, 37.
Svestka, Z. et al., 1982a, Solar Phys., **75**, 305.
Svestka, Z., Dennis, B.R., Pick, M., Raoult, A., Rapley, C.G., Stewart, R.T., and Woodgate, B.E., 1982b, Solar Phys., **80**, 143.
Svestka, Z., Dodson-Prince, H.W., Martin, S.F., Mohler, O.C., Moore, R.L., Nolte, J.T., and Petrasso, R.D., 1982c, Solar Phys., **78**, 271.
Svestka, Z., Schrivjer, H., Somov, B., Dennis, B.R., Woodgate, B.E., Furst, E., Herth, W., Klein, L., and Raoult, A., 1983, Solar Phys., **85**, 313.
Sylwester, J., Schrijver, H., and Mewe, R., 1980, Solar Phys., **67**, 285.
Sylwester, J., Lemen, J.R. and Mewe, R., 1984, Nature, **310**, 665.
Takakura, T. and Scalise, E., 1970, Solar Phys., **26**, 151.
Takakura, T., Degaonkar, S.S., Ohki, K., Kosugi, T., and Enome, S., 1983, Solar Phys., **89**, 379.
Tanaka, K. and Nakagawa, Y., 1973, Solar Phys, **33**, 187.
Tanaka, K., Smith, Z., and Dryer, M., 1980, in M. Dryer and E. Tandberg-Hanssen (eds.): Solar and Interplanetary Dyanmics, Int. Astron. Union Symp., 91, 231.
Tanaka, K., Nitta, N., Akita, K., and Watanabe, T., 1983, Solar Phys., **86**, 91.
Tanaka, K., 1983, IAU Colloquium 71, Activity in Red Dwarf Stars, (P.B. Byrne and M. Rodono, eds.), p.307.
Tanaka, K., Watanabe, T., and Nitta, N., 1984, Ap.J, **282**, 793.
Tandberg-Hanssen, E., Reichmann, E., and Woodgate, B.E., 1983, Solar Phys., **86**, 159.
Tandberg-Hanssen, E., Kaufmann, P., Reichmann, E.J., Teuber, D.L., Moore, R.L., Orwig, L.E., and Zirin, H., 1984, Solar Phys., **90**, 41.
Tsuneta, S., 1983, Proc. of Japan - France Seminar, Active Phenomena in the Outer Atmosphere of the Sun and Stars, Paris, Oct. 1983, (J.-C. Pecker and Y. Uchida, eds.), CNRS and Observatoire de Paris, p. 243.
Tsuneta, S., Nitta, N., Ohki, K., Takakura, T., Tanaka, K., Makashima, K., Murakami, T., Oda, M., and Ogawara, Y., 1984a, Ap.J., **284**, 827.
Tsuneta, S., Takakura, T., Nitta, N., Ohki, K., Tanaka, K., Makishima, K., Murakami, T., Oda, M., Ogawara, Y., and Kondo, I., 1984b, Ap.J., **280**, 887.
Tucker, W.H., 1975, Radiation Processes in Astrophysics, The MIT Press, Cambridge, Mass., p.202.

Underwood, J.H., Antiochos, S.K., Feldman, U., and Dere, K.P., 1978, Ap.J., **224**, 1017.
Van Hoven, G., 1979, Solar Phys. **61**, 115.
Van Hoven, G., 1981, Solar Flare Magnetohydrodynamics, E.R. Priest (ed.), Gordon and Breach, N.Y., p. 217.
Veck, N.J., Strong, K.T., Jordan, C., Simnett, G.M., Cargill, P.J., and Priest, E.R., 1984, MNRAS, in press.
Wagner, W.J., 1984, Ann. Rev. Astron. Astrophys., **22**, 267.
Webb, D.A. et al., 1980, Solar Flares: A Monograph from the Skylab Solar Workshop II, (P.A. Sturrock, ed. Colorado Assoc. Univ. Press, Boulder, p. 471.
Webb, D.R. and Kundu, M.R., 1978, Solar Phys., **57**, 155.
Widing, K.G. and Spicer, D.S., 1980, Ap.J., **242**, 1243.
Wiehl, H.J. and Matzler, C., 1980, Astron. Astrophys., **82**, 93.
Withbroe, G.L., 1975, Solar Phys., **45**, 301.
Wolfson, C.J., Doyle, J.G., Leibacher, J.W., and Phillips, K.J.H, 1983, Ap.J., **269**, 319.
Woodgate, B.E., Shine, R.A., Poland A., and Orwig, L.E., 1983, Ap.J., **265**, 530.
Woodgate, B.E., 1984, Proc. COSPAR Symposium, Graz, Austria, to be published.
Wu, S.T. and Hu, Y.Q., 1982, in V.N. Obridko and E.V. Ivanvov (eds.), Sun-Earth Council, Proceedings of Crimean Solar Maximum Year (SMY) Workshop, Academy of Sciences, IZMIRAN, USSR, Vol. 1, p.148.
Wu, S.T. et al., 1983, Solar Phys. **85**, 351.
Wu, S.T., Hu, Y.Q., Krall, K.R., Hagyard, M.J., and Smith, J.R., 1984, Solar Phys., **90**, 117.
Yang, H.S. and Chang, H.M., 1981, Chinese Astron. & Astrophys., **5**, 77.
Yang, H.S., Chang, J.M. and Harvey, J.W., 1983, Solar Phys., **84**, 139.

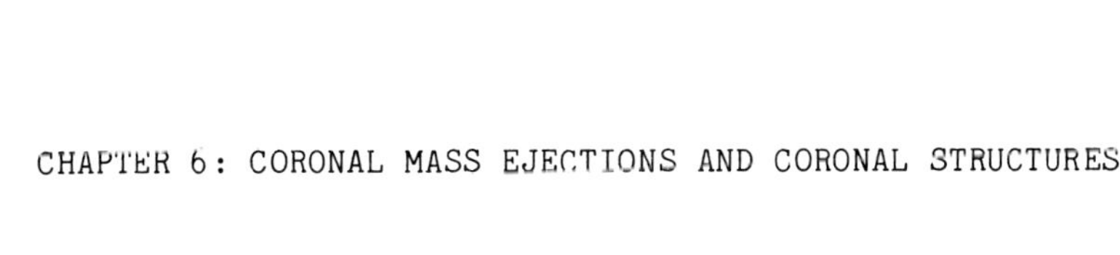

CHAPTER 6: CORONAL MASS EJECTIONS AND CORONAL STRUCTURES

CHAPTER 6: CORONAL MASS EJECTIONS AND CORONAL STRUCTURES

TABLE OF CONTENTS

E. Hildner, J. Bassi, J. L. Bougeret, R. A. Duncan, D. E. Gary, T. E. Gergely, R. A. Harrison, R. A. Howard, R. M. E. Illing, B. V. Jackson, S. W. Kahler, K. Kopp, B. C. Low, P. Lantos, K. J. H. Phillips, G. Poletto, N. R. Sheeley, Jr., R. T. Stewart, Z. Svestka, P. W. Waggett, S. T. Wu.

Page

TABLE OF CONTENTS (continued)

CHAPTER 6: CORONAL MASS EJECTIONS AND CORONAL STRUCTURES

E. Hildner, J. Bassi, J. L. Bougeret, R. A. Duncan, D. E. Gary, T. E. Gergely, R. A. Harrison, R. A. Howard, R. M. E. Illing, B. V. Jackson, S. W. Kahler, R. Kopp, B. C. Low, P. Lantos, K. J. H. Phillips, G. Poletto, N. R. Sheeley, Jr., R. T. Stewart, Z. Svestka, P. W. Waggett, S. T. Wu.

6.1 INTRODUCTION

The coronal portion of the solar atmosphere consists of a wide variety of structures which exhibit a similarly wide variety of dynamical processes and kinds of activity. The launch of the SMM presented an opportunity to study the low and intermediate corona from space with multiple instruments for protracted periods of time. This opportunity had been lacking for years, and it was enthusiastically awaited by those interested in coronal studies. The research performed during the SMM Workshop and reported here shows how successfully the SMM and collaborative observations have been used to advance our knowledge of the corona and how these observations have stimulated our theoretical understanding of why the corona is the way we observe it to be.

This chapter intends to present the work occurring during, and as a result of, the SMM Workshop. While we have made an effort to put this work in context, not all researchers participated in the SMM workshop; it is beyond the scope of this chapter to summarize the entirety of the research performed elsewhere and to summarize the current state of knowledge of coronal mass ejections and coronal structures. For additional information, the interested reader is referred to a recent review of coronal mass ejections by Hundhausen et al. (1984b) and by Wagner (1984) and to the references cited therein.

Early in the Workshop, it was apparent that members of the Coronal Structures Team were interested in a variety of coronal structures and processes; however, the team members' interests centered predominantly on the coronal response to flares and, especially, the phenomenon of coronal mass ejections (CMEs). The research described in this chapter reflects the team's distribution of interests. Modelling of post-flare arches, the reconnection theory of flares, and the slow variation of coronal structure indicate the diversity of topics considered. Some team members were interested in the interplanetary detection, evolution, and consequences of mass ejections after they had propagated through the corona. The remainder of the research was focussed on the origins of CMEs and how they propagate through the corona.

Post-flare arches (Section 6.4.4) are a newly discovered phenomenon, wherein a very large coronal loop appears to undergo energization after a flare, thus allowing it to shine more brightly in X-rays. Some post-flare arches seem to be re-energized in nearly homologous fashion after each of a sequence of flares, without suffering significant disruption. The reconnection theory of flares (Section 6.4.5), in all its variants, has been a mainstay of solar physics for quite some time, but it was pointed out during the Workshop that if one analytically describes the coronal magnetic structures involved in a flare, it is possible to estimate the amount of magnetic energy available for liberation by a two-ribbon flare. Another non-CME topic was the slow variation of coronal density structures (Section 6.6.3) in the context of how the architecture of the corona slowly evolves.

Research on coronal mass ejections took a variety of forms, both observational and theoretical. On the observational side there were: case studies of individual events (Section 6.2.1), in which it was attempted to provide the most complete descriptions possible, using correlative observations in diverse wavelengths; statistical studies of the properties of CMEs (Section 6.2.2) and their associated activity; observations which may tell us about the initiation of mass ejections (Section 6.3); interplanetary observations of associated shocks and energetic particles (Section 6.5.3) - even observations of CMEs traversing interplanetary space (Section 6.5.2); and the beautiful synoptic charts which show to what degree mass ejections affect the background corona and how rapidly (if at all) the corona recovers its predistrubance form (Seftion 6.6.3).

In the five sections which follow, these efforts are described in capsule form with an emphasis on presenting pictures, graphs, and tables so that the reader can form a personal appreciation of the work and its results. The Summary, Section 6.7, highlights some of the notable results contained in earlier sections of the chapter.

6.2 OBSERVATIONS

6.2.1 Case Studies

A number of coronal transient events from the SMM period are individually of sufficient interest to be included here. These range from new observations of transients in the inner corona as low as 1.2 R_0 with the Mauna Loa (MLO) K-coronameter to reconstructions of transient brightness distributions at 0.3 AU with the Helios spacecraft. Two events observed with the HAO coronagraph/polarimeter (C/P) on SMM (MacQueen et al., 1980) show unusual features that may offer insight into the physical structure and processes occurrring in transients. One of these is a "disconnection" event that has been interpreted as a pinching-off of a transient loop, so that the magnetic fields threading the transient no longer connect to the Sun. The other shows features that expand in a self-similar fashion, as predicted by Low (1982). (See Section 6.4.3). For five events, good metric radio observations exist. Such observations are important because they indicate the presence of energetic electrons and/or shock waves at the same range of coronal heights as traversed by the transient. For the first time, it is possible to associate the radio sources with identifiable distinct features within transients, enabling workers to set stricter limits on physical parameters within the sources of metric bursts of Types I, II, and IV.

6.2.1.1 5 August 1980 -- An Event Observed From 1.2 to 6 R_0

Space-based coronagraphs employ an occulting disk to block the overwhelmingly bright light of the solar disk. A larger occulting disk reduces the scattered light in the outer part of the field of view so that coronal features can be observed to great heights, at the price of cutting off the inner edge of the field of view at greater height. The inner edge of the field of view of HAO's C/P was the lowest yet, down to nearly 1.5 R_0. Still, coronal transients observed with SMM appear fully formed as they emerge from below the occulting disk. During the SMM period, however, the question of how transients evolve low in the corona, near their sites of initiation, could be answered for the first time, thanks to the newly completed Mark III K-coronameter at HAO's Mauna Lao Observatory (MLO), Hawaii. The event of 5 August 1980 is one of the best observed with both MLO and SMM. The MLO event, between 1.2 and 2.2 R_0 from Sun center, is discussed by Fisher, Garcia, and Seagraves (1981, henceforth FGS). The synthesis of the MLO and SMM observations, outlined below, is described in detail by Illing, Hundhausen, and Fisher, (1985).

The MLO Mark III K-coronameter is a conventional, internally occulted coronagraph. A full description of the coronameter and associated H-alpha prominence monitor coronagraph may be found in Fisher and Poland (1981) and Fisher et al. (1981).

Figure 6.2.1 shows trios of pictures taken simultaneously from the SMM and MLO coronagraphs at three times early in the event. The C/P direct intensity image, the MLO polarization x brightness (pB) image, and the MLO difference (event minus pre-event) image are shown at each time. MLO's difference image at 1829 UT shows the transient as a deficit of brightness (hence, material); successive differences show the depletion to be moving outward. Although the transient was well underway in the lower corona, the SMM images show no change from the pre-event image at this time. By 1915 UT the outward-moving depletion has developed a "rim" of enhanced brightness at its flanks and top (see Figure 6.2.1b); this enhancement is a combination of additional mass from lower in the corona and pushing aside of the pre-existing streamer. The small bright feature within the depleted region (associated by FGS with the rising prominence) is seen in the MLO difference images as early as 1834 UT, and continues to rise with the surrounding depleted volume.

Figure 6.2.1c shows the event at 1959 UT, when the top has passed out of the MLO field of view but is fully within the C/P's field. Figure 6.2.2a, a schematic drawing of the transient at this time, shows that a bright loop now forms the "front" of the transient. The loop is followed by a conspicuous dark space and central bright region (P in Figure 6.2.2a). Further development of the event is shown in the lower panels of Figure 6.2.2.

The central core, P, is the remnant of the eruptive prominence seen in the prominence monitor. The dark shell seems to be the prominence cavity (as suggested by Low, Munro, and Fisher 1982), while the bright rim in the MLO field of view appears to become the bright outer loop in the C/P field of view.

A time-height plot of the important parts of the transient, as seen by both instruments, is shown in Figure 6.2.3. The lines drawn through

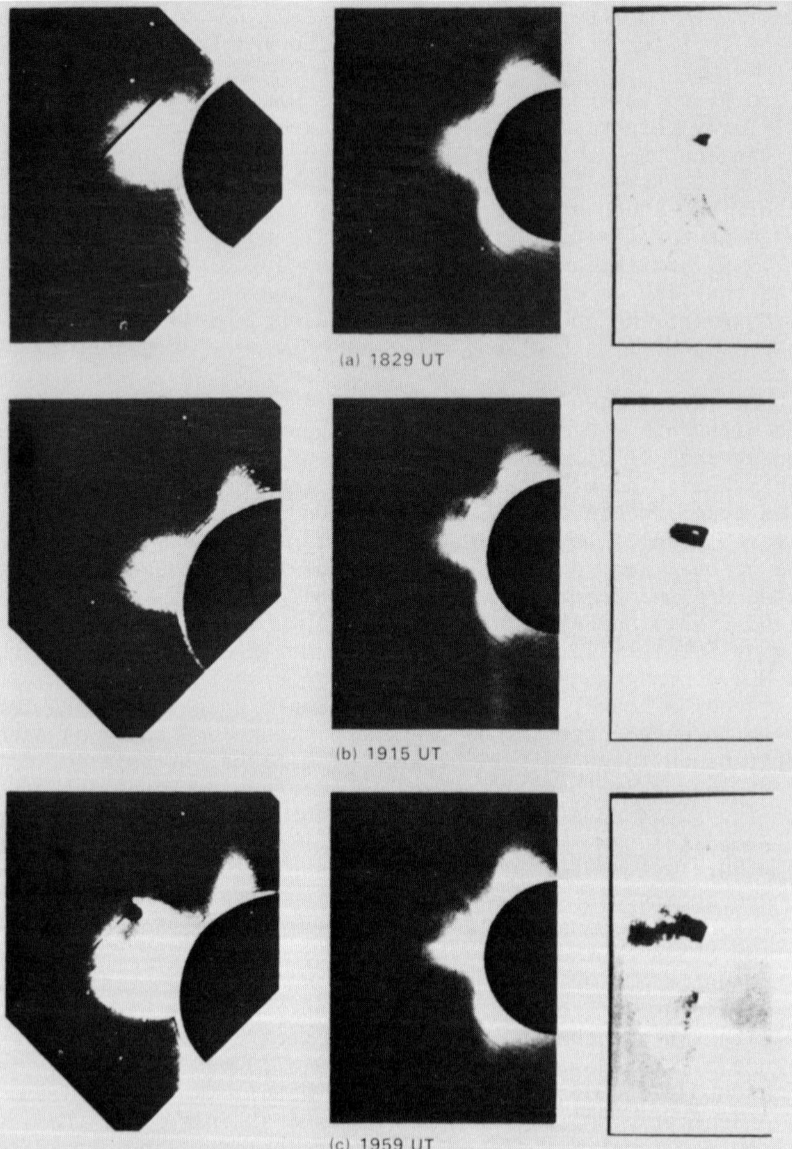

Figure 6.2.1 Early development of the coronal mass ejection of 5 August 1980. Images from the SMM coronagraph appear at the left of each set of three. MLO direct intensity images are given in the center of each triplet. MLO difference images with base frame at 1800 UT are shown in the rightmost of each trio. The time for each set of three images is indicated below the pictures. North is up, east to the left. All images are printed to the same scale. SMM images at 1829 and 1959 UT are through the green broadband filter.

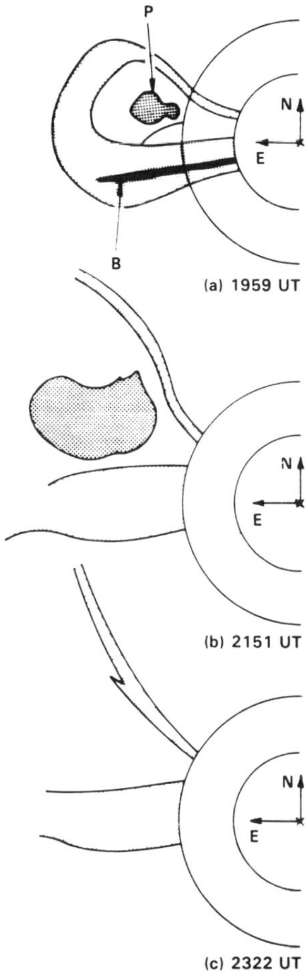

Figure 6.2.2 Schematic representation of the 5 August 1980 coronal transient at several times during the event. P is the remnant of an erupting prominence; B is a narrow, bright spike within the leg of the CME. (a) 1959 UT; (b) 2151 UT; (c) 2322 UT.

the SMM data are least squares fits. The velocities derived from the loop data are 345 km s^{-1} for the leading edge, and 334 km s^{-1} for the trailing edge; within the experimental uncertainties, these are identical. The velocity of the prominence is considerably lower, 194 km s^{-1}. Also shown in Figure 6.2.3 are the times of occurrence of disk activity. No flares were reported in association with the eruptive prominence observed from Mauna Loa.

The initiation of the transient is consistent with the model of Low, Munro, and Fisher (1982) of an underdense volume that has become magnetically buoyant. The bright front is then presumably material swept up from the background corona. We know from the MLO data that the only mass that rises from below the occulting disk is the erupting prominence.

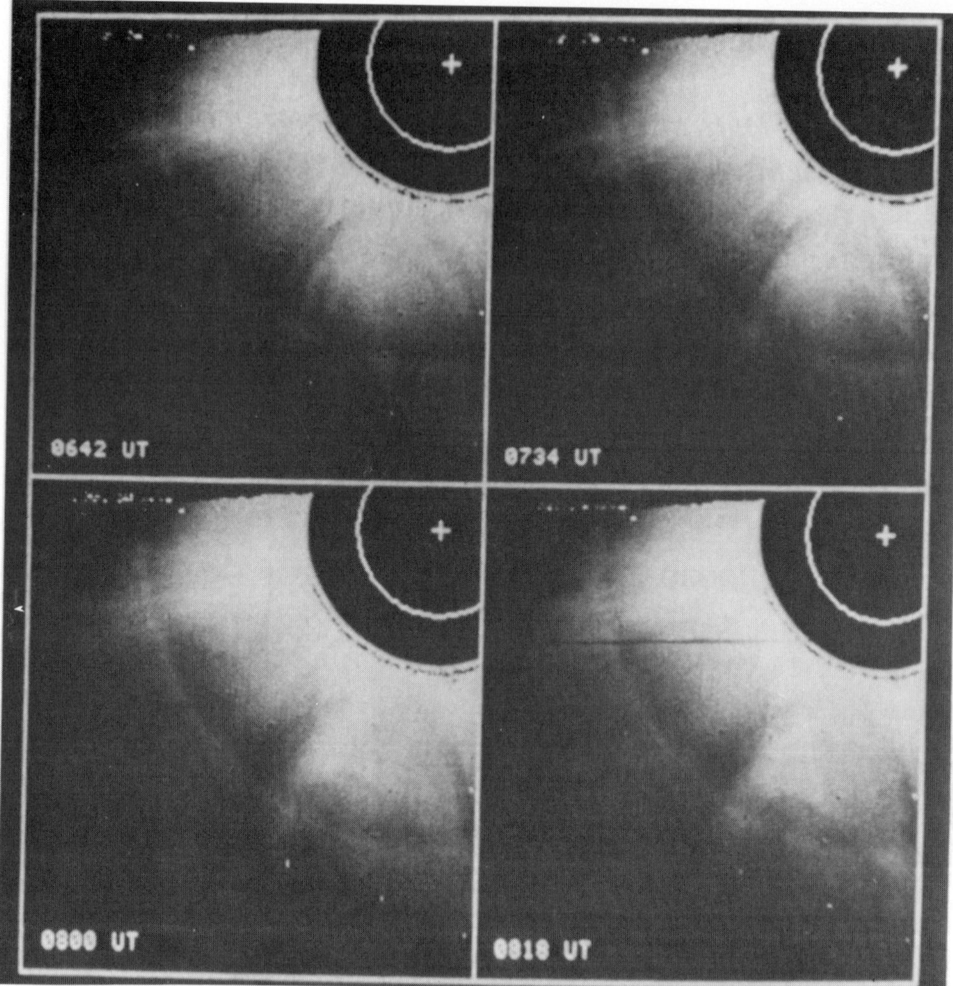

Figure 6.2.4 Coronagraph/Polarimeter images showing the progress of the concave upward structure on 16 March 1980. The time for each frame is indicated at the lower left. North is to the upper left, east to the lower left. The radius of the first, brightest diffraction ring surrounding the occulting disk is 1.61 R_0. The length of one side of a frame is approximately 5.5 R_0. The center and limb of the Sun are surrounded on each image.

Table 6.2.1 Best Fit Lines

Feature	v(km/s)	r(t = 0100) (R_0)	t(1 R_0)
D	200 ± 18	−0.28 ± 0.33	0215 ± 20
F	110 ± 5	0.02 ± 0.10	0237 ± 14
a	73 ± 1	0.54 ± 0.04	0213 ± 6
b	66 ± 3	0.47 ± 0.07	0234 ± 13
c	55 ± 4	0.54 ± 0.12	0237 ± 26

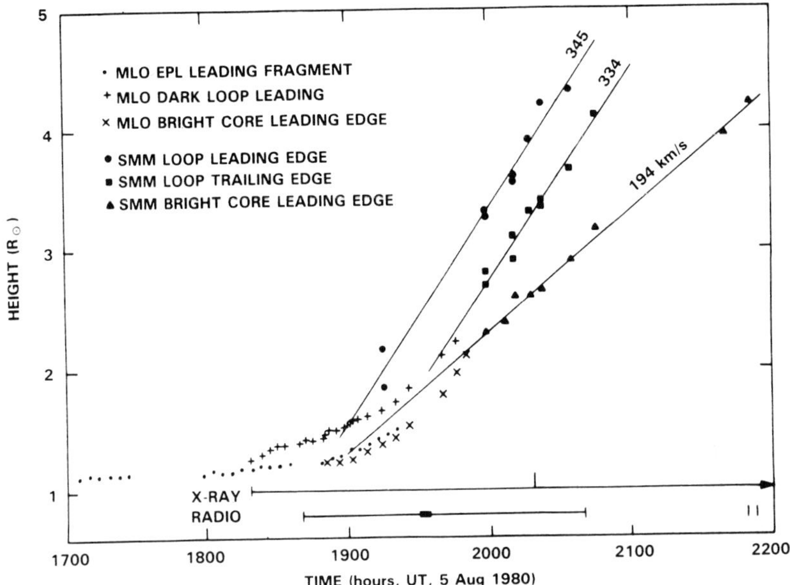

Figure 6.2.3 Height as a function of time for distinct features of the mass ejection as seen in both the C/P and MLO data. The line marked X-ray indicates the duration and peak (vertical mark) of the GOES X-ray event. The line marked Radio shows the duration of the 2800 MHz event; the thick portion indicates the continuum burst, and the two vertical bars the IIIG,W bursts.

Since the bright rim is first seen high in the MLO field of view, it cannot have originated below the occulting disk; that is, the mass must be due only to the material previously present in the background corona. We can estimate the amount of this pre-existing mass from the coronal model of Saito, Poland, and Munro (1977). The amount of mass available in a wedge 30° in latitude, 30° in longitude, and extending from 1.2 to 2 R_0, is 4.6×10^{15}g. Since the loop mass is only 0.95×10^{15}g, there is support for the the suggestion of Hildner et al. (1975a) that the excess material seen being ejected through a spaceborne coronagraph's field of view originated in the corona and was at coronal temperatures when the ejection began (see also Hildner et al., 1975b; Schmahl and Hildner, 1977).

6.2.1.2 15-16 March 1980 -- A "Disconnection" Event

This event, described in detail by Illing and Hundhausen (1983), is the first published observational evidence of an effect of magnetic reconnection suggested by many authors and summarized by MacQueen (1980) -- the "pinching off" of a transient loop or bubble from the Sun. Figure 6.2.4 is a sequence of images, taken with the C/P, of a region within 45° of the solar east limb. Figure 6.2.5 shows the orientation of the images and the outlines of features of interest.

The observations suggest three "phases" in the evolution of the transient and its propagation through the corona.

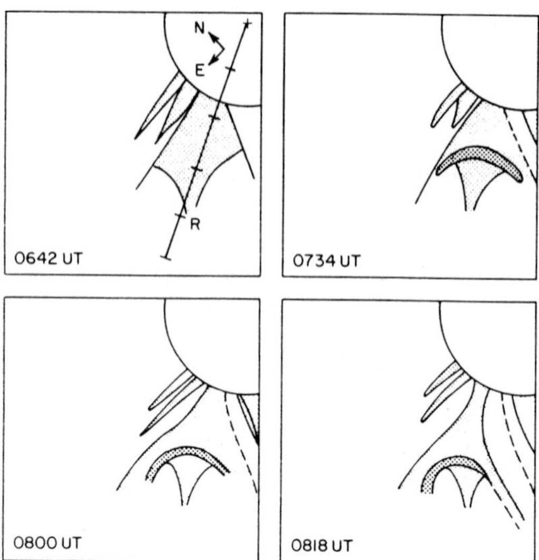

Figure 6.2.5 Schematic diagram of the 16 March 1980 images in Figure 6.2.4, showing the major parts of the event. The position of the front has been measured along the radius indicated R, about 27° south of the projected equator. The center of the occulting disk is again shown by a cross, and radius R is marked in units of R_o.

Phase 1. The ejection of bright (dense) coronal material with loop-like structures visible within the ejecta.

Phase 2. Outward motion of a bright front that is concave away from the Sun. This front evolves from an "inverted arch" shape to a semi-circular annulus as its center moves outward at 175 km s^{-1}, three to four times faster than the radial velocities of structures seen in phase 1.

Phase 3. The rapid contraction of a fan-shaped bright region that appeared beneath the bright front. This contraction seems to occur from the bottom to the top of the structure and leads to a single narrow, bright ray in the region previously filled with bright transient material.

Illing and Hundhausen (1983) interpret these observations as a direct indication of magnetic reconnection at an X-type neutral point. The pinch-off point must be below 1.6 R_0 since no features are seen moving toward the occulting disk (as was expected by MacQueen 1980).

How common are such events within the SMM data set? An initial examination of 68 CME's reveals features similar to that described above in seven events; only additional study of these ejections (most of which are seen on fewer images than the 1980 March 15-16 transient) will reveal whether this interpretation can be applied to them. Detailed examination of SMM data from another two-day interval (27-28 March 1980) has revealed a similar structure that evolves in the manner described above, indicating that this disconnection event is not unique.

6.2.1.3 23 March 1980 -- Self-Similar Expansion

Many CMEs appear to expand into the corona rather like an inflating balloon. This behavior, or more exactly, self-similar expansion, has been discussed theoretically by Low (1982). The 23 March 1980 event, described in detail by Illing (1984), is the first to be examined critically for quantitative evidence of self-similar behavior. Selected frames of the event are shown in Figure 6.2.6. Schematic drawings of these frames are given in Figure 6.2.7, with the major parts of the transient labeled.

Figure 6.2.8 gives the time-height plots for these structures. The lines shown are least squares fits to the data points, with velocities and intercepts (together with standard errors) given in Table 6.2.1. The CME's velocity is rather low, 55 to 220 km s^{-1}. Also given in Table 6.2.1 are the extrapolated times at which the features would have been at 1 R_0, with uncertainties derived from the linear fit.

The apparent divergence of the substructures from a common point, as shown in Figure 6.2.8, allows a test of Low's (1982) theory, described in Section 6.4.3 of this chapter. A dynamical system is said to evolve self-similarly in time if its motion can be described in terms of a variable that relates its spatial and temporal dependences. The system is then coupled in a very particular way; all forces acting on any mass element in the system have constant relative magnitudes throughout all time and space. This restrictive condition is not met, in general, for an arbitrary dynamical system. A result of self-similar motion is that distinct substructures are related such that at a given time their individual velocities are a function only of height. In Figure 6.2.9 we plot the velocities of all the substructures against their respective heights at 0700 UT. The vertical bars represent the velocity uncertainties given in Table 6.2.1. Within the errors, the points appear to lie on a straight line; the line shown is a least squares fit to the points. Since we have selected several distinct substructures in the event, Figure 6.2.9 suggests that the entire structure is a self-similar dynamic system.

6.2.1.4 29 June 1980

a. The Type II association. Many CMEs depart the Sun at speeds exceeding the typical Alfven speed in the corona. From magnetohydrodynamical considerations, such rapidly moving disturbances must be or be preceded by shock waves. That shock waves actually exist in the inner corona is amply shown by the observation of metric Type II bursts, which are due to plasma emission from electrons accelerated locally by the shock. The advantage of simultaneous observations of transients and spatially resolved Type II bursts is obvious -- such observations are the only way to determine the exact spatial relationship between CMEs and shock waves and, given good fortune, to determine the electron density in the Type II-emitting source. All during the Skylab era, however, no simultaneous observation was successfully made. The 29 June 1980 event, described by Gary et al. (1985), is one of only four events observed during the SMM period accompanied by a spatially resolved Type II burst.

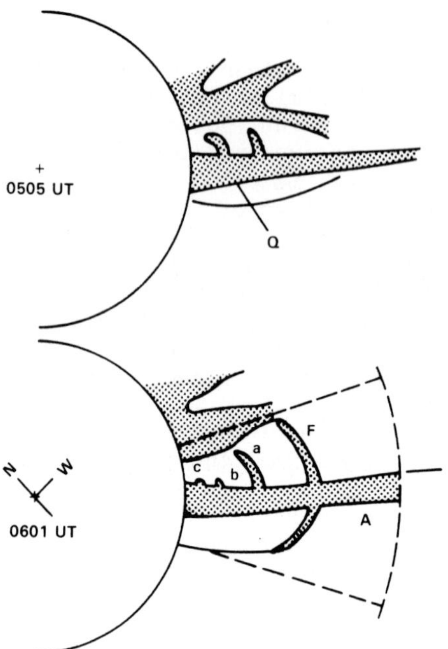

Figure 6.2.7 Schematic representation of the 23 March 1980 event shown in Figure 6.2.6. Streamers and substructures are labeled as they are referred to in the text. The dotted line in the drawing for 0601 UT indicates the area covered by the excess mass calculation for the entire event. The radial line protruding from the outer dotted line marks position angle 230°.

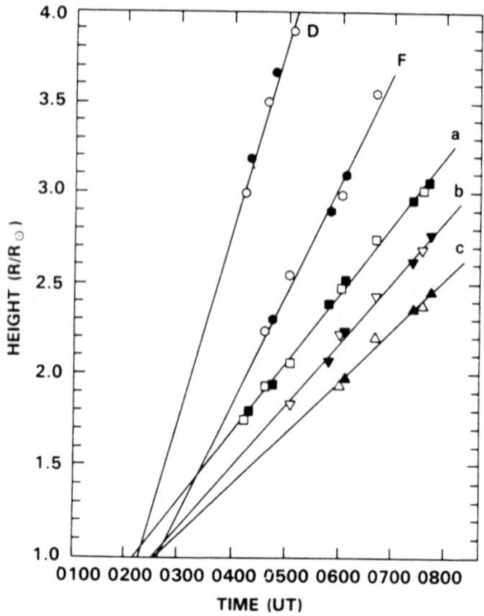

Figure 6.2.8 Height-time plots for several substructures in the 23 March 1980 event shown in Figures 6.2.6 and 6.2.7. Positions for features F, a, b, and c are measured from scans at position angle 229°, verified by inspection of the direct frames. Positions of the diffuse front D are measured from radial scans at position angle 242.5°, where D is seen moving through a relatively dim background. Open symbols indicate measurements from a south sector frame, filled symbols, from a west sector frame. Parameters of the least-squares fit lines shown are given in Table 6.2.1.

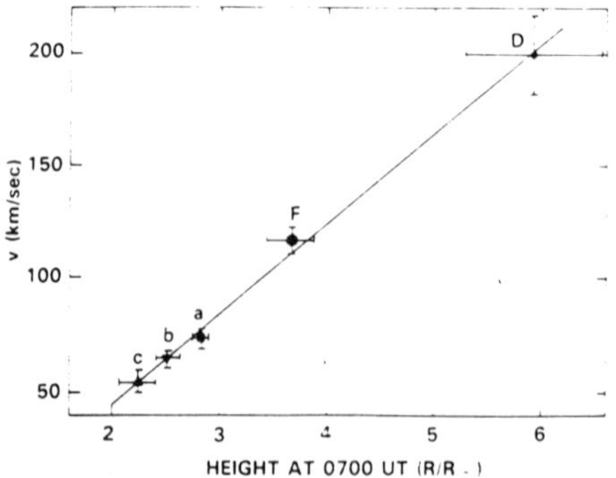

Figure 6.2.9 Velocity of substructures as a function of radial distance at 0700 UT 23 March 1980. Symbols and labels correspond to similarly labeled features in Figures 6.2.7 and 6.2.8 and in Table 6.2.1.

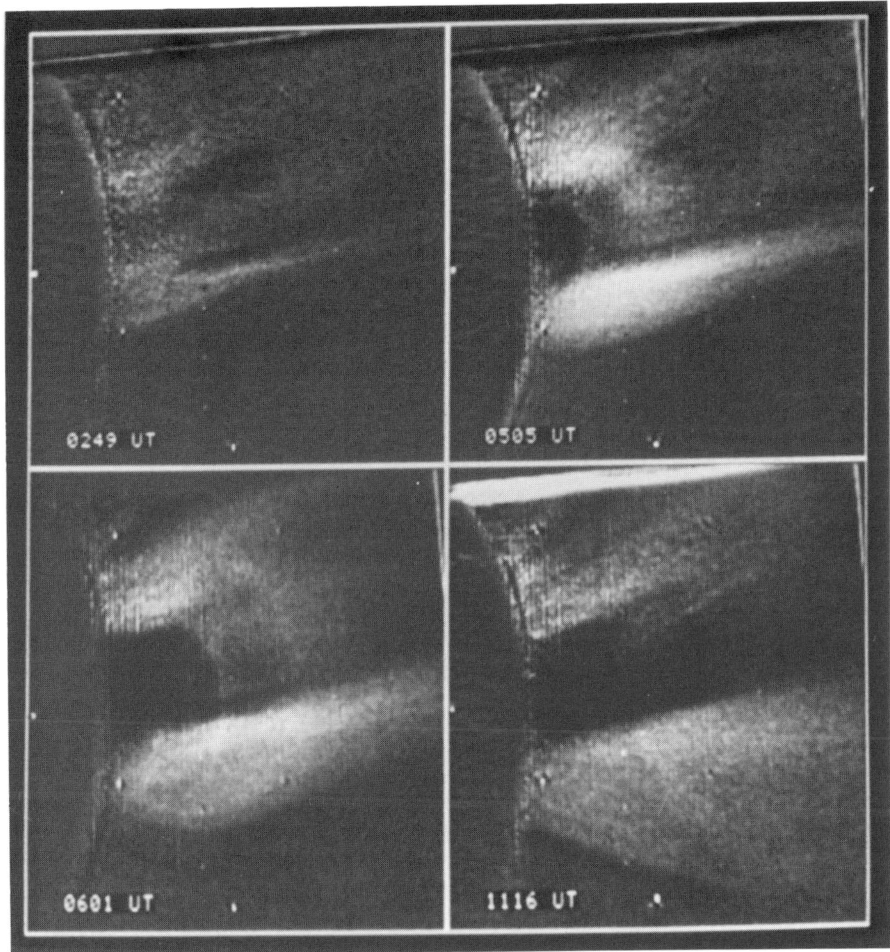

Figure 6.2.6 SMM Coronagraph/Polarimeter difference images showing the development of the 23 March 1980 depletion CME. Frames are oriented with north to the upper left and west to the upper right. The brightest diffraction ring of the occulting disk occurs at 1.61 R_0. All times are UT. The reference frame was observed at 0113 UT; scaling is identical for all frames.

The CME was associated with the first of three limb flares that occurred in Hale Region 16923 on 29 June 1980. The initiation of the coronal mass ejection and the timing of the X-ray bursts associated with it are discussed in Section 6.3.2.

The CME appears initially as two loops that move radially outward at about the same speed, ~ 600 km s^{-1} (see Figure 6.2.10). A third loop is located northward from the original loops, a faint halo leads all three loops and a remarkable arc-like feature stands off to the north of the loop complex, extendng from about PA 250° to PA 280°. (Each of these features is shown schematically in Figure 6.2.10c.) The faint arc does not resemble the loops of the CME, but appears as a circular arc whose center of curvature is very near the position of the flare. The arc moves at an approximately constant velocity of about 900 km s^{-1}. Note that the faint arc is quite easily seen in the SOLWIND image taken at 0322 UT and presented as Figure 6.5.1b.

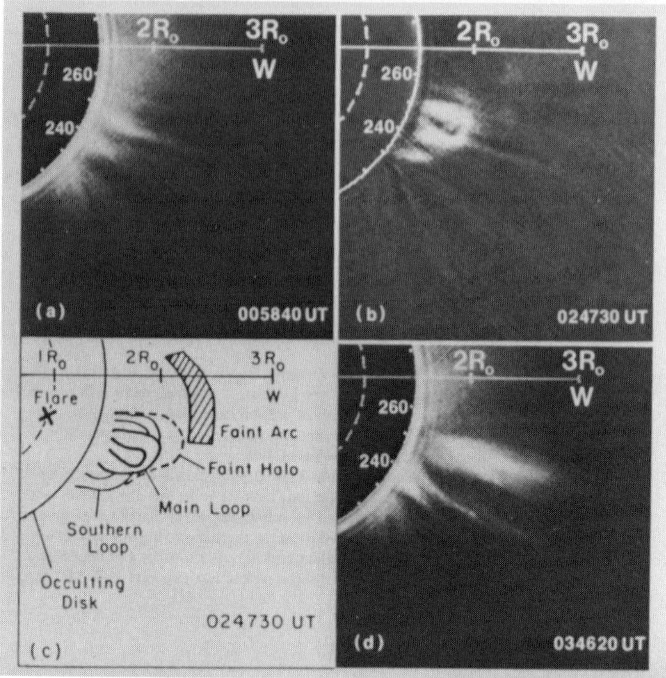

Figure 6.2.10 (a) Pre-event C/P image taken 0059 UT 29 June 1980, before the beginning of the transient. (b) A subtracted image in which the image shown in (a) was subtracted from an image taken during the transient at 0248 UT. (c) A schematic drawing of the image of (b) illustrating the main features discussed in the text. (d) the appearance of the corona at 0346 UT after the transient, showing the bright central streamer flanked by thin legs.

Simultaneous observations of radio spectra, polarization and positions were obtained with the instruments of the CSIRO Division of Radiophysics at Culgoora, Australia. Type III/V bursts mark the start of the flare at 0233 UT, and a strong fundamental/harmonic Type II (shock wave related) burst begins at about 0241 UT.

It is possible that the faint arc marks the density enhancement (compression region) immediately behind a shock wave. Its position ahead of the loops of the transient, its faster speed, and its circular arc shape are all suggestive of this. To see whether the arc is the site of Type II emission, we determine the electron density within the enhancement, averaged over the length of the arc, and display it in Figure 6.2.11 as a function of height for two cases. The maximum density suggests that emission at the second harmonic of the plasma frequency should have been visible near 20 MHz, and higher frequency emission should have been visible at earlier times when the arc was traversing higher density layers lower in the corona. No such emission is seen. We conclude that the faint arc was not a source of radio emission. The observed Type II emission was, instead, associated with the loops, as demonstrated in Figure 6.2.12.

To determine whether the faint halo of enhanced density shown in Figure 6.2.10 is a forerunner, we adopt the Jackson and Hildner (1978) definition of the leading edge of the forerunner as the point where the

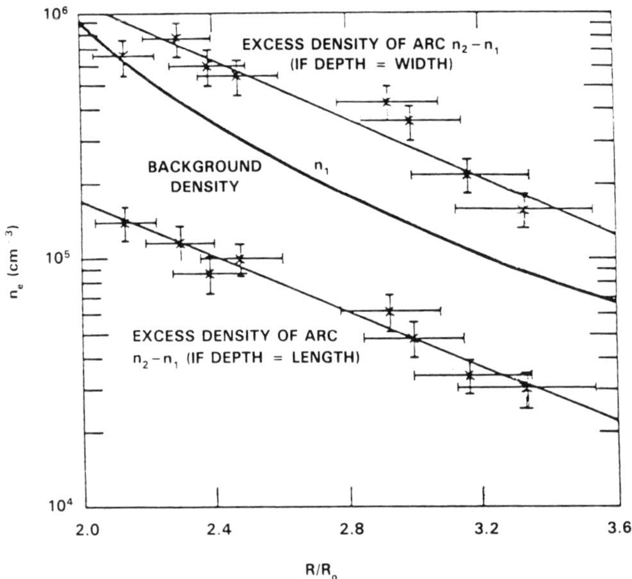

Figure 6.2.11 Comparison of the excess density of the arc n_2-n_1 with background density n_1 (heavy line) for two cases: (1) depth of the arc same as arc width (upper points) and (2) depth of the arc same as arc length (lower points). Horizontal bars represent the width of the front. Vertical bars represent measurement uncertainties. Uncertainty in n_1 due to our choice of F-coronal brightness ranges from 8% at 2 R_0 to 40% at 3.6 R_0.

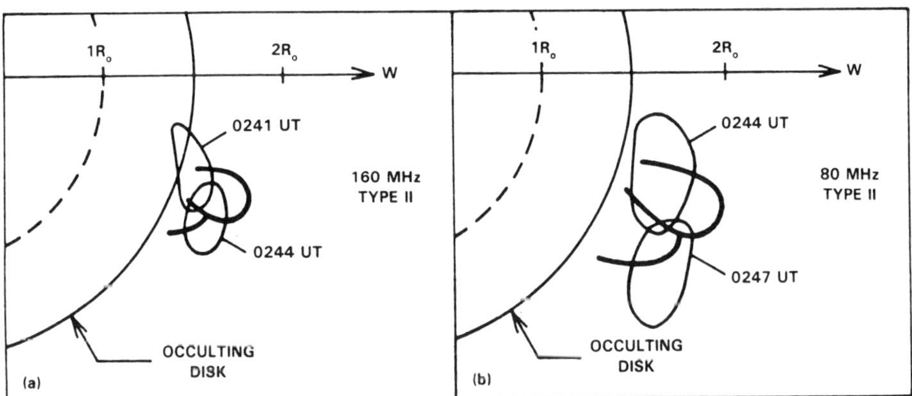

Figure 6.2.12 Schematic diagrams of the early 29 June 1980 CME at two times (a) 0244 UT and (b) 0248 UT, and comparison with the 160 and 80 MHz Type II burst positions.

excess columnar density falls below the 2σ noise level (1.15×10^{-8} g cm^{-2}); we set the second contour level at 1.3×10^{-7} g cm^{-2}, to delineate the boundary between the transient and the forerunner. The similar appearance of the forerunner in Figure 6.2.13 and the ATM forerunners shown by Jackson and Hildner (1978) leads us to conclude that this is the same phenomenon.

Also visible in Figure 6.2.13 are portions of the faint arc ahead of both the CME and forerunner. The arc extends to position angles far to the north of the forerunner and there is no indication of a forerunner behind that portion of the arc. However, at other position angles the density enhancement seems to extend from the arc to the loop transient.

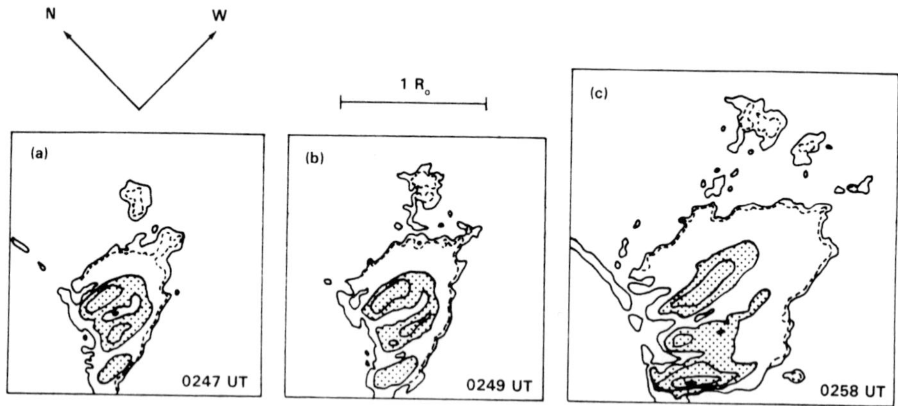

Figure 6.2.13 Contour plot of loops and forerunner at three times on 29 June 1980: (a) 0248 UT, (b) 0249 UT, and (c) 0257 UT. The lowest contour corresponds to the 2σ noise level of 1.15×10^{-8} g cm^{-2}. The hatched area denotes the brightest (densest) parts of the transient above 1.3×10^{-7} g cm^{-2}. The areas of slightly enhanced density outside the forerunner are parts of the faint arc. The dashed contour at 2×10^{-8} g cm^{-2} is included to show the faintness of the arc.

The relationships among the faint arc, forerunner and CME loops shown in Figure 6.2.13 find a natural explanation if the faint arc and the forerunner are the compression region behind a shock front. In this picture, the loop-shaped CME is the piston that drives the shock; a piston that consists of material originating in the low corona. This is the same picture that was proposed by Dulk et al. (1976) to explain the CME and forerunner of 14-15 September 1973 (although the term forerunner was not used by them).

If we assume that the density enhancement in the faint arc is due to compression of material passing through a shock front, we can estimate the range of Mach numbers for the shock from the two extremes of desity given in Figure 6.2.11. At R = 3 R_0, Figure 6.2.11 shows the excess density $\Delta n = n_2 - n_1$ of the shock front to be $0.3n_1 < n < 2n_1$ where n_1 is the background density also shown in the figure. The density ratio across the shock is then $1.3 < n_2/n_1 < 3$, giving a range in Mach number [from the Rankine-Hugoniot jump relation $n_2/n_1 = 4M_A{}^2/(3 + M_M{}^2)$] of $1.2 \leq M_A \leq 3$. From the observed speed of the arc (~ 900 km s^{-1}) and assuming an ambient solar wind speed of ~ 150 km s^{-1} at 3 R_0 (Parker, 1958), the corresponding range of Alfven speed and magnetic field strength in the ambient medium can be obtained. The results are in Table 6.2.2. The parameters for d = 1.5 R_0 seem to be most consistent with other observations.

Table 6.2.2 Parameters for shock wave (at R = 3 R_0)

	Line of Sight Depth	
	d = 0.2 R_0	d = 1.5 R_0
n_2/n_1	3.0	1.3
Mach Number	3.0	1.2
v_A (km s)	250	625
B (gauss)	0.05	0.12

An idealized sequence of events that is consistent with the data is as follows: The magnetic configuration in the corona above the active region becomes unstable and begins to expand, eventually forming a loop-shaped CME. A quasi-parallel, piston-driven shock wave develops in the upper corona above the transient, which does not generate radio emission. Accompanying the expansion are alterations to the low coronal environment that lead, a few minutes later, to the impulsive flare. The resultant thermal pulse initiates a blast wave that evolves into a shock, generating Type II radiation as it moves quasiperpendicularly to the magnetic field lines in the lower corona and in the ejected material of the transient. As it progresses farther into the confused corona - into faster moving material - its velocity relative to the surroundings becomes smaller, finally becoming sub-Alfvenic, ceasing to be a shock and quenching the Type II emission. We note that Cane (1984) presents further evidence of the existence of two shocks in some CME events, based on the continuation of Type II bursts to low frequencies observed by the ISEE-3 spacecraft.

b. The Type IV association. The discussion above concerned the radio emission produced at the leading portions of the CME. The interior region behind the transient is also of interest in understanding what becomes of the electrons accelerated during the disruption, and how the corona relaxes into its post-ejection state. The study of metric Type I and Type IV sources is potentially a direct way to quantify conditions within CMEs, but only when the mechanism responsible for these emissions are known. The 29 June 1980 event, described in detail by Gary et al. (1984), offers a chance to determine the emission mechanism for at least one kind of Type IV burst, a burst observed at Culgoora at 80 and 43 MHz.

The appearances of the CME at 0258 UT during the transient, and at 0346 UT after the CME reaches the edge of the field of view of the C/P, are shown in Figure 6.2.14. By 0346 UT, a bright streamer remains where the northern leg of the loops had been. Subsequent images show this streamer decreases slowly in brightness. It is with this streamer that the late Type IV sources are associated. The time behavior of the transient and the Type IV sources at 80 and 43 MHz are shown schematically in Figure 6.2.15.

There are two possible emission mechanisms that can account for Type

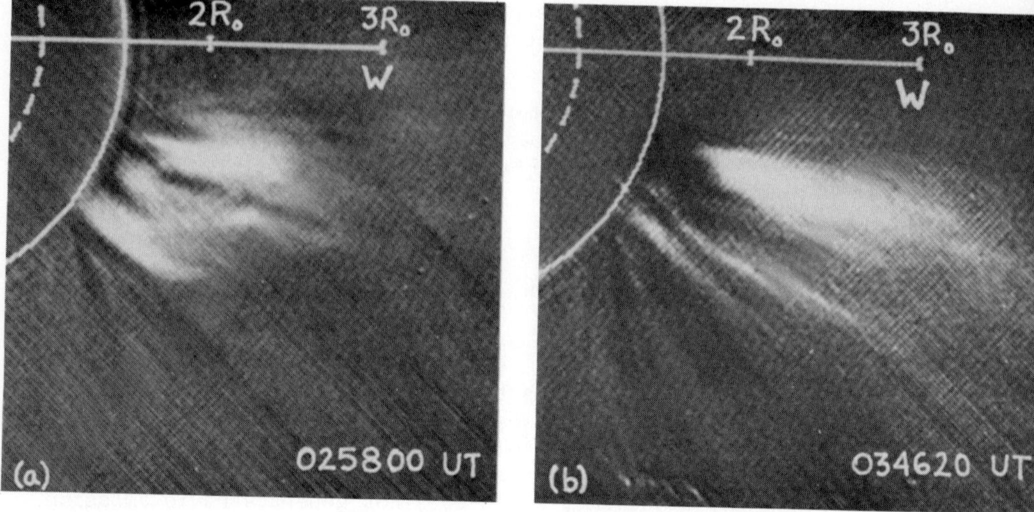

Figure 6.2.14 Subtracted C/P images of the 29 June 1980 transient at (a) 0258 UT and (b) 0346 UT, identifying the northern leg of the loop in (a) with the remnant streamer in (b).

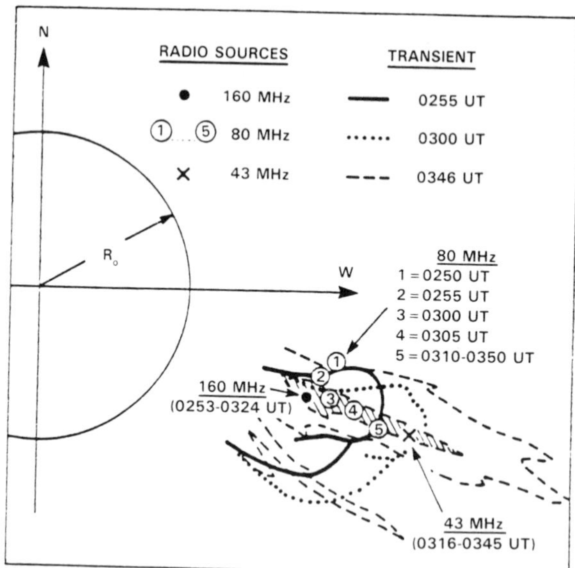

Figure 6.2.15 Comparison of corrected radio source positions with visible-light features in the C/P images of 29 June 1980. The 80 MHz Type IV source (circled numbers 1-5) is always associated with the densest part of the transient.

IV radio emission -- plasma emission (either fundamental or harmonic) and gyrosynchrotron emission. These were examined in detail by Gary et al. (1984), with the following results:

The sources at 80 and 43 MHz showed characteristics expected of harmonic plasma emission:

1. The 80 MHz source moves along the densest part of the transient in conjunction with rising of the relevant (40 MHz) plasma level.
2. Both 80 and 43 MHz sources are associated at all times with features whose density is probably high enough to account for the emission as harmonic plasma emission.
3. The polarization of the 80 MHz source is consistent with harmonic plasma emission.
4. At times the spectrograph shows weak Type II-like bursts that imply acceleration of electrons from lower in the corona to energies of a few keV. The source of these electrons could be the source of the electrons producing the continuum emission.

Gyrosynchrotron emission is found to be less likely, especially when Razin-Tsytovich suppression is taken into account; about 10% of the electrons in the source region would need to have energy greater than 10 keV, with the average energy of emitting particles being ~ 40 keV. Another requirement, which seems unlikely to be met, is that the magnetic field strength at 2.5 R_0 be ~ 2.8 gauss.

It is also found that the emission at 80 and 43 MHz is probably not due to optically thick gyroresonance emission at low harmonics of the gyrofrequency. If it were, the highest harmonic that is optically thick for reasonable numbers of energetic electrons is s = 5, for which a magnetic field strength of 5.7 gauss is required.

In conclusion, it seems more likely that the 80 and 43 MHz sources are due to harmonic plasma emission than to gyrosynchrotron emission. If so, the radiation could be due to ~ 1 to 3 keV electrons continuously accelerated lower in the corona and having a plateau distribution in velocity space.

6.2.1.5 30 March 1980 -- A Type I Noise Storm Associated Event

Another type of continuum emission localized below some transients is the metric Type I noise storm. Type I emission differs from Type IV emission in that, although its intensity can increase during a coronal transient, it is generally independent of the CME itself, existing both prior to and after the event. This CME, described in a series of papers by Lantos et al. (1981), Lantos and Kordraon (1984) and Lantos (1984), is a loop-shaped event that occurs in a region where there already existed a weak Type I source, observed at 169 MHz by the Nancay Radioheliograph.

On 30 March, active region 2363 was situated at 25° N and 25° E and consisted of two main spots of the same polarity bordered on their east and south sides by a chain of H-alpha filaments. The southern filament began to dissipate before 0938 UT, but the radio and X-ray events began only around 1300 UT. Region A in Figure 6.2.16 was a source of Type I continuum emission from before 0510 UT. Beginning at 1310 UT, the brightest point of the emission left source location A and followed the trajectory indicated in Figure 6.2.16, again becoming stationary at location B after 1330 UT.

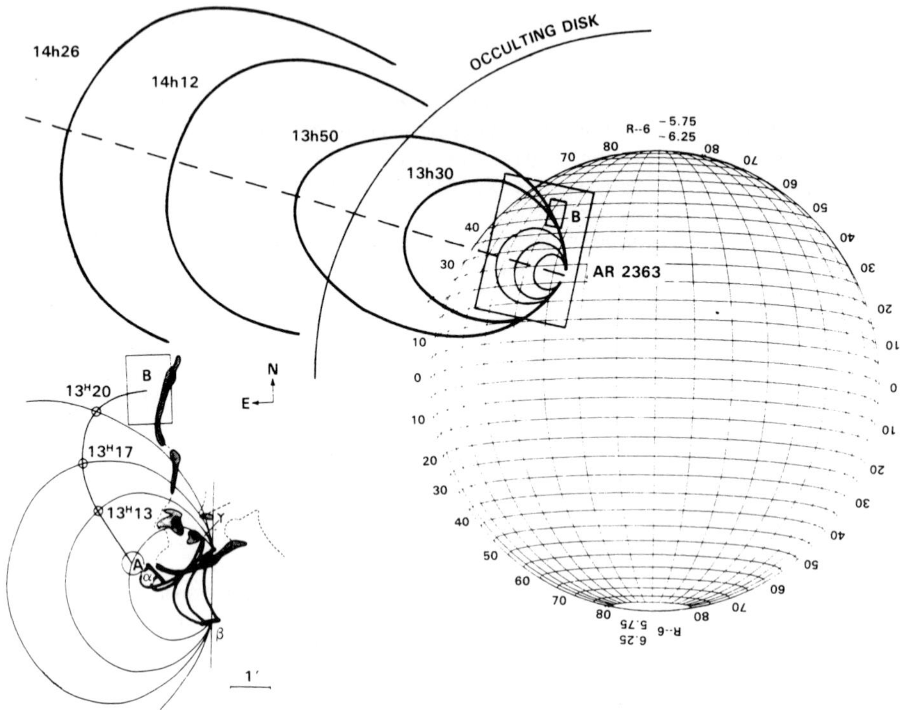

Figure 6.2.16 The coronal mass ejection of 30 March 1980. Inset is a schematic of the active region. The loop features at lowest height are derived from soft X-ray data (after Lantos *et al.,* 1981). The line joining A and B is the path of a type I metric noise storm. Based on this, Lantos and Kerdraon, (1984), constructed the ascending loops as shown in the small figure; these expand into the corona in a way which is consistent with the coronal transient observations shown in the larger figure. H-alpha ribbons are connected as shown α with δ, β with γ.

The source at point A is large (about 4' x 5'), with at least two components, indicating that the continuum emission probably originates from an arcade of loops rather than from a singleloop. The apparent velocity of the 169 MHz source from point A to B is ~ 300 km s^{-1}. A CME (sketched in Figure 6.2.16) is observed at 1412 UT and 1426 UT with the C/P instrument. The ejection appears loop-like, with a lateral width about half the radial height above active region 2363. If we assume radial motion of the transient, the observed velocity corresponds to a true speed of 600-800 km s^{-1}. This seems too high for a CME associated with a disappearing filament in the absence of an H-alpha flare (Gosling et al., 1976). Thus, it is likely (and we shall assume) that the motion was not radial but inclined somewhat ~ 30°) from the local meridian toward the east.

The emission mechanism for the noise storm in this event will be assumed to be due to plasma waves excited by ~ 10-30 keV electrons (Melrose 1980, Benz and Wentzel, 1981). According to plasma wave theories, the radio emission occurs near the plasma frequency level defined by

$$\omega_p^2 = \frac{4\pi\, n_e\, e^2}{m_e} \quad .$$

The radio emission at a given frequency takes place at a constant electron density which, for 169 MHz, is $n_e = 3.5 \times 10^8\ cm^{-3}$. The motion of the noise storm is thus a trace of this density level during the evolution of the involved coronal structure.

The event can be modeled with a simple loop geometry to determine whether the radio emission could have occurred in the CME itself. The initial loop is taken to be circular, isothermal, in hydrostatic equilibrium, and with a top altitude of 10^5 km. The foot points are anchored (like the X-ray emitting loops) on H-alpha ribbons β and δ (see Figure 6.2.16). The top of the loop in the model rises with a velocity of 370 km s^{-1} as suggested by the coronagraph observations. We increase the ellipticity of the model loop during its expansion to fit the CME observations. Comparison of the model with observation gives reasonable agreement when the initial loop has a density of $3.1 \times 10^8\ cm^{-3}$ at the top and a temperature of 4.5×10^4 K. The density gradient inside the initial loop is the main parameter that determines the trajectory of the constant density level during the motion. Thus, temperature in the model is directly dependent on the hydrostatic equilibrium assumption.

The noise storm motion could be fit with a model assuming a loop velocity of 370 km s^{-1}; extrapolation of the motion until 1412 and 1426 UT coincides with the location of the CME observed with the C/P instrument (Figure 6.2.16). Thus, our model of the displacement of the source is consistent with the idea that the emission arose in the CME itself during its early phases. Placing the Type I source in the leg of a CME recalls a similar feature of stationary type IV bursts such as the 29 June 1980 event we have just seen. We will see this same phenomenon in the next two events as well.

6.2.1.6 7 April 1980 -- A Type IV Associated Event

This event, described in a preliminary way by Wagner et al. (1981), contains a moving Type IV radio burst -- that is, a source of continuum emission that moves outward. The event also contains a more typical stationary Type IV burst, similar to the event of 29 June 1980 just described. The stationary Type IV in this event differs from that of the previous event, however, because the 43 MHz source early in this event is associated not with a density enhancement, but rather with a void. The following description is based on work by Bassi, Dulk, and Wagner (1985).

The CME appears in visible light as a large loop extending from near the solar equator to the north pole. The associated meter wave radio sources cover a similar range of positions. At the time of the first SMM coronagraph image (0405 UT), the ejection is already well developed with a very bright leg to the west and a smoothly northward curving arc, as shown in Figure 6.2.17. During the loop's expansion, blobs of brighter material part way up the loop move non-radially in a direction perpendicular to the loop with speeds on the order of 450 km s^{-1}. In contrast, a "kink" feature (a discontinuity in the curvature of the loop) moves radially outward with a speed of about 550 km s^{-1}, i.e., it does not participate in the whip-like motion evidenced by the blobs.

Radio observations with the Culgoora radioheliograph show both moving and stationary Type IV sources at 80 MHz and 43 MHz. No Type II burst was observed at any time during the transient.

The 80 MHz source moves almost radially outward from the Sun (Figure 6.2.17), with the last observed position of the source centroid being coincident with the kink rather than with a bright blob proposed by Wagner et al. (1981). Analysis of subsequent images shows that the kink movement is at approximately the same speed and direction as the previously observed 80 MHz Type IV source. This spacial coincidence between the kink and the moving source leads us to speculate that the dynamic phenomenon that produces the kink also is an exciting mechanism for the 80 MHz emission. It is possible that the radio source is due to plasma emission as the kink moved up the leg, and that the emission stopped at 0405 UT because the kink travelled to a region of density that is too low to support 80 MHz emission. In support of this idea, we find that the density in the loop below the kink at 0405 UT is 10^7 cm^{-3}, resulting in a plasma frequency of ~ 30 MHz. The inferred density is somewhat greater if the brightness contribution from the background corona is taken into account. Thus, the result appears consistent with the hypothesis that the observed radiation at 80 MHz was produced by second-harmonic plasma emission, which requires a plasma frequency ~ 40 MHz (density ~ 2 x 10^7 cm^{-3}).

The 43 MHz emission at 0405 UT could be interpreted either as coming from a single large, elongated source or from a double source. The more northerly part of the source lies over the void below the loop; the other part is coincident with the west leg of the CME. As time progresses, the part of the source coincident with the void fades, while the other part becomes coincident with the leg (Figure 6.2.18) and elongated in the direction of the leg. The association of stationary Type IV sources with the legs of transient loops seems a common feature (as in the 29 June 1980 event of Section 6.2.1.4b).

6.2.1.7 9 April 1980 -- A Type II (Shock Wave) and Type IVM Associated Event

This event is described by Gergely et al. (1984). The event was comprehensibly observed; in H-alpha at the Haleakala Observatory of the University of Hawaii, in visible light by the Coronagraph/Polarimeter (C/P) experiment aboard SMM, as well as the coronagraph aboard the US Air Force P78-1 satellite, and at meter-decameter radio wavelengths by Clark Lake (CLRO) and Culgoora Radio Observatories. As for the 29 Jue 1980 event, both Type II and Type IV radio emissions were observed with this event. However, the first C/P image was after the Type II burst ceased at

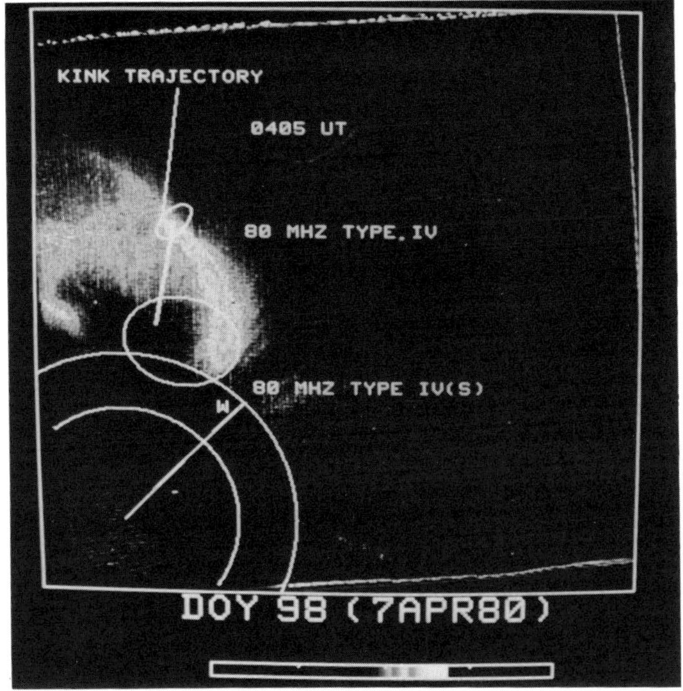

Figure 6.2.17 A difference image showing the south leg of the loop-shaped CME of 7 April 1980, the trajectory of the 80 MHz Type IVM radio source until 0405 UT, and the trajectory of the kink in the loop after 0405 UT. Also shown are the positions and sizes of the 80 MHz Type IV stationary and moving sources at 0405 UT.

CLRO, so no firm conclusions about the relative locations of the transient and the shock can be reached. The moving Type IV burst (Type IVM), however, can be associated with an overdense feature in the transient.

Observations at Haleakala show the development of an eruptive prominence in the low corona in association with the event. The evolution of the CME observed by the C/P is shown in the sequence of Figure 6.2.19. The complex structure suggests multiple loops with some similarity to the arcade of loops seen at the same position angle ten hours earlier.

A bright blob, which we believe to be a genuine density enhancement, moves with a velocity similar to that of the other features; it appears to be situated within a loop, and is most clearly seen in Figure 6.2.19c as the brightest part of the outer loop.

CLRO observed a Type II (shock related) burst that lasted from about 2236 to 2243 UT. The Culgoora radioheliograph went into operation at 2255 UT, after the Type II burst had ceased. From 2255 UT until ~ 0000 UT, a complex of six distinct sources were observed at positions shown in

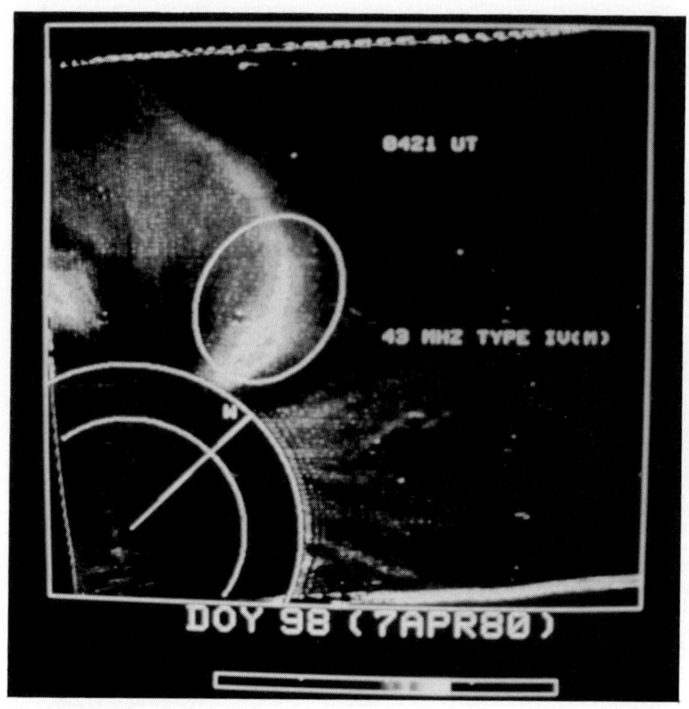

Figure 6.2.18 The position of the later 43 MHz Type IV source at 0421 UT. The evolution of the source at 43 MHz suggests that the component in the void below the loop has faded, and a new source has appeared over the dense leg of the transient.

Figure 6.2.20. After correcting for ionospheric effects by using the known Type I noise storm source A as a reference, it is found that all of the sources remain stationary to within a source diameter, except for one (source E), which moves outward from ~ 2.4 to ~ 3.0 R_0 with a projected speed 500 km s^{-1}. We thus identify the sources as follows: A and B are persistent sources, most likely of Type I; C and D are very intermittent and may be Type I sources also; E is a moving Type IV source; the nature of source F is not clear.

The Type II burst, observed with the CLRO arrays and with the Culgoora spectrograph, evolves as shown in Figure 6.2.20. At any given frequency, the source moves toward the disk; the lower frequencies are displaced further to the south.

The Type II burst had large tangential source motions and its frequency drifted at a very low rate. The joint occurrence of low drift rates and large tangential motions in some Type II bursts was first pointed out by Weiss (1963). The slow drift and large tangential motion of the burst sources may be attributable to a shock wave propagating transverse to open magnetic field lines in the corona, possibly along the axis of the arcade of loops seen later in whitelight.

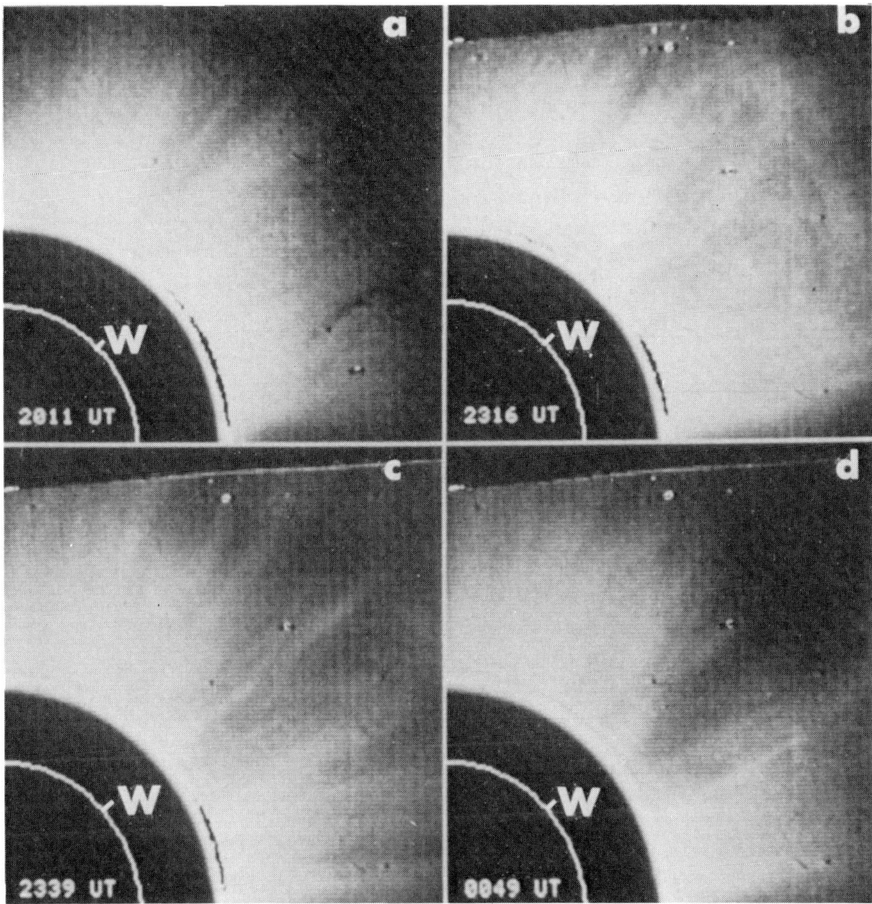

Figure 6.2.19 The evolution of the 9 April 1980 CME coronal transient as obsérved by the C/P aboard SMM. The images have been obtained at: (a) 2011 UT; (b) 2316 UT; (c) 2320 UT; (d) 2339 UT.

The moving Type IV burst at 80 MHz at 2252 and 2307 UT is also shown in Figure 6.2.20 as source E, associated with the bright blob observed by the C/P in visible light at 2316 UT. We estimate for the blob a plasma frequency of ~ 23 MHz. On the basis of Figure 6.2.20, we associate the moving Type IV source with the excess matter contained in the bright blob. Unfortunately, simultaneous observations of the white light CME and moving Type IV source are not available.

We discuss now possible radiation mechanisms for the Type IVM source E. The high degree of polarization of the source (~ 30% RH) rules out second harmonic plasma emission (Dulk and Suzuki 1980). This leaves the possibility of fundamental plasma or gyrosynchrotron radiation. The source was not bright enough to rule out the latter. However, if the

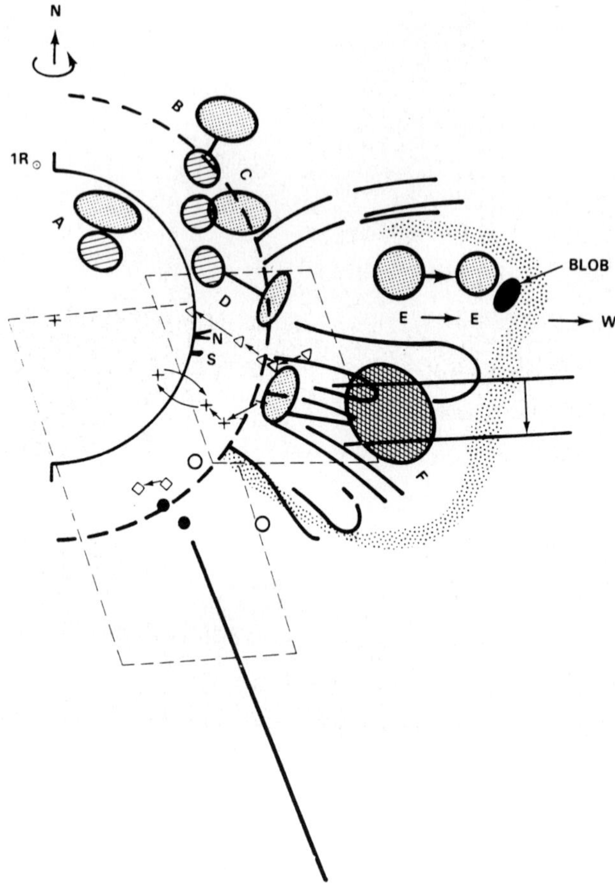

Figure 6.2.20 Composite sketch of the visible light features and radio sources observed during the 9 April 1980 transient. The stationary radio sources A, B, C, and D observed at 160 and 80 MHz with the Culgoora heliograph are shown. The motion of the centroid of the Type II burst at frequencies between 60-50, 50-40, and 40-30 MHz is represented by the triangles, crosses, and squares, respectively. We also show typical source sizes in the 60-50 and 40-30 MHz ranges and the position of the moving source, E (at 2252 and 2307 UT) and F observed at Culgoora and at Clark Lake. The location of one prominence on the limb (N and S), of the whitelight loops, the white light blob (at 2316 UT), the outer edge of the transient, and the streamer south of the transient are all indicated. The open and full circles show the approximate location of the Type II bursts which started at 44 and 55 MHz, respectively.

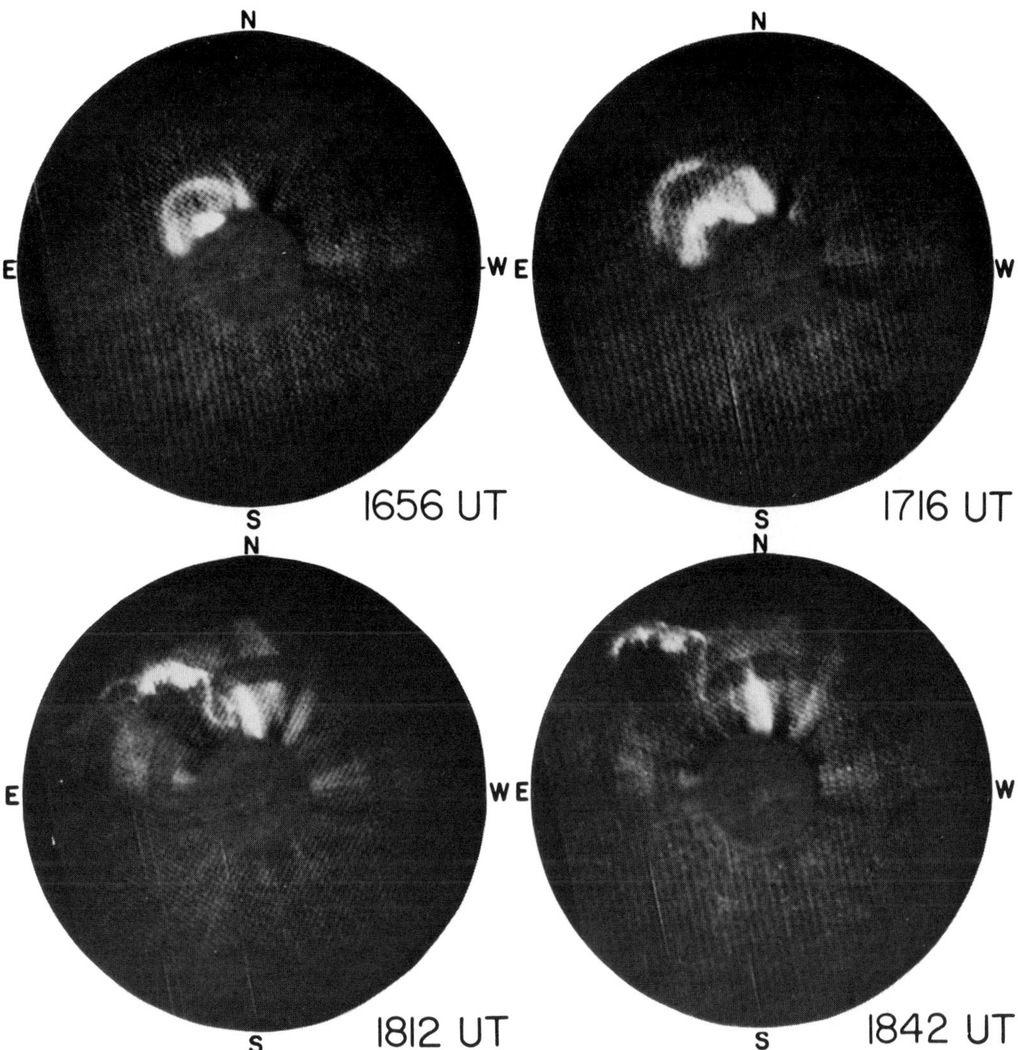

Figure 6.2.21 SOLWIND coronagraph observations of the great solar mass ejection of 24 May 1979. Coronal material forms the loop-like structure ahead of the erupting prominence. The coronagraph's occulting disk is at 2.6 R_o, the outer field of view at 8 R_o. In these difference images a dark void trails the prominence material. These features accelerate to 900 km s^{-1} before leaving the field. A narrow, bright structure moves out very slowly to the north.

plasma frequency was ≥ 40 MHz, then Razin suppression would apply. Since the white light source may have expanded and the true size of the IVM source might be smaller than the 60% contour size, it is reasonable to conclude that the plasma frequency must have exceeded 40 MHz at the time of the 80 MHz observations. Hence, the source may have been due to emission at the fundamental plasma frequency.

6.2.1.8. Helios Spacecraft Observations of Mass Ejection Transients

Prior to the SMM era, when a coronal transient left the outer field of view of a coronagraph, its subsequent evolution could only be guessed at. Yet, knowledge of this evolution is crucial if we are to relate CMEs at the Sun to interplanetary disturbances measured with in situ spacecraft, or to the interaction of the ejected plasma with the magnetosphere of Earth. Fortunately, Jackson and Leinert (1985) recognized that the zodiacal light photometers on board the Helios spacecraft, although constructed for another purpose, can be used to "image" mass ejection transients far from the Sun. Because the two Helios spacecrafts orbit the Sun, rather than Earth, comparison of their observations with those from Earth-orbiting coronagraphs allows stereoscopic views of the outer solar atmosphere. At least three major CMEs with loop-like characteristics -- as observed in Earth-orbiting coronagraphs -- were observed with the Helios photometers. An example is shown in Figures 6.2.21-6.2.22 and 6.5.1-6.5.2. The observational technique for Helios is explained in Section 6.5.2.

Figure 6.2.22a is a Helios 2 contour plot obtained at approximately 0936 UT, 25 May 1979. The spacecraft was then 60 degrees behind the west limb of the Sun as seen from Earth. The ejection, which traveled northeast (Figure 6.2.21) as seen from Earth, is observed to the solar north and northwest in Figure 6.2.22a. Helios observations confirm that a depletion followed the ejection. The 15-hour Sun-spacecraft transit time of the leading material gives an outward speed of approximately 750 km s^{-1}. In Figure 6.2.22b, 1.8 days later, more mass is observed to move outward in two features of fairly narrow angular extent -- possibly the legs of the loop. The speed of this material is approximately 400 km s^{-1}.

This example is one of three CMEs later observed from Helios that had loop-like characteristics as viewed from Earth. From the Helios observations, we find that either the three ejections have large dimensions along the line-of-sight from Earth (i.e., large extent in heliocentric longitude), or the Helios spacecraft did not observe these features edge on. The shapes of ejections evolve as they move outward from the Sun. For instance, in the 24 May 1979 ejection, general features of what was once a prominence can be distinguished, while the outermost portion of the event was probably not observed in the Helios photometers. There is little evidence of structure in the loop legs even though there are several Helios photometer positions within each leg. The complex loop ejection of 29 June 1980 (shown in Figures 6.5.1 and 6.5.2) can be discerned as at least two outward-moving features when viewed from Helios, although from an Earthly perspective the two overlap and thus do not give rise to an apparently broad angular extent.

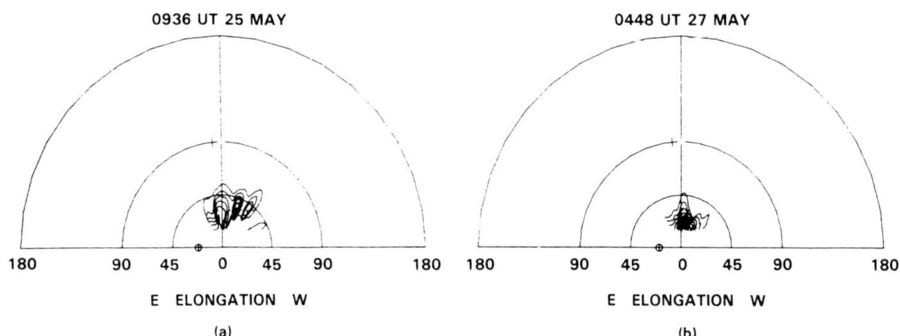

Figure 6.2.22 Helios 2 observations of the 24 May 1980 mass ejection. Visual intensities are converted to excess mass contours which form a coarse image of the ejection (contour interval is 6 × 10^{12} g deg^{-2}). In this presentation the Sun is centered, and the point behind the observer on the spacecraft, i.e., at 180° elongation is represented as the outer circle. The position of the earth (⊕) and the solar pole are indicated relative to the ecliptic plane (horizontal line). (a) Observations of the loop at 0936 UT 25 May. (b) Observations of the legs of the ejection at 0447 UT 27 May. At this late stage of the ejection the high-speed material is well beyond the outer bins of the 31° photometers.

6.2.2 Statistical Studies

6.2.2.1 Introduction

Since the discovery of optical manifestations of coronal mass ejections (CMEs), they have been intensively studied to determine their significance, their cause, and their effects on the solar wind. The whitelight coronagraphs on OSO-7 and Skylab observed and documented many properties of CMEs; both operated during the declining phase near the minimum of solar cycle 20. Thus their measurements of, for example, the frequency of occurrence, speeds, and associations to eruptive prominences and/or flares may have been influenced by phase in the solar cycle.

More recently, the solar corona has been monitored nearly continuously by a combination of two coronagraphs and a coronameter for the period prior to the maximum of sunspot cycle 21, through the maximum, and on into the declining phase. The SOLWIND coronagraph on the P78-1 satellite began routine observations in April 1979, followed a year later by the C/P instrument on the SMM satellite. In the summer of 1980, the ground-based Mark-III coronameter on Mauna Lao in Hawaii began routine observations of the inner corona. The combination of these three instruments has produced far more observations of CMEs than were obtained during the previous solar cycle. In addition to studies of individual CMEs, a number of statistical studies have produced some very general conclusions about CMEs. These studies can be divided into three areas: (1) properties of CMEs themselves, (2) properties of the associations of CMEs with solar radio events, and (3) properties of the associations of CMEs with solar X-ray events. In the next three sections, we summarize the results of team members in these three areas.

6.2.2.2 Properties of CMEs

The major properties -- speed, mass, and energy -- of CMEs derived from the Skylab era observations are summarized in a review by Rust, Hildner et al. (1980). On average, these properties of CMEs during the solar maximum era have remained the same -- the only change appears to be that the maximum and minimum values appear to be more extreme. This difference is presumably due to the increased numbers of CMEs observed rather than any solar cycle dependence. In Figure 6.2.23, from Howard et al. (1984), we present distributions of the various CME properties during the period 1979-1981, a period spanning solar activity maximum.

Perhaps the most striking difference between the low-activity and the maximum epochs is the occurrence of CMEs at high latitudes during solar maximum; during the 1973-1974 period, only 10% of the CMEs were observed above 30° latitude and no CME was observed above 48° latitude (Munro et al., 1979). In contrast, both SOLWIND observations (Sheeley et al., 1980; Howard et al., 1985) and SMM observations (Hundhausen et al., 1984a; Wagner, 1984) reveal CMEs appearing at position angles excluded during the earlier epoch. In the SOLWIND set of observations, the average angular span or extent in the plane of the sky is 45° compared to 28° for the Skylab CMEs. In comparing the absolute numbers, it should be noted that the SOLWIND observations of angular span were made higher in the corona than were the Skylab and SMM observations. Nonetheless, the trend toward larger CMEs during the solar maximum is clearly present. We speculate that the presence of the large polar coronal holes excluded CME emission from high-latitude position angles during the near-minimum Skylab and OSO-7 missions.

The most extreme example of CMEs with large angular spans is the class of "halo" CMEs (Howard et al., 1982). This class of CMEs, about 2% of all of the SOLWIND CMEs, is distinguished by emission surrounding the occulting disk. The interpretation of these events is that the brightness enhancements seen at nearly all position angles around the occulting disk are due to CMEs emitted nearly Earthward from somewhere near the center of the solar disk; the excess brightness is projected onto a large range of position angles in the plane of the sky.

The frequencies of occurrence of CMEs inferred from the C/P and the SOLWIND coronagraphs differ from the solar minimum value and from each other. From an analysis of C/P data, Hundhausen et al. (1984a) obtain an average rate of 0.9 CMEs/day for 1980, whereas Howard et al. (1985) analyze SOLWIND data to obtain an average rate of 1.8 CMEs/day for 1979-1981. Recall that Hildner et al. (1976) obtain a rate of 0.34 CME/day for the Skylab data. Hundhausen et al. recompute the Skylab rate as 0.7 CME/day, using the same criteria used for the SMM analysis. From this, they conclude that the CME rate had not increased in direct proportion to the change in Zurich sunspot number. Howard et al., while finding a higher rate of CME occurrences, also do not find an obvious correlation between fluctuations in the sunspot number and fluctuations in the CME rate.

Though a detailed comparison of the rates inferred from the C/P and SOLWIND instruments during the period of overlapping observations has not been completed, useful comments can be made about the difference between

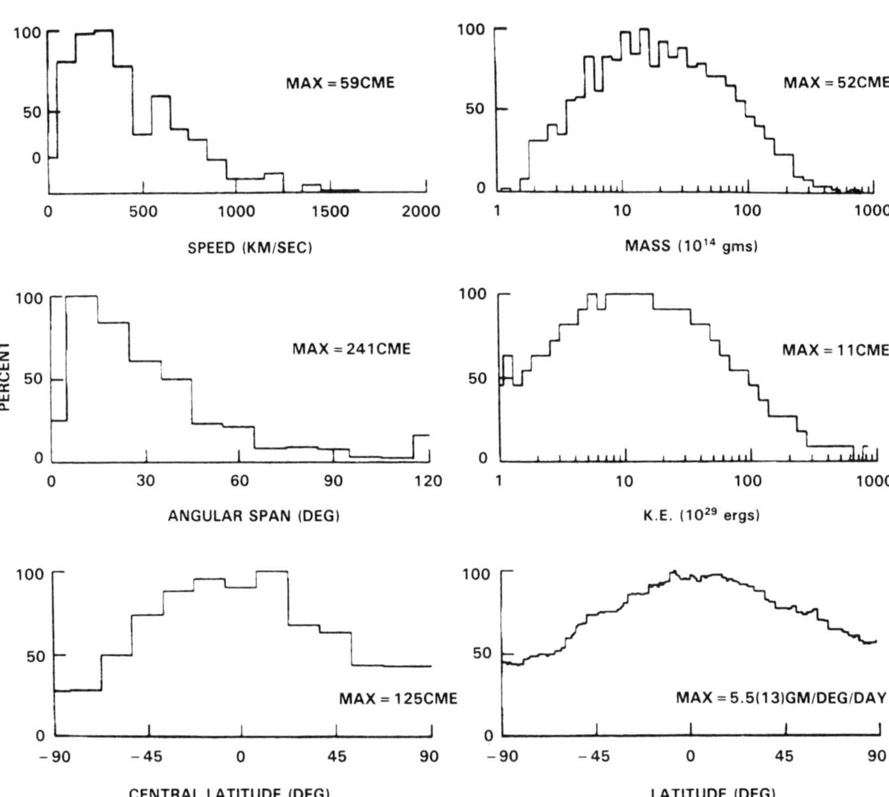

Figure 6.2.23 Properties of all CMEs during the interval surrounding the maximum of the solar cycle, 1979-1981. The distributions of speed, spread, central latitude, mass and kinetic energy are plotted as histograms. Each plot has been normalized to 100% of the maximum number of CMEs. The maximum values used in the normalizations are indicated on the plots. The sixth distribution gives the average mass ejected into each degree of latitude. Also, the plot is normalized to an average daily ejected mass of 5.5×10^{13} gm/deg/day. To derive the daily ejected mass distribution, the correction for instrumental duty cycle has been applied. Note that all angular measurements are made projected in the plane of the sky.

the rates. SOLWIND observers used a differencing technique, subtracting an earlier image from the one being considered, to search systematically through the data and identify the occurrence of CMEs; C/P observers used the direct coronal images to identify the occurrence or non-occurrence of a CME. The direct images were also used by the Skylab observers. Thus, SOLWIND results are biased to include fainter events than might be included in C/P and Skylab result. On the other hand, since the SOLWIND observations extend from 2.5 to 10.0 R_0, one could envision that, because CMEs fade with height, the contrast of a CME above the background K + F coronal emission for SOLWIND might be less than the contrast of the same CME in the Skylab and C/P observations which extend from approximately 1.6 R_0. Thus, the Skylab and C/P analyses using direct images of the inner corona might include fainter CMEs than detected in undifferenced SOLWIND images covering the outer corona.

6.2.2.3 Association of CMEs with Type II and IV Solar Radio Bursts

A meter-wave Type II radio burst is generally believed to indicate the presence of a shock front in the corona. The speeds of many CMEs in the corona exceed the local Alfven velocity and therefore are expected to drive shocks. Prior to the recent observations at solar maximum, the shocks were generally assumed to occur at appropriate standoff distances ahead of the leading edges of CMEs (e.g., see reviews by MacQueen, 1980; Maxwell and Dryer, 1982). For Skylab-era CMEs, Gosling et al. (1976) find that about 85% of the CMEs with speeds greater than 500 km s^{-1} are associated with Type II and/or Type IV radio bursts, and that these bursts are associated only with those CMEs whose speeds are greater than 400 km s^{-1}. From another statistical study of the associations of Skylab CMEs to other solar phenomena, Munro et al. (1979) find that nearly all Type II or Type IV bursts occurring within 45° of the limb are associated with CMEs.

Interestingly, these close statistical associations found for the Skylab era are not confirmed by the recent observations (Sheeley et al., 1984 and Kahler et al., 1984a). Thus, the pre-maximum picture of the relationship of CMEs to Type II and IV radio bursts is, at least, confused. Sheeley et al. (1984) and Kahler et al. (1984a, 1985a), in comparing SOLWIND CMEs with metric Type II bursts, find that about 60-70% of Type II bursts were associated with CMEs and 30-40% were not. On the other hand, about 40% of all CMEs whose speeds exceeded 450 km s^{-1} had no associated Type II. Thus, a class of Type II bursts exists for which no visible CME was produced, and, conversely, a class of fast CMEs exists for which no associated metric Type II emission was observed. One is tempted to speculate that backside events might obscure about half of the Type II bursts. However, Sheeley et al. (1984) find 5 cases in the Culgoora time zone and Kahler et al. (1985a) find 15 cases in all time zones, in which an H-alpha and 1-8A flare can be associated with the CME, but for which no Type II was observed.

The association of Type II bursts with fast CMEs argues for a piston-driven model of Type II shocks. Both Sheeley et al. and Kahler et al. suggest that variations in the ambient magnetic field might give rise to variations in local characteristic speeds; such variations might enable a fast CME to be travelling locally sub-Alfvenic, and, therefore, be ineffective in producing Type II emission. However, as Kahler et al. point out, in the piston-driven model, CMEs of larger angular width would pass through larger regions of the corona and thus be more effective than smaller CMEs in finding local regions in the corona where the Alfven speed was appropriate for producing Type II emission. No correlation between CME size and Type II emission is found, though a correlation is found between CME brightness and Type II burst association.

Arguing against a model which suggests that Type II bursts come from shocks driven by CMEs are several studies of speeds, timing, and other associations. For at least two events for which simultaneous, or near simultaneous, observations of a Type II burst and its associated CME are available (Gergely, 1984; Gary et al., 1984) the speed of the CME is lower than that of the Type II burst source. Wagner and MacQueen (1983) note that in some events the Type II is seen well below the top of the CME; they propose to account for this disparity by suggesting that the CME starts to rise first. Then, somewhat later, the impulsive phase of energy liberation generates a shock which propagates up through the CME causing Type II emission when the shock encounters the high-density regions associated with the legs of the CME. Cane (1984) extended the Wagner-MacQueen hypothesis to account for the observations of interplanetary Type II bursts and interplanetary shocks. Kahler et al. (1985) find the peak strength of the accompanying 3-cm burst is as important as the CME speed in determining the Type II burst association. These studies support the hypothesis of Wagner and MacQueen that the CME does not drive the shock which, in turn, is responsible for any associated Type II burst. In this view, the CME is simply a result of the big flare syndrome (Kahler, 1982) in which large, energetic evens are more likely to be associated with other large, energetic phenomena, but without a direct cause and effect relationship.

Additional evidence is offered by Gergely (1984), who compares the velocity distributions of Type II associated CMEs observed with the Skylab and P78-1 coronagraphs (Rust, Hildner et al., 1980; Sheeley et al., 1984) to the velocity distribution of all Type IIs observed during the period 1968-1982 (Gergely, 1984). Figure 6.2.24 shows the distributions for CMEs from Skylab (shaded) and for CMEs from SOLWIND, while Figure 6.2.25 shows the distribution of speeds of Type II bursts. Some representative parameters of the distributions are given in Table 6.2.3.

From a comparison of these two figures, one sees that the velocity distribution of Type II associated CMEs is similar at the declining and maximum phases of the solar cycle. Note that the distribution of all CMEs (Gosling et al., 1976; Howard et al., 1985) is also similar for these two phases. The average speed of the Type II bursts exceeds the average speed of the CMEs by about 70%, whereas the rms dispersion of Type II speeds is considerably higher than the rms dispersion of CME speeds. These results should be qualified with the following caveats. Firstly, most Type II speeds have been derived from drift rates and, therefore, depend on the density model (most often the 2x Newkirk streamer model) adopted for the corona. Secondly, the speeds of Type IIs were determined in the 1.5-2.5

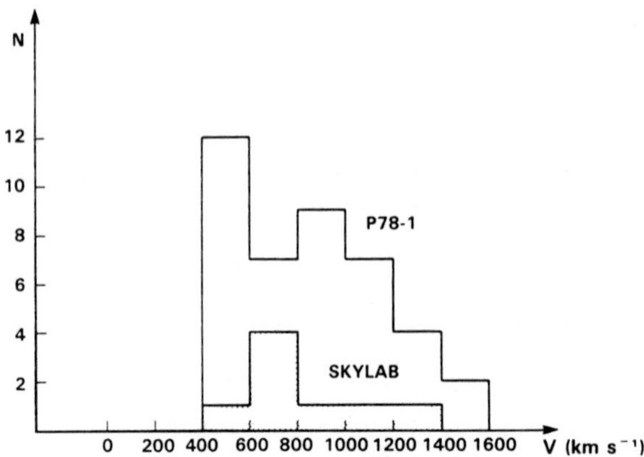

Figure 6.2.24 Speed distributions of the leading edge of CMEs accompanied by Type II bursts. The shaded distribution indicates the speeds of CMEs observed with the Skylab coronagraph; the unshaded distribution indicates the speeds of CMEs observed with the P78-1 coronagraph. Note the absence of any speeds slower than 400 km/s and the similarity of the distributions during solar minimum and solar maximum.

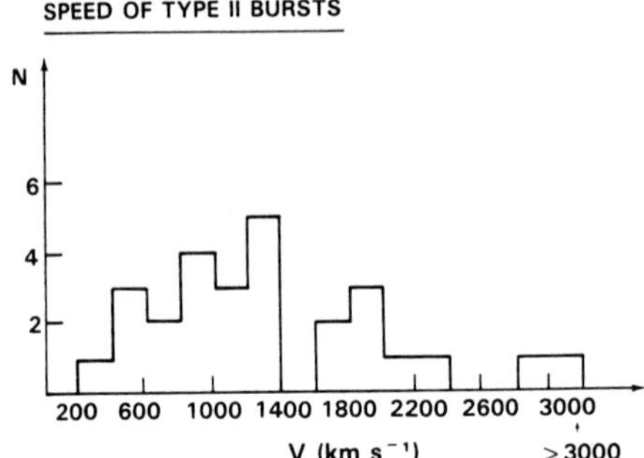

Figure 6.2.25 Distribution of Type II velocities. The Type II speeds were compiled from a literature search of all Type II speeds. Note the very large spread in speeds, and the maximum in the 1200-1400 km/s bin.

Table 6.2.3 Comparison of Speeds of CMEs, Type II and Type IV

	CMEs Observed With		Metric Radio Bursts	
	Skylab	P78-1	II	IV
Number of Events	8	40	26	46
Lowest Speed (km s^{-1})	450	400	370	100
Highest Speed (km s^{-1})	>1200	1500	4900	1300
Mean Speed (km s^{-1})	795	825	1380	515
RMS Dispersion (km s^{-1})	255	293	720	286

R_0 height range, whereas the speeds of CMEs were determined at greater heights. Thirdly, the Type II speed determinations are very inhomogeneous in terms of method, instrument used, phase of the solar cycle, etc., as contrasted with the CME speed determinations which were very homogeneous.

Stewart (1984) compared the positions of Type II bursts observed with the Culgoora radioheliograph at 160 MHz with the positions of filaments and optical flares for the years 1980-1982. Figures 6.2.26 and 6.2.27 show Carrington plots of the Type II radio positions and the filament channels for 1980 and 1981. The circles give the approximate positional error of the Type II burst (5°). Stewart finds that 93% of the Type II bursts occurring above the limb lie within 5° of a filament channel and that 51% lie within 1°. For disk events, 86% of the Type II positions lie within 5°. In a similar comparison with optical flare positions, he finds that 83% of the Type II burst positions occurred further than 10° from the flare site. Since the local Alfven speed is expected to be lower in a streamer than outside it, it is likely that an MHD wave will steepen into a shock at the streamer. Thus, Stewart's results are consistent with the concept of the Type II emission being generated by the interaction of an MHD wave with a helmet streamer overlying the magnetic neutral line (Stewart, 1984).

1980 TYPE II BURSTS (CULGOORA)

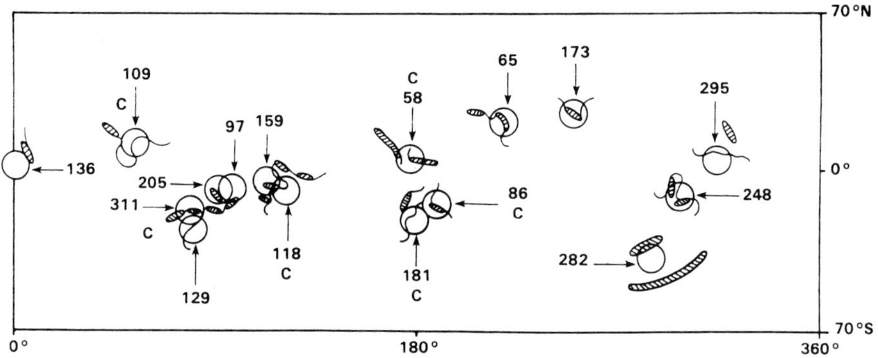

Figure 6.2.26 Carrington plots of the positions of Type II bursts and H-alpha filament channels for 1980. The position of the Type II bursts were derived from the Culgoora 160 MHz radioheliograph and are indicated with circles, the size of which indicates the estimated positional error. The positions of the H-alpha filament channels are indicated by the hatched regions and thin lines. The numbers are the day of year on which the Type II burst occurred. The letter "C" indicates those events associated with a CME observed either with the SMM C/P or the P78-1 SOLWIND coronagraph. Note the close association of the positions of Type IIs and the filament channels.

1981 TYPE II BURSTS (CULGOORA)

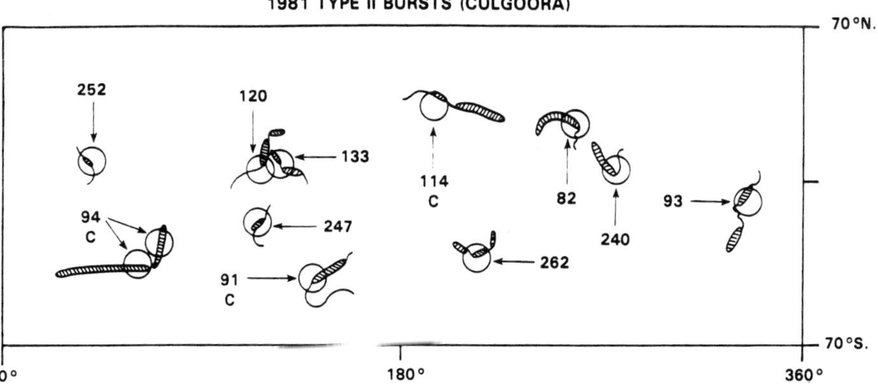

Figure 6.2.27 Carrington plot of the position of Type II bursts and H-alpha filament channels for 1981. The format is exactly the same as for Figure 6.2.26. Again, note the close association of the positions of Type IIs and the filament channels.

Kahler et al. (1985) question whether the Type II bursts associated with H-alha flares but not with CMEs could be pure blast waves. For pure blast waves, the Type II bursts should correlate with the impulsive release of energy as measured by the size of the 3-cm peak flux. However, for a thick target model, Kahler et al. estimate that small Type II bursts have only one-tenth to one-hundredth the threshold energies necessary for shock formation in large flares. Also, only 20% of the large 3-cm bursts were associated with reported Type II bursts. Thus, Kahler et al. conclude that the relationship between impulsive phase energy releases and shocks identified by Type II bursts is poor.

At least two Type IV bursts have been identified with dense plasmoids observed within, or near the leading edge of loops or bubble-shaped CMEs (Gergely et al., 1984; Stewart et al., 1982). Gergely (1984) examined the statistical distribution of Type IV burst velocities for the period 1968-1982, as shown in Figure 6.2.28. Some parameters of the distribution are given in Table 6.2.3. Comparison of the Type IV speed distribution with the distribution of speeds for all CMEs implies

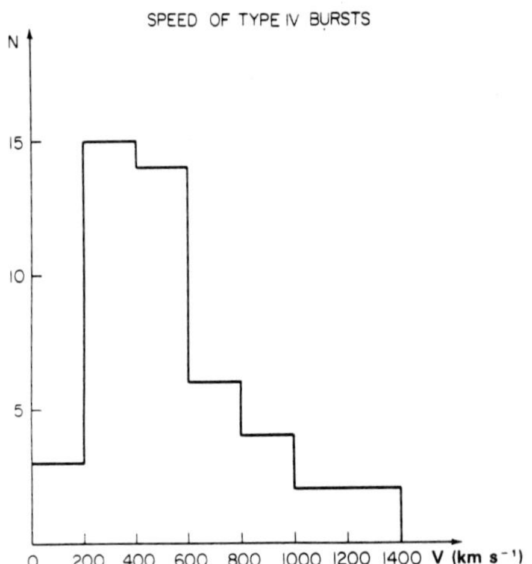

Figure 6.2.28 Speeds of Type IV bursts during 1968-1982. Note the broad maximum of speeds from 200-600 km/s.

that (a) the mean speed of the Type IVs is less than the mean speed for all CMEs, and (b) the rms dispersions of the speed distributions of CMEs and Type IVs are identical. Thus, the moving Type IV bursts appear to be moving outward (sometimes to as high as 5 or 6 R_0 before fading) either slightly behind, or with the leading edge of the CME. In contrast with those inferred for Type II bursts, the speeds of Type IVs are not dependent on coronal density models; consequently they are more reliable than the Type II speeds. Since only a small number of instruments have been used to determine their speeds, the Type IV observations are also more homogeneous.

6.2.2.4 Association With Soft X-Rays

The long-duration, soft, X-ray events during the Skylab mission were usually accompanied by coronal mass ejections (see the review by Rust, Hildner et al., 1980). Sheeley et al. (1983a) examined the relationship of the SOLWIND CMEs to the 1-8 A flux from the SMS-GOES satellite for the period 1979-1981. They find that the probability for a CME to be associated with an X-ray long-duration event increases as the duration of the X-ray event increases -- from about 25% for X-ray events less than 2 hours long to 100% for X-ray events lasting longer than 6 hours. The distribution of X-ray durations of those X-ray events with CMEs has a broad peak centered at about 3.5 hours, with a median value of about 4.5 hours. Thus, the longer an X-ray event lasts, the more likely that a CME accompanies it.

6.3 The INITIATION OF CORONAL MASS EJECTIONS

6.3.1 Introduction

Surprisingly little is known about the initiation of coronal mass ejections. Although the first flare observation took place over a century ago (Carrington, 1859), and we have studied the passage of material through the solar corona for several decades (e.g., Payne-Scott, Yabsley, and Bolton, 1947-Type II; Wild, Roberts, and Murray, 1954-Type III; Tousey, 1973-coronal transients), their relationship is poorly understood. In asking how a mass ejection is initiated, we are requiring to know (a) the original location of the mass which is to be ejected, (b) when the onset is triggered, (c) the time-height profiles for the various components of the ejection, and (d) their relationship to other solar activity. With these observational questions answered we may then confidently postulate the basic mechanism behind these events.

A coronal transient is characterized by changes in brightness of regions of the corona with time-scales of tens of minutes. The majority of such events are thought to be indicative of CMEs, outward mass motions with velocities of < 1000 km s^{-1}. These events are most often associated with eruptive filaments whether or not they are flare associated.

In this section, we discuss the relative timing of the various components of ejecta and compare the sequence of events to surface activity. Of particular interest is the pre-flare role of ejecta and their association with pre-flare surface activity. We then describe some recent observations relating to below-coronagraph images of ejecta and, finally, due to the quality and quantity of data obtained for an event on 30 March 1980, we describe this event in some detail, emphasizing different aspects of the event than those treated in Section 6.2.1.5.

6.3.2 Relative Timing

The relationship between a solar active region, its flare activity, and related mass ejections cannot be sensibly modelled in the absence of detailed studies into the relative timing of the various active phenomena. There are problems, however, in obtaining suitable measurements. The use of externally occulted whitelight coronagraphs, which obscure the low corona, has tended to detach the study of the coronal response to mass motions from the study of solar surface phenomena. Furthermore, the usual plane-of-sky projection view introduces complications in matching a CME to its associated surface activity and produces a selection effect in favor of limb events which, due to their location, are not ideally suited to other types of investigations.

6.3.2.1 Eruptive Filaments

Filaments often exhibit pre-flare disturbances (Martin and Ramsey, 1972; Webb, Krieger, and Rust, 1976), typically showing some upward motion and internal material motion tens of minutes prior to flare onset. Heyvaerts, Priest, and Rust (1977) noted that at the time of a flare's impulsive phase, "...twisting motions in the filament are often observed. Occasionally the filament remains, though disturbed somewhat, but usually it rises (with a much greater acceleration than before), untwists or flies apart, and disappears completely."

An ongoing study begun at the SMM Workshop by Kahler, Moore, Kane, and Zirin is an investigation into the relationship between the flare impulsive phase and filament eruption. A close temporal relatioship has long been known (Svestka and Fritzova-Svestkova, 1974) between the onset of Type II bursts, indicative of shocks, and the impulsive phase as measured in microwaves or hard X-rays. This has led to suggestions that an impulsive phase "explosion", perhaps generated by a rapid chromospheric deposition of energy carried by electrons of $E > 10$ keV, may initiate the shock. In the model of Lin and Hudson (1976) for large flares, this explosion also produces the mass ejection. This concept has also been incorporated in numerical hydrodynamical models (see Dryer, 1982). However, filament eruptions have often been thought to be magnetically controlled, and, as mentioned, filament activity tens of minutes before the impulsive phase is well known. Kahler et al. seek to resolve the question of just what is the activity in the filament (taken to be indicative of non-potential magnetic fields) at the time of the impulsive phase.

From a four-event subset of flares well-observed both with the University of California at Berkeley's ISEE-3 hard X-ray detector and in H-alpha at the Big Bear Solar Observatory (26 April 1979, 28 May 1980, 25 June 1980 and 27 July 1981), preliminary results show that the filaments rapidly accelerate during the impulsive phase. This acceleration does not seem to be a result or effect of the impulsive phase, nor does it seem to drive the impulsive phase. It appears that a catastrophic magnetic action (presumably reconnection) results in the impulsive phase, and both the particle acceleration and rapid filament eruption are manifestations of this causal action. Kahler et al. hope to show the temporal and spacial relationships of the filaments and impulsive phase energetic electrons as clearly as possible in the near future.

If the eruptive filament displays pre-flare activity and then exhibits a violent reaction to events during the impulsive phase, how does this relate to the main body of the CME? The fact that pre-flare filament activity is observed, combined with the fact that coronal material commonly leads the ascending filamentary material (Schmahl and Hildner, 1977; Maxwell and Dryer, 1982), begs the question: what is the relative timing between the main mass ejection onset and the flare onset?

6.3.2.2 Coronal Mass Ejections

One feature of the CME which exhibits pre-flare motion is the so-called forerunner (Jackson and Hildner, 1978; Jackson, 1981), a tenuous, broad envelope of material which leads the main component of the mass ejection. According to Jackson, the entire mass ejection process begins with acceleration of a high altitude, pre-existing structure which becomes the forerunner and ends with the onset of surface eruptions and flare activity. Using Skylab coronagraph data from 15 ejection events, he was able to show that 7 events showed outward motion of tenuous material prior to any surface H-alpha eruption and another 5 exhibited excess mass in the corona prior to an ejection. The other 3 events were rejected for various reasons. No event showed an H-alpha eruption prior to the motion of coronal mass.

In recent years it has been suggested that the main body of the transient also has a pre-flare activation (e.g., Gary, 1982; Wagner, 1983, 1984 and references therein). A thorough investigation into the onset times of several CMEs has been made during the SMM Workshop by Harison et al. (1985). They have been able to confirm, for seven flares observed by the SMM observatory, that the mass ejection onset was prior to the associated flare. Perhaps the most significant fact is that in all of these events the ejection onset appears to be closely associated with a soft X-ray flare precursor. This is illustrated in Figure 6.3.1. Figure 6.3.1a shows the coronal mass ejection altitude history as recorded by the Coronagraph/Polarimeter for an event early on 29 June 1980 (Gary, 1982; Harrison et al., 1985). A projection from a best fit to the data points, assuming no acceleration, indicates that the ejection left the vicinity of the low corona at about 0228 UT, some 5 minutes prior to the soft X-ray burst. The 3.5-5.5 keV intensity, as recorded with HXIS (van Beek et al., 1980) for the flare event (post-0233 UT), and its precursor (pre-0233 UT) associated with this mass ejection, is superimposed onto the figure. The correlation between the mass ejection onset and the precursor is even closer if we assume continuing acceleration in the early stages of the ejection.

The same scenario is found for the 1823 UT flare on 29 June 1980, illustrated in Figure 6.3.1b. The solid line labelled T is the best fit to the mass ejection's altitude as observed with a K-coronameter (Wu et al., 1983; Harrison et al., 1984); projection to the limb region, assuming no acceleration, suggests an onset time of 1811 UT. Harrison et al. have attempted to describe the acceleration phase of this ejection by linking the K-coronameter data to ejection heights estimated from soft X-ray images (described later). Again, the HXIS 3.5 5.5 keV record (a, b, c, d) for this flare is superimposed onto the figure. As with the previous flare, the mass ejection is associated with a small event preceding a larger one. In this case the precursory event is a relatively large burst.

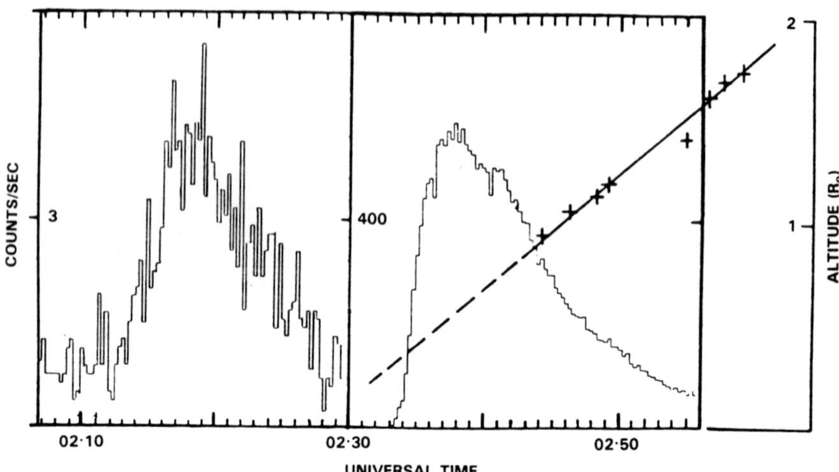

Figure 6.3.1a The relationship between the 0233 UT 29 June 1980 flare and an associated whitelight CME. The 3.5-5.5 keV intensity-time profile is shown for the flare (post-0233 UT) and its precursor (pre-0233 UT), along with the trajectory of the ejection. The solid line is a best fit to the observed leading edge, and the dashed path is a projection assuming no acceleration. The 3.5-5.5 keV background level is 0.05 count s^{-1}.

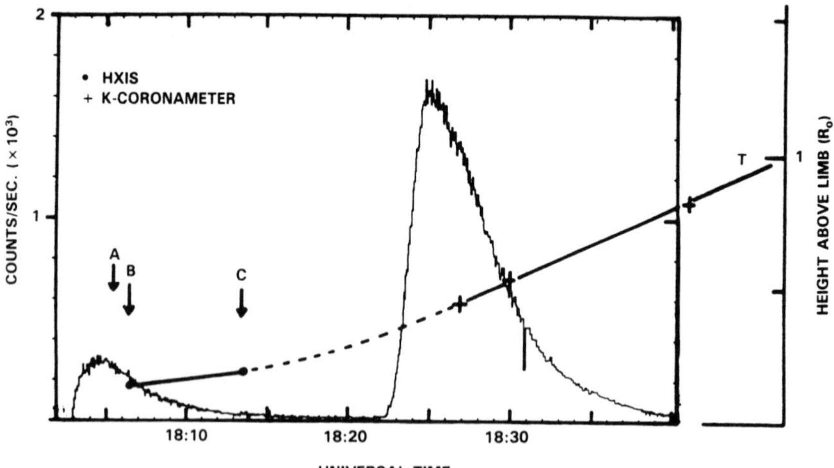

Figure 6.3.1b The relationship between the 1800 to 1900 UT 29 June 1980 flares and associated coronal activity. The 3.5 to 5.5 keV intensity-time profile is shown for the 1823 UT flare and the smaller 1803 UT event. Superimposed onto this are the altitude histories of a whitelight coronal mass ejection (T) and an X-ray transient as shown in Figure 6.3.3.

These analyses depend upon height vs time plots derived from coronagraph images, and such plots are susceptible to a number of uncertainties. Although the time at which each image is obtained is precisely known, the low contrast of CME features ensures uncertainty of 0.1 R_0 in most height measurements. Furthermore, the variation of a CME's brightness contrast as it moves through the field of view has the potential to introduce a systematic error in measurements of the heights of the CME's leading edge on successive images. Finally, a small error in the inferred speed of a slow CME produces a large change in where the trajectory line intercepts the time axis; that is, a change of five or so minutes in the inferred time of CME start, in the absence of acceleration or deceleration, could result from measurement uncertainties. However, as there is no accepted way to reduce these uncertainties, the data are taken to be as measured.

In addition to showing that for the seven events studied the mass ejections appeared to originate at the time of the flare precursor, Harrison et al. have attempted to show that all coronal mass ejections are accompanied by a soft X-ray burst and that a flare may follow, usually, some tens of minutes later. They describe an event on 27 June 1980, when a mass ejection, not associated with a flare, left the low corona at the time of a "lone precursor".

Harrison et al. point out that the coronal mass ejection/ soft X-ray precursor association wuld be vitiated by the existence of either deceleration of the ejection or high altitude onset. They reject both, the latter on the grounds that the average onset altitude for the seven events analyzed would be 0.66 solar radii above the photosphere, and the former on the grounds that acceleration is frequently detected in coronagraph images -- deceleration is rare to non-existent.

6.3.2.3 Summary

To summarize the work of Kahler et al., Jackson, Harrison et al., and others (Webb, Krieger, and Rust, 1976; Dryer, 1982; Schmahl and Hildner, 1977), we make use of a schematic time-plot as in Figure 6.3.2. The lower half of this figure is devoted to the intensity-time curves of the HXIS 3.5-5.5 keV and 22-30 keV energy bands, for a "typical" flare event. A soft X-ray precursor is seen from time A; the hard X-ray burst is seen from time B, just after the onset of the soft X-ray component of the main flare which maximizes at C. Of course, the separation in time between A and B may range from a few to tens of minutes. The top half of the figure shows three height-time plots. The lowest curve represents the location of the eruptive filament, which begins to rise at the time of the precursor but displays more rapid acceleration during the impulsive phase. The middle curve represents the height of the CME's leading edge. We assume that the mass to become part of the ejection is stored in low coronal loops (see, e.g., the model due to Simnett and Harrison, 1985) prior to onset. The structure entraining this mass starts rising at the time of the X-ray precursor, reaching an altitude of about 1 R_0 soon after flare onset. It travels, at any moment, at least as fast as the following filamentary material. The highest curve represents the forerunner's trajectory. The forerunner may well start several solar radii above the photosphere and is accelerated well before B, although the actual timing of the acceleration stage is uncertain.

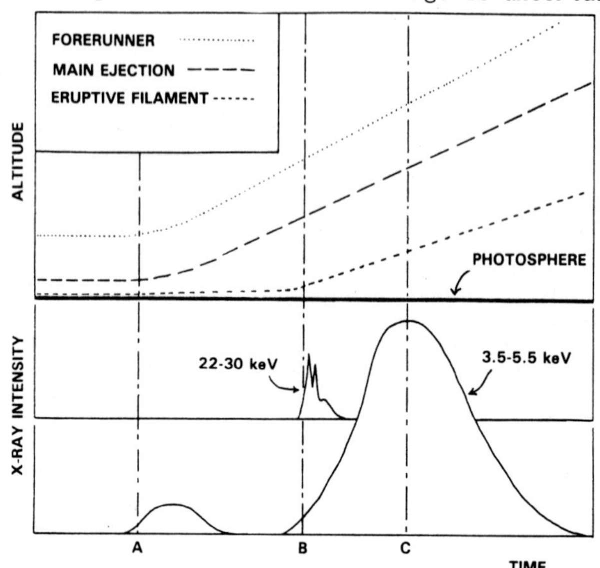

Figure 6.3.2 The relative timing between the X-ray signatures of a solar flare and the mass ejection onsets, shown schematically. The curves in the lower half of the figure represent "typical" soft and hard X-ray intensity profiles for a flare event. The energy bands shown are the highest and lowest bands of HXIS. A soft X-ray precursor is evident from A, the hard X-ray burst onset is at B, and the maximum intensity of the soft X-ray flare is at C. The upper plots show the heights of the eruptive filament, the main mass ejection, and the forerunner on the same time axis.

6.3.3 Low-Height Observations Relevant to Mass Ejections

We define low-height observations as those which view the atmosphere at altitudes below the occulting discs of whitelight coronagraphs, i.e., below about 150,000 km. In this category we include observations of sprays, surges and filamentary eruptions (Rust, Hildner, et al., 1980). Whereas many surges are neither flare related nor result in the escape of material into the high corona, sprays and filamentary eruptions are commonly associated with coronal mass ejections (Munro et al., 1979), sprays originate in flares, and it has been suggested that sprays and filamentary eruptions are not fundamentally different (Rust, Hildner, et al., 1980). These phenomena are associated with the lowest altitude profile of Figure 6.3.2. What observations may be made to observe the low-height behavior of the main coronal mass ejection?

6.3.3.1 X-ray Coronal Transients

Using images from the HXIS in the 3.5-5.5 keV energy range, Harrison, Bentley, Phillips, and their colleagues (Harrison et al., 1984) believe they have identified X-ray-emitting counterparts of white light CMEs (cf., Rust and Hildner, 1976). For three limb flares which were well observed by the SMM instruments on 29 June 1980, they examined temporal changes in the X-ray intensity of the low corona. Evidence was found for two classes of X-ray coronal transients.

Figure 6.3.3 shows an X-ray transient observed from about 1806 UT on 29 June. Panel a is a 7.7s exposure from 1806 UT, during the precursor event illustrated in Figure 6.3.1b. The subsequent panels show exposures from 1806, 1813, and 1818 UT, respectively. These images clearly show an X-ray coronal transient ascending at < 40 km s^{-1}. The altitudes of the X-ray disturbance (the crosses in Figure 6.3.3) are plotted in Figure 6.3.1b. Under the assumption that the material producing the X-ray enhancement is part of the subsequent whitelight CME, the altitudes estimated from HXIS and C/P data are suggestive of acceleration in the early stages of the ejection.

A second X-ray transient is illustrated in Figure 6.3.4 for the flare of 1041 UT on 29 June; it is seen to be much more intense than the first example. The first image is accumulated during 9.2s after 1043 UT. The subsequent images are from 1046 and 1053 UT, respectively. They reveal a transient which is travelling into the corona at 60 km s^{-1}, during the main phase of the flare (see Figure 6.3.1a).

Harrison et al. believe these X-ray transients are due to material motion. In support of this, they demonstrate that as the intensity enhancements move outward, the variation in the intensity levels of the two lowest energy bands of HXIS suggests that the emission is coming from regions of significant density enhancement.

The X-ray transients analyzed by Harrison et al. fall into two classes. Events in the first class, like the 1806 UT 29 June event are observed during the precursor stage and are associated with the onsets of the main coronal transients. Events in the second class, like the 1041 UT 29 June event, occur during the main phases of flares and are not associated with whitelight transients or fresh energy releases. It is thought that this second class is due to hot material rising with the reconnecting magnetic fields above a flare site.

6.3.3.2 High-Velocity X-ray Ejecta in Flares

A study, principally made by Bentley and Phillips, has recently focused on the identification of high-velocity, low-altitude, X-ray emitting ejections which occur at flare onset times. Observations with the Bent Crystal Spectrometer (BCS) on SMM have revealed spectrally discrete, short-lived emission line features near intense parent lines in solar flare X-ray spectra for three large flares: 29 June 1980 at 1041 and 1823 UT and 21 May 1980 at 2100 UT. A preliminary account of them has already been given by Bentley et al. (1984). The line displacements, assumed to be Doppler shifts from the parent lines, indicate line-of-sight velocities of 300-400 km s^{-1}. The observed shifts imply surprisingly large speeds, especially for the two limb flares of 29 June, when one might think that most material would be ejected radially rather than along the line of sight. For the events mentioned, no high-speed feature can be positively identified in images. Bentley et al. speculate that the X-ray line features are due to fast-moving material connected in some way with the fast-moving material seen as a visible ejection.

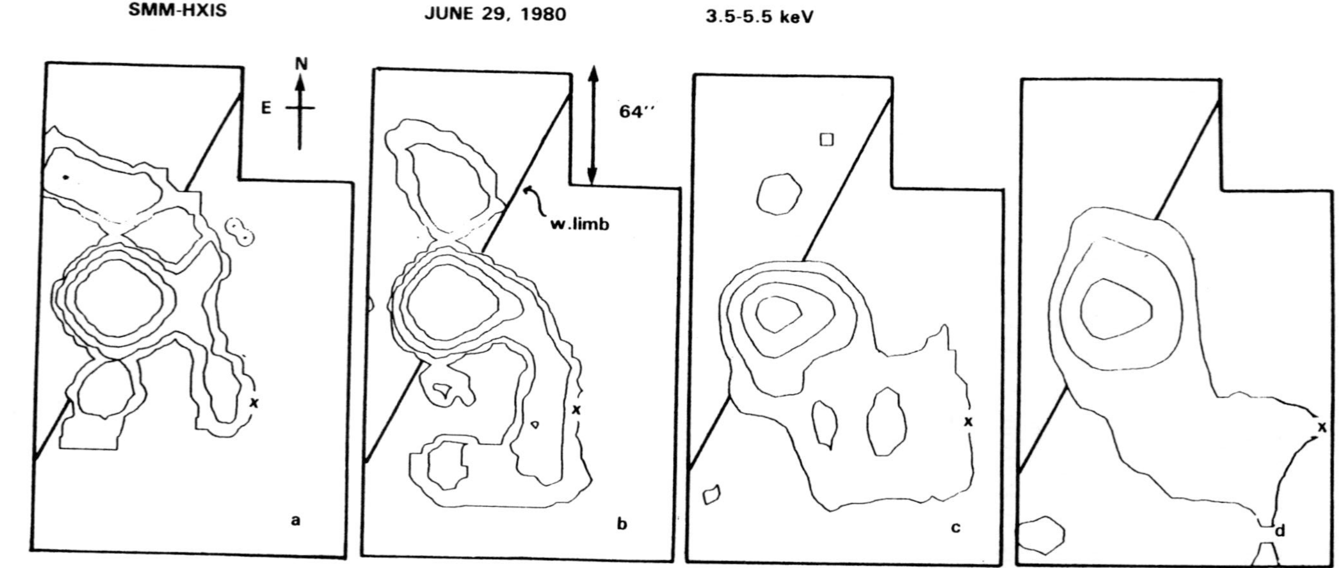

Figure 6.3.3 An X-ray transient seen in the precursor phase of the last 29 June 1980 flare. A series of HXIS 3.5 to 5.5 keV images are shown: (a) 18:05:48 UT, exposure duration 7.7s; (b) 18:06:16 UT, exposure 77s; (c) 18:13:10 UT, exposure 3.33s; and (d) 18:18:07 UT, exposure 32.3s. Intensity contours are plotted at 20, 5, 1 and 0.2 counts/sec. The background level is 0.005 counts/sec per pixel. The altitudes of the crosses are plotted in Figure 6.3.1b.

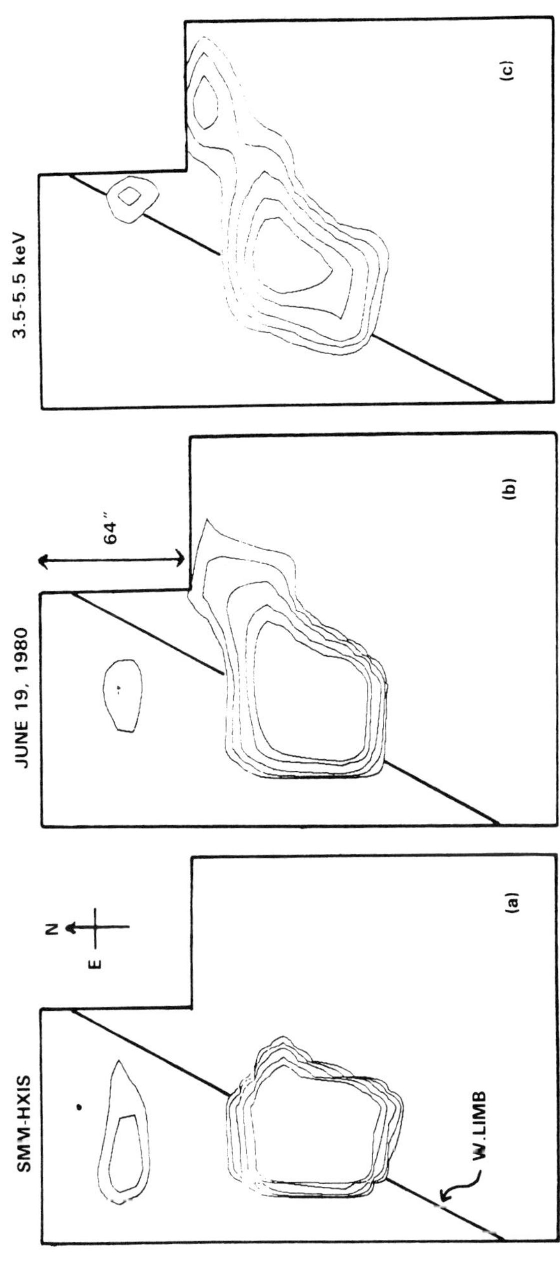

Figure 6.3.4 The X-ray transient seen during the second 29 June 1980 flare. Three HXIS 3.5 to 5.5 keV images are shown; (a) 10:42:50 UT, exposure 9.2s; (b) 10:46:00 UT, exposure 13.9s; and (c) 10:53:34 UT, exposure 20.5s. Intensity contours are plotted at 100, 50, 20, 10 and 5 counts/sec.

For two west-limb flares of 29 June the discrete line features appear to the short wavelength side of the Fe XXV resonance line at 1.85 A (BCS Channel 7) for only ~ 20s. In each case, the time is coincident with a spike-like burst in hard X-rays as seen with the Hard X-ray Burst Spectrometer (HXRBS on SMM). Figure 6.3.5 shows a sequence of BCS channel 7 spectra around the time of the line feature during the 1041 UT event,

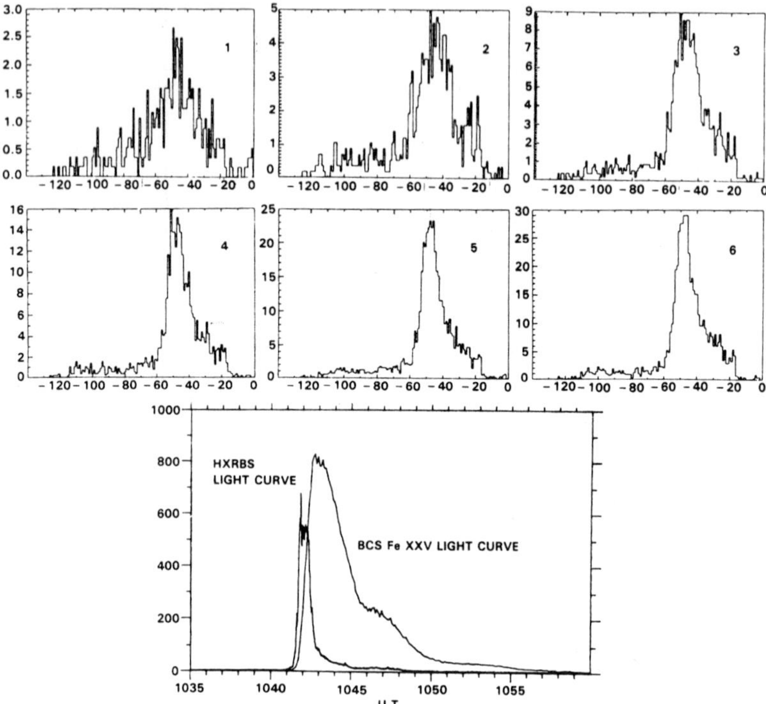

Figure 6.3.5 A. Six sequential BCS spectra of the Fe XXV resonance line ($1^1 - 2^1P$) during the rise of the soft X-ray event on 29 June 1980, at 1040 UT. Each spectrum covers the range 1.842 to 1.855A, and were obtained over an integration time of 5.63 at the following mean times: (1) 1041:45; (2) 1041:41; (3) 1041:57; (4) 1042:02; (5) 1042:08; (6) 1042:14. The vertical scale is in photon counts per wavelength interval per second. The line feature at bin 60 is reproducible in successive spectra starting at spectrum (2). (The structure at bins 20-30 is probably due to the blue-shifted Fe XXV line $1^1S_0 - 23^3P_2$.) B. Hard X-ray flux (HXRBS, all channels summed, energy range 25-200 keV) leading the Fe XXV line flux (BCS channel 7, bins 35-60) for the 1040 UT flare. (Vertical scales not absolute.) HXRBS light-curve from Dr. Allen Kiplinger, GSFC.

together with a light-curve of the parent Fe XV line and all channels (25-500 keV) of the HXRBS. For each event, the line features have at least a 3σ level of significance. The displacement of the features from the centroid of the Fe XXV line is a Doppler effect rather than a spacial displacement within the BCS field of view, since the latter would require flare activity > 5 arcmin from the main flare site, which is not observed. Approach velocities of 290 km s^{-1} for the 1041 UT event and 370 km s^{-1} for the 1823 UT event are indicated. The features are not clearly resolved in the BCS Ca XIX detector which has only half the spectral resolution of channel 7.

The large X-ray velocities and the existence of visible-light ejecta in all three events suggest that hot material is driven out with the whitelight ejection. The line features are narrow, so their spacial extent is small (at least in the direction of dispersion, normally E-W), suggesting that the emitting volumes are small and blob-like rather than, e.g., sheet-like. Though Rust and Webb (1977) reported a fast-moving, soft-X-ray-emitting blob coincident with an H-alpha spray, in the case of an eruptive filament Rust and Hildner (1976) report a large volume of soft X-ray emitting material rising to become, perhaps, the CME itself. It is very suggestive that all three events discussed are spray associated.

The possibility that spray-associated flares give rise to Doppler-shifted discrete, X-ray line features has led Phillips to investigate spray-associated flares listed in Solar Geophysical Data (SGD, U.S. Dept. Commerce) for the SMM operational period. Excluding the events already discussed (SGD lists the 29 June events as surges rather than sprays), of the 93 sprays listed in SGD only 3 were compatible with the SMM pointing and showed Ca XIX brightening (the BCS's most sensitive indicator of flare activity). BCS Ca and Fe spectra were examined for each of the three flares. For the event of 1904 UT 9 September, a discrete line feature appears only in Ca XIX just 2 standard deviations above background. The other two flares show nothing. For periods when sprays were not reported, the SGD was consulted for metric Type II radio bursts occurrence. BCS spectra taken at the times of Type II-associated events were examined, but no clear examples of discrete line features were found.

The absence of identifiable discrete line features in the examined flares seems to arise from poor count statistics at the flare start. Typically, the surveyed events had maximum channel 1 counting rates below 500 s^{-1}, whereas the counting-rate for the three large flares with discrete line events was at least 2000 s^{-1}. The appearance of discrete lines seems to be a phenomenon distinct from the broadening and mild blue-shifting of the parent line, which occurs at the time of the impulsive hard X-ray burst; thus the short-lived discrete features should be most easily visible when the parent line broadening is least. The least broadening of the parent lines occurs for flares on or near the limb, so a search for Doppler-shifted discrete line features of the type discussed here might be profitably directed at strong limb flares. It is hoped that the search will be renewed with the repaired SMM.

6.3.4 The 30 March 1980 Event

The event of 30 March 1980 has been selected for detailed description since an unusually complete set of observations exists. The active region (AR 2363) was at N25E25, well onto the disk, yet C/P observed a coronal transient resulting from activity within this region at about 1245 UT (assuming constant speed of the ejection from the surface to the heights observed later). Thus, the region is ideally suited for studying the initiation of the mass ejection. The activity of AR 2363 has been previously described (Lantos et al., 1981; Lantos, 1984; Lantos and Kerdraon, 1984; and in Section 6.2.1.5). The complementary analysis done by Lantos and Waggett during the Workshop is summarized here.

The low coronal activity associated with the 30 March CME was varied and included a filament eruption, an intense metric noise storm, and an X-ray long duration event (LDE). At metric wavelengths, noise storms with durations of a few hours have been associated with filament eruptions (e.g., Webb and Kundu, 1978). Large amplitude X-ray LDEs are often associated with solar flares distinguished by their large physical sizes, low energy densities, and their association with whitelight coronal transients and centimetric gradual rise and fall bursts (e.g., Sheeley et al., 1975; Pallavicini, Serio, and Vaiana, 1977). Like most others, the 30 March LDE appeared in soft X-rays as a system of large diffuse loops, brighter at the top and crossing a filament channel.

The structure of the low coronal region was imaged by the FCS; it is shown schematically in Figure 6.2.16. Two regions had been steady soft X-ray emitters, one associated with the magnetic inversion line to the northwest of the area labelled A and the other associated with the inversion line to the north of A, as well as various loop arcades. The footpoints of these arcades are denoted by H-alpha ribbons, also shown in the figure. Lantos et al. (1981) concluded that the two main loops are as in Figure 6.2.16., straddling a long filament which is due to erupt.

6.3.4.1 The Precursor Phase

The HXIS and BCS light curves indicate the presence of a small, soft X-ray burst between 1247 and 1250 UT of peak temperature and emission measure 6.4×10^{6} K and 1.3×10^{54} m^{-3}. In Figure 6.2.16., two patches of emission are seen associated with the two areas previously mentioned. HXIS images reveal structure in both, and this - combined with their locations - implies that they are separate loop systems. However, the light curves for the two patches are similar that there must be a connection between them, probably a large loop. This loop must be complex, since a simple link between the two regions would pass directly over the main sunspot group of the active region. This loop could either be passing to the north or south of the sunspots; as there is no evidence in the H-alpha photographs for any loops passing to the north, we will assume that the southern route is the more likely. Lying south of the sunspots implies that the loop will be close to those involved in the main X-ray LDE. This precursory activity could be an indicator of the actual initiation, the X-ray signature being due to loop-top merging to the south of the spot group.

6.3.4.2 The Main Event

The metric noise storm began after 1305 UT, about the time of the southern filament's disappearance. As discussed in Section 6.2.1.5, from about 1305 to 1330 UT, the motion of the noise storm is consistent with the motion of the poleward legs of the loops as they expand and carry the CME into the corona. Lantos and Kerdraon (1984) propose that as this scenario unfolded, the footpoints of some of the loops making up the transient were still attached to the chromosphere at least as late as

1330 UT. The motion discussed in Section 6.2.1.5 suggests a mass ejection onset at ~ 12:50 UT, assuming no acceleration. This implies a precursor/ CME association of the type discussed by Harrison et al. (1985).

C/P observations of the CME were made at 1406 and 1426 UT, from which a relationship between the whitelight transient location and the rising loop structure is inferred and indicated in Figure 6.2.16.

Occurring contemporaneously with the preceding phases is the reconnection phase. During this phase, the main activity in the low corona is the reconnection of ruptured field lines; this powers the main X-ray event as well as the radio noise storm as soon as it is detached from the rising structure. The X-ray and radio intensity profiles are similar, peaking at about 1345 UT, although the noise storm (at B Figure 6.2.16) is several arcmin to the north of the X-ray emitting region, and they are associated with different legs of the ascending loop.

6.3.4.3 Summary

Summarizing the features of this event and the analysis:

1. Apparently a loop feature, between 1305 to 1426 UT, rose from a height of 100,000 km out to several solar radii.
2. This loop, which crossed over a filament involved in the ejection, became a CME.
3. The onset of the ejection was prior to the main X-ray event, perhaps associated with a precursor. The ejection appears to be the result of the merging of loop-tops.
4. The near-surface X-ray and radio signatures were due to reconnection.

6.4 MODELLING OF CORONAL MASS EJECTIONS AND POST-FLARE ARCHES

6.4.1 Introduction

For a comprehensive physical understanding of the CME phenomenon, theorists and modellers must address the following three broad questions and their ramifications: Under what circumstance and by what mechanism can a mass ejection be initiated, or triggered, in the low corona? How is the mass ejection accelerated and propelled dynamically through the corona? What are the manifestations of the mass ejection in interplanetary space? Though this chapter shows that observational data exist, sometimes in abundance, we are quite far from being able to provide answers to these questions in any complete form. Many interesting ideas have been pursued and debated. We refer the readers to recent review articles for surveys of these ideas (Anzer, 1980; Dryer, 1982; Rosner, Low and Holzer, 1984; Hundhausen et al., 1984b). In this section, we report on some specific developments in the theory and modelling of mass ejections and other related coronal eruptions which were presented and discussed at the Workshop.

Existing spaceborne coronagraphs have fields of view covering 1.6 to 10 R_0. A mass ejection observed in these fields of view is already in a fully developed state of motion. Thus, mass ejections in this already-evolved dynamical state tended to dominate the interests of modellers and theorists in the early part of the Skylab era. Gradually, emphasis broadened to include the question of initiation. Data from the present generation of spaceborne coronagraphs cannot be expected to give direct information on the manner and circumstance of the initiation of mass ejections in the low corona. We emphasize two collections of indirect evidence relevant to the question of initiation. Some Skylab-era mass ejections were found to have been preceded, for an hour or more, by broad, faint outflows in the corona called transient forerunners (Jackson and Hildner, 1978; Jackson, 1981). These authors took this result to suggest that, for some mass ejections, the initiation involves a non-impulsive precursory phase. A contrasting view for the initiation of some SMM-era CMEs is presented in Section 6.3.2.2. In at least one case, a forerunner has been identified accompanying an SMM-era mass ejection (Gary et al., 1984). The second group of indirect evidence comes from the statistical association of CMEs with the occurrence of other forms of eruptions in their space-time vicinity. The Skylab CMEs had a significant association with flares and an even more significant association with prominence eruptions (Munro et al., 1979). This pattern of association has been found to be basically unchanged at the current solar maximum (Sawyer et al., 1985; but see also Webb and Hundhausen, 1985). Unlike the SMM era, the prominences that erupted in association with the Skylab CMEs were found to tend to have north-south orientations before their eruptions (Trottet and MacQueen, 1980).

Recently, direct observation of the low corona in whitelight became possible with the operation of HAO's K-coronameter at Mauna Loa, which has an annular field of view of 1.2 to 2.2 R_0 (Fisher et al., 1981). Observations of mass ejections in their early dynamical development show distinctly different kinematics for flare- and prominence-associated mass ejections in the low corona (MacQueen and Fisher, 1983). The former tend to be already at high speeds (> 300 km s^{-1}), when first observed, and have little discernible acceleration thereafter, whereas the latter tend to be at low speeds (< 300 km s^{-1}) when first observed and often show detectable accelerations. Above 2 R_0 in the fields of view of spaceborne coronagraphs, mass ejections tend to move at approximately constant speeds, making their acceleration profiles nearly indistinguishable. These observations suggest that different mass ejections may be subject to different types of acceleration process in the low corona.

The good association between some mass ejections and flares suggests that the initiation processes of mass ejections and of flares may be coupled. Numerical MHD models developed thus far have taken as a starting point that the flare energy initiates and drives the mass ejection by way of impulsively generated nonlinear MHD waves (e.g., Dryer, 1982). It has also been suggested that mass ejections may be magnetically driven as a consequence of the magnetic reconnection postulated for the liberation of the flare energy (Pneuman, 1980). On the other hand, there are CMEs not associated with flares which appear without any impulsive signature in the chromosphere. For these ejections, it has been suggested that they result from the transition of a highly stressed, slowly evolving structure into a state of non-equilibrium (Low, 1981).

Over the length and time scales of concern, ideal magnetohydrodynamics (MHD) provides an appropriate, lowest order description for CMEs. The contention between different theoretical ideas suggested for the initiation and propulsion of mass ejections has led, among other things, to the recognition that we need to develop our basic intuition and understanding of the magnetohydrodynamic medium in the presence of gravity (e.g., Rosner, Low, and Holzer, 1984). We seem to be at a stage where the questions posed by observation on mass-ejection initiation and dynamics are sufficiently focused to warrant intensifying theoretical efforts to resolve basic theoretical issues and to develop the next generation of models. The question concerning the interplanetary manifestations of CMEs is relatively new. The observational picture is, at the present, incomplete, and the theoretical ideas on this question are much less developed.

With the above overview, we present the following five projects developed at the Workshop. In Sections 6.4.2 and 6.4.3, we report on two projects aimed at understanding basic MHD. In Section 6.4.2, Hildner and Wu present a parametric study of an ideal MHD numerical model to investigate the range of physical characteristics exhibited by the impulsively perturbed atmosphere. In Section 6.4.3, Low reports his recent discovery of analytic, self-similar solutions to the time-dependent equations of MHD, which opens up the possibility of building analytic models and testing numerical MHD codes. The projects in Sections 6.4.4, 6.4.5, and 6.4.6 are concerned with various aspects of mass-ejection initiation. In Section 6.4.4, Svestka interprets the formation of post-flare coronal arches observed with the HXIS instrument on the SMM satellite. In Svestka's study, the flares are studied not only for their own sakes, but also for their possible relationships with the initiation of associated mass ejections observed with NRL's SOLWIND coronagraph. In Section 6.4.5, Poletto and Kopp treat a magnetic reconnection model to refine the method of estimating the energy extractable from the magnetic fields in a two-ribbon flare. Finally, in Section 6.4.6, Low demonstrates, theoretically, the linear MHD-stability of coronal structures.

6.4.2 Parametric Study with a Numerical MHD Model

Numerical models have the technical advantage that within the numerical code developed for a particular model, one can freely prescribe the details of the initial and boundary conditions. Numerical modelling has s far concentrated on pulse-initiated disturbances in an ambient atmosphere (Dryer, 1982). Despite the existence of numerical modelling studies with a rich diversity of initial conditions and initial perturbing pulse characteristics, till now there has not been a systematic attempt to learn how the simulated coronal response is affected by varying the initial conditions (e.g., the initial magnetic field strength). In this section, we present the results of a systematic study of this type. We used a single numerical model (Wu et al., 1983) incorporating ideal MHD equations to calculate the response of the solar atmosphere to a standard pressure-pulse perturbation, introducing the pulse into a different member of a family of magnetic configurations for each calculation. The members of the family of magnetic configurations differ in their multipole number and in their base strength. The standard pressure pulse was introduced at one or the other of two places in each initial configuration and strength, as explained below. The results of the 12 calculations allow us to gauge how initial field strength at the coronal base, overall field configuration -- and the fall-off of field strength with height, and the location of the pulse affect the response of the corona.

An isothermal ($T = 2 \times 10^6$ K) hydrostatic corona of base density $n_e = 10^{8.5}$ cm^{-3} is permeated by a potential field. Because the calculation is two-dimensional and axisymmetric, there is no variation of any parameter in heliocentric longitude.

Three initial field configurations are considered: a dipole (180 deg symmetry), a quadrupole (90 deg symmetry), and a hexapole (60 deg symmetry), as shown in Figure 6.4.1. Of course, the magnetic field strength decreases more rapidly with height for the higher multipole fields. Coronal mass ejections often arise in active regions, only rarely

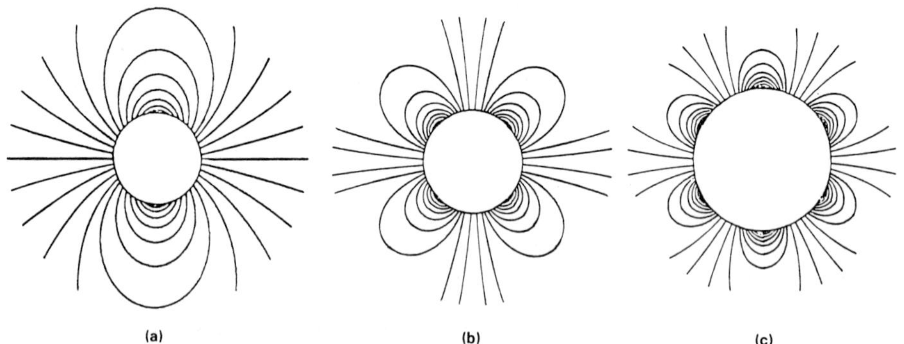

Figure 6.4.1 Meridional cut through the magnetic field configurations for parametric study. Each configuration employed in the study has rotational symmetry about an axis in the plane of the paper; the orientation of the symmetry axis is chosen to make the equatorial magnetic field either open or closed. (a) Dipole configuration. (b) Quadrupole. (c) Hexapole.

subtending as much as 60 deg of heliographic angle; even the hexapole configuration may span too large an angle to be realistic, but we are interested in discerning trends in coronal response as we examine magnetic structures of various scales. Since the gas pressure in the initial, hydrostatic corona varies only with height, the initial plasma beta, the ratio of gas to magnetic pressure, varies with latitude and height. In our calculations, we use two magnetic field strengths for each configuration; the pre-perturbation beta at the site of the perturbing pulse is either 0.1 or 1.0 for a perturbation placed on the open-field symmetry axis. These field strengths imply beta is 0.45 for 4.5 for a perturbation on the closed-field symmetry axis.

The side boundaries of the computational domain are defined by radial rays along symmetry axes of the initial magnetic configuration; no fluid flows through these boundaries, but the pressure and density evolve in response to the perturbation. Though the computational domain covers only one half of one lobe in each magnetic configuration, we take the boundary conditions as periodic, with a perturbing pulse identical to the one in the calculational domain being placed similarly in each lobe of the magnetic field around the sun. The top and bottom boundaries are transparent to fluid flow. After the perturbation is established, the temperature and density are maintained constant on the lower boundary, and the compatibility relations determine the flow; temperature and density are allowed to evolve no the upper boundary. The lower boundary is at 1.0 R_0, and the upper boundary is at 5 R_0 for the dipole, 3.5 R_0 for the quadrupole, and 3.0 R_0 for the hexapole, respectively. Once the calculation begins, the magnetic field lines are not "tied" to a particular location nor required to have a specified orientation on any boundary.

To start each calculation, the gas pressure on three mesh points at the base of the mesh is raised, in a fast linear-ramp fashion, to a value five times the initial value. To achieve the pressure increase, the temperature is increased by 3 and the density by 1.67. At the perturbation site, the temperature and density values are maintained at their elevated values for the duration of each calculation. The perturbing pulse is placed at the base of the corona on a symmetry axis, either the open or the closed axis.

For this ideal MHD calculation, the governing equations are given by Wu et al. (1985) and Wang et al. (1982).

Comparisons among the 12 calculations are facilitated by Table 6.4.1, which emphasizes the outward flow of plasma engendered by the perturbing pulse. A stronger pressure pulse would be expected to increase the flow speed somewhat, especially in the open field configuration and for the weak magnetic field case ($\beta = 1.0$). Other tables could be drawn to contrast other aspects of the way the corona responds to the standard pulse, of course. In Table 6.4.1, at two times for each configuration, we have listed an estimate of the on-axis speed of the contour which represents a 100 percent local density enhancement and the maximum relative density enhancement.

Perusal of the table allows us to state a number of comparative results which are verified by close examination of the more detailed plots of the coronal evolution simulated by the calculations.

1. In the weak-field cases, those in which beta = 1.0 on the open-field symmetry axis at the lower boundary, the outward speed of strong density enhancement is nearly the same for the open- and closed-field trials of a particular configuration. Similarly, the magnitude of the maximum density enhancement differs little between trials when the disturbance is placed in the open or closed portion of a configuration. Even between configurations the variation is fairly small.
2. In sharp contrast to result 1, in the strong-field case, those in which beta = 0.1 (0.45) at the base of the open (closed) field, there is a considerable difference between the open- and closed-field trials of each configuration, especially in the magnitude of the maximum density enhancement which occurs.
3. The chosen excess density contour rises at speeds within the range of observed coronal mass ejections speeds.
4. The ascent of the contour appears to be slowing with time,

generally. This contrasts sharply with the observed behavior of coronal mass ejections in the lower corona; MacQueen and Fisher (1983) report that CMEs typically exhibit steady or increasing speeds in this height range.

Table 6.4.1 Coronal Response for Various Magnetic Fields

	Speed[b] (km s^{-1}) Density Enhancement[c] STRENGTH			
	β = 1.0[a]		β = 0.1[a]	
CONFIGURATION	**OPEN**	**CLOSED**	**OPEN**	**CLOSED**
DIPOLE				
t = 1200 s	340[b]	390	380	560
	6.5[c]	5.2	8.4	4.6
t = 2400 s	275[d]	360[d]	350	530
	12.0[d]	7.8[d]	17	9.8
QUADRUPOLE				
t = 1200 s	410	380	360	480
	5.6	4.9	7.1	2.5
t = 2400 s	265	310	300	250
	8.7	8.9	13	4.4
HEXAPOLE				
t = 1200 s	220[e]	240[e]	310	230
	4.4[e]	3.5[e]	5.8	2.1
t = 2400 s	240[f]	280[f]	240	180
	6.8[f]	8.0[f]	9.5	3.3

Notes: a. The ratio of gas to magnetic pressure at the base of the open field. Pre-perturbation beta is ~4.5 times greater at the base of the closed-field symmetry axis.
b. The estimated, on-axis height of the $n_e(r,t)=2n_e(r,o)$ contour, divided by the elapsed time; see text.
c. The maximum density enhancement attained anywhere in the computational domain at the noted time.
d. At t = 2800 s
e. At t = 1050 s
f. At t = 2250 s

From these results we may draw the following interpretations and conclusions:

1. The effect of the magnetic field is small when beta = 1.0 at the base of the open-field corona. This is understandable in light of the fact that for our assumed initial field (and for any non-pathological, realistic solar magnetic field) beta increases rapidly with height; for beta = 1.0 (4.5) at the base of the open (closed) corona, therefore, the gas pressure dominates the magnetic pressure everywhere in the initial configuration.
2. When the magnetic field is sufficiently strong, the character of the flow is significantly different for the perturbation introduced on the open-field axis of the configuration than it

is for the perturbation introduced on the closed-field axis. The introduction of the perturbation sets up a pressure gradient in its vicinity. In response, the plasma can flow outward (along the field lines) if the site of the high-pressure perturbation is on the open axis of the magnetic field configuration. In these cases, the outflow from the perturbation site adds mass to the computational domain, and dense material rising from below into the more tenuous, ambient upper corona causes ever-increasing, strong, local density enhancements. By contrast, outward fluid flow -- which must produce deformation of the overlying magnetic field -- is strongly retarded in the closed-field cases. Because fluid flows away from the perturbation less readily, less mass is added to the computational domain per unit of time, and there is a much smaller maximum density enhancement than in the open-field cases.

Although in the calculation waves propagate away from the site of the perturbation, their density contrast ($\Delta n_e/n_e$ < one percent) is insufficient for them to be identified with observed CMEs, whose much higher brightness contrast with the background ($(B-B_o)/B_o > 0.6$ according to Sime, MacQueen and Hundhausen (1984)) implies a higher density contrast. The shape of the wave front is affected by the choice of the site of the perturbation, since MHD fast waves propagate more rapidly across than along field lines.

In summary, it is difficult to see how a reasonable pressure pulse at the bottom of a potential magnetic field configuration can give rise to rapid outflow of large amounts of mass, as seen in CMEs. In the calculations reported here, for the more realistic low-beta cases, a significant fraction of the energy given to the fluid by the excess gradient of pressure in the vicinity of the perturbation site is absorbed in the deformation of the magnetic field, initially in its lowest-energy state. Thus, the field is retarding, rather than driving the mass flow. Nevertheless, wave motion propagates significant energy away from the perturbation site even though the bulk flow is small; this wave motion alters local plasma conditions as it goes.

It appears that an appropriate next step in attempts to simulate CMEs via numerical modelling is to start from an equilibrium configuration in which the magnetic field is stressed, i.e., contains stored energy, so that magnetic energy can be transferred to the fluid once the perturbed flow begins (Low, 1982). Whether this initial state is best simulated as a magnetohydrodynamic steady state (more realistic) or as a magnetohydrostatic one (easier to calculate) is not yet clear.

6.4.3 Self-Similar MHD Modelling of Mass Ejections

In reality, mass ejections are three-dimensional objects. Since time-dependent MHD flows with variations in three-dimensional space pose exceedingly difficult mathematical problems, modellers have so far concentrated on two-dimensional systems, in particular, the axisymmetric atmosphere. It is generally acepted that a spacially one-dimensional system is of little use in the study of mass ejections, because many of the interesting, anisotropic magnetic effects in the MHD medium take trivial forms in such a system. Fortunately, a two-dimensional system is adequate to capture the essence of these effects; two-dimensional, time-dependent MHD flows are very complicated, and numerical methods provide the only practical means of building models. Still, a strictly numerical approach is not often useful. There is a need to build up our intuition and knowledge of the behavior of the highly nonlinear MHD system to enable ourselves to question and to interpret numerical results. Analytic solutions, whenever they can be found, serve this need well. We are fortunate that analytic, time-dependent solutions to the MHD equations exist (Low, 1982; 1984a). In the following, we describe briefly what these solutions are and point out some of their physical implications for the mass ejection phenomenon.

As demonstrated in Low (1984a), the ideal MHD equations admit time-dependent solutions describing a radial global velocity field of the form

$$\underline{v} = \frac{r}{\Phi}\frac{d\Phi}{dt}\hat{r} \tag{6.4.1}$$

where Φ is a strict function of time satisfying

$$\left(\frac{d\Phi}{dt}\right)^2 = \eta - \frac{2\alpha}{\Phi} \tag{6.4.2}$$

with η and α being arbitrary constants. In such a velocity field, the plasma and magnetic field evolve in a self-similar manner as described by

$$p(\underline{r},t) = \Phi^{-4}\, P(\zeta,\theta,\phi), \tag{6.4.3}$$

$$\rho(\underline{r},t) = \Phi^{-3}\, D(\zeta,\theta,\phi), \tag{6.4.4}$$

$$\underline{B}(\underline{r},t) = \Phi^{-2}\, \underline{H}(\zeta,\theta,\phi), \tag{6.4.5}$$

where the similarity variable

$$\zeta = \frac{r}{\Phi} \tag{6.4.6}$$

has been introduced. In fact, if Equations (6.4.3)-(6.4.6) are substituted into the MHD equations, the conservation laws for mass, magnetic flux, and entropy are trivially satisfied for all functional forms of P, D, and H, provided we set $\gamma = 4/3$. Explicit time-dependence can then be transformed away from the momentum equation to give rise to

$$\alpha\,\zeta\, D\hat{r} = \frac{1}{4\pi}\,(\underline{\nabla}^* \times \underline{H}) \times \underline{H} - \underline{\nabla}P - \frac{DGM_0}{\zeta^2}\, r, \tag{6.4.7}$$

where ∇^* is the usual operator ∇, but with r replaced by ζ. In this final procedure, we have transformed the time-dependent MHD problem to a "static" problem cast in the (ζ,Θ,ϕ) space. To construct a solution, we solve the pseudo-magnetostatic problem posed by Equation (6.4.7), seeking equilibrium states in which the Lorentz force balances a pressure gradient and a body force made up of Newtonian gravity and an inertial force arising from the non-Galilean similarity transformation. To every solution so constructed and a solution Φ of Equation (6.4.2), we can generate a time-dependent MHD solution by transforming from the (ζ,Θ,ϕ) space back to real space-time. It is important to note that this formulation does not require the physical system to be two-dimensional. Both axisymmetric and fully three-dimensional solutions are admissible. Axisymmetric solutions are the ones that received immediate attention for reasons of tractability. However, we should bear in mind that this formulation has opened up the feasibility of building three-dimensional solutions (Low, Hundhausen, an Hu, 1985).

Figure 6.4.2 displays the time-development of an axisymmetic solution which simulates a mass ejection in the form of an expanding loop structure. This type of mass ejection is commonly observed (Munro et al., 1979). The mathematical construction of the solution displayed in Figure 6.4.2 is given in Low (1984a). In the figure, we are looking at a global outflow carrying an axisymmetric magnetic field. This outflow plows into an ambient atmosphere having no magnetic field. A spherical contact surface, marked $r = R_c$, forms between the two fluids and drives a strong gasdynamic shock propagating into the ambient atmosphere. The top panel shows the magnetic field lines projected onto the meridional plane at three successive instants of time. In this particular example, there is a ϕ-component of the magnetic field. Thus, the set of nested, closed loops of field-line projection correspond to a twisted toroidal magnetic flux rope in three-dimensional space.

The free parameters of the solution and the density profile of the ambient atmosphere have been chosen to produce the following dynamical behaviors. At some initial moment, say $t = t_o$, the outflow has been established with $R_c = 2 \times 10^{11}$ cm and the shock created right on $r = R_c$. The ambient atmosphere is such that the shock immediately separates and propagates with a speed which increases as the 1/7th power of the shock radius. At the same time, the contact surface moves outward at a constant speed of 540 km s^{-1}. All this is achieved with an ambient atmosphere whose density falls with radius approximately like $r^{-26/7}$, a profile which is steeper than r^{-2} drop for a solar wind with a constant speed. Thus, this ambient density profile would correspond to a region of accelerating solar wind in the corona. The three instants of time indicated in Figure 6.4.2 are reckoned from the initial moment $t = t_o$.

Consider the second panel of Figure 6.4.2 showing the time development of density in the outflow. We find low density in the equatorial region, flanked by higher densities at higher latitudes. Let us define $\Delta\rho$ to be the excess density above the spherically-symmetric reference density given by the run of the initial density along the polar axis. The departure from spherical symmetry, $\Delta\rho$, is analogous to the result of subtracting a standard pre-event photograph from a coronagraph photograph of a mass ejection, a procedure commonly used in data

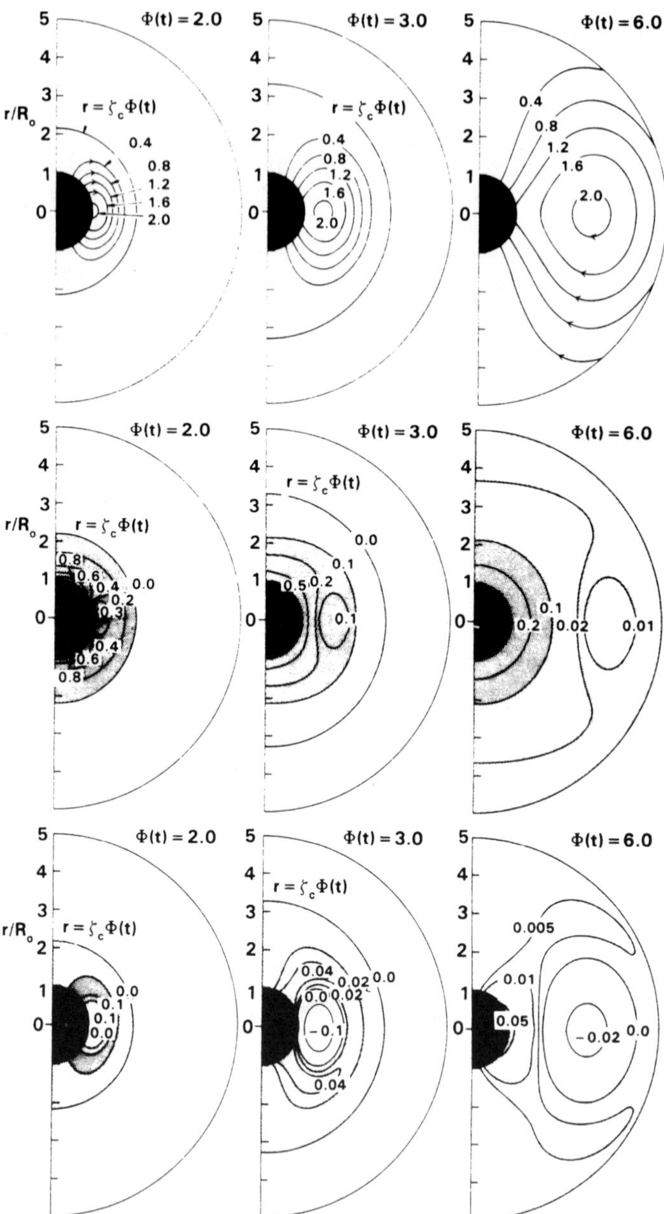

Figure 6.4.2 A particular axisymmetric MHD solution simulating a loop-like mass ejection. The top panel shows magnetic field lines projected on the meridional plane at three instants of time. The boundary $r = \zeta_c \Phi$ is a contact surface. The lower two panels show the distributions, at the same respective instants of time, of the density ρ and density excess $\Delta\rho$, as described in the text.

analysis. The time-development of $\Delta\rho$ is shown in the third panel in Figure 6.4.2. We identify a density enhancement ($\Delta\rho > 0$) in the form of a loop structure in the r, θ plane, enclosing a cavity of density depletion ($\Delta\rho < 0$). In the time development, the top of the "loop" fades away, leaving the appearance of a pair of radial "legs". The legs tend towards a fixed angular displacement between them, a phenomenon commonly observed for loop-like mass ejections. Eventually, the density depletion cavity rises. and a density enhancement takes its place. This and other illustrative MHD solutions are constructed and analyzed in Low (1984a).

Self-similar MHD solutions are being generated both for comparison with specific mass ejection observations and for studying their basic physics. Since these solutions can be constructed analyticlly, they also provide an opportunity to test the accuracy of existing numerical codes. In the following, we point out two interesting physical implications which have emerged from the study of self-similar MHD. The form of radial velocity given by Equation 6.4.1 is specialized. Realistic mass ejections exhibit small motions in the θ-direction (e.g., Hildner, 1977). The radial velocity field of Equation (6.4.1) must be regarded as a lowest-order description. In fact, self-similar solutions often are lowest-order descriptions of general, non-self-similar solutions in regions of space-time which are not sensitive to the influence of boundary and initial conditions (Barenblatt and Zel'dovich, 1972). It seems worthwhile to look quantitatively for self-similar behaviors in observed mass ejections; see the study by Illing (1984).

The time development of self-similar flows is dictated by the function fixed through Equation (6.4.2) by prescribing the constants η and α. Only positive values of η are admissible for unbounded expansion flows. For these flows the sign of α dictates an accelerating ($\alpha > 0$) or decelerating ($\alpha < 0$) flow. In all cases, the flow rapidly becomes inertial so that each plasma parcel sees no net force and cruises at constant speed. There is a spacially and temporally varying (Eulerian) velocity field, since different plasma parcels may move with different inertial speeds. The value of α then controls the magnitude of the terminal speeds of the cruising plasma parcels. The value of α is not bounded below so that, in principle, the expansion flow can have extremely low inertial speeds. These properties allow us to understand for the first time why most mass ejections are observed to have constant speeds above 2 R_0 and why their constant speeds can cover a broad range, from below 100 km s^{-1} to about 1000 km s^{-1} (Gosling et al., 1976; Rust, Hildner, et al., 1980; MacQueen, 1980). It appears that such inertial states are preferred asymptotic states, irrespective of initial conditions. In the inertial states, Lorentz forces and pressure gradients act to balance gravity so that ballistic deceleration of mass ejections is seldom observed. Notice that the construction of a self-similar MHD solution involves solving Equation (6.4.7) the pseudo-magnetostatic problem and Equation (6.4.5) the time-evolution problem, as two independent steps. Thus, an infinite variety of plasma and magnetic structures, generated by Equation (6.4.7), can evolve from the same velocity field given by Equation (6.4.5). This suggests that the dominant effect in a mass ejection may be a global outflow due to gravitational instability (Low, 1984a). In such an outflow, an entire coronal structure expands and is carried outward. That the corona is gravitationally unstable is not new. We know from the study of solar wind that efficient heating and thermal conduction, in the absence of adequate interstellar pressure to confine the corona, are natural causes for an expanding corona (Parker, 1963). Magnetic fields provide the only means to halt this expansion for localized low coronal regions through the tension force of field lines anchored in the dense photosphere. Various effects, for example the occurrence of a flare, can cause the magnetic tension force to be inadequate to hold down a coronal structure. The coronal structure then expands outward. In such a picture of the mass ejection, density-enhanced and density-depleted features are different parts of a single expanding structure and can have a different dynamical relationship than might be postulated for a compressional wave and its trailing rarefaction wave.

6.4.4 Post-Flare Coronal Arches Imaged in > 3.5 keV X-rays

The Hard X-ray Imaging Spectrometer aboard the SMM (van Beek et al., 1980) detected gigantic post-flare arches in the solar corona that are seen in X-rays above 3.5 keV and extend along the $B_{\parallel} = 0$ line to altitudes between 10^5 km and 2×10^5 km. So far we have detected a number of arches of this kind, of which four have been studied in detail: on 22 May 1980, following the flare of 2055 UT on 21 May; on 6 November 1980, following flares at 0329 and 1444 UT; and on 7 November 1980, following the flare at 0444 UT. The relationship between these arches and the similarly-sized loops which brighten just prior to flares discussed in Sections 6.3.3.1 and 6.3.4 for two events, is not clear. In this section, the post-flare behavior of large arches is considered.

The basic characteristics of the arches can be summarized as follows:

(a) All flares that produced or revived a coronal arch were flares associated with a Type IV radio burst (mostly accompanied by Type II). Flares of this kind have the form of two bright ribbons in H-alpha at footpoints of a growing system of loops (e.g., Svestka, 1976). The existence of bright ribbons and/or growing loops has been confirmed for all flares that produced the arches and for which H-alpha pictures were available. Other flares in the same active regions, even of H-alpha importance 2B in one case (6 November, 1726 UT), did not produce or revive the arch.

As the radio Type IV burst decays, the continuum becomes noisy, and the burst changes into a long-lasting type I noise storm. Thus, the arch can be considered to be an X-ray image of the lowest part of the (stationary) Type IV burst in its early phase, and of a Type I noise storm in the later phase of its development (examples in Svestka, 1983).

(b) The arch observed after the flare of 21 May 1980 (Svestka et al., 1982a) was stationary, its brightness maximum staying at a constant projected distance of 10^5 km from the $B_{\parallel} = 0$ line (corresponding to an altitude of $\sim 1.5 \times 10^5$ km). There was no other two-ribbon flare in that region: the arch was an isolated feature formed during the flare, and it ceased to be visible 11 hours later.

(c) In contrast to that, in the three consecutive arches on 6 and 7 November 1980, shown in Figure 6.4.3 (Svestka, 1984), brightness maxima

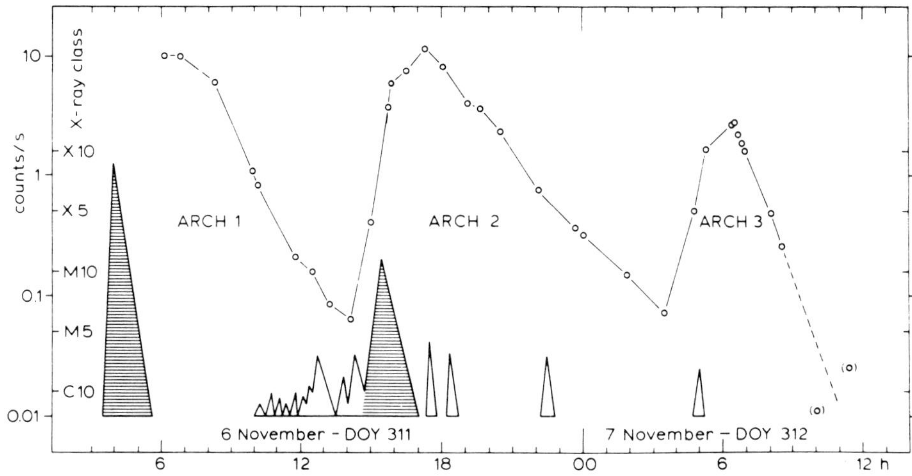

Figure 6.4.3 Time variations of the brightness of the coronal arches on 6 and 7 November 1980: maximum counts per second in one pixel of HXIS coarse field of view (32″ × 32″) in the energy range from 3.5 to 5.5 keV. The triangles below indicate 1-8 A X-ray variations in the active regions below the arch (GOES-2 data, scale C6-X10). The parent flares of the arches are hatched.

moved upwards, eventually disappearing from HXIS field of view (Figure 6.4.4). Arches 2 and 3 were clearly revivals of the preceding arch following new two-ribbon flares in the active region (cf., the shaded triangles in Figure 6.4.3). However, there was another major two-ribbon flare at 1341 UT on November 5. We surmise that this flare produced the first arch of the series (while HXIS looked at another region) so that even arch 1 in Figure 6.4.3 was probably a revival. This supposition is supported by the striking homology of arches 1 and 2 in Figure 6.4.4 (arch 3 was significantly weaker).

(d) A detailed analysis of HXIS images reveals that there were two velocity components in the arches (Figure 6.4.5): a slow one, with 8-12 km s^{-1} in projection on the plane of sky, and a fast one, with 35 km s^{-1} in projection. The slower speed is related to the rise of the brightness maxima, whereas the source of the fast component remains unknown.

Two other kinds of motion are also present: the 1-10 km s^{-1} growth of post-flare loops below the arch, and a possible coronal transient (no observations available) with a speed of several hundred km s^{-1}. Thus, the post-flare velocity pattern in the corona is extremely complex.

(e) In arch 2, which HXIS observed from its beginning, temperature (~ 14 x 10^6 K) peaked ~ 65 minutes, X-ray counts peaked about 2 hours, and emission measure maximized ~ 3.5 hours after the onset of the arch revival. Thus, the coronal arch looks like a magnified flare, with scales both in time and size increased by an order of magnitude.

(f) At the altitude of 10^5 km (mean of six integrated HXIS pixels) the maximum electron density was n_e ~ 2.5 x 10^9 cm^{-3} and energy density E ≈ 11.2 erg cm^{-3}. This leads to a total energy content of the arch of 1.2

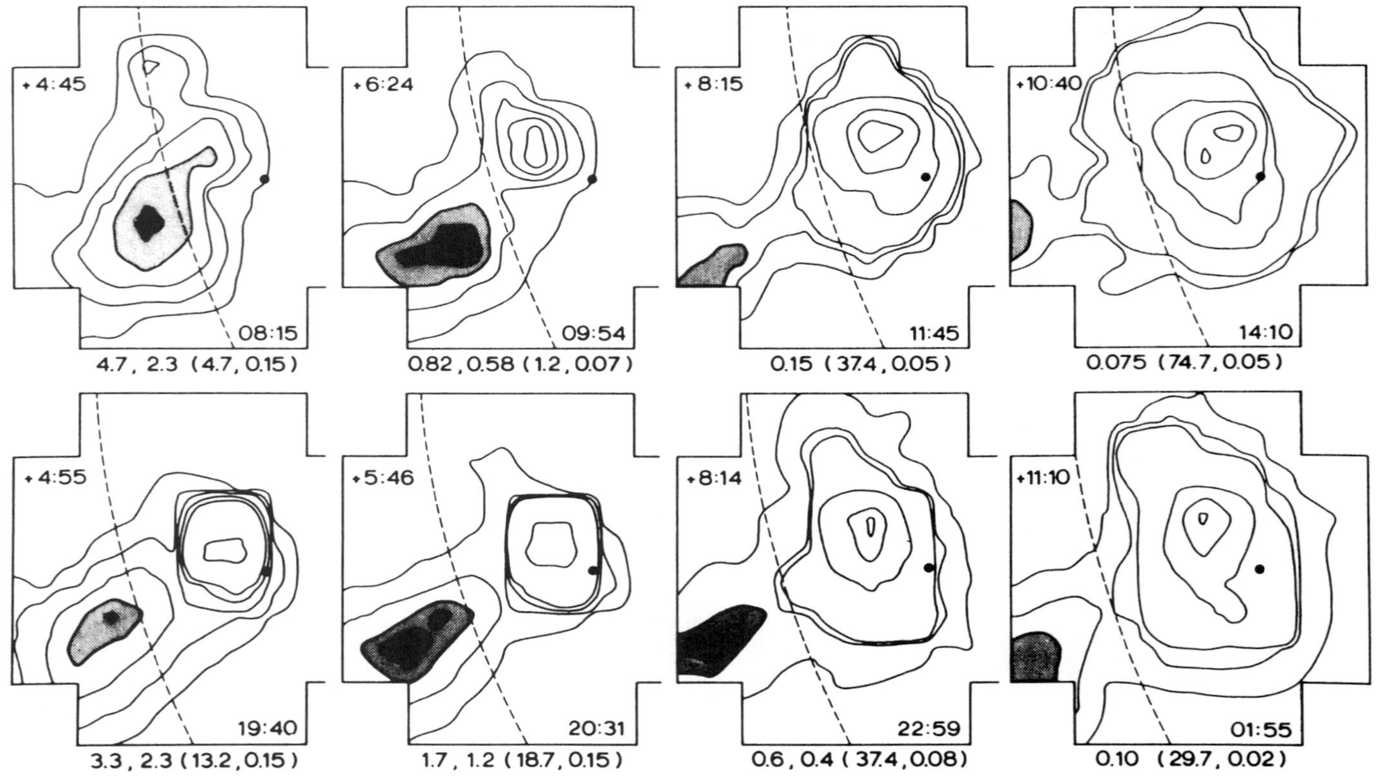

Figure 6.4.4 3.5-5.5 keV images of the arches 1 and 2 of Figure 6.4.3. The dashed line is the solar limb; a black dot marks the big spot in the parent active region. In each image the lower-right number is the time of imaging, while the upper-left number is the time elapsed from the onset of the parent flare. The numbers below each image are counts/second per pixel of HXIS coarse field of view corresponding to the brightest (shaded) contour(s) in the arch and (in brackets) the maximum and minimum contours in the image.

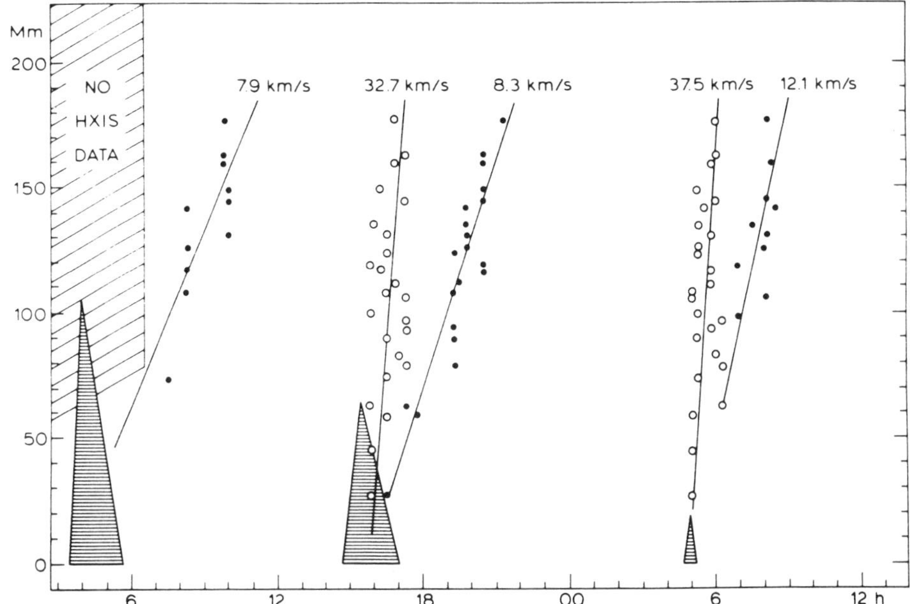

Figure 6.4.5 Velocities on 6 and 7 November 1980, deduced from a set of 25 pixels of HXIS coarse field of view. Each point (or circle) shows the time at which the arch brightness reached maximum in a HXIS pixel located at a given distance (in Mm) from the active region. The triangles (parent flares in X-rays) have been taken from Figure 6.4.3.

x 10^{31} erg at the time of observed maximum energy density and to a total mass of 4.4×10^{15} g.

The arch must cool through radiation, with conduction cooling inhibited. If uninhibited conduction cooling occurred for the duration of the arch, then the arch's total energy content would exceed that of the parent flare below it.

(g) Even at the altitude of 10^5 km, the temperature began to rise at the very onset of the flare that revived the arch. This is evidence that the whole extent of the pre-existing arch was fed with energy from the beginning of its revival. Thus, the revival apparently implies a refilling of the pre-existing arch with heated plasma.

In addition to the revived arch, the images in Figure 6.4.4 and the velocity patterns in Figure 6.4.5 reveal that a new feature was formed in the low corona and propagated upwards. An analysis of the emission measure and temperature reveals that the travelling disturbance was predominantly a temperature enhancement.

(h) Maximum temperature in the arch is always observed slightly above the site of the maximum brightness and rises with a similar speed. This indicates a decrease in density with increasing altitude.

In temperature maps, the rising temperature wave can be studied even in the early phase of the arch development, when the image of the arch cannot be easily separated from the image of the nearby flare. For arch 2 on 6 November we get $v \approx 7.4$ km s^{-1} as the speed of the thermal wave during the period 40 to 80 min after the flare's impulsive peak (compared to ~ 8.3 km s^{-1} for the later period shown in Figure 6.4.5). In a similar way, for the same period of time after the flare, we find a thermal wave with v ≈ 6.3 km s^{-1} for the flare of 4 June 1980. HXIS looked at this flare for one orbit only; still, the existence of a temperature wave moving upwards with a speed similar to the 6 November event indicates strongly that this flare was also the source of a post-flare arch. This offers an attractive possibility to infer the existence of more arches from HXIS data, even when the arch itself was not imaged as HXIS looked elsewhere on the Sun after the flare was over.

Characteristic a. shows clearly that the arch must be a natural component of a two-ribbon flare, formed or revived during the flare formation; at the same time as flare loops are formed below (sometimes for hours after the flare onset), the arch originates in the corona above them. This can be explained well by the Kopp and Pneuman (1976) model of the post-flare loop formation in Sturrock and Smith's (1968) configuration, through sequential reconnection of distended field lines. Each field-line reconnection produces a flare loop below and an elliptical closed field formation above. If the reconnecting field is sheared, the upper formations become interconnected and form a coiled structure over the $B_{\parallel}$ = 0 line (Anzer and Pneuman, 1982; Svestka et al., 1982a; Pneuman, 1983). There is a long sequence of such reconnecting coils that merge, mix, partly reconnect, and eventually give rise to a very complex magnetic field above the flare site; this field confines plasma, excited in the reconnection process, which is seen in X-rays as the post-flare arch. The magnetic field complexity also explains why the conductive cooling is inhibited (characteristic (f) noted above).

If the arch still exists when a new flare of the same type occurs below it, the arch field structure becomes an obstacle for the newly distending field lines. The loop distention at the onset of the flare is stopped, and at least some loops reconnect with the coils in the arch (Svestka, 1984). Thus, particles accelerated in the reconnection process and heated plasma begin to have free access into the pre-existing arch, raising its temperature: the arch revival begins. However, the field lines that reconnect first with the old arch are the outer field lines (the highest ones), most distant from the $B_{\parallel}$ = 0 line. Only later does reconnection of the lower loops occur; and only then can the Kopp and Pneuman mechanism begin to work, since the lowest lines reconnect back first. This really seems to be confirmed in arch 2 on 6 November, where no loop system was seen in the flare during the first 33 - 40 minutes of its development, though temperature in the arch was increasing (Svestka, 1984). A new arch begins to be formed, perhaps, as the upper product of the loop-reconnection process, after the first post-flare loop appeared and the velocity fields found in Figure 6.4.5 may reflect this formation.

A difficult problem to be solved is the relation of these stationary or semi-stationary coronal arches to whitelight coronal transients and coronal mass ejections (CMEs). On 21 May 1980, the preflare filament was greatly activated, but it did not rise and never fully disappeared. Instead, a powerful outburst of dark material was seen to leak southwards from its eastern end. The whitelight coronagraph of NRL observed a CME in the extension of this dark outburst, but it was regarded as an atypical event that could not be included in any well-known category of CMEs (McCabe, Howard, and Svestka, 1985). On the other dates when HXIS recorded the arches, outer coronal images were not available. However, the SOLWIND coronagraph imaged a typical post-CME coronal structure above the eastern limb a few hours after the flare that had produced arch 1 on 6 November (Howard, 1984). This flare produced one of the most extensive Type IV bursts ever seen at Culgoora, as well as a strong Type II burst (Stewart, 1983). Thus, it looks very likely that a CME was associated with this flare. Apart from it, Type II bursts accompanying four out of the six parent flares producing arches signify shocks moving through the corona and are generally associated with CMEs. Thus, though the evidence is only circumstantial, both a post-flare arch and a CME can apparently originate in one and the same flare. There also seems to be radio evidence for associated CMEs: in many flares one can see both moving and stationary Type IV bursts. The moving burst may be related to a mass ejection, whereas the stationary burst is imaged as an arch in X-rays while gradually changing into a Type I noise storm lasting (like the X-ray arch) for hours.

The possible co-existence of mass ejections and semi-stationary post-flare arches is intriguing, particularly in light of the pre-flare brightening of large coronal loops in association with CMEs, as discussed in Section 6.3. We first thought that a transient may be ejected first and only then may an arch begin to form. However, if the revived arches are associated with transients, which seems to be indicated, this explanation cannot be true; the transient and the arch must involve different parts of the magnetic field above the active region.

6.4.5 Extension of the Reconnection Theory of Two-ribbon Solar Flares

The magnetic reconnection theory for the "decay phase" of two-ribbon flares, as developed originally by Kopp and Pneuman (1976) and subsequently by Pneuman (1980, 1982), Cargill and Priest (1982), and others, is generally regarded (Svestka et al., 1980; Pallavicini and Vaiana, 1980) as providing a comprehensive and self-consistent description of the relationships between a wide variety of flare-associated phenomena -- filament eruptions, H-alpha-ribbon brightenings and separations, hot (X-ray) and cool (H-alpha) flare-loop growth, and nonthermal particle generation and storage. Briefly, the theory hypothesizes that a two-ribbon flare is the visible manifestation of magnetic reconnection in the corona above the flare site, the stressed open-field structure within which this reconnection occurs having been created immediately beforehand by a filament activation/disruption and coronal transient. The excess magnetic energy of the distended field is released (rapidly at first and more gradually as the flare progresses) as reconnection allows a lower energy configuration containing closed magnetic loops to form. The field-annihilation process supplies the diverse energy requirements of the flare itself -- enhanced optical emissions, dynamical mass motions, energetic particle releases, etc.

In the following, we extend the analytical basis for the reconnection model by developing a more general representation of the time-dependent magnetic configuration in the flare region than has heretofore been attempted. We will take explicit account of the fact, largely neglected in previous analyses, that the physical size of the hot flare loops generally increases as the flare progresses. The modified theory yields an analytic expression for the temporal variation of the average thermal energy density of the plasma on these closed loops. To compare with earlier studies, we apply the formalism developed here to the large two-ribbon flare of 29 July 1973, for which Skylab-era observations are available.

In previous studies the reconnecting field geometry was assumed to be dipolar near the Sun, but to undergo a smooth transition to a radial field at some radius $r_1(t)$ identified as the height of the neutral point. The dipole was located either at the solar center (Kopp and Pneuman, 1976) or at some specified distance beneath the solar surface (Pneuman, 1980, 1982). In neither case is the detailed analysis adequate: in the first case the global scale of the field is simply too large, whereas in the second an application of the axisymmetric dipole equations to a displaced (from sun center) dipole yields erroneous analytical results.

Large two-ribbon flares generally occur in mature or decaying active regions containing a magnetic neutral line oriented roughly in the east--west direction; but see Trottet and MacQueen (1980). For this reason, we choose a field representation which retains axial symmetry about the solar rotation axis ($\partial/\partial\phi = 0$), but which possesses a high degree of spacial structure in the latitudinal direction. Specifically, we seek a field which at any time t:

(a) is potential between the solar surface ($r = r_0$) and the neutral point level ($r = r_1$);
(b) extends radially outward beyond $r = r_1$; and
(c) has always the same magnetic flux distribution at the solar surface (i.e., an invariant $B_r(\theta)$ at $r = r_0$), since observations indicate that major field rearrangements do not occur at the photospheric level during large flares.

The field that satisfies conditions (a)-(c) is found by solving Laplace's equation $\nabla^2\Phi = 0$, where $B = \nabla\Phi$ in the region (r_0, r_1), subject to the condition that $B = \frac{1}{r}\frac{d\Phi}{d\theta} = 0 \quad r = r_1$ ($B_\phi = 0$ automatically by the assumed symmetry). For $r > r_1$, $B_\theta = 0$, and B_r declines outward as r^{-2} from its value at $r = r_1$. It suffices to consider fields proportional to a single Legendre polynomial $P_n(\theta)$, giving rise to n "lobes" of field lines between $\theta = 0°$ and $\theta = 180°$, each lobe being bounded latitudinally by radial lines along which $({}^{dP}n/d\,\theta)$ $(\sim B_\theta) = 0$. For appropriate n, one lobe will span the latitudes covered by a given active region. For example, Figure 6.4.6 shows selected field lines for the sixth lobe (from the north pole) of a P_{17} field, for several values of the neutral point height $y = r_1/r$. This particular lobe is centered at colatitude 59.1° and has a latitudinal width of 10.3°.

The total magnetostatic energy $\int(B^2/8\pi)\,dV$ of a single lobe, per radian of longitude, as a function of the neutral point height is:

$$E^{(n)}(y) = \frac{\beta^2 r_o^3 y^{2n+1}}{8\pi} \{(n(n+1) + \tag{6.4.8}$$

$$(n+1)y^{2n+1} - n/y^{2n+1}\} I_{12}(n)$$

$$I_{12}(n) = \int_{x_1}^{x_2} P_n^2(x)dx \tag{6.4.9}$$

$$\beta = B_o/\{n + (n+1)(r_1/r_0)^{2n+1}\} \tag{6.4.10}$$

where the limits of integration x_1 and x_2 correspond to the latitudinal boundaries of the lobe (i.e., the points at which $dP_n/d\theta = 0$) and B_o is the field strength at the poles ($\theta = 0°, 180°$). This stored energy decreases as the neutral point rises higher in the corona:

$$\frac{dE}{dy} = -2n(n+1)(2n+1)^2 b^2 y^{2n}(y^{2n+1}-1)\ \{n + (n+1)y^{2n+1}\}^{-3}, \tag{6.4.11}$$

where $b^2 = r_o^3 B_o^2 I_{12}(n)/8\pi$. We identify $-dE/dt = -(v_n/r_o)dE/dy$, where $v_n = dr_1/dt$ is the upward velocity of the neutral point, as the rate at which energy "reappears" in the flare plasma in an observable form. If we assume that some fraction, f, of this liberated energy is used to supply the thermal energy density, e, of newly formed flare loops, then we can write

$$e = 3N_e kT = -f\frac{dE}{dr_1} \left(\frac{dV}{dr_1}\right)^{-1}, \tag{6.4.12}$$

where dV/dr_1 is the rate at which the volume of the loop system, V (per radian of longitude), grows as the neutral point rises. Let L denote a characteristic arc length over which the enhanced thermal energy of a newly formed loop is significant. Then we can write $dV \approx L\, dr_1\, r_1 \sin\langle\theta\rangle$, where $\langle\theta\rangle$ is the mean latitude of the loop system. For the reasonable assumption that L is twice the loop height, i.e., $L = 2r_0(y-1)$, Equations (6.4.11) and (6.4.12) yield

$$e = \frac{n(n+1)(2n+1)^2 B_o^2 I_{12}(n) f}{8\pi \sin\langle\theta\rangle} \left(\frac{y^{2n-1}}{y-1}\right) \left(\frac{y^{2n+1} - 1}{\{n+(n+1)y^{2n+1}\}^3}\right). \tag{6.4.13}$$

To apply Equation (6.4.13) to a particular flare, we have to know the variation of neutral point height with time. For flares near the solar limb, this information may be derived directly by measuring the height of the highest (i.e., most recently formed) X-ray loops, since the

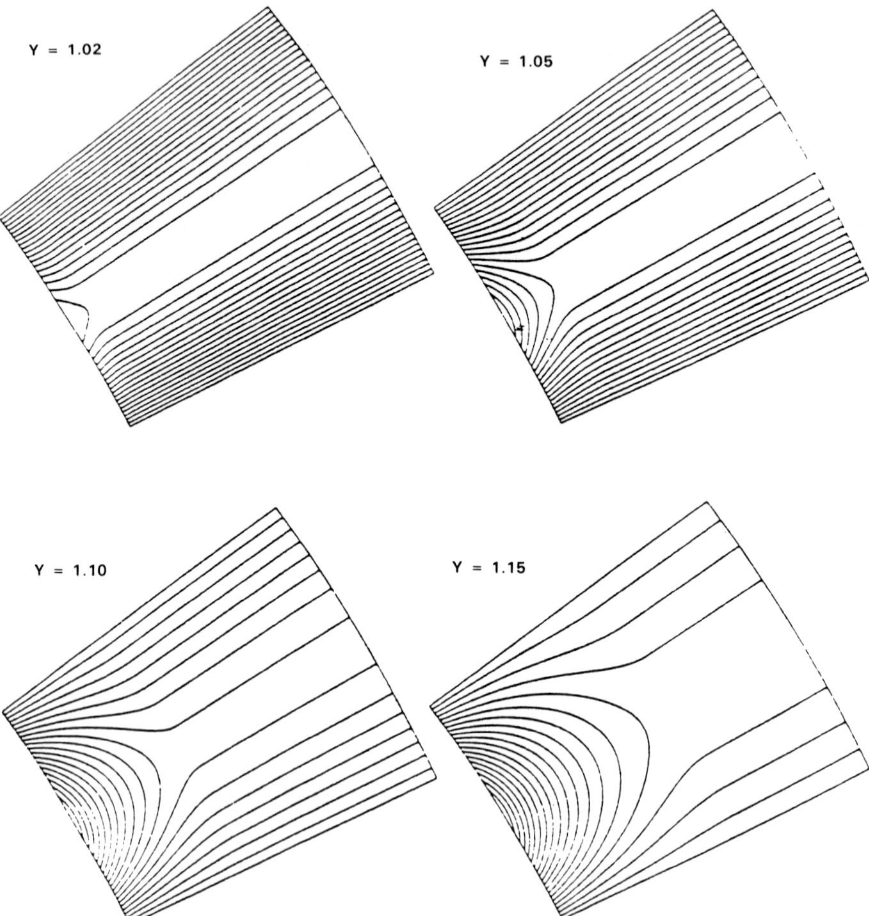

Figure 6.4.6 Selected field lines for the sixth lobe of a P_{17} field as a function of the height of the neutral point (source surface).

neutral point supposedly lies immediately above. However, for flares near disk center we recall that the highest X-ray loops visible at any time appear to have their footpoints anchored in the leading edges of the expanding H-alpha ribbons. Using the field representation developed above, we can calculate numerically the neutral point height, y, which gives a magnetic separatrix which intersects the solar surface with a footpoint separation corresponding to that of the ribbon's leading edges at any time. As a test of this procedure, we present the results of such a calculation for the well-observed disk flare of 29 July 1973; this classical two-ribbon flare occurred sufficiently far away from disk center that both ribbon separations and loop heights were simultaneously measurable during the later stages of the event and are available in the published literature (Moore et al., 1980).

The latitude and size of this flare suggests that the eighth lobe of a P_{18} field is appropriate. We adopt the values of Moore et al. (1980) for the average separation of the leading edges of the H-alpha ribbons (cf., their Figure 8.4) and for the height of the brightest X-ray emission (their Figure 8.5). From these data we derive the empirical relationship between hot-loop height and ribbon separation shown by the filled circles in Figure 6.4.7 (error bars from Nolte et al., 1979). During the time interval 16:43-21:41 UT, the loop-height observations are fit well by the theoretical run of neutral point height versus separatrix-footpoint separation.

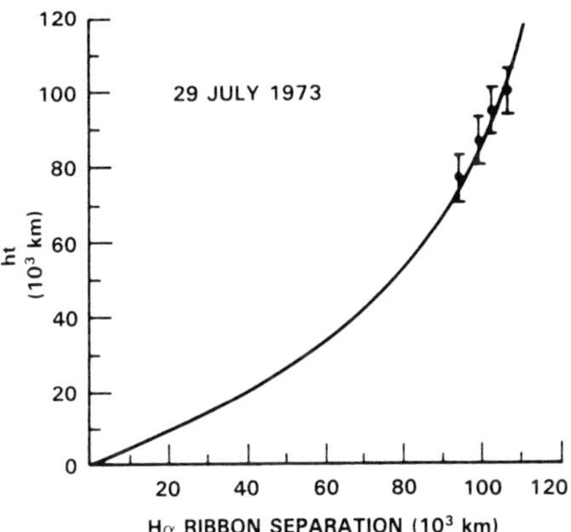

Figure 6.4.7 The heights of the hottest loops seen in X-rays appear to incresae as the H-alpha ribbons separate; the filled circles correspond (left to right) to the observed loop heights at 1643, 1821, 1958, and 2141 UT, respectively. The solid curve depicts the height of the neutral point as a function of the separatrix-footpoint separation for the magnetic field configuration described in the text.

The goodness of fit thus obtained suggests using the theoretical curve to extrapolate the empirical relationship to earlier times in the flare, when soft X-ray images are not available. Combining this extrapolation with the ribbon-separation observations of Moore et al. (1980) yields the neutral point history shown by the filled circles in

Figure 6.4.8; the smooth curve represents an analytical fit to these results, namely

$$y(t) = 1.06704 + 2.927 \times 10^{-2} \log (t-t_o), \qquad (6.4.14)$$

where t is in hours (UT) and $t_o = 13.078$.

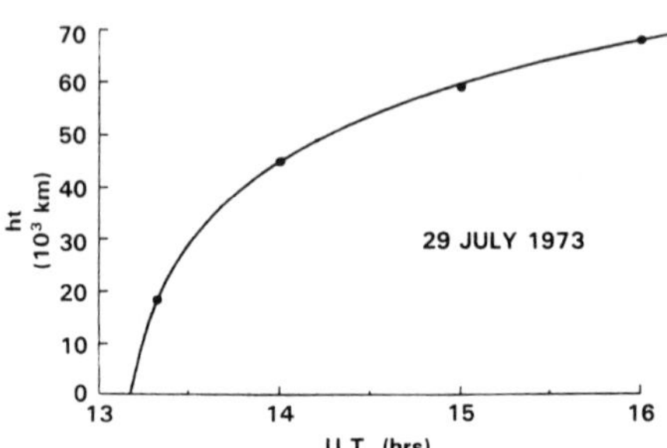

Figure 6.4.8 Temporal variation of the height of the neutral point for the 29 July 1973 flare, as calculated from Equation (6.4.22).

The solid curve in Figure 6.4.9 shows the temporal variation of plasma energy density as calculated from this fit and Equation (6.4.13). The parameter values assumed were B_o = 1600 G (which gives a peak field strength in the magnetic lobe representing the flare region of about 300 G; cf. Michalitsianos and Kupferman (1974)) and f = 0.003. This value of f was chosen to fit the energy density profile (vertical bars in Figure 6.4.9) calculated from the flare-plasma temperature and density values as determined from SOLRAD full-disk data as revised by Svestka et al. (1982). The small value thus derived implies that only a tiny fraction of the total magnetic energy stored in the active region corona apparently need be used to account for the presence of hot plasma on flare loops.

In this project we emphasize two major points pertaining to the reconnection theory for two-ribbon flares. The first is that the theoretical magnetic field model should have a spacial scale and orientation corresponding to that of the active region within which the flare occurs. Most large two-ribbon flares occur in active regions characterized by a bipolar field configuration oriented more-or-less in the north-south direction. To meet this requirement, we have chosen to represent the observed photospheric field distribution in and around the flare site by a single high-degree term of the Legendre-polynomial series expansion that comprises the general solution of Laplace's equation. Also, we have imposed the additional boundary condition that the field lines become radial at the level of the neutral point (more correctly, neutral "line") which rises into the corona as reconnection proceeds, and that they remain radial beyond this level (source-surface model). The

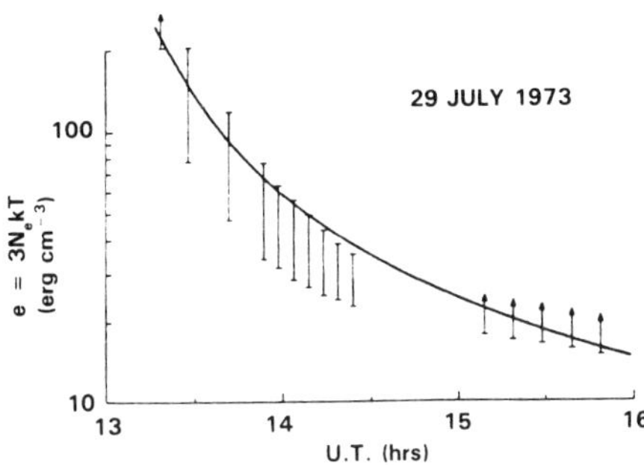

Figure 6.4.9 Thermal energy density of plasma on closed loops for the 29 July 1973 flare. The data (vertical bars) were taken from Svestka *et al.* (1982), whereas the smooth curve was calculated from Equations (6.4.13) for the parameter values given in the text.

resulting field contains volume currents everywhere above the source surface and a current sheet extending radially outward from the neutral point itself. There is no necessity to align the axis of symmetry with the rotational axis since we confine ourselves to a finite sector in longitude. For H-alpha ribbons not in an east-west orientation, we can relocate the axis of symmetry so that the field lobe of interest runs parallel to the observed H-alpha ribbons.

The second point is that, as the flare progresses, the new closed loops which are being formed continually via reconnection represent an increase of total volume of the loop system with time. A proper accounting of this growth is necessary if one is to relate the net rate of reconnection-liberated energy to the energy density of the plasma trapped on these closed field lines. Our approximate treatment of the increasing loop dimensions leads to an expression for the hot-plasma energy density which declines monotonically with time, once the reconnection phase has commenced. This is in good agreement with observations for the 29 July 1973 flare, thereby representing a major improvement over the results of earlier analyses (e.g., Pneuman, 1980) which indicated that the energy density should increase rapidly from zero at flare onset to a maximum value a short time after the onset of reconnection.

Finally, only a very small fraction (0.003) of reconnection-liberated magnetic energy was needed to account for the thermal energy density on the hot loops of the 29 July 1973 flare. This shows that the postulated field configuration contains much more energy than is actually needed to supply these losses. From Equation (6.4.8), when the neutral point rises from the solar surface ($y = 1$) to infinity, the total energy released per radian of longitude is found to be

$$\Delta E_{12}(n) = \frac{r_o^3 \, I_{12}(n) \, B_o^2}{8\pi} \left(\frac{n}{n+1}\right) , \qquad (6.4.15)$$

and for the parameters n = 18, B_o = 1600 G, and $I_{12}^{(n)} = 2.93 \times 10^{-3}$ representing the 29 July 1973 flare, it follows that $\Delta E_{21}^{(n)} \cong 10^{35}$ erg/radian. The longitudinal extent of the active region was about 2 x 10^5 km, so that the net available magnetostatic energy is about 3 x 10^{34} erg. In actuality, the field at flare onset may not be opened up all the way down to the solar surface, and the neutral point will not rise to infinity. Both of these effects tend to reduce the energy actually accessible by reconnection. Even if the range of neutral point heights is limited to that shown in Figure 6.4.8. (i.e., 18,000-70,000 km), it can be shown from Equation (6.4.8) that about two-thirds of the above energy, or 2 x 10^{34} erg, will be liberated. This rather large energy -- two orders of magnitude higher than the observed flare losses -- is consistent with the small fraction found necessary to account for the hot-loop plasma.

We deem it unlikely that future refinements in the potential-field/source-surface model presented in this paper will be of a sufficient magnitude to remove the energy disparity noted above. A more promising possibility is that the rate of reconnection is physically limited to values lower than those otherwise desired by external inflow conditions; i.e., the reconnection is "forced". A restricted merging rate could produce a post-reconnection, three-dimensional, field configuration (e.g., force-free) after only a small fraction of the energy of the original open field has been shed.

6.4.6 Linear Stability of Magnetostatic Coronal Structures

In the low corona, where the solar wind may be neglected, long-lived structures are in approximate magnetostatic equilibrium. These structures may evolve quasi-steadily. Calculations have been presented to suggest that a quasi-steady evolution may terminate in an unstable equilibrium state, whereupon the coronal structure appears to break spontaneously into a dynamical state (Low, 1981; Wolfson, 1982). Such a process may be the origin of some coronal mass ejections, particularly those associated with eruptive prominences without flares (e.g., Low, Munro, and Fisher, 1982). It is worthwhile, therefore, to study what kind of magnetostatic states are stable, and what others are not, in the low corona. Problems in MHD stability are formidable, and our theoretical knowledge is limited because the stability analyses can be carried out only for equilibrium states of the simplest geometries. There are two main technical difficulties. Equilibrium states of complex, but realistic, geometries usually cannot be written down in analytic forms to allow the usual expansion for the linearized perturbation equations. Even if the linearized perturbational equations were available, the equilibrium state description usually introduces such spacially-varying coefficients into the equations as to make the mathematical problem intractable. Low (1984b) presented a theoretical study in which these two difficulties are overcome and several analytic magnetostatic equilibrium states with known stability properties are made available. Magnetostatic equilibrium is described by the equation:

$$\frac{1}{4\pi}\left(\underline{\nabla} \times \underline{B} \times\right) \underline{B} - \underline{\nabla} p = \rho \frac{GM_o}{r^2}\hat{r} = 0 \,. \qquad (6.4.16)$$

Assuming axisymmetry, this equation can be reduced to a non-linear, elliptic, partial differential equation in the r-θ plane (Hundhausen, Hundhausen, and Zweibel, 1981; Uchida and Low, 1981). The following solution describing $\Upsilon = 6/5$ polytropic, axisymmetric atmosphere can be constructed by solving this elliptic equation:

$$\underline{B} = B_o \left(\left(\frac{\cos\theta}{r^4}\right)\hat{r} + \left(\frac{\sin\theta}{r^4}\right)\hat{\theta}\right) , \qquad (6.4.17)$$

$$p = \frac{1}{6}\left(W_o - W_1 \frac{B_o}{2} \frac{\sin^2\theta}{r^2}\right)\left(\frac{GM_o}{r}\right)^6 , \qquad (6.4.18)$$

$$\rho = \left(W_o - W_1 \frac{B_o}{2} \frac{\sin^2\theta}{r^2}\right)\left(\frac{GM_o}{r}\right)^5 , \qquad (6.4.19)$$

where W_o and W_1 are free constants and $B_o = \pi W_1 (GM_o)^6/3$. This analytic solution is displayed in Figure 6.4.10; the geometry is shown by the solid lines in panel (a). Superposed is a set of broken lines representing the potential magnetic field having the same normal flux distribution at the reference level $r = R_0$. Panel (b) shows contours of constant density. The axisymmetric density distribution is characterized by a density depletion low in the equatorial region. This equilibrium state may be visualized to have been formed in the following manner. Consider the initial equilibrium state in which an atmosphere is in spherically-symmetric hydrostatic equilibrium with the potential magnetic field shown in panel (a) of Figure 6.4.10. If we slowly remove plasma from the low equatorial region, the reduced pressure in this locality will not be able to support the weight of the upper atmosphere. Adjustment to new equilibrium states results, and the atmosphere weighs upon the magnetic field, deforming it into the geometry of the non-potential field shown.

The global force balance of the equilibrium state in Figure 6.4.10 may be described in the following terms. The density depletion in the low equatorial region is buoyant in the stratified atmosphere. The lower pressure of the depletion region does not result in its collapse because of its locally large magnetic pressure. The buoyant depletion region is prevented from rising by the downward-acting magnetic tension force. In this equilibrium, gravity has its role to play through the buoyancy force.

The equilibrium state just described can be subjected to a vigorous stability analysis, using the energy principle of Bernstein et al. (1958), assuming rigid boundary conditions at $r = R_0$ and linear perturbations. The mathematical analysis in Low (1984b) establishes the following results. The equilibrium state contains two free parameters, W_o and W_1 in Equations (6.4.17)-(6.4.19). For all values of W_o and W_1 the equilibrium is unstable. However, instability appears only for perturbations with variations in three-dimensional space. The equilibrium state is stable to all axisymmetric perturbations. The detailed analysis shows that to

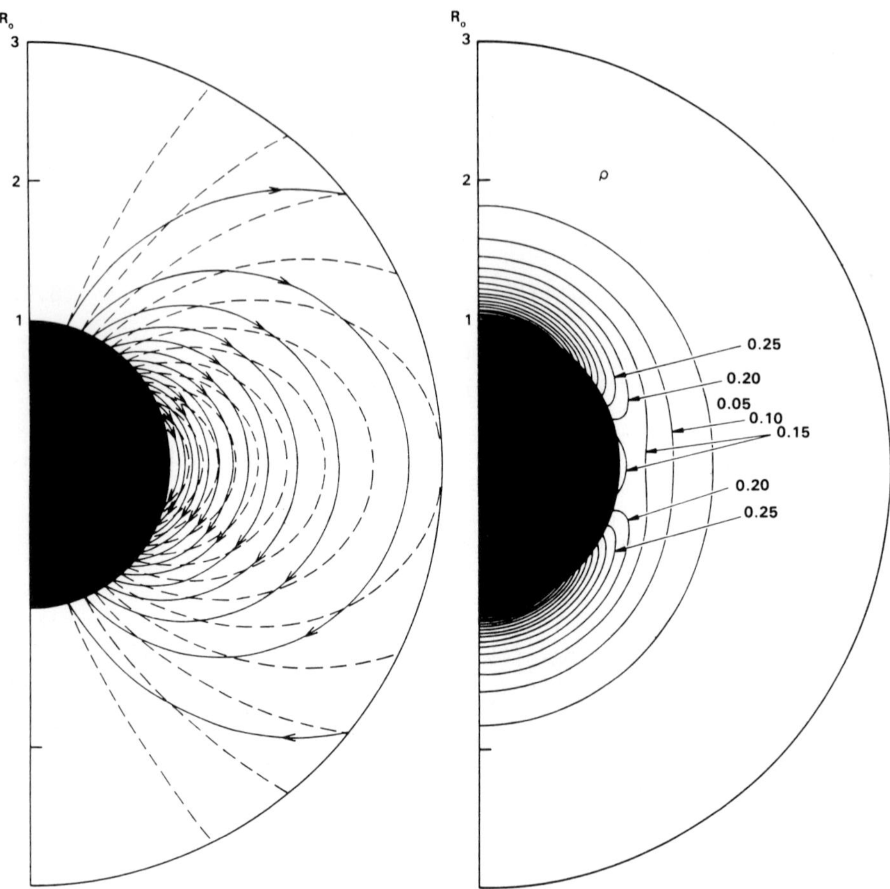

Figure 6.4.10 A particular axisymmetric magnetostatic solution. The left figure shows the magnetostatic field in solid curves and the associated potential magnetic field in broken curves. The right figure shows the distribution of the density ρ.

excite an instability, the perturbation must involve a displacement perpendicular to the magnetic field in the r-θ plane combined with a ϕ-dependent displacement out of the r-θ plane. An obvious example of this type of perturbation is one due to a particular magnetic flux tube rising outward in its r-θ plane. In so doing, the flux tube expands and pushes aside the neighboring magnetic flux tubes. In the three-dimensional development, we find strands of magnetic flux pushing their way out, destroying the axisymmetry of the initial equilibrium state. Finally, this instability is confined to magnetic field lines which subtend angles at $r=R_0$, between their pairs of footpoints, exceeding about 45°. Hence, if the equilibrium is perturbed, instability sets in for the larger magnetic arches that extend high up in the atmosphere. Other magnetostatic equilibria and their stability, as well as a discussion of their relation to mass ejections, can be found in Low (1984b).

Stability analyses establish properties on a firm basis whenever they can be carried out completely. Obviously, many more studies are needed before we can synthesize a broad understanding of stability properties of relevance to coronal structures. From this synthesis, we can hope that a theory will emerge as to why certain coronal structures may become unstable and develop into large scale outflows to be identified with the mass ejection phenomenon.

6.5 INTERPLANETARY EFFECTS OF CORONAL MASS EJECTIONS

6.5.1 Introduction

The mass, energy, and volume observed in the Skylab coronal mass ejections implied that the consequent interplanetary effects of CMEs should be similarly dramatic and readily observed at 1 AU. The association of interplanetary (Gosling et al., 1975) and coronal (Gosling et al., 1976) shocks with fast CMEs seemed to confirm this point of view. The realization that the mechanical energy released in mass motions dominates the easily observed radiative output of associated large flares (Webb et al., 1980) further enhanced this expectation.

The results to date have been rather surprising. The principal effects of CMEs -- interplanetary shocks and energetic particle events -- appear well associated with major CMEs, but the detection of CMEs themselves at 1 AU has remained elusive. Gosling et al. (1977) suggested that non-compressive density enhancements (NCDEs) observed in the solar wind were the interplanetary signatures of CMEs. Another attempt to detect CMEs at 1 AU has been carried out by Klein and Burlaga (1982), who observed "magnetic clouds", regions in which the magnetic field has rotated nearly in a plane, perhaps indicating a loop-like structure. A third approach has been to associate solar wind helium abundance enhancements (HAEs) with CMEs (Borrini et al., 1982). The lack of correlation among NCDEs, magnetic clouds, and HAEs, however, suggests that the interplanetary signatures of CMEs may show considerable variations (Borrini et al., 1982).

In the remainder of this section, we discuss the results of team members engaged in efforts to clarify the interplanetary signatures of CMEs and follow that with a review of work on associating CMEs, interplanetary shocks, and energetic particles.

6.5.2 Direct Detection of CMEs in Interplanetary Space

Klein and Burlaga (1982) surveyed hourly averages of interplanetary magnetic field and plasma data from 1967 to 1978 to search for "magnetic clouds." The clouds are defined as regions with radial dimensions of ~ 0.25 AU in which the magnetic field strength is high and varies so that the measured field direction rotates substantially, nearly parallel to a plane as the cloud passes over a spacecraft. These criteria sieved out 45 identified clouds, which Klein and Burlaga listed in three classes, depending whether each cloud followed a shock, preceded a stream interface, or was a cold magnetic enhancement. Each class had about the same number of clouds, and it was suggested that all clouds arose from the same basic physical cause but found themselves in different environments in the solar wind. The sums of the magnetic and ion pressures in the clouds were higher than the ion pressure outside, implying that the clouds were expanding. Klein and Burlaga suggested that these magnetic clouds might be the 1 AU manifestations of CMEs.

The Klein and Burlaga analysis did not include a search for the solar origin for each of these magnetic clouds. Wilson and Hildner (1984) undertook to search for the solar origins of the 35 clouds observed from 1971-1978. For each cloud they determined a temporal window, based on the maximum and minimum wind speeds observed within each cloud, during which the cloud must have departed the Sun. They then looked for solar activity which might serve as proxy indicators of the occurrence of a CME within each window. As proxy solar events for CMEs they examined H-alpha flares, Type II and IV radio events, radio GRF events and soft X-ray events listed in Solar Geophysical Data. The same proxy phenomena were also examined for temporal windows when no clouds were observed, i.e., for control periods. No significant proxy events were found, relative to the control periods, for the clouds associated with stream interfaces and cold magnetic enhancements. However, for six of the nine clouds associated with shocks, Wilson and Hildner found a Type II burst originating within 49° of central meridian during the event window. Of nine control windows, only three contained a Type II burst and each of those was at least 63° from central meridian. On the assumption that Type II bursts are well correlated with fast CMEs, which are likely to result in interplanetary shocks, Wilson and Hildner (1984) find support for the idea that fast CMEs are expelled nearly radially from the Sun to become shock associated magnetic clouds at 1 AU. Caution about the hypothesis is urged because of the very sparse statistics involved (six of nine cases) and because for three of the six good cloud associations the Type II bursts were associated with H-alpha subflares. The recent study of Kahler et al. (1984a), shows that about 2/3 of all Type II bursts associated with subflares are not associated with CMEs. It is discouraging that the solar signature of the large group of magnetic clouds not associated with shocks is still undetermined.

The attempts to observe CMEs at 1 AU through their particle or magnetic field signatures, as discussed above, lack the global perspective with which CMEs are first observed in coronagraphs. However, as mentioned in Section 6.2.1.8, Jackson has shown that there is a way to observe CMEs globally beyond 10 R_0. As first discussed by Richter, Leinert, and Planck (1982) the zodiacal light photometers on the two solar-orbiting Helios spacecrafts can be used to observe interplanetary plasma clouds and to determine their velocities. Each Helios experiment consists of three photometers pointed at ecliptic latitudes of 16°, 31°, and 90°, with Helios 1 viewing the southern hemisphere and Helios 2 the northern. The spin axes point to the ecliptic poles, and the 16° and 31° photometers sample 32 positions of azimuth through various color filters and polarizers (Leinert et al., 1975; 1981).

Jackson has shown that several CMEs observed with the SOLWIND and SMM coronagraphs can be tracked well into the interplanetary medium using the Helios data (Jackson et al., 1985a,b), and he has used a program to plot brightness contours (and excess density contours deduced from them) in a "fisheye" lens view that shows the shape and position of the CME as seen from Helios, relative to the Sun and Earth. For some CMEs seen in projection with coronagraphs in near-Earth orbit, this offers the possibility of a "stereoscopic" view of their shape and evolution. Figure pairs 6.2.21-6.2.22 and 6.5.1-6.5.2 show coronagraph and Helios views of the 24 May 1979 and 29 June 1980 mass ejections, respectively.

The Helios photometer data do have several important limitations which must be considered. CMEs cannot be observed until they are at least 15 R_0 away from the Sun, even under the most favorable circumstances. The best observations are obtained when the CMEs are several tenths of an AU distant from the Sun. This makes detailed comparisons with coronagraph observations difficult (Jackson et al., 1985a). In addition, the brightness variation due to the Thomson scattering angle and the distance of the CME electrons from the Sun is substantial, which leads to a considerable uncertainty in deconvolving the CME shape from the Helios observations. It is also necessary to assume the distributions of material along the photometer line of sight to determine the total mass in the field of view. Nevertheless, the Helios data allow one to determine the general size and position of a CME as it moves through the interplanetary medium. Further work should help to elucidate the hitherto elusive nature of CMEs at 1 AU.

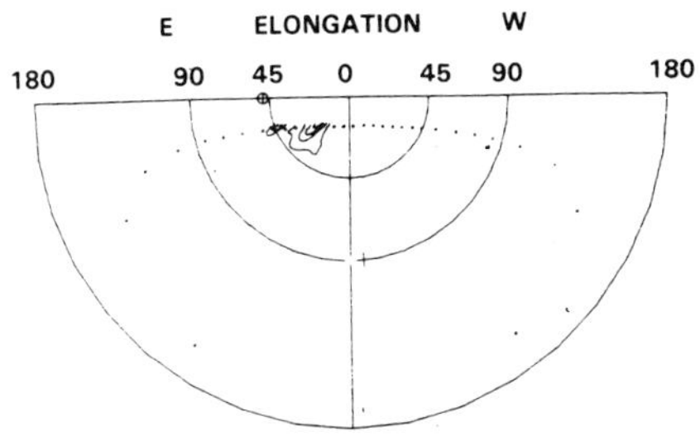

Figure 6.5.2 Helios 1 contour image of the 29 June 1980 mass ejection at 1800 UT 30 June 1980 in levels of 6×10^{12} g deg^{-2}. The Sun is centered, and the point behind the observer on the spacecraft, i.e., at 180° elongation is represented as the outer circle. The position of the Earth (⊕) and the solar pole are indicated relative to the ecliptic plane (horizontal line).

6.5.3 Interplanetary Shocks and Energetic Particles Associated with CMEs.

The large ($\gtrsim$ 1000 CMEs) SOLWIND data set has provided an excellent base for studying CME associations with interplanetary shocks and energetic particles. Using the Max Planck Institut's plasma detector on the Helios 1 spacecraft to detect interplanetary shocks, Sheeley et al. (1983b, 1985) carried out a direct comparison of CMEs and interplanetary shocks. For 80 shocks detected while Helios 1 was within 30° of the Sun's east or west limb and for which there were complementary SOLWIND observations, 40 appropriately-timed, well-associated, major CMEs were found.

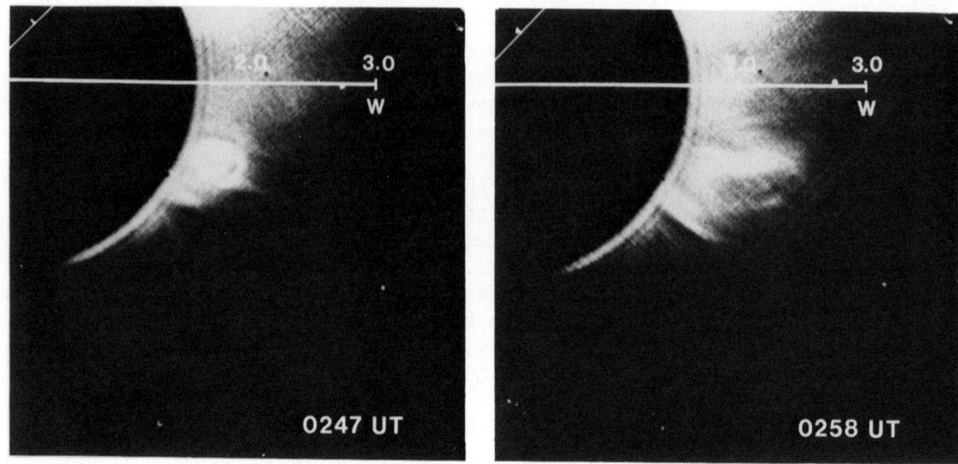

Figure 6.5.1a Coronagraph observations of the early 29 June 1980 mass ejection. (a) C/P observations. In these views of the low corona the ejection is clearly observed as a complex loop system. The coronagraph's occulting disk is at 1.6 R_0 and the field of view extends outward as indicated.

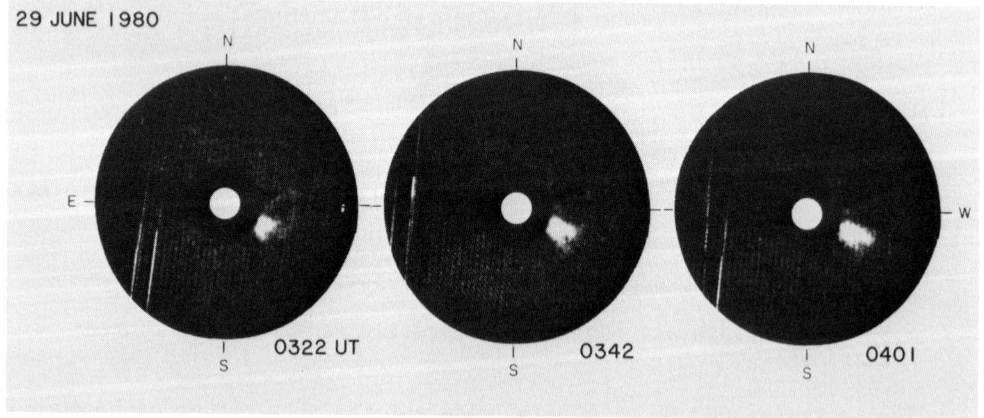

Figure 6.5.1b Coronagraph observations of the early 29 June 1980 mass ejection. (b) SOLWIND coronagraph difference images later in the event.

For the 40 remaining shocks, 19 "possible" associations were found, "possible" for a variety of reasons, and for 20 shocks associations with CMEs were "indeterminate". The associated CMEs generally encompassed the ecliptic plane and usually (but not always) had speeds exceeding 400 km s^{-1}. If one starts with CMEs and asks how many had Helios-observed shocks associated with them, the following answers emerge. There were: 27 major CMEs in Helios' hemisphere not initially associated with Helios shocks; 17 of these did not encompass the equator, i.e., no part of the CME went in the direction of Helios; the remaining 10 grazed or spanned the solar equator, and 7 of these failed to be associated with shocks, though 5 of these 7 were associated with disturbed flows, including NCDEs.

Although this initial association study suggested a strong CME/shock correlation, several questions remained. The first was the origin of the "possible" category of events, which consisted of one-quarter of the sample for which adequate data existed. Second, although the CME occurrence rate was 0.9 major events per day, (Howard et al., 1985), the rate for large, low-latitude CMEs on Helios' limb was only 0.15 per day, making unlikely the occurrence of many random, chance associations. As Sheeley et al. (1983b) pointed out, the shock-associated CMEs were generally bright, fast, low-latitude events not typical of CMEs in general.

As a test of their CME associations Sheeley and his colleagues (Sheeley et al., 1985) updated their original list of Helios shocks to 99 events and matched them and (for control) a comparable set of randomly generated "shock" times with their CME data (see Table 6.5.1). As the percentages in Table 6.5.1 show, when the indeterminate and poor Helios geometry cases are eliminated, there was close temporal association with a CME for 72% of all observed shocks, but an apparent association for only 16% of the random periods or control "shocks". In addition, where half the random "shocks" had no CME association, only one (2%) Helios shock was similarly unassociated. This result establishes that about 70% of the Helios shocks are, in fact, associated with CMEs, although the origin of the remaining 30% of the Helios shocks is unclear. In addition,

Table 6.5.1 CME Associations with Helios Shocks and Random Times

	Helios Shocks	Random "Shocks"
Yes	49 (72%)	7 (16%)
Possible	18 (26%)	15 (34%)
No or Doubtful	1 (2%)	22 (50%)
Indeterminate	21	40
Helios More Than 30° Away From Limb	10	15
	99	99

the relationship between coronal shocks, observed as metric Type II bursts, and interplanetary shocks is also unclear, despite the work of

Cane (1984); considerable work is still required to obtain a comprehensive view of these shocks. Results obtained by Kahler and his colleagues (Kahler et al., 1984b) suggest that the earlier picture that all prompt, i.e., flare associated, proton events are associated with CMEs (Kahler, Hildner, and van Hollebeke, 1978) needs some revision. They found that 26 of the 27 prompt proton events with identified H-alpha flares also had associated CMEs. The one exception, on 7 June 1980, is one of what may be a rare class of events which are associated with short duration, well-connected (W50-W90) flares with no CMEs. These unusual events tend to have γ-ray emission and strong Type III bursts, suggesting a prominent role for the impulsive phase in the production of interplanetary energetic protons despite their lack of CMEs.

A further complication has been found in the sources of the ^{3}He-rich energetic particle events. In these events the ratio ^{3}He/^{4}He exceeds 0.2 at 1.5 MeV/nucleon (Reames and von Rosenvinge, 1983). Sixty-six events have been found in the GSFC ISEE-3 experiment data from 1979 to 1982. Comparing these events with metric Type II bursts and with CMEs, Kahler et al. (1985b) found that there was no statistically significant association of these particle events with the radio bursts nor with CMEs. The accepted view that a preheating phase selectively energizes ^{3}He ions, which are then accelerated in a second, conventional particle acceleration process, implies that ^{3}He-rich events should be significantly associated with CMEs and Type II bursts through the second acceleration process. The scenario for ^{3}He ion acceleration now appears distinctly different than for the conventional proton events and perhaps different than the proton events from the short-duration well-connected flares. These results suggest that particle acceleration processes in the solar corona may be more diverse than previously supposed.

6.6 THE SLOWLY VARYING CORONA NEAR SOLAR ACTIVITY MAXIMUM

6.6.1 Introduction

The background against which we see the continual activity of the solar maximum corona is a large scale structure, coherent over times of the order of weeks. Whether this global pattern of activity has true "time scales" and "spacial scales" in the sense of peaks in a power spectrum is as yet unknown. In this section, we present a brief view of the slowly varying component of the SMM corona, and we emphasize its interaction with the complex phenomena below it.

Since we are concerned with the aspects of long-term stability of coronal structures, we avoid any detailed investigation of the phenomena known collectively as coronal mass ejections. Thus, we explicitly restrict our discussion to time scales of about a day and longer, and spacial scales (in the upper corona) of a solar radius and greater. A short discussion of the inner corona (i.e., below the Coronagraph/Polarimeter occulting disk at 1.6 R_O) as seen by the HXIS instrument and meter-wavelength radio telescopes is given first. Then we present preliminary results of analysis of a synoptic map of the upper corona derived from SMM coronagraph observations during Carrington rotation 1693. We conclude with a glimpse of the global stability of the SMM-era corona on the time scale of multiple solar rotations, presenting pictures of the corona taken at intervals of one Carrington rotation.

6.6.2 Inner Corona

It seems obvious that the inner corona will react to the rapid evolution seen dramatically in observations of flares. However, some coronal loops at the limb, seen in X-rays, appear to have a large scale component which is almost impervious to the disruptive effects of flares, as the examples in Section 6.4.4 show. Observations of radio events also seem to provide evidence that in spite of bursts here and noise storm enchancements there, the underlying magnetic structures are fairly long-lived, evolving on time scales of a day. We discuss the joint radio and whitelight coronagraph observations of Duncan and of Pick and Trottet which bear on this interpretation.

Harrison et al. (1984) have studied a large coronal feature which extended above the west solar limb on 29 June 1980 (cf. Section 6.3.3.1). This feature was observed with HXIS throughout the day. The west limb was exceptionally active on this day, apparently because active regions 2522 and 2530 were near limb passage; three M-class flares occurred in these regions, and major coronal mass ejections were seen in conjunction with the first and last flares, as noted in Sections 6.2.1.6 and 6.3.3.

The coronal feature seen in 3.5-5.5 keV X-rays has been interpreted by Harrison et al. as a set of three loops (or loop systems) interconnecting active regions at the limb. The loop systems were seen on the first and all subsequent orbits of the day, except when the HXIS was in its flare mode, a mode in which sensitivity to low-surface-brightness features is much diminished. Figure 6.6.1 shows the appearance of the loops at 1337 UT. The loop system brightens significantly at various times during the day, often associated with the flares in the nearby active regions. The loops' brightness contours also expand upward, perhaps due to injection of material. However, despite the large energy releases during this period (one flare occurred at a footpoint of one of the loops), the same basic components are visible at the end of the day as at its beginning. This alone argues that the large-scale magnetic structure of the interconnected active regions has not been disrupted substantially by the flares, or, at least, that the large-scale configuration is re-established in a few hours or less after any disruption. The CMEs associated with the first and last X-ray flares (class M4.6 and M5, respectively) exhibited complex loop systems; see Figure 6.6.2 for representative coronagraph frames. It is unfortunate that the HXIS flare observing mode does not allow observation of the X-ray loops during the flare; we cannot know whether the X-ray-emitting loops seen before the flare are the same loops seen later, high in the corona. It may be that here we are seeing the "shedding" of loops by the active region as it readjusts its magnetic structure to nearly that existing before the flare. However we may speculate about the details, one striking fact remains: even in the presence of repeated large energy releases, there is a persistent skeleton of large-scale coronal features lasting at least a day.

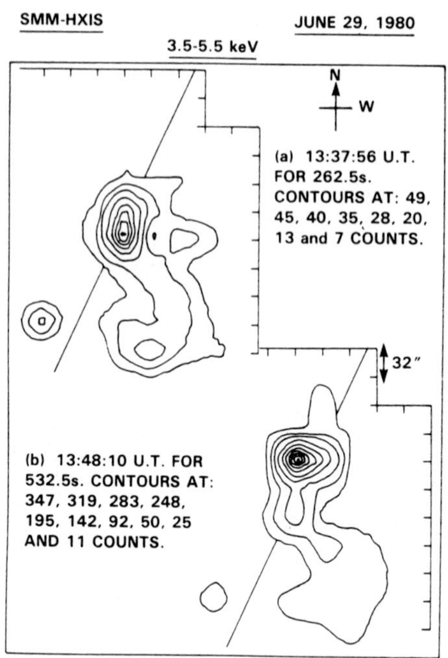

Figure 6.6.1 Contour plots showing the largest of the coronal loops seen by HXIS at the limb. The solar disk is to the left of the diagonal line, and the corona to the right (cf., Figures 6.3.3 and 6.3.4).

The persistence is not absolute, however. The magnetic structure of an active region certainly changes on time scales of a day. Catastrophic changes sometimes are associated with CMEs; slower, more subtle changes also affect large-scale coronal structure. To investigate these slow changes, we turn to joint observations of the corona in whitelight and meter wavelength radio. Duncan (1983) analyzed the relation between the slow injection of mass into the corona and the appearance of radio activity (a Type I storm) under the coronal region. Duncan used a digital technique -- closely akin to unsharp masking in photographic processing -- to reduce the background streamer brightness levels in SMM coronagraph images. In the pictures thus processed, he followed the evolution of several coronal rays. Over several hours, the rays brightened, and a new ray was formed. The brightenings began at the ray's bases and moved upward, as if new material were being forced into the corona along thin, open flux tubes. Simultaneously with the appearance of the new ray, the Culgoora radio spectrograph recorded a Type III burst at an unknown position. Duncan suggested that the ray and Type III bursts were related. In keeping with this result, Pick et al. (1980) have shown that the coronal structure overlying Type III-producing regions is highly structured, with thin dense rays whose densities vary on time scales of a few hours. Similarly, Trottet et al. (1982) located more than 100 Type

III bursts on SMM coronagraph images with positional uncertainties of 2° in latitude and 0.05 R_0 in radius. They found that Type III electron beams tend to occur where the corona is composed of small discrete rays, and that the Type III structure tends to follow that of the corona.

Even more remarkable than the appearance of the new ray in Duncan's study period was the behavior of the system of rays as a whole. Simultaneously with the appearance of the ray brightenings, the entire complex of rays began to expand latitudinally, and continued to do so for 24 hours. The centroid of the ray system remained at nearly the same position angle throughout. The Type I noise storm was motionless near this centroid. This situation was an example of the more general finding of Kerdraon et al. (1983), who found that noise storm onsets or enhancements are systematically associated with the appearance of additional material in the corona, often as thin rays or amorphous

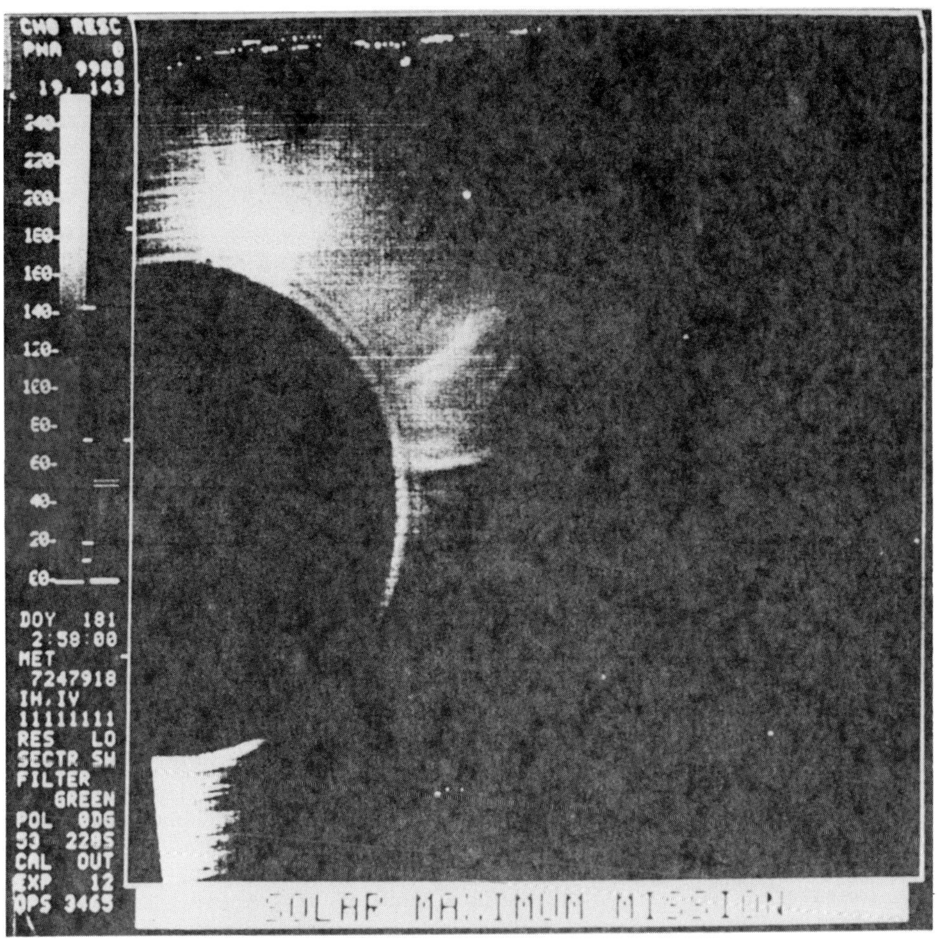

Figure 6.6.2a SMM coronagraph frames showing the 29 June 1980 CMEs associated with the flares seen with HXIS. North is to the upper left, and west to the upper right. The brightest diffraction ring around the occulting disk has a radius of 1.61 solar radii. (a) At 0258 UT.

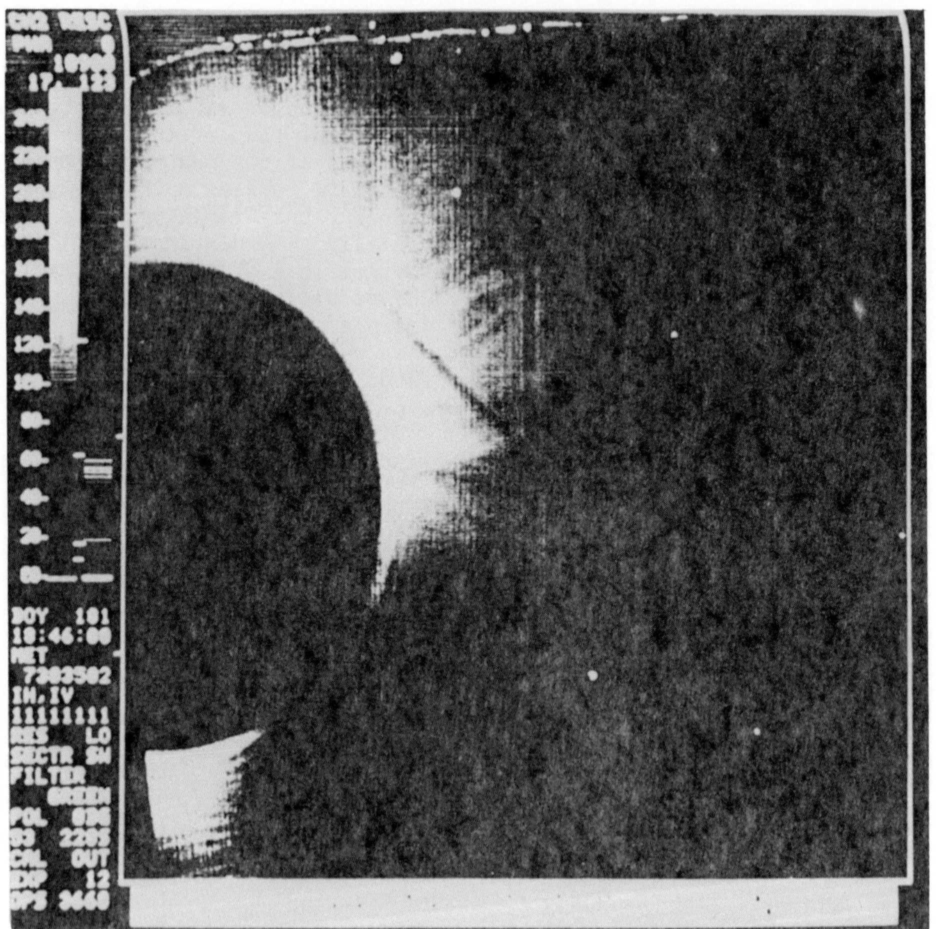

Figure 6.6.2b SMM coronagraph frames showing the 19 June 1980 CMEs associated with the flares seen with HXIS. North is to the upper left, and west to the upper right. The brightest diffraction ring around the occulting disk has a radius of 1.61 solar radii. (b) At 1846 UT.

structures with time scales of brightness increase one hour or less. In all cases, the radio sources were cospacial with regions of coronal mass enhancements. Duncan reached the conclusion that the noise storm is likely caused by the emergence of new flux at the coronal base, precisely at the center of symmetry of the expanding ray system. This interpetation of the observations suggests that the slow evolution of the photospheric magnetic structure is indeed mirrored in the high corona. The effect in the corona is subtle, so that it may be occurring continuously but is usually overlooked because of the small amount of brightness slowly added to existing structures.

6.6.3 Outer Corona

In this section, we discuss the longest temporal and largest spacial scales accessible with the coronagraphic data from the SMM. We proceed in two steps from the scale of a day -- discussed above -- to the scale of several solar rotations. For the first step, we present a detailed synoptic map of the corona prepared by Illing and House for Carrington rotation 1693, and we consider what it shows regarding the large-scale characteristics of the corona. For this purpose we use the map's information about large scales, considering its high-resolution details only as they relate to global phenomena. For the second step, we remark on the variability of the corona at solar maximum from one rotation to the next. The analyses of both the one-rotation synoptic map and the rotation-to-rotation variations are in an extremely preliminary state.

6.6.3.1 Coronal Variations During Rotation 1693

Figure 6.6.3 shows the intensity distribution of the corona during Carrington rotation 1693. This synoptic map was built up from vertical strips, each strip a constant-height scan of coronagraph images. The images were separated, on average, by 96 min (one spacecraft orbit), corresponding to a change of approximately 1° in Carrington longitude between images. However, integration along the line of sight through the optically thin corona sets the true longitude resolution of coronal features, more like 15°. The latitude resolution was chosen to be 1° for convenience of presentation. Since we have not removed the tilt of the solar rotation axis, the latitude scale shown in the figure is slightly incorrect. The Carrington longitude for each image (or scan) is the longitude of central meridian at the time of the image plus (for west limb images) or minus (for east limb images) 90°. We have thus assumed that the coronal features are in the plane of the sky in each image, and that the images represent the state of the corona at the Carrington longitude of the underlying photospheric and chromospheric structures. We have used only north, east, and west quadrants for the maps presented here, since a large portion of the south images is blocked by the support for the C/P's external occulting disk. The north images have been split at position angle zero into east and west sections. The Sun rotates once every 410 SMM orbits, making room for 410 constant-height latitude scans in the synoptic map; operational constraints and instrumental considerations (corrupted images, down time, special observing programs, etc.) reduced coverage by about 25%. In total, the map contains scans of over 900 quadrantal images. East limb observations of rotation 1693 are limited by the beginning of satellite operations on 13 March 1980 (DOY73). In both east and west projections there are no data from 3-4 April 1980 (DOY 94-95) because of instrumental problems. Otherwise, missing data occur more or less at random throughout the observing interval. To minimize visual interference due to data gaps, we have interpolated the intensity distribution across the gaps in the longitudinal direction only.

Investigation of the vertical "streaks" in the synoptic map shows thatabout 75% are due to CMEs; the rest are due to coronal rearrangement, uncorrected photometry shifts, etc. Of the real, abrupt coronal brightenings on one limb or the other, 34 CMEs can be described as loop-like or as major disruptions of streamers. This "major event" group (for only one of five solar rotations observed in 1980) is half as numerous as the total used by Hundhausen et al. (1984a) in their determination of the CME occurrence rate (0.9 ± 0.15 per day, on average) for all of 1980. Hundhausen et al., before Figure 6.6.3 was prepared, counted 21 CMEs in rotation 1693. The discrepancy between the formerly-identified events and the new count of 34 events arises from more-detailed examination of the data using image-difference and blink-comparison techniques on the images suggested by the streaks in Figure 6.6.3. This type of "second look" analysis was not used by Hundhausen et al., intentionally; to retain consistency with the Skylab coronagraph analyses, they examined only the direct C/P images.

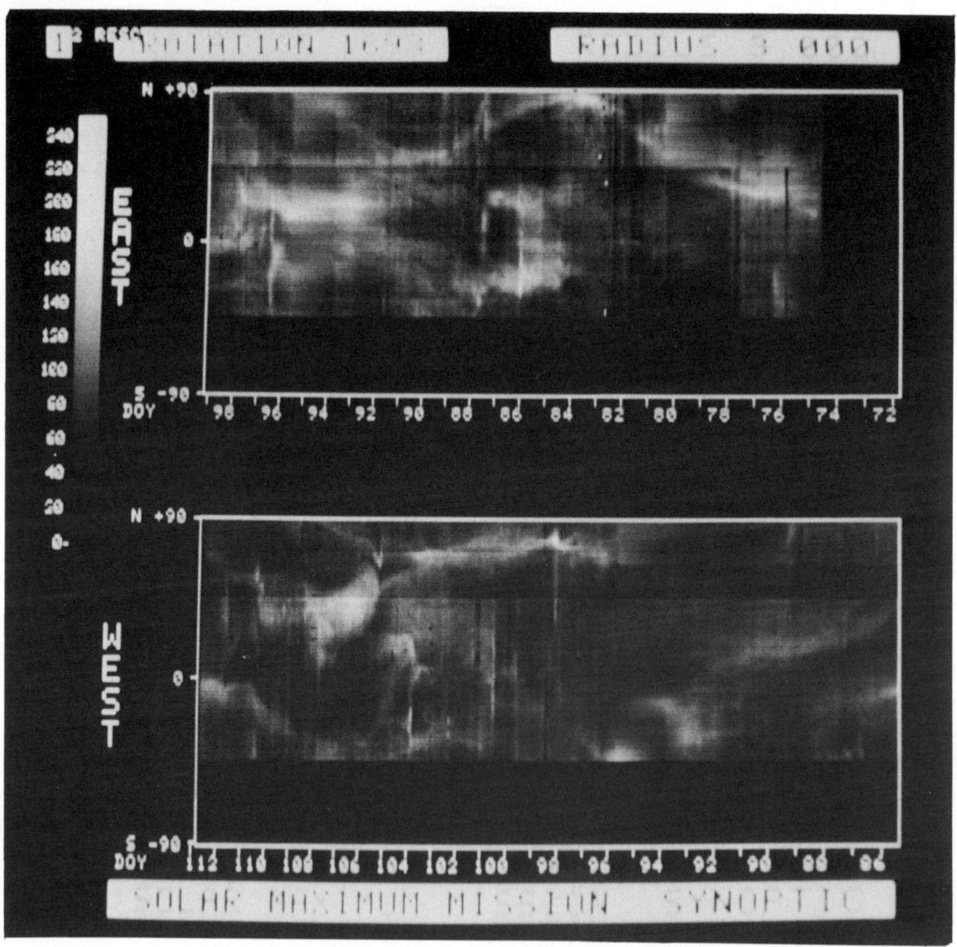

Figure 6.6.3 The detailed synoptic map of Carrington rotation 1693, at a height of 3 R_o. Time increases to the left, measured in day of year, 1980. See the text for an explanation of the construction of this map.

The tendency for transients to be clustered in time is not an artifact of the cadence of data acquisition; we used one coronagraph image (per sector) for each spacecraft orbit, even if more images are available. A notable example of temporal (ergo, spacial) clustering can be found on the west panel, near the equator, between days 102 and 106; Figure 6.6.4 shows a blowup of this area. The original pictures confirm the impression given by the map; there is a great concentration of coronal activity in this area, including at least five major mass ejections and much intermediary restructuring. This concentration of extreme coronal activity occurred above active regions which produced a large number of flares.

In Figure 6.6.3, coronal streamers are clearly visible as longitudinally extended brightness features. Comparison of the map with the heliospheric current sheet inferred by Hoeksema, Wilcox, and Scherrer (1983) suggests that the streamer-like features seen at 3 R_0 approximately outline the current sheet; further investigations of this correlation are in progress. The interaction of a CME with a streamer typically produces mostly a short-term change; the map shows that the large-scale structures basically return to their pre-event forms in about 6 hours or less. Even in such apparently violent events as that seen in the east frame at day 87, an event which extends over the entire east limb and strongly bends and displaces streamers, the long-term disruption is nil. It may be useful to distinguish two ways in which the brightness of observed coronal structures recovers to approximately its pre-event configuration. In the case of neighboring streamers which are obviously bent and displaced, recovery must be due to the relaxation of the CME-strained magnetic fields. The time scales involved are likely to be several hours, the approximate Alfven transit time across a coronal feature of length 1 R_0. Alternatively, in the case of longitudinally extended streamers, we are looking through long arcades; a part of the arcade may erupt, leaving the remainder of the overall structure nearly unaffected. In this case, the time scale for coronal recovery is more difficult to assess, since there can be a larger brightening due to rotation of the whole arcade blended with the smaller brightening due to "refilling" or recovery of the restricted region which erupted.

Of course, there are exceptions to enduring stability. The region of frequent restructuring (days 102-106) mentioned above suffers persistent brightenig or the creation of several small streamers. On day 105, a large eruptive prominence-associated mass ejection occurred (in projection) over the north pole. The southwestern leg of the transient loop impinged on a pre-existing, fairly quiescent streamer; following this interaction, the streamer expanded in latitude and height and was the site of numerous additional coronal mass ejections over the next 5 days, as shown in Figure 6.6.4. Indeed, the original frames indicate that the streamer, in addition to the major coronal transients, underwent almost continous activity in the form of material outflows and frequent small changes of shape. This "activation" of the coronal (magnetic) structure persisted for about 3 days before the streamer "deflated" to nearly its pre-activation, day 105 size. Even after this relaxation, the area was the site of numerous mass ejections.

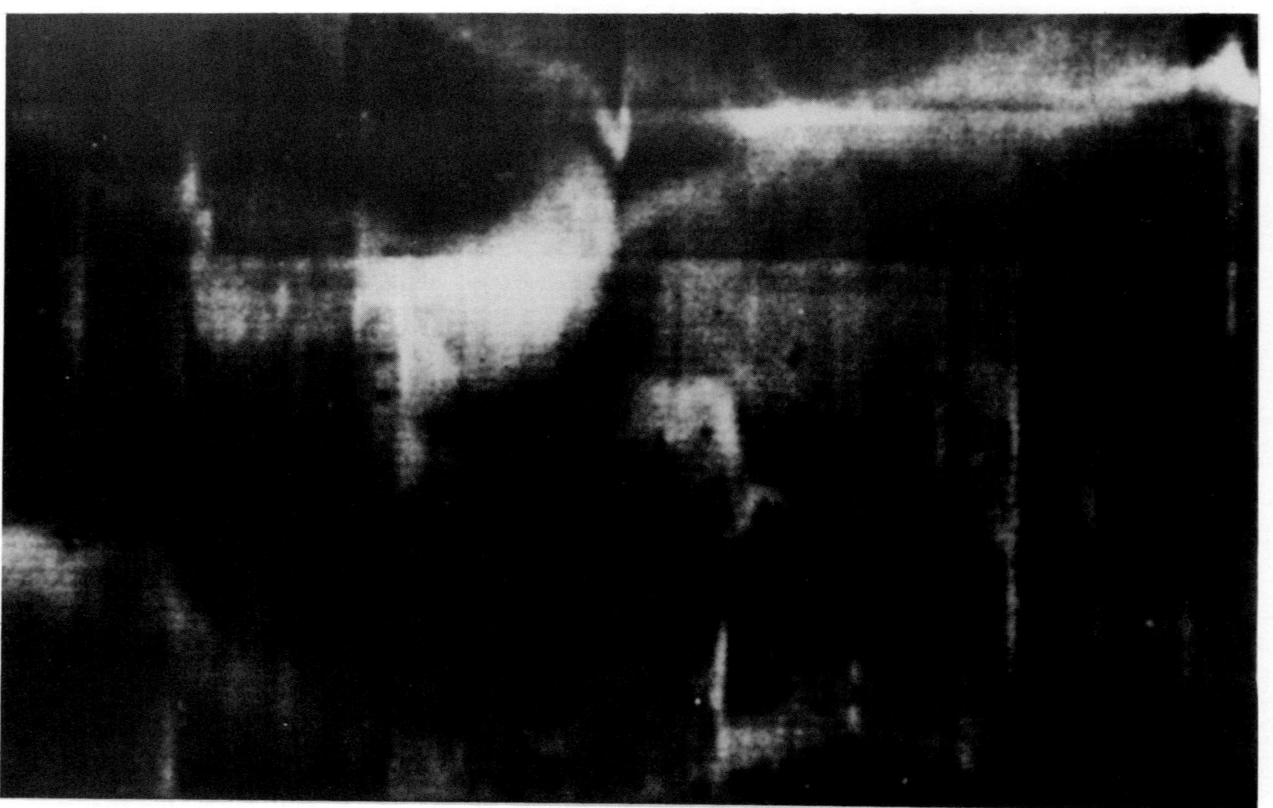

Figure 6.6.4 Expanded view of the coronal activity on days 102 to 106, from the detailed synoptic map of Figure 6.6.3.

The map shows that the upper corona has a remarkable ability to "heal" itself and to recover in a few hours or a few days from the effects of strong disruptions. The day(s)-long recovery from disruption occurs in the context of slower evolution. Comparison between the east and west limb panels in Figure 6.6.3 shows that there is noticeable evolution of the corona during half a rotation, just as MacQueen and Poland (1977) found for the Skylab-era corona. Often, the streamer bands merely shift position somewhat, while keeping their major features intact. In contrast, the bright region on the east near the equator between days 84 and 88 has almost vanished by its west limb passage on days 98 to 102. Comparisons of this type between similar maps of successive rotations give the same conclusion; in 14 to 28 days, one-half to one solar rotation, there takes place considerable movement, appearance and disappearance of major features.

6.6.3.2 Coronal Variation From Rotation to Rotation

The detailed synoptic map of Figure 6.6.3 is not the most appropriate way to present coronal observations for study of the longest time scales of coronal evolution. The welter of detail tends to obscure those trends which are the object of investigation. In this section, we give only a glimpse of these trends.

Figure 6.6.5 shows a set of six SMM coronagraph frames taken at the central meridian passages of Carrington longitude 090° during the SMM mission in 1980. Each frame is a composite of three images of the north, east, and west quadrants. As usual, solar north is to the upper left, and east is to the lower left. The individual images were chosen to be free of CMEs. The dark stripes and spots are instrumental artifacts and should be ignored. There is no picture for rotation 1695.

Though coronal structure varies from rotation to rotation, there is also correlation from rotation to rotation. In each rotation, there is some bright streamer structure on the east limb; though less evident on rotations 1694 and 1696, it becomes strong again in the following rotations. In each rotation, there is fairly bright structure near the north pole. The west is the most variable region, with prominent streamers disappearing and being replaced by other streamers at considerably different positions. Even though the detailed structure of the corona changes from one rotation to the next in Figure 6.6.5, the correlation of bright structure from rotation to rotation implies that the major, determining, magnetic features underlying the corona remain, at least on the north and east limbs.

The large-scale structure does not always correlate well in successive rotations. Figure 6.6.6 shows the corona as it appeared at central meridian passage of Carrington longitude 000° and as it appeared one rotation later. The north and east have almost no features in common between the two rotations and we interpret this as evidence for the occurrence of major change or disruption in the underlying magnetic structure. Comparative analysis of coronal changes with magnetic changes inferred from contemporaneous magnetograms is necessary to achieve complete understanding of these large scale phenomena.

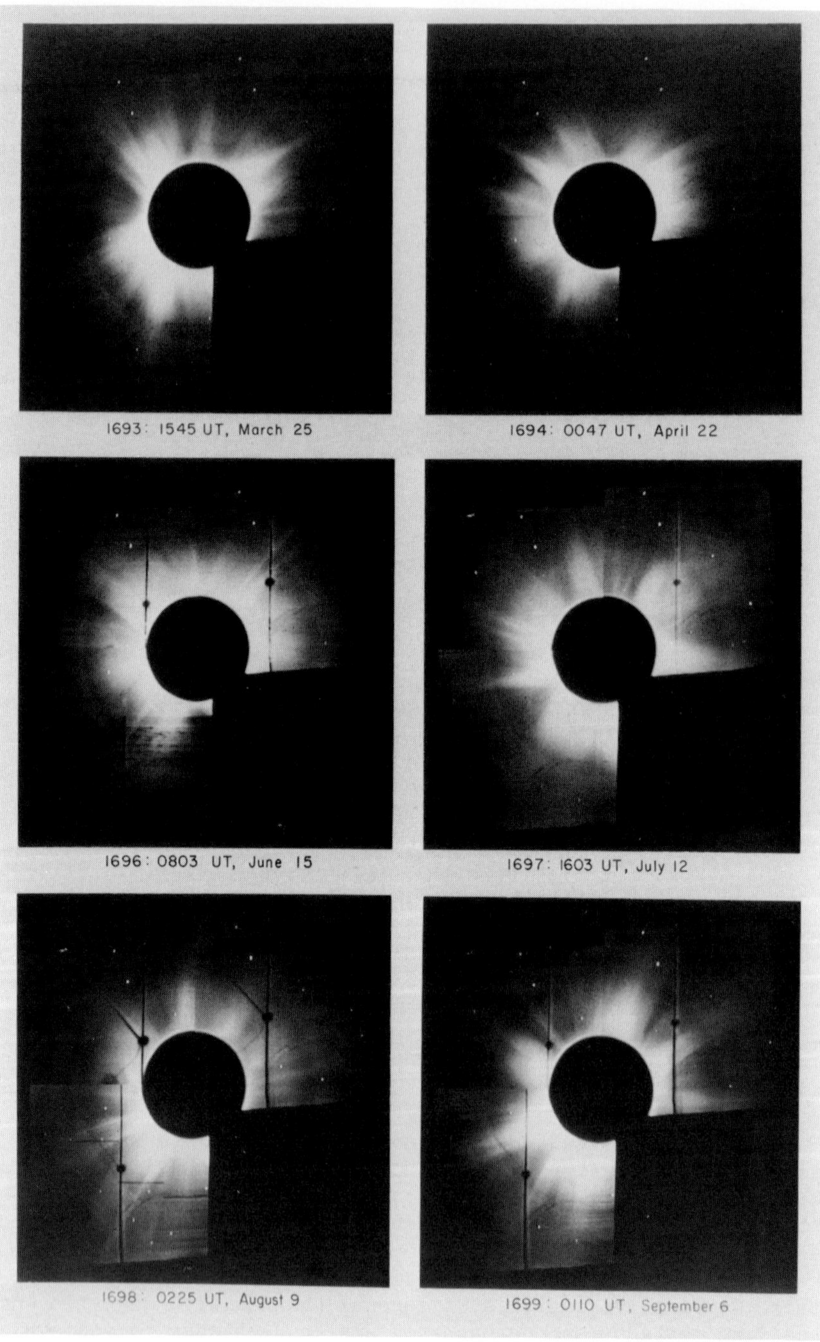

Figure 6.6.5 SMM coronagraph frames taken at 090° disk center Carrington longitude for five of the six rotations included in the 1980 SMM data.

From our still preliminary analyses we can draw the following conclusions about evolution of the SMM-era corona on temporal scales of a day or longer. The corona showed remarkable ability to recover to near pre-event configuration over times of a few days after violent disruption; this at the epoch of maximum solar activity. However, the coherence time for many of the largest features was short, approximately one solar rotation. Both these results are consistent with the results found by Hundhausen, Hansen, and Hansen (1981) in their analysis of K-coronameter observations of the lower corona over almost two full solar cycles.

1696: 1743 UT. June 7

1697: 2243 UT, July 5

Figure 6.6.6 SMM coronagraph frames taken at 000° disk center Carrington longitude for rotations 1696 and 1697.

6.7 SUMMARY

6.7.1 Introduction

The quantity and diversity of results presented in the preceding five sections of this chapter almost defy summarization. Yet, they are so voluminous as to demand an outline of some of the major achievements. Inevitably such a list is idiosyncratic and slights excellent work of extreme interest to some. In Section 6.7.2, we list -- in very abbreviated form -- some of the major conclusions and observations presented during the Workshop. Readers interested in more detail are referred to the appropriate subsection of this chapter. In the final portion of this section, we suggest future work to advance our understanding of the corona and its (slow and rapid) evolution.

6.7.2 Major Observations and Conclusions

The list which follows is a selection of the major observations and conclusions presented during the SMM Workshop. The reader is referred to the appropriate subsection of the chapter for more details and for references to published work regarding the listed items; the reader is referred to the entire chapter for many results which are not in this list of highlights. That the selected items preponderantly deal with coronal mass ejections reflects the interest of the majority of the team members in that phenomenon as observed in the corona. Not all of the members of the Team were exclusively interested in CMEs, so some results on other coronal topics are listed as well.

1. We have gathered together in one place comprehensive, multi-instrument descriptions of seven coronal mass ejection events. Detailed conclusions which follow from these collaboratively observed events are (Section 6.2.1.1-6.2.1.7):

(a) An example of the class of dark or depletion (at 1.2 R_0) transients was reported to acquire a bright rim as it rose through the corona. Thus, this depletion transient (perhaps the ascent of the cavity often seen around a prominence at the limb) became a bright coronal mass ejection in an orbiting coronagraph's field-of view. In this case, it is quite clear that the exess mass ejected from the orbiting coronagraph's field-of-view was not raised from the chromosphere but, rather, started in the corona (Section 6.2.1.1);

(b) In one event there is a strong suggestion of magnetic disconnection from the Sun; that is, there is a rising, outwardly concave, intensity-enhanced structure. Less obvious, but still suggestive, similar structures were seen in perhaps as many as 10 percent of all SMM events (Section 6.2.1.2);

(c) A CME which possibly resulted from the eruption of an arcade rather than a single loop was reported; a moving Type IV radio burst was associated with a denser blob of rising plasma; soft X-ray emission enhancement commenced 17 minutes before the earliest H-alpha activity; a forerunner similar to those reported for Skylab-era events surrounded the ejection (Section 6.2.1.7).

2. Observations with the Helios satellites' zodiacal light photometers enabled us to obtain "stereoscopic" views of three coronal mass ejections and to resolve some of the ambiguities which are inherent in observations from a single vantage point. The observations show that coronal mass ejections typically have a complicated 3-dimensional structure (Section 6.2.1.8; see also 6.5.2), a result supported by the SOLWIND coronagraph observations of "halo" CMEs which sometimes appear to surround the occulting disk completely (Section 6.2.2.2).

3. The long observing period of the SOLWIND coronagraph aboard the P78-1 satellite has enabled statistics to be gathered on more than a thousand CMEs occurring around the time of solar activity maximum (Figure 6.2.23). Aside from the broader range that might be expected when the sample size is increased by more than 100 times, the values of speed, mass, and energy are similar to those reported for Skylab-era CMEs, though the latitude of occurrences spread much higher (Section 6.2.2).

4. It was suggested that at least some coronal mass ejection events -- with or without accompanying flares -- start at a time coincident with weak, soft X-ray bursts. Such "precursory bursts" occur some minutes before the associated flare onset, if any (Section 6.3.2.2).

5. In some flares, both near disk center and at the limb, doppler shifts of Ca XIX or Fe XXV X-ray emission lines were seen briefly during the flares' impulsive phase. The observed magnitude of the doppler shift implies motion of 300-400 km s^{-1} toward the observer during the impulsive phases. The doppler shift and apparent velocity were too brief to be seen in low-cadence X-ray images, so the location of the phenomenon could not be related in detail to flare site morphologies (Section 6.3.3.2).

6. Rising, X-ray-emitting counterparts of whitelight coronal mass ejections were detected at low heights (e.g., 1.14 R_0). In one well-observed case, the X-ray emitting structure started to rise before the main X-ray flare occurred; it may have started at the time of an X-ray "precursory burst" as mentioned in item 3 (Sections 6.3.3.1 and 6.3.4).

7. Characteristically, coronal mass ejections associated with flares show rapid initial acceleration, usually followed by constant speed or deceleration. By contrast, coronal mass ejections associated with prominence eruptions tend to move more slowly and often are still accelerating at great heights in orbiting coronagraphs' fields-of-view (Section 6.4.1).

8. From a family of numerical model calculations, it was concluded that it is impossible to simulate a realistic coronal mass ejection by calculating the response of an atmosphere in hydrostatic equilibrium, permeated by an initially potential magnetic field, to a perturbing pressure pulse at its base (Section 6.4.2).

9. Self-similar solutions of the MHD equations were suggested as a good way to model the asymptotic behavior of coronal mass ejections far from their initiating sites (Section 6.4.3). Appropriate choice of the free parameters in the models appears to give good agreement with some observations (Section 6.2.1.2).

10. Gigantic post-flare coronal arches, emitting soft X-rays for several hours, were discovered after two-ribbon flares. Such flares

typically have Type IV (and, often, Type II) radio bursts. In one case, the arch was stationary, but in another case of subsequent flares from the same active region, the post-flare coronal arches revived, brightening in place in X-ray emission, and rising at different speeds. The arches probably cooled radiatively, with inhibition of cooling by conduction. The relation of these arches to coronal mass ejection is unclear, due to a lack of data (Section 6.4.4).

11. The energy expected to be liberated by magnetic field reconnection and realignment to a potential configuration over a two-ribbon flare was examined analytically. Conservative assumptions about magnetic field strength and geometry yield an energy liberation some 300 times greater than what is needed to power an observed flare taken as an example. Therefore, it may be concluded that the magnetic field in the volume above and around a flare does not relax completely, that is, to a potential field; rather, it remains stressed after the flare is over (Section 6.4.5).

12. An axisymmetric, hydrostatic, equilibrium configuration (with two free configuration-changing parameters) was presented which allowed analytical examination of stability. Though stable against all axisymmetric perturbations, the equilibrium was -- in every case -- unstable against perturbations with variations along all three axes (Section 6.4.6).

13. Based on the observed association between meter-wave Type II bursts at the appropriate time and place and the later observations of interplanetary magnetic clouds, it was suggested that interplanetary magnetic clouds are the manifestations of coronal mass ejections at 1 AU (Section 6.5.2).

14. Observations (obtained with instruments aboard the Helios and other satellites) of the interplanetary consequences of coronal mass ejections were particularly rich (Section 6.5.3):

(a) Of 80 interplanetary shocks observed when Helios was within 30° of a solar limb and when the SOLWIND coronagraph was observing, 40 had good associations with coronal mass ejections, and another 19 of the shocks had possible coronal mass ejections. The mass ejections associated with the shocks at Helios were generally faster, brighter, and at lower latitude than the typical CME; they tended to fill an arc of heliocentric latitude in the corona which encompassed the Helios -- Sun line;

(b) Of 27 prompt proton events with H-alpha flares which were observed, 26 had associated coronal mass ejections. The 27th event appears to be a member of a new class of events associated with short-lived flares having γ-ray and radio Type III bursts but no CME;

(c) By contrast, energetic particle events which were rich in ^{3}He (^{3}He/^{4}He > 0.2 at ~ 1.5 MeV/nucleon) were not well associated with radio bursts or coronal mass ejections.

15. Though the topic was not intensively studied during the Workshop, it is possible to say a few words about coronal evolution

during SMM (Section 6.6.2):

(a) On occasion, there were permanent changes in coronal structures due to coronal mass ejections;
(b) Sometimes, coronal structures endured for more than a day, even in the presence of repeated flaring below them;
(c) Often, there were changes which occurred over a period of hours. These are difficult to identify as coronal mass ejections, but they clearly involved at least the rearrangement and, possibly, the gradual expulsion of material;
(d) Type III storms tended to occur where the coronal structure appeared to be bundles of small, discrete rays;
(e) Generally, the corona was brighter after a radio noise storm than before; it was suggested that the eruption of new flux at the base of the corona created the proper conditions for the noise storm and carried relatively denser plasma laterally, and upward to greater heights as well, thereby brightening the corona.

16. The preparation of synoptic charts, which showed coronal brightness at one limb as an entire rotation brought 360 deg of longitude to the limb, gave a pseudo-image presentation of global coronal structure. This format allowed for the study of evolution and the association of coronal structures with other solar features such as the base of the heliospheric current sheet. The analysis of such displays was only begun during the Workshop (Section 6.6.3.1).

6.7.3 Suggestions for Future Research

As always happens in a vigorously advancing field, the observations and analyses reported by the Coronal Structures team members raised as many questions as they answered. The evolution of the corona and the relation between its structures and the underlying photospheric and chromospheric features has been a neglected area of study. The long series of SOLWIND coronagraph observations, the accumulating Mauna Loa Mark III K-Coronameter data, and the repaired SMM Coronagraph/Polarimeter instrument (now repaired and operating again at fairly rapid cadence) have given us the necessary observations to explore these questions. Another area begging for future investigation is the nature of the large-scale X-ray emitting structures, both the moving structures associated with visible light coronal mass ejections and the gigantic post-flare arches. The energy sources, the magnetic configurations, and the relation of these features to other forms of solar activity will lead to a substantially improved understanding of solar activity in the low corona.

The major issue still confronting us, an issue pointed up by the list of results given in Section 6.7.2, is the inadequacy of our present physical understanding of why and how coronal mass ejections occur. Both analytical and numerical modelling are as toddlers, ready to grow, to improve, and to become more realistic. Basically, we lack insight into the physics of the initiating processes of coronal mass ejection. Though we can be relatively successful at modelling a particular event, we do not possess an over-arching understanding of the causes and mechanisms which give rise to coronal mass ejctions in the first place. During the Workshop, phenomena associated with CMEs were discovered which must be integrated into our conceptual picture. The relationship of post-flare arches must be considered. To be satisfactory, an overall picture must also accommodate the growing body of observations that coronal reconfiguration occurs on all time scales; should CMEs be distinguished from slightly slower reconfigurations, or are they just the more energetic members of the reconfiguration spectrum? Also, the discovery of soft X-ray pulses coincident with CME onsets, prior to flares, will affect our understanding of the initiation of coronal mass ejections. If the rapid movement of hot (Ca XIX and Fe XXV-emitting) plasma during the impulsive phase of flares is related to coronal mass ejections, that too will lead to deeper understanding.

It is widely believed that the magnetic field holds the key to understanding the observed structure of the corona and its variation. The radio observations give us our best hope to infer the strength of the magnetic field in the corona. These observations and their analyses should be pursued vigorously. Another type of observation, whose enormous value is only glimpsed as yet, is the "stereoscopic" examination of coronal mass ejections. The zodiacal light photometers aboard Helios were never intended for coronal mass ejection observations, but the CME-related information wrested from their records with great difficulty is intriguing and important, for it suggests that coronal mass ejections are not loop-like. Indeed, they are not even symmetric about a central axis, despite their frequent loop-like appearance when viewed in projection from ground-based and near-Earth vantage points. An instrument designed specifically to view coronal mass ejections traversing interplanetary space, sited to give a perspective far removed from Earth's, would yield data of exceptional value to our understanding of CME morphology and evolution. The interplanetary manifestations and consequences of coronal mass ejections are still a puzzle. Coronal mass ejections are often high-contrast disruptions of the pre-event corona, but in most cases they appear to be low-contrast perturbations of the 1 AU solar wind. The association between energetic coronal mass ejections and interplanetary shocks at Helios is encouraging, but it leaves unanswered the question of the appearance of the slower coronal mass ejections in the interplanetary medium. The association between coronal mass ejections and prompt proton events is understood not at all, nor is the lack of association between coronal mass ejections and ^{3}He-rich energetic particle events understood. It is tempting to think that the association (or lack of it) between CMEs and interplanetary particle events is a clue to the nature of energetic particle acceleration and release on the one hand and to the nature of CME initiation on the other.

6.8 ACKNOWLEDGEMENTS

With great pleasure, the editor of this chapter acknowledges and thanks the members of the Coronal Structures Team for their enthusiastic participation in the SMM Workshop and for their contributions which are reported here. The organizers of the chapter have been very helpful, even though it cost some of them considerable labor. The editor thanks S. W. Kahler and R. M. MacQueen for their reviews of an earlier draft of the chapter; their comments led to numerous improvements and modifications. To Ms. Leila Reed goes the editor's deepest appreciation for her exceptionally competent and cheerful clerical assistance through the long process of entry, revision, and subsequent re-revisions of the text of this chapter. Finally, thanks go to the SMM Principal Investigators, to the Workshop organizers, and to the Workshop's financial supporters; they brought the Workshop into existence and created the opportunity for such stimulating and fruitful interaction with colleagues engaged in similar research.

6.9 REFERENCES

Anzer, U.: 1980, in Solar and Interplanetary Dynamics, IAU Symp. No. 91, (eds., M. Dryer and E. Tandberg-Hanssen) (Reidel Publ. Co.: Dordrecht), p. 263.

Anzer, U., and Pneuman, G. W.: 1982, Solar Phys. **79**, 129.

Barenblatt, G. I., and Zel'dovich, Ya. B.: 1972, Ann. Rev. Fluid Mech. **4**, 285.

Bassi, J., Dulk, G. A., and Wagner, W. J.: 1985, in preparation.

Bentley, R. D., Lemen, J. R., Phillips. K. J. H., and Culhane, J. L.: 1984, Solar Phys. submitted.

Benz, A., and Wentzel, D. G.: 1981, Astron. Astrophys. **94**, 100.

Bernstein, I. B., Friedman, E. A., Kruskal, M. D., and Kulsrud, R. M.: 1958, Proc. Roy. Soc. Lon. **A244**, 17.

Borrini, G., Gosling, J. T., Bame, S. J., and Feldman, W. C.: 1982, J. Geophys. Res. **87**, 7370.

Cane, H. V.: 1984, Astron. Astrophys. **140**, 205.

Cargill, P. J., and Priest, E. R.: 1982, Solar Phys. **76**, 357.

Carrington, R. C.: 1859, Monthly Not. Roy. Astron. Soc. **20**, 13.

Dryer, M.: 1982 Space Sci. Rev. **33**, 233.

Dulk, G. A., Smerd, S. F., MacQueen, R. M., Gosling, J. T., Magun, A., Stewart, R. T., Sheridan, K. V., Robinson, R. D., and Jacques, S.: 1976, Solar Phys. **49**, 369.

Dulk, G. A., and Suzuki, S.: 1980, Astron. Astrophys. **88**, 203.

Duncan, R. A.: 1983, Solar Phys. **89**, 63.

Fisher, R. R., Garcia, C. J., and Seagraves, P.: 1981, Astrophys. J. (Letters) **246**, L161.

Fisher, R. R., and Poland, A. I.: 1981, Astrophys. J. **246**, 1004.

Fisher, R. R., Lee, R. H., MacQueen, R. M., and Poland, A. I.: 1981, Appl. Opt. **20**, 1094.

Gary, D. E.: Radio emission from solar and stellar coronae. Ph.D. thesis, University of Colorado, 1982.

Gary, D. E., Dulk, G. A., House, L. L., Illing, R. M. E., Sawyer, C., Wagner, W. J., McLean, D. J., and Hildner, E.: 1984, Astron. Astrophys., **134**, 222.

Gary, D. E., Dulk, G. A., House, L. L., Illing, R. M. E., Wagner, W. J., and McLean, D. J.: 1985, Astron. Astrophys., submitted.

Gergely, T. E.: 1984, this workshop.

Gergely, T. E., Kundu, M. R., Erskine, F. T., III, Sawyer, C., Wagner, W. J., Illing, R. M. E., House, L. L., McCabe, M. K., Stewart, R. T., Nelson, G. J., Koomen, M. J., Michels, D., Howard, R., and Sheeley, N.: 1984, Solar Phys. **90**, 161.
Gosling, J. T., Hildner, E., MacQueen, R. M., Munro, R. H., Poland, A. I., and Ross, C. L.: 1975, Solar Phys. **40**, 439.
Gosling, J. T., Hildner, E., MacQueen, R. M., Munro, R. H., Poland, A. I. and Ross, C. L.: 1976, Solar Phys. **48**, 389.
Gosling, J. T., Hildner, E., Asbridge, J. R., Bame, S. J., and Feldman, W. C.: 1977, J. Geophys. Res. **82**, 5005.
Harrison, R. A., Simnett, G. M., Hoyng, P., LaFleur, H., and van Beek, H. F.: 1984, in Proc. STIP Symposium on Solar-Interplanetary Intervals, p. 287.
Harrison, R. A., Waggett, P. W., Bentley, R. D., Phillips, K. J. H., Bruner, M., Dryer, M., and Simnett, G. M.: 1985, Solar Phys., submitted.
Heyvaerts, J., Priest, E. R., and Rust, D. M.: 1977, Astrophys. J. **216**, 123.
Hildner, E.: 1977, in Studies of Travelling Interplanetary Phenomena, (eds., M. A. Shea et al.) p. 3.
Hildner, E., Gosling, J. T., MacQueen, R. M., Munro, R. H., Poland, A. I., and Ross, C. L.: 1975a, Solar Phys. **42**, 163.
Hildner, E., Gosling, J. T., Hansen, T. R., Bohlin, J. D.: 975b, Solar Phys. **45**, 363.
Hildner, E., Gosling, J. T., MacQueen, R. M., Munro, R. H., Poland, A. I., Ross, C. L.: 1976, Solar Phys. **48**, 127.
Hoeksema, J. T., Wilcox, J. M., Scherrer, P. H.: 1983, J. Geophys. Res. **88**, 9910.
Howard, R. A., Michels, D. J., Sheeley, N. R., Jr., Koomen, M. J.: 1982, Astrophys. J. **263**, L101.
Howard, R. A., Sheeley, N. R., Jr., Koomen, M. J., Michels, D. J.: 1984, in Advances in Space Research (Pergamon Press), ed. P. Simon, in press.
Howard, R. A.: 1984, private communication.
Howard, R. A., Sheeley, N. R., Jr., Koomen, M. J., and Michels, D. J.: 1985, J. Geophys. Res., in press.
Hundhausen, J. R., Hundhausen, A. J. and Zweibel, E. G. 1981: J. Geophys. Res. **86**, 11117.
Hundhausen, A. J., Hansen, R. T., Hansen, S. F.: 1981, J. Geophys. Res. **86**, 2079.
Hundhausen, A. J., Sawyer, C. B., House, L., Illing, R. M. E., Wagner, W. J.: 1984a, J. Geophys Res., **89**, 2639.
Hundhausen, A. J. et. al.: 1984b, in Solar-Terrestrial Physics: Present and Future, NASA Ref. Publ. 1120, eds. D. M. Butler and K. Papadopoulos, Chapter 6.
Illing, R. M. E.: 1984, Astrophys. J. **280**, 399.
Illing, R. M. E., and Hundhausen, A. J.: 1983, J. Geophys. Res. **88**, 10210.
Illing, R. M. E., Hundhausen, A. J., and Fisher, R. R.: 1985, in preparation.
Jackson, B. V.: 1981, Solar Phys. **73**, 133.
Jackson, B. V., and Hildner, E.: 1978, Solar Phys. **60**, 155.

Jackson, B. V., Howard, R. A., Sheeley, N. R., Jr., Michels, D. J., and Koomen, M. J.: 1985a, Solar Phys., submitted.
Jackson, B. V., Howard, R. A., Sheeley, N. R., Jr., Michels, D. J., Koomen, M. J., and Illing, R. M. E.: 1985b, in preparation.
Jackson, B. V., and Leinert, C.: 1985, submitted to JGR.
Kahler, S. W.: 1982, J. Geophys. Res. **87**, 3439.
Kahler, S. W., Hildner, E., van Hollebeke, M. A. I.: 1978, Solar Phys. **57**, 429.
Kahler, S. W., Sheeley, N. R., Jr., Howard, R. A., Koomen, M. J., and Michels, D. J.: 1984a, Solar Phys. **93**, 133.
Kahler, S. W., Sheeley, N. R., Jr., Howard, R. A., Koomen, M. J., Michels, D. J., McGuire, R. E., von Rosenvinge, T. T., and Reames, D. V.: 1984b, J. Geophys. Res. **89**, 9683.
Kahler, S., Cliver, E. W., Sheeley, N. R., Jr., Howard, R. A., Koomen, M. J., Michels, D. J.: 1985a, J. Geophys. Res. **90**, 177.
Kahler, S. W., Reames, D. V., Sheeley, N. R., Jr., Howard, R. A., Koomen, M. J., and Michels, D. J.: 1985b, Astrophys. J., in press.
Kerdraon, A., Pick, M., Trottet, G., Sawyer, C., Illing, R., Wagner, W., House, L.: 1983, Astrophys. J. **265**, L19.
Klein, L. W., and Burlaga, L. F.: 1982, J. Geophys. Res. **87**, 613.
Kopp, R. A., and Pneuman, G. W.: 1976, Solar Phys. **50**, 85.
Lantos, P.: 1984, in Proc. International Workshop on Solar Physics and Interplanetary Travelling Phenomena, Kunming, China, November 1983, in press.
Lantos, P., Kerdraon, A., Rapley, G. G., and Bentley, R. D.: 1981, Astron. Astrophys. **101**, 33.
Lantos, P., and Kerdraon, A.: 1984, Solar Phys., submitted.
Leinert, C., Link, H, Pitz, E., Salm, N., and Kuppelberg, D.: 1975, Raumfahrtforschung **19**, 264.
Leinert, C., Pitz, E., Link, H., and Salm, N.: 1981, J. Space Sci. Instr. **5**, 257.
Lin, R. P., and Hudson, H. S.: 1976, Solar Phys. **50**, 153.
Low, B. C.: 1981, Astrophys. J. **251**, 352.
Low, B. C.: 1982, Astrophys. J. **254**, 796.
Low, B. C.: 1984a, Astrophys. J., **281**, 392.
Low, B. C.: 1984b, Astrophys. J., **286**, 772.
Low, B. C.: Mnro, R. H., and Fisher, R. R. 1982, Astrophys. J. **255**, 335.
Low, B. C., Munro, R. H., and Fisher, R. R.: 1982, Astrophys. J. **254**, 335.
Low, B. C., Hundhausen, A. J., and Hu, Y. Q.: 1985, J. Geophys. Res., submitted.
MacQueen, R. M.: 1980, Phil. Trans. Roy. Soc. Lon. **A297**, 605.
MacQueen, R. M., and Poland, A. I.: 1977, Solar Phys. **55**, 143.
MacQueen, R. M., Csoeke-Poeckh, A., Hildner, E., House, L., Reynolds, R., Stanger, A., Tepoel, H., and Wagner, W.: 1980, Solar Phys. **65**, 91.
MacQueen, R. M., and Fisher, R. R.: 1983, Solar Phys. **89**, 89.
Martin, S. F., and Ramsey, H. E.: 1972, Prog. Astronaut. Aeronaut. **30**, 371.
Maxwell, A., and Dryer, M.: 1982, Space Sci. Rev., **32**, 11.
McCabe, M. K., Svestka, Z., Howard, R. A., Jackson, B. V., and Sheeley, N. R., submitted to Solar Phys.

Melrose, D. B.: 1980, Solar Phys. **67**, 357.

Michalitsanos, A. G., and Kupferman, P.: 1974, Solar Phys. **36**, 304.

Moore, R. L., McKenzie, D. L., Svestka, Z., Widing, K. G., Krieger, A. S., Mason, H. E., Petrasso, R. D., Pneuman, G. W., Silk, J. K., Vorpahl, J. A., and Withbroe, G. L.: 1980, in Solar Flares, Skylab Solar Workshop II Monograph (ed., P. A. Sturrock) (Colorado Assoc. University Press: Boulder), p. 341.

Munro, R. H., Gosling, J. T., Hildner, E., MacQueen, R. M., Poland, A. I., and Ross, C. L.: 1979, Solar Phys. **61**, 201.

Nolte, J. T., Gerassimenko, M. Krieger, A. S., Petrasso, R. D., and Svestka, Z.: 1979, Solar Phys. **62**, 123.

Pallavicini, R., Serio, S., and Vaiana, G. S.: 1977 Astrophys. J. **216**, 108.

Pallavicini, R., and Vaiana, G. S.: 1980, Solar Phys. **67**, 127.

Parker, E. N.: 1958, Astrophys. J. **128**, 664.

Parker, E. N.: 1963, Interplanetary Dynamical Process (Interscience: New York).

Payne-Scott, R., Yabsley, D. E., and Bolton, J. G.: 1947, Nature **160**, 697.

Pick, M., Trottett, G., MacQueen, R. M.: 1980, Solar Phys. **63**, 369.

Pneuman, G. W.: 1980, in Solar Flare Magnetohydrodynamics (ed., E. Priest).

Pneuman, G. W.: 1982, Solar Phys. **78**, 229.

Pneuman, G. W.: 1983, Solar Phys. **88**, 219.

Reames, D. V., and von Rosenvinge, T. T.: 1983, in Proc. 18th Int. Cosmic Ray Conf. (Bangalore) **4**, p. 48.

Richter, I., Leinert, C., and Planck, B.: 1982, Astron. Astrophys. **110**, 115.

Rosner, R., Low, B. C. and Holzer, T. E.: 1984, in Physics of the Sun (eds., P. A. Sturrock, T. E. Holzer, and D. Michels) (to be published by the National Academy of Science).

Rust, D., and Hildner, E.: 1976, Solar Phys. **48**, 381.

Rust, D. M., and Webb, D. F.: 1977, Solar Phys. **54**, 403.

Rust, D. M., Hildner, E., Dryer, M., Hansen, R. T., McClymont, A. N., McKenna-Lawlor, S. M. P., McLean, D. J., Schmahl, E. J., Steinolfson, R. S., Tandberg-Hanssen, E., Tousey, R., Webb, D. F. and Wu, S. T.: 1980, in Solar Flares, Skylab Solar Workshop II Monograph (ed., P. A. Sturrock) (Colorado Assoc. University Press: Boulder), p. 273.

Saito, K., Poland, A. I., and Munro, R. H.: 1977, Solar Phys. **55**, 121.

Sawyer, C., Wagner, W. J., House, L. H., and Illing, R. W.: 1984, J. Geophys. Res., in preparation.

Schmahl, E., and Hildner, E.: 1977, Solar Phys. **55**, 473.

Sheeley, N. R., Bohlin, J. D., Brueckner, G. E., Purcell, J. D., Scherrer, V. E., Tousey, T., Smith, J. B., Speich, D. M., Tandberg-Hanssen, E., Wilson, R. M., deLoach, A. C., Hoover, R. B., and McGuire, J. P.: 1975, Solar Phys. **45**, 377.

Sheeley, N. R., Jr., Howard, R. A., Mihels, D. J., Koomen, D. J.: 1980, in IAU Symp. 91, Solar and Interplanetary Dynamics, Dryer and Tandberg-Hanssen, eds., D.-Reidel.

Sheeley, N. R., Jr., Stewart, R. T., Robinson, R. D., Howard, R. A., Koomen, M. J., Michels, D. J.: 1984, Astrophys. J., **279**, 839.

Sheeley, N. R., Jr., Howard, R. A., Koomen, M. J., Michels, D. J.: 1983a, Astrophys. J., **272**, 349.
Sheeley, N. R., Jr., Howard, R. A., Koomen, M. J., Michels, D. J., Schwenn., R., Muhlhauser, K. H., and Rosenbauer, H.: 1983b, in Solar Wind Five, (ed., M. Neugebauer) NASA Conf. Publ. 2280, p. 693.
Sheeley, N. R., Jr., Howard, R. A., Koomen, M. J., Michels, D. J., Schwenn, R., Muhlhauser, K. H., and Rosenbauer, H.: 1985, J. Geophys. Res. **90**, 163.
Sime, D. G., MacQueen, R. M., and Hundhausen, A. J.: 1984, J. Geophys. Res. **89**, 2113.
Simnett, G. M., and Harrison, R. A.: 1985, Solar Phys., in press.
Stewart, R. T., Dulk, G. A., Sheridan, K. V., House, L. L., Wagner, W. J., Sawyer, C., Illing, R.: 1982, Astron. Astrophys., **116**, 217.
Stewart, R. T.: 1983, private communication.
Stewart, R. T.: 1984, Solar Phys., **94**, 379.
Sturrock, P. A., and Smith, S. M.: 1968, Solar Phys. **5**, 87.
Svestka, Z.: 1976, in Solar Flares, (Reidel Publ. Co.: Dordrecht) p. 241.
Svestka, Z.: 1983, Space Sci. Rev. **35**, 259.
Svestka, Z.: 1984, Solar Phys. **94**, 171.
Svestka, Z., and Fritzova-Svestkova, L.: 1974, Solar Phys. **36**, 417.
Svestka, Z., Martin, S. F., and Kopp, R. A.: 1980, in Solar and Interplanetary Dynamics, IAU Symp. 91, (eds., M. Dryer and E. Tandberg-Hanssen), p. 216.
Svestka, Z., Stewart, R. T., Hoyng, P., van Tend, W., Acton, L. W., Gabriel, A. H., Rapley, C. G., and 8 co-authors: 1982a, Solar Phys. **75**, 305.
Svestka, Z., Dodson-Prince, H. W., Martin, S. F., Mohler, O. C., Moore, R. L., Nolte, J. T., and Petrasso, R. D.: 1982b, Solar Phys. **78**, 271.
Tousey, R.: 1973, The solar corona, in Space Research XIII, (eds., M. J. Rycroft and S. K. Runcorn) (Akademie-Verlag: Berlin) p. 713.
Trottet, G., and MacQueen, R. M.: 1980, Solar Phys. **68**, 177.
Trottet, G., Pick, M., House, L., Illing, R., Sawyer, C., Wagner, W.: 1982 Astron. Astrophys. **111**, 306.
Uchida, Y., and Low, B. C.: 1981, J. Ap. Astr. **2**, 405.
van Beek, H. F., Hoyng, P., LaFleur, H., and Simnett, G. M.: 1980, Solar Phys. **65**, 39.
Wagner, W. J.: 1983, Adv. Space Sci. **2**, 203.
Wagner, W. J.: 1984, Ann. Rev. of Astron. and Astrophys., **22**, 267.
Wagner, W. J., Hildner, E., House, L. L., Sawyer, C., Sheridan, K. V., and Dulk, G. A.: 1981, Astrophys. J. Letters **244**, L123.
Wagner, W. J., and MacQueen, R. M.: 1983, Astron. Astrophys., **120**, 36.
Wang, S., Hu, Y, Q, and Wu, S. T.: 1982, Scientia Sinica (Series A) **25**, 1305.
Webb, D., Krieger, A. S., and Rust, D. M.: 1976, Solar Phys. **48**, 159.
Webb, D. F., and Kundu, M. R.: 1978, Solar Phys. **57**, 155.
Webb, D. F., Cheng, C.-C., Dulk, G. A., Edberg, S. J., Martin, S. F., McKenna-Lawlor, S., and McLean, D. J.: 1980, in Solar Flares, (ed., P. A. Sturrock) (Colorado Assoc. University Press: Boulder) p. 471.
Webb, D. F., and Hundhausen, A. J.: 1985, JGR, in preparation.
Weiss, A., A.: 1963, Australian J. Phys. **16**, 240.
Wild, J. P., Roberts, J. A., and Murray, J. D.: 1954, Nature **173**, 532.

Wilson, R. M., and Hildner, E.: 1984, Solar Phys. **91**, 169.
Wolfson, R. L. T.: 1982, Astrophys. J. **255**, 774.
Wu, S. T., Wang, S., Dryer, M., Poland, A. I., Sime, D. G., Wolfson, C. J., Orwig, L. E., and Maxwell, A.: 1983, Solar Phys. **85**, 351.
Wu, S. T., Wang, S., Hu, Y. Q., Michels, D. J., Howard, R., and Sheeley, N.: 1985. Astrophys. J., to be submitted.

CHAPTER 7: INTERCOMPARISON OF NUMERICAL MODELS OF FLARING CORONAL LOOPS

CHAPTER 7: INTERCOMPARISON OF NUMERICAL MODELS OF FLARING CORONAL LOOPS

TABLE OF CONTENTS

R. A. Kopp, G. H. Fisher, P. MacNiece, R. W. P. McWhirter and G. Peres

Page

CHAPTER 7: INTERCOMPARISON OF NUMERICAL MODELS OF FLARING CORONAL LOOPS

R.A. Kopp, G.H. Fisher, P. MacNeice,
R.W.P. McWhirter, and G. Peres

7.1 INTRODUCTION

The past decade of solar observations from space, which has seen the extension of high spatial resolution and temporal cadence into the XUV spectral regime, has demonstrated convincingly that the corona is pervaded at all times by loop-shaped features that appear to be closely aligned with ambient magnetic field lines. This structuring is most strikingly evident in solar active regions. Together with photospheric magnetograms, the orientations of these loop structures allow the magnetic topology of much of the corona to be mapped out in detail. Whereas some fraction of the coronal volume may be permeated by open field lines or may even be virtually field-free, the plasma contained therein is of a lower density than that contained in the loops and is thereby more difficult to diagnose. Consequently, most interpretative studies have concentrated on the development and application of loop models to active region phenomena. Especially important to devise, but nonetheless difficult, are quantitative models for describing temporal changes of the plasma resulting from, for example, fluctuations of loop energy input. The extreme case is that which occurs during a solar flare, when the energy supply of a given magnetic loop may vary by many orders of magnitude on a time scale of a few seconds or less.

At the first meeting of the SMM Flare Workshop (23-28 January, 1983) it became apparent that, among the many efforts currently underway around the world to simulate various flare-related phenomena using large computer codes, several groups seemed to be in a fairly advanced state in their capabilities to carry out calculations relevant to the problem mentioned above--namely, the hydrodynamic and radiative response of a single magnetic flux tube to a sudden release of energy within it. This led rather quickly to the idea of using the SMM Flare Workshop Series as a forum for the intercomparison of code calculations on a specific standardized loop-heating problem. Such an opportunity, whereby a large group of numerical analysts from the international solar physics community gathers simultaneously, seldom presents itself in the everyday course of events. Even on the rare occasion when it does, the atmosphere is usually not conducive to an honest self-appraisal of code capabilities and, perhaps more importantly, of inherent limitations. The work to be summarized here represents the principal activity of the Flare Dynamical Modeling Group (FDMG) of the SMM Workshop--the seventh group organized for the purpose of studying various aspects of the flare process. This Modeling Group, however, was unique in that its members were drawn from each of the other six groups. Whereas this posed severe constraints on the amount of time available for the required discussions, it carried the advantage of representing the expertise of all segments of the Workshop at large.

7.2 THE ORIGINAL BENCHMARK PROBLEM

The physical configuration which the FDMG participants agreed to consider was chosen to be simple enough that all of the applicable computer codes could be used with only minor modifications, yet sufficiently complex that the basic nonlinear processes believed to govern the physics of real loops (radiaton, thermal conduction, compressible hydrodynamics, gravity, nonthermal heating) could be incorporated with some degree of realism. However, it should be kept in mind that the so-called "Benchmark Model" which resulted is nothing more than an attempt to establish a baseline calculation against which the performance of a given code might be judged. It was hoped thereby to provide a means for quantitatively comparing one code calculation with another. The model, on the other hand, was specifically not meant to represent either an average or a particular flare loop observed by the SMM or by anyone else, although of course to keep the problem interesting the choice of physical parameters was dictated by the conditions believed to typify flare loops in general.

The proposed Benchmark Problem consists of an infinitesimal magnetic flux tube containing a low-beta plasma. The field strength is assumed to be so large that the plasma can move only along the flux tube, whose shape remains invariant with time (i.e., the fluid motion is essentially one-dimensional). The flux tube cross section is taken to be constant over its entire length. In planar view the flux tube has a semi-circular shape, symmetric about its midpoint $s = s_{max}$ and intersecting the chromosphere-corona interface (CCI) perpendicularly at each footpoint; see Figure 7.2.1a. The arc length from the loop apex to the CCI is 10,000 km. The flux tube extends an additional 2000 km below the CCI to include the chromosphere, which initially has a uniform temperature of 8000 K. The temperature at the top of the loop was fixed initially at 2×10^6 K. The plasma is assumed to be a perfect gas ($\gamma = 5/3$), consisting of pure hydrogen which is considered to be fully ionized at all temperatures. For simplicity, moreover, the electron and ion temperatures are taken to be everywhere equal at all times (corresponding to an artificially enhanced electron-ion collisional coupling).

The equations describing the one-dimensional temporal evolution of the loop plasma are:

$$\frac{\partial \rho}{\partial t} + \frac{\partial}{\partial s}(\rho v) = 0 \qquad (1)$$

$$\rho \frac{Dv}{Dt} = - \frac{\partial P}{\partial s} - \rho g_{\|} \qquad (2)$$

$$\frac{3}{2}\frac{DP}{Dt} = H(s,t) - N_e^2\, \Lambda(T) + \frac{\partial}{\partial s}\{K(T)\, \frac{\partial T}{\partial s}\} - \frac{5}{2}\, P\, \frac{\partial v}{\partial s}\ , \qquad (3)$$

where N_e is the electron density, $P = 2N_ekT$ is the gas pressure, $\rho = m_pN_e$ is the plasma density, and v is the fluid velocity; g is the component of gravity tangential to the loop. Moreover, s denotes arc length from the loop footpoint and D/Dt represents the time derivative following a fluid element as it moves. H(s,t) denotes the nonthermal heating per unit volume (see below), $N_e^2\,\Lambda(T)$ is the (optically thin) radiative loss function, and K(T) is the plasma thermal conductivity.

For the radiative losses, Λ(T) was chosen to approximate the form for a plasma with normal solar abundances, viz.,

$$\Lambda(T) = 3 \times 10^{-22}\ (T/(2/10^4))^3 \ \ldots\ldots\ (T \leqq 2 \times 10^4\ K) \quad (4a)$$

$$\Lambda(T) = 3 \times 10^{-22} \ \ldots\ldots\ (2 \times 10^4 \leqq T \leqq 2 \times 10^5\ K) \quad (4b)$$

$$\Lambda(T) = 3 \times 10^{-22}\ (T/(2 \times 10^5))^{-1/2} + 2 \times 10^{-23}\ (T/10^8)^{1/2} \ \ldots\ldots\ (T \geqq 2 \times 10^5\ K)\ . \quad (4c)$$

The initially steady preflare state of the loop is maintained by a nonthermal heating function $H_o(s)$ with the following properties:

i. in the chromosphere, that necessary to balance radiative losses at each point, thereby maintaining isothermal conditions;

ii. above the chromosphere, a constant value such that the net energy input exactly balances the integrated radiative losses from the corona and transition region.

The thermal conductivity was taken to be of the classical Spitzer form--$K(T) = 9.203 \times 10^{-7}T^{5/2}$--and the resulting heat flow was assumed not to be flux-limited.

Along with the various solar constants (gravity, radius, etc.), the above details prescribe the quiescent thermodynamical state of the loop plasma that persists up to, say, t = 0. To initiate a transient response an additional input of energy was invoked at later times; the net heating function was taken to be of the form

$$H(s,t) = H_o(s) + H_F(s,t), \quad (5)$$

where

$$H_F(s,t) = \begin{cases} E \exp\,[-(s - s_{max})^2/\sigma^2] & (0 \leqq t \leqq 5s) \\ 0. & (t \geqq 5s) \end{cases} \quad (6)$$

The gaussian width of deposition, σ, was chosen to have the value 5000 km, and the constant E was to be determined by the condition that the integral of H_F over either half of the flux tube correspond to a transient energy flux of 10^{11} erg cm^{-2} s^{-1}, a not flare-unlike value.

The anticipated dynamical response of the loop atmosphere to this transient heating function is depicted by the general scenario of Figure 7.2.1b and can be described qualitatively as follows. For the assumed value of σ, nearly all of the flare energy is deposited in the corona. This leads to a rapid rise in the coronal temperature from its initial value on a time scale given by $\tau_F = 3N_e kT/E \approx 0.02$ s, which is much less than the acoustic transit time for the loop: $\tau_a\ s_{max}/c \approx 50$ s. Thus, much of the heating takes place before substantial mass motions can occur. The temperature rise is most rapid near the loop top (where the heating is

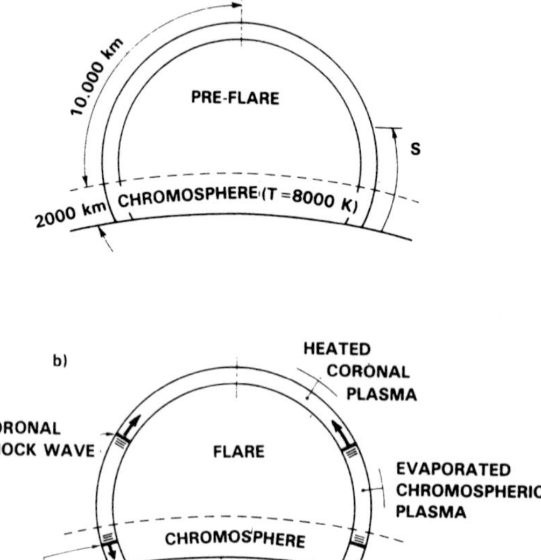

Figure 7.2.1 Schematic plan view of the loop configuration used in the Benchmark Problem (a) before and (b) after the initiation of the flare energy release.

strongest), and this drives a supersonic thermal wave downward along the loop (cf. Figure 7.2.2). When this conduction front reaches the top of the chromosphere, the resultant sudden heating of the cool plasma there causes an expansion in both directions along the flux tube. The downward-propagating pressure wave rapidly steepens to form a shock, which ultimately overtakes the thermal wave as both move deeper into the chromosphere. At the same time, the upward-moving (evaporated) chromospheric plasma pushes a weaker pressure wave ahead of it into the corona. This wave may or may not have time to steepen into a shock before reaching the top of the loop. In any case, the loop soon becomes filled with hot and dense matter. It was agreed by all participants to try to follow the dynamical history of the loop plasma for a period of ten seconds following the switch-on of the transient heating function.

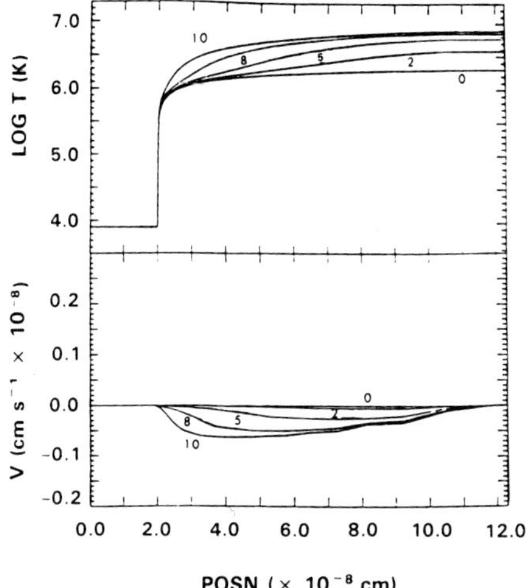

Figure 7.2.2 Loop temperature and velocity profiles at various times during the first 10 s of the transient flare heating, as calculated by P. MacNeice (CAMB) for the final Benchmark Problem parameters. The density profile changes but little from that of the initial state during this time and thus is not shown here.

7.3 INITIAL SOLUTION COMPARISONS

Originally eight groups or individuals expressed confidence in their ability to perform the above calculation in advance of the second SMM Workshop (9-14 June 1983). These are listed alphabetically by name of the principal worker in Table 7.3.1. By the time of this second gathering an initial comparison of the solution curves had been assembled at Rutherford Laboratory, from which it became immediately apparent that large discrepancies existed among the various calculations. In fact, while there was more-or-less unanimous agreement as to certain global properties of the system behavior (peak temperature reached, thermal-wave time scales, etc.), no two groups could claim satisfactory accord when a more detailed comparison of solutions was attempted.

Table 7.3.1 Benchmark Calculation Participants

Name	Institute	Identifier	Type of Code
C.C. Cheng	Naval Research Laboratory	NRL	Eulerian (FCT)
G.H. Fisher	Univ. of Calif., San Diego	UCSD	Lagrangian
R.A. Kopp	Los Alamos National Laboratory	LANL	Lagrangian
P. MacNeice	Cambridge University	CAMB	Eulerian (FCT)
F. Nagai	Marshall Space Flight Center	MSFC	Eulerian
G. Peres	Osserv. Astronomico, Palermo	PLRMO	Eulerian
A.I. Poland	NASA Goddard Space Flight Center	GSFC	Eulerian (FCT)
D.F. Smith	Berkeley Research Associates	BRA	Eulerian

As a result of discussions held during the second Workshop, it was concluded that some of the differences between solutions could be accounted for by the realization that, even though purportedly agreed to in advance, no two groups had actually solved exactly the same problem. For example, MSFC had used a thermal conduction flux limiter; PLRMO had inadvertently spread the transient heat input over too much of the loop; and LANL, GSFC, and NRL had secretly "modified" the radiative loss function at low temperatures (albeit in different ways) from that given above, to stabilize the initial atmosphere against an apparent tendency to heat up almost explosively (a behavior explained subsequently by PLRMO and UCSD as being caused by an unrealistic prescription of the quiescent heating function; see below). However, not all of the discrepancies could be accounted for by these differences. It was generally felt by the FDMG participants that none of the numerical solutions which had been carried to completion were to be wholly trusted, because each had failed to resolve adequately the structure of the thermal wave front once it enters the chromosphere; as was pointed out by Fisher (UCSD), for the adopted flare heat input one would expect this front to have a thickness of only about 10 cm! Even were it possible to resolve the front region, say by means of an automatic dynamic rezoner (only the UCSD code had this capability), the small time steps which would be required to achieve reasonable solution accuracy would have rendered the calculation impractical on present-day computers.

The issue of numerical resolution of the thermal wave front is a complex one. If it could be demonstrated that the structure of this front is not important to the global dynamics of the loop plasma, then one might be able to introduce a numerical thermal conductivity to spread out the front artificially over a few mesh points, analogous to the use of an artificial viscosity for shockwave problems. On the other hand, if the detailed structure of the thermal front turns out to be important for determining the global behavior (e.g., the net evaporation rate or the peak coronal temperature), then it is essential to resolve the front even if one is interested only in the global properties of the loop. Studies by McClymont and Fisher (UCSD) indicate that, if the total energy flux into the corona is large compared with coronal radiative losses, then it is not essential to resolve the thermal front to get the correct evaporation rate and coronal temperature. However, once sufficient evaporation has taken place that the total coronal radiative loss rate is of the same magnitude as the total coronal energy flux, the subsequent numerical solution of the evaporation problem is grossly incorrect unless the thermal front is resolved.

But these considerations aside, it was recognized that the posed problem is also physically unrealistic for various reasons. For example, linear heat flow theory (as exemplified by the use of the Spitzer conductivity) does not even apply in regions of strong thermal gradients; instead, the actual heat flow will be flux-limited and, ideally, a description of heat flow based on "non-local" properties of the atmosphere should be used. Lacking such, one can rightfully question the significance of numerical heat-flow simulations for which the mesh size required for numerical accuracy/stability is much smaller than an electron mean free path. However, the primary goal of the Benchmark Model was to intercompare code calculations on a standardized, although hypothetical, problem, rather than to establish the best possible physical model. To this end, it was decided to repeat the basic Benchmark Model calculation for the third and final Workshop meeting (13-17 February, 1984), using a transient energy flux reduced by two orders of magnitude from the original value. This would have the effect of driving a much gentler thermal wave into the chromosphere; since the thickness of the wave front increases with decreasing thermal energy flux, adequate numerical resolution of this front was now expected to pose less of a constraint on obtaining a solution. It was moreover agreed to leave the transient heating on for the duration of the problem, which was extended to 100 s from the original 10 s (thus, the total "flare" input was smaller than that of the original problem by only a factor of 5). Although the thermal wave front in the problem as redefined is still quite thin, it was nevertheless felt that the codes with dynamic rezoners might at least have a chance of running to completion with the finite computer resources available.

Interestingly, whereas the intercomparison of this second round of calculations showed a modest improvement in agreement, the improvement was not as marked as expected. Still there were found to be large differences in the velocity of the evaporated chromospheric plasma, and even the temperature at the top of the loop--perhaps the least sensitive parameter used in the comparison--varied considerably from one solution to the next. Herein was realized a second major problem--this one associated with the basic definition of the quiescent heating function for the pre-flare atmosphere. Recall that this was chosen arbitrarily to be a time-independent "volumetric" heating function (i.e., the units are erg cm^{-3} s^{-1}) of position s along the flux tube, and that it was to be left on at all times. A plot of this function against height shows a steep exponential decrease up through the chromosphere--a direct result of the small density scale height there--followed by a discontinuous drop by an order of magnitude to its (assumed) uniform coronal value for $s \geq$ 2000 km. It is this sudden jump that plays havoc with numerical codes. For, even without a transient heat input, finite-difference errors will cause the pre-flare atmosphere t be slightly out of hydrostatic equilibrium (by a varying degree with each code) and some initial readjustment of the density structure will inevitably occur. Consider, for example, the result of such an adjustment by which the corona settles downward ever so slightly, compressing the chromosphere below. The coronal plasma which was originally just above 2000 km now finds itself in the region of strong chromospheric heating, but being of much lower density than the chromosphere is incapable of radiating away this increased heat input. Consequently, the temperature of this region begins to rise; the resulting localized pressure enhancement initiates an outward expansion of the plasma, causing the density to decrease still further and the heating to become even more unbalanced. This unstable situation rapidly leads to expansion of an almost explosive nature.

The same argument can be used to show that, were finite-difference errors nonexistent and the pre-flare atmosphere perfectly in equilibrium, the initiation of transient heating would still cause an unrealistically violent expansion of the upper chromosphere to take place immediately upon arrival of the leading edge of the thermal wave in these layers. As was originally suggested by George Fisher (UCSD), this unphysical behavior can be largely avoided by using a quiescent heating function which is defined (and kept constant) per unit mass rather than per unit volume. Then, when the plasma starts to expand as the result of a heating imbalance, the amount of (quiescent) heating of each mass element will remain constant and radiative losses will be better able to restrict a further temperature rise.

The extreme sensitivity of the plasma behavior to the adopted definition of the quiescent heating function was illustrated most vividly via a calculation carried out by Giovanni Peres (PLRMO). Therein the original volumetric heating function was divided by the mass density at each point to give an equivalent heating function per unit mass, and the dependence of this function on position was kept constant throughout the calculation (the simplest procedure when using an Eulerian code). In the original (volumetric heating) case the temperature at the top of the loop continued to increase at late times as chromospheric evaporation, supplied ever-faster by the quiescent heating function, grew rapidly in intensity, whereas with the revised definition the temperature approached a well-defined limit. Other properties of the solution changed by even greater amounts. Thus a seemingly minor change of definition of the quiescent heating function was shown to have a dramatic effect on the temporal evolution of the loop plasma.

Why didn't the other groups experience the same difficulties with "explosive" evaporation as did PLRMO? Confronted with the above results, it turned out that nearly all had. For example, UCSD, being probably the first to identify the problem but unable to convince others of its importance, had decided early on to abandon the volumetric heating function in favor of one defined per unit mass. And, as mentioned earlier, several groups had independently chosen to "modify" their radiative loss function so that it vanished at chromospheric temperatures; since the magnitude of the chromospheric heating function is determined by the condition that it balance radiative losses at each point, this procedure clearly eliminates the problem, although in an artificial manner.

7.4 FINAL BENCHMARK PROBLEM AND SOLUTION COMPARISON

With these facts in hand, it was decided at the final Workshop that those participants who hadn't already done so, and who were willing and able, would perform a final calculation using a quiescent heating function properly defined per unit mass. Even so, one realized by now that certain intrinsic differences between Lagrangian and Eulerian code logistics would make impractical a precise comparison of results; a quiescent heating function that "remains constant in time" is interpreted differently by the two approaches. Nevertheless, it was expected that even a rough comparison of results would yield substantially closer agreement than had been obtained previously.

This expectation was in fact borne out. Figures 7.4.1-7.4.4 show some of the results of the final comparison, completed some time after the end of the third Workshop. We note first that the number of participants has diminished markedly, from the original eight to now only four. The quantities displayed are arranged roughly in order of increasing discrepancies between solutions. For example, all calculations are now in close accord as to the time history of temperature at the loop top; this quantity is primarily a function of the total loop heat input. On the other hand, appreciable differences are seen to persist in the maximum electron density reached in the chromospheric shock wave, a result which is not too surprising in view of the fact that this quantity

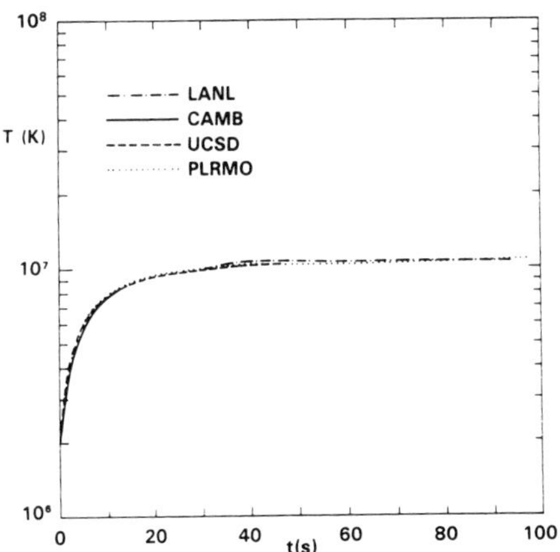

Figure 7.4.1 Temperature versus time at the top of the loop, as predicted by the four codes used for the final Benchmark Model calculation.

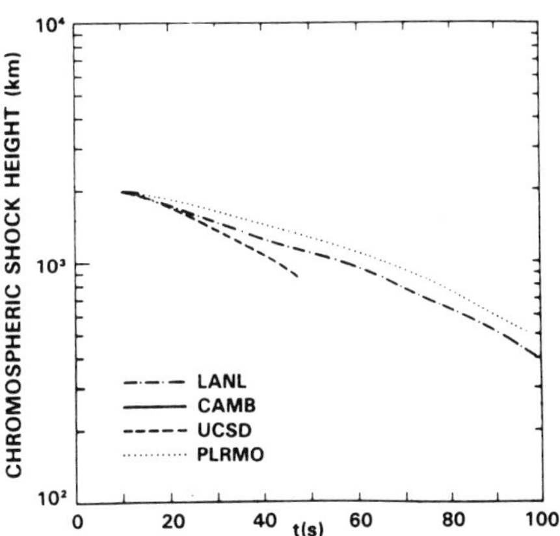

Figure 7.4.3 Position of the downward-propagating shock wave in the chromosphere. Note that this shock first appears at about 10 s in all the calculations, i.e., when the thermal wave first reaches the top of the chromosphere.

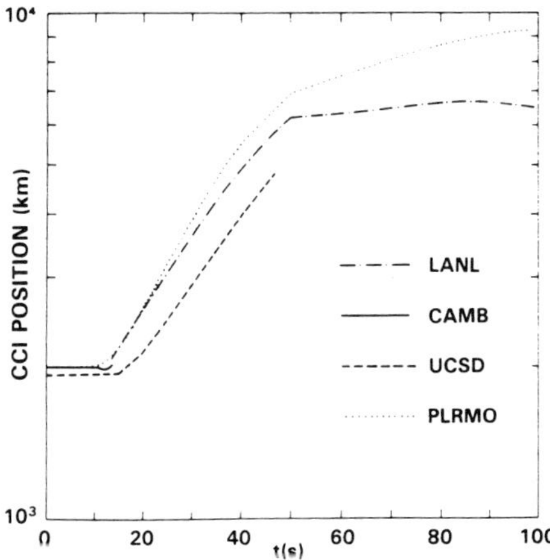

Figure 7.4.2. Height of the original chromosphere-corona interface (CCI) as a function of time.

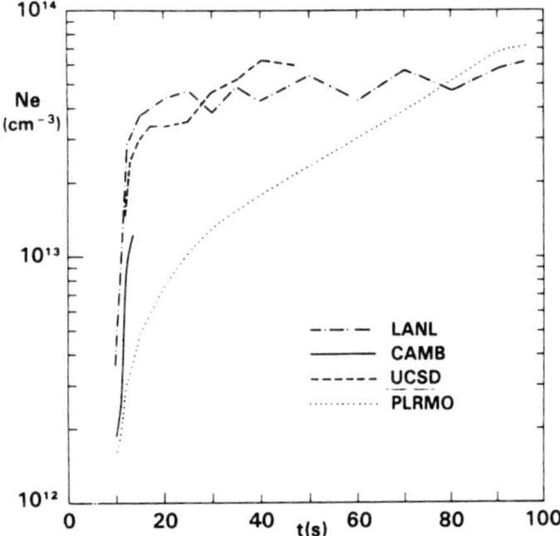

Figure 7.4.4 Maximum electron density reached behind (i.e., just above) the chromospheric shock front.

is quite sensitive to the particular grid spacing used to resolve the shock wave (which varied considerably from one calculation to another).

Thus, the results shown in Figures 7.4.1-7.4.4 comprise in fact a "Benchmark Model" against which other flare-loop codes can be tested for the loop heating problem described here. The specific parameters used for this problem are collected in Table 7.4.1.

Nevertheless, it is important to note that significant discrepancies remain between the code results, as shown in Figures 7.4.2-7.4.4. Whereas this is not particularly comforting to those individuals who performed the calculations, it is important that these results be presented in the form shown here rather than for example, as an "averaged" solution, because the adopted format conveys some notion of the inherent uncertainties that still exist in the Benchmark Model. Future numerical solutions of the Benchmark Problem, either by the original participants or by others, should be aimed at resolving these remaining discrepancies.

Table 7.4.1 Benchmark Problem — Definition and Parameters

1. Symmetric semi-circular loop of uniform cross section—cf. Figure 7.2.1a

   s_{max} = 2000 km (chromosphere) + 10,000 km (corona).

2. Fully ionized hydrogen atmosphere (perfect gas)—$N_i = N_e$; $T_i = T_e$.

3. Optically thin radiative losses—cf. Equations (4a-c).

4. Spitzer thermal conductivity (no flux limit imposed)

   $K(T) = 9.203 \times 10^{-7}\, T^{5/2}$ erg cm^{-1} s^{-1} K^{-1}.

5. Initial atmosphere

   a. Hydrostatic equilibrium ($g = 2.738 \times 10^4$ cm s^{-2}):

   $$\frac{\partial}{\partial t} = 0;\ v = 0;\ \frac{dP}{ds} = -\varrho g_{\parallel}.$$

   b. Isothermal chromosphere ($0 \le s \le 2000$ km):

   $T = 8000$ K; $H_0(s) = N_e^2\, \Lambda(T)$.

   c. Corona (2000 km $\le s \le s_{max}$):

   $T = 8000$ K at chromospheric boundary ($s = 2000$ km);

   $T = 2 \times 10^6$ K and $\frac{dT}{ds} = 0$ at loop top ($s = s_{max}$);

   $H_0(s)$ = a constant such that $H_0 \int ds = \int N_e^2\, \Lambda(T)\, ds$, where the integrals extend over $(2000, s_{max})$.

6. Flare heating function

   $H_F(s,t) = H_F(s) = E \exp[-(s - s_{max})^2/\sigma^2]$, where $\sigma = 5000$ km and $E = 2 \times 10^9/(\sigma\sqrt{\pi})$ erg cm^{-3} s^{-1};

   Flare heating left on for 100 s, the problem run time.

7.5 CONCLUSIONS

The flare modeling activity at the SMM Flare Workshop Series represented the dedicated efforts of several individuals and substantial computer resources of their respective institutions. Whereas these have not yet converged upon a unique Benchmark Model, we have succeeded in defining a well posed Benchmark Problem, and this is a necessary first step. Future modeling work directed toward this problem will no doubt produce the desired unique solution. Perhaps more valuable than this was the general recognition within the flare-modeling community of certain pitfalls and difficulties associated with trying to model the energetic flare process numerically. In particular, we call attention once more to the importance in any calculation of confronting directly the difficult numerical problems associated with the rapid motion of a very steep thermal wave front through the chromosphere, as well as that of the extremely thin, dense compression wave that runs ahead of it.

In addition, we have learned how difficult it is to intercompare the results obtained with diverse and highly complex computer codes. This is partly due to the intrinsic differences in mathematical formulation used by various codes (Eulerian versus Lagrangian hydrodynamics, fixed-nonuniform zoning versus adaptive rezoning, etc.), which render a detailed comparison of many quantities impossible without making fairly major code modifications or extensions. But this is not an insoluble dilemma. Perhaps the most important lesson learned is how careful one must be to define a meaningful problem in the first place, the solution of which will provide a viable test of real simulational capabilities and not just magnify seemingly insignificant differences of problem definition to the point where these dominate the results.

Finally, it was felt by all participants that, although an attempt had been made to reduce the Benchmark Problem to its bare essentials, it would nevertheless be useful to have available a selection of even simpler test problems, for the purpose of verifying the mechanics of a given code for each elementary physical process by itself. The following list is by no means all inclusive. We simply reference here a few situations for which analytical solutions (e.g., similarity solutions) are known, without elaborating on any of them. Of course, the value of these solutions is limited to codes which (a) possess the capability of switching off all physical processes except that which is being checked, and (b) can accommodate the required boundary conditions.

1. Thermal-wavefront dynamics (no hydrodynamics or radiation). In a medium with nonlinear heat conduction, the presence of transient energy sources can give rise to steep thermal fronts (Marshak waves) which propagate away from the source regions and transport energy to other parts of the problem. Useul similarity solutions can be found in References 1 and 2.

2. Continuum hydrodynamics (i.e, no shock waves) with gravity. For the case of one-dimensional time-dependent flow with a prescribed (fixed) temperature profile and flowtube geometry, a type of similarity solution has been given by Reference 3; the flow velocity at each point is time-invariant, but the density grows or decays exponentially with time.

3. Shockwave dynamics (no heat conduction or radiation). Besides the obvious requirement that the Rankine-Hugoniot conditions be satisfied across a shock front, a simple test case that addresses the interaction of a flow discontinuity with a problem boundary is that of reflection off a rigid wall of a piston-driven shock wave (Ref. 4).

4. Optically thin radiation (no fluid motions or heat conduction). The obvious test here would be to simulate numerically the analytical form of the radiative cooling curve:

$$T(t) = \{T_o^{1-n} + \frac{(n-1)A}{3k} N_e t\}^{\frac{1}{1-n}} ,$$

which results from solving the energy equation for the case that $\Lambda(T) = A T^n$, where A and n ($\neq$ 1) are prescribed constants.

7.6 REFERENCES

1. Richtmeyer, R. D. and Morton, K. W.: 1967, Difference Methods for Initial Value Problems, Interscience, New York, pp. 201-206.
2. Zel'dovich, Ya. B. and Raizer, Yu. P.: 1967, Physics of Shock Waves and High-Temperature Hydrodynamic Phenomena, Vol. II, Academic Press, New York, pp. 652-684.
3. Kopp R. A.: 1980, Solar Phys. 68, 307.
4. Landau, L. D. and Lifshitz, E. M.: 1959, Fluid Mechanics, Pergamon Press, London, p. 365.

SUBJECT INDEX

SMM PARTICIPANTS

SMM PARTICIPANTS

Name	Institution	Country
Costas Alissandrakis	U. of Athens	GREECE
Spiro Antiochos	Naval Research Laboratory	USA
Ester Antonucci	Univ. di Torino	ITALY
A. Aydemir	U. of Texas, Austin	USA
Thomas R. Ayres	U. of Colorado, JILA	USA
Taeil Bai	Stanford University	USA
Joseph Bassi	University of Colorado	USA
David Batchelor	Johns Hopkins Applied Phys. Lab.	USA
F. Bely-Dubau	Observatoire de Nice	FRANCE
Robert Bentley	Mullard Space Science Lab.	ENGLAND
A.O. Benz	ETH-Zentrum, Zurich	SWITZERLAND
Patricia Bornmann	NOAA Space Environment Lab.	USA
J.L. Bougeret	Meudon	FRANCE
John C. Brown	University of Glasgow	IRELAND
Francois Brunel	U. of Texas, Austin	USA
Marilyn Bruner	Lockheed Palo Alto Research Lab.	USA
Richard C. Canfield	University of Hawaii	USA
Peter Cargill	University of Maryland	USA
Gary Chanan	University of California	USA
Hou Mei Chang	U. of Alabama, Huntsville	USA
Gary A. Chapman	California State University	USA
Robert D. Chapman	NASA/GSFC	USA
Chung-Chieh Cheng	Naval Research Lab.	USA
F. Chiuderi-Drago	Observatoire de Paris	FRANCE
Edward L. Chupp	University of New Hampshire	USA
E.W. Cliver	Air Force Geophysis Lab.	USA
Carol J. Crannel	NASA/GSFC	USA
J. Leonard Culhane	Mullard Space Science Lab.	ENGLAND
A deLoach	NASA/GSFC	USA
B.R. Dennis	NASA/GSFC	USA
George Doschek	Naval Research Lab.	USA
J.G. Doyle	Armagh Observatory	N. IRELAND
George Dulk	University of Colorado	USA
R. Duncan	CSIRO	AUSTRALIA
D. Ellison	NASA/GSFC	USA
A. Gordon Emslie	Univ. of Alabama, Huntsville	USA
S. Enome	Nagoya University	JAPAN
Paul Evenson	University of Delaware	USA
Roberto Falciani	Arcetri Astrophysical Obs.	ITALY
George Fisher	IGPP/LLNL, Livermore, CA	USA
Terry Forbes	University of New Hampshire	USA
David J. Forrest	University of New Hampshire	USA
A.H. Gabriel	Rutherford Appleton Labs.	ENGLAND
V. Gaizauskas	Hertzberg Inst. of Astrophysics	CANADA
G. Alen Gary	NASA/GSFC	USA
Dale Gary	Caltech	USA
Katherine B. Gebbie	University of Colorado	USA

SMM PARTICIPANTS

Name	Institution	Country
Thomas E. Gergely	NASA/GSFC	USA
R.S. Gill	Oxford Unifersity	ENGLAND
J.B. Gurman	NASA/GSFC	USA
M.J. Hagyard	NASA/GSFC	USA
Bernhard M. Haisch	Lockheed Palo Alto Res. Lab.	USA
R. Harrison	High Altitude Observatory	USA
J. Hyaward	Napier College, Edinburgh	SCOTLAND
J.C. Henoux	Observatoire de Paris	FRANCE
William Henze, Jr.	NASA/MSFC	USA
E. Hiei	Tokyo Astronomy Observatory	JAPAN
Ernest Hildner	High Altitude Observatory	USA
Gordon Holman	NASA/GSFC	USA
Alan Hood	University of St. Andrews	SCOTLAND
Robert Howard	Hale Observatories	USA
Peter Hoyng	Space Research Laboratory	THE NETHERLANDS
Hugh Hudson	Univ. of Calif., San Diego	USA
Gordon Hurford	Calif. Institute of Technology	USA
Charles Hyder	High Altitude Observatory	USA
Rainer Illing	High Altitude Observatory	USA
Bernard Jackson	Univ. of Calif., San Diego	USA
C. de Jager	Space Research Laboratory	THE NETHERLANDS
Harrison P. Jones	NASA/GSFC/KPNO	USA
Carole Jordan	Oxford University	ENGLAND
Steven Kahler	Hanscom Air Force Base	USA
Sharad Kane	Univ. of Calif., Berkeley	USA
Judy Karpen	Naval Research Laboratory	USA
Pierre Kaufmann	INPE:CRAAM	BRAZIL
A. Kiplinger	NASA/GSFC	USA
Roger Kopp	Los Alamos National Lab.	USA
Takeo Kosugi	University of Tokyo	JAPAN
J. Kroger	Max Planck	GERMANY
M.R. Kundu	University of Maryland	USA
K. Lang	Tufts University	USA
P. Lantos	Observatoire de Meudon	FRANCE
John Leibacher	Kitt Peak Observatory	USA
James Lemen	Mullard Space Science Lab.	ENGLAND
R.P. Lin	Univ. of Calif., Berkeley	USA
L. Loiacono	University of Florence	ITALY
Boon Chye Low	High Altitude Observatory	USA
A.L.MacKinnon	University of Glasgow	IRELAND
Peter MacNeice	Rutherford Appleton Labs	ENGLAND
Andreas Magun	Inst. of Applied Phys., Bern	SWITZERLAND
John Mariska	Naval Research Laboratory	USA
Piet Martens	NASA/GSFC	USA
Sara Martin	Calif. Institute of Technology	USA
M.J. Martres	Observatoire de Paris	FRANCE
Robert McGuire		

SMM PARTICIPANTS

Name	Institution	Country
R.W.P. McWhirter	Rutherford Appleton Labs	ENGLAND
D.B. Melrose	University of Sydney	AUSTRALIA
Y. Mok	University of California	USA
Ron Moore	NASA/MSFC	USA
Ron Murphy	NASA/GSFC	USA
F. Nagai	NASA/MSFC	USA
Yoshinari Nakagawa	Sagami Inst. of Technology	JAPAN
H. Nakajima	Nobeyama Solar Radio Obs.	JAPAN
Donald Neidig	Sacramento Peak Observatory	USA
Werner Neupert	NASA/GSFC	USA
K. Ohki	Tokyo Astronomical Observatory	JAPAN
Larry E. Orwig	NASA/GSFC	USA
Roberto Pallavicini	Osservatorio Astrofisico di Arcetri	ITALY
Giovanni Peres	Palermo	ITALY
Mark Pesses	Applied Research Corporation	USA
Vahe Petrosian	Stanford University	USA
Kenneth Phillips	Rutherford Appleton Labs.	ENGLAND
Monique Pick	Observatoire de Paris	FRANCE
Gerald Pneuman	High Altitude Observatory	USA
Arthur I. Poland	NASA/GSFC	USA
Giannina Poletto	Osservatorio Astrofisico di Arcetri	ITALY
Jason Porter	NASA/MSFC	USA
Eric R. Priest	University of St. Andrews	SCOTLAND
Douglas Rabin	National Solar Observatory	USA
Reuven Ramaty	NASA/GSFC	USA
A. Raoult	Observatoire de Paris	FRANCE
Donald Reames	NASA/GSFC	USA
Eric Reichmann	NASA/MSFC	USA
P.J. Ricchiazzi	University of Hawaii	USA
E. Reiger	Max Planck	GERMANY
H. Rowland	University of Maryland	USA
David M. Rust	Johns Hopkins Applied Phys. Lab.	USA
J. Ryan	University of New Hampshire	USA
A. Schadee	Space Research Laboratory	THE NETHERLANDS
E.J. Schmahl	University of Maryland	USA
Brigitte Schmeider	Observatoire de Paris	FRANCE
J. Schrijver	Space Research Laboratory	THE NETHERLANDS
R.A. Schwartz	NASA/GSFC	USA
S. Serio	Palermo	ITALY
Gerald Share	Naval Research Laboratory	USA
Neil R. Sheeley, Jr.	Naval Research Laboratory	USA
John Sherman	Rutherford Appleton Labs.	ENGLAND
R.K. Shevgaonkar	University of Maryland	USA
Richard A. Shine	Lockheed Palo Alto Res. Lab.	USA
George Simnett	University of Birmingham	ENGLAND

SMM PARTICIPANTS

Name	Institution	Country
G. Simon	Observatoire de Paris	FRANCE
Dean Smith	Berkeley Research Association	USA
Jesse B. Smith, Jr.	NASA/MSFC	USA
David Speich	NASA/GSFC	USA
Richard Steinolfson	University of California	USA
R. Stewart	CSIRO	AUSTRALIA
Keith Strong	NASA/GSFC	USA
Peter Sturrock	Stanford University	USA
Zdenek Svestka	Space Research Laboratory	THE NETHERLANDS
J. Sylwester	Instutut Astronomiczny	POLAND
K. Tanaka	Tokyo University	JAPAN
Juri Toomre	University of Colorado	USA
Gerard Trottet	Observatoire de Paris	FRANCE
Saku Tsuneta	University of Tokyo	JAPAN
Gerard vanHoven	University of California	USA
N.J. Veck	Marconi Research Center	ENGLAND
L. Vlahos	University of Maryland	USA
Peter Waggett	Marconi Research Center	ENGLAND
David F. Webb	American Science & Engineering	USA
H.J. Wiehl	Inst. of Applied Physics, Bern	SWITZERLAND
Bruce Woodgate	NASA/GSFC	USA
S.T. Wu	Univ. of Alabama, Huntsville	USA
Harold Zirin	Calif. Inst. of Technology	USA